AF765240

Bayesian Econometric Modelling for Big Data

This book delves into scalable Bayesian statistical methods designed to tackle the challenges posed by big data. It explores a variety of divide-and-conquer and subsampling techniques, seamlessly integrating these scalable methods into a broad spectrum of econometric models.

In addition to its focus on big data, the book introduces novel concepts within traditional statistics, such as the summation, subtraction, and multiplication of conjugate distributions. These arithmetic operators conceptualize pseudo data in the conjugate prior, sufficient statistics that determine the likelihood, and the posterior as a balance between data and prior information, adding an intriguing dimension to Bayesian analysis. This book also offers a deep dive into Bayesian computation. Given the intricacies of the floating-point representation of real numbers, computer programs can sometimes yield unexpected or theoretically impossible results. Drawing from his experience as a senior statistical software developer, the author shares valuable strategies for designing numerically stable algorithms.

The book is an essential resource for a diverse audience: graduate students seeking foundational knowledge in Bayesian econometric models, early-career statisticians eager to explore cutting-edge advancements in scalable Bayesian methods, data analysts struggling with out-of-memory challenges in large datasets, and statistical software users and developers striving to program with efficiency and numerical stability.

Hang Qian is the principal engineer of the Econometrics Toolbox for MATLAB and has been dedicated to statistical software development at MathWorks since 2012. He earned his PhD in economics, specializing in Bayesian statistics, big data analysis, and computational finance. His research has been published in journals such as Bayesian Analysis, Journal of Business & Economic Statistics, and Journal of Econometrics.

CHAPMAN & HALL/CRC Series on Statistics in Business and Economics

Recently Published Titles

Empirical Research in Accounting
Tools and Methods
Ian D. Gow and Tongqing Ding

For more information about this series, please visit: https://www.routledge.com/Chapman-and-HallCRC-Series-on-Statistics-in-Business-and-Economics/book-series/CHSBE

Bayesian Econometric Modelling for Big Data

Hang Qian

CRC Press
Taylor & Francis Group
Boca Raton London New York

CRC Press is an imprint of the
Taylor & Francis Group, an **informa** business
A CHAPMAN & HALL BOOK

Designed cover image: Shutterstock_ 1871497255

MATLAB® and Simulink® are trademarks of The MathWorks, Inc. and are used with permission. The MathWorks does not warrant the accuracy of the text or exercises in this book. This book's use or discussion of MATLAB® or Simulink® software or related products does not constitute endorsement or sponsorship by The MathWorks of a particular pedagogical approach or particular use of the MATLAB® and Simulink® software.

First edition published 2025
by CRC Press
2385 NW Executive Center Drive, Suite 320, Boca Raton, FL 33431

and by CRC Press
4 Park Square, Milton Park, Abingdon, Oxon, OX14 4RN

CRC Press is an imprint of Taylor & Francis Group, LLC

© 2025 Hang Qian

Reasonable efforts have been made to publish reliable data and information, but the author and publisher cannot assume responsibility for the validity of all materials or the consequences of their use. The authors and publishers have attempted to trace the copyright holders of all material reproduced in this publication and apologize to copyright holders if permission to publish in this form has not been obtained. If any copyright material has not been acknowledged, please write and let us know so we may rectify in any future reprint.

Except as permitted under U.S. Copyright Law, no part of this book may be reprinted, reproduced, transmitted, or utilized in any form by any electronic, mechanical, or other means, now known or hereafter invented, including photocopying, microfilming, and recording, or in any information storage or retrieval system, without written permission from the publishers.

For permission to photocopy or use material electronically from this work, access www.copyright.com or contact the Copyright Clearance Center, Inc. (CCC), 222 Rosewood Drive, Danvers, MA 01923, 978-750-8400. For works that are not available on CCC, please contact mpkbookspermissions@tandf.co.uk

Trademark notice: Product or corporate names may be trademarks or registered trademarks and are used only for identification and explanation without intent to infringe.

Library of Congress Cataloging-in-Publication Data
Names: Qian, Hang (Economist), author.
Title: Bayesian econometric modelling for big data / Hang Qian.
Description: First edition. | Boca Raton, FL : CRC Press, 2025. | Series: Chapman and Hall/CRC series on statistics in business and economics | Includes bibliographical references and index. | Summary: "This book delves into scalable Bayesian statistical methods designed to tackle the challenges posed by big data. It explores a variety of divide-and-conquer and subsampling techniques, seamlessly integrating these scalable methods into a broad spectrum of econometric models. In addition to its focus on big data, the book introduces novel concepts within traditional statistics, such as the summation, subtraction, and multiplication of conjugate distributions. These arithmetic operators conceptualize pseudo data in the conjugate prior, sufficient statistics that determine the likelihood, and the posterior as a balance between data and prior information, adding an intriguing dimension to Bayesian analysis. This book also offers a deep dive into Bayesian computation. Given the intricacies of floating-point representation of real numbers, computer programs can sometimes yield unexpected or theoretically impossible results. Drawing from his experience as a senior statistical software developer, the author shares valuable strategies for designing numerically stable algorithms. The book is an essential resource for a diverse audience: graduate students seeking foundational knowledge in Bayesian econometric models, early-career statisticians eager to explore cutting-edge advancements in scalable Bayesian methods, data analysts struggling with out-of-memory challenges in large datasets, and statistical software users and developers striving to program with efficiency and numerical stability"-- Provided by publisher.
Identifiers: LCCN 2024057171 (print) | LCCN 2024057172 (ebook) | ISBN 9781032915258 (hardback) | ISBN 9781032915838 (paperback) | ISBN 9781003564027 (ebook)
Subjects: LCSH: Bayesian statistical decision theory. | Big data.
Classification: LCC QA279.5 .Q24 2025 (print) | LCC QA279.5 (ebook) | DDC 519.5/42--dc23/eng/20250408
LC record available at https://lccn.loc.gov/2024057171
LC ebook record available at https://lccn.loc.gov/2024057172
ISBN: 978-1-032-91525-8 (hbk)
ISBN: 978-1-032-91583-8 (pbk)
ISBN: 978-1-003-56402-7 (ebk)

DOI: 10.1201/9781003564027

Typeset in CMR10 font
by KnowledgeWorks Global Ltd.

Publisher's note: This book has been prepared from camera-ready copy provided by the authors.

To my mother Xiaoping Wang

Contents

Preface

The subject of the book is scalable Bayesian statistical methods designed to address the challenges posed by big data in econometric models. Simulation-based Bayesian inference becomes computationally intensive when dealing with large datasets. It is imperative that statistical methods evolve to accommodate the surge in data volume.

The text explores a variety of strategies for scalable Bayesian inference. Some approaches are general and broadly applicable, while others are tailored to specific model structures. For example, Chapter 2 discusses divide-and-conquer strategies and subsampling methods that can be applied to the Metropolis-Hastings sampler in general. In contrast, Chapters 4 and 5 introduce double marginalized subsampling methods that exploit the conditionally conjugate structure embedded in micro-econometric models with data augmentation.

Scalable Bayesian methods cannot be developed in isolation. The book seamlessly integrates them with traditional statistical methods. A section typically begins by presenting self-contained traditional methods, followed by extensions for big data scenarios. The exposition is accessible, akin to a graduate-level textbook. Even if readers choose to bypass the sections on big data analysis, they will encounter a concise treatment of Bayesian econometrics, covering most traditional models and methods.

Even in the realm of traditional econometrics, the book introduces novel concepts. For instance, conjugate distributions serve as an entry point to Bayesian analysis. Instead of the conventional approach of analyzing the prior and posterior densities, we define arithmetic operators for conjugate distributions. While linear regressions are standard material in econometric textbooks, we present them through summation, subtraction, and multiplication of conjugate distributions. These arithmetic operators conceptualize pseudo data in the conjugate prior, sufficient statistics that determine the likelihood, and the posterior as a balance between data and prior information.

The book places significant emphasis on Bayesian computation. Due to the floating-point representation of real numbers, computer programs can sometimes yield unexpected or theoretically impossible results. Drawing from his experience as a senior statistical software developer, the author shares insights into overcoming numerical pitfalls and designing stable algorithms. These discussions are particularly highlighted in the "Software and Computational Considerations" section found in each chapter.

The book aims to be accessible to readers with some prior knowledge of elementary statistics and matrix algebra. The target audience includes 1) graduate students seeking foundational insights into Bayesian econometric models, 2) early-career statisticians interested in cutting-edge advancements in scalable Bayesian methods, 3) data analysts contending with out-of-memory problems caused by large datasets, and 4) statistical software users and developers striving to program with efficiency and numerical stability.

Prior to diving into the pool of econometric models, some readers doubt that the entry to the pool is deep. To ease the readers into the book, we provide a gentle introduction to Bayesian inference and offer a big-picture view of the models and methods covered in Chapters 1 to 8. Let us address five frequently asked questions below.

The first question is what Bayesian econometrics is and how it differs from classical (frequentist) statistics.

Econometrics involves using probability models to explain and predict economic data. In a parametric probability model, the likelihood function $p(Y|\theta)$ describes the data generating process, with parameters θ indexing the distribution of data Y. Classical statistics treats parameters as fixed but unknown quantities. In contrast, Bayesian analysts adopt a subjective view of probability to measure uncertainty about parameters. These parameters are treated as random variables, and their prior $p(\theta)$ and posterior $p(\theta|Y)$ are connected by Bayes' rule:

$$p(\theta|Y) = \frac{p(\theta)\,p(Y|\theta)}{p(Y)},$$

where $p(Y) = \int p(\theta)\,p(Y|\theta)\,d\theta$ is the marginal likelihood. As an integral over parameters, the marginal likelihood does not depend on θ and can be put into the proportionality operator $\propto$ such that

$$p(\theta|Y) \propto p(\theta)\,p(Y|\theta).$$

The marginal likelihood plays a key role in Bayesian model comparison, analogous to classical hypothesis testing. The Bayes factor, as a ratio of the marginal likelihood under two competing models (e.g., null hypothesis vs. alternative), provides evidence in favor of one model over the other. The marginal likelihood is an average of the likelihood weighted by the prior, while the posterior predictive distribution averages the likelihood weighted by the posterior:

$$p(Y_f|Y) = \int p(\theta|Y)\,p(Y_f|\theta)\,d\theta,$$

which provides Bayesian prediction on future data Y_f.

The maximum likelihood estimator (MLE) is widely used in classical econometrics. Bayes' rule establishes a connection between the MLE and Bayesian estimators: the maximizer of the posterior density (also known as the posterior mode or maximum a posteriori estimator) coincides with the maximizer of the likelihood function under a diffuse prior $p(\theta) \propto 1$. In simple models, point estimators obtained via classical and Bayesian methods tend to be numerically close. However, in complex models, Bayesian methods are computationally favorable if we only have access to an unbiased estimator by the simulated likelihood (i.e., simulated MLE).

A further link between the MLE and Bayesian estimators under the diffuse prior can be seen by the second-order Taylor approximation of the log-likelihood function:

$$p(\theta|Y) \propto \exp\left[\ln p\left(Y\middle|\hat{\theta}\right) + \frac{\partial \ln p\left(Y\middle|\hat{\theta}\right)}{\partial \hat{\theta}'}\left(\theta-\hat{\theta}\right) + \frac{1}{2}\left(\theta-\hat{\theta}\right)'\frac{\partial^2 \ln p\left(Y\middle|\hat{\theta}\right)}{\partial \hat{\theta}\partial \hat{\theta}'}\left(\theta-\hat{\theta}\right)\right].$$

If we select $\hat{\theta}$ as the MLE, the first-order term equals zero: $\frac{\partial \ln p(Y|\hat{\theta})}{\partial \hat{\theta}} = 0$. It follows that

$$p(\theta|Y) \propto \exp\left[\frac{1}{2}\left(\theta-\hat{\theta}\right)'\frac{\partial^2 \ln p\left(Y\middle|\hat{\theta}\right)}{\partial \hat{\theta}\partial \hat{\theta}'}\left(\theta-\hat{\theta}\right)\right],$$

which indicates that the posterior distribution is asymptotically normal with the mean/mode $\hat{\theta}$ and the covariance matrix $\left[-\frac{\partial^2 \ln p(Y|\hat{\theta})}{\partial \hat{\theta}\partial \hat{\theta}'}\right]^{-1}$. The negative inverse Hessian corresponds to the MLE covariance matrix. Consequently, the posterior standard deviation provides a Bayesian counterpart of the classical MLE standard error.

In summary, a brief answer to the first question is that Bayesian data analysis follows a unified approach for parameter estimation, prediction, and model comparison by Bayes' rule. The differences between classical and Bayesian inference are overshadowed by their similarities. From a practical perspective, most users resort to Bayesian estimation out of computational advantages rather than philosophically sound foundations. Even a non-Bayesian user can leverage Bayesian methods for replicating the MLE and its standard error for statistical inference. Bayes' rule is used extensively throughout the book, in computing the posterior distribution (see Chapter 1), in specifying the target distribution of the Markov Chain for posterior simulation (see Chapter 2), in double marginalized subsampling (see Chapter 4), in deriving the Kalman filter (see Chapter 6), in designing efficient particle filtering algorithms (see Chapter 7), and so on.

The second question is what priors are and how they are selected.

Bayesian analysis begins with a probability statement about the prior, which measures parameter uncertainty, reflects personal belief on weights given to each possible parameter value, and summarizes the existing knowledge about parameters before seeing data. The appropriate methods for prior specification can vary across different models. We suggest five practical methods tailored to econometric modeling.

First, incorporating domain knowledge in the prior. Econometrics, at its core, combines economic theory and mathematical statistics. Econometric analysis of Dynamic Stochastic General Equilibrium (DSGE) in Section 8.8 exemplifies prior elicitation by expert opinion, common knowledge, and previous economic studies. DSGE models typically involve many interpretable parameters familiar to economists, while the sample size of macroeconomic data is relatively small. Tight priors are clearly documented in empirical studies. For instance, Smets and Wouters (2007) specify a normal prior for the intertemporal elasticity of substitution at 1.5 with a standard deviation of 0.375. Although tight priors may face criticism, they are valuable for regularizing parameter estimation to prevent unreasonable results. Section 5.2 provides another example of Bayesian regularization in addressing ill-behaved MLE in logistic regressions with separation problems.

Second, imposing hard and soft parameter constraints by priors. A hard constraint must be satisfied at all times, whereas a soft constraint is preferable if the enforcement cost is not prohibitive. The prior distribution inherently incorporates soft constraints by assigning weights to possible parameter values. As an example of hard constraints, many time series models discussed in Chapter 8 have vector autoregressive components, of which eigenvalues must be in the unit circle for stationarity. A simple method of imposing nonlinear inequality constraints on parameters is to truncate the prior distribution:

$$p(\theta) \propto \tilde{p}(\theta) I(\theta \in \Theta),$$

where $\tilde{p}(\theta)$ is a prior without truncation, and the indicator function $I(\theta \in \Theta)$ constrains parameters in an appropriate set. By Bayes' rule, truncation is transferred to the posterior:

$$p(\theta|Y) \propto \tilde{p}(\theta) p(Y|\theta) I(\theta \in \Theta).$$

Third, assimilating priors into modeling assumptions. Chapters 6 and 7 are devoted to linear and nonlinear state space models, of which initialization is an indispensable part of model assumptions. The initial state distribution serves as a prior on state variables. The Kalman filter provides an excellent demonstration of prior washout as observations are sequentially incorporated. Another example of assimilating priors into model specification is panel data regression (see Section 8.6). The distinction between random and fixed effects models (in econometric terminology) stems from different priors on unobserved effects: a hierarchical prior for random effects and a diffuse prior for fixed effects. The diffuse prior allows for arbitrary dependence between unobserved effects and observed regressors. Additional examples of priors versus model assumptions can be found in Chapter 3. Different regularization methods correspond to distinct prior specifications: normal priors lead to ridge regressions (see Section 3.2), while double exponential priors lead to lasso regressions (see Section 3.3).

Fourth, avoiding subjectivity by non-informative priors. It may be tempting to specify $p(\theta) \propto 1$ so that the classical MLE aligns with Bayesian estimation. Another approach is constructing a non-informative prior using the Fisher information matrix (Jeffreys, 1946). A novel interpretation of the non-informative prior is presented in Chapter 1, where summation operations are defined over conjugate distributions. The non-informative prior serves as a zero in summation.

Fifth, eliciting priors in terms of pseudo data. A prominent example is Bayesian vector autoregressions (see Section 8.3), which can be implemented without Bayesian software by concatenating observations with pseudo data extracted from the prior for standard linear regressions. A novelty of Chapter 1 is to show pseudo data extracted from various distributions in the exponential family. The process of concatenating pseudo and genuine data corresponds to summation operation of conjugate distributions.

In summary, a brief answer to the second question is that an informative prior serves as a regularization tool to prevent unreasonable results, while a non-informative prior makes Bayesian estimation competitive and comparable to the MLE, as it can replicate the latter at a lower computational cost. A method of prior elicitation involves casting priors as pseudo data that supplement observations for statistical inference. There is no clear dividing line between priors and model assumptions, as priors can be interpreted as honest statements on assumptions of a probability model.

The third question is how the posterior is used to obtain point and interval estimates.

In Bayesian statistics, parameters are conceptualized as random variables characterized by prior and posterior distributions, which reflect the uncertainty surrounding unknown parameters. While it may be appealing to estimate the random variable θ by a specific point estimate $\hat{\theta}$, estimation error and utility loss are inevitable. A Bayesian point estimator is defined as the minimizer of an expected loss function: $E\left[f\left(\theta-\hat{\theta}\right)|Y\right]$. The most prevalent loss function takes a quadratic form: $f\left(\theta-\hat{\theta}\right)=\left(\theta-\hat{\theta}\right)^2$, for which the minimizer is the posterior mean: $\hat{\theta}=E(\theta|Y)$. Alternatively, when employing an absolute-value loss function $f\left(\theta-\hat{\theta}\right)=\left|\theta-\hat{\theta}\right|$, the posterior median becomes the minimizer. It is also feasible to specify an all-or-nothing loss function $f\left(\theta-\hat{\theta}\right)=I\left(\left|\theta-\hat{\theta}\right|>k\right)$, where the posterior mode tends to be the minimizer as $k\to 0$.

In classical statistics, a confidence interval is interpreted in terms of coverage: a fixed parameter is covered by random interval endpoints with a certain probability. Conversely, in Bayesian inference, the credible interval offers a more intuitive interpretation: a random parameter falls within a fixed interval with a specific probability. Two primary types of credible intervals are frequently utilized in practice. The equal-tail credible interval (a,b) is defined by the condition:

$$p(\theta<a|Y)=p(\theta>b|Y)=\frac{\alpha}{2},$$

where α is a significance level, commonly set at 0.05. The highest posterior density (HPD) credible set is defined as the smallest set with the probability of $1-\alpha$. Notably, the HPD set can comprise a union of disjoint intervals, with points included having higher density than those excluded.

In summary, a brief answer to the third question is that the posterior mean, median, and mode can serve as point estimators, contingent upon the specification of the loss function. Bayesian interval estimates, whether equal-tail or HPD intervals, provide a more intuitive alternative compared to confidence intervals in classical statistics.

The fourth question is what tools are used in posterior inference.

Posterior distributions are analytically tractable in a few conjugate models, such as linear regressions, Beta-Bernoulli trials, and Gamma-Poisson sampling. Section 1.9 addresses these models within a unified framework of arithmetic operations on conjugate distributions in the exponential family.

For most econometric models discussed in Chapters 3 to 8, posteriors do not conform to common distributions. Thus, contemporary Bayesian statistics heavily relies on Markov Chain Monte Carlo (MCMC) methods. Chapter 2 explores a variety of MCMC techniques, including the Metropolis-Hastings, Gibbs and slice samplers, Hamiltonian Monte Carlo (including Langevin dynamics), multiple-try Metropolis sampler for parallel computing, trans-dimensional methods, and pseudo-marginal methods. A significant portion of Chapter 2 is dedicated to big data MCMC methods, utilizing divide-and-conquer strategies and subsampling.

In time series econometrics, Sequential Monte Carlo (SMC) provides a powerful tool for state and parameter estimation. Chapter 7 presents SMC methods based on importance sampling and resampling. MCMC kernels can be incorporated into SMC proposals for rejuvenating particles (Gilks and Berzuini, 2001; Chopin, 2002), and SMC can be integrated into MCMC for evaluating an intractable target density by an unbiased estimator (Andrieu and Roberts, 2009; Andrieu et al., 2010).

In summary, a brief answer to the fourth question is that MCMC and SMC have emerged as main tools for Bayesian data analysis by sampling from high dimensional distributions. However, a complex model often has conditionally conjugate sub-models, in which posterior conditional distributions are analytically tractable. Analytical results should be consulted for marginalization (i.e., integrating out some variables in closed form) to reduce computational burden of Monte Carlo simulation.

The fifth question concerns the mathematical background required to study Bayesian econometrics.

The book is intended to be accessible to readers with some prior knowledge on calculus, matrix algebra, and elementary statistics (measure-theoretic treatment of probability is not required). Throughout the text, we avoid making propositions without proofs, which are compiled in a separate section at the end of each chapter.

There are different styles of presenting Bayesian statistical methods. For instance, pseudo-marginal MCMC methods offer advantages over simulated MLE. The seminal papers by Andrieu and Roberts (2009) and Andrieu et al. (2010) provide formal justifications for particle MCMC theory, though they may be challenging for some readers. Flury and Shephard (2011) present the same topic in a more accessible manner, focusing directly on the core issues. If readers find the presentation in Flury and Shephard (2011) approachable, they should encounter little technical difficulty in engaging with this book.

My gratitude is due to David Grubbs and Curtis Hill for their excellent editorial work and helpful suggestions that greatly improve the book quality. I am also grateful to Rick Baker, Bill Mueller, Kevin Shea, and David Resendes for their generous help in toolbox development. If readers have questions and comments, please send emails to matlabist@gmail.com.

Hang Qian

Natick, Massachusetts

November 2024

1

Linear Regressions

Consider the scenario where the response variable Y_i and predictors (i.e., explanatory variables) X_i are random samples from a distribution $p(X_i, Y_i | \theta)$, with θ representing a set of parameters that have a prior $p(\theta)$. The notation $p(\cdot)$ is employed in a generic context, rather than referring to particular functions. Econometric models vary in assumptions on the conditional distribution of the response variable $p(Y_i | \theta, X_i)$, contingent upon the data types being analyzed (e.g., skewed and fat-tailed data in Chapter 4; binary, multinomial, truncated, censored, count, and duration data in Chapter 5). This chapter focuses on Gaussian linear regressions with continuous and unconstrained data, assuming the conditional distribution:

$$p(Y_i | \theta, X_i) = \frac{1}{\sqrt{2\pi}\sigma} e^{-\frac{(Y_i - X_i\beta)^2}{2\sigma^2}}, \tag{1.1}$$

where $\theta = (\beta, \sigma^2)$.

Since $X_i\beta$ and σ are the location and scale parameters of the normal distribution, standardization by defining $\varepsilon_i = \sigma^{-1}(Y_i - X_i\beta)$ yields the regression form of Equation (1.1):

$$Y_i = X_i\beta + \sigma\varepsilon_i, \tag{1.2}$$

where the disturbance term ε_i follows a standard normal distribution.

Linear regressions serve two primary purposes: predicting the response variable and analyzing the marginal effects of explanatory variables. The linear regression functions as a conditional mean model:

$$E(Y_i | \theta, X_i) = X_i\beta,$$

where the marginal effects, represented as $\frac{\partial E(Y_i|\theta, X_i)}{\partial X_i}$, are captured by the coefficients β. If the coefficients were known, our prediction would be $\hat{Y}_i = X_i\beta$. However, it is essential to incorporate parameter uncertainty into posterior inference. This chapter will elucidate the process of marginalizing over parameters to derive the posterior predictive distribution.

The linear regression model is a fundamental component of every Bayesian econometrics textbook. We approach it from a novel perspective, emphasizing the arithmetic operations of summation, subtraction, and multiplication applied to conjugate distributions. Utilizing arithmetic operators simplifies Bayesian analysis, even with large datasets, and adds an intriguing dimension to our exploration of Bayesian econometric models.

1.1 Conjugate Distributions

Stacking n observations ($i = 1, \ldots, n$) in Equation (1.2) results in a vector/matrix representation of the linear regression:

$$Y = X\beta + \sigma\varepsilon,$$

DOI: 10.1201/9781003564027-1

where $Y = \begin{pmatrix} Y_1 \\ \cdots \\ Y_n \end{pmatrix}$, $X = \begin{pmatrix} X_1 \\ \cdots \\ X_n \end{pmatrix}$, and $\varepsilon = \begin{pmatrix} \varepsilon_1 \\ \cdots \\ \varepsilon_n \end{pmatrix}$.

1.1.1 The Prior

The likelihood function represents the joint probability density of observations, given the parameters of a model. Assuming random sampling, Equation (1.1) provides the normal likelihood function:

$$p\left(Y \mid \beta, \sigma^2\right) \propto \left(\sigma^2\right)^{-\frac{n}{2}} e^{-\frac{1}{2}\sigma^{-2}(Y-X\beta)'(Y-X\beta)}, \tag{1.3}$$

without conditioning on X explicitly denoted hereafter. Bayesian model specification is completed by incorporating a prior distribution, balancing flexibility and tractability. The natural conjugate prior to the normal likelihood is the normal-inverse-gamma (NIG) distribution.

Definition 1.1 (NIG) *In linear regressions with d predictors, the coefficients β and the disturbance variance σ^2 follow the $NIG\left(\mu, \Lambda, a, b\right)$ distribution if*

$$p\left(\beta, \sigma^2\right) \propto \left(\sigma^2\right)^{-\left(a+\frac{d}{2}+1\right)} e^{-\sigma^{-2}\left[b+\frac{1}{2}(\beta-\mu)'\Lambda(\beta-\mu)\right]}. \tag{1.4}$$

As the name suggests, the NIG distribution is a combination of multivariate normal and inverse gamma distributions. Marginal-conditional decomposition yields $p\left(\beta, \sigma^2\right) = p\left(\beta \mid \sigma^2\right) p\left(\sigma^2\right)$, where

$$p\left(\beta \mid \sigma^2\right) \propto \left(\sigma^2\right)^{-\frac{d}{2}} e^{-\frac{1}{2}\sigma^{-2}(\beta-\mu)'\Lambda(\beta-\mu)},$$

$$p\left(\sigma^2\right) \propto \left(\sigma^2\right)^{-(a+1)} e^{-\sigma^{-2}b}.$$

The former represents a multivariate normal distribution with mean μ and precision $\sigma^{-2}\Lambda$, where the covariance matrix is $\sigma^2\Lambda^{-1}$. The latter corresponds to an inverse gamma distribution with hyperparameters a and b. Exercise 1.1 provides a detailed exploration of the NIG distribution. The hyperparameter μ represents the prior mean of coefficients β, and the hyperparameter Λ quantifies parameter uncertainty: a smaller Λ indicates weaker prior belief regarding the coefficients. In the limiting case where $\Lambda = 0$, the prior density curve becomes flat, implying a lack of prior knowledge.

A common question arises as to why one might not specify the normal and inverse gamma distributions independently. The straightforward answer is the desire for a closed-form posterior that retains the NIG form. While it might be tempting to assume an independent prior $p\left(\beta, \sigma^2\right) = p\left(\beta\right) p\left(\sigma^2\right)$, this only achieves conditional conjugacy: the posterior conditional distributions remain normal and inverse gamma, but the posterior joint distribution does not take a recognizable form (see Exercise 1.2).

1.1.2 The Posterior

By Bayes' rule, the posterior is proportional to the likelihood function multiplied by the prior:

$$p\left(\beta, \sigma^2 \mid Y\right) = \frac{p\left(\beta, \sigma^2\right) p\left(Y \mid \beta, \sigma^2\right)}{\int\int p\left(\beta, \sigma^2\right) p\left(Y \mid \beta, \sigma^2\right) d\beta d\sigma^2},$$

where the integral in the denominator does not depend on β and σ^2, thus it need not be computed. The numerator is the product of Equations (1.3) and (1.4):

$$p\left(\beta,\sigma^2\left|Y\right.\right)\propto\left(\sigma^2\right)^{-\left(a+\frac{n+d}{2}+1\right)}e^{-\sigma^{-2}\left[b+\frac{1}{2}(\beta-\mu)'\Lambda(\beta-\mu)+\frac{1}{2}(Y-X\beta)'(Y-X\beta)\right]}.$$

Completing the square for the quadratic function of β yields

$$(\beta-\mu)'\Lambda(\beta-\mu)+(Y-X\beta)'(Y-X\beta)=(\beta-\bar{\mu})'\bar{\Lambda}(\beta-\bar{\mu})-\bar{\mu}'\bar{\Lambda}\bar{\mu}+\mu'\Lambda\mu+Y'Y,$$

where $\bar{\mu}$ and $\bar{\Lambda}$ are given by Equations (1.5) and (1.6). This completes our derivation of the posterior distribution, which preserves the NIG form as summarized in Proposition 1.1.

Proposition 1.1 (Conjugate Bayesian linear regression) *If β and σ^2 have the prior $NIG(\mu,\Lambda,a,b)$, then the posterior is $NIG(\bar{\mu},\bar{\Lambda},\bar{a},\bar{b})$, where*

$$\bar{\mu}=(\Lambda+X'X)^{-1}(\Lambda\mu+X'Y), \tag{1.5}$$

$$\bar{\Lambda}=\Lambda+X'X, \tag{1.6}$$

$$\bar{a}=a+\frac{n}{2}, \tag{1.7}$$

$$\bar{b}=b+\frac{1}{2}Y'Y+\frac{1}{2}\mu'\Lambda\mu-\frac{1}{2}\bar{\mu}'\bar{\Lambda}\bar{\mu}. \tag{1.8}$$

Some researchers prefer a non-informative prior to maintain objectivity in statistical inference, allowing the data to "speak for themselves" and rationalizing maximum likelihood estimation within the Bayesian framework (Jeffreys, 1946; Bernardo, 1979; Berger, 1985). A typical non-informative prior takes the form $p(\beta,\sigma^2)\propto\sigma^{-2}$, represented by an improper distribution $NIG\left(\mu_0,0_{d\times d},-\frac{d}{2},0\right)$, for an arbitrary $d\times 1$ vector μ_0. Readers may wonder about the term $-\frac{d}{2}$ in the non-informative prior, which stems from Definition 1.1, where β and σ^2 are not independent and $p\left(\beta\left|\sigma^2\right.\right)$ involves the term $\left(\sigma^2\right)^{-\frac{d}{2}}$.

Under the non-informative prior, the posterior is given by Proposition 1.2, as a corollary of Proposition 1.1. The posterior mean, median, and mode of β coincide with the ordinary least squares (OLS) and the maximum likelihood estimators, and the posterior standard deviation nearly matches the classical standard error.

Proposition 1.2 (Regression with non-informative prior) *If β and σ^2 have the non-informative prior $p(\beta,\sigma^2)\propto\sigma^{-2}$, then the posterior is $NIG\left(\tilde{\mu},\tilde{\Lambda},\tilde{a},\tilde{b}\right)$, where*

$$\tilde{\mu}=(X'X)^{-1}X'Y,$$

$$\tilde{\Lambda}=X'X,$$

$$\tilde{a}=\frac{n-d}{2},$$

$$\tilde{b}=\frac{1}{2}Y'Y-\frac{1}{2}Y'X(X'X)^{-1}X'Y.$$

Throughout this chapter, upper-bar variables (e.g., $\bar{\mu}$) denote the posterior under an informative (proper) prior, and tilde-form variables (e.g., $\tilde{\mu}$) represent the posterior under a non-informative prior.

1.1.3 Pseudo Data

The NIG distribution can be viewed as a repository of compressed data in the form of sufficient statistics, with the non-informative distribution $NIG\left(\mu_0, 0_{d\times d}, -\frac{d}{2}, 0\right)$ as an empty warehouse. Once data X, Y are introduced, they are compressed into the sufficient statistics $X'X$, $X'Y$, and $Y'Y$, and saved in the tilde-form posterior $NIG\left(\tilde{\mu}, \tilde{\Lambda}, \tilde{a}, \tilde{b}\right)$. By compression, we mean the dimensions of the sufficient statistics do not increase with the number of observations. The sizes of X and Y are $n \times d$ and $n \times 1$, respectively, while the sizes of the sufficient statistics are at most $d \times d$, regardless of the sample size n.

Proposition 1.2 poses an inverse problem that extracts data saved in the tilde-form posterior. Due to data compression, the extracted data are unique up to the sufficient statistics, hence, the name pseudo data in Proposition 1.3.

Proposition 1.3 (Pseudo data in NIG) *Consider d-dimensional* $NIG\left(\tilde{\mu}, \tilde{\Lambda}, \tilde{a}, \tilde{b}\right)$, *where* $\tilde{\Lambda}$ *is positive definite,* $\tilde{a} > 0$, $\tilde{b} > 0$ *and* $n = 2\tilde{a} + d$ *is a positive integer. The pseudo data take the form*

$$X = \begin{pmatrix} \tilde{\Lambda}^{1/2} \\ 0_{(n-d)\times d} \end{pmatrix}, Y = \begin{pmatrix} \tilde{\Lambda}^{1/2}\tilde{\mu} \\ \sqrt{\tilde{b}/\tilde{a}} \cdot 1_{(n-d)\times 1} \end{pmatrix},$$

where $\tilde{\Lambda}^{1/2}$ *is the upper triangular Cholesky-like factor that satisfies* $\tilde{\Lambda}^{1/2\prime}\tilde{\Lambda}^{1/2} = \tilde{\Lambda}$. *Under the non-informative prior and pseudo data, the posterior distribution aligns with* $NIG\left(\tilde{\mu}, \tilde{\Lambda}, \tilde{a}, \tilde{b}\right)$. *Moreover, if another dataset* X^*, Y^* *yields the same posterior distribution, the data must satisfy* $X^{*\prime}X^* = \tilde{\Lambda}$, $X^{*\prime}Y^* = \tilde{\Lambda}\tilde{\mu}$ *and* $Y^{*\prime}Y^* = 2\tilde{b} + \tilde{\mu}'\tilde{\Lambda}\tilde{\mu}$.

Prior elicitation is a critical aspect of practical Bayesian analysis (Garthwaite et al., 2005), requiring the specification of hyperparameters μ, Λ, a, b before performing regressions. The pseudo-data interpretation of the NIG distribution shifts our focus from plausible hyperparameter values to pseudo data underlying the NIG distribution. Specifying hyperparameters becomes a process of recalling subjective data prior to using actual observations.

1.2 NIG Summation

The pseudo-data interpretation of conjugate distributions motivates the definition of the NIG summation operator. The sum of two NIGs equals the tilde-form posterior obtained by concatenating the data extracted from both distributions. Alternatively, one NIG can serve as the prior, with data extracted from the other. The upper-bar posterior corresponds to the sum of NIGs.

Definition 1.2 (NIG summation) *Consider two d-dimensional* $NIG(\mu_1, \Lambda_1, a_1, b_1)$ *and* $NIG(\mu_2, \Lambda_2, a_2, b_2)$. *If a distribution* $NIG(\mu, \Lambda, a, b)$ *satisfies*

1. $\mu = (\Lambda_1 + \Lambda_2)^{-1}(\Lambda_1\mu_1 + \Lambda_2\mu_2)$,
2. $\Lambda = \Lambda_1 + \Lambda_2$,
3. $a = a_1 + a_2 + \frac{d}{2}$,
4. $b = b_1 + b_2 + \frac{1}{2}(\mu_1 - \mu)'\Lambda_1(\mu_1 - \mu) + \frac{1}{2}(\mu_2 - \mu)'\Lambda_2(\mu_2 - \mu)$,

it is the sum of NIG distributions, denoted by

$$NIG\left(\mu, \Lambda, a, b\right) = NIG\left(\mu_1, \Lambda_1, a_1, b_1\right) + NIG\left(\mu_2, \Lambda_2, a_2, b_2\right),$$

or more compactly,

$$NIG\left(\mu, \Lambda, a, b\right) = \sum_{i=1}^{2} NIG\left(\mu_i, \Lambda_i, a_i, b_i\right).$$

1.2.1 Zero in Summation

The sum of integers results in the total value combined, and adding zero leaves the total unchanged. The NIG summation operator exhibits similar properties.

Proposition 1.4 (Properties of NIG summation) *The NIG summation operator satisfies commutativity, associativity, and additive identity.*

1. *Commutativity:*

$$NIG\left(\mu_1, \Lambda_1, a_1, b_1\right) + NIG\left(\mu_2, \Lambda_2, a_2, b_2\right) = NIG\left(\mu_2, \Lambda_2, a_2, b_2\right) + NIG\left(\mu_1, \Lambda_1, a_1, b_1\right).$$

2. *Associativity:*

$$\sum_{i=1}^{3} NIG\left(\mu_i, \Lambda_i, a_i, b_i\right) = NIG\left(\mu_1, \Lambda_1, a_1, b_1\right) + \sum_{i=2}^{3} NIG\left(\mu_i, \Lambda_i, a_i, b_i\right).$$

3. *Additive identity:*

$$NIG\left(\mu, \Lambda, a, b\right) + NIG\left(\mu_0, 0_{d\times d}, -\frac{d}{2}, 0\right) = NIG\left(\mu, \Lambda, a, b\right). \tag{1.9}$$

Proposition 1.4 indicates that the operator satisfies the usual properties of summation in mathematics. A striking feature is that the non-informative distribution $NIG\left(\mu_0, 0_{d\times d}, -\frac{d}{2}, 0\right)$ fulfills the central role of zero in the NIG summation. Three interpretations of Equation (1.9) are available. First, extract data from $NIG\left(\mu, \Lambda, a, b\right)$ and perform Bayesian linear regression under the non-informative prior; the posterior remains $NIG\left(\mu, \Lambda, a, b\right)$. Second, extract trivial data (i.e., empty matrices with zero sample size) from $NIG\left(\mu_0, 0_{d\times d}, -\frac{d}{2}, 0\right)$ and perform Bayesian linear regression with the prior $NIG\left(\mu, \Lambda, a, b\right)$. By Proposition 1.1, the posterior remains unchanged. Third, extract data from both NIGs, concatenate them, and run Bayesian linear regression under the non-informative prior; the posterior is $NIG\left(\mu, \Lambda, a, b\right)$.

1.2.2 Bridge for Bayesian and Classical Estimators

A frequent critique of Bayesian statistics is its reliance on subjective priors in statistical inference. Proposition 1.5 clarifies the distinct roles of the prior and the data in the posterior NIG distribution, bridging the gap between Bayesian and classical estimators. The upper-bar posterior mean represents a typical Bayesian estimator, while the tilde-form posterior mean is analogous to the classical OLS estimator. Proposition 1.5 can be informally interpreted as a bridge: Bayesian estimator = prior + classical estimator.

Proposition 1.5 (Separation of prior and data in posterior NIG) *The priors and posteriors specified in Propositions 1.1 and 1.2 are connected by the equation:*

$$NIG\left(\bar{\mu}, \bar{\Lambda}, \bar{a}, \bar{b}\right) = NIG\left(\mu, \Lambda, a, b\right) + NIG\left(\tilde{\mu}, \tilde{\Lambda}, \tilde{a}, \tilde{b}\right).$$

Algorithm 1.1 Divide-and-conquer big data regression

Input: data X, Y and prior $NIG(\mu, \Lambda, a, b)$.
Output: posterior $NIG(\bar{\mu}, \bar{\Lambda}, \bar{a}, \bar{b})$ in Proposition 1.1.
1 Partition big data into J subsets X_j, Y_j, $j = 1, \ldots, J$.
2 Compute subset tilde-form posteriors $NIG\left(\tilde{\mu}_j, \tilde{\Lambda}_j, \tilde{a}_j, \tilde{b}_j\right)$.
3 Sum up subset tilde-form posteriors

$$NIG\left(\tilde{\mu}, \tilde{\Lambda}, \tilde{a}, \tilde{b}\right) = \sum_{j=1}^{J} NIG\left(\tilde{\mu}_j, \tilde{\Lambda}_j, \tilde{a}_j, \tilde{b}_j\right).$$

4 Sum up prior and tilde-form posterior for upper-bar posterior

$$NIG(\bar{\mu}, \bar{\Lambda}, \bar{a}, \bar{b}) = NIG(\mu, \Lambda, a, b) + NIG\left(\tilde{\mu}, \tilde{\Lambda}, \tilde{a}, \tilde{b}\right).$$

1.2.3 Big Data Regressions

A notable feature of the NIG summation is its scalability for big-data Bayesian linear regressions and variable selection via a divide-and-conquer algorithm, which can be processed independently in parallel. Proposition 1.6 decomposes the tilde-form posteriors. Data are divided into multiple subsets for separate regressions, and the subset regression results are merged using the NIG summation operator.

Proposition 1.6 (Subset posteriors of NIG) *Partition data X, Y into J subsets X_j, Y_j, $j = 1, \ldots, J$. Let the subset tilde-form posterior distributions be $NIG\left(\tilde{\mu}_j, \tilde{\Lambda}_j, \tilde{a}_j, \tilde{b}_j\right)$. The full-sample tilde-form posterior $NIG\left(\tilde{\mu}, \tilde{\Lambda}, \tilde{a}, \tilde{b}\right)$ can be computed as*

$$NIG\left(\tilde{\mu}, \tilde{\Lambda}, \tilde{a}, \tilde{b}\right) = \sum_{j=1}^{J} NIG\left(\tilde{\mu}_j, \tilde{\Lambda}_j, \tilde{a}_j, \tilde{b}_j\right).$$

Propositions 1.5 and 1.6 lead to the divide-and-conquer (also known as split-and-merge) Algorithm 1.1 for big data Bayesian linear regressions. The methodology involves two-stage regressions. In the first stage, the big data are split into subsets for distributed processing. The subset tilde-form posterior distributions are summed. The prior information is not considered until the second stage, where the prior and the combined subset posterior distributions are merged using the NIG summation operator.

Several important aspects of Algorithm 1.1 warrant discussion. First, the subset data can vary in size, yet the summation results remain unaffected by the partition methods employed. Second, there is no dependency between subset regressions, allowing them to be processed independently across multiple computational threads without requiring the exchange of intermediate variables. Third, the NIG summation operator is associative, enabling Step 3 to be processed recursively in pairs, which aligns with a divide-and-conquer approach. For instance, $\sum_{i=1}^{4} NIG(\mu_i, \Lambda_i, a_i, b_i)$ can be processed by computing the subtotals $\sum_{i=1}^{2} NIG(\mu_i, \Lambda_i, a_i, b_i)$ and $\sum_{i=3}^{4} NIG(\mu_i, \Lambda_i, a_i, b_i)$, and then adding up the subtotals. Fourth, commutativity and associativity of the summation operator justify online updates for streaming data. When new data X_{J+1}, Y_{J+1} arrive, an update is made:

$$NIG(\bar{\mu}, \bar{\Lambda}, \bar{a}, \bar{b}) + NIG\left(\tilde{\mu}_{J+1}, \tilde{\Lambda}_{J+1}, \tilde{a}_{J+1}, \tilde{b}_{J+1}\right),$$

equivalent to the regression results under the concatenated data

$$NIG(\mu, \Lambda, a, b) + \sum_{j=1}^{J+1} NIG\left(\tilde{\mu}_j, \tilde{\Lambda}_j, \tilde{a}_j, \tilde{b}_j\right).$$

Additionally, Algorithm 1.1 facilitates secure information exchange. After proprietary data are processed by distributed units, the aggregated regression results can be obtained by exchanging the subset regression outputs without revealing the private data. Refer to Ghosh and Reiter (2013) for secure Bayesian model averaging of linear regressions.

An application of Algorithm 1.1 involves big data ridge regression (Hoerl and Kennard, 1970), which can be interpreted as Bayesian linear regression under a conjugate prior $NIG(0, \Lambda, a, b)$, where $\Lambda = diag(\lambda_1, \ldots, \lambda_d)$ comprises regularization parameters. As observed from the posterior density

$$p\left(\beta, \sigma^2 | Y\right) \propto \left(\sigma^2\right)^{-\left(a+\frac{n+d}{2}+1\right)} e^{-\frac{1}{2}\sigma^{-2}\left[2b+(Y-X\beta)'(Y-X\beta)+\sum_{j=1}^{d}\lambda_j\beta_j^2\right]},$$

the posterior mode of β minimizes the objective function in the classical ridge regression

$$\min_{\beta} (Y - X\beta)' (Y - X\beta) + \sum_{j=1}^{d} \lambda_j \beta_j^2.$$

For the normal distribution, the posterior mode coincides with the posterior mean $(\Lambda + X'X)^{-1} X'Y$, which corresponds to the classical ridge point estimator, regardless of the hyperparameters a and b. Further discussion on Bayesian shrinkage will be provided in Chapter 3.

1.3 NIG Subtraction

The NIG distribution functions as a repository of compressed data in the form of sufficient statistics. The summation operator adds data to this repository, while the subtraction operator extracts data from it.

1.3.1 Subtraction as Reversed Summation

The NIG subtraction operator is defined by reverting Definition 1.2.

Definition 1.3 (NIG subtraction) *Consider d-dimensional distributions* $NIG(\mu, \Lambda, a, b)$ *and* $NIG(\mu_1, \Lambda_1, a_1, b_1)$*. If a distribution* $NIG(\mu_2, \Lambda_2, a_2, b_2)$ *satisfies*

1. $\mu_2 = (\Lambda - \Lambda_1)^{-1} (\Lambda\mu - \Lambda_1\mu_1)$,
2. $\Lambda_2 = \Lambda - \Lambda_1$,
3. $a_2 = a - a_1 - \frac{d}{2}$,
4. $b_2 = b - b_1 - \frac{1}{2}(\mu_1 - \mu)' \Lambda_1 (\mu_1 - \mu) - \frac{1}{2}(\mu_2 - \mu)' \Lambda_2 (\mu_2 - \mu)$,

it is the difference of NIG distributions, denoted by

$$NIG(\mu_2, \Lambda_2, a_2, b_2) = NIG(\mu, \Lambda, a, b) - NIG(\mu_1, \Lambda_1, a_1, b_1). \quad (1.10)$$

Definition 1.3 indicates that subtraction is the inverse operation of summation, as Equation (1.10) is equivalent to

$$NIG(\mu, \Lambda, a, b) = NIG(\mu_1, \Lambda_1, a_1, b_1) + NIG(\mu_2, \Lambda_2, a_2, b_2).$$

Subtraction also has a pseudo-data interpretation. By Proposition 1.3, pseudo data X, Y can be extracted from $NIG(\mu, \Lambda, a, b)$, and similarly X_1, Y_1 from $NIG(\mu_1, \Lambda_1, a_1, b_1)$. For any pseudo data X_2, Y_2 with sufficient statistics $X_2'X_2 = X'X - X_1'X_1$, $X_2'Y_2 = X'Y - X_1'Y_1$, and $Y_2'Y_2 = Y'Y - Y_1'Y_1$, regressions using pseudo data X_2, Y_2 under the non-informative prior yields the posterior $NIG(\mu_2, \Lambda_2, a_2, b_2)$.

1.3.2 Applications of Subtraction

The NIG subtraction operator finds broad applicability. We briefly discuss its use in cross-validation (Hastie et al., 2009) and rolling-window regressions (Swanson, 1998) with multicollinearity (Farrar and Glauber, 1967).

The NIG subtraction operator facilitates k-fold cross-validation in Bayesian linear regressions. Data are partitioned into k subsets: X_j, Y_j, $j = 1, \ldots, k$; one reserved for testing and the remainder for parameter estimation. The process is repeated k times, with each subset used once as validation data. Instead of repeated estimation, an efficient approach is to estimate the model once with the full-sample data for $NIG(\bar{\mu}, \bar{\Lambda}, \bar{a}, \bar{b})$ by Algorithm 1.1. The NIG subtraction operation $NIG(\bar{\mu}, \bar{\Lambda}, \bar{a}, \bar{b}) - NIG\left(\tilde{\mu}_j, \tilde{\Lambda}_j, \tilde{a}_j, \tilde{b}_j\right)$ fulfills the goal of excluding the jth subset from the training data.

Another application of the NIG subtraction operator is in rolling-window regressions. Suppose we have daily data and update regressions monthly with the most recent data of $m + 1$ months. Let $NIG(\mu, \Lambda, a, b)$ represent the prior and $NIG\left(\tilde{\mu}_j, \tilde{\Lambda}_j, \tilde{a}_j, \tilde{b}_j\right)$ be the subset posterior by the jth month data alone. The posterior in period t is given by

$$NIG(\bar{\mu}_t, \bar{\Lambda}_t, \bar{a}_t, \bar{b}_t) = NIG(\mu, \Lambda, a, b) + \sum_{j=t-m}^{t} NIG\left(\tilde{\mu}_j, \tilde{\Lambda}_j, \tilde{a}_j, \tilde{b}_j\right).$$

In the subsequent month, the rolling-window regression updates by incorporating new data and removing old data:

$$\begin{aligned} NIG(\bar{\mu}_{t+1}, \bar{\Lambda}_{t+1}, \bar{a}_{t+1}, \bar{b}_{t+1}) &= NIG(\bar{\mu}_t, \bar{\Lambda}_t, \bar{a}_t, \bar{b}_t) \\ &+ NIG\left(\tilde{\mu}_{t+1}, \tilde{\Lambda}_{t+1}, \tilde{a}_{t+1}, \tilde{b}_{t+1}\right) \\ &- NIG\left(\tilde{\mu}_{t-m}, \tilde{\Lambda}_{t-m}, \tilde{a}_{t-m}, \tilde{b}_{t-m}\right). \end{aligned}$$

Additionally, the NIG subtraction operator can address perfect collinearity in linear regressions. Suppose predictors include an intercept and monthly dummy variables. Subset regressions using j^{th} month data alone suffer from perfect collinearity (though not in the full-sample regression). This issue can be addressed by borrowing a prior for each subset regression:

$$NIG\left(\ddot{\mu}_j, \ddot{\Lambda}_j, \ddot{a}_j, \ddot{b}_j\right) = NIG(\mu, \Lambda, a, b) + NIG\left(\tilde{\mu}_j, \tilde{\Lambda}_j, \tilde{a}_j, \tilde{b}_j\right).$$

Naturally, borrowed priors must be repaid. The m duplicated priors should be subtracted from the posterior:

$$NIG(\bar{\mu}_t, \bar{\Lambda}_t, \bar{a}_t, \bar{b}_t) = \sum_{j=t-m}^{t} NIG\left(\ddot{\mu}_j, \ddot{\Lambda}_j, \ddot{a}_j, \ddot{b}_j\right) - \sum_{j=1}^{m} NIG(\mu, \Lambda, a, b).$$

1.4 NIG Scalar Multiplication

Multiplication is a fundamental arithmetic operation. We extend the NIG summation to multiplication and discuss its applications.

1.4.1 Multiplication as Repeated Summation

In the previous section, borrowed priors amounted to $\sum_{j=1}^{m} NIG(\mu, \Lambda, a, b)$, which can be defined as m times $NIG(\mu, \Lambda, a, b)$ to simplify repeated summation. Moreover, the multiplier is not limited to an integer. When the NIG distribution is multiplied by a real number between zero and one, the multiplier serves as a discount factor for the NIG distribution.

Definition 1.4 (NIG scalar multiplication) *Consider d-dimensional $NIG(\mu, \Lambda, a, b)$. Let δ be a nonnegative scalar. If a distribution $NIG\left(\hat{\mu}, \hat{\Lambda}, \hat{a}, \hat{b}\right)$ satisfies*

1. $\hat{\mu} = \mu$,
2. $\hat{\Lambda} = \delta\Lambda$,
3. $\hat{a} = \delta\left(a + \frac{d}{2}\right) - \frac{d}{2}$,
4. $\hat{b} = \delta b$,

it is the scalar multiplication of the NIG distribution, denoted by

$$NIG\left(\hat{\mu}, \hat{\Lambda}, \hat{a}, \hat{b}\right) = \delta \cdot NIG(\mu, \Lambda, a, b).$$

An interpretation of the NIG scalar multiplication is that δ copies of data are used in the regression. Definition 1.4 indicates the multiplication effects on the posterior distribution of the coefficient β and the disturbance variance σ^2: 1) the mean of β remains μ, as the weighted average of identical NIGs leaves the mean unchanged; 2) the precision of β is scaled by δ, as multiple NIG copies bring more pseudo data, which are concatenated to increase the precision; 3) the parameters a and b in the NIG distribution determine the distribution of σ^2. Both a and b increase linearly with δ. Since $E\left(\sigma^2\right) = \frac{b}{a-1}$, $Var\left(\sigma^2\right) = \frac{b^2}{(a-1)^2(a-2)}$, multiplication has little impact on the mean, but the variance decreases as δ increases, implying reduced uncertainty in σ^2.

Proposition 1.7 (Properties of NIG scalar multiplication) *The NIG scalar multiplication operator satisfies distributivity, compatibility, and identity/zero elements.*

1. *Distributivity:*

$$\sum_{i=1}^{n} \delta_i \cdot NIG(\mu, \Lambda, a, b) = \left(\sum_{i=1}^{n} \delta_i\right) \cdot NIG(\mu, \Lambda, a, b),$$

$$\sum_{i=1}^{n} \delta \cdot NIG(\mu_i, \Lambda_i, a_i, b_i) = \delta \cdot \sum_{i=1}^{n} NIG(\mu_i, \Lambda_i, a_i, b_i).$$

2. *Compatibility:*

$$\delta_1\delta_2 \cdot NIG(\mu, \Lambda, a, b) = \delta_1 \cdot \left[\delta_2 \cdot NIG(\mu, \Lambda, a, b)\right].$$

3. *Identity/zero element:*

$$1 \cdot NIG(\mu, \Lambda, a, b) = NIG(\mu, \Lambda, a, b),$$

$$0 \cdot NIG(\mu, \Lambda, a, b) = NIG\left(\mu, 0_{d\times d}, -\frac{d}{2}, 0\right).$$

Equivalence between scalar multiplication and repeated summation follows directly from Proposition 1.7. For an integer δ, the distributive law of scalar multiplication under $\delta = \sum_{j=1}^{\delta} 1$ indicates that

$$\delta \cdot NIG(\mu, \Lambda, a, b) = \sum_{j=1}^{\delta} NIG(\mu, \Lambda, a, b).$$

1.4.2 Applications of Multiplication

Weighted regressions are prevalent in financial time series analysis, where it is desirable to down-weight past observations as time elapses. The NIG multiplication operator facilitates weighted regressions, with the real-valued $\delta \in [0, 1]$ serving as a discount factor (also known as a forgetting or smoothing factor). Just as geometrically weighted least squares extend OLS, weighted NIG summation extends equal-weight summation.

Weighted summation satisfies the equation

$$\sum_{i=1}^{t} \delta^{t-i} \cdot NIG(\mu_i, \Lambda_i, a_i, b_i) = NIG\left(\hat{\mu}_t, \hat{\Lambda}_t, \hat{a}_t, \hat{b}_t\right),$$

where

$$\hat{\mu}_t = \left(\sum_{i=1}^{t} \delta^{t-i}\Lambda_i\right)^{-1}\left(\sum_{i=1}^{t} \delta^{t-i}\Lambda_i\mu_i\right),$$

$$\hat{\Lambda}_t = \sum_{i=1}^{t} \delta^{t-i}\Lambda_i,$$

$$\hat{a}_t = -\frac{d}{2} + \sum_{i=1}^{t} \delta^{t-i}\left(a_i + \frac{d}{2}\right),$$

$$\hat{b}_t = \sum_{i=1}^{t} \delta^{t-i} b_i + \frac{1}{2}\sum_{i=1}^{t} \delta^{t-i}(\mu_i - \hat{\mu}_t)' \Lambda_i (\mu_i - \hat{\mu}_t).$$

Let $NIG(\mu_i, \Lambda_i, a_i, b_i)$ represent the posterior distribution obtained by regressing Y_i on X_i under the non-informative prior. The sufficient statistics $\Lambda_i = X_i'X_i$ and $\Lambda_i\mu_i = X_i'Y_i$ indicate that

$$\hat{\mu}_t = \left(\sum_{i=1}^{t} \delta^{t-i}X_i'X_i\right)^{-1}\left(\sum_{i=1}^{t} \delta^{t-i}X_i'Y_i\right),$$

which is recognized as the geometrically weighted least squares estimator if $\delta < 1$. The OLS estimator corresponds to a special case under $\delta = 1$.

Weighted regressions support online updates via a recursive formula:

$$NIG\left(\hat{\mu}_t, \hat{\Lambda}_t, \hat{a}_t, \hat{b}_t\right) = NIG(\mu_t, \Lambda_t, a_t, b_t) + \delta \cdot NIG\left(\hat{\mu}_{t-1}, \hat{\Lambda}_{t-1}, \hat{a}_{t-1}, \hat{b}_{t-1}\right).$$

In cases of perfect collinearity or inadequate sample size in subset regressions, a solution is to borrow a prior $NIG(\mu, \Lambda, a, b)$ and apply the distributive law of NIG scalar multiplication:

$$\begin{aligned} NIG\left(\bar{\mu}_t, \bar{\Lambda}_t, \bar{a}_t, \bar{b}_t\right) &= \sum_{i=1}^{t} \delta^{t-i} \cdot \left[NIG\left(\mu_i, \Lambda_i, a_i, b_i\right) + NIG\left(\mu, \Lambda, a, b\right)\right] \\ &= NIG\left(\hat{\mu}_t, \hat{\Lambda}_t, \hat{a}_t, \hat{b}_t\right) + \frac{1-\delta^t}{1-\delta} NIG\left(\mu, \Lambda, a, b\right). \end{aligned}$$

Finally, apply NIG subtraction to offset the effects of the borrowed prior:

$$NIG\left(\hat{\mu}_t, \hat{\Lambda}_t, \hat{a}_t, \hat{b}_t\right) = NIG\left(\bar{\mu}_t, \bar{\Lambda}_t, \bar{a}_t, \bar{b}_t\right) - \frac{1-\delta^t}{1-\delta} NIG\left(\mu, \Lambda, a, b\right).$$

1.5 Marginal Likelihood

A model is an abstraction of reality, designed to simplify and encapsulate the essential features of a complex system. Model selection (Kadane and Lazar, 2004) involves comparing a set of candidate models to evaluate whether they adequately describe real-world data. In classical inference, a null hypothesis regarding model specification is tested against an alternative hypothesis. A small p-value, typically less than 0.05, indicates that the statistic falls in the tail of the sampling distribution, suggesting that the data do not support the null hypothesis. In Bayesian statistics, the null and alternative hypotheses are viewed as distinct models, compared by the marginal likelihood. In Bayes' rule,

$$p(\theta \,|Y) = \frac{p(\theta)\, p(Y\,|\theta)}{p(Y)},$$

the denominator

$$p(Y) = \int p(\theta)\, p(Y\,|\theta)\, d\theta$$

is termed the marginal likelihood, which integrates the likelihood function over the parameter space. Thus, the marginal likelihood is the weighted average of the likelihood function by the prior density.

1.5.1 Bayesian Model Comparison

While often treated as a constant when computing the posterior distribution of parameters under a specific model, the marginal likelihood plays a pivotal role in Bayesian model selection and averaging, quantifying the evidence in favor of each model (Raftery et al., 1997; Wasserman, 2000).

Consider a collection of candidate models $M \in \{1, \ldots, m\}$. To emphasize model dependency, the marginal likelihood under the jth model is denoted as $p(Y\,|M = j)$. Let $p(M = j)$ represent the prior probability of the model. The posterior probability of the model is derived via Bayes' rule

$$p(M = j\,|Y) = \frac{p(M = j)\, p(Y\,|M = j)}{\sum_{i=1}^{m} p(M = i)\, p(Y\,|M = i)}.$$

Two competing models can be compared by their posterior probability ratio

$$\frac{p(M = i\,|Y)}{p(M = j\,|Y)} = \frac{p(M = i)\, p(Y\,|M = i)}{p(M = j)\, p(Y\,|M = j)},$$

where $\frac{p(Y|M=i)}{p(Y|M=j)}$ is known as the Bayes factor (Kass and Raftery, 1995), which is determined by the marginal likelihood of the competing models.

1.5.2 Computational Methods of Marginal Likelihood

Evaluating the marginal likelihood for most models is challenging due to the intractability of integration over parameters. However, it is feasible to compute the marginal likelihood of the linear regression model under the conjugate NIG prior specified in Equation (1.4). The marginal likelihood can be derived via a two-step approach by examining the equation $Y = X\beta + \sigma\varepsilon$. First, conditional on σ^2, the data Y follows a multivariate normal distribution with mean $X\mu$ and covariance matrix $\sigma^2 V$, where $V = I_n + X\Lambda^{-1}X'$, since the prior of β is normal with mean μ and covariance matrix $\sigma^2\Lambda^{-1}$. Second, marginalizing over the inverse gamma distributed σ^2 yields a scale mixture that follows the multivariate t distribution (see Exercise 1.5). The marginal likelihood is the density of $t\left(X\mu, \frac{b}{a}V, 2a\right)$:

$$p(Y) = \frac{\Gamma\left(a+\frac{n}{2}\right) b^a}{\Gamma(a)(2\pi)^{\frac{n}{2}}} |V|^{-\frac{1}{2}} \left[b + \frac{1}{2}(Y - X\mu)' V^{-1} (Y - X\mu)\right]^{-\left(a+\frac{n}{2}\right)},$$

without conditioning on the model explicitly denoted.

Two technical challenges arise when applying the above formula with a large sample size n. First, an out-of-memory error may occur when loading and saving an $n \times n$ matrix in computer memory. Second, computing the matrix determinant and inversion is cost-prohibitive, as the complexity amounts to $O\left(n^3\right)$. Exercise 1.6 offers a workaround using the linear algebra toolkit, including the Woodbury inversion, Sylvester determinant, and Cholesky decomposition. Alternatively, the NIG summation operator provides a straightforward method for evaluating the marginal likelihood via Bayes' rule

$$p(Y) = \frac{p\left(\beta,\sigma^2\right) p\left(Y\left|\beta,\sigma^2\right.\right)}{p\left(\beta,\sigma^2\left|Y\right.\right)},$$

where $p\left(\beta,\sigma^2\right)$ and $p\left(\beta,\sigma^2\left|Y\right.\right)$ are densities of $NIG(\mu,\Lambda,a,b)$ and $NIG\left(\bar{\mu},\bar{\Lambda},\bar{a},\bar{b}\right)$ respectively. It is feasible to evaluate densities even with large sample sizes, as the posterior distribution can be computed using the divide-and-conquer NIG summation described in Algorithm 1.1. Notably, NIG densities can be evaluated at arbitrary β and σ^2 without altering the marginal likelihood results. It is convenient to evaluate densities at $\beta = 0$ and $\sigma^2 = 1$, under which the marginal likelihood is simplified as follows:

$$p(Y) = \frac{p\left(\sigma^2\right) p\left(\beta\left|\sigma^2\right.\right) p\left(Y\left|\beta,\sigma^2\right.\right)}{p\left(\sigma^2\left|Y\right.\right) p\left(\beta\left|Y,\sigma^2\right.\right)},$$

where

$$\begin{aligned}
p\left(\sigma^2\right) &= \frac{b^a}{\Gamma(a)} e^{-b}, \\
p\left(\beta\left|\sigma^2\right.\right) &= (2\pi)^{-\frac{d}{2}} |\Lambda|^{\frac{1}{2}} e^{-\frac{1}{2}\mu'\Lambda\mu}, \\
p\left(Y\left|\beta,\sigma^2\right.\right) &= (2\pi)^{-\frac{n}{2}} e^{-\frac{1}{2}Y'Y}, \\
p\left(\sigma^2\left|Y\right.\right) &= \frac{\bar{b}^{\bar{a}}}{\Gamma(\bar{a})} e^{-\bar{b}}, \\
p\left(\beta\left|Y,\sigma^2\right.\right) &= (2\pi)^{-\frac{d}{2}} |\bar{\Lambda}|^{\frac{1}{2}} e^{-\frac{1}{2}\bar{\mu}'\bar{\Lambda}\bar{\mu}}.
\end{aligned}$$

All of these expressions are determined by the sufficient statistics $X'X$, $X'Y$ and $Y'Y$. Assuming $n > d$, the computational complexity reduces to $O\left(d^2 n\right)$ instead of $O\left(n^3\right)$.

1.6 Prediction

Bayesian prediction focuses on the posterior predictive distribution of the response variables Y_f, conditional on the observed data Y. This distribution is a weighted average of the predicted likelihood function, with weights given by the posterior density of the parameters:

$$p(Y_f|Y) = \int p(\theta|Y)\, p(Y_f|\theta, Y)\, d\theta,$$

where $p(Y_f|\theta, Y) = p(Y_f|\theta)$, as the observed data provide no additional information about future data conditional on parameters under the random sampling assumption. By integrating out the unknown parameters, Bayesian prediction incorporates parameter uncertainty.

1.6.1 Point Prediction

The density function $p(Y_f|Y)$ provides a full picture of the predictive distribution, but it is of interest to characterize the distribution by a point and an interval, similar to the point estimator and the confidence interval in classical inference. Point prediction depends on the loss function specification. As shown in Proposition 1.8, the optimal point is the mean of the posterior predictive distribution under the quadratic loss function, the median under the absolute-value loss function, and the mode under the all-or-nothing loss function.

Proposition 1.8 (Point prediction) *Let $p(Y_f|Y)$ be the posterior predictive distribution, where Y_f is a scalar. Let Y_f^* be a point that minimizes the expected loss function $L\left(Y_f^*, Y_f\right)$ such that*

$$\min_{Y_f^*} \int L(Y_f^*, Y_f)\, p(Y_f|Y)\, dY_f.$$

1. *The optimum is the mean under the quadratic loss function:*
if $L\left(Y_f^*, Y_f\right) = \left(Y_f^* - Y_f\right)^2$,
then $Y_f^* = E(Y_f|Y)$.
2. *The optimum is a quantile under the asymmetric loss function:*
$$\text{if } L\left(Y_f^*, Y_f\right) = \begin{cases} k_1\left(Y_f - Y_f^*\right) & Y_f \geq Y_f^* \\ k_2\left(Y_f^* - Y_f\right) & Y_f < Y_f^* \end{cases},$$
then $Y_f^* = F^{-1}\left(\frac{k_1}{k_1+k_2}\right)$, *where* $F(c) = \int_{-\infty}^{c} p(Y_f|Y)\, dY_f$.
3. *The optimum is the median under the absolute-value loss function:*
if $L\left(Y_f^*, Y_f\right) = \left|Y_f^* - Y_f\right|$,
then $Y_f^* = F^{-1}\left(\frac{1}{2}\right)$.
4. *The optimum is the mode under the all-or-nothing loss function:*
if $L\left(Y_f^*, Y_f\right) = I\left(Y_f^* \neq Y_f\right)$ *and* $p(Y_f|Y)$ *is unimodal,*
then $Y_f^* = \arg\max_{Y_f} p(Y_f|Y)$.

1.6.2 Interval Prediction

Credible intervals are the Bayesian counterparts of confidence intervals, but frequentist and Bayesian interpretations differ. Bayesian intervals are straightforward: a random variable falls within an interval with the specified probability. In contrast, confidence intervals are interpreted in terms of coverage: an interval with two random endpoints covers a point with the specified probability.

Two types of credible intervals are commonly used: equal-tail and highest posterior density (HPD) intervals. Let α denote a significance level. The equal-tail interval is determined by two thresholds at $\frac{\alpha}{2}$ and $1-\frac{\alpha}{2}$ quantiles, ensuring that the random variable falls within the credible interval with probability $1-\alpha$. As the name suggests, HPD is defined as an interval where the density inside is uniformly higher than the density outside the interval. HPD is the shortest among intervals with the specified probability. For a unimodal and symmetric distribution, HPD and equal-tail intervals overlap. Otherwise, HPD is not necessarily symmetric or even connected. A key property of HPD intervals is that the density at the endpoints is equal. This property suggests computing HPD intervals by a horizontal scanning algorithm, where a horizontal line moves from the peak of the density function downward, marking intervals of increasing width until the cumulative probability reaches $1-\alpha$. See Chen and Shao (1999) and Eberly and Casella (2003) for credible interval estimation.

1.6.3 Computational Methods of Posterior Predictive distribution

For Bayesian linear regressions, the response variable is generated by

$$Y_f = X_f\beta + \sigma\varepsilon_f.$$

The posterior predictive distribution

$$p\left(Y_f\,|Y\right) = \int\int p\left(\beta,\sigma^2\,|Y\right)p\left(Y_f\,|\beta,\sigma^2,Y\right)d\beta d\sigma^2$$

is analytically tractable under the NIG conjugate prior. By Proposition 1.1, $p\left(\beta,\sigma^2\,|Y\right)$ follows $NIG\left(\bar{\mu},\bar{\Lambda},\bar{a},\bar{b}\right)$. The posterior predictive distribution can be derived via a two-step approach. First, conditional on σ^2, future data Y_f follows a multivariate normal distribution with mean $X_f\bar{\mu}$ and covariance matrix σ^2V_f, where $V_f = I + X_f\bar{\Lambda}^{-1}X_f'$, as the posterior mean and variance of β are $\bar{\mu}$ and $\sigma^2\bar{\Lambda}^{-1}$, respectively. Second, marginalization over the posterior distribution of σ^2, namely $IG\left(\bar{a},\bar{b}\right)$, yields the posterior predictive distribution as the multivariate t:

$$Y_f\,|Y \sim t\left(X_f\bar{\mu},\frac{\bar{b}}{\bar{a}}V_f,2\bar{a}\right).$$

The posterior predictive mean and variance are

$$E\left(Y_f\,|Y\right) = X_f\bar{\mu},$$

$$Var\left(Y_f\,|Y\right) = \frac{\bar{b}}{\bar{a}-1}V_f.$$

The NIG summation operator offers an alternative method for evaluating the posterior predictive density. Bayes' rule indicates that

$$p\left(Y_f\,|Y\right) = \frac{p\left(\beta,\sigma^2\,|Y\right)p\left(Y_f\,|\beta,\sigma^2,Y\right)}{p\left(\beta,\sigma^2\,|Y_f,Y\right)},$$

where $p\left(\beta,\sigma^2\,|Y\right)$ represents the density of $NIG\left(\bar{\mu},\bar{\Lambda},\bar{a},\bar{b}\right)$, while $p\left(\beta,\sigma^2\,|Y_f,Y\right)$ is the posterior distribution by concatenating Y and Y_f, corresponding to the NIG summation

$$NIG\left(\bar{\mu},\bar{\Lambda},\bar{a},\bar{b}\right) + NIG\left(\tilde{\mu}_f,\tilde{\Lambda}_f,\tilde{a}_f,\tilde{b}_f\right),$$

where $NIG\left(\tilde{\mu}_f,\tilde{\Lambda}_f,\tilde{a}_f,\tilde{b}_f\right)$ denotes the posterior updated by X_f, Y_f under the non-informative prior.

1.7 Posterior Predictive Assessment

In addition to out-of-sample forecasting, the posterior predictive distribution can also be used for goodness-of-fit assessment of the model (Gelman et al., 1996). Future data Y_f are generalized as replicated data that could have been observed if the data-generating process for the observed data Y were replicated. If the model fits the data well, the observed and replicated data should look similar. A significant discrepancy, where the observed data fall in the extreme tail of the posterior predictive distribution, indicates evidence of misfit.

1.7.1 Cross-validation

Consider cross-validation (Shao, 1993; Arlot and Celisse, 2010) that reserves the ith observation Y_i as the validation data. The remaining observations, denoted by Y_{-i}, are used for parameter estimation. The NIG summation and subtraction can be used for computing the posterior $p\left(\beta, \sigma^2 \,|Y_{-i}\right)$ as $NIG\left(\bar{\mu}_{-i}, \bar{\Lambda}_{-i}, \bar{a}_{-i}, \bar{b}_{-i}\right)$. Let Y_f be the replicated counterpart of Y_i. The latter is generated by $Y_i = X_i\beta + \sigma\varepsilon_i$, while the former is generated by $Y_f = X_i\beta + \sigma\varepsilon_f$, where ε_f and ε_i have the same distribution but different realizations. The leaving-one-out posterior predictive distribution is

$$Y_f \,|Y_{-i} \sim t\left(X_i\bar{\mu}_{-i}, \frac{\bar{b}_{-i}}{\bar{a}_{-i}}V_i, 2\bar{a}_{-i}\right),$$

where $V_i = 1 + X_i\bar{\Lambda}_{-i}^{-1}X_i'$.

Let q_α be the α quantile of the preceding t distribution. A validation data falls in the tail of the posterior predictive distribution if it is too small ($Y_i < q_{\alpha/2}$) or too large ($Y_i > q_{1-\alpha/2}$), suggesting that the current model is unlikely to generate Y_i. Of course, an outlier may be produced by chance. Cross-validation cycles through $p\left(Y_f \,|Y_{-i}\right), i = 1, \ldots, n$. If a substantial portion of observed data are in the tails, the model should be revised to improve fitting.

1.7.2 Bayesian P-value

Posterior predictive checking supports generalized test statistics $T\left(Y, \theta\right)$, as a function of both data and parameters. In the context of linear regressions, the parameters are $\theta = \left(\beta, \sigma^2\right)$. Let Y_f be the replicated counterpart of Y, sharing the same set of predictors X. The observed data are generated by $Y = X\beta + \sigma\varepsilon$, while the replicated data are generated by $Y_f = X\beta + \sigma\varepsilon_f$, where the identically distributed ε_f and ε have different realizations. Gelman et al. (1996) define Bayesian p-value as the posterior predictive probability that the replicated data statistics $T\left(Y_f, \theta\right)$ could be more extreme than the observed data statistics $T\left(Y, \theta\right)$:

$$P\left[T\left(Y_f, \theta\right) > T\left(Y, \theta\right) |Y\right] = \int\int I\left[T\left(Y_f, \theta\right) > T\left(Y, \theta\right)\right] p\left(\theta\,|Y\right) p\left(Y_f\,|\theta\right) dY_f d\theta.$$

Although the integral generally lacks a closed-form solution, Bayesian p-value can be computed via Monte Carlo simulation. For the linear regression model, Proposition 1.1 indicates that $p\left(\theta\,|Y\right)$ follows $NIG\left(\bar{\mu}, \bar{\Lambda}, \bar{a}, \bar{b}\right)$, from which the draws $\theta_t = \left(\beta_t, \sigma_t^2\right)$, $t = 1, \ldots, T$ can be simulated. Conditional on each θ_t, the replicated data are simulated via $Y_{f,t} = X\beta_t + \sigma_t\varepsilon_{f,t}$. By computing the test statistics $T\left(Y_{f,t}, \theta_t\right)$ and $T\left(Y, \theta_t\right)$, the sample analogue of Bayesian p-value is given by

$$\frac{1}{T}\sum_{t=1}^{T} I\left[T\left(Y_{f,t},\theta_t\right) > T\left(Y,\theta_t\right)\right].$$

Extreme values close to zero or one suggest potential model misspecification.

1.8 Multivariate Extension

The NIG summation operator can be extended to vector regressions with m response variables and d predictors such that

$$Y_i = X_i B + \varepsilon_i,$$

where Y_i is a $1\times m$ response vector, X_i is a $1\times d$ predictor vector, B is a $d\times m$ coefficient matrix, and ε_i follows the multivariate normal distribution with mean zero and covariance matrix Σ. The matrix form is obtained by stacking the observations:

$$Y = XB + \varepsilon,$$

where $Y = \begin{pmatrix} Y_1 \\ \cdots \\ Y_n \end{pmatrix}$ and $X = \begin{pmatrix} X_1 \\ \cdots \\ X_n \end{pmatrix}$. Their sizes are $n\times m$ and $n\times d$, respectively. The sufficient statistics take the form $X'X = \sum_{i=1}^{n} X_i'X_i$, $X'Y = \sum_{i=1}^{n} X_i'Y_i$ and $Y'Y = \sum_{i=1}^{n} Y_i'Y_i$.

1.8.1 Normal-Inverse-Wishart Distribution

Conjugate Bayesian model specification is completed by incorporating the normal-inverse-Wishart (NIW) prior, as shown in Definition 1.5.

Definition 1.5 (NIW) *Let B be a $d\times m$ coefficient matrix and Σ be a $m\times m$ covariance matrix. They jointly follow $NIW(U,\Lambda,\Omega,v)$ distribution if*

$$p(B,\Sigma) \propto |\Sigma|^{-\frac{v+d+m+1}{2}} \exp\left\{-\frac{1}{2}tr\left[(B-U)'\Lambda(B-U)\Sigma^{-1} + \Omega\Sigma^{-1}\right]\right\},$$

where dimensions of hyperparameters U,Λ,Ω,v are $d\times m$, $d\times d$, $m\times m$, 1×1, respectively.

The NIW is a concatenation of normal and inverse Wishart distributions. By marginal-conditional decomposition, the marginal distribution $p(\Sigma)$ is inverse Wishart, and the conditional distribution $p(B|\Sigma)$ is multivariate normal with a Kronecker structure in the covariance matrix (Exercise 1.7). The NIW distribution is the natural conjugate prior for the Gaussian likelihood (Exercise 1.8)

$$p(Y|B,\Sigma) \propto |\Sigma|^{-\frac{n}{2}} \exp\left\{-\frac{1}{2}tr\left[(Y-XB)'(Y-XB)\Sigma^{-1}\right]\right\}.$$

The posterior $p(B,\Sigma|Y) \propto p(B,\Sigma)\,p(Y|B,\Sigma)$ takes the form

$$p(B,\Sigma|Y) \propto |\Sigma|^{-\frac{n+v+d+m+1}{2}} \exp\left[-\frac{1}{2}tr\left(M\Sigma^{-1}\right)\right],$$

where $M = \Omega + (B-U)'\Lambda(B-U) + (Y-XB)'(Y-XB)$. Since M is a quadratic function with respect to B, the posterior maintains the NIW form shown in Proposition 1.9, serving as a multivariate extension to Proposition 1.1.

Proposition 1.9 (Posterior NIW under conjugate prior) *If B and Σ have the prior $NIW(U,\Lambda,\Omega,v)$, then the posterior is $NIW(\bar{U},\bar{\Lambda},\bar{\Omega},\bar{v})$, where*

$$\bar{U} = (\Lambda + X'X)^{-1}(\Lambda U + X'Y),$$
$$\bar{\Lambda} = \Lambda + X'X,$$
$$\bar{\Omega} = \Omega + Y'Y + U'\Lambda U - \bar{U}'\bar{\Lambda}\bar{U},$$
$$\bar{v} = n + v.$$

Just as NIG distributions admit the non-informative prior $NIG(\mu_0, 0_{d\times d}, -\frac{d}{2}, 0)$ as the additive identity in arithmetic operations, the multivariate counterpart is $NIW(U_0, 0_{d\times d}, 0_{m\times m}, -d)$, where U_0 is an arbitrary $d\times m$ matrix. This is an improper density such that $p(B,\Sigma) \propto |\Sigma|^{-\frac{m+1}{2}}$, serving as a non-informative prior in the sense that the posterior mean coincides with the OLS estimator. As a corollary to Proposition 1.9, the posterior distribution under the non-informative prior is given by Proposition 1.10.

Proposition 1.10 (Posterior NIW under non-informative prior) *If B and Σ have a non-informative prior $p(B,\Sigma) \propto |\Sigma|^{-\frac{m+1}{2}}$, then the posterior is $NIW(\tilde{U},\tilde{\Lambda},\tilde{\Omega},\tilde{v})$ where*

$$\tilde{U} = (X'X)^{-1}X'Y,$$
$$\tilde{\Lambda} = X'X,$$
$$\tilde{\Omega} = Y'Y - \tilde{U}'\tilde{\Lambda}\tilde{U},$$
$$\tilde{v} = n - d.$$

Subjectivists elicit a prior that reflects non-sample information such as belief, expert opinion, parameter constraint and regularization. Prior elicitation on multitudinous hyperparameters in the NIW distribution could be a daunting task. A workaround is to conceive fictitious samples and infer hyperparameters of the prior distribution. This strategy is prevalent in Bayesian vector autoregression literature, a prominent application of conjugate NIW analysis. Refer to Sims and Zha (1998) for prior elicitation by dummy observations in dynamic multivariate models.

Proposition 1.11 demonstrates pseudo data in the NIW distributions. The non-informative distribution $NIW(U_0, 0_{d\times d}, 0_{m\times m}, -d)$ contains no data, while $NIW(\tilde{U},\tilde{\Lambda},\tilde{\Omega},\tilde{v})$ saves data X, Y compressed as sufficient statistics $X'X$, $X'Y$ and $Y'Y$. Pseudo data are unique up to sufficient statistics.

Proposition 1.11 (NIW pseudo data) *Consider $NIW(\tilde{U},\tilde{\Lambda},\tilde{\Omega},\tilde{v})$, where $\tilde{\Lambda},\tilde{\Omega}$ are $d\times d$ and $m\times m$ positive definite matrices, with upper triangular Cholesky-like factors $\tilde{\Lambda}^{1/2}$ and $\tilde{\Omega}^{1/2}$, respectively. A pseudo dataset of the size $n = \tilde{v} + d$ can be extracted in three steps:*

1. *Construct d data points as $X_{(1)} = \tilde{\Lambda}^{1/2}$, $Y_{(1)} = \tilde{\Lambda}^{1/2}\tilde{U}$.*
2. *Construct m data points as $X_{(2)} = 0_{m\times d}$, $Y_{(2)} = \tilde{\Omega}^{1/2}$.*
3. *Construct $n-d-m$ data points as $X_{(3)} = 0_{(n-d-m)\times d}$, $Y_{(3)} = 0_{(n-d-m)\times m}$.*

With pseudo data, the posterior distribution under the non-informative prior is $NIW(\tilde{U},\tilde{\Lambda},\tilde{\Omega},\tilde{v})$. Moreover, if another dataset X^*, Y^* yields the same posterior distribution, the data must satisfy $X^{*\prime}X^* = \tilde{\Lambda}$, $X^{*\prime}Y^* = \tilde{\Lambda}\tilde{U}$, $Y^{*\prime}Y^* = \tilde{\Omega} + \tilde{U}'\tilde{\Lambda}\tilde{U}$.

1.8.2 Summation Operator

The NIW summation operator is a multivariate extension of the NIG summation. An interpretation of the sum is the posterior distribution obtained by concatenating NIW pseudo data under the non-informative prior. Another interpretation is to extract data from one NIW while the other serves as the prior.

Definition 1.6 (NIW summation) *Consider distributions* $NIW(U_1, \Lambda_1, \Omega_1, v_1)$ *and* $NIW(U_2, \Lambda_2, \Omega_2, v_2)$. *If a distribution* $NIW(U, \Lambda, \Omega, v)$ *satisfies*

1. $U = (\Lambda_1 + \Lambda_2)^{-1}(\Lambda_1 U_1 + \Lambda_2 U_2)$,
2. $\Lambda = \Lambda_1 + \Lambda_2$,
3. $\Omega = \Omega_1 + \Omega_2 + (U_1 - U)' \Lambda_1 (U_1 - U) + (U_2 - U)' \Lambda_2 (U_2 - U)$,
4. $v = v_1 + v_2 + d$,

it is the sum of NIW distributions, denoted by

$$NIW(U, \Lambda, \Omega, v) = NIW(U_1, \Lambda_1, \Omega_1, v_1) + NIW(U_2, \Lambda_2, \Omega_2, v_2),$$

or more compactly,

$$NIW(U, \Lambda, \Omega, v) = \sum_{i=1}^{2} NIW(U_i, \Lambda_i, \Omega_i, v_i).$$

Proposition 1.12 (Properties of NIW summation) *The NIW summation operator satisfies commutativity, associativity and additive identity.*

1. *Commutativity:*

$$NIW(U_1, \Lambda_1, \Omega_1, v_1) + NIW(U_2, \Lambda_2, \Omega_2, v_2) = NIW(U_2, \Lambda_2, \Omega_2, v_2) + NIW(U_1, \Lambda_1, \Omega_1, v_1).$$

2. *Associativity:*

$$\sum_{i=1}^{3} NIW(U_i, \Lambda_i, \Omega_i, v_i) = NIW(U_1, \Lambda_1, \Omega_1, v_1) + \sum_{i=2}^{3} NIW(U_i, \Lambda_i, \Omega_i, v_i).$$

3. *Additive identity:*

$$NIW(U, \Lambda, \Omega, v) + NIW(U_0, 0_{d\times d}, 0_{m\times m}, -d) = NIW(U, \Lambda, \Omega, v).$$

Given an arbitrary $d \times m$ matrix U_0, the non-informative distribution $NIW(U_0, 0_{d\times d}, 0_{m\times m}, -d)$ serves as zero in the NIW summation. The non-informative distribution has no impact on the sum due to null data (as empty matrices) in its repository.

1.8.3 Big Data Regressions

The NIW summation disentangles the roles of the prior and data in the posterior distribution and disintegrates big datasets for subset regressions. Propositions 1.13 and 1.14 are multivariate extensions of Propositions 1.5 and 1.6.

Proposition 1.13 (Separation of prior and data in posterior NIW) *The priors and posteriors specified in Propositions 1.9 and 1.10 are connected by the equation:*

$$NIW(\bar{U}, \bar{\Lambda}, \bar{\Omega}, \bar{v}) = NIW(U, \Lambda, \Omega, v) + NIW\left(\tilde{U}, \tilde{\Lambda}, \tilde{\Omega}, \tilde{v}\right).$$

Algorithm 1.2 Divide-and-conquer big data vector regression

Input: data X, Y and prior $NIW(U, \Lambda, \Omega, v)$.
Output: posterior $NIW(\bar{U}, \bar{\Lambda}, \bar{\Omega}, \bar{v})$ in Proposition 1.9.
1 Partition big data into J subsets X_j, Y_j, $j = 1, \ldots, J$.
2 Perform subset regressions separately for $NIW\left(\tilde{U}_j, \tilde{\Lambda}_j, \tilde{\Omega}_j, \tilde{v}_j\right)$.
3 Sum up subset posterior distributions

$$NIW\left(\tilde{U}, \tilde{\Lambda}, \tilde{\Omega}, \tilde{v}\right) = \sum_{j=1}^{J} NIW\left(\tilde{U}_j, \tilde{\Lambda}_j, \tilde{\Omega}_j, \tilde{v}_j\right).$$

4 Sum up the prior and combined subset posterior distributions

$$NIW(\bar{U}, \bar{\Lambda}, \bar{\Omega}, \bar{v}) = NIW(U, \Lambda, \Omega, v) + NIW\left(\tilde{U}, \tilde{\Lambda}, \tilde{\Omega}, \tilde{v}\right).$$

Proposition 1.14 (Subset posteriors of NIW) *Partition data X, Y into J subsets X_j, Y_j, $j = 1, \ldots, J$. Let the subset tilde-form posterior distributions be $NIW\left(\tilde{U}_j, \tilde{\Lambda}_j, \tilde{\Omega}_j, \tilde{v}_j\right)$. The full-sample tilde-form posterior $NIW\left(\tilde{U}, \tilde{\Lambda}, \tilde{\Omega}, \tilde{v}\right)$ can be computed as*

$$NIW\left(\tilde{U}, \tilde{\Lambda}, \tilde{\Omega}, \tilde{v}\right) = \sum_{j=1}^{J} NIW\left(\tilde{U}_j, \tilde{\Lambda}_j, \tilde{\Omega}_j, \tilde{v}_j\right).$$

Propositions 1.13 and 1.14 are fundamental to big data regressions, where the sample size is too large to be processed by a single machine. The massive data are divided into subsets, and regressions are performed with subset data separately under the non-informative prior. Subset regressions support distributed and parallel computing. Proposition 1.14 indicates that the sum of subset posteriors aligns with the full-sample posterior under the non-informative prior. Summing up the prior and combined subset posteriors leads to the posterior under the conjugate prior, as indicated by Proposition 1.13. Algorithm 1.2 is the multivariate extension of Algorithm 1.1.

1.8.4 Marginal Likelihood

The NIW summation operator facilitates evaluating the marginal likelihood, which is the key to Bayesian model selection and averaging. Bayes' rule ensures that the marginal likelihood is analytically tractable whenever the posterior is tractable:

$$p(Y) = \frac{p(B, \Sigma)\, p(Y\,|B, \Sigma)}{p(B, \Sigma\,|Y)},$$

where $p(B, \Sigma)$ and $p(B, \Sigma\,|Y)$ are densities of $NIW(U, \Lambda, \Omega, v)$ and $NIW(\bar{U}, \bar{\Lambda}, \bar{\Omega}, \bar{v})$, respectively. The latter can be computed by the NIW summation under Propositions 1.13 and 1.14, especially when the sample size is large. As $p(Y)$ is invariant to the evaluation points of B and Σ, a convenient choice is to evaluate the prior and posterior densities at $B = 0$ and $\Sigma = I_m$. The results are given by Proposition 1.15.

Proposition 1.15 (Marginal likelihood) *For the multivariate/vector regression model $Y = XB + \varepsilon$ under the prior $NIW(U, \Lambda, \Omega, v)$, the marginal likelihood is given by*

$$p(Y) = (2\pi)^{-\frac{mn}{2}} \frac{f(\Lambda, \Omega, v)}{f(\bar{\Lambda}, \bar{\Omega}, \bar{v})},$$

where

$$f(\Lambda, \Omega, v) = \left[2^{\frac{mv}{2}} \Gamma_m\left(\frac{v}{2}\right)\right]^{-1} |\Omega|^{\frac{v}{2}} |\Lambda|^{\frac{m}{2}},$$

and Γ_m is the multivariate gamma function.

1.9 Extension to Exponential Family

It is instructive to place the NIG summation within a broader framework, where the NIG distribution is a member of the exponential family, accommodating various common distributions such as normal, gamma, Poisson, beta, and binomial distributions. Diaconis and Ylvisaker (1979) analyze conjugate priors for the exponential family.

1.9.1 Conjugate Prior and Pseudo Data

Let $Y_i, i = 1, \ldots, n$ be random samples from an exponential family distribution

$$p(Y_i | \theta) = e^{h(Y_i) - A(\theta) + \eta(\theta)' T(Y_i)},$$

where $T(Y_i)$ serves as sufficient statistics, $\eta(\theta)$ is a mapping to natural parameters, and $A(\theta)$ is a normalization factor. These triple functions are abbreviated as $T\eta A$ in the subsequent presentations. Gelman et al. (2014, p.37) and Bernardo and Smith (2000, p.266) show that the conjugate prior distribution, denoted by $\epsilon_{T\eta A}(\tau, v)$, for the exponential family takes the form

$$p(\theta) \propto e^{\eta(\theta)'\tau - vA(\theta)},$$

where τ and v are hyperparameters in the prior. The posterior distribution $\epsilon_{T\eta A}(\bar{\tau}, \bar{v})$ is obtained by adding τ to the sufficient statistics and adding v to the sample size:

$$p(\theta | Y) \propto p(\theta) \prod_{i=1}^{n} p(Y_i | \theta) \propto e^{\eta(\theta)'\bar{\tau} - \bar{v}A(\theta)},$$

$$\bar{\tau} = \tau + \sum_{i=1}^{n} T(Y_i),$$
$$\bar{v} = v + n.$$

Specifying $\tau = 0$ and $v = 0$ leads to a flat and diffuse prior $p(\theta) \propto 1$, as a heuristic representation of a non-informative prior. Some researchers prefer a different specification of τ and v for alternative prior forms, such as the Jeffreys prior (Jeffreys, 1946). We denote $\epsilon_{T\eta A}(\tau_0, v_0)$ as a non-informative prior, from which pseudo data can be extracted, as shown in Proposition 1.16. Notably, the definition of pseudo data depends on the specification of the non-informative prior, which is merely a name for some benchmark distribution with hyperparameters τ_0 and v_0. It does not matter whether a non-informative prior conveys any information about the parameter θ.

Proposition 1.16 (Pseudo data in the prior) *Let $\epsilon_{T\eta A}(\tau, v)$ denote the conjugate prior for the exponential family and $\epsilon_{T\eta A}(\bar{\tau}, \bar{v})$ represent the posterior obtained by (genuine) data $Y_i, i = 1, \ldots, n$. With a non-informative prior $\epsilon_{T\eta A}(\tau_0, v_0)$ serving as a benchmark, define pseudo data $\tilde{Y}_i, i = 1, \ldots, n_0$ from the prior $\epsilon_{T\eta A}(\tau, v)$ such that*

$$\sum_{i=1}^{n_0} T\left(\tilde{Y}_i\right) = \tau - \tau_0,$$

$$n_0 = v - v_0.$$

1. *Using pseudo data to update the non-informative prior, the posterior equals $\epsilon_{T\eta A}(\tau, v)$.*
2. *Using the concatenated pseudo and genuine data to update the non-informative prior, the posterior equals $\epsilon_{T\eta A}(\bar{\tau}, \bar{v})$.*

Proposition 1.16 indicates that pseudo data are analogous to genuine data. First, if pseudo data were extracted from $\epsilon_{T\eta A}(\tau, v)$ and treated as if they were actual observations, the posterior would replicate $\epsilon_{T\eta A}(\tau, v)$. Second, if pseudo data were extracted from the prior and concatenated with genuine data, the posterior would algin with $\epsilon_{T\eta A}(\bar{\tau}, \bar{v})$.

1.9.2 Summation Operator in the Exponential Family

The pseudo data interpretation of the conjugate prior motivates Definition 1.7, where summation rules are defined over conjugate distributions in the exponential family. The sum of two distributions can be viewed as the posterior obtained by concatenating pseudo data extracted from two distributions under the non-informative prior. Alternatively, one of the distributions can serve as the prior, with pseudo data extracted from the other; the posterior aligns with the sum of the two distributions.

Definition 1.7 (Summation in exponential family) *Given a non-informative prior $\epsilon_{T\eta A}(\tau_0, v_0)$ as a benchmark, consider distributions $\epsilon_{T\eta A}(\tau_1, v_1)$ and $\epsilon_{T\eta A}(\tau_2, v_2)$. If a distribution $\epsilon_{T\eta A}(\tau, v)$ satisfies*

$$\tau = \tau_1 + \tau_2 - \tau_0,$$

$$v = v_1 + v_2 - v_0,$$

it is the sum of two distributions, denoted by

$$\epsilon_{T\eta A}(\tau, v) = \epsilon_{T\eta A}(\tau_1, v_1) + \epsilon_{T\eta A}(\tau_2, v_2).$$

The summation operator satisfies commutativity and associativity. Most importantly, Definition 1.7 indicates that the non-informative prior $\epsilon_{T\eta A}(\tau_0, v_0)$ serves as the zero in the summation:

$$\epsilon_{T\eta A}(\tau, v) + \epsilon_{T\eta A}(\tau_0, v_0) = \epsilon_{T\eta A}(\tau, v).$$

The summation operator separates the roles of the prior and data in the posterior, allowing for a simple equation to connect Bayesian and classical estimators, as shown in Proposition 1.17.

Proposition 1.17 (Separation of prior and data in posterior) *Let $\epsilon_{T\eta A}(\bar{\tau}, \bar{v})$ be the posterior under a conjugate prior $\epsilon_{T\eta A}(\tau, v)$. Let $\epsilon_{T\eta A}(\tilde{\tau}, \tilde{v})$ be the posterior under a non-informative prior $\epsilon_{T\eta A}(\tau_0, v_0)$ with the same data. These distributions in the exponential family are connected by*

$$\epsilon_{T\eta A}(\bar{\tau}, \bar{v}) = \epsilon_{T\eta A}(\tau, v) + \epsilon_{T\eta A}(\tilde{\tau}, \tilde{v}).$$

1.9.3 Beta-Bernoulli Conjugate Distributions

In introductory courses on Bayesian statistics, the Beta-Bernoulli model serves as a standard example of conjugate Bayesian analysis.

Let $Y_i, i = 1, \ldots, n$ be random samples from the Bernoulli distribution with the success probability θ. The likelihood function is

$$p(Y_i | \theta) = \theta^{Y_i} (1 - \theta)^{1 - Y_i},$$

or in the log-likelihood form:

$$\ln p(Y_i | \theta) = Y_i \ln \frac{\theta}{1 - \theta} + \ln (1 - \theta),$$

or as a member in the exponential family:

$$T(Y_i) = Y_i,$$

$$\eta(\theta) = \ln \frac{\theta}{1 - \theta},$$

$$A(\theta) = -\ln (1 - \theta).$$

The conjugate prior for Bernoulli sampling is constructed as

$$p(\theta) \propto e^{\eta(\theta)'\tau - vA(\theta)} = \theta^{\tau} (1 - \theta)^{v - \tau},$$

which is recognized as a Beta distribution, parametrized as $Beta(\tau + 1, v - \tau + 1)$. This is a concrete instance of the abstract distribution $\epsilon_{T\eta A}(\tau, v)$. By defining $a = \tau + 1$ and $b = v - \tau + 1$, the prior can be reparametrized in a standard form $Beta(a, b)$. The posterior remains a Beta distribution: $Beta(a + \sum_{i=1}^{n} Y_i, b + n - \sum_{i=1}^{n} Y_i)$.

An intuitive specification of the non-informative prior is the uniform distribution $p(\theta) \propto 1, \forall \theta \in (0, 1)$, corresponding to $Beta(1, 1)$, namely the benchmark distribution $\epsilon_{T\eta A}(\tau_0, v_0)$ with $\tau_0 = 0$ and $v_0 = 0$. By Definition 1.7, summation rules of Beta distributions are given by

$$Beta(a, b) = Beta(a_1, b_1) + Beta(a_2, b_2),$$

$$a = a_1 + a_2 - 1,$$

$$b = b_1 + b_2 - 1.$$

By Proposition 1.16, pseudo data extracted from $Beta(a, b)$ are $a - 1$ successes and $b - 1$ failures under the uniform prior. The sum of $Beta(a_1, b_1)$ and $Beta(a_2, b_2)$ involves concatenating pseudo data as $a_1 - 1 + a_2 - 1$ successes and $b_1 - 1 + b_2 - 1$ failures. Under the uniform prior $Beta(1, 1)$ and concatenated pseudo data, the posterior is $Beta(a_1 + a_2 - 1, b_1 + b_2 - 1)$, aligning with the sum of two Beta distributions.

An alternative specification of the non-informative distribution is the Jeffreys (1946) prior $p(\theta) \propto \theta^{-1/2} (1 - \theta)^{-1/2}$, which corresponds to $Beta(1/2, 1/2)$, or $\epsilon_{T\eta A}(\tau_0, v_0)$ with $\tau_0 = -1/2$ and $v_0 = -1$. Pseudo data and summation rules depend on the non-informative prior specification. Pseudo data extracted from $Beta(a, b)$ adjust to $a - 1/2$ successes and $b - 1/2$ failures. Under the Jeffreys prior, summation rules are modified as follows:

$$a = a_1 + a_2 - \frac{1}{2},$$

$$b = b_1 + b_2 - \frac{1}{2}.$$

Despite these changes, the sum of Beta distributions can still be interpreted as the posterior distribution obtained by concatenating pseudo data under the new definition of the non-informative distribution.

Another specification is the Haldane (1932) prior $p(\theta) \propto \theta^{-1}(1-\theta)^{-1}$, corresponding to $Beta(0,0)$, or $\epsilon_{T\eta A}(\tau_0, \upsilon_0)$ with $\tau_0 = -1$ and $\upsilon_0 = -2$. Under the Haldane prior, summation rules are modified as follows:

$$a = a_1 + a_2,$$

$$b = b_1 + b_2.$$

1.9.4 Gamma-Poisson Conjugate Distributions

In introductory Bayesian statistics, Gamma-Poisson sampling is another example of conjugate distributions.

Let $Y_i, i = 1, \ldots, n$ be random samples from the Poisson distribution with intensity parameter θ. The likelihood function is

$$p(Y_i|\theta) = \frac{\theta^{Y_i}}{Y_i!}e^{-\theta},$$

or in the log-likelihood form:

$$\ln p(Y_i|\theta) = Y_i \ln\theta - \theta - \ln Y_i!,$$

or as a member in the exponential family:

$$T(Y_i) = Y_i,$$

$$\eta(\theta) = \ln\theta,$$

$$A(\theta) = \theta.$$

The conjugate prior for Poisson sampling is constructed as

$$p(\theta) \propto e^{\eta(\theta)'\tau - \upsilon A(\theta)} = \theta^{\tau} e^{-\upsilon\theta},$$

which is recognized as a Gamma distribution, parametrized as $Gamma(\tau+1, \upsilon)$. This is a concrete instance of the abstract distribution $\epsilon_{T\eta A}(\tau, \upsilon)$. By defining $a = \tau + 1$ and $b = \upsilon$, the prior is reparametrized as $Gamma(a, b)$. The posterior remains a Gamma distribution: $Gamma\left(a + \sum_{i=1}^{n} Y_i, b+n\right)$.

An intuitive specification of the non-informative prior is the positive uniform distribution $p(\theta) \propto 1$, $\forall\theta > 0$, corresponding to $Gamma(1,0)$, namely the benchmark distribution $\epsilon_{T\eta A}(\tau_0, \upsilon_0)$ with $\tau_0 = 0$ and $\upsilon_0 = 0$. By Definition 1.7, summation rules of Gamma distributions are given by

$$Gamma(a,b) = Gamma(a_1,b_1) + Gamma(a_2,b_2),$$

$$a = a_1 + a_2 - 1,$$

$$b = b_1 + b_2.$$

By Proposition 1.16, pseudo data extracted from $Gamma(a,b)$ are b data points that sum to $a-1$. Summing $Gamma(a_1,b_1)$ and $Gamma(a_2,b_2)$ involves concatenating b_1+b_2 data points that sum to $a_1 - 1 + a_2 - 1$. As the non-informative prior $Gamma(1,0)$ is updated by concatenated data, the posterior is $Gamma(a_1+a_2-1, b_1+b_2)$, aligning with the sum of two Gamma distributions.

An alternative specification of the non-informative distribution is the Jeffreys (1946) prior $p(\theta) \propto \theta^{-1/2}$, which corresponds to $Gamma\,(1/2, 0)$, or $\epsilon_{T\eta A}(\tau_0, v_0)$ with $\tau_0 = -1/2$ and $v_0 = 0$. Under the Jeffreys prior, summation rules are modified as follows:

$$a = a_1 + a_2 - \frac{1}{2},$$

$$b = b_1 + b_2.$$

1.9.5 Normal Conjugate Distributions

The normal distribution is perhaps the most well-known example of conjugate distributions in Bayesian statistics.

Let $Y_i, i = 1, \ldots, n$ be random samples from a normal distribution with mean θ, while the variance is fixed at σ^2. The likelihood function is

$$p(Y_i\,|\theta) = \frac{1}{\sqrt{2\pi}\sigma} e^{-\frac{(Y_i-\theta)^2}{2\sigma^2}},$$

or in the log-likelihood form:

$$\ln p(Y_i\,|\theta) = Y_i\theta\sigma^{-2} - \frac{1}{2}\sigma^{-2}\theta^2 - \frac{1}{2}\ln 2\pi - \frac{1}{2}\ln\sigma^2 - \frac{1}{2}Y_i^2\sigma^{-2},$$

or as a member in the exponential family:

$$T(Y_i) = Y_i,$$

$$\eta(\theta) = \theta\sigma^{-2},$$

$$A(\theta) = \frac{1}{2}\sigma^{-2}\theta^2.$$

The conjugate prior for normal sampling is constructed as

$$p(\theta) \propto e^{\eta(\theta)'\tau - vA(\theta)} = e^{\theta\sigma^{-2}\tau - \frac{1}{2}\sigma^{-2}\theta^2 v}.$$

Completing the square with respect to θ yields

$$p(\theta) \propto e^{-\frac{1}{2}\sigma^{-2}v\left(\theta - \frac{\tau}{v}\right)^2},$$

which is recognized as a normal distribution, parametrized as $N\left(\frac{\tau}{v}, \frac{\sigma^2}{v}\right)$. This is a concrete instance of the abstract distribution $\epsilon_{T\eta A}(\tau, v)$. By defining $\mu = \frac{\tau}{v}$ and $V = \frac{\sigma^2}{v}$, the prior can be reparametrized in a familiar form $N(\mu, V)$. The posterior remains a normal distribution $N(\bar{\mu}, \bar{V})$, where

$$\bar{\mu} = \left(V^{-1} + n\sigma^{-2}\right)^{-1}\left(V^{-1}\mu + \sigma^{-2}\sum_{i=1}^{n} Y_i\right),$$

$$\bar{V} = \left(V^{-1} + n\sigma^{-2}\right)^{-1}.$$

The non-informative prior can be specified as $p(\theta) \propto 1$, corresponding to $N(0, \infty)$, namely the benchmark distribution $\epsilon_{T\eta A}(\tau_0, v_0)$ with $\tau_0 = 0$ and $v_0 = 0$. By Definition 1.7, summation rules of normal distributions are given by

$$N(\mu, V) = N(\mu_1, V_1) + N(\mu_2, V_2),$$

$$V^{-1} = V_1^{-1} + V_2^{-1},$$
$$V^{-1}\mu = V_1^{-1}\mu_1 + V_2^{-1}\mu_2.$$

By Proposition 1.16, pseudo data extracted from $N(\mu, V)$ are $V^{-1}\sigma^2$ data points that sum to $V^{-1}\mu\sigma^2$. Summing $N(\mu_1, V_1)$ and $N(\mu_2, V_2)$ involves concatenating $\left(V_1^{-1} + V_2^{-1}\right)\sigma^2$ data points that sum to $\left(V_1^{-1}\mu_1 + V_2^{-1}\mu_2\right)\sigma^2$. Using concatenated data for updating the non-informative prior, the posterior equals the sum of two normal distributions.

1.9.6 NIG Conjugate Distributions

We revisit linear regressions, in which the data $Y_i, i = 1, \ldots, n$ are generated by

$$Y_i = X_i\beta + \sigma\varepsilon_i,$$

where X_i is a $1 \times d$ vector of regressors and ε_i follows the standard normal distribution. The unknown parameters are $\theta = (\beta, \sigma^2)$. The likelihood function is

$$p(Y_i | \theta) = \frac{1}{\sqrt{2\pi}\sigma} e^{-\frac{(Y_i - X_i\beta)^2}{2\sigma^2}},$$

or in the log-likelihood form:

$$\ln p(Y_i | \theta) = -\frac{1}{2}\ln 2\pi - \frac{1}{2}\ln\sigma^2 - \frac{1}{2}\sigma^{-2}Y_i^2 + \sigma^{-2}\beta' X_i' Y_i - \frac{1}{2}\sigma^{-2}\beta' X_i' X_i\beta.$$

Notably, the scalar term $\beta' X_i' X_i \beta$ functions as the inner product of two vectors $vec(\beta\beta')'\, vec(X_i' X_i)$, which is identified as a member of the exponential family:

$$T(Y_i) = \begin{pmatrix} vec(X_i' X_i) \\ X_i' Y_i \\ Y_i^2 \end{pmatrix},$$

$$\eta(\theta) = -\frac{1}{2}\sigma^{-2}\begin{pmatrix} vec(\beta\beta')' \\ -2\beta \\ 1 \end{pmatrix},$$

$$A(\theta) = \frac{1}{2}\ln\sigma^2.$$

The conjugate prior is constructed as

$$p(\theta) \propto e^{\eta(\theta)'\tau - vA(\theta)} = e^{\eta(\theta)'\tau}\left(\sigma^2\right)^{-v/2},$$

in which the hyperparameter vector τ consists of three components: $d^2 \times 1$ vector τ_A, $d \times 1$ vector τ_B and a scalar τ_C, comprising a linear expression:

$$\eta(\theta)'\tau = -\frac{1}{2}\sigma^{-2} vec(\beta\beta')'\,\tau_A + \sigma^{-2}\beta'\tau_B - \frac{1}{2}\sigma^{-2}\tau_C.$$

By Definition 1.1, $p(\theta)$ represents $NIG(\mu, \Lambda, a, b)$, with a one-to-one mapping between parameters $(\tau_A, \tau_B, \tau_C, v)$ and (μ, Λ, a, b):

$$\tau_A = vec(\Lambda),$$

$$\tau_B = \Lambda\mu,$$

$$\tau_C = 2b + \mu' \Lambda \mu,$$

$$v = 2a + d + 2.$$

Equivalently, the conjugate prior is $NIG\left(\tilde{\tau}_A^{-1}\tau_B, \tilde{\tau}_A, \frac{v-d-2}{2}, \frac{\tau_C - \tau_B' \tilde{\tau}_A^{-1} \tau_B}{2}\right)$, where $\tilde{\tau}_A$ reshapes the $d^2 \times 1$ vector τ_A as a $d \times d$ matrix: $vec\,(\tilde{\tau}_A) = \tau_A$.

A non-informative prior is specified as $p\,(\theta) \propto \sigma^{-2}$, corresponding to $NIG\left(0, 0, -\frac{d}{2}, 0\right)$, namely the benchmark distribution $\epsilon_{T\eta A}\,(\tau_0, v_0)$ with $\tau_0 = 0$ and $v_0 = 2$. Definitions 1.2 and 1.7 provide the same summation rules for the NIG distributions:

$$NIG\,(\mu, \Lambda, a, b) = NIG\,(\mu_1, \Lambda_1, a_1, b_1) + NIG\,(\mu_2, \Lambda_2, a_2, b_2)\,,$$

$$vec\,(\Lambda) = vec\,(\Lambda_1) + vec\,(\Lambda_2)\,,$$

$$\Lambda\mu = \Lambda_1\mu_1 + \Lambda_2\mu_2,$$

$$b = b_1 + b_2 + \frac{1}{2}\mu_1'\Lambda_1\mu_1 + \frac{1}{2}\mu_2'\Lambda_2\mu_2 - \frac{1}{2}\mu'\Lambda\mu,$$

$$a = a_1 + a_2 + \frac{d}{2}.$$

By Proposition 1.16, pseudo data $\tilde{X}, \tilde{Y}$ extracted from $NIG\,(\mu, \Lambda, a, b)$ are $2a + d$ data points with sufficient statistics $\tilde{X}'\tilde{X} = \Lambda$, $\tilde{X}'\tilde{Y} = \Lambda\mu$, $\tilde{Y}'\tilde{Y} = 2b + \mu'\Lambda\mu$, aligning with the results in Proposition 1.3. The summation of NIG distributions is the process of concatenating pseudo data by adding up sufficient statistics.

1.10 Software and Computational Considerations

The NIG arithmetic is not only a mathematical abstraction, but also implementable on a scientific computing platform such as MATLAB.

1.10.1 Operator Overloading

From the perspective of software development, the summation operator `A+B` is equivalent to the built-in function `plus(A,B)` for computing the sum of `A` and `B`. Operator overloading refers to implementing arithmetic operators that are appropriate for user-defined data types. In an object-oriented programing language, a developer can define a custom `NIG` class with properties like μ, Λ, a, b and class methods like `plus` (+), `minus` (-), and `times` (*) by Definitions 1.2, 1.3 and 1.4, respectively. This extends the meaning of arithmetic operators for customized data types beyond their built-in functionality.

We encourage readers to implement NIG arithmetic operators in their preferred software or use the toolbox accompanying this book to verify the propositions in this chapter. For example, the following code snippet validates the additive identity: adding the non-informative distribution to any NIG distribution leaves the sum unchanged.

```
NIG1 = NIG(mu1,Lambda1,a1,b1);
NIG0 = NIG(zeros(d,1),zeros(d),-d/2,0);
NIG10 = NIG1 + NIG0;
```

Similarly, pseudo data exacted from NIG distributions can be validated by the following code:

```
NIG2 = NIG(mu2,Lambda2,a2,b2);
```

```
NIG12 = NIG1 + NIG2;
[X,y] = extract(NIG2);
EstNIG = estimate(NIG1,X,y);
```

In theory, `NIG1==NIG10` and `NIG12==EstNIG`. However, an exact equality is unattainable on a computer, where real numbers are approximated by double-precision floating-point data with 15–16 digits of precision. Although rounding errors are unavoidable, computational pitfalls that lead to catastrophic cancellation errors can be avoided by employing numerically stable algorithms.

1.10.2 Computational Methods of OLS

Computational issues are a recurring theme throughout the book. In this section, we focus on the OLS estimator, specifically the variable $\tilde{\mu} = (X'X)^{-1} X'Y$ in Proposition 1.2. A naive implementation like `inv(X'*X)*X'*Y` is neither computationally efficient nor numerically stable.

Users may be surprised at the performance improvement with a seemingly innocuous addition of a parenthesis: `inv(X'*X)*(X'*Y)`. The revised code runs faster, as it reduces about $2d^2n$ floating-point operations. Furthermore, matrix decomposition routines by QR, SVD, or Cholesky algorithms are recommended for solving linear equations instead of direct matrix inversion `inv(X'*X)`.

The QR solver decomposes the predictor matrix as $X = QR$, where Q is orthonormal and R is upper triangular, leading to

$$(X'X)^{-1} X'Y = R^{-1}Q'Y.$$

The backslash operator `X\Y` is likely to invoke the QR solver for $(X'X)^{-1} X'Y$ if $n > d$.

The SVD solver decomposes the $n \times d$ predictor matrix as $X = USV'$. Assuming $n > d$, a compact (economy-size) version of SVD is available: S is a $d \times d$ diagonal matrix of singular values, V is a $d \times d$ orthonormal matrix, and U is a $n \times d$ matrix of orthogonal columns. It follows that

$$(X'X)^{-1} X'Y = VS^{-1}U'Y.$$

If the NIG prior hyperparameters are specified as $\mu = 0$ and $\Lambda = \delta I$, Proposition 1.1 provides the ridge estimator $\bar{\mu} = (X'X + \delta I)^{-1} X'Y$. The SVD solver can neatly handle the ridge regression

$$(X'X + \delta I)^{-1} X'Y = V\left(S + \delta S^{-1}\right)^{-1} U'Y.$$

The Cholesky solver decomposes the positive definite precision matrix as $X'X = LL'$, where L is lower triangular, resulting in

$$(X'X)^{-1} X'Y = L^{-1\prime}L^{-1}X'Y.$$

The Cholesky solver boasts excellent run-time performance compared to other matrix decomposition routines in some optimized linear algebra libraries (e.g., LAPACK). For Bayesian linear regressions, it is of interest to compute the intermediate variable $X'X$, as it represents the precision matrix. The backslash operator `(X'*X)\(X'*Y)` is likely to invoke the Cholesky solver. The Cholesky decomposition requires a positive definite matrix. If the precision matrix is of reduced rank (positive semi-definite), an alternative decomposition scheme resorts to eigenvalues such that $X'X = VDV'$, where D is a diagonal matrix of eigenvalues (squared singular values), and V is an orthonormal matrix of eigenvectors (the same as the SVD output).

1.11 An Application on Mortgage Shopping

Mortgage shoppers seek quotes from different banks to find the best available mortgage rate. Home buyers could save thousands of dollars by obtaining an additional rate quote, providing a significant economic incentive (Damen and Buyst, 2017). However, comparing offers from multiple mortgage lenders can be costly, as applying for credit is considered a hard inquiry, which can negatively impact the credit score of the applicant. The high search costs imply incomplete information and a market of imperfect competition, even if the commodity (i.e., borrowed money) is homogeneous. Big data can provide insights into multiple-bank interest rates for mortgage shoppers without making hard inquiries.

1.11.1 Data Description

Under the U.S. Home Mortgage Disclosure Act (HMDA), financial institutions are mandated to make loan-level data publicly available. See Munnell et al. (1996) and Avery et al. (2007) for HMDA studies. In 2021, over four thousand institutions reported data on 26 million loans to the HMDA database, which includes detailed information on loans, properties, applicants, and lenders. Approximately 57% of these loans were originated, with the rest being denied, withdrawn, or purchased on the secondary market. The purposes of these loans varied, with home purchases accounting for 35%, refinancing for 37%, cash-out refinancing for 20%, and home improvements for 4%.

To study the factors influencing mortgage rates for home buyers, we extracted records of originated loans for single-family home purchases with fixed interest rates ranging from 10 to 30 years. We created dummy variables for 100 largest lenders by their Legal Entity Identifiers. Our analysis also considers a variety of other factors, including the debt-to-income ratio, loan-to-property-value ratio, estimated market interest rate, loan size, loan term, popular term (indicating 15 or 30-year fixed-rate mortgages), down payment, discount points, origination charge, and indicators for non-primary residence, commercial property, applicant demographics, special loan types, regional incomes, and U.S. state variables. In total, we included 172 predictors in our model. To facilitate interpretation, level variables are expressed in logarithmic form, allowing us to interpret coefficients in terms of percentage changes.

The response variable in our study is the mortgage interest rate, quoted in percentage points. In 2021, the median interest rate hovered around 3%. Our parameter estimation is based on a sample size of approximately 4.3 million loans, with 1% of randomly sampled observations set aside for out-of-sample forecast validation.

1.11.2 Full-sample Estimation

Bayesian linear regression results under the non-informative prior are presented in Table 1.1. For brevity, only the first 10 lenders are displayed. The regression results demonstrate the payoff of mortgage shopping. While borrowed money may be homogeneous, lenders are not. Larger banks tend to offer lower interest rates, as evidenced by significantly negative coefficients. For example, Lender 5 offers a competitive mortgage rate that is 0.234 percentage points lower compared to smaller banks. If a home buyer borrows $500,000 with a 30-year fixed rate of 3.0, a reduction of 0.234 translates to a saving of $63 on the monthly payment.

The regression model has good interpretability, as the coefficients of most predictors have expected signs and significant values. Under the non-informative prior, the posterior

TABLE 1.1
Full and subset posterior means and merged results. For brevity, 5 out of 50 states and 10 out of 100 lenders are displayed in the table.

	Full	AL	AK	AZ	AR	CA	Merged
debt/Income	0.0039	0.0041	0.0030	0.0041	0.0035	0.0038	0.0039
loan/Value	0.0049	0.0049	0.0033	0.0060	0.0050	0.0061	0.0049
marketRate	0.3807	0.4751	0.3285	0.4071	0.4563	0.3693	0.3807
loanSize	−0.0015	−0.0022	−0.0019	−0.0012	−0.0023	−0.0010	−0.0015
downPayment	−0.0053	−0.0100	−0.0028	−0.0040	−0.0036	−0.0024	−0.0053
loanTerm	0.0022	0.0018	0.0020	0.0022	0.0018	0.0022	0.0022
popularTerm	−0.2964	−0.3201	−0.1379	−0.2044	−0.3639	−0.1650	−0.2964
discountPoints	−0.1491	−0.0927	−0.0698	−0.2453	−0.0501	−0.2093	−0.1491
originationFee	0.1440	0.0837	−0.0407	0.2169	0.0427	0.1936	0.1440
investment	0.4839	0.5230	0.7469	0.4937	0.5666	0.4272	0.4839
commercial	0.1344	0.0656	0.1066	0.0130	−0.1194	0.1088	0.1344
income	0.0010	0.0009	0.0014	0.0009	0.0008	0.0011	0.0010
age	0.0001	−0.0004	0.0007	−0.0007	−0.0006	0.0007	0.0001
female	−0.0066	−0.0044	−0.0202	−0.0078	−0.0002	0.0007	−0.0066
jointFilers	0.0004	−0.0052	0.0016	0.0189	0.0042	−0.0126	0.0004
asian	−0.0414	−0.0556	−0.0084	−0.0964	−0.0812	−0.0210	−0.0414
african	0.0847	0.1153	0.0195	0.0719	0.1503	0.0607	0.0847
hispanic	0.0841	0.0802	−0.0034	0.0749	0.0338	0.0358	0.0841
FHA	0.1570	0.1346	−0.0553	0.2701	0.1598	0.0335	0.1570
VA	−0.3902	−0.4378	−0.4244	−0.4350	−0.3376	−0.4716	−0.3902
FSA	−0.1626	−0.2248	−0.2264	−0.2544	−0.1289	−0.3570	−0.1626
regionIncome	−0.0004	0.0002	0.0002	0.0008	−0.0001	−0.0003	−0.0004
AL	2.1781	2.4838	0.0000	0.0000	0.0000	0.0000	2.1781
AK	2.0616	0.0000	1.6665	0.0000	0.0000	0.0000	2.0616
AZ	2.2737	0.0000	0.0000	0.4221	0.0000	0.0000	2.2737
AR	2.1292	0.0000	0.0000	0.0000	2.9863	0.0000	2.1292
CA	2.2297	0.0000	0.0000	0.0000	0.0000	1.0970	2.2297
lender1	−0.1728	−0.1421	0.0985	−0.2163	−0.0919	−0.2453	−0.1728
lender2	−0.1428	−0.0480	0.0279	−0.1675	−0.0199	−0.3326	−0.1428
lender3	−0.1753	−0.1679	0.0255	−0.2344	−0.1610	−0.2282	−0.1753
lender4	−0.1555	−0.0977	0.0853	−0.2205	−0.1423	−0.1385	−0.1555
lender5	−0.2344	−0.2070	−0.0657	−0.2513	−0.0976	−0.2743	−0.2344
lender6	−0.0263	0.0000	0.0050	0.0206	0.0606	−0.1406	−0.0263
lender7	−0.1267	−0.1140	0.0156	−0.1284	−0.0921	−0.2308	−0.1267
lender8	−0.0565	0.8286	0.0000	−0.0936	1.3838	−0.1209	−0.0565
lender9	−0.1510	−0.1096	0.0016	−0.1523	−0.1102	−0.3187	−0.1510
lender10	−0.1256	−0.0580	0.1098	−0.2166	0.0531	−0.2803	−0.1256
sigma2	0.1553	0.1510	0.0901	0.1556	0.1512	0.1354	0.1553

mean corresponds to the OLS estimator. The empirical results are the following. 1) Both the debt-to-income ratio and the loan-to-property-value ratio are positively correlated with the mortgage interest rate. If a home buyer decreases her debt-to-income ratio, she is expected to get a lower mortgage rate. Additionally, making a larger down payment contributes to a better interest rate. 2) Traditionally, the jumbo loan interest rate was higher than the standard loan rate, but in recent years, the trend has reversed. Data suggest that loan size is negatively correlated with the interest rate. 3) The coefficient of the loan term is positive, implying an upward-sloping yield curve. On the mortgage market, the most popular choices are 15- and 30-year fixed-rate mortgages, and their rates are indeed better than loans with less common terms. 4) Discount points are negatively correlated with the interest rate, as they are designed to reduce it. 5) Mortgages for primary residences are cheaper than those for investments, and loans for commercial purposes are more expensive. 6) Our empirical results suggest no obvious gender discrimination. Female applicants tend to get a better rate. 7) Asian applicants may obtain a good mortgage, possibly due to the high saving rate of Asian families influenced by Confucian culture.

Based on the fitted model, we use the reserved test data for out-of-sample prediction. The root mean squared error between the predicted and actual values is 0.396, which is slightly larger than the estimated $\sigma = 0.394$. The parameter uncertainty is low, while the noise term of the regression is relatively large. For data privacy and security, the public-use HMDA data do not disclose applicant credit scores, which can be a major cause of the discrepancy between the predicted and actual values.

1.11.3 Divide-and-conquer NIG Arithmetic

For brevity, Table 1.1 displays 5 out of 50 state dummy variables. The mortgage rates in some states are significantly lower than those in other states. It is of interest to run the regression for each state, which is most relevant to home buyers living in that state.

Consider a divide-and-conquer strategy that splits the big data into subsets by the states, runs the state-specific regressions, and then combines the subset posterior distributions. By Algorithm 1.1, the combined results should reproduce the full-sample results. However, the state dummy variables are problematic for the state-specific regressions under the non-informative prior, as the OLS estimator is not uniquely determined due to multicollinearity (some data are all zeros or ones). To resolve the problem, we specify a weakly informative ridge prior $NIG(0, I, 3, 1)$ for each state-specific regression. The subset posterior mean is identical to the prior mean for the state dummy variable coefficients, because the likelihood function does not change with those parameters when the subset observations are all zeros.

When we combine the 50 subset posterior distributions by NIG summation, the prior is repeated for 50 times. Therefore, we resort to the NIG subtraction and scalar multiplication operators, deducting $50 \cdot NIG(0, I, 3, 1)$ from the merged posteriors. The last column of the table confirms that the merged subset posteriors are identical to the full-sample posterior.

1.12 An Application on Large Bayesian Vector Autoregression

Vector Autoregression (VAR) has been a workhorse model in macroeconomics since the pioneering work of Sims (1980). Until recently, most empirical work employs small VARs that rarely include more than ten series. Although a high-dimensional VAR model can capture complex dynamic relationships among macroeconomic variables, the large number of parameters poses challenges for classical inference. Bayesian VAR addresses the curse of dimensionality by imposing a tight prior on the parameters. Litterman (1986) proposes a regularization method known as the Minnesota prior, where higher-order lagged variables are assigned smaller prior variances, resulting in shrinkage toward univariate random walk or autoregressive processes. Banbura et al. (2010) demonstrate a large Bayesian VAR with over a hundred variables. Koop (2013) examines the forecast performance of large Bayesian VAR versus factor models in a data-rich context. These promising findings generate a surge of interest in large Bayesian VAR applications, including Carriero et al. (2012), Ellahie and Ricco (2017), Huber and Feldkircher (2019), Carriero et al. (2019), Chan (2020), Cross et al. (2020), Kastner and Huber (2020), Cimadomo et al. (2022), among others. For more discussions of shrinkage methods in large Bayesian VAR, refer to Section 8.3.

1.12.1 Model Specification

The VAR model specification is conformable to the multivariate/vector regressions specified in Section 1.8:

$$Y_t = X_t B + \varepsilon_t,$$

where Y_t is a $1 \times m$ vector of endogenous variables, $X_t = (Y_{t-1}, \ldots, Y_{t-p})$ is a $1 \times pm$ vector of lagged variables and p indicates the VAR order. The disturbances ε_t follow a multivariate normal distribution with mean zero and covariance matrix Σ.

The VAR model can be analyzed from the perspective of m linear regressions, each having the same set of predictors, namely $Y_{t-1}, \ldots, Y_{t-p}$. Symmetric predictors enable the use of a conjugate NIW prior on the VAR coefficients B, Σ. Alternative views on the VAR as a state space model will be elaborated in Section 8.3.

We consider a VAR(4) model with 38 endogenous variables. The number of predictors is large relative to the number of observations, necessitating an informative prior $NIW(U, \Lambda, \Omega, v)$ for parameter regularization. We compare shrinkage toward random walk versus stationary AR(1) by specifying $\phi = \{1, \frac{1}{2}\}$. The hyperparameters are specified as follows:

$$U = \begin{pmatrix} \phi I_{38} \\ 0_{116\times 38} \end{pmatrix}, \Lambda = \begin{pmatrix} I_{38} & & & & \\ & 2I_{38} & & & \\ & & 3I_{38} & & \\ & & & 4I_{38} & \\ & & & & 0_2 \end{pmatrix}, \Omega = 10^{-4} I_{38}, v = 40.$$

It is possible to implement the conjugate Bayesian VAR using the regression functionality of standard statistical software. The idea is to exact pseudo data from the NIW prior by Proposition 1.11 and concatenate actual and pseudo data for standard regressions. Our prior specification corresponds to 194 pseudo data points consisting of 1) 154 dense data obtained by the Cholesky factor of Λ, 2) 38 sparse data obtained by the Cholesky factor of Ω, and 3) 2 points of zeros.

1.12.2 Data Description

We consider a moderately large VAR with 38 macroeconomic variables, including national accounts, business activities, labor markets, prices, money, banking, and financial indicators. The VAR series are the following:

- Real Gross Domestic Product
- Real Personal Consumption Expenditures

- Real Gross Private Domestic Investment
- Real Government Consumption Expenditures and Gross Investment
- Real Exports of Goods and Services
- Real Imports of Goods and Services
- Real Disposable Personal Income
- Industrial Production: Total Index
- Consumer Price Index for All Urban Consumers
- Consumer Price Index: All Items Less Food and Energy
- Producer Price Index by Commodity: All Commodities
- Gross Domestic Product: Chain-type Price Index
- Personal Consumption Expenditures: Chain-type Price Index
- Unemployment Rate
- Average Weeks Unemployed
- All Employees, Total Nonfarm
- Nonfarm Business Sector: Hourly Compensation for All Workers
- Capacity Utilization: Manufacturing
- New Privately-Owned Housing Units Started: Total Units
- Real Manufacturing and Trade Industries Sales
- Commercial and Industrial Loans
- University of Michigan Consumer Sentiment
- M2, Billions of Dollars
- Reserves of Depository Institutions: Total
- Federal Funds Effective Rate
- Treasury Securities at 1-Year Constant Maturity
- Treasury Securities at 3-Year Constant Maturity
- Treasury Securities at 5-Year Constant Maturity
- Treasury Securities at 7-Year Constant Maturity
- Treasury Securities at 10-Year Constant Maturity
- Treasury Securities at 20-Year Constant Maturity
- Moody's Seasoned AAA Corporate Bond Yield
- Moody's Seasoned BAA Corporate Bond Yield

- Wilshire 5000 Total Market Index
- NASDAQ Composite Index
- Spot Crude Oil Price: West Texas Intermediate
- U.S. Dollars to U.K. Pound Sterling Spot Exchange Rate
- Japanese Yen to U.S. Dollar Spot Exchange Rate

Quarterly data from 1975Q1 to 2023Q3 are available at Federal Reserve Economic Data database (fred.stlouisfed.org). Data are seasonally adjusted and log-transformed for most series. An intercept and linear trend are included in regressions.

1.12.3 Empirical Results

Figures 1.1 and 1.2 depict key macroeconomic variables over the past decade and compare out-of-sample forecast results for the subsequent 12 quarters under different priors. The notable outliers from 2020 to 2022 reflect the pandemic shock on the economy. Both models provide similar predictions regarding the growth momentum of output, consumption, investment, price, and stock market conditions. However, predictions differ for unemployment, manufacturing capacity utilization, and bond yields. Under the prior of $\phi = 1$, current trends persist over the next three years, and the time series exhibit random walk behavior.

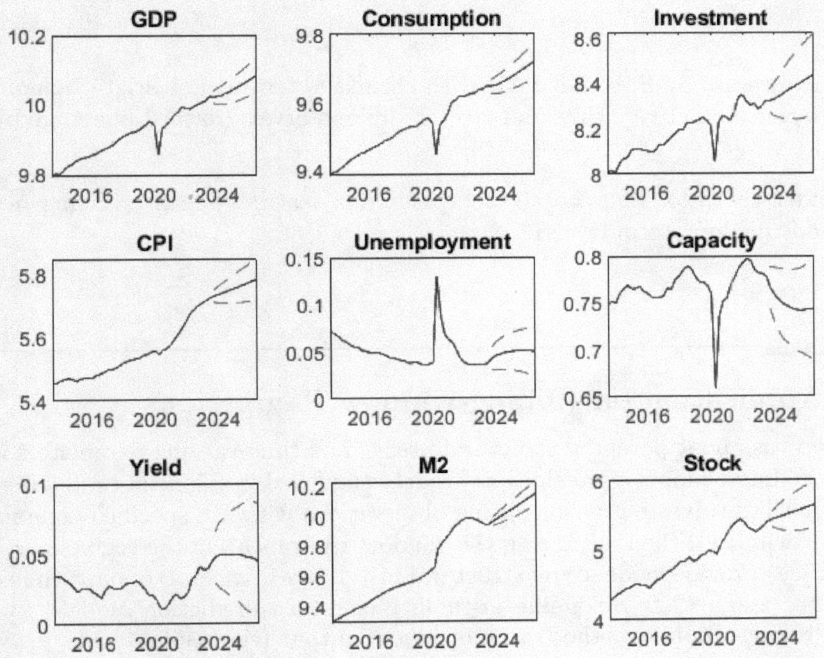

FIGURE 1.1
Macroeconomic forecast by Bayesian VAR with shrinkage toward univariate autoregressive processes. Posterior predictive means and 90% credible intervals (dashed lines) are plotted.

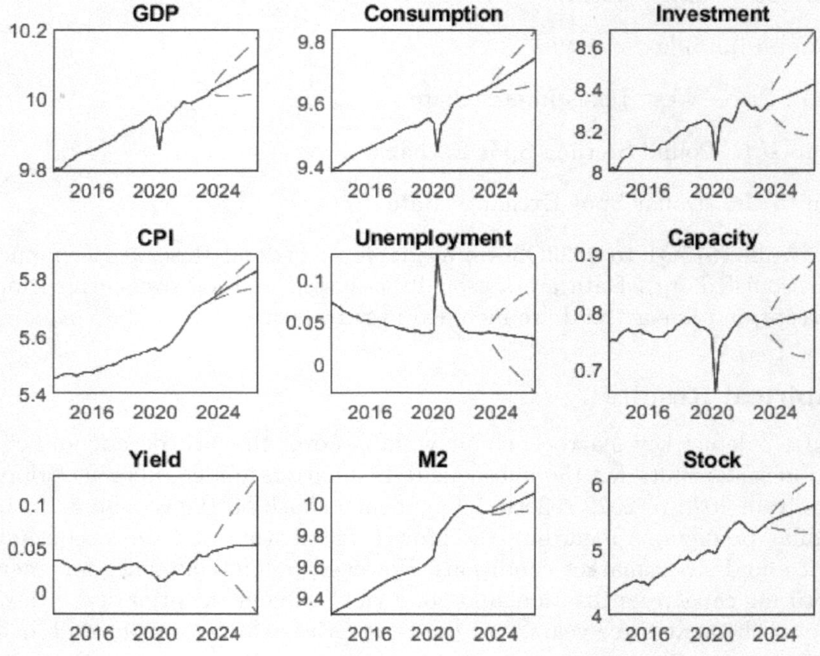

FIGURE 1.2
Macroeconomic forecast by Bayesian VAR with shrinkage toward univariate random walk processes. Posterior predictive means and 90% credible intervals (dashed lines) are plotted.

Conversely, under the prior of $\phi = 0.5$, the prediction features mean-reverting behavior toward pre-pandemic long-term levels.

1.13 An Application on Rolling-window Regressions

To address concerns about potential structural breaks and time-varying parameters in economic models, rolling-window regressions are widely employed in financial time series analysis. This approach involves regressions using observations within a specified time interval (e.g., a ten-year window), then advancing the window and repeating the regressions. Stock and Watson (1996) provide evidence on structural instability in macroeconomic time series. Paye and Timmermann (2006) examine instability of return prediction models. Swanson (1998) applies rolling-window methods to money and output regressions.

1.13.1 Five-factor Asset Pricing Model

To illustrate the NIG summation and subtraction operators, we consider rolling-window regressions based on the five-factor model of Fama and French (2015), which extends the influential three-factor asset pricing model (Fama and French, 1993). The factors capture the effects of size, value, profitability, investment, and market excess returns in the following regression form:

$$R_{it} - R_{ft} = \alpha + \beta_1 (R_{mt} - R_{ft}) + \beta_2 SMB_t + \beta_3 HML_t + \beta_4 RMW_t + \beta_5 CMA_t + \varepsilon_{it},$$

where R_{it} denotes the return of the ith asset or portfolio, R_{ft} represents the risk-free rate, and R_{mt} is the market return. SMB_t refers to the spread between small and big portfolio returns. HML_t refers to the spread between high and low book-to-market portfolio returns. RMW_t refers to the spread between robust and weak profitability portfolio returns. CMA_t refers to the spread between conservative and aggressive investment portfolio returns. If the factor exposures (i.e., the terms β_1, β_2, β_3, β_4, β_5) capture all variation in the expected excess returns, the intercept (i.e., the term α) equals zero.

1.13.2 Empirical Results

In this application, we analyze the daily returns of five industry portfolios spanning from July 1963 to June 2023, with data sourced from Professor French's website.

Assuming that regression coefficients remain relatively stable over a decade, we employ ten-year rolling windows, each comprising 2520 observations, to capture the enduring characteristics of the portfolios. The initial window covers the period from July 1963 to June 1973, with subsequent windows advancing monthly. In total, 600 rolling-window regressions are performed. According to Proposition 1.2, the posterior mean under a non-informative prior provides the OLS estimator. The estimated beta coefficients for the industry portfolios are illustrated in Figure 1.3.

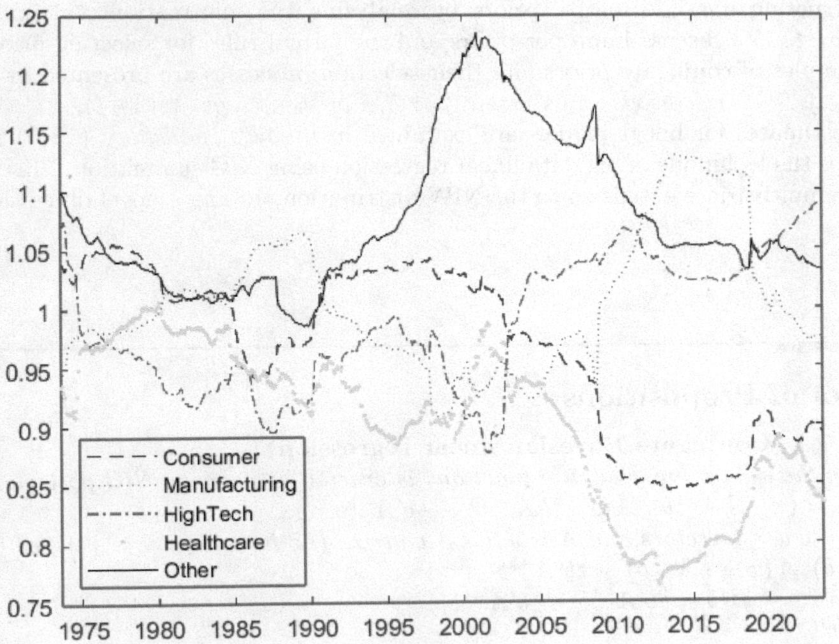

FIGURE 1.3
Rolling-window regressions of Fama-French five-factor asset pricing model.

The NIG summation and subtraction operators significantly streamline the computational process of rolling-window regressions. By incorporating new data with the NIG summation and phasing out old data with the NIG subtraction, we efficiently update our estimations as each window progresses by one month, adding approximately 21 observations. This method requires only 6000 floating-point operations (flops), a stark contrast to 210,000 flops needed for a direct regression. The use of NIG arithmetic thus offers a substantial reduction in computational load, enabling faster analysis and more efficient use of computer resources.

1.14 Bibliographic Notes

As a workhorse model of econometrics, linear regression is extensively covered in many Bayesian econometrics textbooks, including Koop (2003), Lancaster (2004), Geweke (2005), Greenberg (2013). These texts are excellent for easing readers into topics covered in this book. The comparative approach used in Poirier (1995) sheds light on how Bayesian inference differs from classical (frequentist) statistics. Poirier (1995) focuses on Bayesian theory and methodology (rather than Bayesian computation), making it a useful supplement to this book. Koop et al. (2007) and Chan et al. (2019) offer invaluable resources organized through a series of worked exercises, enabling hands-on learning and application of econometric principles.

Prior distributions are fundamental in Bayesian analysis. Diaconis and Ylvisaker (1979) delve into conjugate priors for the exponential family, a topic further explored by Bernardo and Smith (2000) with in-depth theory and applications. Jeffreys (1946) introduces a general rule for constructing non-informative priors by analyzing the information matrix. Kass and Wasserman (1996) discuss improper priors and the formal rules for selecting priors. A variety of examples of conjugate priors and their selection processes are presented in texts by Gelman et al. (2014), Bolstad and Curran (2017), Christensen et al. (2011).

Bayesian estimates for linear models are examined in Lindley and Smith (1972). Qian (2018) develops the technique of big data linear regression using NIG summation. This book introduces the multivariate extension to the NIW distribution, adding a novel dimension to the field.

1.15 Proof of Propositions

Proposition 1.1 (Conjugate Bayesian linear regression)

Completing the square for quadratic functions is crucial for deriving NIG posteriors:
$x'Ax + x'b = \left(x + \frac{1}{2}A^{-1}b\right)' A \left(x + \frac{1}{2}A^{-1}b\right) - \frac{1}{4}b'A^{-1}b,$
where x, b are $d \times 1$ vectors and A is a $d \times d$ matrix. The formula can be verified by
$\left(x + \frac{1}{2}A^{-1}b\right)' A \left(x + \frac{1}{2}A^{-1}b\right) - \frac{1}{4}b'A^{-1}b$
$= x'Ax + \frac{1}{2}b'x + \frac{1}{2}x'b + \frac{1}{4}b'A^{-1}b - \frac{1}{4}b'A^{-1}b$
$= x'Ax + x'b.$
For Bayesian linear regressions, the posterior density takes the form
$p\left(\beta, \sigma^2 \left| Y\right.\right) \propto \left(\sigma^2\right)^{-\left(a + \frac{n+d}{2} + 1\right)} e^{-\sigma^{-2}\left[b + \frac{1}{2}(\beta-\mu)'\Lambda(\beta-\mu) + \frac{1}{2}(Y - X\beta)'(Y - X\beta)\right]}.$
Completing the square for the quadratic function of β yields

$(\beta-\mu)'\Lambda(\beta-\mu)+(Y-X\beta)'(Y-X\beta)$
$=\beta'(\Lambda+X'X)\beta-2\beta'(\Lambda\mu+X'Y)+\mu'\Lambda\mu+Y'Y$
$=(\beta-\bar{\mu})'\bar{\Lambda}(\beta-\bar{\mu})-\bar{\mu}'\bar{\Lambda}\bar{\mu}+\mu'\Lambda\mu+Y'Y,$
where $\bar{\mu}=(\Lambda+X'X)^{-1}(\Lambda\mu+X'Y)$, *and* $\bar{\Lambda}=\Lambda+X'X$.
It follows that
$p(\beta,\sigma^2|Y)\propto(\sigma^2)^{-\left(a+\frac{n+d}{2}+1\right)}e^{-\frac{1}{2}\sigma^{-2}\left[2b+(\beta-\bar{\mu})'\bar{\Lambda}(\beta-\bar{\mu})-\bar{\mu}'\bar{\Lambda}\bar{\mu}+\mu'\Lambda\mu+Y'Y\right]}.$
This is the density of $NIG(\bar{\mu},\bar{\Lambda},\bar{a},\bar{b})$,
where $\bar{a}=a+\frac{n}{2}$ *and* $\bar{b}=b+\frac{1}{2}Y'Y+\frac{1}{2}\mu'\Lambda\mu-\frac{1}{2}\bar{\mu}'\bar{\Lambda}\bar{\mu}$.

Proposition 1.2 (Regression with non-informative prior)
The posterior density takes the form
$p(\beta,\sigma^2|Y)\propto(\sigma^2)^{-\left(\frac{n}{2}+1\right)}e^{-\frac{1}{2}\sigma^{-2}(Y-X\beta)'(Y-X\beta)}.$
Let $\tilde{\mu}=(X'X)^{-1}X'Y$ *be the OLS estimator. Expand the quadratic function as*
$(Y-X\beta)'(Y-X\beta)$
$=(Y-X\tilde{\mu}+X\tilde{\mu}-X\beta)'(Y-X\tilde{\mu}+X\tilde{\mu}-X\beta)$
$=(Y-X\tilde{\mu})'(Y-X\tilde{\mu})+(\beta-\tilde{\mu})'X'X(\beta-\tilde{\mu})-2(\beta-\tilde{\mu})'X'(Y-X\tilde{\mu}).$
Residuals are orthogonal to regressors, implying $X'(Y-X\tilde{\mu})=0$ *and*
$p(\beta,\sigma^2|Y)\propto(\sigma^2)^{-\left(\frac{n}{2}+1\right)}e^{-\frac{1}{2}\sigma^{-2}\left[(Y-X\tilde{\mu})'(Y-X\tilde{\mu})+(\beta-\tilde{\mu})'X'X(\beta-\tilde{\mu})\right]}.$
This is the density of $NIG\left(\tilde{\mu},\tilde{\Lambda},\tilde{a},\tilde{b}\right)$,
where $\tilde{\Lambda}=X'X$, $\tilde{a}=\frac{n-d}{2}$ *and* $\tilde{b}=\frac{1}{2}Y'Y-\frac{1}{2}Y'X(X'X)^{-1}X'Y$.

Proposition 1.3 (NIG pseudo data)
For notation simplicity, we drop tildes and focus on $NIG(\mu,\Lambda,a,b)$.
The pseudo data provide the sufficient statistics:
$X'X=(\Lambda^{1/2})'\Lambda^{1/2}=\Lambda,$
$X'Y=\Lambda\mu,$
$Y'Y=\mu'\Lambda\mu+2b.$
It follows that
$(X'X)^{-1}X'Y=\mu,$
$\frac{1}{2}\left[Y'Y-Y'X(X'X)^{-1}X'Y\right]=b.$
By Proposition 1.2, pseudo data lead to the posterior distribution $NIG(\mu,\Lambda,a,b)$.
Proposition 1.2 also indicates that the posterior under X^*,Y^* *satisfies*
$(X^{*\prime}X^*)^{-1}X^{*\prime}Y^*=\mu,$
$X^{*\prime}X^*=\Lambda,$
$\frac{1}{2}Y^{*\prime}Y^*-\frac{1}{2}(X^{*\prime}Y^*)'(X^{*\prime}X^*)^{-1}X^{*\prime}Y^*=b.$
Solving the three equations results in
$X^{*\prime}X^*=\Lambda$, $X^{*\prime}Y^*=\Lambda\mu$, $Y^{*\prime}Y^*=2b+\mu'\Lambda\mu$.

Proposition 1.4 (Properties of NIG summation)
Commutativity follows from the symmetry of two NIG distributions in Definition 1.2.
The sum $NIG(\mu,\Lambda,a,b)=\sum_{j=1}^J NIG(\mu_j,\Lambda_j,a_j,b_j)$ *is defined by*
$\mu=\left(\sum_{j=1}^J\Lambda_j\right)^{-1}\left(\sum_{j=1}^J\Lambda_j\mu_j\right),$
$\Lambda=\sum_{j=1}^J\Lambda_j,$
$a+\frac{d}{2}=\sum_{j=1}^J\left(a_j+\frac{d}{2}\right),$

$b = \sum_{j=1}^{J} b_j + \frac{1}{2}\sum_{j=1}^{J} (\mu_j - \mu)' \Lambda_j (\mu_j - \mu)$.

The sum does not depend on the order of participating NIGs, hence associativity.

For an arbitrary μ_0, *Definition 1.2 indicates that*

$(\Lambda + 0_{d\times d})^{-1} (\Lambda\mu + 0_{d\times d}\mu_0) = \mu$,

$\Lambda + 0_{d\times d} = \Lambda$,

$a - \frac{d}{2} + \frac{d}{2} = a$,

$b + 0 + \frac{1}{2}(\mu - \mu)' \Lambda (\mu - \mu) + \frac{1}{2}(\mu_0 - \mu)' 0_{d\times d} (\mu_0 - \mu) = b$.

It follows that $NIG\left(\mu_0, 0_{d\times d}, -\frac{d}{2}, 0\right)$ *serves as a zero in the summation.*

Proposition 1.5 (Separation of prior and data in posterior NIG)

By Proposition 1.2,

$\widetilde{\mu} = (X'X)^{-1} X'Y$, $\widetilde{\Lambda} = X'X$, $\widetilde{a} = \frac{n-d}{2}$, $\widetilde{b} = \frac{1}{2}Y'Y - \frac{1}{2}\widetilde{\mu}'\widetilde{\Lambda}\widetilde{\mu}$.

The sum $NIG\left(\widehat{\mu}, \widehat{\Lambda}, \widehat{a}, \widehat{b}\right) = NIG(\mu, \Lambda, a, b) + NIG\left(\widetilde{\mu}, \widetilde{\Lambda}, \widetilde{a}, \widetilde{b}\right)$ *is defined by*

$$\widehat{\mu} = \left(\Lambda + \widetilde{\Lambda}\right)^{-1} \left(\Lambda\mu + \widetilde{\Lambda}\widetilde{\mu}\right) = (\Lambda + X'X)^{-1} (\Lambda\mu + X'Y),$$

$$\widehat{\Lambda} = \Lambda + \widetilde{\Lambda} = \Lambda + X'X,$$

$$\widehat{a} = a + \widetilde{a} + \frac{d}{2} = a + \frac{n}{2},$$

$$\begin{aligned}\widehat{b} &= b + \widetilde{b} + \frac{1}{2}(\mu - \widehat{\mu})' \Lambda (\mu - \widehat{\mu}) + \frac{1}{2}(\widetilde{\mu} - \widehat{\mu})' \widetilde{\Lambda} (\widetilde{\mu} - \widehat{\mu}) \\ &= b + \frac{1}{2}Y'Y + \frac{1}{2}\mu'\Lambda\mu - \frac{1}{2}\widehat{\mu}'\widehat{\Lambda}\widehat{\mu}.\end{aligned}$$

The results align with Proposition 1.1, thus $NIG\left(\widehat{\mu}, \widehat{\Lambda}, \widehat{a}, \widehat{b}\right) = NIG\left(\overline{\mu}, \overline{\Lambda}, \overline{a}, \overline{b}\right)$.

Proposition 1.6 (NIG subset posteriors)

By Proposition 1.2,

$\widetilde{\mu}_j = \left(X_j'X_j\right)^{-1} X_j'Y_j$, $\widetilde{\Lambda}_j = X_j'X_j$, $\widetilde{a}_j = \frac{n_j-d}{2}$, $\widetilde{b}_j = \frac{1}{2}Y_j'Y_j - \frac{1}{2}\widetilde{\mu}_j'\widetilde{\Lambda}_j\widetilde{\mu}_j$.

The sufficient statistics satisfy

$X'X = \sum_{j=1}^{J} X_j'X_j$, $X'Y = \sum_{j=1}^{J} X_j'Y_j$, $Y'Y = \sum_{j=1}^{J} Y_j'Y_j$, $n = \sum_{j=1}^{J} n_j$.

The sum $NIG\left(\widehat{\mu}, \widehat{\Lambda}, \widehat{a}, \widehat{b}\right) = \sum_{j=1}^{J} NIG\left(\widetilde{\mu}_j, \widetilde{\Lambda}_j, \widetilde{a}_j, \widetilde{b}_j\right)$ *is defined by*

$$\widehat{\mu} = \left(\sum_{j=1}^{J} X_j'X_j\right)^{-1} \left(\sum_{j=1}^{J} X_j'Y_j\right) = (X'X)^{-1} X'Y,$$

$$\widehat{\Lambda} = \sum_{j=1}^{J} X_j'X_j = X'X,$$

$$\widehat{a} = \frac{n-d}{2},$$

$$\begin{aligned}\widehat{b} &= \sum_{j=1}^{J} \widetilde{b}_j + \frac{1}{2}\sum_{j=1}^{J} (\widetilde{\mu}_j - \widehat{\mu})' \widetilde{\Lambda}_j (\widetilde{\mu}_j - \widehat{\mu}) \\ &= \frac{1}{2}Y'Y - \widehat{\mu}'\left(\sum_{j=1}^{J} \widetilde{\Lambda}_j\widetilde{\mu}_j\right) + \frac{1}{2}\widehat{\mu}'\left(\sum_{j=1}^{J} \widetilde{\Lambda}_j\right)\widehat{\mu} \\ &= \frac{1}{2}Y'Y - \frac{1}{2}\widehat{\mu}'\widehat{\Lambda}\widehat{\mu}.\end{aligned}$$

The results align with Proposition 1.2, thus $NIG\left(\widehat{\mu}, \widehat{\Lambda}, \widehat{a}, \widehat{b}\right) = NIG\left(\widetilde{\mu}, \widetilde{\Lambda}, \widetilde{a}, \widetilde{b}\right)$.

Proposition 1.7 (Properties of NIG scalar multiplication)

We will show the case of $n = 2$. *Results for* $n > 2$ *can be obtained by induction.*

To show distributivity,

$$\begin{aligned}&\delta_1 \cdot NIG(\mu, \Lambda, a, b) + \delta_2 \cdot NIG(\mu, \Lambda, a, b) \\ &= NIG\left[\mu, \delta_1\Lambda, \delta_1\left(a + \tfrac{d}{2}\right) - \tfrac{d}{2}, \delta_1 b\right] + NIG\left[\mu, \delta_2\Lambda, \delta_2\left(a + \tfrac{d}{2}\right) - \tfrac{d}{2}, \delta_2 b\right] \\ &= NIG\left[\mu, (\delta_1 + \delta_2)\Lambda, (\delta_1 + \delta_2)\left(a + \tfrac{d}{2}\right) - \tfrac{d}{2}, (\delta_1 + \delta_2) b\right] \\ &= (\delta_1 + \delta_2) \cdot NIG(\mu, \Lambda, a, b).\end{aligned}$$

$\delta \cdot NIG(\mu_1, \Lambda_1, a_1, b_1) + \delta \cdot NIG(\mu_2, \Lambda_2, a_2, b_2)$

$= NIG\left[\mu_1, \delta\Lambda_1, \delta\left(a_1 + \frac{d}{2}\right) - \frac{d}{2}, \delta b_1\right] + NIG\left[\mu_2, \delta\Lambda_2, \delta\left(a_2 + \frac{d}{2}\right) - \frac{d}{2}, \delta b_2\right]$
$= NIG\left[\mu, \delta\Lambda, \delta\left(a + \frac{d}{2}\right) - \frac{d}{2}, \delta b\right]$
$= \delta \cdot NIG\left(\mu, \Lambda, a, b\right),$
where $\mu = \left(\Lambda_1 + \Lambda_2\right)^{-1}\left(\Lambda_1\mu_1 + \Lambda_2\mu_2\right)$, $\Lambda = \Lambda_1 + \Lambda_2$, $a = a_1 + a_2 + \frac{d}{2}$,
$b = b_1 + b_2 + \frac{1}{2}\left(\mu_1 - \mu\right)'\Lambda_1\left(\mu_1 - \mu\right) + \frac{1}{2}\left(\mu_2 - \mu\right)'\Lambda_2\left(\mu_2 - \mu\right)$.
By Definition 1.2, this is the sum of two NIGs.
To show compatibility,
$\delta_1 \cdot \left[\delta_2 \cdot NIG\left(\mu, \Lambda, a, b\right)\right]$
$= \delta_1 \cdot \left[NIG\left(\mu, \delta_2\Lambda, \delta_2\left(a + \frac{d}{2}\right) - \frac{d}{2}, \delta_2 b\right)\right]$
$= NIG\left(\mu, \delta_1\delta_2\Lambda, \delta_1\delta_2\left(a + \frac{d}{2}\right) - \frac{d}{2}, \delta_1\delta_2 b\right)$
$= \delta_1\delta_2 \cdot NIG\left(\mu, \Lambda, a, b\right).$
To show identity element,
$1 \cdot NIG\left(\mu, \Lambda, a, b\right) = NIG\left(\mu, 1 \cdot \Lambda, 1 \cdot \left(a + \frac{d}{2}\right) - \frac{d}{2}, 1 \cdot b\right) = NIG\left(\mu, \Lambda, a, b\right).$
$0 \cdot NIG\left(\mu, \Lambda, a, b\right) = NIG\left(\mu, 0 \cdot \Lambda, 0 \cdot \left(a + \frac{d}{2}\right) - \frac{d}{2}, 0 \cdot b\right) = NIG\left(\mu, 0_{d\times d}, -\frac{d}{2}, 0\right).$

Proposition 1.8 (Point prediction)
1. Let $\bar{Y}_f = E\left(Y_f \,|Y\right)$ *be the mean of the posterior predictive distribution.*
For brevity, denote $E\left[L\left(Y_f^*, Y_f\right)\right] = \int L\left(Y_f^*, Y_f\right) p\left(Y_f \,|Y\right) dY_f$.
It is understood that Y_f *is the random variable, to which the expectation operator is applied (with respect to the posterior predictive distribution), but* Y_f^* *and* $\bar{Y}_f$ *are predetermined (i.e., not random variables). The expectation takes the form:*
$E\left[\left(Y_f - Y_f^*\right)^2\right] = E\left[\left(Y_f - \bar{Y}_f + \bar{Y}_f - Y_f^*\right)^2\right] = E\left[\left(Y_f - \bar{Y}_f\right)^2\right] + \left(\bar{Y}_f - Y_f^*\right)^2.$
$E\left[\left(Y_f - \bar{Y}_f\right)^2\right]$ *is the variance of posterior predictive distribution, not a function of* Y_f^*.
Also, $\left(\bar{Y}_f - Y_f^*\right)^2$ *is non-negative and* $Y_f^* = \bar{Y}_f$ *is the unique minimizer.*
2. Under the asymmetric loss function, the expected loss is given by
$\int L\left(Y_f^*, Y_f\right) p\left(Y_f \,|Y\right) dY_f$
$= \int_{-\infty}^{Y_f^*} k_2\left(Y_f^* - Y_f\right) p\left(Y_f \,|Y\right) dY_f + \int_{Y_f^*}^{\infty} k_1\left(Y_f - Y_f^*\right) p\left(Y_f \,|Y\right) dY_f.$
Taking derivatives with respect to Y_f^* *yields the first-order condition*
$\int_{-\infty}^{Y_f^*} k_2 p\left(Y_f \,|Y\right) dY_f - \int_{Y_f^*}^{\infty} k_1 p\left(Y_f \,|Y\right) dY_f = 0.$
It follows that $k_2 F\left(Y_f^*\right) - k_1\left[1 - F\left(Y_f^*\right)\right] = 0,$
where $F\left(Y_f^*\right) = \int_{-\infty}^{Y_f^*} p\left(Y_f \,|Y\right) dY_f$ *is the cumulative distributional function.*
The solution is the quantile function $Y_f^* = F^{-1}\left(\frac{k_1}{k_1+k_2}\right)$.
3. This is a special case of the asymmetric loss: $k_1 = k_2 = 1$.
The optimal forecast is the median of posterior predictive distribution $Y_f^* = F^{-1}\left(\frac{1}{2}\right)$.
4. Instead of all-or-nothing loss function, consider a soft version with a tolerance level

$$L\left(Y_f^*, Y_f\right) = \begin{cases} 1 & \left|Y_f - Y_f^*\right| > \tau \\ 0 & \left|Y_f - Y_f^*\right| \leq \tau \end{cases}.$$

The expected loss is given by
$\int L\left(Y_f^*, Y_f\right) p\left(Y_f \,|Y\right) dY_f$
$= \int_{-\infty}^{Y_f^* - \tau} p\left(Y_f \,|Y\right) dY_f + \int_{Y_f^* + \tau}^{\infty} p\left(Y_f \,|Y\right) dY_f.$
Taking derivatives with respect to Y_f^* *yields the first-order condition*

$p\left(Y_f^* - \tau \,|Y\right) = p\left(Y_f^* + \tau \,|Y\right)$,

implying the slope of the line connecting two points $Y_f^* - \tau$ *and* $Y_f^* + \tau$ *equals zero.*

Taking the limit $\tau \to 0$ *yields* $p'\left(Y_f^* \,|Y\right) = 0$.

It follows that Y_f^* *is a local mode of the posterior predictive distribution.*

For a unimodal distribution, Y_f^* *is also the global mode:* $Y_f^* = \arg\max_{Y_f} p\left(Y_f \,|Y\right)$.

Proposition 1.9 (Posterior NIW under conjugate prior)

Completing the square for quadratic functions is crucial for deriving NIW posteriors:

$x'Ax + x'b + b'x = \left(x + A^{-1}b\right)' A \left(x + A^{-1}b\right) - b'A^{-1}b$,

where x, b *are* $d \times m$ *matrices and* A *is* $d \times d$ *matrix. The formula can be verified by*

$$\begin{aligned}&\left(x + A^{-1}b\right)' A \left(x + A^{-1}b\right) - b'A^{-1}b\\ &= x'Ax + b'x + x'b + b'A^{-1}b - b'A^{-1}b\\ &= x'Ax + x'b + b'x.\end{aligned}$$

For Bayesian linear regressions, the posterior density takes the form

$p\left(B, \Sigma \,|Y\right) \propto |\Sigma|^{-\frac{n+v+d+m+1}{2}} \exp\left[-\frac{1}{2} tr\left(M\Sigma^{-1}\right)\right]$,

where $M = \Omega + (B-U)' \Lambda (B-U) + (Y - XB)'(Y - XB)$.

Completing the square for the quadratic function of B *yields*

$$\begin{aligned}&(B-U)' \Lambda (B-U) + (Y-XB)'(Y-XB)\\ &= B'\left(\Lambda + X'X\right)B - B'\left(\Lambda U + X'Y\right) - \left(\Lambda U + X'Y\right)'B + U'\Lambda U + Y'Y\\ &= \left(B - \bar{U}\right)' \bar{\Lambda} \left(B - \bar{U}\right) - \bar{U}'\bar{\Lambda}\bar{U} + U'\Lambda U + Y'Y,\end{aligned}$$

where $\bar{U} = \left(\Lambda + X'X\right)^{-1}\left(\Lambda U + X'Y\right)$, *and* $\bar{\Lambda} = \Lambda + X'X$.

It follows that

$p\left(B, \Sigma \,|Y\right) \propto |\Sigma|^{-\frac{\bar{v}+d+m+1}{2}} \exp\left\{-\frac{1}{2} tr\left[\left(B - \bar{U}\right)' \bar{\Lambda} \left(B - \bar{U}\right)\Sigma^{-1} + \bar{\Omega}\Sigma^{-1}\right]\right\}$.

This is the density of $NIW\left(\bar{U}, \bar{\Lambda}, \bar{\Omega}, \bar{v}\right)$,

where $\bar{\Omega} = \Omega + Y'Y + U'\Lambda U - \bar{U}'\bar{\Lambda}\bar{U}$ *and* $\bar{v} = n + v$.

Proposition 1.10 (Posterior NIW under non-informative prior)

The posterior density takes the form

$p\left(B, \Sigma \,|Y\right) \propto |\Sigma|^{-\frac{n+m+1}{2}} \exp\left\{-\frac{1}{2} tr\left[(Y-XB)'(Y-XB)\Sigma^{-1}\right]\right\}$,

where

$$\begin{aligned}&(Y-XB)'(Y-XB)\\ &= B'X'XB - B'X'Y - Y'XB + Y'Y\\ &= \left(B - \tilde{U}\right)' \tilde{\Lambda} \left(B - \tilde{U}\right) + \tilde{\Omega}.\end{aligned}$$

We recognize that it is the density of $NIW\left(\tilde{U}, \tilde{\Lambda}, \tilde{\Omega}, \tilde{v}\right)$,

where $\tilde{U} = \left(X'X\right)^{-1}X'Y$, $\tilde{\Lambda} = X'X$, $\tilde{\Omega} = Y'Y - \tilde{U}'\tilde{\Lambda}\tilde{U}$, $\tilde{v} = n - d$.

Proposition 1.11 (NIW pseudo data)

The sufficient statistics are

$$\begin{aligned}X'X &= X'_{(1)}X_{(1)} + X'_{(2)}X_{(2)} + X'_{(3)}X_{(3)} = \tilde{\Lambda},\\ X'Y &= X'_{(1)}Y_{(1)} + X'_{(2)}Y_{(2)} + X'_{(3)}Y_{(3)} = \tilde{\Lambda}\tilde{U},\\ Y'Y &= Y'_{(1)}Y_{(1)} + Y'_{(2)}Y_{(2)} + Y'_{(3)}Y_{(3)} = \tilde{\Omega} + \tilde{U}'\tilde{\Lambda}\tilde{U}.\end{aligned}$$

It follows that the posterior distribution is $NIW\left(\tilde{U}, \tilde{\Lambda}, \tilde{\Omega}, \tilde{v}\right)$.

Proposition 1.10 indicates that X^*, Y^* *satisfies*

$(X^{*\prime}X^{*})^{-1}X^{*\prime}Y^{*}=\tilde{U}$,
$X^{*\prime}X^{*}=\tilde{\Lambda}$,
$Y^{*\prime}Y^{*}-\tilde{U}^{\prime}\tilde{\Lambda}\tilde{U}=\tilde{\Omega}$.
Solving these equations results in
$X^{*\prime}X^{*}=\tilde{\Lambda}$, $X^{*\prime}Y^{*}=\tilde{\Lambda}\tilde{U}$, $Y^{*\prime}Y^{*}=\tilde{\Omega}+\tilde{U}^{\prime}\tilde{\Lambda}\tilde{U}$.

Proposition 1.12 (Properties of NIW summation)
Commutativity, associativity, and additive identity follow from the definition of NIW summation. See the proof of Proposition 1.4.

Proposition 1.13 (Separation of prior and data in posterior NIW)
By Proposition 1.10,
$\tilde{U}=(X^{\prime}X)^{-1}X^{\prime}Y$, $\tilde{\Lambda}=X^{\prime}X$, $\tilde{\Omega}=Y^{\prime}Y-\tilde{U}^{\prime}\tilde{\Lambda}\tilde{U}$, $\tilde{v}=n-d$.
The sum $NIW\left(\widehat{U},\widehat{\Lambda},\widehat{\Omega},\widehat{v}\right)=NIW\left(U,\Lambda,\Omega,v\right)+NIW\left(\tilde{U},\tilde{\Lambda},\tilde{\Omega},\tilde{v}\right)$ *is defined by*
$\widehat{U}=\left(\Lambda+\widetilde{\Lambda}\right)^{-1}\left(\Lambda U+\widetilde{\Lambda}\widetilde{U}\right)=(\Lambda+X^{\prime}X)^{-1}(\Lambda U+X^{\prime}Y)$,
$\widehat{\Lambda}=\Lambda+\widetilde{\Lambda}=\Lambda+X^{\prime}X$,
$\widehat{\Omega}=\Omega+Y^{\prime}Y+U^{\prime}\Lambda U-\hat{U}^{\prime}\widehat{\Lambda}\widehat{U}$,
$\widehat{v}=n+v$.
The results align with Proposition 1.9, thus $NIW\left(\widehat{U},\widehat{\Lambda},\widehat{\Omega},\widehat{v}\right)=NIW\left(\bar{U},\bar{\Lambda},\bar{\Omega},\bar{v}\right)$.

Proposition 1.14 (NIW subset posteriors)
By Proposition 1.10,
$\tilde{U}_j=\left(X_j^{\prime}X_j\right)^{-1}X_j^{\prime}Y_j$, $\tilde{\Lambda}_j=X_j^{\prime}X_j$, $\tilde{\Omega}_j=Y_j^{\prime}Y_j-\tilde{U}_j^{\prime}\tilde{\Lambda}_j\tilde{U}_j$, $\tilde{v}_j=n_j-d$.
The sufficient statistics satisfy
$X^{\prime}X=\sum_{j=1}^{J}X_j^{\prime}X_j$, $X^{\prime}Y=\sum_{j=1}^{J}X_j^{\prime}Y_j$, $Y^{\prime}Y=\sum_{j=1}^{J}Y_j^{\prime}Y_j$, $n=\sum_{j=1}^{J}n_j$.
The sum $NIW\left(\widehat{U},\widehat{\Lambda},\widehat{\Omega},\widehat{v}\right)=\sum_{j=1}^{J}NIW\left(\tilde{U}_j,\tilde{\Lambda}_j,\tilde{\Omega}_j,\tilde{v}_j\right)$ *is defined by*
$\widehat{U}=\left(\sum_{j=1}^{J}X_j^{\prime}X_j\right)^{-1}\left(\sum_{j=1}^{J}X_j^{\prime}Y_j\right)=(X^{\prime}X)^{-1}X^{\prime}Y$,
$\widehat{\Lambda}=\sum_{j=1}^{J}X_j^{\prime}X_j=X^{\prime}X$,
$\widehat{\Omega}=Y^{\prime}Y-\hat{U}^{\prime}\widehat{\Lambda}\widehat{U}$,
$\tilde{v}=n-d$.
The results align with Proposition 1.10, thus $NIW\left(\widehat{U},\widehat{\Lambda},\widehat{\Omega},\widehat{v}\right)=NIW\left(\tilde{U},\tilde{\Lambda},\tilde{\Omega},\tilde{v}\right)$.

Proposition 1.15 (Marginal likelihood under NIW prior)
By Bayes' rule
$p\left(Y\right)=\frac{p(\Sigma)p(B|\Sigma)p(Y|B,\Sigma)}{p(\Sigma|Y)p(B|\Sigma,Y)}$,
Evaluating each of those densities at $B=0$ *and* $\Sigma=I_m$ *results in*
$p\left(\Sigma\right)=\left[2^{\frac{mv}{2}}\Gamma_m\left(\frac{v}{2}\right)\right]^{-1}\left|\Omega\right|^{\frac{v}{2}}e^{-\frac{1}{2}tr(\Omega)}$,
$p\left(B\left|\Sigma\right.\right)=(2\pi)^{-\frac{dm}{2}}\left|\Lambda\right|^{\frac{m}{2}}e^{-\frac{1}{2}tr\left(U^{\prime}\Lambda U\right)}$,
$p\left(\Sigma\left|Y\right.\right)=\left[2^{\frac{m\bar{v}}{2}}\Gamma_m\left(\frac{\bar{v}}{2}\right)\right]^{-1}\left|\bar{\Omega}\right|^{\frac{\bar{v}}{2}}e^{-\frac{1}{2}tr\left(\bar{\Omega}\right)}$,
$p\left(B\left|\Sigma,Y\right.\right)=(2\pi)^{-\frac{dm}{2}}\left|\bar{\Lambda}\right|^{\frac{m}{2}}e^{-\frac{1}{2}tr\left(\bar{U}^{\prime}\bar{\Lambda}\bar{U}\right)}$,
$p\left(Y\left|B,\Sigma\right.\right)=(2\pi)^{-\frac{mn}{2}}e^{-\frac{1}{2}tr\left(Y^{\prime}Y\right)}$.

Because of $\bar{\Omega} = \Omega + Y'Y + U'\Lambda U - \bar{U}'\bar{\Lambda}\bar{U}$,
all terms involving $e^{-\frac{1}{2}tr(\cdot)}$ *are cancelled, simplifying the marginal likelihood as*
$p(Y) = (2\pi)^{-\frac{mn}{2}} \frac{f(\Lambda,\Omega,v)}{f(\bar{\Lambda},\bar{\Omega},\bar{v})}$,
where $f(\Lambda, \Omega, v) = \left[2^{\frac{mv}{2}} \Gamma_m \left(\frac{v}{2}\right)\right]^{-1} |\Omega|^{\frac{v}{2}} |\Lambda|^{\frac{m}{2}}$,
and Γ_m *is the multivariate gamma function.*

Proposition 1.16 (Pseudo data in the prior)
1. The non-informative prior $\epsilon_{T\eta A}(\tau_0, v_0)$ *is updated by pseudo data* $\tilde{Y}$ *such that*
$p\left(\theta \middle| \tilde{Y}\right) \propto e^{\eta(\theta)'\left[\tau_0 + \sum_{i=1}^{n_0} T(\tilde{Y}_i)\right] - (v_0+n_0)A(\theta)}$.
Pseudo data are defined by $\sum_{i=1}^{n_0} T\left(\tilde{Y}_i\right) = \tau - \tau_0$ *and* $n_0 = v - v_0$.
Therefore, the posterior is $\epsilon_{T\eta A}(\tau, v)$.
2. The non-informative prior $\epsilon_{T\eta A}(\tau_0, v_0)$ *is updated by pseudo and genuine data:*
$p\left(\theta \middle| \tilde{Y}\right) \propto e^{\eta(\theta)'\left[\tau_0 + \sum_{i=1}^{n_0} T(\tilde{Y}_i)\tau_0 + \sum_{i=1}^{n} T(Y_i)\right] - (v_0+n_0+n)A(\theta)}$.
Since $\bar{\tau} = \tau + \sum_{i=1}^{n} T(Y_i)$ *and* $\tau = \tau_0 + \sum_{i=1}^{n_0} T\left(\tilde{Y}_i\right)$,
as well as $\bar{v} = v + n$ *and* $v = v_0 + n_0$, *the posterior is* $\epsilon_{T\eta A}(\bar{\tau}, \bar{v})$.

Proposition 1.17 (Separation of prior and data in posterior)
Let $Y_i, i = 1, \ldots, n$ *be the data. The posterior under the conjugate prior is obtained by*
$\bar{\tau} = \tau + \sum_{i=1}^{n} T(Y_i)$,
$\bar{v} = v + n$.
Similarly, the posterior under the non-informative prior is
$\tilde{\tau} = \tau_0 + \sum_{i=1}^{n} T(Y_i)$,
$\tilde{v} = v_0 + n$.
By the summation rules,
$\epsilon_{T\eta A}(\tau, v) + \epsilon_{T\eta A}(\tilde{\tau}, \tilde{v})$
$= \epsilon_{T\eta A}(\tau + \tilde{\tau} - \tau_0, v + \tilde{v} - v_0)$
$= \epsilon_{T\eta A}\left(\tau + \sum_{i=1}^{n} T(Y_i), v + n\right)$
$= \epsilon_{T\eta A}(\bar{\tau}, \bar{v})$.

1.16 Exercises

Exercise 1.1 (Characteristics of NIG distribution) *The density of* $NIG(\mu, \Lambda, a, b)$ *is defined as*

$$p\left(\beta, \sigma^2\right) \propto \left(\sigma^2\right)^{-\left(a+\frac{d}{2}+1\right)} e^{-\sigma^{-2}\left[b+\frac{1}{2}(\beta-\mu)'\Lambda(\beta-\mu)\right]},$$

where the normalizing constant is concealed by the proportionality operator ($\propto$). *The joint density determines marginal and conditional densities of* β *and* σ^2, *as well as moments like means, variances and covariances. Show the following properties of the NIG distribution.*

1. The normalizing constant is

$$(2\pi)^{-\frac{d}{2}} |\Lambda|^{\frac{1}{2}} \left[\Gamma(a)\right]^{-1} b^a.$$

2. Marginal distributions are

$$\sigma^2 \sim IG\left(a,b\right),$$

$$\beta \sim MVT\left(\mu, \frac{b}{a}\Lambda^{-1}, 2a\right).$$

3. Conditional distributions are

$$\beta\left|\sigma^2\right. \sim N\left(\mu, \sigma^2\Lambda^{-1}\right),$$

$$\sigma^2\left|\beta\right. \sim IG\left[a+\frac{d}{2}, b+\frac{1}{2}\left(\beta-\mu\right)'\Lambda\left(\beta-\mu\right)\right].$$

4. Means and variances are

$$E\left(\beta\right)=\mu, Var\left(\beta\right)=\frac{b}{a-1}\Lambda^{-1},$$

$$E\left(\sigma^2\right)=\frac{b}{a-1}, Var\left(\sigma^2\right)=\frac{b^2}{\left(a-1\right)^2\left(a-2\right)}.$$

5. Random variables β and σ^2 are not independent, but uncorrelated:

$$corr\left(\beta,\sigma^2\right)=0.$$

Exercise 1.2 (Independent prior and conditional conjugacy) *The conjugate NIG prior for the pair $\left(\beta,\sigma^2\right)$ may not be as intuitive as an independent prior such that $\beta \sim N\left(\mu,\Lambda^{-1}\right)$ and $\sigma^2 \sim IG\left(a,b\right)$. We consider Bayesian linear regressions under the independent prior specification.*

1. Show that the posterior distribution $p\left(\beta,\sigma^2\left|Y\right.\right)$ is not the density of a common distribution under the independent prior, but the posterior conditional distributions are of recognizable forms: $\beta\left|\sigma^2\right.,Y \sim N\left(\bar{\mu},\bar{\Lambda}^{-1}\right)$ and $\sigma^2\left|\beta\right.,Y \sim IG\left(\bar{a},\bar{b}\right)$, where

$$\bar{\mu}=\left(\Lambda+\sigma^{-2}X'X\right)^{-1}\left(\Lambda\mu+\sigma^{-2}X'Y\right),$$

$$\bar{\Lambda}=\Lambda+\sigma^{-2}X'X,$$

$$\bar{a}=a+\frac{n}{2},$$

$$\bar{b}=b+\frac{1}{2}\left(Y-X\beta\right)'\left(Y-X\beta\right).$$

2. Show that the posterior conditional distributions are determined by the sufficient statistics $X'X$, $X'Y$ and $Y'Y$.

3. How do you compute the sum of squared residuals $\left(Y-X\beta\right)'\left(Y-X\beta\right)$? Do you expect any numerical issues?

4. How do you specify the hyperparameters μ,Λ,a,b for the independent prior and the conjugate NIG prior?

Exercise 1.3 (Data extraction and NIG summation) *Consider two NIG distributions $NIG(\mu_1, \Lambda_1, a_1, b_1)$ and $NIG(\mu_2, \Lambda_2, a_2, b_2)$. First, we apply Proposition 1.3 and extract data from the NIG distributions. Second, we concatenate extracted data. Third, we apply Proposition 1.2 and run a Bayesian linear regression with concatenated data under the non-informative prior. Show that the posterior distribution is given by the sum of two NIGs.*

Exercise 1.4 (Borrow priors) *Suppose that we split the data X, Y into J subsets X_j, Y_j, $j = 1, \ldots, J$. Instead of running subset regressions under the non-informative prior, we may borrow the original prior $NIG(\mu, \Lambda, a, b)$ for each of the J subset regressions. As J copies of the prior are used by the NIG summation, we must subtract $J-1$ duplicates from the total, leaving only one copy of the prior.*

Consider an alternative scheme that leverages the scalar multiplication operator: we borrow the discounted prior $\frac{1}{J} \cdot NIG(\mu, \Lambda, a, b)$ for each of the J subset regressions. Show that NIG summation of the subset posteriors provides the correct result, without the need of subtracting the duplicated priors.

Exercise 1.5 (Multivariate t as a scalar mixture) *The multivariate t distribution can be represented by a scale mixture of normal distributions, which is helpful for Bayesian linear regressions. Let $\sigma^2 \sim IG(a, b)$ and $\beta \,|\sigma^2 \sim N\left(\mu, \sigma^2 V\right)$. Show that $\beta \sim MVT\left(\mu, \frac{b}{a} V, 2a\right)$, where both normal and t distributions are parametrized by the covariance/scale matrix V.*

Exercise 1.6 (Marginal likelihood computation) *In the linear regression model, the marginal likelihood follows the multivariate t distribution, namely $t\left(X\mu, \frac{b}{a}V, 2a\right)$, where $V = I_n + X\Lambda^{-1}X'$, with the density*

$$p(Y) = \frac{\Gamma\left(a + \frac{n}{2}\right)(2\pi)^{-\frac{n}{2}} b^a}{\Gamma(a)} |V|^{-\frac{1}{2}} \left[b + \frac{1}{2}(Y - X\mu)' V^{-1} (Y - X\mu)\right]^{-\left(a + \frac{n}{2}\right)}.$$

When the sample size is large, it is not feasible to compute the inverse and determinant of V. We resort to linear algebra tools to simplify computation.

1. Consider Woodbury inversion formula:

$$(A + BCD)^{-1} = A^{-1} - A^{-1}B\left(C^{-1} + DA^{-1}B\right)^{-1} DA^{-1}.$$

Use the formula to simplify V^{-1}.

2. Consider Sylvester determinant formula:

$$|I_n + BD| = |I_d + DB|.$$

Use the formula to simplify $|V|^{-\frac{1}{2}}$.

3. Let L be the Cholesky factor of $\Lambda + X'X$, that is, $LL' = \Lambda + X'X$. Does the Cholesky factor simplify the matrix inversion and determinant computation?

4. Simplify the marginal likelihood $p(Y)$ using the above results.

5. Write $p(Y)$ as a function of the sufficient statistics $X'X$, $X'Y$ and $Y'Y$.

Exercise 1.7 (Decomposition of NIW distribution) *The density of* $NIW(U, \Lambda, \Omega, v)$ *is given by*

$$p(B, \Sigma) \propto |\Sigma|^{-\frac{v+d+m+1}{2}} \exp\left\{-\frac{1}{2}tr\left[(B-U)' \Lambda (B-U) \Sigma^{-1} + \Omega\Sigma^{-1}\right]\right\}.$$

We decompose the joint density as

$$p(\Sigma) \propto |\Sigma|^{-\frac{v+m+1}{2}} \exp\left\{-\frac{1}{2}tr\left[\Omega\Sigma^{-1}\right]\right\},$$

$$p(B|\Sigma) \propto |\Sigma|^{-\frac{d}{2}} \exp\left\{-\frac{1}{2}tr\left[(B-U)' \Lambda (B-U) \Sigma^{-1}\right]\right\}.$$

Show that $p(\Sigma)$ *is the density of the inverse Wishart distribution* $IW(\Omega, v)$, *while* $p(B|\Sigma)$ *is the density of the matrix-value normal distribution* $N(U, \Lambda^{-1}, \Sigma)$. *If we reshape the* $d \times m$ *matrix* B *as a* $dm \times 1$ *vector, then* $vec(B)|\Sigma$ *follows the multivariate normal distribution with the mean* $vec(U)$ *and the covariance matrix* $\Sigma \otimes \Lambda^{-1}$. *Explain the covariance matrix.*

Exercise 1.8 (Gaussian likelihood) *Consider a linear regression system with* m *equations, each of which has* d *predictors:* $Y_i = X_i B + \varepsilon_i$, *where* Y_i *is a* $1 \times m$ *response vector,* X_i *is a* $1 \times d$ *predictor vector,* B *is a* $d \times m$ *coefficient matrix, and* ε_i *follows the multivariate normal distribution with the mean zero and the covariance matrix* Σ. *Show that the likelihood function is given by*

$$p(Y|B, \Sigma) \propto |\Sigma|^{-\frac{n}{2}} \exp\left\{-\frac{1}{2}\sum_{i=1}^{n}(Y_i - X_i B)\Sigma^{-1}(Y_i - X_i B)'\right\},$$

or in the matrix form:

$$p(Y|B, \Sigma) \propto |\Sigma|^{-\frac{n}{2}} \exp\left\{-\frac{1}{2}tr\left[(Y-XB)'(Y-XB)\Sigma^{-1}\right]\right\}.$$

Exercise 1.9 (Linear regression with a fixed disturbance variance) *In some applications, we want to fix the disturbance variance* $\bar{\sigma}^2$ *in the linear regression:* $Y = X\beta + \bar{\sigma}\varepsilon$. *Assume that the prior of* β *is multivariate normal with the mean* μ *and the precision* $\bar{\sigma}^{-2}\Lambda$ *(the covariance matrix is* $\bar{\sigma}^2\Lambda^{-1}$*).*

1. Show that the posterior of β *is multivariate normal with the mean* $\bar{\mu}$ *and the precision* $\bar{\sigma}^{-2}\bar{\Lambda}$, *where* $\bar{\mu}$ *and* $\bar{\Lambda}$ *given by Proposition 1.1.*

2. Suppose that we approximate a fixed disturbance variance $\bar{\sigma}^2$ *by an inverse gamma distribution* $IG(\kappa, \bar{\sigma}^2\kappa)$, *where* κ *is a large positive scalar (large-*κ *approximation). Under the prior* $NIG(\mu, \Lambda, \kappa, \bar{\sigma}^2\kappa)$, *we apply Proposition 1.1 for the posterior distribution. Show that the result under a fixed variance is the limiting case of Proposition 1.1 as* $\kappa \to \infty$.

Exercise 1.10 (Exogenous predictors in Gaussian linear regression) *Consider the Gaussian linear regression model*

$$p(Y_i|\theta, X_i) = \frac{1}{\sqrt{2\pi}\sigma}e^{-\frac{(Y_i - X_i\beta)^2}{2\sigma^2}}.$$

Define $\varepsilon_i = \sigma^{-1}(Y_i - X_i\beta)$, *and we have the identity*

$$Y_i = X_i\beta + \sigma\varepsilon_i,$$

1. Show that ε_i *follows the standard normal distribution.*

2. Show that ε_i *is independent to* X_i. *The Gaussian linear regression model implies that the predictors are exogenous.*

Exercise 1.11 (Data Rescaling) *Consider Bayesian vector autoregression (VAR) model formulated as multivariate linear regressions:*

$$Y_t = X_t B + \varepsilon_t,$$

where Y_t *is a* $1 \times m$ *vector of response variables,* $X_t = (Y_{t-1}, \ldots, Y_{t-p})$ *is a* $1 \times pm$ *vector of lagged variables and* p *indicates the VAR order. The disturbances* ε_t *follow the multivariate normal distribution with the mean zero and the covariance matrix* Σ. *Suppose that typical values of the data* Y_t *are so large that we encounter ill-conditioned matrices in computing the posterior. It is desirable to rescale the data as* $\tilde{Y}_t = Y_t/c$ *and* $\tilde{X}_t = X_t/c$, *where* c *is a constant (e.g.,* $c = 100$*).*

1. If we specify a non-informative prior, does the posterior of B *and/or* Σ *change after rescaling the data? Explain the reason.*

2. If we specify a conjugate prior $NIW(U, \Lambda, \Omega, v)$, *does the posterior of* B *and/or* Σ *change after rescaling the data? If yes, discuss how to reproduce the original results by adjusting the prior.*

2

Markov Chain Monte Carlo Methods

Chapter 1 delves into exact Bayesian inference, focusing on scenarios where posterior distributions are analytically tractable. However, for most econometric models discussed in subsequent chapters, the posterior does not conform to a common distribution form. In such cases, the posterior density can only be evaluated up to a normalizing constant. Particularly challenging are models like stochastic volatility, where the likelihood function cannot be evaluated in closed form, necessitating approximation by an unbiased estimator.

Since the 1990s, the Markov chain Monte Carlo (MCMC) method has become a cornerstone of Bayesian computation, rendering Bayesian statistical inference as accessible as classical statistics. The computational efficiency of MCMC appeals to empirical researchers, even those who may not fully embrace subjectivism. Compared to numerical maximum likelihood estimation (MLE), which may falter with poorly specified initial values, MCMC is robust, less sensitive to starting points, and commonly initiated with over-dispersed values. MCMC excels over MLE in complex models with numerous parameters, such as time-varying-parameter vector autoregressions and high-dimensional Gaussian mixtures. In scenarios where the likelihood function is intractable, simulated MLE is hampered by Monte Carlo noise, whereas the correlated pseudo-marginal MCMC method still targets the exact posterior distribution. This approach is exemplified in nonlinear state space models, where the particle filter approximates the likelihood by the sample average of importance weights. Further exploration of state space models is reserved for Chapters 6, 7 and 8.

This chapter provides an overview of MCMC methods, unified by the Metropolis-Hastings (MH) sampling technique, which is powerful, adaptable, and scalable. We begin with the discrete-state MH algorithm, offering a gentle introduction to the topic. Its simplicity allows for elementary probabilistic tools to establish convergence results, fostering an understanding of reversibility, irreducibility, and aperiodicity of the Markov chain. We explore variations of MH sampling, including the Gibbs sampler, slice sampling, and the Metropolis-within-Gibbs algorithm. The multiple-try MH variant facilitates large jumps without compromising acceptance rates, and its suitability for parallel computing enhances efficiency. Hamiltonian Monte Carlo aligns with the standard MH framework by its single-step leapfrog version, known as Langevin Monte Carlo. Trans-dimensional MCMC methods also fit within the MH sampling paradigm.

In the latter half of the chapter, we adapt MH methods to address big data challenges. To alleviate the computational load of evaluating target densities in massive datasets, we consider divide-and-conquer and subsampling MCMC techniques, such as consensus Monte Carlo, MH sampling with delayed acceptance, firefly Monte Carlo, subsampling with control variates, and stochastic gradient Langevin dynamics. The correlated pseudo-marginal method emerges as a promising strategy for designing big data MCMC algorithms.

DOI: 10.1201/9781003564027-2

2.1 Discrete-state MH Sampler

Consider a sequence of random variables with the joint distribution

$$p(\theta_0, \theta_1, \ldots \theta_T) = p(\theta_0) \prod_{t=1}^{t} p(\theta_t | \theta_{t-1}, \ldots, \theta_0).$$

If the probability of transitioning to the next state depends only on the current state, i.e., $p(\theta_t | \theta_{t-1}, \ldots, \theta_0) = p(\theta_t | \theta_{t-1})$, and the functional form is time-invariant, the sequence is termed a homogeneous Markov chain. In such a chain, the future evolution depends solely on the present state, without any path dependency on past states.

When the state space is finite, such as $\theta_t \in \{1, \ldots, m\}$, the sequence is referred to as a discrete-state Markov chain. The law of state evolution $p(\theta_t | \theta_{t-1})$ is characterized by an $m \times m$ transition matrix $Q = \{q_{ij}\}$, where q_{ij} indicates the probability of transitioning from state i to j in one step.

The marginal distribution $p(\theta_t)$, represented as a $1 \times m$ vector $\pi^{(t)}$, can be recursively computed using the relation

$$\pi^{(t)} = \pi^{(t-1)} Q.$$

The stationary distribution (also known as the invariant distribution, equilibrium distribution, etc.) of a discrete-state Markov chain is represented by a $1 \times m$ probability vector π such that

$$\pi = \pi Q,$$

implying that the marginal distributions remain π if the initial state distribution is π. The existence and uniqueness of the stationary distribution are intrinsic properties of the transition matrix.

MCMC methods are designed to construct a Markov chain with a specified stationary distribution, rather than finding the stationary distribution of a given Markov chain. Most MCMC methods utilize the reversibility condition to construct such chains.

Proposition 2.1 (Reversibility condition) *Let $\pi = \{\pi_i\}$ be the target distribution and $Q = \{q_{ij}\}$ be the transition matrix. If they satisfy the reversibility condition (also known as the detailed balance condition):*

$$\pi_i q_{ij} = \pi_j q_{ji}, \forall i, j \in \{1, \ldots, m\},$$

then π is a stationary distribution of the Markov chain Q.

Proposition 2.1 provides a test for stationarity. Given a target distribution π, we construct a Markov chain with transition matrix Q. If the reversibility condition is satisfied, then π must be a stationary distribution of Q. However, failure of the reversibility test does not imply that π is not a stationary distribution. Reversibility is a sufficient but not necessary condition for stationarity. Example 2.1 demonstrates a non-reversible chain.

Example 2.1 (Non-reversible chain) *Let $\pi = \left(\frac{1}{3}, \frac{1}{3}, \frac{1}{3}\right)$ and $Q = \begin{pmatrix} 0.2 & 0.3 & 0.5 \\ 0.5 & 0.2 & 0.3 \\ 0.3 & 0.5 & 0.2 \end{pmatrix}$. Due to $\pi_1 q_{12} \neq \pi_2 q_{21}$, the reversibility condition does not hold. However, π is the unique stationary distribution of the regular Markov chain Q.*

Algorithm 2.1 Discrete-state MH sampler

Input: target distribution $\pi = \{\pi_i\}$, proposal $R = \{r_{ij}\}$, and starting value θ_0.
Output: MCMC draws $\theta_1, \theta_2, \ldots$.
Iterate the following steps for $t = 1, 2, \ldots$
1 Sample θ^* from the proposal $P(\theta^* = j) = r_{ij}$, $j = 1, \ldots, m$, given $i = \theta_{t-1}$.
2 Compute acceptance probability $\alpha_{ij} = \min\left(1, \frac{\pi_j r_{ji}}{\pi_i r_{ij}}\right)$, where $j = \theta^*$.
3 Set $\theta_t = \begin{cases} \theta^* & \textit{with probability } \alpha_{ij} \\ \theta_{t-1} & \textit{with probability } 1 - \alpha_{ij} \end{cases}$.

The reversibility condition is fundamental to the MH sampler (see Algorithm 2.1). Initially, the proposal transition matrix may not satisfy the reversibility condition, but adjustments can be made to achieve it. As shown in Equation (2.1), the modified transition matrix is constructed by discounting the off-diagonal elements and enhancing the diagonal elements of the matrix. An alternative description of this modification involves accepting a proposed transition with a carefully designed probability: if accepted, the process transitions to the new state; if rejected, it remains in the current state. Proposition 2.2 establishes the reversibility of the modified Markov chain, ensuring the target distribution is its stationary distribution.

Proposition 2.2 (Reversibility of MH sampling) *Let $\pi = \{\pi_i\}$ be the target distribution and $R = \{r_{ij}\}$ be a proposal transition matrix. The modified transition matrix $Q = \{q_{ij}\}$ is constructed by*

$$q_{ij} = \alpha_{ij} r_{ij} + I(i = j) \sum_{k=1}^{m} (1 - \alpha_{ik}) r_{ik}, \tag{2.1}$$

where $\alpha_{ij} = \min\left(1, \frac{\pi_j r_{ji}}{\pi_i r_{ij}}\right)$, and $I(i = j)$ is an indicator function that equals one if $i = j$. Then, π is a stationary distribution of the Markov chain Q.

Several remarks on Proposition 2.2 are noteworthy. The proposition guarantees the existence of the stationary distribution for the constructed Markov chain. It ensures that the marginal distributions of the draws generated by Algorithm 2.1 conform to the stationary distribution, provided the initial value is drawn from this distribution. However, it does not address what occurs if the initial value is sampled from a different distribution. Additionally, it offers no insight into whether the sample average of the draws approximates the mean of the target distribution. These questions are intricately linked to the asymptotic theory of Markov chains, as outlined in Proposition 2.3.

The asymptotic behavior of a Markov chain is determined by its irreducibility and aperiodicity. A Markov chain is irreducible if there exists a positive transition probability from one state to any other state in a finite number of steps, ensuring that all states can communicate. The chain is aperiodic if the greatest common divisor of the return times is one, allowing states to return at irregular time intervals. Proposition 2.3 demonstrates the convergence of an irreducible and aperiodic Markov chain. By convergence, we mean that the marginal distribution converges to the stationary distribution, and the empirical distribution of a sample path converges to the stationary distribution. If the MH sampler is run for a sufficiently long time, not only is the last draw approximately generated from the stationary distribution, but the sample path also characterizes the stationary distribution.

Irreducibility and aperiodicity are crucial for the convergence of Markov chains. The MH sampler may be ill-behaved due to reducibility or periodicity, as shown in Examples 2.2 and 2.3.

Proposition 2.3 (Convergence of Markov chains) *Let $\theta_t \in \{1,\ldots,m\}$, $t = 0,1,2,\ldots$ be an irreducible and aperiodic Markov chain with the stationary distribution π. The initial distribution is arbitrary. For all $j = 1,\ldots,m$, we have*

1. $\lim_{t\to\infty} P(\theta_t = j) = \pi_j$.
2. $\frac{1}{T}\sum_{t=1}^{T} I(\theta_t = j) \xrightarrow{a.s.} \pi_j$.

Example 2.2 (Reducible chain) *Consider a target distribution $\pi = \left(\frac{1}{3},\frac{1}{3},\frac{1}{3}\right)$ and a proposal transition matrix $R = \begin{pmatrix} 1/2 & 1/4 & 1/4 \\ 1/4 & 3/4 & 0 \\ 0 & 1/2 & 1/2 \end{pmatrix}$. The MH algorithm modifies the transition matrix as $Q = \begin{pmatrix} 3/4 & 1/4 & 0 \\ 1/4 & 3/4 & 0 \\ 0 & 0 & 1 \end{pmatrix}$. Although R is irreducible, Q is reducible because the transition probability from the third state to any other state is zero, meaning not all states communicate. Even though Q satisfies the reversibility condition and π is a (non-unique) stationary distribution, the Markov chain does not converge to π unless the initial distribution is π. If Algorithm 2.1 is executed, the draws either oscillate between the first two states or get trapped in the third state. Consequently, the large-sample frequencies are $\left(\frac{1}{2},\frac{1}{2},0\right)$ or $(0,0,1)$, depending on the initial state.*

Example 2.3 (Periodic chain) *Consider a target distribution $\pi = \left(\frac{1}{4},\frac{1}{2},\frac{1}{4}\right)$ and a quasi-random walk proposal $R = \begin{pmatrix} 0 & 1 & 0 \\ 1/2 & 0 & 1/2 \\ 0 & 1 & 0 \end{pmatrix}$. Since R satisfies the reversibility condition, the MH sampler accepts all proposals, resulting in $Q = R$. The Markov chain is irreducible, and the unique stationary distribution is $\left(\frac{1}{4},\frac{1}{2},\frac{1}{4}\right)$. However, the chain has a periodicity of 2, as states return only in even periods, not in odd periods. If the initial distribution is (a_1,a_2,a_3), the marginal distribution in an odd period becomes $\left(\frac{a_2}{2}, a_1+a_3, \frac{a_2}{2}\right)$, while in an even period, it is $\left(\frac{a_1+a_3}{2}, a_2, \frac{a_1+a_3}{2}\right)$. These marginal distributions oscillate and do not converge to the stationary distribution, unless the initial distribution satisfies $a_2 = a_1+a_3 = \frac{1}{2}$. Nonetheless, running Algorithm 2.1 results in draws that appear well-behaved, with the large-sample frequency being $\left(\frac{1}{4},\frac{1}{2},\frac{1}{4}\right)$, regardless of the initial state. Although the draws follow the wrong marginal distribution, the sample path still provides a good approximation of the target distribution. It is important to retain the entire path of draws without thinning.*

The discrete-state MH sampler is not only of pedagogical interest but also of practical importance. One application is sampling from a high-dimensional categorical distribution. Consider sampling from a k dimensional binary vector, which corresponds to a target distribution with 2^k states. The curse of dimensionality makes it infeasible to enumerate all components of the categorical distribution. The MH sampler avoids the overwhelming task of enumeration by evaluating π_i and π_j on the fly. A random-walk proposal randomly flips one element of the k dimensional vector, resulting in a simplified MH acceptance probability: $\alpha_{ij} = \min\left(1, \frac{\pi_j}{\pi_i}\right)$. A proposed jump to a higher-probability state is always accepted, whereas a lower-probability proposal is accepted with probability $\frac{\pi_j}{\pi_i}$. Consequently, states with relatively large probabilities are more likely to be sampled. Even if the number of draws is significantly less than 2^k, the draws can still provide a good approximation of the mean or mode of the target distribution.

2.2 MH Sampler in the General Form

We extend the finite-state Markov chain to the real-valued version, which is fundamental to MCMC sampling for multivariate continuous distributions.

2.2.1 Discrete versus Real-valued Markov Chains

A Markov chain in its general form is a sequence of real-valued random variables where the future evolution of the chain depends solely on the current state. We compare discrete-valued states $\theta_t \in \{1, \ldots, m\}$ with real-valued states $\theta_t \in R^d$, while keeping the time steps discrete: $t = 0, 1, 2, \ldots$.

A discrete-state Markov chain is characterized by a transition matrix $Q = \{q_{ij}\}$ such that $q_{ij} \geq 0$ and $\sum_{j=1}^{m} q_{ij} = 1$. In contrast, a real-valued Markov chain features a transition kernel $q(\theta^*; \theta)$ such that $q(\theta^*; \theta) \geq 0$ and $\int q(\theta^*; \theta)\, d\theta^* = 1$.

The marginal distribution of a discrete chain is a probability vector obtained recursively by $\pi^{(t)} = \pi^{(t-1)} Q$. In contrast, the marginal distribution of a real-valued chain is a density function satisfying the recursion

$$\pi^{(t)}(\theta^*) = \int \pi^{(t-1)}(\theta)\, q(\theta^*; \theta)\, d\theta.$$

The stationary distribution of a discrete Markov chain is a probability vector such that $\pi = \pi Q$. In contrast, the stationary distribution of a real-valued chain is a density function $\pi(\cdot)$ that satisfies

$$\pi(\theta^*) = \int \pi(\theta)\, q(\theta^*; \theta)\, d\theta.$$

The existence and uniqueness of the stationary distribution are intrinsic properties of the transition kernel. Examples 2.4 and 2.5 illustrate real-valued Markov chains in time series econometrics.

Example 2.4 (AR series) *An autoregressive process is a continuous-state Markov chain. Let $\theta_t = c + \rho\theta_{t-1} + \sigma\varepsilon_t$, where ε_t follows the standard normal distribution. The transition kernel is $q(\theta^*; \theta) = \phi(\theta^*; c + \rho\theta, \sigma^2)$. If $|\rho| < 1$, the Markov chain has the stationary distribution $\pi(\theta) = \phi\left(\theta; \frac{c}{1-\rho}, \frac{\sigma^2}{1-\rho^2}\right)$. Similarly, vector autoregressions $\theta_t = c + A\theta_{t-1} + B\varepsilon_t$ has the stationary distribution $\pi(\theta) = \phi\left(\theta; (1-A)^{-1} c, V\right)$, where V is the solution to the Lyapunov equation $V = AVA' + BB'$, if eigenvalues of A are in the unit circle.*

Conversely, given a target distribution $\pi(\theta) = \phi(\theta; \mu, V)$, it is possible to construct a Markov chain that has $\pi(\theta)$ as its stationary distribution. The transition kernel can be constructed as $q(\theta^*; \theta) = \phi(\theta^*; (I - A)\mu + A\theta, BB')$, where A is a user-specified autoregressive matrix, and B can be the Cholesky factor of $V - AVA'$.

Example 2.5 (ARCH series) *Consider an autoregressive conditional variance model: $\theta_t = c + v_t$, $v_t = \sigma_t\varepsilon_t$, $\sigma_t^2 = k + \alpha v_{t-1}^2$, where ε_t is the standard normal disturbance and σ_t is the conditional volatility. Rewriting the model as $\theta_t = c + \sqrt{k + \alpha(\theta_{t-1} - c)^2}\,\varepsilon_t$, it forms a continuous-state Markov chain, with the transition kernel $q(\theta^*; \theta) = \phi\left(\theta^*; c, k + \alpha(\theta - c)^2\right)$. The stationary distribution exists if $k > 0$ and $0 < \alpha < 1$.*

2.2.2 Reversible Markov Chain

Instead of seeking the stationary distribution of a given Markov chain, MCMC methods construct a Markov chain that has the target distribution (e.g., the posterior) as its stationary distribution. If the reversibility condition holds, the target distribution must be a stationary distribution of the Markov chain, as demonstrated in Proposition 2.4 (see also Proposition 2.1 for the discrete counterpart). However, if the reversibility condition does not hold, it is still possible for the target distribution to be a stationary distribution of the Markov chain, as illustrated by Example 2.6. Reversibility is a sufficient but not necessary condition for validating that an MCMC method targets the correct distribution.

Proposition 2.4 (Reversibility) *Let $\pi(\cdot)$ be the target distribution and $q(\cdot)$ be the transition kernel of a Markov chain. If they satisfy the reversibility condition (also known as the detailed balance condition):*

$$\pi(\theta)\, q(\theta^*; \theta) = \pi(\theta^*)\, q(\theta; \theta^*),$$

for all points θ, θ^* in the state space, then $\pi(\cdot)$ is a stationary distribution of the Markov chain.

Example 2.6 (Non-reversible chain) *Consider a target distribution*

$$\pi(\theta) = 2(1-\theta)\, I(0 \leq \theta \leq 1).$$

Construct a Markov chain on the state space $[0, 1]$, with the transition kernel

$$q(\theta^*; \theta) = \left[e^{1-\theta} - e^{\theta^*-\theta} I(\theta \leq \theta^*)\right] \cdot I(0 \leq \theta^* \leq 1).$$

This is a proper conditional density that satisfies $q(\theta^*; \theta) \geq 0$ and $\int q(\theta^*; \theta)\, d\theta^* = 1$. We can verify the integral $\int \pi(\theta)\, q(\theta^*; \theta)\, d\theta = \pi(\theta^*)$, which indicates that $\pi(\cdot)$ is a stationary distribution of $q(\cdot)$. However, the reversibility condition does not hold. For example, let $\theta = 0.5$ and $\theta^* = 0.6$. We have $\pi(\theta)\, q(\theta^*; \theta) = 0.54$, while $\pi(\theta^*)\, q(\theta; \theta^*) = 1.19$.

2.2.3 MH Sampling

Let $\pi(\theta)$ be the target distribution and $r(\theta^*; \theta)$ be a proposal transition kernel, from which the MH sampler makes a proposal θ^* conditional on the current state θ. A proposal is accepted with a carefully designed probability to satisfy the reversibility condition. If the proposal is rejected, the chain remains in the current state. The implication is that the modified transition kernel $q(\theta^*; \theta)$ becomes a mixture with some probability mass concentrated on θ, even if $r(\theta^*; \theta)$ is specified as a continuous function, such as a multivariate normal density.

To describe the mixture distribution, we use the Dirac delta function $\delta_\theta(\theta^*) = 0, \forall \theta^* \neq \theta$ and $\int_A \delta_\theta(\theta^*)\, d\theta^* = I(\theta \in A)$. An interpretation of the Dirac delta function is the limit of a density function, such as a normal density with mean θ and an infinitesimal variance, which produces a spike at θ.

The modified transition kernel takes the form

$$q(\theta^*; \theta) = \alpha(\theta^*, \theta)\, r(\theta^*; \theta) + \delta_\theta(\theta^*) \int [1 - \alpha(\theta^*, \theta)]\, r(\theta^*; \theta)\, d\theta^*, \tag{2.2}$$

where

$$\alpha(\theta^*, \theta) = \min\left[1, \frac{\pi(\theta^*)\, r(\theta; \theta^*)}{\pi(\theta)\, r(\theta^*; \theta)}\right].$$

Algorithm 2.2 MH sampler

Input: target distribution $\pi(\cdot)$, proposal $r(\cdot)$, and starting value θ_0.
Output: MCMC draws $\theta_1, \theta_2, \ldots$.
Iterate the following steps for $t = 1, 2, \ldots$
1 Generate θ^* from the proposal $r(\theta^*; \theta)$, where $\theta = \theta_{t-1}$.
2 Compute acceptance probability $\alpha(\theta^*, \theta) = \min\left[1, \frac{\pi(\theta^*) r(\theta; \theta^*)}{\pi(\theta) r(\theta^*; \theta)}\right]$.
3 Set $\theta_t = \begin{cases} \theta^* & \textit{with probability } \alpha(\theta^*, \theta) \\ \theta_{t-1} & \textit{with probability } 1 - \alpha(\theta^*, \theta) \end{cases}$.

Equation (2.2) is analogous to Equation (2.1) (reproduced below for quick reference):

$$q_{ij} = \alpha_{ij} r_{ij} + I(i = j) \sum_{k=1}^{m} (1 - \alpha_{ik}) r_{ik}.$$

Note that $\alpha_{ij} r_{ij}$ is the discrete counterpart of $\alpha(\theta^*, \theta) r(\theta^*; \theta)$, which describes a successful transition from state i to j, or θ to θ^* (that is, a proposal is accepted), while $I(i = j)$ is analogous to $\delta_\theta(\theta^*)$, which describes a rejected transition. The probability that the chain stays in the original state is given by $\sum_{k=1}^{m} (1 - \alpha_{ik}) r_{ik}$ and $\int [1 - \alpha(\theta^*, \theta)] r(\theta^*; \theta) d\theta^*$ for the discrete and continuous cases, respectively. The row-sum of the transition matrix (the integral of the transition kernel) is always equal to one. Apart from some notation differences between discrete and continuous versions, the idea of constructing a reversible Markov chain is the same.

The MH sampler facilitates Bayesian inference in situations where an intractable posterior density can only be evaluated pointwise up to a normalizing constant. Instead of directly sampling from the target distribution, the MH sampler takes an indirect sampling approach by constructing a Markov chain that has the posterior as its stationary distribution. Algorithm 2.2 describes the MH sampler, and the first part of Proposition 2.5 validates stationarity: if the current state follows the target distribution, then the subsequent state follows the same distribution. The second part of the proposition provides an alternative validation method: the reversibility condition holds, and the Markov chain must have the posterior as its stationary distribution.

Proposition 2.5 (Stationary distribution of MH sampling) *Let $\pi(\cdot)$ be the target distribution and $r(\cdot)$ be a proposal transition kernel.*

1. Suppose that θ follows $\pi(\theta)$, and θ^* is generated by $r(\theta^*; \theta)$. Let

$$\tilde{\theta} = \begin{cases} \theta^* & \textit{with probability } \alpha(\theta^*, \theta) \\ \theta & \textit{with probability } 1 - \alpha(\theta^*, \theta) \end{cases}.$$

For an arbitrary square integrable function h, we have $E\left[h\left(\tilde{\theta}\right)\right] = E[h(\theta)]$.

2. $\pi(\cdot)$ is a stationary distribution of the reversible $q(\cdot)$ in Equation (2.2).

2.2.4 Choice of Transition Kernel

The performance of the MH sampler is heavily influenced by the choice of the proposal transition kernel, with popular options including the random walk, independence, and product proposals.

The random walk MH sampler features a symmetric transition kernel $r(\theta^*;\theta) = r(\theta;\theta^*)$. A common specification is the multivariate normal density: $r(\theta^*;\theta) = \phi(\theta^*;\theta,\Sigma)$. Alternatively, a multivariate t density can be used to encourage larger jumps. The symmetry of the kernel simplifies the MH acceptance probability to $\min\left[1, \frac{\pi(\theta^*)}{\pi(\theta)}\right]$. It implies that 1) a proposal must be accepted if $\pi(\theta^*) > \pi(\theta)$, ensuring most draws are sampled from the high-density regions; 2) a proposal must be rejected if $\pi(\theta^*) = 0$, constraining all draws within the support of the target distribution; and 3) a low-density proposal has a positive probability of being accepted, facilitating comprehensive exploration of the entire parameter space.

The independence MH sampler has a transition kernel that does not depend on the current state of the Markov chain $r(\theta^*;\theta) = r(\theta^*)$. Typically, the proposal distribution is specified as a Gaussian approximation of the target distribution: $r(\theta^*;\theta) = \phi(\theta^*;\mu,\Sigma)$, where μ and Σ represent the posterior mode and the inverse Hessian, respectively. A proposal from the independence MH sampler is accepted with probability $\min\left[1, \frac{w(\theta^*)}{w(\theta)}\right]$, where $w(\theta) = \frac{\pi(\theta)}{r(\theta)}$, reminiscent of the importance weight in importance sampling.

The product MH sampler is applicable when the target density is proportional to the product of two terms $\pi(\theta^*) \propto f(\theta^*)\, g(\theta^*)$, with $f(\theta^*)$ being the density function of a common distribution. Proposals are generated from $f(\theta^*)$ and accepted with probability $\min\left[1, \frac{g(\theta^*)}{g(\theta)}\right]$.

Sampler tuning involves the judicious specification of parameters in the proposal distribution, a task that can be challenging, especially in high-dimensional settings. The random walk MH sampler can underperform if the covariance matrix Σ is not appropriately sized. Similarly, the independence MH sampler may fail if the mean μ is far from the posterior mode or if the covariance matrix Σ is not carefully specified.

A common tuning strategy for a tailored transition kernel involves numerical optimization for the posterior mode: $\hat{\theta} = \arg\max_\theta \pi(\theta)$. Consider the second-order Taylor approximation of the log posterior density

$$\ln \pi(\theta) = \ln \pi\left(\hat{\theta}\right) + G\left(\hat{\theta}\right)\left(\theta - \hat{\theta}\right) + \frac{1}{2}\left(\theta - \hat{\theta}\right)' H\left(\hat{\theta}\right)\left(\theta - \hat{\theta}\right),$$

where $G\left(\hat{\theta}\right)$ and $H\left(\hat{\theta}\right)$ represent the gradient and Hessian, respectively. Since $\hat{\theta}$ maximizes the log posterior density, the gradient evaluated at the mode equals zero: $G\left(\hat{\theta}\right) = 0$. Also, the Hessian matrix evaluated at the maximizer should be negative definite, even if the objective function is not globally concave. Thus, the posterior density $\pi(\theta)$ is approximately proportional to $\exp\left[\frac{1}{2}\left(\theta - \hat{\theta}\right)' H\left(\hat{\theta}\right)\left(\theta - \hat{\theta}\right)\right]$, justifying a multivariate normal approximation with mean $\hat{\theta}$ and covariance matrix $\Sigma = \left[-H\left(\hat{\theta}\right)\right]^{-1}$.

Unfortunately, this descent proposal does not always work in practice. First, numerical optimization itself is computationally intensive. Second, there is no guarantee that the optimization program converges to the posterior mode; it may settle at a saddle point, and the Hessian may not be negative definite. A symptom of this issue is difficulty in inverting the Hessian and failure in the Cholesky decomposition of the covariance matrix. Third, even if the posterior mode is identified, the numerical Hessian obtained by finite differencing can be computationally expensive and inaccurate due to catastrophic cancellation errors.

Some workarounds are available. First, the quasi-Newton optimization algorithm has an internal approximation of the inverse Hessian, which is updated with each evaluation of the gradient. This positive definite matrix does not resemble the true Hessian but captures curvature. Such a by-product can be recycled as the proposal for MCMC simulation at no additional computing cost. Second, if the sample size is large and the prior is not tight, we may consider the outer product of gradients of the likelihood function as an approximation of the inverse Hessian. It ignores the prior, but at least the matrix remains positive definite. Third, when computing the full Hessian can be too expensive, it is possible to compute only the diagonals, leaving the off-diagonal elements as zero. The bottom line is that the MH sampler does not require an accurate evaluation of $\left[-H\left(\hat{\theta}\right)\right]^{-1}$. It suffices to have a positive definite approximation of it.

Another tuning strategy involves multi-stage simulation. Starting with an initial guess of the proposal, a trial run of MCMC generates the sample mean and covariance matrix of the draws for updating the proposal for subsequent simulations. This process is repeated until the MH acceptance rate is roughly 1/4 (Gelman et al., 1997). If previous samples are treated as burn-in draws, it still aligns with the MH sampler. Otherwise, the proposal distribution is dynamically modified during the simulation, leading to the adaptive MCMC (Haario et al., 2001; Roberts et al., 2007).

Note that a trial run of the MH sampler generates not only samples but also the target density evaluated at each proposal draw. It is possible to estimate the location and scale of the target distribution using the recycled densities. This recycling strategy is economical when it is computationally expensive to evaluate the target distribution. Let θ_t^*, $t = 1, \ldots, T$ be the proposals, and $\ln \pi\left(\theta_t^*\right)$ be the corresponding log densities. Assume that the log density is approximated by a quadratic function: $\ln \pi(\theta) = c + \theta'\beta + \frac{1}{2}\theta' H \theta$, where the coefficients c, β, H can be estimated by least squares using the "regression data": θ_t^*, $\ln \pi\left(\theta_t^*\right)$, $t = 1, \ldots, T$. If the matrix H is negative definite, the maximizer of the quadratic function is $-H^{-1}\beta$. A reasonable proposal distribution is a multivariate normal or t distribution with mean $-H^{-1}\beta$ and covariance matrix proportional to $-H^{-1}$. The estimated $-H^{-1}\beta$ is usually close to the mode of the target density obtained by computationally expensive numerical optimization. The quadratic approximation strategy does not always make good proposals, but it suggests a way to recycle the densities calculated by the trial MH sampler to estimate the location and scale of the target distribution at minimal computing cost.

2.2.5 Convergence of Markov Chain

Proposition 2.5 shows the existence of an equilibrium such that the constructed Markov chain has the target distribution as its stationary distribution. However, it does not show how to reach the equilibrium from an arbitrary initial state. Irreducibility and aperiodicity are crucial for the convergence of the Markov chain. A sufficient condition for an irreducible MH chain is that the proposal transition kernel is everywhere positive on the support of the target distribution, ensuring that states are communicable. That is, $r\left(\theta^*; \theta\right) > 0$, for all $\left(\theta^*, \theta\right) \in S \times S$, where S is the support of the target distribution. This positivity condition ensures that every measurable set on the support of the target distribution can be reached in a single step. A sufficient condition for aperiodicity is that not all proposals are accepted and the probability of the event $\{\theta_{t+1} = \theta_t\}$ is positive. Since it is possible for a state to depart and return in a single step, the greatest common divisor of the round-trip time equals one, and states return at irregularly spaced time intervals. In addition, sample path convergence requires a recurrent Markov chain.

Under these regularity conditions, the marginal distributions converge to the stationary distribution, and the empirical average of the sample path, such as $\frac{1}{T}\sum_{t=1}^{T} h\left(\theta_t\right)$, converges to the population moment $E\left[h(\theta)\right]$, where $h(\cdot)$ is a function of interest. If the MCMC sampler is run for a sufficiently long time, the draws are approximately sampled from the stationary distribution, and the sample path (excluding burn-in draws) can be used for statistical inference. See Tierney (1994) and Robert and Casella (2004) for rigorous treatments of convergence theorems.

From a practical perspective, it is necessary to assess convergence by MCMC draws in an empirical study. Cowles and Carlin (1996) provide a comparative review of 13 convergence diagnostic methods (see Table 1 in their paper).

In the context of posterior simulation of Bayesian econometric models, we consider a two-step procedure for convergence assessment.

First, generate artificial data of increasing sample sizes (say 500, 5000, and 50000) conformable to the model specification and perform posterior simulation for each artificial dataset under a weak prior. The fundamental difficulty of assessing convergence is that the MCMC chain converges to an unknown distribution. However, as the sample size increases, they degenerate toward the specified parameters that generate the artificial data. The sample means should tend to the true parameters, and the sample variances tend to zero. Dry runs with artificial data are useful for 1) assessing convergence speed, 2) revealing an unidentified model, and 3) eliminating bugs in computer programs.

Second, use real-world data and generate multiple MCMC chains with over-dispersed initial values. Although the MCMC chain converges to an unknown distribution, different chains should converge to the same distribution. Convergence from an arbitrary initial state is a strength of the MCMC sampler, compared to numerical maximum likelihood optimization, which may not converge unless the initial values are carefully specified. For an experienced researcher, it is possible to diagnose convergence by monitoring the trace plots, but it is more reliable to compute summary statistics such as $\hat{R}$ (Gelman and Rubin, 1992), which is based on the variances between and within MCMC chains.

Suppose m chains are simulated, each with n draws. Let ψ_{ij} be the ith draw of the jth chain. The chain-specific sample mean and variance are

$$\bar{\psi}_j = \frac{1}{n}\sum_{i=1}^{n}\psi_{ij},$$

$$s_j^2 = \frac{1}{n}\sum_{i=1}^{n}\left(\psi_{ij} - \bar{\psi}_j\right)^2.$$

The pooled mean and variance are

$$\bar{\psi} = \frac{1}{mn}\sum_{j=1}^{m}\sum_{i=1}^{n}\psi_{ij} = \frac{1}{m}\sum_{j=1}^{m}\bar{\psi}_j,$$

$$s^2 = \frac{1}{mn}\sum_{j=1}^{m}\sum_{i=1}^{n}\left(\psi_{ij} - \bar{\psi}\right)^2 = \frac{1}{m}\sum_{j=1}^{m}s_j^2 + \frac{1}{m}\sum_{j=1}^{m}\left(\bar{\psi}_j - \bar{\psi}\right)^2.$$

To keep the notation simple, the above formulae did not make the sample size adjustment for an unbiased variance estimator. Let $W = \frac{1}{m}\sum_{j=1}^{m} s_j^2$ be the within variance, and $B = \frac{n}{m}\sum_{j=1}^{m}\left(\bar{\psi}_j - \bar{\psi}\right)^2$ be the between variance (scaled by n because the variance of $\bar{\psi}_j$ decreases with n). They satisfy $s^2 = W + \frac{1}{n}B$. The Gelman-Rubin statistics is defined as $\hat{R} = \sqrt{\frac{s^2}{W}}$, which converges to one as the number of draws n increases. A rule of thumb is that $\hat{R} > 1.1$ signals non-convergence.

2.2.6 Autocorrelations and Inefficiency Factor

MCMC draws are inherently correlated. Even when using the independence MH sampler, the positive probability of the event $\{\theta_{t+1} = \theta_t\}$ results in correlated draws. High dependency among draws leads to slow mixing of the chain.

Consider a stationary series $\{\theta_t\}$ with a constant mean $\mu = E(\theta_t)$, $\forall t$. Let $\gamma_j = E[(\theta_t - \mu)(\theta_{t-j} - \mu)]$ denote the jth order autocovariance, and $\rho_j = \frac{\gamma_j}{\gamma_0}$ represent the autocorrelation. Consider the sample-path average $\bar{\theta}_T = \frac{1}{T}\sum_{t=1}^{T}\theta_t$. For a given sample size T, the variance of $\bar{\theta}_T$ is given by (see Exercise 2.4)

$$Var\left(\bar{\theta}_T\right) = \frac{1}{T}\left[1 + 2\sum_{k=1}^{T}\left(1 - \frac{k}{T}\right)\rho_k\right]\gamma_0.$$

Assume that the covariances are absolutely summable: $\sum_{k=1}^{\infty}|\gamma_k| < \infty$. As $T \to \infty$, the Law of Large Numbers ensures that $\bar{\theta}_T \xrightarrow{p} \mu$ (converge in probability to the mean) and

$$Var\left(\sqrt{T}\bar{\theta}_T\right) \to \left(1 + 2\sum_{k=1}^{\infty}\rho_k\right)\gamma_0.$$

The large-sample variance of $\bar{\theta}_T$ is approximated by

$$Var\left(\bar{\theta}_T\right) \approx \frac{IF}{T}\gamma_0,$$

where the inefficiency factor (also known as the integrated autocorrelation time) is defined as

$$IF = 1 + 2\sum_{k=1}^{\infty}\rho_k.$$

The inefficiency factor measures the loss of efficiency against independent samples, under which the variance reduces to $Var\left(\bar{\theta}_T\right) = \frac{\gamma_0}{T}$. In other words, the inefficiency factor measures the number of correlated samples required to produce one independent sample, when evaluating the sample path average. A closely related metric is the effective sample size, which is defined as the number of draws divided by the inefficiency factor.

Estimating the inefficiency factor can be challenging due to sampling variation in high-order autocorrelations. In practice, it is necessary to truncate the sample autocorrelations, either by specifying a cutoff lag (e.g., 100) or by terminating the sum when the sample analogue of ρ_j violates theoretical properties. Geyer (1992) demonstrates that for a stationary, irreducible, and reversible Markov chain, the adjacent autocorrelation pairs $\rho_{2j} + \rho_{2j+1}$ are strictly positive, monotone decreasing, and convex.

Alternatively, the inefficiency factor can be estimated using sample batches (Jones et al., 2006; Flegal and Jones, 2010). This method involves dividing the full sequence of draws into multiple non-overlapping batches. The asymptotic behavior of a batch mirrors that of the full sequence. The batch method approximates the inefficiency factor by examining the variance of batch means.

Specifically, the sequence is divided into m non-overlapping batches, each with a length $n = \frac{T}{m}$ (as an integer). Assuming that the batch length is sufficiently long and that high order autocovariances are small enough, the dependency between batches is negligible. Let

$$\bar{\theta}_{(i)} = \frac{1}{n}\sum_{t=(i-1)n+1}^{ni}\theta_t, i = 1, \ldots, m$$

be the ith batch mean. Because the batch means $\bar{\theta}_{(1)}, \ldots, \bar{\theta}_{(m)}$ are nearly independent, the sample variance

$$V = \frac{1}{m-1}\sum_{i=1}^{m}\left(\bar{\theta}_{(i)} - \bar{\theta}_T\right)^2$$

provides an unbiased estimator of $Var\left(\bar{\theta}_{(i)}\right)$. As $m \to \infty$, V is also a consistent estimator.

Consider the batch mean $\bar{\theta}_{(i)}$ as $n \to \infty$. The asymptotic behavior of a batch is the same as that of the full sequence, except that the batch length is denoted by n instead of T. The Law of Large Numbers ensures that $\bar{\theta}_{(i)} \xrightarrow{p} \mu$ and

$$Var\left(\sqrt{n}\bar{\theta}_{(i)}\right) \to \left(1 + 2\sum_{k=1}^{\infty} \rho_k\right) \gamma_0.$$

The large-sample variance of $\bar{\theta}_{(i)}$ is approximated by

$$Var\left(\bar{\theta}_{(i)}\right) \approx \frac{IF}{n}\gamma_0.$$

With V as an unbiased and consistent estimator of $Var\left(\bar{\theta}_{(i)}\right)$, the inefficiency factor can be inferred from the variance of batch means:

$$IF \approx nV\gamma_0^{-1}.$$

Consequently, the large-sample variance of $\bar{\theta}_T$ is approximated by

$$Var\left(\bar{\theta}_T\right) \approx \frac{n}{T}V.$$

2.3 Gibbs Sampler

For many practitioners, the Gibbs sampler is the preferred MCMC method if the full posterior conditional distributions are of recognizable forms. The Gibbs sampler cycles through the posterior conditionals, converting a high-dimensional sampling problem into a series of univariate or low-dimensional problems. An appeal of the Gibbs sampler is that it is free of tuning parameters, whereas a pain point of implementing the MH sampler is that the proposal transition kernel can be delicate to tune.

The Gibbs sampler was introduced by Geman and Geman (1984) in the context of image processing. The data augmentation technique, developed by Tanner and Wong (1987), makes the Gibbs sampler a fundamental tool for posterior simulation under conditionally conjugate structures. Influential papers such as Gelfand and Smith (1990), Casella and George (1992), Smith and Roberts (1993) are crucial for development and popularization of the Gibbs sampler in statistical modeling.

2.3.1 Conditional Distributions

Assume that variables are partitioned as two blocks, say A and B. Let $p_{A|B}\left(\theta_A;\theta_B\right)$ be the density function of block-A variables conditional on block-B variables, evaluated at θ_A and θ_B, respectively. In this section, we avoid a generic notation like $p\left(\theta_A \left|\theta_B\right.\right)$ because we want to be clear about the distributional form and the points at which the density function is evaluated. Similarly, we define the conditional distribution $p_{B|A}\left(\theta_B;\theta_A\right)$, the joint distribution $p_{A,B}\left(\theta_A,\theta_B\right)$, as well as the marginal distributions $p_A\left(\theta_A\right)$ and $p_B\left(\theta_B\right)$. The joint, marginal and conditional distributions satisfy the equations:

$$p_{A,B}\left(\theta_A,\theta_B\right) = p_A\left(\theta_A\right)p_{B|A}\left(\theta_B;\theta_A\right),$$

$$p_{A,B}\left(\theta_A,\theta_B\right) = p_B\left(\theta_B\right)p_{A|B}\left(\theta_A;\theta_B\right),$$

Algorithm 2.3 Gibbs sampler

Input: conditional distributions $p_{A|B}$ and $p_{B|A}$, starting value θ_{A0}.
Output: MCMC draws $(\theta_{A1}, \theta_{B1}), (\theta_{A2}, \theta_{B2}), \ldots$.
Iterate the following steps for $t = 1, 2, \ldots$
1 Generate θ_{Bt} from $p_{B|A}(\theta_{Bt}; \theta_{A,t-1})$, conditional on $\theta_{A,t-1}$.
2 Generate θ_{At} from $p_{A|B}(\theta_{At}; \theta_{Bt})$, conditional on θ_{Bt}.

$$p_A(\theta_A) = \int p_{A,B}(\theta_A, \theta_B)\, d\theta_B,$$

$$p_B(\theta_B) = \int p_{A,B}(\theta_A, \theta_B)\, d\theta_A.$$

The marginal distributions cannot determine the joint distribution, but the full set of conditional distributions can recover the joint distribution, which is the essence of the Hammersley-Clifford Theorem that justifies the Gibbs sampler. Proposition 2.6 presents the bivariate version, and Exercise 2.6 discusses a multivariate extension via the divide-and-conquer method, which recursively applies Proposition 2.6 to pairs of variables.

Proposition 2.6 (Hammersley-Clifford) *Let $p_{A,B}$, p_A, p_B, $p_{A|B}$, $p_{B|A}$ be the joint, marginal and conditional distributions. If $p_A(\theta_A) > 0$ and $p_B(\theta_B) > 0$ imply $p_{A,B}(\theta_A, \theta_B) > 0$ (the positivity condition), the joint distribution is determined by the conditional distributions such that*

$$p_{A,B}(\theta_A, \theta_B) = \frac{p_{B|A}(\theta_B; \theta_A)}{\int \frac{p_{B|A}(\theta_B; \theta_A)}{p_{A|B}(\theta_A; \theta_B)} d\theta_B}.$$

2.3.2 Gibbs Sampling Algorithm

Although Proposition 2.6 establishes an explicit formula of the joint density as a function of the conditional densities, it is not a computational tool due to the intractable integral. The computational vehicle is the Gibbs sampler, which alternately generates draws from $p_{B|A}$ and $p_{A|B}$, as described in Algorithm 2.3. It corresponds to a Markov chain with the transition kernel that moves from (θ_A, θ_B) to (θ_A^*, θ_B^*):

$$q(\theta_A^*, \theta_B^*; \theta_A, \theta_B) = p_{B|A}(\theta_B^*; \theta_A)\, p_{A|B}(\theta_A^*; \theta_B^*). \tag{2.3}$$

In Equation (2.3), the conditioning variable θ_B is absent in the transition kernel, implying that θ_A itself evolves as a marginal Markov chain with the transition kernel

$$\tilde{q}(\theta_A^*; \theta_A) = \int q(\theta_A^*, \theta_B^*; \theta_A, \theta_B)\, d\theta_B^*. \tag{2.4}$$

Proposition 2.7 shows that 1) the marginal Markov chain $\tilde{q}$ has p_A as its stationary distribution, because the reversibility condition is satisfied, and 2) the joint Markov chain q has $p_{A,B}$ as the stationary distribution, without resorting to the reversibility condition.

Proposition 2.7 (Stationary distribution of the Gibbs sampler) *Let $p_{A,B}$ and p_A be the joint and marginal target distributions. Let q and $\tilde{q}$ be the joint and marginal transition kernels in Equations (2.3) and (2.4).*

1. *The reversible Markov chain $\tilde{q}$ has p_A as a stationary distribution.*
2. *The Markov chain q has $p_{A,B}$ as a stationary distribution.*

2.3.3 Correlation Pattern and Rao-Blackwellization

Draws generated by the Gibbs sampler are correlated. What do we know about the correlation pattern? Liu et al. (1994) provide a clear answer: the autocorrelations are non-negative and monotone decreasing. The covariance structure is determined by the high-order conditional expectations. This is an intriguing concept that deserves careful explanation.

Consider a scalar-valued function of block-A variables $g_0(\theta_A)$. For example, g_0 may extract the first element of block-A variables. The first-order expectation is defined as $g_1(\theta_B) = E[g_0(\theta_A)|\theta_B]$, reflecting an educated guess of $g_0(\theta_A)$ given θ_B. The second-order expectation is defined as $g_2(\theta_A) = E[g_1(\theta_B)|\theta_A]$, representing an educated guess of the first-order expectation given θ_A. Similarly, a higher-order expectation is an educated guess of the previous guess. The odd-order expectations are conditional on θ_B, while the even-order expectations are conditional on θ_A. Consequently, the high-order expectations are random variables, as functions of θ_B for the odd lags and θ_A for the even lags. The high-order expectations are determined by the joint distribution of θ_A and θ_B. Specifically, in the context of multivariate normal distributions, it is possible to derive closed-form high-order expectations (see Exercise 2.7). Interested readers may extend the use cases of high-order expectations to areas beyond econometrics (e.g., the game theory).

Provided that the joint distribution of $(\theta_{At}, \theta_{Bt})$ is time-invariant, the functional forms of the high-order expectations are time-invariant as well. Proposition 2.8 shows that the monotone decreasing autocorrelations of the stationary draws generated by the Gibbs sampler are characterized by the variances of high-order expectations.

Proposition 2.8 (High-order expectations) *Let $(\theta_{At}, \theta_{Bt})$, $t = 1, 2, \ldots$ be samples of the two-block Gibbs sampler. Assume that the chain has reached the stationary distribution and the joint distribution of $(\theta_{At}, \theta_{Bt})$ does not change over time. Let g_0 be a square integrable function of interest. Denote $x_t = g_0(\theta_{At})$. The jth order conditional expectation g_j is defined recursively such that*

$g_j(\theta_{Bt}) = E[g_{j-1}(\theta_{At})|\theta_{Bt}]$, if j is odd,
$g_j(\theta_{At}) = E[g_{j-1}(\theta_{Bt})|\theta_{At}]$, if j is even.
Let $\gamma_j = cov(x_t, x_{t-j})$, $j = 0, 1, 2, \ldots$ be the autocovariances.
Then we have $\gamma_j = Var(g_j)$ and $\gamma_j \geq \gamma_{j+1} \geq 0$.

The Gibbs sampler generates draws $(\theta_{At}, \theta_{Bt})$, $t = 1, 2, \ldots, T$. When approximating $E[g_0(\theta_{At})]$, a straightforward estimator is the sample average

$$s_1 = \frac{1}{T}\sum_{t=1}^{T} g_0(\theta_{At}).$$

However, this may not always be the optimal choice. Monte Carlo simulation should be used only when necessary, and analytical results should be leveraged whenever available. In the Gibbs sampler, the conditional distributions are tractable, suggesting that $E[g_0(\theta_{At})|\theta_{Bt}]$ likely has a closed form. The law of iterated expectations ensures that $E[g_0(\theta_{At})] = E\{E[g_0(\theta_{At})|\theta_{Bt}]\}$. The Rao-Blackwellized estimator takes the form

$$s_2 = \frac{1}{T}\sum_{t=1}^{T} E[g_0(\theta_{At})|\theta_{Bt}].$$

The key difference between s_1 and s_2 is that the former uses draws of θ_{At}, while the latter uses draws of θ_{Bt}. If the draws were independent and identically distributed, the Rao-Blackwellized estimator would be preferable due to its smaller variance. This is because the unconditional variance is the sum of the variance of the conditional mean and the mean of the conditional variance:

$$Var\left[g_0\left(\theta_{At}\right)\right] = Var\left\{E\left[g_0\left(\theta_{At}\right)|\theta_{Bt}\right]\right\} + E\left\{Var\left[g_0\left(\theta_{At}\right)|\theta_{Bt}\right]\right\}.$$

Taking autocorrelations into account, Proposition 2.8 indicates that the Rao-Blackwellized estimator, in the context of the two-block Gibbs sampler, still maintains a smaller variance, because

$$Var\left(s_1\right) = \frac{1}{T^2}\left[T\gamma_0 + 2\sum_{j=1}^{T}\left(T-j\right)\gamma_j\right],$$

$$Var\left(s_2\right) = \frac{1}{T^2}\left[T\gamma_1 + 2\sum_{j=1}^{T}\left(T-j\right)\gamma_{j+1}\right],$$

where the autocovariances γ_j are monotone decreasing as stated in Proposition 2.8.

2.3.4 Slice Sampling

The Gibbs sampler finds broad applicability. First, parameters in some models are naturally split in multiple blocks. For example, the linear regression involves parameters β and σ^2, allowing for iterative sampling from $p\left(\beta\left|\sigma^2, Y\right.\right)$ and $p\left(\sigma^2\left|\beta, Y\right.\right)$. Second, some models contain missing data or latent variables Z, conditional on which the parameter θ is easy to sample. With data augmentation, the Gibbs sampler cycles through $p\left(Z\left|\theta, Y\right.\right)$ and $p\left(\theta\left|Z, Y\right.\right)$. Third, an auxiliary variable Z can be constructed to enable the Gibbs sampler on an extended parameter space for the joint distribution $\pi\left(\theta, Z\right) = \pi\left(\theta\right)p\left(Z\left|\theta\right.\right)$. The auxiliary variable Z can be solely determined by θ, and may not be relevant to the model. An example is the slice sampler, which replaces a complex sampling problem by drawing from a series of uniform distributions.

The slice sampler is an application of the Gibbs sampler. Let $p_A\left(\theta_A\right)$ be the target density and $S = \{\theta_A : p_A\left(\theta_A\right) > 0\}$ be the support of the distribution. Construct an auxiliary variable with a conditionally uniform distribution:

$$p_{B|A}\left(\theta_B;\theta_A\right) = \frac{1}{p_A\left(\theta_A\right)} I\left[0 < \theta_B < p_A\left(\theta_A\right)\right].$$

The joint density is given by

$$p_{A,B}\left(\theta_A,\theta_B\right) = p_A\left(\theta_A\right) p_{B|A}\left(\theta_B;\theta_A\right) = I\left[0 < \theta_B < p_A\left(\theta_A\right)\right].$$

It follows that

$$p_{A|B}\left(\theta_A;\theta_B\right) \propto I\left[p_A\left(\theta_A\right) > \theta_B\right].$$

A basic version of the slice sampler cycles through $p_{A|B}$ and $p_{B|A}$, both of which are uniformly distributed. On the one hand, $p_{B|A}$ is the density of the uniform distribution $U\left[0, p_A\left(\theta_A\right)\right]$. On the other hand, $p_{A|B}$ follows a uniform distribution in the set $\{\theta_A \in S : p_A\left(\theta_A\right) > \theta_B\}$. Sampling θ_A can be challenging, except in two special cases. First, when the inequality $p_A\left(\theta_A\right) > \theta_B$ has an explicit solution, draws can be obtained directly. For example, if θ_A follows an exponential distribution with $p_A\left(\theta_A\right) = e^{-\theta_A} I\left(\theta_A > 0\right)$, then $\theta_A \in \left(0, -\ln\theta_B\right)$. Second, when the support S is a bounded set, rejection sampling may be used. A uniform draw from S is proposed, and if $p_A\left(\theta_A\right) > \theta_B$, it is accepted. Otherwise, new proposals are made until the condition is satisfied.

If the support S is unbounded, a clever workaround involves imposing a bound using an auxiliary variable C such that

$$p_{C|A,B}\left(\theta_C;\theta_A,\theta_B\right)=\frac{1}{w}I\left[\theta_A-w<\theta_C<\theta_A\right],$$

where w is a tuning parameter that determines the window width. As the joint density takes the form

$$p_{A,B,C}\left(\theta_A,\theta_B,\theta_C\right)=I\left[0<\theta_B<p_A\left(\theta_A\right)\right]\cdot\frac{1}{w}I\left[\theta_A-w<\theta_C<\theta_A\right],$$

the full posterior conditional distributions are given by

$$p_{C|A,B}\left(\theta_C;\theta_A,\theta_B\right)\propto I\left[\theta_A-w<\theta_C<\theta_A\right],$$

$$p_{B|A,C}\left(\theta_B;\theta_A,\theta_C\right)\propto I\left[0<\theta_B<p_A\left(\theta_A\right)\right],$$

$$p_{A|B,C}\left(\theta_A;\theta_B,\theta_C\right)\propto I\left[p_A\left(\theta_A\right)>\theta_B\right]\cdot I\left[\theta_C<\theta_A<\theta_C+w\right].$$

The Gibbs sampler cycles through three variable blocks. First, given θ_A,θ_B, we sample θ_C from the uniform distribution $U\left(\theta_A-w,\theta_A\right)$. Second, given θ_A,θ_C,we sample θ_B from the uniform distribution $U\left[0,p_A\left(\theta_A\right)\right]$. Third, given θ_B,θ_C, we sample θ_A by rejection sampling. A candidate is drawn from the uniform distribution $U\left(\theta_C,\theta_C+w\right)$. If $p_A\left(\theta_A\right)>\theta_B$, it is accepted. Otherwise, new candidates are proposed until acceptance. This is feasible because the previous θ_A (obtained from the last iteration of the Gibbs sampler) resides in the interval $\left(\theta_C,\theta_C+w\right)$ and satisfies $p_A\left(\theta_A\right)>\theta_B$. If the density curve of p_A is smooth, a region where the density is higher than θ_B should exist.

The effectiveness of the slice sampler depends on the tuning parameter w. Neal (2003) develops an adaptive slice sampler in which the window width can be self-tuned by automatic stepping out and shrinking.

2.3.5 Gibbs and MH Samplers

The Gibbs sampler is closely related to the MH sampler. The transition kernel of the Gibbs sampler can be decomposed into two sub-kernels:

$$r_1\left(\theta_A^*,\theta_B^*;\theta_A,\theta_B\right)=p_{B|A}\left(\theta_B^*;\theta_A\right)I\left(\theta_A^*=\theta_A\right),$$

$$r_2\left(\theta_A^*,\theta_B^*;\theta_A,\theta_B\right)=p_{A|B}\left(\theta_A^*;\theta_B^*\right)I\left(\theta_B^*=\theta_B\right).$$

The first kernel updates B without changing A, while the second kernel updates A only.

Consider a two-step MH sampler where the transition kernel is the product of two sub-kernels, each constructed using the MH method with the same stationary distribution (see Exercise 2.8). The MH sampler targeting $p_{A,B}$ accepts proposals from r_1 with probability

$$\min\left[1,\frac{p_{A,B}\left(\theta_A^*,\theta_B^*\right)p_{B|A}\left(\theta_B;\theta_A^*\right)I\left(\theta_A^*=\theta_A\right)}{p_{A,B}\left(\theta_A,\theta_B\right)p_{B|A}\left(\theta_B^*;\theta_A\right)I\left(\theta_A^*=\theta_A\right)}\right]=1.$$

This is because r_1 satisfies the reversibility condition, ensuring all proposals from r_1 are accepted. Similarly, the acceptance probability of r_2 equals one as well. Thus, the Gibbs sampler can be interpreted as a special case of a multi-step transition MH sampler that accepts proposals with probability one.

2.3.6 Metropolis-within-Gibbs Sampler

Tuning a high-dimensional MH sampler, where all parameters are updated simultaneously, can be challenging. This issue can be addressed using a blocking strategy, where parameters are partitioned into smaller blocks for sequential updating. This approach is reminiscent of the Gibbs sampler, which cycles through the full posterior conditional distributions. The key difference is that the multi-block MH sampler does not require recognizable forms of the posterior conditional distributions; it only evaluates the conditional densities, which are proportional to the joint density.

For simplicity, assume that variables are partitioned into two blocks A and B. Let $p_{A,B}$, p_A, p_B, $p_{A|B}$, $p_{B|A}$ be the joint, marginal and conditional distributions. To sample from the target distribution $p_{A,B}$, a transition kernel is constructed to move (θ_A, θ_B) to (θ_A^*, θ_B^*) in two steps. First, conditional on θ_B, a partial transition kernel is constructed to move θ_A to θ_A^*, denoted by $q_1(\theta_A^*; \theta_A, \theta_B)$. Second, conditional on θ_A^*, a partial transition kernel is constructed to move θ_B to θ_B^* denoted by $q_2(\theta_B^*; \theta_B, \theta_A^*)$. Provided that the stationary distribution of q_1 is $p_{A|B}$ and that of q_2 is $p_{B|A}$, the joint distribution $p_{A,B}$ is a stationary distribution of the combined transition kernel $q_1 q_2$. See Exercise 2.9 for a derivation.

Regarding the choice of q_1 and q_2, if both $p_{A|B}$ and $p_{B|A}$ are easy to sample, we specify

$$q_1(\theta_A^*; \theta_A, \theta_B) = p_{A|B}(\theta_A^*; \theta_B),$$

$$q_2(\theta_B^*; \theta_B, \theta_A^*) = p_{B|A}(\theta_B^*; \theta_A^*).$$

Consequently, $q = q_1 q_2$ coincides with the transition kernel of the Gibbs sampler. Thus, the Gibbs sampler can be interpreted as a special case of the multi-block MH sampler.

A universal approach for constructing q_1 is the MH method, which evaluates the target density $p_{A|B}$, proportional to the joint density $p_{A,B}$. Similarly, the MH method can be used for constructing q_2 by evaluating the conditional density $p_{B|A}$, also proportional to the joint density $p_{A,B}$. In other words, implementation of the multiple-block MH sampler only requires evaluating the joint density, as in a standard MH sampler.

In summary, the multi-block MH sampler updates each variable block iteratively. Whenever the posterior conditional distribution is analytically tractable, draws can be directly generated. Otherwise, a Metropolis component is inserted. The resulting MCMC sampler is called the Metropolis-within-Gibbs sampler. Chib and Ramamurthy (2010) demonstrate that it is possible to randomize the number of blocks and the composition of blocks in each iteration.

2.4 Hamiltonian Monte Carlo

The random walk MH sampler often generates highly correlated samples. If the proposal distribution has a small variance, the acceptance probability is high, but the chain tends to explore the space locally. Conversely, if the proposal variance is large, most proposals are rejected, causing the chain to become sticky. In contrast, Hamiltonian Monte Carlo (HMC) is a gradient-based sampler that typically generates distant proposals with high acceptance probabilities. HMC samples exhibit lower correlations, which means the target distribution can be well approximated with fewer samples. Originally developed in the context of physics by Duane et al. (1987), HMC was later popularized in statistics by Neal (2011).

2.4.1 Hamiltonian and Leapfrog Move

Assume the target density $\pi(\theta)$ is positive and differentiable everywhere on $\mathbb{R}^d$. Instead of sampling θ alone, HMC operates on an extended state space by introducing a d-dimentional auxiliary variable z with a symmetric density function $f(z) = f(-z)$. A common choice is the normal density: $f(z) \propto e^{-\frac{1}{2}z'z}$. The variables θ and z are independent, with the joint distribution $\Pi(\theta, z) = \pi(\theta) f(z)$. The HMC sampler simulates the pair (θ, z) with the stationary distribution $\Pi(\theta, z)$, and the marginal distribution of interest is $\pi(\theta)$.

The HMC sampler generates proposals using Hamiltonian dynamics. Define the Hamiltonian $H(\theta, z) = -\ln \pi(\theta) - \ln f(z)$, so that $\Pi(\theta, z) = e^{-H(\theta,z)}$. Physically, θ represents position with potential energy $-\ln \pi(\theta)$, and z corresponds to momentum with kinetic energy $-\ln f(z)$. The Hamiltonian $H(\theta, z)$ describes the total energy, which remains constant as θ and z evolve over time in a closed system. The Hamiltonian dynamics requires $\frac{dz}{dt}$ and $\frac{d\theta}{dt}$ satisfy the law of motion:

$$\frac{dz}{dt} = -\frac{\partial H}{\partial \theta} = \frac{d \ln \pi(\theta)}{d\theta},$$

$$\frac{d\theta}{dt} = \frac{\partial H}{\partial z} = -\frac{d \ln f(z)}{dz}.$$

Under that law of motion, the total differentiation is

$$\frac{dH}{dt} = \frac{\partial H}{\partial z}\frac{dz}{dt} + \frac{\partial H}{\partial \theta}\frac{d\theta}{dt} = 0,$$

which implies that $H(\theta, z)$ and $\Pi(\theta, z)$ remain unchanged. Note that the target distribution must be unconstrained, because $\frac{d\ln \pi(\theta)}{d\theta}$ is not well-defined if $\pi(\theta) = 0$.

The differential equations describe the trajectory of (θ, z) in continuous time. To simulate a discrete-time Markov chain, we discretize the Hamiltonian dynamics. It is tempting to approximate the derivatives $\frac{dz}{dt}$, $\frac{d\theta}{dt}$ by finite differencing $\frac{z_{t+\delta}-z_t}{\delta}$, $\frac{\theta_{t+\delta}-\theta_t}{\delta}$, where δ is the step size. This is the Euler discretization:

$$z_{t+\delta} = z_t + \delta \frac{d \ln \pi(\theta_t)}{d\theta_t},$$

$$\theta_{t+\delta} = \theta_t - \delta \frac{d \ln f(z_t)}{dz_t}.$$

The leapfrog method (Definition 2.1) refines the Euler discretization by simulating z in two half-steps with derivatives evaluated at the most recent values.

Definition 2.1 (Leapfrog move) *Let $\pi(\theta)$ be the target density, and $f(z)$ be a symmetric auxiliary density, and δ be the step size. A one-step leapfrog move from the current value (θ, z) to the next value (θ^*, z^*) is a deterministic mapping $(\theta^*, z^*) = g(\theta, z)$ such that*

$$z_{1/2} = z + \frac{1}{2}\delta \frac{d \ln \pi(\theta)}{d\theta},$$

$$\theta^* = \theta - \delta \frac{d \ln f(z_{1/2})}{dz_{1/2}},$$

$$z^* = z_{1/2} + \frac{1}{2}\delta \frac{d \ln \pi(\theta^*)}{d\theta^*}.$$

The multiple-step leapfrog move refers to successive application of the mapping $g(\ldots g(\theta, z))$.

Algorithm 2.4 HMC sampler

Input: target distribution $\pi(\cdot)$, auxiliary distribution $f(\cdot)$, and starting value θ_0.
Output: MCMC draws $(\theta_1, z_1), (\theta_2, z_2), \ldots$.
Iterate the following steps for $t = 1, 2, \ldots$
1 Generate z from the auxiliary distribution $f(z)$.
2 Make leapfrog move from (θ, z) to (θ^*, z^*), where $\theta = \theta_{t-1}$.
3 Compute acceptance probability $\alpha = \min\left[1, \frac{\pi(\theta^*)f(z^*)}{\pi(\theta)f(z)}\right]$.
4 Set $(\theta_t, z_t) = \begin{cases} (\theta^*, -z^*) & \textit{with probability } \alpha \\ (\theta, z) & \textit{with probability } 1-\alpha \end{cases}$.

2.4.2 HMC and Langevin Monte Carlo

The HMC sampler makes proposals by one or more leapfrog moves. Unlike the continuous-time trajectory that conserves the Hamiltonian, discretization induces a slight discrepancy between $\Pi(\theta^*, z^*)$ and $\Pi(\theta, z)$. The HMC sampler compensates such discrepancy by accepting a proposal with the probability $\min\left[1, \frac{\Pi(\theta^*,z^*)}{\Pi(\theta,z)}\right]$, which is usually close to one provided that the step size is not excessively large.

Algorithm 2.4 outlines the HMC sampler. Since $\pi(\theta)$ is the primary distribution of interest, it is possible to simplify the algorithm by not tracking the auxiliary variable z, which is refreshed at the beginning of each iteration without path dependency.

What is the rationale of Algorithm 2.4? The HMC under the one-step leapfrog move can be framed as a standard MH sampler (i.e., Langevin Monte Carlo). However, for the multiple-step leapfrog move, we need to dive deeper on properties of the leapfrog move to prove stationarity of the HMC sampler.

Langevin Monte Carlo is a gradient-based MCMC method conformable to the MH sampling paradigm (Algorithm 2.2). Conditional on the current state θ, the Langevin sampler makes a proposal

$$\theta^* = \theta + \frac{1}{2}\delta^2 \nabla(\theta) + \delta z,$$

where $\nabla(\theta) = \frac{d \ln \pi(\theta)}{d\theta}$ is the gradient of the log density, and z follows the standard normal distribution. The proposal transition kernel is

$$r(\theta^*; \theta) = e^{-\frac{1}{2}\delta^{-2}\left[\theta^* - \theta - \frac{1}{2}\delta^2\nabla(\theta)\right]'\left[\theta^* - \theta - \frac{1}{2}\delta^2\nabla(\theta)\right]}.$$

The Langevin Monte Carlo utilizes the gradient information to make a proposal, which is likely to obtain a higher density compared to the current state. It is reminiscent of numerical optimization by the steepest descent/ascent method (refer to Section 5.8). Following the MH sampling procedure, the proposal is accepted by the probability

$$\alpha(\theta^*, \theta) = \min\left[1, \frac{\pi(\theta^*)\, r(\theta; \theta^*)}{\pi(\theta)\, r(\theta^*; \theta)}\right].$$

Proposition 2.9 shows that the HMC sampler with the one-step leapfrog move is equivalent to the Langevin Monte Carlo, since they make the same proposal and accept it by the same probability.

Proposition 2.9 (Langevin Monte Carlo) *The proposal transition kernel $r(\theta^*; \theta)$ of Langevin Monte Carlo aligns with the one-step leapfrog move. The proposal acceptance probability $\alpha(\theta^*, \theta)$ in Langevin Monte Carlo is identical to that of the HMC sampler.*

2.4.3 Multiple-step Leapfrog Move and Stationarity

In practice, Langevin Monte Carlo can still produce highly correlated samples, indicating that a single leapfrog step may not propose a state sufficiently far from the current one. Successful implementation of HMC often involves multiple leapfrog steps per iteration. In that case, HMC diverges from the traditional MH framework, but it is still possible to use leapfrog properties (Proposition 2.10) to verify directly that $\Pi(\theta, z)$ is the stationary distribution of the Markov chain constructed by HMC.

Proposition 2.10 demonstrates that the leapfrog move represents a one-to-one mapping. The inverse mapping transforms (θ^*, z^*) back to (θ, z), while the original mapping transforms $(\theta^*, -z^*)$ to $(\theta, -z)$, illustrating the reversibility of Hamiltonian dynamics. Physically, negative momentum at the destination propels a round-trip back to the starting position. Another crucial property is volume preservation: a leapfrog move deterministically maps (θ, z) to (θ^*, z^*), with an identity Jacobian matrix. The implication is that the density of (θ^*, z^*) can be evaluated using $\Pi(\theta, z)$, without accounting for the Jacobian term induced by the change of variables. Based on reversibility and volume preservation, Proposition 2.11 confirms that the HMC sampler targets the correct distribution $\Pi(\theta, z)$, implying that $\pi(\theta)$ is the stationary distribution for θ draws.

Proposition 2.10 (Properties of leapfrog move) *The leapfrog move $(\theta^*, z^*) = g(\theta, z)$ in Definition 2.1 has the properties:*

1. *One-to-one mapping: the inverse mapping $(\theta, z) = g^{-1}(\theta^*, z^*)$ is*

$$z_{1/2} = z^* - \frac{1}{2}\delta \frac{d \ln \pi(\theta^*)}{d\theta^*},$$

$$\theta = \theta^* + \delta \frac{d \ln f(z_{1/2})}{dz_{1/2}},$$

$$z = z_{1/2} - \frac{1}{2}\delta \frac{d \ln \pi(\theta)}{d\theta}.$$

2. *Reversibility of the Hamiltonian dynamics:*

$$g(\theta^*, -z^*) = (\theta, -z),$$

$$\Pi\left[g^{-1}(\theta^*, z^*)\right] = \Pi\left[g(\theta^*, -z^*)\right].$$

3. *Volume preservation:*

$$p(\theta^*, z^*) = \Pi(\theta, z).$$

4. *Multiple-step leapfrog move satisfies reversibility and volume preservation.*

Proposition 2.11 (Stationary distribution of HMC sampler) *Consider*

$$\Pi(\theta, z) = \pi(\theta) f(z),$$

where $\pi(\theta)$ is the target density and $f(z)$ is a symmetric auxiliary density. Let (θ, z) be the current state, (θ^*, z^*) be a proposed one-step or multiple-step leapfrog move, and

$$\left(\tilde{\theta}, \tilde{z}\right) = \begin{cases} (\theta^*, -z^*) & \text{with probability } \alpha \\ (\theta, z) & \text{with probability } 1-\alpha \end{cases},$$

where $\alpha = \min\left[1, \frac{\Pi(\theta^*, z^*)}{\Pi(\theta, z)}\right]$. If (θ, z) follows $\Pi(\cdot)$, then $\left(\tilde{\theta}, \tilde{z}\right)$ also follows $\Pi(\cdot)$, hence stationarity.

2.4.4 Implementation Issues

The implementation of the HMC sampler is ideally suited for unconstrained distributions where the densities are positive throughout the domain. However, in practice, many distributions are subject to inequality constraints. A common solution is to reparametrize the variables. For instance, the variance σ^2 must be positive, while the reparametrized $\tau = \ln \sigma^2$ is unconstrained. The reparametrized density is $p_\tau(\tau) = p_{\sigma^2}(e^\tau) e^\tau$. Alternatively, heuristic approaches, such as stopping or bouncing rules, could be employed to prevent reaching regions where $\pi(\theta) = 0$.

Moreover, the HMC sampler is designed for continuous distributions. When dealing with discrete variables, it is possible to partition the variables into continuous and discrete blocks. The HMC sampler can then be applied to the continuous variables, targeting the posterior conditional distribution given the discrete variables. Subsequently, the discrete variables are sampled conditional on the continuous ones. This HMC-within-Gibbs approach is reminiscent of the Metropolis-within-Gibbs sampler.

The performance of HMC is sensitive to two critical user-specified parameters: the step size and the number of leapfrog steps. To address this sensitivity, Hoffman and Gelman (2014) introduced the No-U-Turn Sampler, which automatically tunes these parameters by adaptively setting path lengths in HMC.

The HMC sampler has been effectively implemented in statistical software such as Stan (Carpenter et al., 2017). Its advantages and disadvantages can be summarized as follows. First, HMC is an efficient MCMC method that generates distant proposals with high acceptance probabilities, leading to lower correlations among samples compared to the MH or Gibbs samplers. However, the requirement for gradient information can be a significant drawback for users who are hesitant to compute analytic gradients. While numerical derivatives or automatic differentiation are alternatives, they may slow down the computation. In our experience, HMC often requires only a few hundred samples to achieve satisfactory results, whereas the MH sampler might need thousands of samples to reach similar accuracy. Nonetheless, each HMC sample typically requires more computational time than an MH sample, which makes the standard MH sampler a viable competitor to HMC in practical applications.

2.5 Multiple-try MH Sampler

The standard MH sampler generates a single proposal at each iteration. In contrast, the multiple-try MH sampler, introduced by Liu et al. (2000), generates several proposals, selects one, and accepts or rejects it using a carefully defined probability. The strengths of this approach include: 1) multiple proposals allow the sampler to make larger and potentially more effective jumps without reducing the acceptance rate, 2) the standard MH sampler is inherently sequential, but multiple proposals are well-suited for parallel computing, and 3) the multiple-try MH sampler provides insights into the particle marginal MCMC method (Andrieu et al., 2010), which is a powerful technique for parameter estimation in nonlinear state space models (see Section 7.7).

There are two versions of the multiple-try MH sampler. Proposition 2.12 introduces a random-walk version and demonstrates that the resulting Markov chain has the target distribution as its stationary distribution. Proposition 2.13 presents a multiple-try counterpart to the independence MH sampler. For downward compatibility, the random-walk version with $n = 1$ reduces to the standard random-walk MH sampler, while the independence version with $n = 1$ reduces to the standard independence MH sampler. From a computational perspective, the random-walk version requires $2n-1$ evaluations of the target density per iteration, whereas the independence version requires n evaluations without generating auxiliary variables.

In the random-walk multiple-try MH sampler, one of the n proposals is selected based on the target density: $P(\theta^* = \theta_i^*) \propto \pi(\theta_i^*)$, as detailed in step 2 of Proposition 2.12. It implies that proposals with higher densities are more likely to be chosen, prompting the sampler to move toward high-density regions of the target distribution.

In the independence multiple-try MH sampler, one of the n proposals is selected based on the importance weight: $P(\theta^* = \theta_i^*) \propto w(\theta_i^*)$, as outlined in step 2 of Proposition 2.13. This approach is reminiscent of importance sampling and resampling. As $n \to \infty$, the selected proposal approximates a draw from the target distribution. If the proposal is accepted with probability $\min\left[1, \frac{\sum_i w(\theta_i^*)}{w(\theta)-w(\theta^*)+\sum_i w(\theta_i^*)}\right]$, the acceptance probability converges to one, since the term $\sum_i w(\theta_i^*)$ dominates as $n \to \infty$.

Another version of the independence multiple-try MH sampler specifies the acceptance probability as $\min\left[1, \frac{\sum_i w(\theta_i^*)}{\sum_i w(\theta_i)}\right]$, where $\theta_1, \dots, \theta_n$ are recycled from the proposals generated in the previous iteration. This can be interpreted as a standard MH sampler on an extended state space with the joint target distribution

$$\tilde{\pi}(\theta, \tau, \theta_{1:n}) = \frac{1}{n}\pi(\theta) I(\theta_\tau = \theta) \prod_{i \neq \tau} r(\theta_i),$$

where $\tau \in \{1, \dots, n\}$ is an index variable. The target distribution suggests that θ follows $\pi(\theta)$, and independently τ follows the discrete-valued uniform distribution. Conditional on θ and τ, we set $\theta_\tau = \theta$ and sample the remaining variables by $r(\cdot)$. The proposal distribution of θ_i^* is $r(\cdot)$, $i = 1, \dots, n$. The index τ^* indicates the selected proposal based on the importance weight $P(\tau^* = i) \propto w(\theta_i^*)$. We then set $\theta^* = \theta_{\tau^*}^*$. With these extended target and proposal distributions, a standard independence MH sampler results in an acceptance probability $\min\left[1, \frac{\sum_i w(\theta_i^*)}{\sum_i w(\theta_i)}\right]$.

In Chapter 7, we will explore the particle marginal MCMC method (Andrieu et al., 2010), which closely relates to the independence multiple-try MH sampler. Here, multiple proposals are viewed as particles, and selecting one is akin to resampling a particle. The acceptance probability $\frac{\sum_i w(\theta_i^*)}{\sum_i w(\theta_i)}$ corresponds to the ratio of the sample average of importance weights, serving as an unbiased estimator of the likelihood function.

Proposition 2.12 (Random-walk multiple-try MH sampler) *Let $\pi(\theta)$ be the target distribution and $r(\theta^*; \theta) = r(\theta; \theta^*)$ be a symmetric proposal transition kernel, where θ is the current state. The random-walk multiple-try MH sampling iterates the following steps:*

1 Generate proposals θ_i^* by $r(\theta_i^*; \theta)$, $i = 1, \dots, n$.

2 Sample a proposal θ^* by $P(\theta^* = \theta_i^*) = \frac{\pi(\theta_i^*)}{\sum_j \pi(\theta_j^*)}$, $i = 1, \dots, n$.

3 Generate auxiliary variables θ_i by $r(\theta_i; \theta^*)$, $i = 1, \dots, n-1$. Set $\theta_n = \theta$.

4 Set $\tilde{\theta} = \begin{cases} \theta^* & \text{with probability } \alpha \\ \theta & \text{with probability } 1-\alpha \end{cases}$, where $\alpha = \min\left[1, \frac{\sum_i \pi(\theta_i^*)}{\sum_i \pi(\theta_i)}\right]$.

If θ follows $\pi(\cdot)$, then $\tilde{\theta}$ also follows $\pi(\cdot)$, hence stationarity.

Proposition 2.13 (Independence multiple-try MH sampler) *Let $\pi(\theta)$ be the target distribution and $r(\theta^*; \theta) = r(\theta^*)$ be an independence proposal transition kernel, where θ is the current state. Let $w(\theta) = \frac{\pi(\theta)}{r(\theta)}$ be the importance weight. The independence multiple-try MH sampling iterates the following steps:*

1 Generate proposals θ_i^* *by* $r(\theta_i^*)$, $i = 1, \ldots, n$.
2 Sample a proposal θ^* *by* $P(\theta^* = \theta_i^*) = \frac{w(\theta_i^*)}{\sum_j w(\theta_j^*)}$, $i = 1, \ldots, n$.
3 Recycle proposals from the previous iteration, denoted by $\theta_1, \ldots, \theta_n$.
4 Set $\tilde{\theta} = \begin{cases} \theta^* & \text{with probability } \alpha \\ \theta & \text{with probability } 1-\alpha \end{cases}$,
where $\alpha = \min(1, \rho)$, *and two choices of* ρ *specifications are available:*
$\rho = \frac{\sum_i w(\theta_i^*)}{w(\theta) - w(\theta^*) + \sum_i w(\theta_i^*)}$, *or* $\rho = \frac{\sum_i w(\theta_i^*)}{\sum_i w(\theta_i)}$.
If θ *follows* $\pi(\cdot)$, *then* $\tilde{\theta}$ *also follows* $\pi(\cdot)$, *hence stationarity.*

2.6 Trans-dimensional MCMC Methods

Bayesian model selection and averaging address the uncertainty associated with models and their parameters. Consider a collection of candidate models, denoted by $\gamma \in \{1, \ldots, M\}$, each with its own set of parameters θ_γ. Our goal is to determine the posterior model distribution $p(\gamma|Y)$ and the model-specific parameter distribution $p(\theta_\gamma|Y, \gamma)$. Since parameter dimensions can vary between models, this presents a trans-dimensional problem.

The approaches of Carlin and Chib (1995) and Godsill (2001) allow for converting a trans-dimensional problem into a fixed-dimensional one by constructing a composite model. In this model, the concatenated parameters $\theta = (\theta_1, \ldots, \theta_M)$ are shared across all models. The composite model is defined in three stages. First, the model γ is assigned a prior probability $p(\gamma)$. Second, the parameters used by model γ have a genuine prior $p(\theta_\gamma|\gamma)$. Third, unused parameters $\theta_{-\gamma} = (\theta_1, \ldots, \theta_{\gamma-1}, \theta_{\gamma+1}, \ldots, \theta_M)$ are given a pseudo prior $p(\theta_{-\gamma}|\gamma)$. As the name implies, the pseudo prior is irrelevant because the likelihood function $p(Y|\gamma, \theta_\gamma)$ does not depend on unused parameters.

With fixed dimensions in the composite model, standard MCMC methods like the MH and Gibbs samplers can be employed. The joint distribution

$$p(\gamma, \theta, Y) = p(\gamma)\, p(\theta_\gamma|\gamma)\, p(\theta_{-\gamma}|\gamma)\, p(Y|\gamma, \theta_\gamma)$$

indicates that the full posterior conditional distributions are given by

$$p(\theta_\gamma|Y, \gamma, \theta_{-\gamma}) \propto p(\theta_\gamma|\gamma)\, p(Y|\gamma, \theta_\gamma),$$

$$p(\theta_{-\gamma}|Y, \gamma, \theta_\gamma) \propto p(\theta_{-\gamma}|\gamma),$$

$$p(\gamma|Y, \theta) \propto p(\gamma)\, p(\theta_\gamma|\gamma)\, p(\theta_{-\gamma}|\gamma)\, p(Y|\gamma, \theta_\gamma).$$

The Gibbs sampler proposed by Carlin and Chib (1995) cycles through the full posterior conditional distributions. First, conditional on the model γ, the posterior of θ_γ is proportional to the genuine prior multiplied by the likelihood, typically in a conjugate form. Second, conditional on the model γ, unused parameters $\theta_{-\gamma}$ are sampled from the pseudo prior, as they do not affect the likelihood and the data cannot update the pseudo prior. Third, $p(\gamma|Y, \theta)$ follows a categorical distribution, since γ is discrete.

The Gibbs sampler is practical if M is not excessively large. However, real-world model selection tasks often involve a vast number of models, such as $M = 2^d$,where d is the number of predictors in a regression model. Godsill (2001) addresses this curse of dimensionality using the MH sampler with a proposed transition kernel from (γ, θ) to (γ^*, θ^*):

$$r(\gamma^*, \theta^*; \gamma, \theta) = r_1(\gamma^*; \gamma)\, r_2(\theta^*_{\gamma^*}; \gamma^*)\, p(\theta_{-\gamma^*}|\gamma^*).$$

This proposal consists of three components. First, model transition from γ to γ^*. Second, transition of parameter used by model γ^*. Third, transition of unused parameters using the pseudo prior. The target distribution for the MH sampler is $p(\gamma, \theta | Y)$. According to Algorithm 2.2, the acceptance probability is given by $\min(1, \rho)$, where

$$\begin{aligned}\rho &= \frac{p(\gamma^*,\theta^*|Y)r(\gamma,\theta;\gamma^*,\theta^*)}{p(\gamma,\theta|Y)r(\gamma^*,\theta^*;\gamma,\theta)} \\ &= \frac{p(\gamma^*)p\left(\theta^*_{\gamma^*}|\gamma^*\right)p\left(\theta^*_{-\gamma^*}|\gamma^*\right)p\left(Y\middle|\gamma^*,\theta^*_{\gamma^*}\right)r_1(\gamma;\gamma^*)r_2(\theta_\gamma;\gamma)p(\theta_{-\gamma}|\gamma)}{p(\gamma)p(\theta_\gamma|\gamma)p(\theta_{-\gamma}|\gamma)p(Y|\gamma,\theta_\gamma)r_1(\gamma^*;\gamma)r_2\left(\theta^*_{\gamma^*};\gamma^*\right)p\left(\theta_{-\gamma^*}|\gamma^*\right)} \\ &= \frac{p(\gamma^*)p\left(\theta^*_{\gamma^*}|\gamma^*\right)p\left(Y\middle|\gamma^*,\theta^*_{\gamma^*}\right)r_1(\gamma;\gamma^*)r_2(\theta_\gamma;\gamma)}{p(\gamma)p(\theta_\gamma|\gamma)p(Y|\gamma,\theta_\gamma)r_1(\gamma^*;\gamma)r_2\left(\theta^*_{\gamma^*};\gamma^*\right)}.\end{aligned}$$

Interestingly, all terms involving unused parameters and the pseudo prior cancel out in the MH acceptance probability. Thus, transitioning unused parameters is merely conceptual and does not require implementation, nor is it necessary to specify the functional form for the pseudo prior.

The model transition r_1 and the parameter transition r_2 are user-specified. If we choose $r_2\left(\theta^*_{\gamma^*};\gamma^*\right) = p\left(\theta^*_{\gamma^*}|Y,\gamma^*\right)$, which is the posterior distribution of parameters given the model γ^*, then the MH acceptance probability is simplified:

$$\rho = \frac{p(\gamma^*)\,p(Y|\gamma^*)\,r_1(\gamma;\gamma^*)}{p(\gamma)\,p(Y|\gamma)\,r_1(\gamma^*;\gamma)}.$$

This approach is recognized as a marginalized MH sampler targeting $p(\gamma|Y)$, with model parameters integrated out. Evaluating the marginal likelihood is feasible if the model-specific posterior distribution of parameters is analytically tractable.

The pseudo-prior MH sampler aligns with the reversible jump MCMC by Green (1995), a powerful trans-dimensional method for moving between models by drawing dimension-matching variables (u^* and u). In the pseudo-prior MH sampler, the matching variables are specified such that $(\theta_\gamma, u) = \left(u^*, \theta^*_{\gamma^*}\right)$, with the Jacobian term equal to one.

2.7 Pseudo-marginal MH Sampler

The MH sampler requires pointwise evaluation of the target density $\pi(\theta)$. However, in certain models, evaluating the density function in a closed form is not feasible. Instead, we might have access to a Monte Carlo estimator $\hat{\pi}(\theta; Z)$, where Z represents random variables utilized during simulation. Often, this noisy estimator remains unbiased, satisfying $E[\hat{\pi}(\theta; Z)] = \pi(\theta)$. One might be inclined to substitute the target density with its Monte Carlo counterpart in the MH sampler, a concept reminiscent of the maximum simulated likelihood method (Flury and Shephard, 2011).

Remarkably, the noise-affected MH sampler still targets the exact distribution, provided that the estimator is nonnegative and unbiased. This property, known as exact approximation, is justified within the pseudo-marginal framework introduced by Andrieu and Roberts (2009). Pseudo-marginal methods find broad applicability, including nonlinear state space likelihood evaluation via the particle filter, Monte Carlo integration of the marginal likelihood in latent variable models, and subsampling strategies for Bayesian inference with large datasets.

2.7.1 Independent Pseudo-marginal Method

To implement the Monte Carlo estimator $\hat{\pi}(\theta; Z)$, a straightforward approach involves generating new random numbers Z for each evaluation point of θ. When the MH sampler proposes a transition from θ to θ^*, it also generates a new Z^* independent to Z, and evaluates $\hat{\pi}(\theta^*; Z^*)$ as an estimator of $\pi(\theta^*)$. The original random numbers Z are not reused.

As demonstrated in Proposition 2.14, the pseudo-marginal method can be interpreted as a standard MH sampler on an extended state space, targeting the joint distribution of θ and Z. The extended Markov chain admits $\pi(\theta)$ as the marginal stationary distribution.

Proposition 2.14 (Independent pseudo-marginal MH sampler) *Let $\pi(\theta)$ be the target density and $\hat{\pi}(\theta; Z)$ be a nonnegative unbiased estimator such that*

$$\int p(Z)\,\hat{\pi}(\theta; Z)\,dZ = \pi(\theta), \forall\theta,$$

where Z is an auxiliary random vector with the distribution $p(Z)$. The pseudo-marginal MH sampler follows all the steps in Algorithm 2.2, except that the intractable $\pi(\theta)$ is replaced by the unbiased estimator $\hat{\pi}(\theta; Z)$ for evaluating the acceptance probability:

$$\min\left[1, \frac{\hat{\pi}(\theta^*; Z^*)\, r(\theta; \theta^*)}{\hat{\pi}(\theta; Z)\, r(\theta^*; \theta)}\right],$$

where Z^* and Z are independent and identically distributed, and $r(\theta^*; \theta)$ is a proposal transition kernel.

On an extended state space, the target and proposal distributions are $p(Z)\,\hat{\pi}(\theta; Z)$ and $r(\theta^*; \theta)\, p(Z^*)$, respectively. The pseudo-marginal MH sampler has the target distribution as its stationary distribution, and the marginal stationary distribution is $\pi(\theta)$.

Several remarks on Proposition 2.14 are noteworthy. First, both θ and Z are states of the Markov chain, and the pair (θ, Z) should be accepted or rejected together. Second, it is permissible to omit a normalizing constant when evaluating the unbiased estimator of $\pi(\theta)$, as the constant will cancel out in the MH acceptance probability. It facilitates Bayesian inference, where the posterior is proportional to the prior times the likelihood, and the likelihood is approximated by an unbiased estimator. Third, the transition kernel $r(\theta^*; \theta)\, p(Z^*)$ implies that Z^* is sampled independently from Z, so Proposition 2.14 is termed the independent pseudo-marginal method.

The pseudo-marginal MH sampler targets the exact distribution, even if the unbiased estimator is highly noisy. However, there is no free lunch. The Monte Carlo noise can undermine the independent pseudo-marginal method through a prohibitively large variance of $\frac{\hat{\pi}(\theta^*; Z^*)}{\hat{\pi}(\theta; Z)}$. The sampler might become trapped by a spuriously large $\hat{\pi}(\theta; Z)$, which merely reflects sampling variation rather than a high-density $\pi(\theta)$. The root cause is the discrepancy between Z and Z^* as independent draws. Even if the transition from θ to θ^* is small, the change from $\hat{\pi}(\theta; Z)$ to $\hat{\pi}(\theta^*; Z^*)$ can be substantial.

2.7.2 Correlated Pseudo-marginal Method

The correlated pseudo-marginal method by Deligiannidis et al. (2018) mitigates Monte Carlo variation by introducing correlations between Z and Z^*. The intuition is that $\hat{\pi}(\theta; Z)$ and $\hat{\pi}(\theta^*; Z^*)$ should be close if 1) θ and θ^* are close, 2) Z and Z^* are close, and 3) the function is smooth.

Correlations are obtained by a reversible transition kernel $K(Z^*; Z)$ with the stationary distribution $p(Z)$ such that

$$p(Z)\, K(Z^*; Z) = p(Z^*)\, K(Z; Z^*).$$

If Z follows a multivariate normal distribution, the reversible transition kernel has a closed-form solution. For example,

$$p(Z) = \phi(Z; \mu, V),$$
$$K(Z^*; Z) = \phi[Z^*; (I - A)\mu + AZ, V - AVA'],$$

where A is a user-specified matrix with all eigenvalues in the unit circle. A convenient choice is $A = \rho I$, where $0 < \rho < 1$. Specifically, if Z follows the standard normal distribution ($\mu = 0$ and $V = I$), the transition from Z to Z^* is characterized by a Gaussian autoregressive process

$$Z^* = \rho Z + \sqrt{1 - \rho^2}\varepsilon,$$

where ε is the standard normal variate. If the parameter ρ is close to one, there is a strong correlation between Z and Z^*, while both of them are standard normal variates marginally.

Correlations in non-normal distributions can be created by the change of variable approach. A generic recipe is the inverse cumulative distribution function (CDF) method, as illustrated in Exercise 2.10. If U is a standard normal variate with the CDF $\Phi(\cdot)$, then $\Phi(U)$ is uniform and $Z = F^{-1}(\Phi(U))$ has the CDF $F(\cdot)$. The inverse CDF method is routinely used by statistical software for random number generation from common distributions. If $\hat{\pi}(\theta; Z)$ is an unbiased estimator of $\pi(\theta)$, then $\hat{\pi}(\theta; F^{-1}(\Phi(U)))$ is the reparametrized estimator that remains unbiased.

The correlated pseudo-marginal MH sampler follows all steps in Algorithm 2.2, except that $\pi(\theta)$ and $\pi(\theta^*)$ are replaced by $\hat{\pi}(\theta; Z)$ and $\hat{\pi}(\theta^*; Z^*)$ for evaluating the acceptance probability. Proposition 2.15 outlines the algorithm, demonstrating that it can be framed as a standard MH sampler targeting the joint distribution of θ and Z. The marginal stationary distribution remains $\pi(\theta)$, preserving the feature of exact approximation. A comparison between Propositions 2.14 and 2.15 reveals that the independent pseudo-marginal sampler is a special case of the correlated method by specifying $K(Z^*; Z) = p(Z^*)$.

Proposition 2.15 (Correlated pseudo-marginal MH sampler) *Let $\pi(\theta)$ be the target density and $\hat{\pi}(\theta; Z)$ be a nonnegative unbiased estimator such that*

$$\int p(Z)\hat{\pi}(\theta; Z)\,dZ = \pi(\theta), \forall\theta,$$

where Z is an auxiliary random vector with the distribution $p(Z)$. Let $K(Z^*; Z)$ be a reversible transition kernel with the stationary distribution $p(Z)$:

$$p(Z)K(Z^*; Z) = p(Z^*)K(Z; Z^*).$$

Let $r(\theta^*; \theta)$ be a proposal transition kernel. The correlated pseudo-marginal MH sampling involves the following steps, for $t = 1, 2, \ldots$

1. Generate a proposal (θ^*, Z^*) by the transition kernel

$$\hat{r}(\theta^*, Z^*; \theta, Z) = r(\theta^*; \theta)K(Z^*; Z),$$

where $\theta = \theta_{t-1}$ and $Z = Z_{t-1}$.

2. Compute the acceptance probability

$$\min\left[1, \frac{\hat{\pi}(\theta^*; Z^*)r(\theta; \theta^*)}{\hat{\pi}(\theta; Z)r(\theta^*; \theta)}\right].$$

3. Set $(\theta_t, Z_t) = \begin{cases} (\theta^*, Z^*) & \text{with probability } \alpha \\ (\theta, Z) & \text{with probability } 1 - \alpha \end{cases}.$

The correlated pseudo-marginal MH sampler targets the joint distribution $p(Z)\hat{\pi}(\theta; Z)$ as its stationary distribution. The marginal stationary distribution is $\pi(\theta)$.

2.8 Big Data MCMC: Divide and Conquer

Implementing conventional MCMC samplers on massive datasets presents significant computational challenges, as each iteration of these samplers necessitates processing the entire dataset. The computational concern has catalyzed the development of scalable Bayesian inference techniques, which have become an active area of research. Two predominant methodologies have emerged. First, divide-and-conquer methods partition a large dataset into manageable subsets and then combine the estimation results from these subsets. Second, subsampling methods reduce the number of data points for evaluating the likelihood function and/or gradients by randomly selecting a small subset of data at each MCMC iteration. Bardenet et al. (2017) provides a comprehensive review of divide-and-conquer and subsampling MCMC methods for tall data.

This section focuses on divide-and-conquer methods. When the data volume exceeds the capacity of a single machine, a logical approach is to partition the data, compute or simulate subset posterior distributions (termed subposteriors), and then combine these results. The divide-and-conquer methods proposed in the literature differ in the distributional assumptions about subposteriors, and the efficient combination of subposteriors remains an open challenge. For example, the combination method of Scott et al. (2016) would be exact if subposteriors were normally distributed. The subposteriors in Neiswanger et al. (2014) are approximated by the Gaussian kernel density estimator. Wang et al. (2015) assume that each subposterior is approximated by a K-block histogram. Wang and Dunson (2013) approximate the subposterior densities via the Weierstrass transformation. Nemeth and Sherlock (2018) estimate subposterior densities by Gaussian process regressions. The subposteriors can also be combined through their geometric center (Minsker et al., 2017; Srivastava et al., 2018).

Compared to subsampling, the divide-and-conquer methods have the key advantage of supporting distributed storage and computing, akin to the MapReduce scheme (Dean and Ghemawat, 2008). Distributed storage allows data to be split across multiple servers, avoiding the entire dataset loaded into memory at once. Distributed computing enables parallel processing of subposteriors without the need for inter-thread communication. These features render divide-and-conquer methods a scalable solution for Bayesian inference with large datasets.

2.8.1 Exact Divide and Conquer

In the context of Bayesian linear regression with the conjugate priors, it is feasible to design an exact divide-and-conquer algorithm that reconstructs the full posterior by aggregating subposteriors. The process involves data partitioning, prior splitting, distributed computing and subposterior merging via NIG arithmetic operations introduced in Chapter 1.

Divide the data X, Y into subsets $X^j, Y^j, j = 1, \ldots, J$. Propositions 1.5 and 1.6 elucidate the subposterior combination rule:

$$NIG\left(\bar{\mu}, \bar{\Lambda}, \bar{a}, \bar{b}\right) = NIG\left(\mu, \Lambda, a, b\right) + \sum_{j=1}^{J} NIG\left(\tilde{\mu}_j, \tilde{\Lambda}_j, \tilde{a}_j, \tilde{b}_j\right),$$

where $NIG\left(\mu, \Lambda, a, b\right)$ denotes the prior, $NIG\left(\tilde{\mu}_j, \tilde{\Lambda}_j, \tilde{a}_j, \tilde{b}_j\right)$ represents the jth subposterior under the non-informative prior, and $NIG\left(\bar{\mu}, \bar{\Lambda}, \bar{a}, \bar{b}\right)$ is the full posterior.

Split the prior equally into J parts and distribute them across subsets using the NIG scalar multiplication rule:

$$\frac{1}{J} \cdot NIG(\mu, \Lambda, a, b) = NIG\left(\mu, J^{-1}\Lambda, J^{-1}\left(a + \frac{d}{2}\right) - \frac{d}{2}, J^{-1}b\right).$$

Commutativity and associativity of the NIG summation indicate that

$$NIG\left(\bar{\mu}, \bar{\Lambda}, \bar{a}, \bar{b}\right) = \sum_{j=1}^{J} \left[\frac{1}{J} \cdot NIG(\mu, \Lambda, a, b) + NIG\left(\tilde{\mu}_j, \tilde{\Lambda}_j, \tilde{a}_j, \tilde{b}_j\right)\right].$$

This equation justifies an exact divide-and-conquer algorithm for Bayesian linear regressions. The process involves four steps: First, divide the data into subsets. Second, split the prior using the NIG multiplication. Third, compute subposteriors using the NIG summation. Fourth, merge the subposteriors via the NIG summation. The merged subposteriors reconstruct the full posterior.

2.8.2 Fundamental Formula of Subposteriors

An exact divide-and-conquer method is not available in most statistical models beyond Bayesian linear regressions. To tackle the challenge of subposterior recombination, approximate divide-and-conquer methods adhere to a fundamental principle: the full posterior density is proportional to the product of subposterior densities.

Let $p(Y|\theta)$ be the likelihood, where Y denotes the data and θ represents the model parameters with the prior $p(\theta)$. A standard MCMC sampler targets the full posterior

$$p(\theta|Y) \propto p(\theta)\, p(Y|\theta),$$

or in a verbose form $p(\theta|Y) = \frac{p(\theta)p(Y|\theta)}{p(Y)}$, where the marginal likelihood is $p(Y) = \int p(\theta)\, p(Y|\theta)\, d\theta$.

Divide the data into subsets $Y^j, j = 1, \ldots, J$ such that

$$p(Y|\theta) = \prod_{j=1}^{J} p\left(Y^j|\theta\right),$$

and perform subposterior MCMC simulation with the target distribution

$$p\left(\theta|Y^j\right) \propto [p(\theta)]^{1/J}\, p\left(Y^j|\theta\right),$$

for $j = 1, \ldots, J$ separately with distributed and parallel computing. The split prior density $[p(\theta)]^{1/J}$ is proper if the normalizing constant $k = \int [p(\theta)]^{1/J}\, d\theta$ is finite. The subposterior density can be written as

$$p\left(\theta|Y^j\right) = \frac{k^{-1}[p(\theta)]^{1/J}\, p\left(Y^j|\theta\right)}{p\left(Y^j\right)},$$

where the subset marginal likelihood is $p\left(Y^j\right) = \int k^{-1}[p(\theta)]^{1/J}\, p\left(Y^j|\theta\right) d\theta$.

It follows that the full and subposterior densities satisfy the equation

$$p(\theta|Y) = \frac{k^J \prod_{j=1}^{J} p\left(Y^j\right)}{p(Y)} \prod_{j=1}^{J} p\left(\theta|Y^j\right),$$

or in a compact form

$$p(\theta\,|Y) \propto \prod_{j=1}^{J} p(\theta\,|Y^j)\,.$$

This formula will be used extensively in the subsequent discussions, as it underpins various divide-and-conquer approaches for subposterior recombination.

2.8.3 Consensus Monte Carlo

The consensus Monte Carlo method, proposed by Scott et al. (2016), recombines subposterior samples in a manner that would be exact if the subposteriors were normally distributed. One of the appealing features of consensus Monte Carlo is its closed-form formula for merging subposteriors, which incurs a low computational cost.

Suppose that the subposteriors follow multivariate normal distributions:

$$p\left(\theta\,\middle|Y^j\right) \propto e^{-\frac{1}{2}(\theta-\mu_j)'\Sigma_j^{-1}(\theta-\mu_j)},$$

where the subposterior mean μ_j and the covariance matrix Σ_j are estimated by the sample analogue of subposterior MCMC draws. The normality assumption is reasonable in the context of regular models with large sample sizes, owing to the Bernstein-von Mises theorem (Scott, 2017).

The full posterior density is proportional to the product of subposterior densities. Completing the square yields the full posterior that preserves the multivariate normal distribution:

$$p(\theta\,|Y) \propto \prod_{j=1}^{J} p\left(\theta\,\middle|Y^j\right) \propto e^{-\frac{1}{2}(\theta-\bar{\mu})'\bar{\Sigma}^{-1}(\theta-\bar{\mu})},$$

where the aggregated mean $\bar{\mu}$ and the covariance matrix $\bar{\Sigma}$ are given by

$$\bar{\Sigma} = \left(\sum_{j=1}^{J} \Sigma_j^{-1}\right)^{-1},$$

$$\bar{\mu} = \bar{\Sigma}\left(\sum_{j=1}^{J} \Sigma_j^{-1}\mu_j\right),$$

which is reminiscent of the NIG summation operator (see Definition 1.2), where the precision matrices (i.e., inverse covariance matrices) add up, and the full-sample mean is the weighted average of subset means, with the weight proportional to the precision matrix.

An alternative method is to combine subposterior MCMC samples instead of Gaussian densities. Let θ_{jt} be the tth draw generated by the jth subposterior sampler. Consider the weighted average of subposterior draws:

$$\bar{\theta}_t = \bar{\Sigma}\left(\sum_{j=1}^{J} \Sigma_j^{-1}\theta_{jt}\right).$$

If θ_{jt} is a draw from the multivariate normal distribution with the mean μ_j and the covariance matrix Σ_j, then the linear combination $\bar{\theta}_t$ preserves normality, with the mean and covariance matrix given by

$$E\left(\bar{\theta}_t\right) = \bar{\Sigma}\left(\sum_{j=1}^{J} \Sigma_j^{-1}\mu_j\right) = \bar{\mu},$$

$$Var\left(\bar{\theta}_t\right) = \bar{\Sigma}\left(\sum_{j=1}^{J} \Sigma_j^{-1}\Sigma_j\Sigma_j^{-1}\right)\bar{\Sigma} = \bar{\Sigma}.$$

Consequently, $\bar{\theta}_t$ can be viewed as a draw from the aggregated normal distribution.

2.8.4 Kernel Density Approximation

The consensus Monte Carlo method by Scott et al. (2016) performs effectively when subposteriors closely resemble normal distributions. To address deviations from normality, Neiswanger et al. (2014) propose a non-parametric merging algorithm, where each subposterior is approximated using a kernel density estimator of MCMC draws θ_{jt}, $j = 1, \ldots, J$ and $t = 1, \ldots, T$:

$$\hat{p}\left(\theta \left| Y^j \right.\right) \propto \frac{1}{T}\sum_{t=1}^{T} \phi\left(\theta; \theta_{jt}, hI\right),$$

where $\phi(\cdot)$ is the multivariate normal density, and the variance h serves as the bandwidth, a user-specified tuning parameter. The kernel density evaluated at θ is determined by the average distance to the points $\theta_{j1}, \ldots, \theta_{jT}$. A shorter distance indicates a higher density.

The full posterior is proportional to the product of subposteriors:

$$\hat{p}\left(\theta \left| Y \right.\right) \propto \prod_{j=1}^{J} \hat{p}\left(\theta \left| Y^j \right.\right) \propto \sum_{t_1=1}^{T} \cdots \sum_{t_J=1}^{T} \prod_{j=1}^{J} \phi\left(\theta; \theta_{j,t_j}, hI\right).$$

Let $z = (t_1, \ldots t_J)$ be a vector of integers ranging from 1 to T, which selects one draw from each subposterior, namely $(\theta_{1,t_1}, \ldots \theta_{J,t_J})$. The product of multivariate normal densities preserves normality. Completing the square yields the full posterior approximated by a Gaussian mixture distribution:

$$\hat{p}\left(\theta \left| Y \right.\right) \propto \sum_{t_1=1}^{T} \cdots \sum_{t_J=1}^{T} w_z \phi\left(\theta; \bar{\theta}_z, J^{-1}hI\right),$$

where each component of the mixture has the mean

$$\bar{\theta}_z = \frac{1}{J}\sum_{j=1}^{J} \theta_{j,t_j},$$

with the weight

$$w_z \propto \prod_{j=1}^{J} \phi\left(\bar{\theta}_z; \theta_{j,t_j}, hI\right).$$

However, the number of mixture components amounts to T^J, which is too large to be processed analytically. To overcome this curse of dimensionality, MCMC methods, such as the discrete-state MH sampler, can be used. At each iteration, an element in z is updated using the weight w_z. Conditional on z, samples of θ are obtained using the multivariate normal density $\phi\left(\theta; \bar{\theta}_z, J^{-1}hI\right)$. This aggregation method appears intuitive, as samples of θ are centered at $\bar{\theta}_z$, which is formulated by sampling one draw from each subposterior and averaging them.

The kernel density estimator is effective when the dimensionality of θ is low. For high-dimensional problems, adequately characterizing the subset posterior requires a large number of draws (T), which makes the merging algorithm computationally intensive. Additionally, Bardenet et al. (2017) point out that the upper bound of the mean squared error of the estimator grows exponentially with the number of subsets (J).

2.8.5 Gaussian Process Approximations

Recent advancements by Nemeth and Sherlock (2018) have further refined divide-and-conquer methods. Their work addresses a notable gap in existing methodologies, highlighting the underutilization of subposterior densities computed during the MCMC sampling process. The MH or HMC sampler must evaluate target densities to accept or reject proposals. Once MCMC draws are obtained, the densities are typically discarded. These subposterior densities, though by-products of the sampling algorithm, encapsulate crucial information regarding the subposterior distributions. The core challenge addressed by Nemeth and Sherlock (2018) is a probabilistic interpolation problem: how to extrapolate the subposterior distributions across their entire support using information from a finite set of points.

Consider the jth subposterior density function $p\left(\theta \left| Y^j \right.\right) : \mathbb{R}^d \to \mathbb{R}$. Limited information about this function is available from the output of the MCMC sampler, which has evaluated target densities at T discrete points. The goal is to predict the subposterior density at a new, unobserved point. This prediction is approached through non-parametric regression, where the response variable is the subposterior density $p\left(\theta \left| Y^j \right.\right)$, and the predictors are the d-dimensional parameters θ. The subposterior MCMC sampler provides T observations $\left\{\theta_{jt}, p\left(\theta_{jt} \left| Y^j \right.\right)\right\}, t = 1, \ldots, T$, serving as the training data for model fitting and prediction. Nemeth and Sherlock (2018) assume that the logarithm of the subposterior density $\ln p\left(\theta \left| Y^j \right.\right)$ follows a Gaussian process, characterized by a mean function $m(\theta)$ and a covariance function $K\left(\theta, \tilde{\theta}\right)$. Gaussian processes are particularly advantageous because they allow for the prediction of function values at new points based on the multivariate normal distribution. Let observations be $\theta_{1:T} = (\theta_{j1}, \ldots, \theta_{jT})$ and a new point be θ. For the Gaussian process, the joint distribution is multivariate normal:

$$\begin{bmatrix} \ln p\left(\theta \left| Y^j \right.\right) \\ \ln p\left(\theta_{1:T} \left| Y^j \right.\right) \end{bmatrix} \sim N\left\{\begin{bmatrix} m(\theta) \\ m(\theta_{1:T}) \end{bmatrix}, \begin{bmatrix} K(\theta, \theta) & K(\theta, \theta_{1:T}) \\ K(\theta_{1:T}, \theta) & K(\theta_{1:T}, \theta_{1:T}) \end{bmatrix}\right\}.$$

Given the observations for $\ln p\left(\theta_{1:T} \left| Y^j \right.\right)$, the conditional distribution remains normal:

$$\ln p\left(\theta \left| Y^j \right.\right) \left| \ln p\left(\theta_{1:T} \left| Y^j \right.\right) \right. \sim N\left(\bar{\mu}_j, \bar{\Sigma}_j\right),$$

where the conditional mean and variance take the form

$$\bar{\mu}_j = m(\theta) + K(\theta, \theta_{1:T}) \left[K(\theta_{1:T}, \theta_{1:T})\right]^{-1} \left[\ln p\left(\theta_{1:T} \left| Y^j \right.\right) - m(\theta_{1:T})\right],$$

$$\bar{\Sigma}_j = K(\theta, \theta) - K(\theta, \theta_{1:T}) \left[K(\theta_{1:T}, \theta_{1:T})\right]^{-1} K(\theta_{1:T}, \theta).$$

One of the appealing features of the log-normal specification is its additivity, enabling the full sample posterior $p(\theta | Y) \propto \prod_{j=1}^{J} p\left(\theta \left| Y^j \right.\right)$ to follow a log-normal distribution, with parameters $\sum_{j=1}^{J} \bar{\mu}_j$ and $\sum_{j=1}^{J} \bar{\Sigma}_j$. A reasonable approximation of $\ln p(\theta | Y)$ and $p(\theta | Y)$ are

$$\ln \hat{p}(\theta | Y) = \sum_{j=1}^{J} \bar{\mu}_j,$$

$$\hat{p}(\theta | Y) = e^{\sum_{j=1}^{J} \bar{\mu}_j + \frac{1}{2} \sum_{j=1}^{J} \bar{\Sigma}_j}.$$

The approximated full posterior density facilitates the use of MCMC methods for simulating the full posterior distribution. However, the computational complexity can be significant, since constructing the Gaussian process estimator involves inverting a potentially large $T \times T$ covariance matrix. Therefore, the number of draws should be moderate. The HMC sampler is particularly well-suited for this task, as it efficiently explores the target distribution with a relatively small number of samples.

To implement the Gaussian process regression, the mean function $m(\theta)$ and the covariance function $K\left(\theta, \tilde{\theta}\right)$ must be specified. A common choice for the covariance function can be the squared exponential kernel akin to the normal density (Rasmussen and Williams, 2006):

$$K\left(\theta, \tilde{\theta}\right) = \sigma_0^2 e^{-\frac{1}{2}\sigma_1^{-2}(\theta-\tilde{\theta})'(\theta-\tilde{\theta})}.$$

Regarding the mean function, Nemeth and Sherlock (2018) propose a quadratic form

$$m(\theta) = \beta_0 + \theta'\beta_1 + diag(\theta\theta')\beta_2,$$

where $\beta_2 < 0$, reflecting the tail behavior of the log density as θ approaches infinity.

2.8.6 Importance Sampling

Importance sampling uses weighted samples from a proposal distribution to approximate a target distribution. The target density, proportional to $f(\theta)g(\theta)$, can be approximated by proposals from $f(\theta)$ with unnormalized importance weight $g(\theta)$. When applied to divide-and-conquer MCMC methods, the target distribution is the full posterior, which is proportional to the product of subposterior densities: $p(\theta|Y) \propto \prod_{j=1}^{J} p(\theta|Y^j)$. Proposal samples (i.e., particles) are generated by subposterior MCMC samplers from $p(\theta|Y^j)$, and each particle is assigned an unnormalized importance weight

$$\tilde{w}_j(\theta) = p(\theta)^{\frac{J-1}{J}} \prod_{i \neq j} p(Y^i|\theta).$$

These weighted particles collectively approximate the full posterior distribution. The importance sampling can be applied to each subset separately ($j = 1, \ldots, J$), resulting in J estimators of the full posterior.

The importance sampling method has advantages and disadvantages. It provides a consistent estimate of the full posterior expectation without assumptions about the shape of the subposterior distributions being combined. However, it can be computationally intensive to evaluate importance weights, as each subposterior sampler broadcasts particles to other samplers for likelihood evaluation. If each sampler generates T draws, weight evaluation requires $(J-1)T$ passes over massive data, which can be a computational burden. Additionally, importance sampling faces challenges such as weight degeneracy, if some subposteriors differ significantly from the full posterior. A small number of particles may carry heavy weights, while others contribute little to the full posterior approximation, potentially leading to inefficient approximations.

To address the computational issues of importance sampling, we consider matched samples in the subposterior simulation, allowing posterior densities used by subposterior MCMC samplers to be recycled for evaluating importance weights.

Matched samples are constructed using global and local proposals. Global proposals from $q(\theta^*)$ are shared across all subposterior samplers, each of which has a local proposal density $q_j(\theta^*|\theta)$ for the jth subposterior MCMC. Instead of direct sampling from local proposal densities, subposterior samplers employ rejection sampling to select from global proposals, thereby enhancing computational efficiency. For example, $q(\theta^*)$ and $q_j(\theta^*|\theta)$ can be Gaussian densities $\phi(\theta^*; \mu, \Sigma)$ and $\phi(\theta^*; \mu_j(\theta), \Sigma_j)$, respectively. A global proposal from the former is accepted by the probability

$$\frac{\phi(\theta^*; \mu_j, \Sigma_j)}{\phi(\theta^*; \mu, \Sigma) B_j},$$

Algorithm 2.5 MH sampler with matched samples

Input: subset data Y^j, proposal densities $q(\cdot)$ and $q_j(\cdot)$.
Output: MCMC samples $\{\theta_{jt}\}$ from the subposterior $p(\theta | Y^j)$.
1 Generate global proposals $\theta^*_{(n)}$, $n = 1, \ldots, N$ by $q(\cdot)$.
2 Iterate for $t = 1, \ldots, T$:
2.1 Select θ^*_{jt} from $\theta^*_{(n)}$ by rejection sampling with the target $q_j(\cdot)$.
2.2 Compute $\alpha = \min\left[1, \frac{p(\theta^*_{jt})^{1/J} p(Y^j|\theta^*_{jt}) q_j(\theta|\theta^*_{jt})}{p(\theta)^{1/J} p(Y^j|\theta) q_j(\theta^*_{jt}|\theta)}\right]$, where $\theta = \theta_{j,t-1}$.
2.3 Set $\theta_{jt} = \begin{cases} \theta^*_{jt} & \textit{with probability } \alpha \\ \theta & \textit{with probability } 1-\alpha \end{cases}$.

where $\mu_j(\theta)$ is abbreviated as μ_j, and the upper bound of the density ratio is given by

$$B_j = \frac{|\Sigma|^{1/2}}{|\Sigma_j|^{1/2}} e^{-\frac{1}{2}\left[\mu_j'\Sigma_j^{-1}\mu_j - \mu'\Sigma^{-1}\mu - (\Sigma_j^{-1}\mu_j - \Sigma^{-1}\mu)'(\Sigma_j^{-1} - \Sigma^{-1})^{-1}(\Sigma_j^{-1}\mu_j - \Sigma^{-1}\mu)\right]}.$$

A special case of the global-local proposal specification is $q_j(\theta^* | \theta) = q(\theta^*)$, leading to an independence MH sampler, wherein the transition kernel does not depend on the current state of the Markov chain. In this scenario, all global proposals are accepted as local proposals. All proposals for the MH sampler are generated independently before the first iteration of MCMC sampling, instead of making a proposal at each iteration.

In Algorithm 2.5, global proposals are denoted by $\theta^*_{(n)}$, $n = 1, \ldots, N$, local proposals by θ^*_{jt}, $t = 1, \ldots, T$, and MCMC samples by θ_{jt}, $t = 1, \ldots, T$. For example, if $\theta^*_{(1)}, \theta^*_{(3)}, \theta^*_{(5)}, \ldots$ are accepted by the rejection sampling for the jth chain, local proposals are indexed as $\theta^*_{j1} = \theta^*_{(1)}$, $\theta^*_{j2} = \theta^*_{(3)}$, $\theta^*_{j3} = \theta^*_{(5)}$, and so on. Because all MCMC samples are members of local proposals, and all local proposals are members of global proposals, we refer to them as matched samples.

Parallel computing can further enhance the efficiency of Algorithm 2.5, as subposterior MCMC simulations can proceed concurrently without inter-sampler communication. Matched samples are created by fixing the random number seed for generating global proposals $\theta^*_{(n)}, n = 1, \ldots, N$. This is equivalent to broadcasting global proposals to all samplers before the first iteration of the MH sampling, streamlining the likelihood evaluation process. The MH samplers evaluate the likelihood function $p(Y^j | \theta^*_{jt})$, $j = 1, \ldots, J$, $t = 1, \ldots, T$. This output is what we aim to recycle.

A divide-and-conquer MCMC algorithm with importance sampling via matched samples is presented in Algorithm 2.6. The novelty here is the use of recycled densities in importance weight calculations at step 4. As θ_{jt} is a member of global proposals, the likelihood function $p(Y^i | \theta_{jt})$ is likely to be evaluated by the ith subposterior sampler and can be recycled for weight evaluation. The recycle rate is maximized under $q_j(\theta^* | \theta) = q(\theta^*)$, $\forall j$, resulting in the importance weights fully determined by recycled densities. In that case, calculating importance weights in step 4 does not require rescanning the original data, thereby reducing computing time relative to subposterior sampling.

The feature of sample matching does not add complexity to the asymptotic analysis, if each subposterior Markov chain underlying the MH sampler is examined separately. For an ergodic Markov chain, the sample average converges to the expected value. Weighted particles can consistently estimate expectations with respect to the full posterior. Proposition 2.16 shows that Algorithm 2.6 provides J consistent estimators of $\int h(\theta) p(\theta | Y) d\theta$ under mild regularity conditions.

Algorithm 2.6 Divide and conquer by matched samples

Input: data Y, proposal densities $q(\cdot)$ and $q_j(\cdot)$, and a function $h(\theta)$.
Output: a collection of estimators of $\int h(\theta)\, p(\theta\,|Y)\, d\theta$.
1 Divide data Y into subsets $Y^1, \ldots, Y^J$.
2 Generate global proposals $\theta^*_{(n)}$, $n = 1, \ldots, N$ by $q(\cdot)$.
3 Share global proposals and apply Algorithm 2.5 to each subposterior sampler.
4 Compute unnormalized and normalized importance weight

$$\tilde{w}_j(\theta_{jt}) = p(\theta_{jt})^{\frac{J-1}{J}} \prod_{i \neq j} p(Y^i\,|\theta_{jt}),$$

$$w_j(\theta_{jt}) = \frac{\tilde{w}_j(\theta_{jt})}{\sum_{\tau=1}^{T} \tilde{w}_j(\theta_{j\tau})}.$$

5 Approximate $\int h(\theta)\, p(\theta\,|Y)\, d\theta$ by $\sum_{t=1}^{T} h(\theta_{jt})\, w_j(\theta_{jt})$, $j = 1, \ldots, J$.

Proposition 2.16 (Divide-and-conquer importance sampling) *Consider the jth MH sampler targeting the subposterior $p(\theta\,|Y^j)$ in Algorithm 2.6. Suppose that the MCMC samples θ_{jt}, $t = 1, \ldots, T$ are generated by an irreducible, aperiodic and Harris recurrent Markov chain, $\int |h(\theta)\, \tilde{w}_j(\theta)|\, p(\theta\,|Y)\, d\theta < \infty$ and $\int |\tilde{w}_j(\theta)|\, p(\theta\,|Y)\, d\theta < \infty$. As $T \to \infty$, we have*

$$\sum_{t=1}^{T} h(\theta_{jt})\, w_j(\theta_{jt}) \xrightarrow{p} \int h(\theta)\, p(\theta\,|Y)\, d\theta.$$

Asymptotic analysis of this method reveals that while all estimators are consistent, their accuracy may vary.

Let $\bar{H}_j = T^{-1} \sum_{t=1}^{T} h(\theta_{jt})\, \tilde{w}_j(\theta_{jt})$ and $\bar{W}_j = T^{-1} \sum_{t=1}^{T} \tilde{w}_j(\theta_{jt})$. The importance sampling estimator $\hat{\mu}_j$ equals the ratio of $\bar{H}_j$ and $\bar{W}_j$:

$$\hat{\mu}_j = \frac{\bar{H}_j}{\bar{W}_j} = \sum_{t=1}^{T} h(\theta_{jt})\, w_j(\theta_{jt}).$$

Ergodicity of the Markov chain indicates consistency of $\bar{H}_j$ and $\bar{W}_j$ such that $\bar{H}_j \xrightarrow{p} c_j\mu$ and $\bar{W}_j \xrightarrow{p} c_j$, where the constant $c_j = \frac{p(Y)}{p(Y^j)k}$ and $\mu = \int h(\theta)\, p(\theta\,|Y)\, d\theta$. The Central Limit Theorem of the MCMC chain yields asymptotic normality:

$$\sqrt{T} \begin{pmatrix} \bar{H}_j - c_j\mu \\ \bar{W}_j - c_j \end{pmatrix} \xrightarrow{d} N(0, \Gamma_j),$$

where the covariance matrix Γ_j captures the correlation structure of MCMC samples:

$$\Gamma_j = \sum_{i=-\infty}^{\infty} \begin{pmatrix} \alpha_{ji} & \beta_{ji} \\ \beta_{ji} & \gamma_{ji} \end{pmatrix},$$

$$\alpha_{ji} = cov\left[h(\theta_{jt})\, \tilde{w}_j(\theta_{jt}), h(\theta_{j,t-i})\, \tilde{w}_j(\theta_{j,t-i})\right],$$

$$\beta_{ji} = cov\left[h(\theta_{jt})\, \tilde{w}_j(\theta_{jt}), \tilde{w}_j(\theta_{j,t-i})\right],$$

$$\gamma_{ji} = cov\left[\tilde{w}_j(\theta_{jt}), \tilde{w}_j(\theta_{j,t-i})\right].$$

Note that $\frac{\bar{H}_j}{\bar{W}_j}$ is a continuously differentiable function of $\bar{H}_j$ and $\bar{W}_j$. On the one hand, consistency of $\bar{H}_j$ and $\bar{W}_j$ implies that $\hat{\mu}_j \xrightarrow{p} \mu$. On the other hand, the Delta Method provides the asymptotic distribution of the ratio estimator:

$$\sqrt{T}\left(\hat{\mu}_j - \mu\right) \xrightarrow{d} N\left(0, G_j \Gamma_j G_j'\right),$$

where the gradient $G_j = c_j^{-1}(1, -\mu)$.

If the MCMC sampler could generate perfectly independent samples, the covariance terms would be zero: $\alpha_{ji} = 0$, $\beta_{ji} = 0$, $\gamma_{ji} = 0$, $\forall i \neq 0$. In that case, the asymptotic variance $G_j \Gamma_j G_j'$ takes a simple form:

$$G_j \Gamma_j G_j' = c_j^{-2} \begin{pmatrix} 1 & -\mu \end{pmatrix} \begin{pmatrix} \alpha_{j0} & \beta_{j0} \\ \beta_{j0} & \gamma_{j0} \end{pmatrix} \begin{pmatrix} 1 \\ -\mu \end{pmatrix} = c_j^{-2} E\left\{\tilde{w}_j \left(\theta_{jt}\right)^2 \left[h\left(\theta_{jt}\right) - \mu\right]^2\right\}.$$

We approximate the constant c_j and μ by the consistent estimator, namely $\bar{W}_j$ and $\hat{\mu}_j$, so that the approximated asymptotic variance of the ratio estimator is given by

$$Var\left(\hat{\mu}_j\right) = \sum_{t=1}^{T} w_j \left(\theta_{jt}\right)^2 \left[h\left(\theta_{jt}\right) - \hat{\mu}_j\right]^2.$$

To gain some intuition of the estimator variance, we consider the scenario that the subposteriors align perfectly with the full posterior. In that case, $w_j\left(\theta_{jt}\right) = \frac{1}{T}$, $\forall t = 1, \ldots, T$. The variance is estimated by averaging the squared residuals: $Var\left(\hat{\mu}_j\right) = \frac{1}{T^2} \sum_{t=1}^{T} \left[h\left(\theta_{jt}\right) - \hat{\mu}_j\right]^2$, where $\hat{\mu}_j = \frac{1}{T} \sum_{t=1}^{T} h\left(\theta_{jt}\right)$. The variance is akin to that of a simple Monte Carlo estimator: the sample variance divided by T. However, when subposteriors diverge, the variance is estimated using the weighted sum of squared residuals.

2.9 Big Data MCMC: Subsampling

The MH sampler is a cornerstone of MCMC methods, widely used for posterior simulation in Bayesian statistics. However, when applied to large datasets, the MH sampler can become computationally intensive due to the necessity of evaluating the likelihood function for all data points at each proposed parameter update. The computational challenge has led to the development of subsampling techniques, which aim to reduce the number of likelihood evaluations by using only a subset of the data at each iteration.

Subsampling methods can be categorized into exact and approximate approaches. An exact subsampling method ensures that the target distribution of the MCMC sampler remains the full posterior distribution, whereas an approximate method results in a slightly perturbed target distribution. In contrast with divide-and-conquer strategies, which distribute data across multiple machines, subsampling methods typically assume that the full dataset is available in memory, allowing for random selection of a subset of data.

2.9.1 Delayed Acceptance and Early Rejection

A well-tuned MH sampler often maintains an acceptance rate around 25%, implying that three out of four proposals are rejected. Heuristically, even a small random subset of data can often reveal an unpromising proposal that causes significant drop of the likelihood.

Banterle et al. (2019) propose a method to accelerate the MH sampler through delayed acceptance, a two-stage process designed to screen out unpromising proposals at a lower computational cost. The first stage is a qualification test using a small subsample. The procedure follows the standard MH sampler (Algorithm 2.2), except that the likelihood is evaluated by the subsample only. Unpromising proposals with low acceptance probabilities are likely be quickly rejected in the first stage. Promising proposals proceed to the second stage, where the final decision on acceptance is made. The key question is how to compute the second-stage acceptance probability. Should we use the full-sample posterior density, the full-sample likelihood, the remaining-sample posterior, or the remaining-sample likelihood? The answer is the last one, as shown by Proposition 2.17, which demonstrates that the delayed acceptance MH sampler maintains the full posterior distribution as its stationary distribution, thus qualifying as an exact subsampling method.

Proposition 2.17 (Delayed acceptance and early rejection) *Let the target distribution be $\pi(\theta) \propto p(\theta)\prod_{i=1}^{n} p(Y_i|\theta)$, and $r(\theta^*;\theta)$ is a proposal transition kernel. Let $Y_1,\ldots,Y_m$ $(m<n)$ be a subsample. Similar to Equation (2.2), the two-stage MH sampler modifies the transition kernel such that*

$$q(\theta^*;\theta) = \alpha(\theta^*,\theta)\, r(\theta^*;\theta) + \delta_\theta(\theta^*)\int [1-\alpha(\theta^*,\theta)]\, r(\theta^*;\theta)\, d\theta^*,$$

where the acceptance probability takes the form:

$$\begin{aligned}
\alpha(\theta^*,\theta) &= \alpha_1(\theta^*,\theta)\,\alpha_2(\theta^*,\theta),\\
\alpha_1(\theta^*,\theta) &= \min[1,\rho_1(\theta^*,\theta)],\\
\alpha_2(\theta^*,\theta) &= \min[1,\rho_2(\theta^*,\theta)],\\
\rho_1(\theta^*,\theta) &= \frac{p(\theta^*)\prod_{i=1}^{m} p(Y_i|\theta^*)\, r(\theta;\theta^*)}{p(\theta)\prod_{i=1}^{m} p(Y_i|\theta)\, r(\theta^*;\theta)},\\
\rho_2(\theta^*,\theta) &= \frac{\prod_{i=m+1}^{n} p(Y_i|\theta^*)}{\prod_{i=m+1}^{n} p(Y_i|\theta)}.
\end{aligned}$$

The modified transition kernel satisfies the reversibility condition

$$\pi(\theta)\, q(\theta^*;\theta) = \pi(\theta^*)\, q(\theta;\theta^*),$$

ensuring $\pi(\theta)$ as a stationary distribution.

Proposition 2.17 is applicable to an arbitrary subsample, but the efficiency of the two-stage MH sampler is enhanced by using a representative subsample. When tuning the sampler, a larger proposal variance may be beneficial, especially with a random-walk MH sampler, because of the asymmetric treatment of rejected and accepted proposals: most proposals are quickly rejected in the qualification test with a small subset of data, while an accepted proposal requires a sweep over the entire dataset. A low acceptance rate implies a sticky chain with infrequent large jumps, which can be mitigated by thinning MCMC draws to reduce autocorrelations.

Proposition 2.17 is a simplified version of multiple-stage MH samplers with delayed acceptance (Banterle et al., 2019), in which the full-sample acceptance ratio is decomposed as the product of multiple components: $\rho(\theta^*,\theta) = \prod_{j=1}^{k}\rho_j(\theta^*,\theta)$, where the component-wise acceptance ratio satisfies the balance condition: $\rho_j(\theta^*,\theta) = 1/\rho_j(\theta,\theta^*)$. The original MH algorithm uses the acceptance probability $\min[1,\rho(\theta^*,\theta)]$, while the delayed-acceptance algorithm uses $\prod_{j=1}^{k}\alpha_j(\theta^*,\theta)$, where $\alpha_j(\theta^*,\theta) = \min[1,\rho_j(\theta^*,\theta)]$. The approach maintains the reversibility condition and asymptotically generates draws from the exact target distribution.

2.9.2 Firefly Monte Carlo

Firefly Monte Carlo, introduced by Maclaurin and Adams (2014), is an exact MCMC subsampling method. By augmenting the data with an auxiliary variable that determines whether a data point is bright (included in the likelihood evaluation) or dark (excluded from the likelihood evaluation), each iteration of the MCMC sampler only queries the likelihood of a subset data. Maclaurin and Adams (2014) describe this metaphor as evoking the image of fireflies, with individual data points blinking on and off.

Let $p(Y_i|\theta)$, $i = 1, \ldots, n$ denote the likelihood for the data Y_i, given the parameters θ with a prior $p(\theta)$. The target distribution is the posterior density, proportional to $p(\theta)\prod_{i=1}^{n} p(Y_i|\theta)$. A fundamental assumption underlying the Firefly Monte Carlo method is that the likelihood function $p(Y_i|\theta)$ has a positive lower bound $b(Y_i, \theta)$, ensuring a well-defined probability

$$0 \leq \frac{p(Y_i|\theta) - b(Y_i,\theta)}{p(Y_i|\theta)} \leq 1.$$

Each data point is associated with a binary auxiliary variable $Z_i \in \{0, 1\}$, $i = 1, \ldots, n$. Conditional on θ and Y, the probability that $Z_i = 1$ (bright data) is constructed as $\frac{p(Y_i|\theta)-b(Y_i,\theta)}{p(Y_i|\theta)}$, leading to

$$p(Z_i|\theta, Y_i) = \left[\frac{p(Y_i|\theta) - b(Y_i,\theta)}{p(Y_i|\theta)}\right]^{Z_i} \left[\frac{b(Y_i,\theta)}{p(Y_i|\theta)}\right]^{1-Z_i}.$$

The joint posterior distribution $p(\theta, Z|Y) \propto p(\theta)\prod_{i=1}^{n} p(Y_i|\theta)\, p(Z_i|\theta, Y_i)$ takes the form

$$p(\theta, Z|Y) \propto p(\theta)\prod_{i=1}^{n} b(Y_i,\theta) \prod_{i=1}^{n} \left[\frac{p(Y_i|\theta) - b(Y_i,\theta)}{b(Y_i,\theta)}\right]^{Z_i}.$$

Assuming the product $\prod_{i=1}^{n} b(Y_i,\theta)$ is determined by sufficient statistics (rather than individual data points), it can be evaluated without sweeping through the entire dataset. Maclaurin and Adams (2014) demonstrate the bound specification for logistic regression.

Any MCMC sampler can be used for posterior simulation of $p(\theta, Z|Y)$. Maclaurin and Adams (2014) propose a Metropolis-within-Gibbs sampler that alternates between sampling from $p(\theta|Y, Z)$ and $p(Z_i|\theta, Y_i)$. On the one hand, $p(\theta|Y, Z)$ is proportional to $p(\theta, Z|Y)$, which only involves the likelihood of the bright data. The posterior sampler only queries a subset of data where $Z_i = 1$. The computational savings can be substantial if most data are dark. The closer the lower bound approaches the likelihood function, the more data points are dark. On the other hand, $p(Z_i|\theta, Y_i)$ is determined by individual data points. Updating all $Z_1, \ldots, Z_n$ in every iteration would defeat the purpose of subsampling. Maclaurin and Adams (2014) argue that "the bottleneck for mixing is usually the exploration of θ, not Z_n". Therefore, a random subset of the auxiliary vector is updated.

Bardenet et al. (2017) extend the Firefly Monte Carlo with a pseudo-marginal approach. The data generating process indicates that the joint distribution of θ, Y, Z can be decomposed as

$$p(\theta, Y, Z) = p(\theta)\prod_{i=1}^{n} p(Y_i|\theta)\, p(Z_i|\theta, Y_i).$$

An alternative decomposition scheme reverses the order of Y and Z:

$$p(\theta, Y, Z) = p(\theta)\prod_{i=1}^{n} p(Z_i|\theta)\, p(Y_i|\theta, Z_i).$$

The marginal probability of the binary variable Z_i is given by

$$p(Z_i = 1\,|\theta) = \int [p(Y_i\,|\theta) - b(Y_i,\theta)]\, dY_i = 1 - I_\theta,$$

$$p(Z_i = 0\,|\theta) = I_\theta,$$

where $I_\theta = \int b(Y_i,\theta)\, dY_i$. Given the auxiliary variable Z_i, the conditional distribution of Y_i is given by

$$p(Y_i\,|\theta, Z_i = 1) = \frac{p(Y_i\,|\theta) - b(Y_i,\theta)}{1 - I_\theta},$$

$$p(Y_i\,|\theta, Z_i = 0) = \frac{b(Y_i,\theta)}{I_\theta}.$$

Our target distribution is $\pi(\theta) = p(\theta)\prod_{i=1}^{n} p(Y_i\,|\theta)$, which depends on the likelihood of all data points. An unbiased estimator of the target distribution is

$$\hat{\pi}(\theta) = p(\theta)\prod_{i=1}^{n} p(Y_i\,|\theta, Z_i),$$

where Z_i is drawn from a Bernoulli distribution with the probability $1 - I_\theta$, namely $p(Z_i = 1\,|\theta)$. Taking the expectation with respect to Z_i yields $E[\hat{\pi}(\theta)] = \pi(\theta)$. The pseudo-marginal method indicates that running the MH sampler with $\hat{\pi}(\theta)$ in lieu of $\pi(\theta)$ still targets at the exact posterior distribution. The advantage of $\hat{\pi}(\theta)$ is that the MH sampler only computes the likelihood of the bright data.

2.9.3 Subsampling with Control Variates

The pseudo-marginal method (Andrieu and Roberts, 2009) offers a promising direction for scalable Bayesian research. The goal is to develop a non-negative and unbiased estimator of the likelihood function through subsampling, thereby reducing the number of data points required for likelihood evaluation. Although achieving an exactly unbiased estimator is often infeasible, it is possible to reduce the bias to an acceptable level, resulting in an approximate subsampling method. For comprehensive discussions on unbiased estimators, readers are referred to Jacob and Thiery (2015) and Rhee and Glynn (2015).

Consider the scenario where $\ell_i(\theta, Y_i) = \ln p(Y_i\,|\theta)$ denotes the log-likelihood of the ith observation, and $\mathcal{L}(\theta, Y) = \sum_{i=1}^{n} \ell_i(\theta, Y_i)$ represents the log-likelihood of the full dataset, from which a subset can be drawn by subsampling m out of n data points with replacement, where $m < n$. Specifically, uniform random integers $u_j, j = 1, \ldots, m$ are drawn from $\{1, \ldots, n\}$ to construct the subsample $Y_{u_1}, \ldots, Y_{u_m}$. The subsample log-likelihood is then aggregated and rescaled:

$$\hat{\mathcal{L}}(\theta, Y, u) = \frac{n}{m}\sum_{j=1}^{m} \ell_{u_j}(\theta, Y_{u_j}).$$

For a given parameter value θ and the data Y, the rescaled subsample log-likelihood $\hat{\mathcal{L}}(\theta, Y, u)$ serves as an estimator of the full-sample log-likelihood $\mathcal{L}(\theta, Y)$, and estimation uncertainty stems from the random variable u. In this context, the mean and variance of an estimator are considered with respect to u, holding θ and Y fixed. The integer-valued u_i implies ℓ_{u_j} following a categorical distribution:

$$P(\ell_{u_j} = \ell_i) = \frac{1}{n}, \forall i = 1, \ldots, n,$$

with mean and variance

$$E\left(\ell_{u_j}\right) = \frac{1}{n}\sum_{i=1}^{n}\ell_i = \frac{1}{n}\mathcal{L},$$

$$Var\left(\ell_{u_j}\right) = \frac{1}{n}\sum_{i=1}^{n}\ell_i^2 - \left(\frac{1}{n}\sum_{i=1}^{n}\ell_i\right)^2 \equiv \sigma_L^2,$$

where σ_L^2 reflects heterogeneity and volatility of the log-likelihood $\ell_1, \ldots, \ell_n$. The dependency of σ_L^2 on θ and Y is implicit in our notation. The mean and variance

$$E\left(\hat{\mathcal{L}}\right) = \frac{n}{m}\sum_{j=1}^{m}E\left(\ell_{u_j}\right) = \mathcal{L},$$

$$Var\left(\hat{\mathcal{L}}\right) = \frac{n^2}{m^2}\sum_{j=1}^{m}Var\left(\ell_{u_j}\right) = \frac{n^2}{m}\sigma_L^2,$$

indicate that the rescaled subsample log-likelihood is a non-negative and unbiased estimator of the full sample log-likelihood, and the estimator variance increases with the volatility of the log-likelihood.

The MH sampler evaluates the likelihood function on the original scale, rather than the log scale. It is important to note that $\hat{\mathcal{L}}$ is an unbiased estimator of $\mathcal{L}$, but $\exp\left(\hat{\mathcal{L}}\right)$ is not an unbiased estimator of $\exp\left(\mathcal{L}\right)$ due to Jensen's inequality. Assume that asymptotically $\hat{\mathcal{L}}$ follows a normal distribution with mean $\mathcal{L}$ and variance $\frac{n^2}{m}\sigma_L^2$, implying that $\exp\left(\hat{\mathcal{L}}\right)$ follows a log-normal distribution with mean $\exp\left(\mathcal{L} + \frac{n^2}{2m}\sigma_L^2\right)$. Ideally, $\exp\left(\hat{\mathcal{L}} - \frac{n^2}{2m}\sigma_L^2\right)$ could serve as an unbiased estimator of $\exp\left(\mathcal{L}\right)$ if $\hat{\mathcal{L}}$ was normal and σ_L^2 was computed exactly. A bias-corrected estimator of the likelihood function is

$$\hat{\exp}\left(\mathcal{L}\right) = \exp\left(\hat{\mathcal{L}} - \frac{n^2}{2m}\hat{\sigma}_L^2\right),$$

where $\hat{\sigma}_L^2$ is an estimator of σ_L^2. In principle, the exact σ_L^2 could be computed by a sweep over the whole dataset, but it would be too computationally expensive, since σ_L^2 varies with θ at each iteration of the MH sampling. A feasible estimator of σ_L^2 is the subsample counterpart:

$$\hat{\sigma}_L^2 = \frac{1}{m}\sum_{j=1}^{m}\ell_{u_j}^2 - \left(\frac{1}{m}\sum_{j=1}^{m}\ell_{u_j}\right)^2.$$

The inherent bias in $\hat{\exp}\left(\mathcal{L}\right)$ arises from the heterogeneity and volatility of the log-likelihood $\ell_1, \ldots, \ell_n$. The bias could be eliminated if they were homogeneous (i.e., $\sigma_L^2 = 0$). Quiroz et al. (2019) propose an efficient bias-correction method of homogenizing the population by employing control variates, which approximate the likelihood function at a low computing cost.

Let $q_i\left(\theta, Y_i\right)$ be a control variate (the functional form will be discussed below). Let the residual be

$$r_i\left(\theta, Y_i\right) = \ell_i\left(\theta, Y_i\right) - q_i\left(\theta, Y_i\right),$$

which satisfies the equation $\mathcal{L} = \sum_{i=1}^{n} r_i + \sum_{i=1}^{n} q_i$.

Instead of subsampling the log-likelihood, consider subsampling the residual with replacement. Let $u_j, j = 1, \ldots, m$ be uniform random indices. The residual subsample consists of $r_{u_1}, \ldots, r_{u_m}$, and the rescaled subsample log-likelihood with control variates, as an estimator of the full-data log-likelihood, is constructed as

$$\tilde{\mathcal{L}}\left(\theta, Y, u\right) = \frac{n}{m}\sum_{j=1}^{m} r_{u_j} + \sum_{i=1}^{n} q_i.$$

For a given θ and Y, the random variable r_{u_j} follows a categorical distribution

$$P\left(r_{u_j} = r_i\right) = \frac{1}{n}, \forall i = 1, \ldots, n,$$

with mean and variance

$$E\left(r_{u_j}\right) = \frac{1}{n}\sum_{i=1}^{n} r_i,$$

$$Var\left(r_{u_j}\right) = \frac{1}{n}\sum_{i=1}^{n} r_i^2 - \left(\frac{1}{n}\sum_{i=1}^{n} r_i\right)^2 \equiv \sigma_r^2,$$

where σ_r^2 reflects heterogeneity and volatility of the residual $r_1, \ldots, r_n$. It follows that

$$E\left(\tilde{\mathcal{L}}\right) = \frac{n}{m}\sum_{j=1}^{m} E\left(r_{u_j}\right) + \sum_{i=1}^{n} q_i = \mathcal{L},$$

$$Var\left(\tilde{\mathcal{L}}\right) = \frac{n^2}{m^2}\sum_{j=1}^{m} Var\left(r_{u_j}\right) = \frac{n^2}{m}\sigma_r^2.$$

Comparing estimators with and without control variates, it is observed that both $\hat{\mathcal{L}}$ and $\tilde{\mathcal{L}}$ are non-negative and unbiased estimator of the full-sample log-likelihood, but the variances are different: $Var\left(\tilde{\mathcal{L}}\right) < Var\left(\hat{\mathcal{L}}\right)$ if $\sigma_r^2 < \sigma_L^2$. Provided that control variates have predictive power over the log-likelihood using information from θ and Y, the residual variance is reduced, leading to a more accurate bias-corrected estimator of the likelihood:

$$\widetilde{\exp}\left(\mathcal{L}\right) = \exp\left(\tilde{\mathcal{L}} - \frac{n^2}{2m}\tilde{\sigma}_r^2\right),$$

where $\tilde{\sigma}_r^2$ is an estimator of σ_r^2 such as the subsample analogue:

$$\tilde{\sigma}_r^2 = \frac{1}{m}\sum_{j=1}^{m} r_{u_j}^2 - \left(\frac{1}{m}\sum_{j=1}^{m} r_{u_j}\right)^2.$$

The choice of control variates must balance relevance and computational convenience. Relevance ensures a reasonable approximation of the log-likelihood, while convenience demands low computational cost for calculating $\sum_{i=1}^{n} q_i$. If the whole dataset was scanned to compute the sum, it would defeat the purpose of control variates.

Quiroz et al. (2019) explore methods for constructing control variates through parameter expansion or data expansion. Assume the log-likelihood $\ell_i\left(\theta, Y_i\right)$ is differentiable with respect to θ, allowing for the Taylor expansion around a parameter value θ^*. Let the control variates be $q_i\left(\theta, Y_i\right) = \tau\left(\theta, \theta^*, Y_i\right)$, where

$$\tau\left(\theta, \theta^*, Y_i\right) = \ell_i\left(\theta^*, Y_i\right) + \frac{\partial \ell_i\left(\theta^*, Y_i\right)}{\partial \theta^*}\left(\theta - \theta^*\right) + \frac{1}{2}\left(\theta - \theta^*\right)' \frac{\partial^2 \ell_i\left(\theta^*, Y_i\right)}{\partial \theta^* \partial \theta^*}\left(\theta - \theta^*\right).$$

The sum of control variates is evaluated as

$$\sum_{i=1}^{n} q_i = \mathcal{L}(\theta^*, Y_i) + \frac{\partial \mathcal{L}(\theta^*, Y_i)}{\partial \theta^*}(\theta - \theta^*) + \frac{1}{2}(\theta - \theta^*)' \frac{\partial^2 \mathcal{L}(\theta^*, Y_i)}{\partial \theta^* \partial \theta^*}(\theta - \theta^*),$$

where $\mathcal{L}(\theta^*, Y_i)$, $\frac{\partial \mathcal{L}(\theta^*, Y_i)}{\partial \theta^*}$ and $\frac{\partial^2 \mathcal{L}(\theta^*, Y_i)}{\partial \theta^* \partial \theta^*}$ are pre-computed before the MH sampler starts. At each iteration of the parameter update in the MH sampling, computing the sum $\sum_{i=1}^{n} q_i$ does not require sweeping over the whole dataset.

It is possible to specify multiple centroids of parameters using Taylor approximations around θ^* and θ^{**}, respectively, with the control variates defined as

$$q_i(\theta, Y_i) = \begin{cases} \tau(\theta, \theta^*, Y_i) & |\theta - \theta^*| < |\theta - \theta^{**}| \\ \tau(\theta, \theta^{**}, Y_i) & |\theta - \theta^*| \geq |\theta - \theta^{**}| \end{cases}.$$

As an alternative to the parameter expansion of Bardenet et al. (2017), Quiroz et al. (2019) propose data-expanded control variates that evaluate the log-likelihood at the nearest centroid of data. For illustration, the data are classified in two groups with the centroid Y^* and Y^{**}. Define the control variates as

$$q_i(\theta, Y_i) = \begin{cases} \varphi(\theta, Y_i, Y^*) & |Y_i - Y^*| < |Y_i - Y^{**}| \\ \varphi(\theta, Y_i, Y^{**}) & |Y_i - Y^*| \geq |Y_i - Y^{**}| \end{cases},$$

where $\varphi(\theta, Y_i, Y^*)$ is the Taylor approximation of $\ell_i(\theta, Y_i)$ with respect to Y_i around the centroid Y^*:

$$\varphi(\theta, Y_i, Y^*) = \ell(\theta, Y^*) + \frac{\partial \ell(\theta, Y^*)}{\partial Y^*}(Y_i - Y^*) + \frac{1}{2}(Y_i - Y^*)' \frac{\partial^2 \ell(\theta, Y^*)}{\partial Y^* \partial Y^*}(Y_i - Y^*).$$

Denote $\ell(\theta, Y^*) = \ell_i(\theta, Y^*)$, as it does not change with the ith data point, but varies with the parameter θ, implying that at each iteration of the MH sampler the gradient and Hessian are evaluated at the centroids, but not at other data points. Additionally, $\sum_{i=1}^{n} q_i$ can be obtained by sufficient statistics, such as $\sum_{i=1}^{n} Y_i I(|Y_i - Y^*| < |Y_i - Y^{**}|)$ without sweeping over the dataset.

With the control variates q_i and the bias-corrected estimator $\widetilde{\exp}(\mathcal{L})$, the pseudo-marginal MH sampler is performed, wherein $p(\theta)\widetilde{\exp}(\mathcal{L})$ serving as an approximation of the posterior density (up to a normalizing constant). While the bias is largely mitigated, it is not entirely eliminated. Nonetheless, the MCMC sampler remains valid, converging to a slightly perturbed distribution, specifically $p(\theta)E[\widetilde{\exp}(\mathcal{L})]$ instead of the exact posterior $p(\theta)\exp(\mathcal{L})$. The discrepancy is often negligible in practice. By leveraging control variates to reduce the variance of the log-likelihood estimator, the sampler is likely to perform effectively.

To further enhance computational efficiency, correlations among random variables, specifically uniformly distributed integers u, can be introduced. Deligiannidis et al. (2018) propose using a reversible transition kernel, such as the Gaussian AR(1) process, to induce correlations. By applying the inverse CDF method, Gaussian random variables can be transformed into uniformly distributed integers.

2.9.4 Stochastic Gradient Langevin Dynamics

Numerical optimization methods such as steepest descent and quasi-Newton algorithms leverage gradient information to update variables. Evaluating the full-sample gradients necessitates a complete sweep over the dataset. For large-scale problems, this can be computationally prohibitive. Stochastic Gradient Descent (SGD), introduced by Robbins and Monro

(1951), is a well-established method for handling large datasets by randomly sampling a single data point (or a mini batch) to approximate the gradient at each iteration. These methods will be explored in Section 5.8.

Stochastic Gradient Langevin Dynamics (SGLD), proposed by Teh et al. (2016), is an approximate subsampling MCMC method, which adapts the Metropolis Adjusted Langevin Algorithm (MALA) for big data. As demonstrated in Proposition 2.9, MALA can be interpreted as the Hamiltonian Monte Carlo (HMC) sampler with a single-step leapfrog move. In continuous-time dynamics, the Hamiltonian is conserved, leading to proposals being accepted with probability one if the step size is sufficiently small. By discretizing Langevin diffusion with a suitably small and decreasing sequence of step sizes, SGLD eliminates the need to compute the MH acceptance probability, always accepting proposals generated by the discretized Langevin equation:

$$\theta^* = \theta + \frac{1}{2}\delta^2 \nabla(\theta) + \delta z,$$

where the step size $\delta > 0$ and z is a standard normal variate. The gradient of the log posterior density $\nabla(\theta)$ is computed by subsampling m out of n data points with replacement, where $m < n$. Let $u_j, j = 1, \ldots, m$ be uniform random integers drawn from $\{1, \ldots, n\}$. The stochastic gradient approximates the full-sample gradient by

$$\nabla(\theta) = \frac{\partial \ln p(\theta)}{\partial \theta} + \frac{n}{m} \sum_{j=1}^{m} \frac{\partial \ln p\left(Y_{u_j} \mid \theta\right)}{\partial \theta}.$$

Although SGLD results in a non-homogeneous Markov chain due to the decreasing step sizes, Teh et al. (2016) demonstrate that these decreasing step sizes enable convergence to the exact posterior distribution.

In conclusion, a wide array of divide-and-conquer and subsampling methods discussed in Sections 2.8 and 2.9 offer powerful tools for scalable Bayesian inference in the presence of large datasets. These methods effectively balance computational efficiency with statistical accuracy, making them indispensable in contemporary Bayesian analysis.

2.10 Software and Computational Considerations

MCMC methods are powerful tools for posterior simulation in Bayesian inference, yet their implementation poses several challenges. This section delves into key implementation issues, including sampler validation, numerical pitfalls, and the opportunities presented by parallel computing.

2.10.1 Software Validation

Posterior simulations are often performed using custom computer code, which is susceptible to programming errors that can yield inaccurate or incorrect results. For instance, when implementing the Gibbs sampler for complex Bayesian models, it is not uncommon for some posterior conditional distributions to be incorrectly coded. Therefore, validating the software through formal statistical tests is crucial.

Samples from the joint distribution $p(\theta, Y)$ can be generated in two ways:

$$p(\theta)\, p(Y \mid \theta) = p(Y)\, p(\theta \mid Y).$$

The left-hand side is managed by the data-generating software, which is assumed to be error-free due to its simplicity. In contrast, the right-hand side is handled by the model-fitting software, which may be flawed due to the complexity of sampling from the posterior using MCMC methods.

Cook et al. (2006) develop Bayesian software validation methods using posterior quantiles. The algorithm involves four steps. First, generate a draw θ_0 from the prior $p(\theta)$ and artificial data Y from the likelihood $p(Y|\theta = \theta_0)$ using the data-generating software. The number of observations can be arbitrary. Second, given the artificial data, generate MCMC draws $\theta_1, \ldots, \theta_T$ from the posterior $p(\theta|Y)$ using the model-fitting software. Third, compute the sample quantile $\frac{1}{T}\sum_{t=1}^{T} I(\theta_t < \theta_0)$, reflecting the ranking of θ_0 among the posterior draws $\theta_1, \ldots, \theta_T$. For vector-valued θ, compute the sample quantile for each element. Fourth, repeat these steps to obtain multiple independent replications of the quantile realizations.

The key insight from Cook et al. (2006) is that the sample quantile follows a uniform distribution on the unit interval, under the null hypothesis that the model-fitting software correctly produces MCMC draws from the posterior distribution asymptotically. Statistical tests, such as the Kolmogorov test, goodness-of-fit test by bin counts, and chi-square test, can be employed to validate the uniform distribution.

The intuition is that the ranking of θ_0 among the posterior draws $\theta_1, \ldots, \theta_T$ is uniform if they are sampled from the same distribution. Although θ_0 is generated from the prior and $\theta_1, \ldots, \theta_T$ are generated from the posterior, the equation $p(\theta_0)\, p(Y|\theta_0) = p(Y)\, p(\theta_0|Y)$ indicates that the pair (θ_0, Y) can be viewed as a hypothetical draw from $p(Y)$, followed by another hypothetical draw from $p(\theta_0|Y)$, despite the actual data-generating process being in reverse order. Conditional on the same data Y, we interpret θ_0 and $\theta_1, \ldots, \theta_T$ as samples from the same posterior distribution. An alternative interpretation is that the pairs (θ_0, Y), $(\theta_1, Y), \ldots, (\theta_T, Y)$ are correlated samples from the joint distribution $p(\theta, Y)$. Nevertheless, the marginal distribution of θ_0 and $\theta_1, \ldots, \theta_T$ are the same. A third interpretation is that θ_0 is the true value of the parameters, and the posterior inference should be correct in the sense that the true parameter should be contained in a 95% interval with probability 0.95. Proposition 2.18 provides mathematical rigor to our intuitive arguments for the uniform distribution.

Proposition 2.18 (Uniform posterior quantile) *Suppose that θ_0, Y and $\theta_1, \ldots, \theta_T$ are generated in the following order: generate a draw θ_0 from the prior $p(\theta)$, generate artificial data Y from the likelihood $p(Y|\theta)$, and generate $\theta_1, \ldots, \theta_T$ by an ergodic MCMC sampler targeting $p(\theta|Y)$. Let θ_{jt} and θ_{j0} be the jth element of θ_t and θ_0. Let $q_t = I(\theta_{jt} < \theta_{j0})$ and $Q = \frac{1}{T}\sum_{t=1}^{T} q_t$. As $T \to \infty$, Q follows the uniform distribution $U(0,1)$.*

Geweke (2004) proposes an alternative validation method using joint distribution tests of posterior simulators. The sampler involves cycling through the conditional distributions $p(\theta|Y)$ and $p(Y|\theta)$, akin to the Gibbs sampler that produces MCMC draws from the joint distribution $p(\theta, Y)$. Meanwhile, the data-generating software can directly sample from $p(\theta, Y) = p(\theta)\, p(Y|\theta)$, allowing for statistical tests to determine if the two samples originate from the same distribution, such as the t-statistic for comparing the means.

The primary distinction between Geweke (2004) and Cook et al. (2006) is whether the artificial data Y is updated during each iteration of MCMC sampling. If updated, the parameter θ is drawn from the prior; otherwise, it is drawn from the posterior.

2.10.2 Numerical Pitfall

If statistical tests reject the null hypothesis that model-fitting software produces correct results, potential causes include programming errors and numerical pitfalls. Programming errors deviate from the mathematical formulae underlying the algorithm, while numerical pitfalls occur due to finite precision in representing real numbers on computers.

The MH sampler evaluates the posterior density up to a proportionality constant. Evaluating the posterior density in logarithmic form is recommended, as it mitigates the risk of underflow when the product of probabilities falls below the double-precision floating-point lower bound, which is approximately 2.2×10^{-308}.

For numerical stability, avoid function calls like `sum(log(normpdf(y)))`, because the standard normal density underflows if the evaluation point is larger than 38.5, resulting in `normpdf(39)=0`. As an alternative, using `-y'*y/2` is more computationally efficient and numerically stable: 1) the proportionality constant is omitted, 2) `log()` and `exp()` are cancelled, and 3) `y'*y` is faster than `sum(y.^2)` on some computing platforms like MATLAB. Transposing a large matrix should be avoided due to its impact on memory, but transposing a vector is usually harmless. Additionally, dividing a real number by 2 is acceptable if the software treats integers as double floating-point numbers; otherwise, dividing by 2.0 is a safe operation.

The MH sampler accepts a proposal with probability $\alpha(\theta^*, \theta)$. An accepted proposal leads to $\theta_t = \theta^*$, while a rejected proposal results in $\theta_t = \theta$. Let u be a uniform random number. Compare two implementations:

if $\alpha(\theta^*, \theta) > u$; $\theta_t = \theta^*$; else $\theta_t = \theta$;
if $\alpha(\theta^*, \theta) < u$; $\theta_t = \theta$; else $\theta_t = \theta^*$;

From a defensive programming perspective, the former is preferable, as it is robust to exceptional situations. As the MH sampler evaluates posterior densities, it may occasionally encounter out-of-range evaluation points or parameters. For example, $\sigma^2 \sim IG(a, b)$ requires $\sigma^2 > 0$, $a > 0$ and $b > 0$. We expect $\alpha(\theta^*, \theta) = 0$ for an out-of-range proposal, but in practice, edge cases may not be perfectly handled in a casually written program. What if $\alpha(\theta^*, \theta)$ produces an unexpected `NaN`? The former implementation rejects the proposal, while the latter accepts it. Defensive programming anticipates the unexpected.

The following program illustrates the random-walk MH sampler with a multivariate normal proposal. To save space, the demo only shows the main steps of the sampler, and some variable definitions are omitted.

```
% Initialization
x = initialState; y = logPDF(x); SigmaHalf = chol(Sigma)';

% MCMC iterations
for i = 1:numDraws
  xNew = x + SigmaHalf * randn(dim,1); % Random-walk proposal
  yNew = logPDF(xNew); % Log density evaluated at the proposal
  logRatio = yNew - y; % MH acceptance ratio
  if logRatio > 0 || exp(logRatio) > rand % Accept the proposal
    x = xNew; y = yNew;
  end
  if i > burnIn; Draws(:, i-burnIn) = x; end % Save draws
end
```

2.10.3 Parallel Computing

The advancement of multi-core CPUs, GPUs, and Beowulf clusters facilitates parallel computing. Meanwhile, software enhancements make high-level parallel programming accessible to average users. For example, MATLAB users can simply replace `FOR` with `PARFOR` if the loop is suitable for parallel processing. Additionally, users are encouraged to leverage vector/matrix operations, which are automatically handled by multiple computing threads if the matrix size is large enough.

Parallel Bayesian computation can be implemented through various approaches. First, multiple MCMC chains can be run independently in parallel, without message passing between computational threads. Second, all divide-and-conquer MCMC methods are amenable to parallel computing. The subposterior MCMC samplers proceed separately without communication. Third, some models have conditionally independent components that are suitable for parallel processing. For example, latent variables in the Probit model satisfy $p(z|Y,\beta) = \prod_{i=1}^{n} p(z_i|Y,\beta)$, allowing the vector z to be sampled in parallel. Fourth, the independence MH sampler supports parallel computing because all proposals can be generated before parameter updates. Fifth, batch processing of a few iterations of the MH sampler is possible through prefetching (Brockwell, 2006). Sixth, the multiple-try MH sampler (Liu et al., 2000) generates multiple proposals in parallel.

The conventional wisdom is that MCMC simulation is inherently sequential, as the state transition depends on the current state. However, the independence MH sampler is an exception. The proposals are drawn from a transition kernel that does not depend on the current state: $r(\theta^*, \theta) = r(\theta^*)$, allowing parallel simulation of the entire sequence of proposals $\theta_t^*, t = 1, \ldots, T$ and evaluating the target densities $\pi(\theta_1^*), \ldots, \pi(\theta_n^*)$ in parallel. The MH sampler just uses a sequence of uniform random numbers to accept or reject these proposals. Jacob et al. (2011) advocate permuting the proposal sequence, allowing for, at no additional cost, multiple MCMC chains, each with the target distribution as its stationary distribution.

Brockwell (2006) proposes a prefetching method for the MH sampler. The idea is that there are two outcomes at each iteration: either accept the proposal or reject it. As there are 2^k outcomes after k iterations, it is possible to process those outcomes in parallel to speed up the sampler by $\log_2 P$, where P is the number of processors. However, prefetching may not be computationally efficient, as most of the cached outcomes are wasted.

The multiple-try MH sampler developed by (Liu et al., 2000) is well-suited for parallel computing. The standard MH sampler generates one proposal at each iteration, while the multiple-try MH sampler generates multiple proposals in parallel, selecting one with a carefully specified probability to maintain reversibility.

2.11 An Application on Fertility Choice

In this section, we illustrate Bayesian parameter estimation of a fertility choice model using a large micro dataset and compare alternative MCMC sampling methods.

2.11.1 Model Specification

We consider an overlapping generation model, in which a parent optimizes her adult-age consumption $c_{y,t}$, old-age consumption $c_{o,t+1}$, the number of children n_t and their education e_t that builds up the human capital h_{t+1}:

$$\max_{c_{y,t}, c_{o,t+1}, n_t, e_t} \ln c_{y,t} + \rho \ln c_{o,t+1} + \ln(n_t h_{t+1}).$$

Endowed with 1 unit of time, the parent could earn wh_t if she worked full time, but a fraction of time $\phi \in (0, 1)$ could be spent on raising a child. In addition, the educational cost of children is $w\bar{h}_t e_t n_t$, where the tuition $w\bar{h}_t$ depends on the national average income. The resource constraint is given by

$$c_{y,t} + R^{-1} c_{o,t+1} + w\bar{h}_t e_t n_t = wh_t \left(1 - \phi n_t\right).$$

Also, we assume that the human capital is accumulated by

$$h_{t+1} = \left(a + e_t\right)^b h_t.$$

The interior solution of the optimal fertility choice is

$$n_t = \frac{1-b}{2+\rho} \left[\phi - a \left(h_t / \bar{h}_t\right)^{-1}\right]^{-1},$$

and the optimal education choice is

$$e_t = \frac{\phi b}{1-b} \frac{h_t}{\bar{h}_t} - \frac{a}{1-b}.$$

The theoretic model predicts that rich and well-educated parents raise fewer children (because n_t decreases with h_t) and provide better education for children (because e_t increases with h_t), assuming $b < 1$ (diminishing marginal returns to schooling).

From the perspective of Bayesian analysis, we are interested in estimating the unknown parameters in the model by real-world data. Our presentation focuses on implementation issues of MCMC methods, and we keep the model simple by assuming the discount factor $\rho = 0.9$ and the child-raising cost $\phi = 0.4$. Our aim is to estimate the unknown parameters (i.e., returns-to-schooling parameters a and b, as well as the regression coefficients γ and the noise variance σ^2) by an econometric version of the fertility choice equation:

$$n_t = \frac{1-b}{2+\rho} \left[\phi - a \left(h_t / \bar{h}_t\right)^{-1}\right]^{-1} + x_t \gamma + \sigma \varepsilon_t,$$

where x_t represents exogenous predictors on maternal behaviors and health conditions, such as the mother age, parent age difference, body mass index and medical payment methods. In this section, we assume that ε_t follows the standard normal distribution. Alternatively, we may consider the Poisson regression, as n_t represents the count data.

The model is ideal for illustrating MCMC methods for two reasons. First, it is nonlinear with respect to the parameter a, which induces a multi-modal likelihood curve. The posterior is not a common distribution and we resort to MCMC methods. Second, it contains a conditionally linear structure: conditional on the parameter a, the model reduces to a linear regression

$$n_t = X_t (a) \beta + \sigma \varepsilon_t,$$

where

$$X_t (a) = \left\{ \left[\phi - a \left(h_t / \bar{h}_t\right)^{-1}\right]^{-1}, x_t \right\},$$

$$\beta = \begin{pmatrix} \frac{1-b}{2+\rho} \\ \gamma \end{pmatrix}.$$

The Metropolis-within-Gibbs sampler leverages the conditional conjugacy, reducing a high-dimensional sampling problem to cycling through low-dimensional posterior conditional distributions.

2.11.2 Data Description

In the U.S., birth certificates are required to be completed for all births, and federal law mandates the collection and publication of vital statistics data. In 2021, the public-use National Vital Statistics System (NVSS) micro-data contained 3.67 million birth records, detailing infant and maternal behaviors and health conditions, demographic characteristics, and medical services utilization.

The dataset contains an entry "Live Birth Order" that indicates the total number of live births including the 2021 birth, from which we infer the variable n_t. Additionally, the entries "Mother's Education" and "Father's Education" indicate the educational attainment of mothers and fathers, allowing us to construct a proxy for the variable h_t, and $\bar{h}_t$ is obtained by the sample average of h_t. Data for exogenous predictors x_t are available from the corresponding entries in the NVSS dataset.

After removing missing values, the sample size for parameter estimation is about 3.1 million. Due to the large sample size, we specify a flat prior for all parameters, so the posterior density is proportional to the likelihood.

2.11.3 Profiling and Maximum Likelihood Estimation

It is helpful to analyze the likelihood function before we run MCMC samplers. Understanding the likelihood function facilitates the design of posterior simulation strategies and the tuning of MCMC samplers. It is common to estimate the maximum of the posterior density and use the Hessian evaluated at the log posterior mode to construct a proposal. Because of the flat prior specification, the maximum-a-posteriori estimator is also the maximum likelihood estimator in this application.

The maximum likelihood estimator for the nonlinear regression does not have a closed-form solution. In practice, we resort to numerical optimization techniques, such as the quasi-Newton algorithm, to maximize the posterior density. However, high-dimensional numerical optimization is computationally intensive, and convergence is not guaranteed.

An estimation strategy for our fertility model is the likelihood profiling. Conditional on the parameter a, the model reduces to a linear regression with the log-likelihood function:

$$\mathcal{L}\left(\beta, \sigma^2, a\right) = -\frac{T}{2}\ln 2\pi - \frac{T}{2}\ln \sigma^2 - \frac{1}{2\sigma^2}\sum_{t=1}^{T}\left[n_t - X_t\left(a\right)\beta\right]^2.$$

First, we concentrate the likelihood function by maximizing σ^2 conditional on β and a. The maximizer is

$$\sigma^{2*}\left(\beta, a\right) = \frac{1}{T}\sum_{t=1}^{T}\left[n_t - X_t\left(a\right)\beta\right]^2,$$

and the concentrated likelihood function is

$$\mathcal{L}\left(\beta, \sigma^{2*}, a\right) = -\frac{T}{2}\ln 2\pi + \frac{T}{2}\ln T - \frac{T}{2} - \frac{T}{2}\ln\sum_{t=1}^{T}\left[n_t - X_t\left(a\right)\beta\right]^2.$$

Second, we further concentrate the likelihood function by maximizing β conditional on a. The maximizer is the ordinary least squares estimator

$$\beta^*\left(a\right) = \left[\sum_{t=1}^{T}X_t'\left(a\right)X_t\left(a\right)\right]^{-1}\left[\sum_{t=1}^{T}X_t'\left(a\right)n_t\right],$$

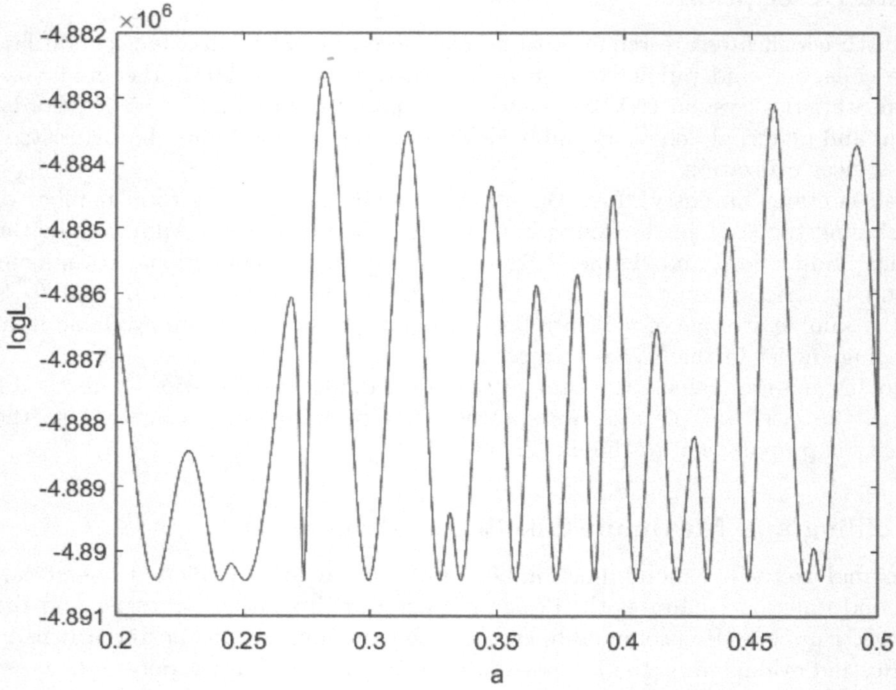

FIGURE 2.1
Profile log-likelihood as a univariate function of the parameter a, with other parameters being concentrated out.

and the concentrated likelihood function is

$$\mathcal{L}\left(\beta^*, \sigma^{2*}, a\right) = -\frac{T}{2}\ln 2\pi + \frac{T}{2}\ln T - \frac{T}{2} - \frac{T}{2}\ln \sum_{t=1}^{T}\left[n_t - X_t\left(a\right)\beta^*\left(a\right)\right]^2.$$

Third, we maximize the concentrated likelihood function with respect to a. Although a closed-form solution is not available, there are various univariate numerical optimization algorithms. Figure 2.1 indicates that an appropriate algorithm for our problem is the grid search, because of the multi-modal objective function. By the likelihood profiling, we find the maximum at $a = 0.282$, and the corresponding maximum likelihood estimator for the returns-to-schooling parameter is $b = 0.987$.

The profiling results can be validated by multivariate maximization of β, σ^2, a simultaneously using a generic numerical optimization program. If the starting value is near the profile estimator, we may obtain the same results. If the results are different, it is likely that that the program gets stuck in a local maximum or saddle point, which is so common that we should be aware of the limitation of numerical optimization.

2.11.4 MH and HMC Samplers

A challenge of implementing the multivariate MH sampler is specifying a good proposal distribution. Since we have obtained the maximum of the posterior density, it is tempting to specify a multivariate normal or t proposal distribution with the covariance matrix proportional to the negative inverse Hessian matrix evaluated at the posterior mode. This strategy works well if we can provide analytical gradients. Otherwise, the numerical Hessian can be computationally intensive and inaccurate due to catastrophic cancellation errors of finite differencing.

In this application, it is straightforward to compute analytical gradients. As an alternative to the tedious Hessian computation, we use an approximation by the outer product of gradients (OPG), which also provides an estimator of the information matrix. The inverse of OPG corresponds to the covariance matrix of the maximum likelihood estimator. A strength of OPG versus the Hessian is that the former is always positive definite, whereas the latter may lose definiteness in practical computing.

To implement the random-walk MH sampler, we specify a multivariate t proposal distribution with 5 degrees of freedom to encourage the sampler to make dispersed proposals. The covariance matrix is proportional to the inverse of OPG. The proportionality constant is a tuning parameter of the MH sampler. A smaller value leads to a higher acceptance rate of proposals. We tune the value to 0.5, resulting in about 30% of proposals being accepted.

Since we have access to analytical gradients, it is worthwhile to try the gradient-based HMC sampler. The step size and the number of leapfrog steps are key tuning parameters of the HMC sampler. Our tuning strategy is to start with the Langevin Monte Carlo (i.e., one-step leapfrog move) with a small enough step size such that nearly all proposals are accepted, as the Hamiltonian should be preserved if the step size tends to zero. To tune the sampler, we gradually increase the step size and the number of leapfrog steps to encourage the HMC sampler to make more distant proposals. During the tuning process, the acceptance rate drops steadily. After some experiments, we set the step size as 2.6e-5 with 5 leapfrog moves. The corresponding acceptance rate of proposals is about 69%.

The empirical results are summarized in Table 2.1. Although we do not have a ground truth about the posterior distribution, we believe that the maximum likelihood result is reliable, as likelihood profiling greatly reduces the numerical risk of non-convergence and ill-conditioned matrices. The results of the MH and HMC samplers closely match the maximum likelihood estimator: the posterior mean is close to the classical point estimator, and the posterior standard deviation is close to the classical standard error.

2.11.5 Metropolis-within-Gibbs Sampler

The multivariate MH and HMC samplers update all parameters simultaneously. For our model, the parameters can be naturally divided into two blocks: a and b, γ, σ^2. The posterior conditional distribution of the latter is analytically tractable, if we reparametrize b as $\frac{1-b}{2+\rho}$ and denote $\beta = \left(\frac{1-b}{2+\rho}, \gamma'\right)'$. Under the conjugate NIG prior $NIG(\mu, \Lambda, A, B)$, the posterior conditional distribution of β, σ^2 is $NIG(\bar{\mu}, \bar{\Lambda}, \bar{A}, \bar{B})$, where

$$\bar{\mu} = \left[\Lambda + \sum_{t=1}^{T} X_t'(a) X_t(a)\right]^{-1} \left[\Lambda\mu + \sum_{t=1}^{T} X_t'(a) n_t\right],$$

$$\bar{\Lambda} = \Lambda + \sum_{t=1}^{T} X_t'(a) X_t(a),$$

$$\bar{A} = A + \frac{T}{2},$$

$$\bar{B} = B + \frac{1}{2}\sum_{t=1}^{T} n_t^2 + \frac{1}{2}\mu'\Lambda\mu - \frac{1}{2}\bar{\mu}'\bar{\Lambda}\bar{\mu}.$$

TABLE 2.1
Maximum likelihood and Bayesian estimation by MH, HMC, Gibbs and CMC samplers for the fertility choice model.

	MLE	MH	HMC	Gibbs	CMC	IS
a	0.2818	0.2817	0.2817	0.2817	0.2818	0.2817
	(0.0001)	(0.0001)	(0.0001)	(0.0001)	(0.0001)	(0.0001)
b	0.9866	0.9865	0.9866	0.9865	0.9866	0.9865
	(0.0002)	(0.0002)	(0.0002)	(0.0002)	(0.0001)	(0.0002)
Age	0.0915	0.0916	0.0914	0.0916	0.0916	0.0916
	(0.0001)	(0.0001)	(0.0001)	(0.0001)	(0.0001)	(0.0001)
ParentsAgeDiff	0.0110	0.0109	0.0109	0.0110	0.0109	0.0109
	(0.0001)	(0.0001)	(0.0001)	(0.0002)	(0.0001)	(0.0001)
BodyMassIndex	0.0122	0.0122	0.0122	0.0122	0.0121	0.0122
	(0.0001)	(0.0001)	(0.0001)	(0.0001)	(0.0001)	(0.0001)
PrivateInsurance	−0.7410	−0.7410	−0.7401	−0.7410	−0.7405	−0.7411
	(0.0014)	(0.0014)	(0.0011)	(0.0014)	(0.0009)	(0.0013)
Intercept	−0.5968	−0.5973	−0.5926	−0.5973	−0.5959	−0.5968
	(0.0050)	(0.0047)	(0.0022)	(0.0046)	(0.0031)	(0.0041)
σ^2	1.3569	1.3568	1.3565	1.3569	1.3573	1.3570
	(0.0008)	(0.0012)	(0.0012)	(0.0011)	(0.0005)	(0.0009)

Although the posterior conditional distribution of a is not in a recognizable form, it is feasible to sample from a scalar-valued distribution by discretization. We may create a grid of discrete points and their weights are proportional to the Gaussian likelihood evaluated at those points. Then we sample from the categorical distribution. Alternatively, we may insert a Metropolis component in the Gibbs sampler to update a. For example, we make a random walk proposal from a normal distribution (its variance is a tuning parameter of the sampler), and we accept the proposal by the MH algorithm. The simulation results of the Metropolis-within-Gibbs sampler are shown in Table 2.1. Both the posterior means and standard deviations are similar to the maximum likelihood point estimators and standard errors.

2.11.6 Divide-and-conquer MCMC

The NVSS micro data contain millions of observations. When the sample size is large, it is computationally intensive to scan the full data at each iteration of MCMC samplers. We consider a divide-and-conquer strategy: 1) divide the data into ten subsets, 2) run MH samplers on each subset, and 3) combine the subposterior simulation results.

The consensus Monte Carlo (CMC) by Scott et al. (2016) combines subposterior samples in a way that would be exact if they were Gaussian. As shown in Table 2.1, the posterior means obtained by CMC are close to those from the full-sample MH sampler, but the posterior standard deviations seem to be underestimated. In practice, the CMC method is still valuable if the point estimator is of primary interest, and a strength of CMC is its low computing cost.

Alternatively, we may combine the subposterior samples using importance sampling methods, although the computing cost increases substantially. To reduce the computational burden, we generate matched samples in subposterior sampling, allowing the subposterior densities to be recycled for evaluating the importance weights. Table 2.1 reports the importance sampling estimator. The results are close to the full-sample MCMC results.

2.12 Bibliographic Notes

The evolution of MCMC methods has kept pace with advances in computational power. MCMC sampling was pioneered by Metropolis et al. (1953), as physicists had access to the first generation of computers. Over the next three decades, influential MCMC works included the generalization by Hastings (1970), simulated annealing in combinatorial optimization by Kirkpatrick et al. (1983), the Gibbs sampler for image restoration by Geman and Geman (1984), and data augmentation by Tanner and Wong (1987). In the 1990s, as personal computers became widespread, Gelfand and Smith (1990) heralded the arrival of MCMC methods in statistics, followed by excellent introductory papers for statisticians such as Chib and Greenberg (1995). MCMC methods have since been indispensable to computational Bayesian analysis. Meanwhile, theoretical advances have also been made. Tierney (1994) provided a rigorous treatment of the convergence conditions of MCMC methods. Haario et al. (2001) introduced an adaptive Metropolis algorithm that adjusts the proposal distribution based on previous draws. Roberts and Rosenthal (2001) discussed the optimal scaling of MH algorithms, which is crucial for efficient simulation in high-dimensional spaces. The book of Robert and Casella (2004) provided an in-depth exploration of MCMC methods, making it a comprehensive resource in the field.

Advances in computation and storage technologies have led to exponential growth in data volume, velocity, and variety, presenting challenges in scaling up MCMC methods to handle big data. Recent years have seen a burgeoning literature on novel methods, including Firefly Monte Carlo by Maclaurin and Adams (2014), adaptive subsampling with MH tests by Korattikara et al. (2014), stochastic gradient Langevin dynamics by Teh et al. (2016), consensus Monte Carlo by Scott et al. (2016), the zig-zag process with subsampling by Bierkens et al. (2019), bias-corrected subsampling with control variates by Quiroz et al. (2019), and MH sampling with delayed acceptance by Banterle et al. (2019), among others. For an excellent review, see Bardenet et al. (2017).

Among the novel MCMC methods, the pseudo-marginal MH sampler by Andrieu and Roberts (2009) stands out, as it makes Bayesian inference feasible for a large collection of statistical models that are difficult to estimate by classical methods. Exact approximation by an unbiased estimator facilitates parameter estimation when the likelihood function is intractable. The pseudo-marginal approach also provides a promising direction for subsampling MCMC methods in Bayesian big data analysis.

2.13 Proof of Propositions

Proposition 2.1 (Reversibility condition)

Consider the jth element of πQ:
$(\pi Q)_j = \sum_{i=1}^{m} \pi_i q_{ij} = \sum_{i=1}^{m} \pi_j q_{ji} = \pi_j$.
It follows that $\pi Q = \pi$.

Proposition 2.2 (Reversibility of MH sampling)

The modified matrix Q remains a well-defined transition matrix, because $q_{ij} \geq 0$ and $\sum_{j=1}^{m} q_{ij} = 1$. We test the reversibility condition $\pi_i q_{ij} = \pi_j q_{ji}$ by considering three cases.

Case 1: $i = j$.

In that case, the reversibility condition holds trivially.
Case 2: $i \neq j$ *and* $\pi_j r_{ji} \leq \pi_i r_{ij}$.
In that case, $\alpha_{ij} = \frac{\pi_j r_{ji}}{\pi_i r_{ij}}$, *while* $\alpha_{ji} = 1$.
It follows that $\pi_i q_{ij} = \pi_i \alpha_{ij} r_{ij} = \pi_j r_{ji} = \pi_j q_{ji}$.
Case 3: $i \neq j$ *and* $\pi_j r_{ji} > \pi_i r_{ij}$.
In that case, $\alpha_{ij} = 1$, *while* $\alpha_{ji} = \frac{\pi_i r_{ij}}{\pi_j r_{ji}}$.
It follows that $\pi_i q_{ij} = \pi_i r_{ij} = \alpha_{ji} \pi_j r_{ji} = \pi_j q_{ji}$.
In summary, the reversibility condition holds in all cases.

Proposition 2.3 (Convergence of Markov chains)
The proof is based on the coupling arguments.

Let $\{\theta_t\}$ *and* $\left\{\tilde{\theta}_t\right\}$ *be two independent Markov chains with the same transition matrix. The initial distribution of the former is arbitrary, but that of the latter coincides with the stationary distribution* π. *Consider the couple* $\left\{\theta_t, \tilde{\theta}_t\right\}$, *which is a two-dimensional Markov chain with the stationary distribution* $\pi_{i,k} = \pi_i \pi_k$. *Because* $\{\theta_t\}$ *and* $\left\{\tilde{\theta}_t\right\}$ *are irreducible and aperiodic Markov chains, the couple chain* $\left\{\theta_t, \tilde{\theta}_t\right\}$ *is irreducible as well. Note that aperiodicity cannot be dropped, and Exercise 2.2 provides an example of a reducible chain when aperiodicity is violated.*

Let $\Gamma = \inf\left\{t \geq 1 : \theta_t = \tilde{\theta}_t = 1\right\}$ *be the first time when the two chains meet at the first state. Because of irreducibility, they will meet eventually:* $P(\Gamma < \infty) = 1$. *Construct a hybrid chain:*

$$\bar{\theta}_t = \begin{cases} \theta_t & t < \Gamma \\ \tilde{\theta}_t & t \geq \Gamma \end{cases}.$$

We have an identity: $P\left(\bar{\theta}_t = j\right) = P(\theta_t = j), \forall t, j$. *To see this, we decompose the marginal distribution as*

$$\begin{aligned}
&P\left(\bar{\theta}_t = j\right) \\
&= P\left(\bar{\theta}_t = j, t < \Gamma\right) + P\left(\bar{\theta}_t = j, t \geq \Gamma\right) \\
&= P(\theta_t = j, t < \Gamma) + P\left(\tilde{\theta}_t = j, t \geq \Gamma\right) \\
&= P(\theta_t = j, t < \Gamma) + \textstyle\sum_{\tilde{\theta}_{t-1}, \ldots, \tilde{\theta}_\Gamma} P\left(\tilde{\theta}_t = j, \tilde{\theta}_{t-1}, \ldots, \tilde{\theta}_\Gamma, t \geq \Gamma\right) \\
&= P(\theta_t = j, t < \Gamma) + \textstyle\sum_{\theta_{t-1}, \ldots, \theta_\Gamma} P(\theta_t = j, \theta_{t-1}, \ldots, \theta_\Gamma, t \geq \Gamma) \\
&= P(\theta_t = j)
\end{aligned}$$

The idea is that $\bar{\theta}_t$ *itself is identical to* θ_t *before the chains meet. When the two chains meet, two new chains are born by the strong Markov property. They have the same initial distribution (fixed at the first state) and the same transition matrix. Therefore, the probability distributions are the same as well.*

Next, consider the discrepancy between the marginal and stationary distributions:

$$\begin{aligned}
&|P(\theta_t = j) - \pi_j| \\
&= \left|P\left(\bar{\theta}_t = j\right) - P\left(\tilde{\theta}_t = j\right)\right| \\
&= \left|P\left(\bar{\theta}_t = j, \Gamma > t\right) + P\left(\bar{\theta}_t = j, \Gamma \leq t\right) - P\left(\tilde{\theta}_t = j\right)\right| \\
&= \left|P\left(\bar{\theta}_t = j, \Gamma > t\right) + P\left(\tilde{\theta}_t = j, \Gamma \leq t\right) - P\left(\tilde{\theta}_t = j\right)\right| \\
&= \left|P\left(\bar{\theta}_t = j, \Gamma > t\right) - P\left(\tilde{\theta}_t = j, \Gamma > t\right)\right| \\
&\leq P\left(\bar{\theta}_t = j, \Gamma > t\right) + P\left(\tilde{\theta}_t = j, \Gamma > t\right)
\end{aligned}$$

$\leq 2P(\Gamma > t)$

Because $\lim_{t\to\infty} P(\Gamma > t) = 0$, *we have* $\lim_{t\to\infty} |P(\theta_t = j) - \pi_j| = 0$. *It indicates that the marginal distribution converges to the stationary distribution.*

Regarding ergodicity, we first define some variables. Let T_j *be the first passage time of the* j^{th} *state:*

$T_j = \inf\{t \geq 1 : \theta_t = j\}$.

Let $\bar{T}_{jj}$ *be the expected first passage time of the* j^{th} *state when the chain starts from the same state:*

$\bar{T}_{jj} = E(T_j | \theta_0 = j)$.

For an irreducible and positive recurrent chain, the unique stationary distribution equals the reciprocal of the first passage time:

$\pi_j = \bar{T}_{jj}^{-1}$.

Let $N_j^{(T)}$ *be the number of visits to the* j^{th} *state in T periods:*

$N_j^{(T)} = \sum_{t=1}^{T} I(\theta_t = j)$.

Let $\Delta_j^{(1)}, \Delta_j^{(2)}, \ldots, \Delta_j^{\left(N_j^{(T)}\right)}$ *be the time intervals between two adjacent visits to the* j^{th} *state. Due to the strong Markov property, those stochastic time intervals are independent and identically distributed, with the expectation* $\bar{T}_{jj}$. *Heuristically, every passage of the* j^{th} *state gives birth to a new Markov chain, which starts from and ends at the same state. The lengths of those stochastic time intervals have the same distribution as that of* T_j. *The expected length is* $\bar{T}_{jj}$.

Regarding the sample path property, we consider $\frac{1}{T}\sum_{t=1}^{T} I(\theta_t = j)$, *namely* $\frac{N_j^{(T)}}{T}$. *Note that T must lies between the time of* $N_j^{(T)}$ *and* $N_j^{(T)} + 1$ *visits. We have the inequality*

$\frac{\Delta_j^{(1)}+\cdots+\Delta_j^{\left(N_j^{(T)}\right)}}{N_j^{(T)}} \leq \frac{T}{N_j^{(T)}} \leq \frac{\Delta_j^{(1)}+\cdots+\Delta_j^{\left(N_j^{(T)}+1\right)}}{N_j^{(T)}}$.

As $T \to \infty$, *both the lower and upper bounds converge to* $\bar{T}_{jj}$, *by the Law of Large Numbers. It follows that*

$\frac{T}{N_j^{(T)}} \xrightarrow{a.s.} \bar{T}_{jj}$.

So we have $\frac{1}{T}\sum_{t=1}^{T} I(\theta_t = j) \xrightarrow{a.s.} \pi_j$.

Proposition 2.4 (Reversibility)

Consider the marginal distribution:

$\int \pi(\theta) q(\theta^*;\theta) d\theta = \int \pi(\theta^*) q(\theta;\theta^*) d\theta = \pi(\theta^*) \int q(\theta;\theta^*) d\theta = \pi(\theta^*)$.

Proposition 2.5 (Stationary distribution of MH sampling)

Regarding the first part of the proposition, by the law of iterated expectations,

$E\left[h\left(\tilde{\theta}\right)\right] = E\left\{E\left[h\left(\tilde{\theta}\right)|\theta,\theta^*\right]\right\}$.

Because $\tilde{\theta} = \begin{cases} \theta^* & \textit{with probability } \alpha(\theta^*;\theta) \\ \theta & \textit{with probability } 1-\alpha(\theta^*;\theta) \end{cases}$, *we have*

$E\left[h\left(\tilde{\theta}\right)|\theta,\theta^*\right] = \alpha(\theta^*;\theta) h(\theta^*) + [1-\alpha(\theta^*;\theta)] h(\theta)$.

The joint distribution of (θ,θ^*) *is* $\pi(\theta) r(\theta^*;\theta)$, *so the expectation takes the form:*

$E\left[h\left(\tilde{\theta}\right)\right] = \int\int \{\alpha(\theta^*;\theta) h(\theta^*) + [1-\alpha(\theta^*;\theta)] h(\theta)\} \pi(\theta) r(\theta^*;\theta) d\theta d\theta^*$

$= A_0 + \int\int h(\theta) \pi(\theta) r(\theta^*;\theta) d\theta d\theta^*$

$= A_0 + E[h(\theta)]$,

where
$A_0 = \int \int [h(\theta^*) - h(\theta)] \alpha(\theta^*; \theta) \pi(\theta) r(\theta^*; \theta) d\theta d\theta^*$
$= \int \int [h(\theta^*) - h(\theta)] \min [\pi(\theta) r(\theta^*; \theta), \pi(\theta^*) r(\theta; \theta^*)] d\theta d\theta^*$.
Switching the labels of θ *and* θ^*, *in the above integral, we note that*
$A_0 = \int \int [h(\theta) - h(\theta^*)] \min [\pi(\theta) r(\theta^*; \theta), \pi(\theta^*) r(\theta; \theta^*)] d\theta d\theta^*$.
We note that $A_0 = -A_0$, *hence* $A_0 = 0$. *It follows that* $E\left[h\left(\tilde{\theta}\right)\right] = E[h(\theta)]$. *Because* $h(\cdot)$ *is arbitrary, we conclude that* θ *and* $\tilde{\theta}$ *follow the same distribution.*

Regarding the second part of the proposition, note that $q(\theta^*; \theta) \geq 0$ *and* $\int q(\theta^*; \theta) d\theta^* = 1$, *hence a proper transition kernel. Denote* $\rho(\theta^*; \theta) = \frac{\pi(\theta^*) r(\theta; \theta^*)}{\pi(\theta) r(\theta^*; \theta)}$. *We validate the reversibility condition* $\pi(\theta) q(\theta^*; \theta) = \pi(\theta^*) q(\theta; \theta^*)$ *by enumerating three cases.*
Case 1: $\theta^* = \theta$. *In that case, the reversibility condition holds trivially.*
Case 2: $\theta^* \neq \theta$ *and* $\rho(\theta^*; \theta) \leq 1$. *In that case,* $\alpha(\theta^*; \theta) = \rho(\theta^*; \theta)$, *while* $\alpha(\theta; \theta^*) = 1$. *Also,* $\theta^* \neq \theta$ *indicates that* $\delta_\theta(\theta^*) = 0$. *It follows that*
$\pi(\theta) q(\theta^*; \theta) = \pi(\theta) \alpha(\theta^*; \theta) r(\theta^*; \theta) = \pi(\theta^*) r(\theta; \theta^*) = \pi(\theta^*) q(\theta; \theta^*)$.
Case 3: $\theta^* \neq \theta$ *and* $\rho(\theta^*; \theta) > 1$. *In that case,* $\alpha(\theta^*; \theta) = 1$, *while* $\alpha(\theta; \theta^*) = 1/\rho(\theta^*; \theta)$. *Also,* $\theta^* \neq \theta$ *indicates that* $\delta_\theta(\theta^*) = 0$. *It follows that*
$\pi(\theta) q(\theta^*; \theta) = \pi(\theta) r(\theta^*; \theta) = \alpha(\theta; \theta^*) \pi(\theta^*) r(\theta; \theta^*) = \pi(\theta^*) q(\theta; \theta^*)$.
In summary, the reversibility condition is satisfied in all cases, and thus $\pi(\cdot)$ *is a stationary distribution of* $q(\cdot)$.

Proposition 2.6 (Hammersley-Clifford)
Under the positivity condition, we write the integral as
$\int \frac{p_{B|A}(\theta_B; \theta_A)}{p_{A|B}(\theta_A; \theta_B)} d\theta_B = \int \frac{p_B(\theta_B)}{p_A(\theta_A)} d\theta_B = \frac{1}{p_A(\theta_A)} = \frac{p_{B|A}(\theta_B; \theta_A)}{p_{A,B}(\theta_A, \theta_B)}$.

Proposition 2.7 (Stationary distribution of the Gibbs sampler)
Regarding the first part of the proposition, $\tilde{q}$ *is a well-defined transition kernel, because* $\tilde{q}(\theta_A^*; \theta_A) \geq 0$ *and* $\int \tilde{q}(\theta_A^*; \theta_A) d\theta_A^* = 1$.
We validate the reversibility condition $p_A(\theta_A) \tilde{q}(\theta_A^*; \theta_A) = p_A(\theta_A^*) \tilde{q}(\theta_A; \theta_A^*)$.
$LHS = \int p_A(\theta_A) p_{B|A}(\theta_B^*; \theta_A) p_{A|B}(\theta_A^*; \theta_B^*) d\theta_B^*$
$= \int p_B(\theta_B^*) p_{A|B}(\theta_A; \theta_B^*) p_{A|B}(\theta_A^*; \theta_B^*) d\theta_B^*$
$RHS = \int p_A(\theta_A^*) p_{B|A}(\theta_B^*; \theta_A^*) p_{A|B}(\theta_A; \theta_B^*) d\theta_B^*$
$= \int p_B(\theta_B^*) p_{A|B}(\theta_A^*; \theta_B^*) p_{A|B}(\theta_A; \theta_B^*) d\theta_B^*$
Therefore, $LHS = RHS$, *the reversibility condition holds, and* p_A *is a stationary distribution of the Markov chain* $\tilde{q}$.

Regarding the second part of the proposition, q *is a well-defined transition kernel, because* $q(\theta_A^*, \theta_B^*; \theta_A, \theta_B) \geq 0$ *and* $\int q(\theta_A^*, \theta_B^*; \theta_A, \theta_B) d\theta_A^* d\theta_B^* = 1$.
We validate stationarity by its definition:
$\int \int p_{A,B}(\theta_A, \theta_B) q(\theta_A^*, \theta_B^*; \theta_A, \theta_B) d\theta_A d\theta_B$
$= \int \left[\int p_{A,B}(\theta_A, \theta_B) d\theta_B\right] p_{B|A}(\theta_B^*; \theta_A) p_{A|B}(\theta_A^*; \theta_B^*) d\theta_A$
$= \int p_A(\theta_A) p_{B|A}(\theta_B^*; \theta_A) p_{A|B}(\theta_A^*; \theta_B^*) d\theta_A$
$= \left[\int p_{A,B}(\theta_A, \theta_B^*) d\theta_A\right] p_{A|B}(\theta_A^*; \theta_B^*)$
$= p_B(\theta_B^*) p_{A|B}(\theta_A^*; \theta_B^*)$
$= p_{A,B}(\theta_A^*, \theta_B^*)$
It follows that $p_{A,B}$ *is the stationary distribution of the Markov chain* q.

Proposition 2.8 (High-order expectations)

The directed acyclic graph $\theta_{A,t-1} \to \theta_{Bt} \to \theta_{At}$ indicates that $\theta_{A,t-1}$ and θ_{At} are independent conditional on θ_{Bt}. To be specific,

$$p(\theta_{A,t-1}, \theta_{At} | \theta_{Bt}) = p(\theta_{A,t-1} | \theta_{Bt}) \, p(\theta_{At} | \theta_{Bt}, \theta_{A,t-1}) = p(\theta_{A,t-1} | \theta_{Bt}) \, p(\theta_{At} | \theta_{Bt}).$$

Regarding the distributional forms, the Gibbs transition kernel is constructed such that $p(\theta_{At} | \theta_{Bt})$ follows $p_{A|B}$. Also, by Bayes' rule

$$p(\theta_{A,t-1} | \theta_{Bt}) = \frac{p(\theta_{A,t-1}) p(\theta_{Bt} | \theta_{A,t-1})}{p(\theta_{Bt})}.$$

As the chain reaches the stationary distribution, $p(\theta_{A,t-1})$ follows p_A, and $p(\theta_{Bt})$ follows p_B. Also, the Gibbs transition kernel is constructed such that $p(\theta_{Bt} | \theta_{A,t-1})$ follows $p_{B|A}$. Bayes' rule indicates that $p(\theta_{A,t-1} | \theta_{Bt})$ follows $p_{A|B}$, which is the same as $p(\theta_{At} | \theta_{Bt})$.

We have established a major feature of the two-block Gibbs sampler: $\theta_{A,t-1}$ and θ_{At} are independent and identically distributed (IID) conditional on θ_{Bt}, if the Gibbs sampler reaches the stationary distribution. Furthermore, $x_t = g_0(\theta_{At})$ is a function of the states, thus x_{t-1} and x_t are IID conditional on θ_{Bt}. The IID feature is crucial for understanding high-order expectations.

Let the stationary mean (unconditional expectation) be $\mu = E(x_t)$, and the first-order conditional expectation be $g_1(\theta_{Bt}) = E(x_t | \theta_{Bt})$. By the law of iterated expectations, we have

$$E[g_1(\theta_{Bt})] = \mu,$$
$$Var[g_1(\theta_{Bt})] = E[g_1(\theta_{Bt}) g_1(\theta_{Bt})] - \mu^2.$$

Regarding the covariances, $cov(x_t, x_{t-1}) = E(x_t x_{t-1}) - \mu^2$, so we concentrate on $E(x_t x_{t-1})$. By the law of iterated expectations and the IID feature we have established, we have

$$\begin{aligned} &E(x_t x_{t-1}) \\ &= E[E(x_t x_{t-1} | \theta_{Bt})] \\ &= E[E(x_t | \theta_{Bt}) \cdot E(x_{t-1} | \theta_{Bt})] \\ &= E[g_1(\theta_{Bt}) g_1(\theta_{Bt})] \end{aligned}$$

It follows that $cov(x_t, x_{t-1}) = var[g_1(\theta_{Bt})]$.

As for the second-order correlations, we have the directed acyclic graph

$$\theta_{A,t-2} \to \theta_{B,t-1} \to \theta_{A,t-1} \to \theta_{Bt} \to \theta_{At}.$$

Also, x_t and x_{t-2} are functions of θ_{At} and $\theta_{A,t-2}$, respectively. So we have

$$\begin{aligned} &E(x_t x_{t-2}) \\ &= E[E(x_t x_{t-2} | \theta_{B,t-1}, \theta_{A,t-1}, \theta_{Bt})] \text{ (law of iterated expectations)} \\ &= E[E(x_t | \theta_{B,t-1}, \theta_{A,t-1}, \theta_{Bt}) E(x_{t-2} | \theta_{B,t-1}, \theta_{A,t-1}, \theta_{Bt})] \text{ (conditional independency)} \\ &= E[E(x_t | \theta_{Bt}) E(x_{t-2} | \theta_{B,t-1})] \text{ (Markov property)} \\ &= E[E(x_t | \theta_{Bt}) E(x_{t-1} | \theta_{B,t-1})] \text{ (IID feature)} \\ &= E[g_1(\theta_{Bt}) g_1(\theta_{B,t-1})] \text{ (functional form of conditional mean)} \\ &= E\{E[g_1(\theta_{Bt}) g_1(\theta_{B,t-1}) | \theta_{A,t-1}]\} \text{ (law of iterated expectations)} \\ &= E\{E[g_1(\theta_{Bt}) | \theta_{A,t-1}] E[g_1(\theta_{B,t-1}) | \theta_{A,t-1}]\} \text{ (conditional independency)} \\ &= E\{E[g_1(\theta_{B,t-1}) | \theta_{A,t-1}] E[g_1(\theta_{B,t-1}) | \theta_{A,t-1}]\} \text{ (IID feature)} \\ &= E\{E[g_1(\theta_{Bt}) | \theta_{At}] E[g_1(\theta_{Bt}) | \theta_{At}]\} \text{ (stationarity)} \end{aligned}$$

The second-order conditional expectation is defined as $g_2(\theta_{At}) = E[g_1(\theta_{Bt}) | \theta_{At}]$, so we have $E(x_t x_{t-2}) = E[g_2(\theta_{At}) g_2(\theta_{At})]$. Also, by the law of iterated expectations, $E[g_2(\theta_{At})] = \mu$. It follows that $cov(x_t, x_{t-2}) = Var[g_2(\theta_{At})]$.

By induction, higher-order correlations are formulated similarly, by higher-order expectations conditional on θ_{At} and θ_{Bt} alternately. They are non-negative because the variance of any random variable is non-negative. They are monotone decreasing, because the unconditional variance equals the variance of the conditional mean plus the mean of the conditional variance:

$$Var(x_t) = Var[E(x_t | \theta_{Bt})] + E[Var(x_t | \theta_{Bt})].$$

We conclude that $Var(x_t)$ is the largest, and $cov(x_t, x_{t-1})$ is the second largest, and $cov(x_t, x_{t-j})$ decreases as we take higher-order expectations.

Proposition 2.9 (Langevin Monte Carlo)

Given the auxiliary density $f(z) \propto e^{-\frac{1}{2}z'z}$, we have $\frac{d \ln f(z)}{dz} = -z$.

By Definition 2.1, a single step of the leapfrog move is given by

$z_{1/2} = z + \frac{1}{2}\delta\nabla(\theta)$,

$\theta^* = \theta - \delta(-z_{1/2})$,

$z^* = z_{1/2} + \frac{1}{2}\delta\nabla(\theta^*)$.

That is,

$\theta^* = \theta + \frac{1}{2}\delta^2\nabla(\theta) + \delta z$,

$z^* = z + \frac{1}{2}\delta\nabla(\theta) + \frac{1}{2}\delta\nabla(\theta^*)$.

Regarding the transition from θ to θ^*, the leapfrog move is exactly the same as the Langevin proposal. Also, we can solve z, z^* as a function of θ, θ^*:

$z = \frac{1}{\delta}\left[\theta^* - \theta - \frac{1}{2}\delta^2\nabla(\theta)\right]$,

$z^* = -\frac{1}{\delta}\left[\theta - \theta^* - \frac{1}{2}\delta^2\nabla(\theta^*)\right]$.

It follows that

$r(\theta^*; \theta) \propto e^{-\frac{1}{2}\delta^{-2}\left[\theta^* - \theta - \frac{1}{2}\delta^2\nabla(\theta)\right]'\left[\theta^* - \theta - \frac{1}{2}\delta^2\nabla(\theta)\right]} = e^{-\frac{1}{2}z'z} \propto f(z)$.

Similarly, $r(\theta, \theta^*) \propto f(z^*)$.

The HMC sampler accepts the proposal by the probability $\min\left[1, \frac{\Pi(\theta^*, z^*)}{\Pi(\theta, z)}\right]$, *where* $\frac{\Pi(\theta^*, z^*)}{\Pi(\theta, z)} = \frac{\pi(\theta^*)f(z^*)}{\pi(\theta)f(z)} = \frac{\pi(\theta^*)r(\theta;\theta^*)}{\pi(\theta)r(\theta^*;\theta)}$.

We conclude that the acceptance probability of the HMC and Langevin sampler are the same. The HMC sampler with one leapfrog move is equivalent to the Langevin Monte Carlo, which is a special case of the MH sampler.

Proposition 2.10 (Properties of leapfrog move)

Regarding the first property, by Definition 2.1, the leapfrog move is a mapping from (θ, z) to (θ^*, z^*) such that

$z_{1/2} = z + \frac{1}{2}\delta\frac{d \ln \pi(\theta)}{d\theta}$,

$\theta^* = \theta - \delta\frac{d \ln f(z_{1/2})}{dz_{1/2}}$,

$z^* = z_{1/2} + \frac{1}{2}\delta\frac{d \ln \pi(\theta^*)}{d\theta^*}$.

Examining the three equations in the reverse order, we obtain the inverse mapping from (θ^*, z^*) to (θ, z):

$z_{1/2} = z^* - \frac{1}{2}\delta\frac{d \ln \pi(\theta^*)}{d\theta^*}$,

$\theta = \theta^* + \delta\frac{d \ln f(z_{1/2})}{dz_{1/2}}$,

$z = z_{1/2} - \frac{1}{2}\delta\frac{d \ln \pi(\theta)}{d\theta}$.

Regarding the second property, by Definition 2.1, the original mapping takes $(\theta^*, -z^*)$ to $\left(\tilde{\theta}, \tilde{z}\right)$ such that

$\tilde{z}_{1/2} = -z^* + \frac{1}{2}\delta\frac{d \ln \pi(\theta^*)}{d\theta^*}$,

$\tilde{\theta} = \theta^* - \delta\frac{d \ln f(\tilde{z}_{1/2})}{d\tilde{z}_{1/2}}$,

$\tilde{z} = \tilde{z}_{1/2} + \frac{1}{2}\delta\frac{d \ln \pi(\tilde{\theta})}{d\tilde{\theta}}$.

The first equation indicates that $\tilde{z}_{1/2} = -z_{1/2}$. *Due to symmetry,* $f(\tilde{z}_{1/2}) = f(z_{1/2})$, *so we have* $\frac{d \ln f(\tilde{z}_{1/2})}{d\tilde{z}_{1/2}} = \frac{d \ln f(z_{1/2})}{dz_{1/2}}$. *Therefore, the second equation indicates that* $\tilde{\theta} = \theta$. *The third equation indicates that* $\tilde{z} = -z$. *If follows that* $(\theta, -z) = g(\theta^*, -z^*)$.

Note that $f(z)$ *is symmetric, which implies that* $\Pi(\theta, z) = \Pi(\theta, -z)$. *Consequently,* $\Pi\left[g^{-1}(\theta^*, z^*)\right] = \Pi\left[g(\theta^*, -z^*)\right]$.

Regarding the third property, as (θ^*, z^*) *is a deterministic function of* (θ, z), *we find the joint distribution of* (θ^*, z^*) *by the change of variables:*

$p(\theta^*, z^*) = \Pi\left[g^{-1}(\theta^*, z^*)\right] \frac{1}{|\det(J)|} = \Pi(\theta, z) \frac{1}{|\det(J)|}$.

To compute the Jacobian J, *as the derivatives of* (θ^*, z^*) *with respect to* (θ, z), *in a leapfrog move, we note that*

$\frac{\partial \theta^*}{\partial \theta} = I - \frac{1}{2}\delta^2 \frac{d^2 \ln f(z_{1/2})}{dz_{1/2}dz_{1/2}} \frac{d^2 \ln \pi(\theta)}{d\theta d\theta}$,

$\frac{\partial \theta^*}{\partial z} = -\delta \frac{d^2 \ln f(z_{1/2})}{dz_{1/2}dz_{1/2}}$,

$\frac{\partial z^*}{\partial \theta} = \frac{1}{2}\delta \frac{d^2 \ln \pi(\theta)}{d\theta d\theta} + \frac{1}{2}\delta \frac{d^2 \ln \pi(\theta^*)}{d\theta^* d\theta^*} \left[I - \frac{1}{2}\delta^2 \frac{d^2 \ln f(z_{1/2})}{dz_{1/2}dz_{1/2}} \frac{d^2 \ln \pi(\theta)}{d\theta d\theta}\right]$,

$\frac{\partial z^*}{\partial z} = I + \frac{1}{2}\delta \frac{d^2 \ln \pi(\theta^*)}{d\theta^* d\theta^*} \left[-\delta \frac{d^2 \ln f(z_{1/2})}{dz_{1/2}dz_{1/2}}\right]$.

It follows that $\det(J) = \frac{\partial \theta^*}{\partial \theta}\frac{\partial z^*}{\partial z} - \frac{\partial \theta^*}{\partial z}\frac{\partial z^*}{\partial \theta} = 1$. *In other words, the density of* (θ^*, z^*) *can be evaluated by* $\Pi\left[g^{-1}(\theta^*, z^*)\right]$, *namely* $\Pi(\theta, z)$ *without accounting for the Jacobian term induced by the change of variables. Note that* $p(\theta^*, z^*)$ *equals* $\Pi(\theta, z)$ *rather than* $\Pi(\theta^*, z^*)$, *so the HMC acceptance probability* $\min\left[1, \frac{\Pi(\theta^*, z^*)}{\Pi(\theta, z)}\right]$ *is not equal to one.*

Regarding the fourth property, recursive application of second and third properties leads to reversibility and volume preservation for the multiple-step leapfrog move.

Proposition 2.11 (Stationary distribution of HMC sampler)

We will show that for an arbitrary square integrable function h, *we have* $E\left[h\left(\tilde{\theta}, \tilde{z}\right)\right] = E\left[h(\theta, z)\right]$.

By the law of iterated expectations,

$E\left[h\left(\tilde{\theta}, \tilde{z}\right)\right] = E\left\{E\left[h\left(\tilde{\theta}, \tilde{z}\right) | \theta, z, \theta^*, z^*\right]\right\}$.

Conditional on θ, z, θ^*, z^*, *we have two scenarios: first, with the probability* $\alpha = \min\left[1, \frac{\Pi(\theta^*, z^*)}{\Pi(\theta, z)}\right]$, *we have an accepted proposal* $\left(\tilde{\theta}, \tilde{z}\right) = (\theta^*, -z^*)$. *Also,* $(\theta, z) = g^{-1}(\theta^*, z^*)$. *In that case, we can rewrite the conditional expectation as an integral with respect to* (θ^*, z^*). *Second, with the probability* $1 - \alpha$, *we have a rejected proposal* $\left(\tilde{\theta}, \tilde{z}\right) = (\theta, z)$. *Also,* $(\theta^*, z^*) = g(\theta, z)$. *In that case, we can rewrite the conditional expectation as an integral of* (θ, z). *To be specific,*

$E\left[h\left(\tilde{\theta}, \tilde{z}\right)\right]$

$= \int\int h(\theta^*, -z^*) \Pi\left[g^{-1}(\theta^*, z^*)\right] \frac{1}{|\det(J)|} \min\left[1, \frac{\Pi(\theta^*, z^*)}{\Pi[g^{-1}(\theta^*, z^*)]}\right] d\theta^* dz^*$

$+ \int\int h(\theta, z) \Pi(\theta, z) \left\{1 - \min\left[1, \frac{\Pi[g(\theta, z)]}{\Pi(\theta, z)}\right]\right\} d\theta dz$.

Due to volume preservation, the Jacobian term equals one $\frac{1}{|\det(J)|}$. *Also, reversibility implies that* $\Pi\left[g^{-1}(\theta^*, z^*)\right] = \Pi\left[g(\theta^*, -z^*)\right]$. *It follows that*

$E\left[h\left(\tilde{\theta}, \tilde{z}\right)\right]$

$= \int\int h(\theta^*, -z^*) \min\left[\Pi\left[g(\theta^*, -z^*)\right], \Pi(\theta^*, z^*)\right] d\theta^* dz^*$

$- \int\int h(\theta, z) \min\left[\Pi(\theta, z), \Pi\left[g(\theta, z)\right]\right] d\theta dz$

$+ \int\int h(\theta, z) \Pi(\theta, z) d\theta dz$.

The first and second integrals are equal, because we can apply the change of variables $(\theta^*, z^*) \to (\theta, -z)$ *to the first integral (with the Jacobian being one), and use the fact that* $\Pi(\theta^*, -z^*) = \Pi(\theta^*, z^*)$. *Consequently, we have*

$$E\left[h\left(\tilde{\theta}, \tilde{z}\right)\right] = \int\int h(\theta, z)\,\Pi(\theta, z)\,d\theta dz = E[h(\theta, z)].$$

Proposition 2.12 (Random-walk multiple-try MH sampler)

We will show that for an arbitrary square integrable function h, *we have* $E\left[h\left(\tilde{\theta}\right)\right] = E[h(\theta)]$.

Denote $\theta_{1:n} = (\theta_1, \ldots, \theta_n)$ *and* $\theta^*_{1:n} = (\theta^*_1, \ldots, \theta^*_n)$.

By the law of iterated expectations,

$$E\left[h\left(\tilde{\theta}\right)\right] = E\left\{E\left[h\left(\tilde{\theta}\right)|\theta, \theta^*_{1:n}, \theta^*, \theta_{1:n}\right]\right\}.$$

Since $\tilde{\theta} = \begin{cases} \theta^* & \text{with probability } \alpha \\ \theta & \text{with probability } 1-\alpha \end{cases}$, *we have*

$$E\left[h\left(\tilde{\theta}\right)|\theta, \theta^*_{1:n}, \theta^*, \theta_{1:n}\right] = \alpha h(\theta^*) + (1-\alpha)\, h(\theta).$$

It follows that

$$E\left[h\left(\tilde{\theta}\right)\right] = E[h(\theta)] + E[\alpha h(\theta^*) - \alpha h(\theta)].$$

The joint distribution of $\theta, \theta^*_{1:n}, \theta^*, \theta_{1:n}$ *is*

$$\pi(\theta)\prod_{i=1}^n r(\theta^*_i; \theta)\left[\sum_{j=1}^n \frac{\pi(\theta^*_j)}{\sum_i \pi(\theta^*_i)} I\left(\theta^* = \theta^*_j\right)\prod_{i=1}^{n-1} r\left(\theta_i; \theta^*_j\right)\right] I(\theta_n = \theta).$$

The expectation operator $E(\cdot)$ *integrates out all of those random variables. We will use indicator functions* $I(\theta_n = \theta)$ *and* $I\left(\theta^* = \theta^*_j\right)$ *to drop* θ *and* θ^* *in the expressions, which appear symmetric between* $\theta_{1:n}$ *and* $\theta^*_{1:n}$:

$$E[\alpha h(\theta^*) - \alpha h(\theta)]$$
$$= \int\int \sum_{j=1}^n \alpha\left[h\left(\theta^*_j\right) - h(\theta_n)\right]\pi(\theta_n)\prod_{i=1}^n r(\theta^*_i; \theta_n)\frac{\pi(\theta^*_j)}{\sum_i \pi(\theta^*_i)}\prod_{i=1}^{n-1} r\left(\theta_i; \theta^*_j\right) d\theta_{1:n} d\theta^*_{1:n}$$
$$= n\int\int \tilde{\alpha}\left[h(\theta^*_n) - h(\theta_n)\right]\pi(\theta_n)\,\pi(\theta^*_n)\prod_{i=1}^n r(\theta^*_i; \theta_n)\prod_{i=1}^{n-1} r(\theta_i; \theta^*_n)\, d\theta_{1:n} d\theta^*_{1:n},$$

where $\tilde{\alpha} = \min\left[\frac{1}{\sum_i \pi(\theta^*_i)}, \frac{1}{\sum_i \pi(\theta_i)}\right]$. *In the above derivations, we leveraged symmetry and the sum* $\sum_{j=1}^n \cdots$ *are the same for* θ^*_j, $j = 1, \ldots, n$, *so that we just replaced* θ^*_j *by* θ^*_n.

The integral remains unchanged if we switch the labels of $\theta_{1:n}$ *and* $\theta^*_{1:n}$:

$$n\int\int \tilde{\alpha}\left[h(\theta_n) - h(\theta^*_n)\right]\pi(\theta^*_n)\,\pi(\theta_n)\prod_{i=1}^n r(\theta_i; \theta^*_n)\prod_{i=1}^{n-1} r(\theta^*_i; \theta_n)\, d\theta_{1:n} d\theta^*_{1:n}.$$

The proposal is symmetric $r(\theta_n; \theta^*_n) = r(\theta^*_n; \theta_n)$. *It follows that*

$$E[\alpha h(\theta^*) - \alpha h(\theta)] = -E[\alpha h(\theta^*) - \alpha h(\theta)].$$

We have $E[\alpha h(\theta^*) - \alpha h(\theta)] = 0$. *It follows that* $E\left[h\left(\tilde{\theta}\right)\right] = E[h(\theta)]$.

Proposition 2.13 (Independence multiple-try MH sampler)

We will show that for an arbitrary square integrable function h, *we have* $E\left[h\left(\tilde{\theta}\right)\right] = E[h(\theta)]$.

Denote $\theta_{1:n} = (\theta_1, \ldots, \theta_n)$ *and* $\theta^*_{1:n} = (\theta^*_1, \ldots, \theta^*_n)$.

First, we consider the sampler with the acceptance probability ratio specified as $\rho = \frac{\sum_i w(\theta^*_i)}{w(\theta) - w(\theta^*) + \sum_i w(\theta^*_i)}$.

By the law of iterated expectations,

$$E\left[h\left(\tilde{\theta}\right)\right] = E\left\{E\left[h\left(\tilde{\theta}\right)|\theta, \theta^*_{1:n}, \theta^*, \theta_{1:n}\right]\right\}.$$

Since $\tilde{\theta} = \begin{cases} \theta^* & \text{with probability } \alpha \\ \theta & \text{with probability } 1-\alpha \end{cases}$, *we have*

$E\left[h\left(\tilde{\theta}\right)|\theta,\theta^*_{1:n},\theta^*,\theta_{1:n}\right] = \alpha h\left(\theta^*\right) + \left(1-\alpha\right)h\left(\theta\right).$

It follows that

$E\left[h\left(\tilde{\theta}\right)\right] = E\left[h\left(\theta\right)\right] + E\left[\alpha h\left(\theta^*\right) - \alpha h\left(\theta\right)\right].$

The joint distribution of $\theta, \theta^*_{1:n}, \theta^*$ *is*

$\pi\left(\theta\right)\prod_{i=1}^{n} r\left(\theta_i^*\right)\left[\sum_{j=1}^{n}\frac{w\left(\theta_j^*\right)}{\sum_i w\left(\theta_i^*\right)} I\left(\theta^* = \theta_j^*\right)\right].$

The expectation operator $E\left(\cdot\right)$ *integrates out all of those random variables.*

$E\left[\alpha h\left(\theta^*\right) - \alpha h\left(\theta\right)\right]$

$= \int\int\sum_{j=1}^{n}\alpha\left[h\left(\theta_j^*\right) - h\left(\theta\right)\right]\pi\left(\theta\right)\prod_{i=1}^{n} r\left(\theta_i^*\right)\frac{w\left(\theta_j^*\right)}{\sum_i w\left(\theta_i^*\right)}d\theta^*_{1:n}d\theta$

$= n\int\int\tilde{\alpha}\left[h\left(\theta_n^*\right) - h\left(\theta\right)\right]\pi\left(\theta\right)\pi\left(\theta_n^*\right)\prod_{i=1}^{n-1} r\left(\theta_i^*\right)d\theta^*_{1:n}d\theta,$

where $\tilde{\alpha} = \min\left[\frac{1}{w\left(\theta_n^*\right)+\sum_{i=1}^{n-1}w\left(\theta_i^*\right)}, \frac{1}{w\left(\theta\right)+\sum_{i=1}^{n-1}w\left(\theta_i^*\right)}\right]$. *In the above derivations, we leveraged symmetry and the sum* $\sum_{j=1}^{n}\cdots$ *are the same for* θ_j^*, $j = 1,\ldots,n$, *so that we just replaced* θ_j^* *by* θ_n^*.

The integral remains unchanged if we switch the labels of θ *and* θ_n^*:

$n\int\int\tilde{\alpha}\left[h\left(\theta\right) - h\left(\theta_n^*\right)\right]\pi\left(\theta_n^*\right)\pi\left(\theta\right)\prod_{i=1}^{n} r\left(\theta_i^*\right)d\theta^*_{1:n}d\theta.$

It follows that

$E\left[\alpha h\left(\theta^*\right) - \alpha h\left(\theta\right)\right] = -E\left[\alpha h\left(\theta^*\right) - \alpha h\left(\theta\right)\right].$

We have $E\left[\alpha h\left(\theta^*\right) - \alpha h\left(\theta\right)\right] = 0$. *It follows that* $E\left[h\left(\tilde{\theta}\right)\right] = E\left[h\left(\theta\right)\right]$.

As for the alternative specification of $\rho = \frac{\sum_i w\left(\theta_i^*\right)}{\sum_i w\left(\theta_i\right)}$, *we cast the problem as a standard MH sampler on an extended state space with* $\left(\theta,\tau,\theta_{1:n}\right)$, *where* τ *is a random integer between 1 and* n.

On the extended state space, we define the target and proposal distributions. The target distribution is a joint distribution of $\left(\theta,\tau,\theta_{1:n}\right)$ *that can be described as follows:* θ *follows* $\pi\left(\theta\right)$, *and* $\tau \in \{1,\ldots,n\}$ *follows the discrete-valued uniform distribution. They are independent. Conditional on* θ *and* τ, *we have* $\theta_\tau = \theta$ *and* $\theta_i, i \neq \tau$ *are* $n-1$ *independent draws from* $r\left(\cdot\right)$.

The target density takes the form:

$\hat{\pi}\left(\theta,\tau,\theta_{1:n}\right) = \frac{1}{n}\pi\left(\theta\right)I\left(\theta_\tau = \theta\right)\prod_{i\neq\tau} r\left(\theta_i\right).$

The proposal is a transition kernel from $\left(\theta,\tau,\theta_{1:n}\right)$ *to* $\left(\theta^*,\tau^*,\theta^*_{1:n}\right)$. *It is an independent proposal (i.e., the distributional form does not depend on* $\theta,\tau,\theta_{1:n}$*). The proposal can be described as follows:* θ_i^*, $i = 1,\ldots,n$ *are independent draws from* $r\left(\cdot\right)$. *We compute the importance weight* $w\left(\theta_i^*\right) = \frac{\pi\left(\theta_i^*\right)}{r\left(\theta_i^*\right)}$ *and generate the discrete-valued* τ^* *with the probability* $P\left(\tau^* = i\right) = \frac{w\left(\theta_i^*\right)}{\sum_j w\left(\theta_j^*\right)}$. *Finally, we set* $\theta^* = \theta^*_{\tau^*}$. *In other words, we select* θ^* *among* $\{\theta_1^*,\ldots,\theta_n^*\}$ *by the importance weight, as described in the multiple-try sampler.*

The proposal transition kernel takes the form:

$\hat{r}\left(\theta^*,\tau^*,\theta^*_{1:n}\right) = \prod_{i=1}^{n} r\left(\theta_i^*\right)\frac{w\left(\theta^*_{\tau^*}\right)}{\sum_j w\left(\theta_j^*\right)}I\left(\theta^* = \theta^*_{\tau^*}\right).$

As we are unable to generate draws directly from the target distribution $\hat{\pi}$, *we employ the standard MH sampler, with the independent transition kernel* $\hat{r}$. *By Algorithm 2.2, the MH acceptance probability ratio takes the form*

$\frac{\hat{\pi}\left(\theta^*,\tau^*,\theta^*_{1:n}\right)\hat{r}\left(\theta,\tau,\theta_{1:n}\right)}{\hat{\pi}\left(\theta,\tau,\theta_{1:n}\right)\hat{r}\left(\theta^*,\tau^*,\theta^*_{1:n}\right)}.$

After some algebra, that ratio can be greatly simplified as $\frac{\sum_i w\left(\theta_i^*\right)}{\sum_i w\left(\theta_i\right)}$.

In summary, the independence multiple-try MH sampler is a standard MH sampler on an extended state space, with the target distribution $\hat{\pi}$ as its invariant distribution, and the marginal stationary distribution is $\pi(\theta)$.

Proposition 2.14 (Independent pseudo-marginal MH sampler)

First, we validate that $f(\theta; Z) = p(Z)\hat{\pi}(\theta; Z)$ is a proper joint density of the pair (θ, Z), and the marginal density is $\pi(\theta)$:

$f(\theta; Z) \geq 0$ *(as a nonnegative estimator),*

$\int f(\theta; Z)\, dZ = \int p(Z)\hat{\pi}(\theta; Z)\, dZ = \pi(\theta)$ *(as an unbiased estimator),*

$\int\int f(\theta; Z)\, dZ d\theta = \int \pi(\theta)\, d\theta = 1.$

Now we show equivalency between the pseudo-marginal MH sampler and a standard MH sampler on the extended state space, targeting the joint distribution $f(\theta; Z)$. Consider the proposed transition from the pair (θ, Z) to (θ^*, Z^*):

$\hat{r}(\theta^*, Z^*; \theta, Z) = r(\theta^*; \theta)\, p(Z^*).$

According to Algorithm 2.2, a proposal is accepted with the probability $\min(1, \rho)$, *where*

$$\rho = \frac{f(\theta^*;Z^*)\hat{r}(\theta,Z;\theta^*,Z^*)}{f(\theta;Z)\hat{r}(\theta^*,Z^*;\theta,Z)} = \frac{p(Z^*)\hat{\pi}(\theta^*;Z^*)r(\theta;\theta^*)p(Z)}{p(Z)\hat{\pi}(\theta;Z)r(\theta^*;\theta)p(Z^*)} = \frac{\hat{\pi}(\theta^*;Z^*)r(\theta;\theta^*)}{\hat{\pi}(\theta;Z)r(\theta^*;\theta)}.$$

As we can see, $p(Z)$ and $p(Z^*)$ are cancelled in the acceptance probability. Therefore, the pseudo-marginal sampler is a standard MH sampler on the extended state space, with the target distribution $p(Z)\hat{\pi}(\theta; Z)$ as its stationary distribution, and the marginal stationary distribution is $\pi(\theta)$.

Proposition 2.15 (Correlated pseudo-marginal MH sampler)

First, we validate that $f(\theta; Z) = p(Z)\hat{\pi}(\theta; Z)$ is a proper joint density of the pair (θ, Z), and the marginal density is $\pi(\theta)$:

$f(\theta; Z) \geq 0$ *(as a nonnegative estimator),*

$\int f(\theta; Z)\, dZ = \int p(Z)\hat{\pi}(\theta; Z)\, dZ = \pi(\theta)$ *(as an unbiased estimator),*

$\int\int f(\theta; Z)\, dZ d\theta = \int \pi(\theta)\, d\theta = 1.$

Now we show equivalency between the correlated pseudo-marginal MH sampler and a standard MH sampler on the extended state space, targeting the joint distribution $f(\theta; Z)$. Consider the proposed transition from the pair (θ, Z) to (θ^*, Z^*):

$\hat{r}(\theta^*, Z^*; \theta, Z) = r(\theta^*; \theta)\, K(Z^*; Z).$

According to Algorithm 2.2, a proposal is accepted with the probability $\min(1, \rho)$, *where*

$$\rho = \frac{f(\theta^*;Z^*)\hat{r}(\theta,Z;\theta^*,Z^*)}{f(\theta;Z)\hat{r}(\theta^*,Z^*;\theta,Z)} = \frac{p(Z^*)\hat{\pi}(\theta^*;Z^*)r(\theta;\theta^*)K(Z;Z^*)}{p(Z)\hat{\pi}(\theta;Z)r(\theta^*;\theta)K(Z^*;Z)} = \frac{\hat{\pi}(\theta^*;Z^*)r(\theta;\theta^*)}{\hat{\pi}(\theta;Z)r(\theta^*;\theta)}.$$

In the above derivation, we resorted to the identity $p(Z)K(Z^*; Z) = p(Z^*)K(Z; Z^*)$ for simplifying the acceptance probability, which looks as if the intractable $\pi(\theta)$ was replaced by the unbiased estimator $\hat{\pi}(\theta; Z)$.

Since we followed the standard MH sampling procedure on the extended state space, the target distribution $p(Z)\hat{\pi}(\theta; Z)$ is the stationary distribution of the Markov chain, and $\pi(\theta)$ is the marginal stationary distribution.

Proposition 2.16 (Divide-and-conquer importance sampling)

We write $p(\theta|Y) \propto p(\theta|Y^j)\,\tilde{w}_j(\theta)$ in a verbose form:

$$p(\theta|Y) = p(\theta|Y^j)\,\tilde{w}_j(\theta)\,\frac{p(Y^j)\,k}{p(Y)},$$

where $k = \int [p(\theta)]^{1/J}\,d\theta$, $p(Y^j) = \int k^{-1}[p(\theta)]^{1/J}\,p(Y^j|\theta)\,d\theta$, and $p(Y) = \int p(\theta)\,p(Y|\theta)\,d\theta$.

The weighted sample average is

$$\sum_{t=1}^{T} h(\theta_{jt})\,w_j(\theta_{jt}) = \frac{T^{-1}\sum_{t=1}^{T} h(\theta_{jt})\,\tilde{w}_j(\theta_{jt})}{T^{-1}\sum_{t=1}^{T}\tilde{w}_j(\theta_{jt})}.$$

As the Markov chain is irreducible, aperiodic and Harris recurrent, the sample average converges to the expected value. That is,

$$T^{-1}\sum_{t=1}^{T} h(\theta_{jt})\,\tilde{w}_j(\theta_{jt}) \xrightarrow{p} \int h(\theta)\,\tilde{w}_j(\theta)\,p(\theta|Y^j)\,d\theta = \frac{p(Y)\int h(\theta)\,p(\theta|Y)\,d\theta}{p(Y^j)\,k},$$

$$T^{-1}\sum_{t=1}^{T}\tilde{w}_j(\theta_{jt}) \xrightarrow{p} \int \tilde{w}_j(\theta)\,p(\theta|Y^j)\,d\theta = \frac{p(Y)}{p(Y^j)\,k}.$$

It follows that $\sum_{t=1}^{T} h(\theta_{jt})\,w_j(\theta_{jt}) \xrightarrow{p} \int h(\theta)\,p(\theta|Y)\,d\theta$.

Proposition 2.17 (Delayed acceptance and early rejection)

The modified transition kernel satisfies $q(\theta^*;\theta) \geq 0$ and $\int q(\theta^*;\theta)\,d\theta^* = 1$, hence a proper transition kernel.

We validate the reversibility condition $\pi(\theta)\,q(\theta^*;\theta) = \pi(\theta^*)\,q(\theta;\theta^*)$ by enumerating all the possible cases:

Case 1: $\theta^* = \theta$. In that case, the reversibility condition holds trivially.

Case 2: $\theta^* \neq \theta$, $\rho_1(\theta^*,\theta) \leq 1$, $\rho_2(\theta^*,\theta) \leq 1$. In that case, $\alpha_1(\theta^*,\theta) = \rho_1(\theta^*,\theta)$, $\alpha_2(\theta^*,\theta) = \rho_2(\theta^*,\theta)$, while $\alpha_1(\theta,\theta^*) = 1$, $\alpha_2(\theta,\theta^*) = 1$. Also, $\theta^* \neq \theta$ indicates that $\delta_\theta(\theta^*) = 0$. It follows that

$\pi(\theta)\,q(\theta^*;\theta)$

$= p(\theta)\prod_{i=1}^{n} p(Y_i|\theta)\,\frac{p(\theta^*)\prod_{i=1}^{m} p(Y_i|\theta^*)\,r(\theta;\theta^*)}{p(\theta)\prod_{i=1}^{m} p(Y_i|\theta)\,r(\theta^*;\theta)}\,\frac{\prod_{i=m+1}^{n} p(Y_i|\theta^*)}{\prod_{i=m+1}^{n} p(Y_i|\theta)}\,r(\theta^*;\theta)$

$= p(\theta^*)\prod_{i=1}^{n} p(Y_i|\theta^*)\,r(\theta;\theta^*)$

$= \pi(\theta^*)\,q(\theta;\theta^*)$

Case 3: $\theta^* \neq \theta$, $\rho_1(\theta^*,\theta) > 1$, $\rho_2(\theta^*,\theta) \leq 1$. In that case, $\alpha_1(\theta^*,\theta) = 1$, $\alpha_2(\theta^*,\theta) = \rho_2(\theta^*,\theta)$, while $\alpha_1(\theta,\theta^*) = 1/\rho_1(\theta^*,\theta)$, $\alpha_2(\theta,\theta^*) = 1$. Also, $\theta^* \neq \theta$ indicates that $\delta_\theta(\theta^*) = 0$. It follows that

$\pi(\theta)\,q(\theta^*;\theta)$

$= p(\theta)\prod_{i=1}^{n} p(Y_i|\theta)\,\frac{\prod_{i=m+1}^{n} p(Y_i|\theta^*)}{\prod_{i=m+1}^{n} p(Y_i|\theta)}\,r(\theta^*;\theta)$

$= p(\theta)\prod_{i=1}^{m} p(Y_i|\theta)\prod_{i=m+1}^{n} p(Y_i|\theta^*)\,r(\theta^*;\theta)$

$\pi(\theta^*)\,q(\theta;\theta^*)$

$= p(\theta^*)\prod_{i=1}^{n} p(Y_i|\theta^*)\,\frac{p(\theta)\prod_{i=1}^{m} p(Y_i|\theta)\,r(\theta^*;\theta)}{p(\theta^*)\prod_{i=1}^{m} p(Y_i|\theta^*)\,r(\theta;\theta^*)}\,r(\theta;\theta^*)$

$= \prod_{i=m+1}^{n} p(Y_i|\theta^*)\,p(\theta)\prod_{i=1}^{m} p(Y_i|\theta)\,r(\theta^*;\theta)$

So we still have $\pi(\theta)\,q(\theta^*;\theta) = \pi(\theta^*)\,q(\theta;\theta^*)$.

Similarly, we can validate the reversibility for the remaining two cases:

Case 4: $\theta^* \neq \theta$, $\rho_1(\theta^*, \theta) \leq 1$, $\rho_2(\theta^*, \theta) > 1$.

Case 5: $\theta^* \neq \theta$, $\rho_1(\theta^*, \theta) > 1$, $\rho_2(\theta^*, \theta) > 1$.

In summary, the reversibly condition holds in all cases. Therefore, $\pi(\cdot)$ *is a stationary distribution of* $q(\cdot)$.

Proposition 2.18 (Uniform posterior quantile)

For notation simplicity, we drop the subscript j *that indicates an element of a vector. Instead, we just treat* θ *as a scalar.*

Note that $q_t = I(\theta_t < \theta_0)$ *is a binary random variable, determined by the random variables* θ_0, Y *and* θ_t. *First, consider the asymptotic distribution conditional on* θ_0 *and* Y:

$P(q_t = 1 | \theta_0, Y) = F_{\theta|Y}(\theta_0)$,

where $F_{\theta|Y}(\theta_0)$ *denote the c.d.f. of the posterior distribution, evaluated at* θ_0.

$F_{\theta|Y}(\theta_0) = \int_{-\infty}^{\theta_0} \frac{p(\theta)p(Y|\theta)}{p(Y)} d\theta$.

It follows that $E[q_t | \theta_0, Y] = F_{\theta|Y}(\theta_0)$.

If the Markov chain is ergodic (irreducible, aperiodic and recurrent), the sample path average of MCMC draws converges to the mean (again, conditional on θ_0 *and* Y*):* $\frac{1}{T}\sum_{t=1}^{T} h(\theta_t) \xrightarrow{a.s.} E[h(\theta_t)]$, *where* $h(\theta_t) = I(\theta_t < \theta_0)$. *So we have* $Q \xrightarrow{a.s.} F_{\theta|Y}(\theta_0)$.

As $T \to \infty$, *we established an asymptotically deterministic mapping from* θ_0, Y *to* Q *such that* $Q = F_{\theta|Y}(\theta_0)$. *The randomness of* Q *inherits from the random variables* θ_0 *and* Y, *but not from* $\theta_1, \ldots, \theta_T$ *because of the law of large numbers. The distribution of* Q *is fully determined by the joint distribution of* θ_0 *and* Y.

By Bayes' rule, $p(\theta)p(Y|\theta) = p(Y)p(\theta|Y)$. *The data generating process can be interpreted as a draw of* Y *from* $p(Y)$, *and then a draw of* θ_0 *from* $p(\theta|Y)$. *Given such interpretation, we consider the distribution of* Q *conditional on* Y. *Let* $c \in [0,1]$ *be a point of evaluation of the c.d.f.,*

$P(Q < c|Y) = P(F_{\theta|Y}(\theta_0) < c|Y) = P\left(\theta_0 < F_{\theta|Y}^{-1}(c)|Y\right) = F_{\theta|Y}\left(F_{\theta|Y}^{-1}(c)\right) = c,$

where $F_{\theta|Y}^{-1}$ *is the inverse function of* $F_{\theta|Y}$ *(that is the reason of operating on a scalar element of the vector-valued* θ*). It indicates that* Q *follows the uniform distribution on the unit interval, conditional on* Y. *Note that the conditioning variable* Y *is absent in the conditional distribution. Therefore, the unconditional distribution is the same:* $P(Q < c) = c$. *In summary,* Q *follows the uniform distribution on the unit interval as* $T \to \infty$.

2.14 Exercises

Exercise 2.1 (Chapman-Kolmogorov equation) *Let* Q *be the one-step transition matrix. Show that* Q^t *is the* t*-step transition matrix that satisfies*

$$\pi^{(t)} = \pi^{(0)} Q^t.$$

Exercise 2.2 (Irreducibility and aperiodicity) *For a discrete-state Markov chain, irreducibility and aperiodicity are intrinsic properties of the transition matrix. Let* $\{\theta_t\}$ *and* $\{\tilde{\theta}_t\}$ *be couple chains with the same transition matrix* $Q = \begin{pmatrix} 0 & 1 \\ 1 & 0 \end{pmatrix}$, *but their initial distributions are different. Consider the two-dimensional Markov chain* $\{\theta_t, \tilde{\theta}_t\}$.

1. *Write down the transition matrix of the two-dimensional chain.*
2. *Show that $\{\theta_t\}$ has the periodicity of 2.*
3. *Show that both $\{\theta_t\}$ and $\left\{\tilde{\theta}_t\right\}$ are irreducible individually, but $\left\{\theta_t, \tilde{\theta}_t\right\}$ is reducible.*

Exercise 2.3 (Reducible chain) *Consider the target distribution $\pi(\theta) \propto e^{-\frac{1}{2}\theta^2}$, namely the standard normal density. Suppose that we adopt the independence MH sampler and propose a half-normal transition kernel:*

$$r(\theta^*; \theta) \propto e^{-\frac{1}{2}\theta^{*2}} I(\theta^* > 0).$$

Does the MH sampler work? why?

Exercise 2.4 (Autocorrelations) *Let $\{\theta_t\}$ be a stationary series with the mean $\mu = E(\theta_t), \forall t$. We consider the sample-path average $\bar{\theta}_T = \frac{1}{T}\sum_{t=1}^{T}\theta_t$. Let $\gamma_j = E[(\theta_t - \mu)(\theta_{t-j} - \mu)]$ be the jth order autocovariance, and $\rho_j = \frac{\gamma_j}{\gamma_0}$ be the autocorrelation. Because of stationarity, the covariance matrix of $\theta_1, \ldots, \theta_T$ takes the form*

$$\Gamma_T = \begin{pmatrix} \gamma_0 & \gamma_1 & \gamma_2 & \cdots & \gamma_{T-1} \\ \gamma_1 & \gamma_0 & \gamma_1 & \cdots & \gamma_{T-2} \\ \gamma_2 & \gamma_1 & \gamma_0 & \cdots & \gamma_{T-3} \\ \cdots & \cdots & \cdots & \ddots & \cdots \\ \gamma_{T-1} & \gamma_{T-2} & \gamma_{T-3} & \cdots & \gamma_0 \end{pmatrix}.$$

1. *For a given sample size T, show that the variance of $\bar{\theta}_T$ is given by*

$$Var\left(\bar{\theta}_T\right) = \frac{1}{T}\left[1 + 2\sum_{k=1}^{T}\left(1 - \frac{k}{T}\right)\rho_k\right]\gamma_0.$$

2. *Is it possible that $Var\left(\bar{\theta}_T\right)$ under correlated draws is smaller than that of the independent draws?*

Exercise 2.5 (MH acceptance probability) *In the standard MH sampler, a proposed transition from the point θ to θ^* is accepted with probability*

$$\alpha(\theta^*, \theta) = \min\left[1, \frac{\pi(\theta^*) r(\theta; \theta^*)}{\pi(\theta) r(\theta^*; \theta)}\right].$$

Consider an alternative acceptance probability

$$\tilde{\alpha}(\theta^*, \theta) = \frac{\pi(\theta^*) r(\theta; \theta^*)}{\pi(\theta) r(\theta^*; \theta) + \pi(\theta^*) r(\theta; \theta^*)}.$$

Show that the reversibility condition still holds, and thus $\tilde{\alpha}(\theta^*, \theta)$ can be used in lieu of $\alpha(\theta^*, \theta)$ by the MH sampler.

Exercise 2.6 (Hammersley-Clifford) *Proposition 2.6 demonstrates that the joint distribution is determined by conditional distributions, when variables are partitioned in two blocks A and B. Now we extend the results to the multiple-block case.*

1. Suppose that variables are partitioned as three blocks: A, B and C, with fully specified conditional distributions $p_{A|B,C}$, $p_{B|A,C}$, $p_{C|A,B}$. Show that p_{ABC} is determined by three conditional distributions. Hint: we could consider a two-step method: first, apply Proposition 2.6 conditional on C, and obtain $p_{A,B|C}$; second, combine A and B as one block, and apply Proposition 2.6 again to obtain p_{ABC}.

2. For the multiple-block case in general, show that the joint density can be obtained by a divide-and-conquer method, with recursive applications of Proposition 2.6.

Exercise 2.7 (Gibbs sampler of bivariate normal) *Instead of direct sampling from a bivariate normal distribution, we consider MCMC sampling with the target distribution*

$$p_{A,B}(\theta_A, \theta_B) = \phi\left[\begin{pmatrix} \theta_A \\ \theta_B \end{pmatrix}; \begin{pmatrix} \mu_A \\ \mu_B \end{pmatrix}, \begin{pmatrix} \Omega_{AA} & \Omega_{AB} \\ \Omega_{BA} & \Omega_{BB} \end{pmatrix}\right].$$

1. Show that the full posterior conditional distributions are

$$p_{A|B}(\theta_A; \theta_B) = \phi\left(\theta_A; \bar{\mu}_A, \bar{\Omega}_{AA}\right),$$

$$p_{B|A}(\theta_B; \theta_A) = \phi\left(\theta_B; \bar{\mu}_B, \bar{\Omega}_{BB}\right),$$

$$\bar{\mu}_A = \mu_A + \Omega_{AB}\Omega_{BB}^{-1}(\theta_B - \mu_B),$$

$$\bar{\mu}_B = \mu_B + \Omega_{BA}\Omega_{AA}^{-1}(\theta_A - \mu_A),$$

$$\bar{\Omega}_{AA} = \Omega_{AA} - \Omega_{AB}\Omega_{BB}^{-1}\Omega_{BA},$$

$$\bar{\Omega}_{BB} = \Omega_{BB} - \Omega_{BA}\Omega_{AA}^{-1}\Omega_{AB}.$$

2. Consider two partial transition kernels:

$$r_1(\theta_A^*, \theta_B^*; \theta_A, \theta_B) = p_{A|B}(\theta_A^*; \theta_B)\, I(\theta_B^* = \theta_B),$$

$$r_2(\theta_A^*, \theta_B^*; \theta_A, \theta_B) = p_{B|A}(\theta_B^*; \theta_A^*)\, I(\theta_A^* = \theta_A).$$

Show that both r_1 and r_2 satisfies the reversibility condition:

$$p_{A,B}(\theta_A, \theta_B)\, r_1(\theta_A^*, \theta_B^*; \theta_A, \theta_B) = p_{A,B}(\theta_A^*, \theta_B^*)\, r_1(\theta_A, \theta_B; \theta_A^*, \theta_B^*),$$

$$p_{A,B}(\theta_A, \theta_B)\, r_2(\theta_A^*, \theta_B^*; \theta_A, \theta_B) = p_{A,B}(\theta_A^*, \theta_B^*)\, r_2(\theta_A, \theta_B; \theta_A^*, \theta_B^*).$$

Consequently, $p_{A,B}$ is the stationary distribution of both r_1 and r_2.

3. Consider the transition kernel of the Gibbs sampler:

$$q(\theta_A^*, \theta_B^*; \theta_A, \theta_B) = p_{A|B}(\theta_A^*; \theta_B)\, p_{B|A}(\theta_B^*; \theta_A^*).$$

Assign some numerical values to the variables, say
$\theta_A = 1, \theta_B = 2, \theta_A^* = 3, \theta_B^* = 4.$
$\mu_A = 1, \mu_B = 2, \Omega_{AA} = 3, \Omega_{AB} = \Omega_{BA} = 4, \Omega_{BB} = 9.$

Show that reversibility condition does not hold:

$$p_{A,B}(\theta_A, \theta_B)\, q(\theta_A^*, \theta_B^*; \theta_A, \theta_B) \neq p_{A,B}(\theta_A^*, \theta_B^*)\, q(\theta_A, \theta_B; \theta_A^*, \theta_B^*).$$

4. Critically evaluate the statement: "the Gibbs sampler is a special case of the MH sampler".

5. Consider a scenario: $\mu_A = \mu_B = 0$, $\Omega_{AA} = \sigma_A^2$, $\Omega_{BB} = \sigma_B^2$, $\Omega_{AB} = \rho\sigma_A\sigma_B$. *Run the Gibbs sampler that cycles through* $p_{A|B}$ *and* $p_{B|A}$. *Compute the sample autocorrelations of* θ_A *draws. Try some different values of* $\sigma_A^2, \sigma_B^2, \rho$. *How do the sample autocorrelations change with* $\sigma_A^2, \sigma_B^2, \rho$*? Also, apply Proposition 2.8 and show that the theoretic autocorrelations are*

$$corr(\theta_{At}, \theta_{A,t-j}) = \rho^{2j}, j = 1, 2, \ldots$$

Exercise 2.8 (Multiple-transition MH sampler) *Consider a multiple-transition MH sampler whose transition kernel is the product of two sub-kernels, each of which is constructed by the MH method with the same stationary distribution. The algorithm of the multiple-transition MH sampler is described as follows:*

Input: target distribution $\pi(\cdot)$, *proposals* $r_1(\cdot), r_2(\cdot)$, *starting value* θ_0.

Output: MCMC draws $\theta_1, \theta_2, \ldots$.

Iterate the following steps for $t = 1, 2, \ldots$

1. Generate θ^* *from the proposal* $r_1(\theta^*; \theta)$, *where* $\theta = \theta_{t-1}$.

2. Compute acceptance probability $\alpha_1(\theta^*, \theta) = \min\left[1, \frac{\pi(\theta^*) r_1(\theta; \theta^*)}{\pi(\theta) r_1(\theta^*; \theta)}\right]$.

3. Reset $\theta^* = \begin{cases} \theta^* & \text{with probability } \alpha_1(\theta^*, \theta) \\ \theta_{t-1} & \text{with probability } 1 - \alpha_1(\theta^*, \theta) \end{cases}$.

4. Generate θ^{**} *from the proposal* $r_2(\theta^{**}; \theta^*)$.

5. Compute acceptance probability $\alpha_2(\theta^{**}, \theta^*) = \min\left[1, \frac{\pi(\theta^{**}) r_2(\theta^*; \theta^{**})}{\pi(\theta^*) r_2(\theta^{**}; \theta^*)}\right]$.

6. Set $\theta_t = \begin{cases} \theta^{**} & \text{with probability } \alpha_2(\theta^{**}, \theta^*) \\ \theta^* & \text{with probability } 1 - \alpha_2(\theta^{**}, \theta^*) \end{cases}$.

Show that $\pi(\cdot)$ *is a stationary distribution of the constructed Markov chain.*

Exercise 2.9 (Multiple-block MCMC sampler) *Let* $p_{A,B}$, p_A, p_B, $p_{A|B}$, $p_{B|A}$ *be the joint, marginal and conditional distributions. If* q_1 *and* q_2 *are partial transition kernels with the stationary distributions* $p_{A|B}$ *and* $p_{B|A}$, *respectively, then* $p_{A,B}$ *is a stationary distribution of the transition kernel* $q_1 q_2$. *To be specific, consider a proposed transition from* (θ_A, θ_B) *to* (θ_A^*, θ_B^*). *Let the transition kernel be*

$$q(\theta_A^*, \theta_B^*; \theta_A, \theta_B) = q_1(\theta_A^*; \theta_A, \theta_B)\, q_2(\theta_B^*; \theta_B, \theta_A^*),$$

with the stationarity conditions

$$p_{A|B}(\theta_A^*; \theta_B) = \int p_{A|B}(\theta_A; \theta_B)\, q_1(\theta_A^*; \theta_A, \theta_B)\, d\theta_A,$$

$$p_{B|A}(\theta_B^*; \theta_A^*) = \int p_{B|A}(\theta_B; \theta_A^*)\, q_2(\theta_B^*; \theta_B, \theta_A^*)\, d\theta_B,$$

show that $p_{A,B}$ *is a stationary distribution:*

$$p_{A,B}(\theta_A^*, \theta_B^*) = \int\int p_{A,B}(\theta_A, \theta_B)\, q(\theta_A^*, \theta_B^*; \theta_A, \theta_B)\, d\theta_A d\theta_B.$$

Exercise 2.10 (Reparametrized unbiased estimator) *Let $\hat{\pi}(\theta; Z)$ be a non-negative unbiased estimator of $\pi(\theta)$. For simplicity, assume that Z is a scalar-valued random variable with the density $p(\cdot)$ and the CDF $F(\cdot)$. By an unbiased estimator, we mean $\int p(Z)\hat{\pi}(\theta; Z)\,dZ = \pi(\theta)$. Let U be a standard normal variate with the density $\phi(\cdot)$ and CDF $\Phi(\cdot)$. Consider the reparametrized estimator $\hat{\pi}\left(\theta; F^{-1}(\Phi(U))\right)$. Show that it remains an unbiased estimator:*

$$\int \phi(U)\,\hat{\pi}\left(\theta; F^{-1}(\Phi(U))\right) dU = \pi(\theta).$$

Exercise 2.11 (Dividing the NIG prior) *Consider the divide-and-conquer method for Bayesian linear regressions with the conjugate prior $NIG(\mu, \Lambda, a, b)$. We divide the data into subsets and divide the prior as well. Suppose that the subposterior has the split prior density proportional to $[p(\theta)]^{1/J}$ Is it still a NIG density? Is it the same as $\frac{1}{J}NIG(\mu, \Lambda, a, b)$?*

Exercise 2.12 (Software validation statistics) *Consider the software validation statistics:*

$$Q = \frac{1}{T}\sum_{t=1}^{T} I(\theta_t < \theta_0),$$

where θ_0, Y and $\theta_1, \ldots, \theta_T$ are generated in the following order: generate a draw θ_0 from the prior $p(\theta)$, generate artificial data Y from the likelihood $p(Y|\theta)$, and generate $\theta_1, \ldots, \theta_T$ by an ergodic MCMC sampler targeting $p(\theta|Y)$. Note that Q takes the form of the sample average, which is reminiscent of the Law of Large Numbers and Central Limit Theorem. Why is Q uniformly distributed, rather than normally distributed asymptotically?

Exercise 2.13 (Cryptanalysis by MCMC) *Consider an application of the discrete-state MH sampler for cryptanalysis. Alice sent Bob a message*

```
the markov chain monte carlo (mcmc) simulation
has revolutionized bayesian computation.
```

in a ciphertext

```
uh&~(b! @y~chbj)~(@)u&~cb!!@~[(c(c]~tj(x!buj@)~
hbt~!&y@!xuj@)ju&&~abv&tjb)~c@(pxubuj@)#
```

without informing him of the encoding key:

```
From: abcdefghijklmnopqrstuvwxyz ()-,.
To  : bac&&fhhji !()@pq!tuxyzuvw~[]%##
```

Bob guessed that the coded symbols represent the English alphabet, of which the transition matrix can be obtained by counting the transition frequency of a large text file. He interpreted the decoding problem as Bayesian statistical inference in a parametric model. Let the ciphertext be the observations $Y_1, \ldots, Y_T$, and f_θ be a mapping from the code space to the English alphabet. The mapping is indexed by a discrete-valued parameter $\theta \in \{1, 2, \ldots, n^m\}$, where n is number of unique symbols in the ciphertext and m is the numbers of states (including English letters and punctuations). Conditional on a specific parameter value, the likelihood function is given by

$$p(Y|\theta) = \pi(f_\theta(Y_1))\prod_{t=2}^{T} Q(f_\theta(Y_{t-1}), f_\theta(Y_t)),$$

where the transition matrix $Q(i,j)$ indicates the transition probability from the state i to j, and $\pi(i)$ indicates the ith element of the stationary distribution of Q.

1. Bayesian model specification is completed by adding a prior $p(\theta)$. What do you think about the prior specification?

2. Is it feasible to compute the posterior $p(\theta|Y) \propto p(\theta)p(Y|\theta)$ in a closed form? Why?

3. Consider the MCMC sampler that targeting the posterior $p(\theta|Y) \propto p(\theta)p(Y|\theta)$. How do you specify the proposal?

4. Implement your MCMC sampler on a computer. Does your sampler successfully decode the ciphertext?

Exercise 2.14 (Parameter expansion for data augmentation) *Suppose that the posterior distribution $p(\theta|Y)$ is not analytically tractable, but by data augmentation of a latent variable Z, the posterior conditional distributions $p(\theta|Y,Z)$ and $p(Z|Y,\theta)$ are of recognizable forms. The Gibbs sampler with data augmentation cycles through the full posterior conditional distributions.*

Liu and Wu (1999) propose parameter expansion for data augmentation (PX-DA) by a low-dimensional (e.g., scalar) auxiliary parameter α, which reparametrizes the latent variable such that $Z = t_\alpha(W)$ or $W = t_\alpha^{-1}(Z)$, where t_α is a one-to-one continuous mapping indexed by α. For example, $W = Z+\alpha$ or $W = Z/\alpha$. Assume that α and θ are independent a priori: $p(\alpha|\theta) = p(\alpha)$. In the reparametrized model, the PX-DA sampler cycles through $p(\alpha, W|Y,\theta)$ and $p(\alpha,\theta|Y,W)$.

1. Let the complete-data likelihood be $p(Y,Z|\theta)$ in the original model. Show that in the reparametrized model, the complete-data likelihood (conditional on α and θ) takes the form:

$$p(Y,W|\theta,\alpha) = p(Y,t_\alpha(W)|\theta)\cdot\left|\frac{\partial t_\alpha(W)}{\partial W}\right|,$$

where $p(Y,t_\alpha(W)|\theta)$ refers to the density function $p(Y,Z|\theta)$ evaluated at $Z = t_\alpha(W)$.

2. Show that the expansion parameter α is identifiable only in the complete-data likelihood, but not identified by the observed data (hence over-parameterization):

$$\int p(Y,W|\theta,\alpha)\,dW = p(Y|\theta).$$

3. Show that the first stage of the PX-DA algorithm, namely sampling from

$$p(\alpha,W|Y,\theta) = p(\alpha|Y,\theta)\,p(W|Y,\theta,\alpha)$$

can be implemented by the following steps:

Step 1: draw α from the prior distribution $p(\alpha)$.

Step 2: draw Z from the posterior conditional distribution $p(Z|Y,\theta)$.

Step 3: reparametrize the latent variable $W = t_\alpha^{-1}(Z)$.

4. Show that the second stage of the PX-DA algorithm, namely sampling from

$$p(\alpha,\theta|Y,W) = p(\alpha|Y,W)\,p(\theta|Y,W,\alpha)$$

can be implemented by the following steps:

Step 4: draw α^* from the posterior marginal distribution $p(\alpha^*|Y,W)$, if possible.

Step 5: reparametrize the latent variable $Z^* = t_{\alpha^*}(W)$, where W is obtained in step 3.

Step 6: draw θ^* from the posterior conditional distribution $p(\theta^*|Y,Z^*)$.

How to handle Step 4?

5. It is possible to rearrange the order the above 6 steps to make the transition like

$$\theta \to Z \to Z^* \to \theta^*,$$

where the transition involves sampling from the familiar distributions $p(\theta\,|Y, Z)$, $p(Z\,|Y, \theta)$, $p(\alpha)$ *as well as a low-dimensional distribution* $p(\alpha\,|Y, W)$*?*

6. Suppose that Z *follows* $p(Z\,|Y)$*. Show that the transition* $Z \to Z^*$ *(defined by Steps 1, 3, 4, 5) makes* Z^* *follow* $p(Z\,|Y)$ *as well.*

7. Show that cycling through $p(\alpha, W\,|Y, \theta)$ *and* $p(\alpha, \theta\,|Y, W)$ *is equivalent to a marginalized Gibbs sampler that cycles through* $p(W\,|Y, \theta)$ *and* $p(\theta\,|Y, W)$*, with* α *integrated out. As the marginalized Gibbs sampler has the stationary distribution* $p(W, \theta\,|Y)$*, draws of* θ *are asymptotically samples from* $p(\theta\,|Y)$*. In short, the PX-DA algorithm targets the correct distribution.*

8. What is the appeal of the PX-DA algorithm?

3

Shrinkage and Variable Selection

A non-informative prior in Bayesian linear regressions allows the data to speak for itself. The predictor data X ideally would be orthogonal (mutually uncorrelated). However, social scientists often face the challenge of working with imperfect observations: 1) predictors with little variability, 2) highly correlated predictors, 3) combinations of predictors that nearly replicate others, and 4) a multitude of predictors that yield good in-sample fits but poor out-of-sample predictions.

This chapter highlights the importance of incorporating prior information when data alone are insufficient for estimating model parameters accurately. Shrinkage and variable selection methods can be interpreted in a Bayesian framework through specific prior distributions. Ridge regression (Hoerl and Kennard, 1970) is a form of conjugate linear regression under the NIG prior. The Least Absolute Shrinkage and Selection Operator (lasso), introduced by Tibshirani (1996), employs a double exponential prior. Stochastic Search Variable Selection (SSVS), developed by George and McCulloch (1993), uses a mixture normal prior with large and small variance components, while MCMC model composition (Madigan et al., 1995) incorporates degenerate components in the mixture prior for excluding a subset of variables from predictors.

Regularization methods penalize large coefficients, resulting in variance reduction and shrinkage toward zero. This variance reduction often outweighs any introduced bias, leading to a favorable trade-off with smaller mean squared errors for out-of-sample predictions. Variable selection methods simplify the model by setting certain coefficients to zero, effectively removing the associated predictors from the regression. Such parsimonious models mitigate overfitting risks, streamlines out-of-sample predictions, and improve model interpretability.

The precision matrix, denoted as $X'X$, serves as a crucial link between multicollinearity, shrinkage, and variable selection. Its eigenvalues and eigenvectors reveal multicollinearity, determine the degree of coefficient shrinkage, and identify variables that are most impacted by the selection process.

Most Bayesian shrinkage and variable selection methods do not have closed-form solutions, necessitating the use of MCMC methods such as the Gibbs and MH samplers. The NIG summation operator, introduced in Chapter 1, facilitates MCMC simulations in big data contexts. It allows for a single pass over the observations, after which the MCMC sampler iteratively combines NIG distributions, bypassing the need to revisit the original dataset.

3.1 Multicollinearity

Empirical researchers often ascribe insignificant regression results to multicollinearity, which vaguely refers to the data problem that a predictor is highly correlated with other predictors, or a group of predictors have a linear relationship. This section addresses the definition, detection, and consequences of multicollinearity.

DOI: 10.1201/9781003564027-3

3.1.1 Perfect and Inexact Multicollinearity

Multicollinearity is fundamentally an issue inherent in the data (rather than an issue of statistical model specification), defined with respect to the realized data matrix.

Definition 3.1 (Multicollinearity) *Let X be a $n \times d$ predictor matrix and v be a $d \times 1$ unit-length vector such that $\|v\| = \sqrt{v'v} = 1$. Let $\varepsilon > 0$ denote a tolerance level.*

1. *If $Xv = 0$, then X has perfect multicollinearity caused by v.*
2. *If $\|Xv\| < \varepsilon$, then X has inexact multicollinearity caused by v.*

According to Definition 3.1, multicollinearity is characterized by the linear combination coefficients v. It is conceivable for the data matrix to exhibit different types of multicollinearity if multiple selections of v render Xv close to zero.

A classic example of perfect multicollinearity is the dummy variable trap, where, for instance, twelve monthly dummy variables sum to one and replicate the intercept term. Another instance occurs in high-dimensional regression when $d > n$. Given that the rank of X is at most n, the linear equation $Xv = 0$ has a non-zero solution, thereby indicating perfect multicollinearity.

Inexact multicollinearity can arise with low-variance predictors whose data are nearly constant. When the intercept term is included in the regression, it results in inexact multicollinearity. Another example is time series autoregression, where predictors are lagged response variables. In the presence of strong autocorrelations, the regression may suffer from severe inexact multicollinearity.

3.1.2 Multicollinearity Detection

In large datasets with numerous variables, a practical question arises: can a machine detect multicollinearity and identify the linear combination coefficients using only the data matrix, without relying on non-sample information such as variable names and descriptions? Proposition 3.1 affirms this possibility: the eigenvalues and eigenvectors of the precision matrix $X'X$ reveal multicollinearity. For a real, symmetric, and positive semi-definite matrix, all eigenvalues must be non-negative. Zero eigenvalues indicate perfect multicollinearity, while small eigenvalues suggest inexact multicollinearity. The corresponding eigenvectors provide the linear combination coefficients.

Proposition 3.1 (Multicollinearity and precision matrix) *Let the eigenvalues of the precision matrix $X'X$ be $\lambda_1, \ldots, \lambda_d$, with the corresponding orthonormal eigenvectors $v_1, \ldots, v_d$. Let $\varepsilon > 0$ be a tolerance level.*

1. *Zero eigenvalue $\lambda_j = 0$ causes perfect multicollinearity: $Xv_j = 0$.*
2. *Tiny eigenvalue $\lambda_j < \varepsilon^2$ causes inexact multicollinearity: $\|Xv_j\| = \sqrt{\lambda_j}$.*

As demonstrated by Proposition 3.1, eigenvalues and eigenvectors are the key to the multicollinearity analysis. Since eigenvalues are scale-dependent, normalizing the data facilitates comparison. A numerically robust method for computing eigenvalues is through singular value decomposition (SVD): $X = USV'$, where the singular values of X are the square roots of the eigenvalues of $X'X$, and the orthonormal matrix V comprises the eigenvectors.

A practical workflow for detecting multicollinearity includes the following steps: 1) rescale variables if data are poorly scaled, 2) compute eigenvalues via SVD, 3) identify eigenvalues close to the machine-epsilon, which indicates perfect multicollinearity, 4) find eigenvalues below a user-specified tolerance level, which suggests inexact multicollinearity, and 5) take the corresponding eigenvectors as the linear combination coefficients that induce multicollinearity.

Multicollinearity is closely related to principal component analysis. For centered predictors, the sample covariance matrix is $\frac{1}{n}X'X$, which shares the same eigenvectors $v_1, \ldots, v_d$ as $X'X$, with eigenvalues rescaled by $\frac{1}{n}\lambda_1, \ldots, \frac{1}{n}\lambda_d$. Arrange eigenvalues in decreasing order. The first principal component Xv_1 has the largest sample variance $\frac{1}{n}\lambda_1$. The second principal component Xv_2 has the second largest sample variance $\frac{1}{n}\lambda_2$, and so forth. The principal components are orthogonal by construction. A principal component with no variability (i.e. $\lambda_j = 0$) indicates perfect multicollinearity along the direction of v_j. If a principal component has minimal variability, it suggests inexact multicollinearity in the direction indicated by the corresponding eigenvectors. For further details, refer to Exercise 3.1.

3.1.3 Consequences of Multicollinearity

The classical ordinary least squares (OLS) estimator can be replicated by Bayesian linear regressions under a non-informative prior. By Proposition 1.2, the posterior mean coincides with the OLS estimator, and the posterior standard deviation is analogous to the OLS standard error.

A consequence of perfect multicollinearity is that the OLS estimator is not uniquely determined, due to non-invertible $X'X$ (the rank of a real symmetric matrix equals the number of non-zero eigenvalues). Data alone cannot separate the marginal effects of predictors, as some predictors could be replicated by a linear combination of other predictors. Although there are infinitely many solutions of an indetermined linear system, two of them are most likely to be returned by statistical software. The first is the minimum-norm least squares solution by the pseudo inverse (e.g., `pinv(X)*Y`). SVD yields zero singular values of the form

$$X = \begin{pmatrix} U_1 & U_2 \end{pmatrix} \begin{pmatrix} S_1 & 0 \\ 0 & 0 \end{pmatrix} \begin{pmatrix} V_1' \\ V_2' \end{pmatrix} = U_1 S_1 V_1',$$

and the pseudo inverse is given by $V_1 S_1^{-1} U_1'$. The second is a sparse solution where the number of non-zero elements matches the rank of the precision matrix (e.g., `X\Y`).

A consequence of inexact multicollinearity is inflation of the estimator variance, which can be seen by diagonalization (i.e., decomposition of a real symmetric matrix by eigenvalues and eigenvectors):

$$X'X = VDV',$$

where the diagonal matrix $D = diag(\lambda_1, \ldots, \lambda_d)$ stacks eigenvalues, and the orthonormal matrix $V = (v_1, \ldots, v_d)$ piles eigenvectors. Diagonalization provides the matrix inversion in the form of eigenvalues and eigenvectors:

$$(X'X)^{-1} = VD^{-1}V' = \sum_{j=1}^{d} \lambda_j^{-1} v_j v_j'.$$

By Proposition 1.2, $Var(\beta | Y, \sigma^2) = \sigma^2 (X'X)^{-1}$, so the variance of the kth predictor coefficient is given by

$$Var(\beta_k | Y, \sigma^2) = \sigma^2 \sum_{j=1}^{d} \lambda_j^{-1} v_{jk}^2,$$

where v_{jk} is the kth element of v_j, A small λ_j inflates the variance, provided that $v_{jk} \neq 0$. The special case $v_{jk} = 0$ implies that the kth predictor is not involved in the multicollinearity caused by the linear combination Xv_j, and it is not affected by the variance inflation.

Regarding the effect of multicollinearity on out-of-sample prediction, consider future data generated by $Y_f = X_f\beta + \sigma\varepsilon_f$. The posterior predictive variance is given by

$$Var\left(Y_f \left|Y, \sigma^2\right.\right) = \sigma^2 \left[1 + X_f \left(X'X\right)^{-1} X_f'\right] = \sigma^2 \left[1 + \sum_{j=1}^{d} \lambda_j^{-1} \left(X_f v_j\right)\left(X_f v_j\right)'\right].$$

Proposition 3.1 shows that $\|Xv_j\| = \sqrt{\lambda_j}$, implying that a small λ_j is accompanied by a small Xv_j. If the multicollinearity pattern carries over to future data, a small $X_f v_j$ may offset a large λ_j^{-1}, and the predictive variance does not necessarily inflate.

3.2 Ridge Regression

Multicollinearity presents a significant challenge when observations lack sufficient information to precisely estimate model parameters. Bayesian methods address this issue by incorporating non-sample information through priors. These priors can originate from various sources: previous studies using different datasets, domain knowledge regarding plausible parameter values or ranges, and the principle of parsimony, which emphasizes maintaining model simplicity and penalizing complexity.

Ridge regression was introduced by Hoerl and Kennard (1970) as a biased estimator with smaller mean-squared error for non-orthogonal problems. From a Bayesian perspective, ridge regression is not a novel model but rather a specific instance of Bayesian linear regression under a conjugate prior: $NIG(0, \delta I, a, b)$, where δ serves as a regularization parameter that dictates the degree of shrinkage. The prior mean is set to zero, reflecting a concern about overfitting models with more parameters than the data can justify. In the absence of prior knowledge about which predictors might be redundant, the preference is for the null model unless the data strongly suggest including certain predictors. According to the pseudo data interpretation of the conjugate prior (Proposition 1.3), the ridge regression augments observations X, Y by the pseudo data $\tilde{X} = \sqrt{\delta}I$ and $\tilde{Y} = 0$.

3.2.1 Ridge Estimator

The ridge estimator is defined as the posterior mean $E(\beta|Y)$ under the conjugate prior $NIG(0, \delta I, a, b)$, given the regression data X, Y. By Proposition 1.1, the ridge estimator adds a small δ to the diagonal of the precision matrix $X'X$ such that

$$\hat{\beta}_\delta = \left(X'X + \delta I\right)^{-1} X'Y. \tag{3.1}$$

The ridge estimator is the also the maximum a posteriori (MAP) estimator, since $E(\beta|Y) = E\left(\beta\left|Y, \sigma^2\right.\right)$ and the latter is the multivariate normal mean, which coincides with the mode (i.e., the maximum of the posterior density function):

$$\hat{\beta}_\delta = \arg\max_\beta p\left(\beta\left|Y, \sigma^2\right.\right).$$

The ridge estimator can also be formulated as the solution of a penalized least squares optimization problem, as the log density of $p\left(\beta\left|Y, \sigma^2\right.\right)$ is proportional to $(Y - X\beta)'(Y - X\beta) + \delta\beta'\beta$, omitting some constants:

$$\hat{\beta}_\delta = \arg\min_\beta (Y - X\beta)'(Y - X\beta) + \delta\beta'\beta.$$

The OLS estimator is a special case of the ridge estimator under $\delta = 0$, if $X'X$ is invertible:

$$\hat{\beta}_0 = (X'X)^{-1} X'Y,$$

which is the minimizer of the sum of squared residuals $(Y - X\beta)' (Y - X\beta)$.

Instead of applying the same δ to all predictors, it is possible to assign a unique regularization parameter to each predictor by the prior $NIG(0, \Lambda, a, b)$, where $\Lambda = diag(\delta_1, \ldots, \delta_d)$. The extended ridge estimator takes the form $(X'X + \Lambda)^{-1} X'Y$, as the minimizer of $(Y - X\beta)' (Y - X\beta) + \sum_{j=1}^{d} \delta_j \beta_j^2$. This approach offers flexibility in handling heterogeneous predictors. If a predictor is rescaled, it is desirable to adjust the regularization parameter accordingly. Additionally, it is customary to include an intercept in the regression. If the intercept is not intended to be penalized, its regularization parameter can be set to zero. Alternatively, standardizing predictors and centering the response variable before applying ridge regression with a single regularization parameter can be an effective strategy.

3.2.2 Shrinkage

As a shrinkage method, the ridge estimator is a weighted average of the OLS estimator and the prior mean, which equals zero under the prior $NIG(0_{d\times 1}, \delta I, a, b)$:

$$\hat{\beta}_\delta = w\hat{\beta}_0 + (I - w)\, 0_{d\times 1},$$

where the weight is determined by the precision matrix and the regularization parameter:

$$w = (X'X + \delta I)^{-1} X'X.$$

As $\delta \to 0$, the ridge estimator converges to the OLS estimator. Conversely, as $\delta \to \infty$, the ridge estimator converges to zero. However, no coefficient is reduced to zero exactly, meaning no predictor is strictly excluded from the ridge regression.

Additional insights into the shrinkage effect can be gained by diagonalizing the positive semi-definite matrix:

$$X'X = VDV',$$

where $D = diag(\lambda_1, \ldots, \lambda_d)$ is a diagonal matrix consisting of non-negative eigenvalues, and $V = (v_1, \ldots, v_d)$ is an orthonormal matrix (i.e., $VV' = I$) that concatenates eigenvectors. A scalar-valued δ enables the decomposition

$$X'X + \delta I = V (D + \delta I) V',$$

indicating that $X'X$ and $X'X + \delta I$ share the same set of eigenvectors, while the eigenvalues increase from λ_j to $\lambda_j + \delta$, $j = 1, \ldots, d$. As all eigenvalues are positive, the matrix is always invertible. Consequently, the ridge estimator is uniquely determined even if X suffers from perfect multicollinearity caused by some $\lambda_j = 0$.

Diagonalization provides a linear connection between the ridge and OLS estimators: $\hat{\beta}_\delta = w\hat{\beta}_0$, where

$$w = V (D + \delta I)^{-1} DV' = \sum_{j=1}^{d} \frac{\lambda_j}{\lambda_j + \delta} v_j v_j'.$$

Small eigenvalues cause variance inflation in the OLS estimators, rendering them statistically insignificant. Ridge regression addresses this by applying greater shrinkage to these insignificant predictors. The ratio $\frac{\lambda_j}{\lambda_j + \delta}$ indicates that shrinkage is more pronounced in the direction of smaller eigenvalues.

To illustrate the heterogeneous effects of shrinkage, consider a special case where the predictors are orthogonal: $X'X = diag\left(\sum_{i=1}^n X_{i1}^2, \ldots, \sum_{i=1}^n X_{id}^2\right)$. The ridge estimator for the kth predictor is given by $\left(\sum_{i=1}^n X_{ik}^2 + \delta\right)^{-1} \sum_{i=1}^n X_{ik}Y_i$, where $\sum_{i=1}^n X_{ik}^2$ reflects the predictor precision. If the precision is low, the OLS estimator is likely statistically insignificant, and the ridge estimator can be substantially smaller than the OLS estimator.

Shrinkage is not merely a technical artifact of the NIG prior; it is a feature of Bayesian linear regression under a broad class of priors whose densities peak at zero, reflecting a preference for the null model by the principle of parsimony. Consider the MAP estimator (as the posterior mode):

$$\begin{aligned}
&\arg\max_\beta p\left(\beta \,|Y, \sigma^2\right) \\
&= \arg\max_\beta \ln p(\beta) - \tfrac{1}{2}\sigma^{-2}(Y - X\beta)'(Y - X\beta) \\
&= \arg\max_\beta \ln p(\beta) - \tfrac{1}{2}\sigma^{-2}\left(\beta - \hat{\beta}_0\right)' X'X\left(\beta - \hat{\beta}_0\right).
\end{aligned}$$

The prior log density is maximized at zero, while the log-likelihood term is maximized at $\hat{\beta}_0$. Therefore, the MAP estimator is a compromise (weighted average) between the OLS estimator and zero, with the weight determined by the precision matrix $X'X$.

3.2.3 Bias and Variance Tradeoff

From a frequentist perspective, we compare the ridge and OLS estimators. The OLS estimator is unbiased, consistent, and efficient, but these attributes do not ensure it outperforms the ridge estimator in out-of-sample prediction. Ultimately, the value of a statistical method is determined by its predictive accuracy.

Assume that a true parameter β_0 generates the data. The mean and variance of the ridge estimator are given by

$$E\left(\hat{\beta}_\delta\right) = (X'X + \delta I)^{-1} X'X\beta_0,$$

$$Var\left(\hat{\beta}_\delta\right) = \sigma^2 (X'X + \delta I)^{-1} X'X (X'X + \delta I)^{-1}.$$

The ridge estimator exemplifies the bias-variance tradeoff. On the one hand, a positive regularization parameter introduces bias:

$$E\left(\hat{\beta}_\delta\right) - \beta_0 = -\delta (X'X + \delta I)^{-1} \beta_0.$$

The bias is avoided if $\delta = 0$ and $n > d$ (though unavoidable if $d > n$). On the other hand, the variance of the ridge estimator is smaller than that of the OLS estimator if $n > d$. The eigenvalues and eigenvectors of the precision matrix $X'X$ quantify variance reduction:

$$Var\left(\hat{\beta}_\delta\right) = \sigma^2 \sum_{j=1}^d \frac{\lambda_j}{(\lambda_j + \delta)^2} v_j v_j'.$$

For an arbitrary $1 \times d$ vector α, we have

$$\alpha Var\left(\hat{\beta}_0\right)\alpha' = \sigma^2 \sum_{j=1}^d \lambda_j^{-1} z_j^2 \geq \sigma^2 \sum_{j=1}^d \frac{\lambda_j}{(\lambda_j + \delta)^2} z_j^2 = \alpha Var\left(\hat{\beta}_\delta\right)\alpha',$$

where $z_j = \alpha v_j$. It follows that $Var\left(\hat{\beta}_\delta\right) \leq Var\left(\hat{\beta}_0\right)$.

The bias-variance tradeoff is captured by the mean-squared error (MSE), defined as $MSE\left(\hat{\beta}_\delta\right) = E\left[\left(\hat{\beta}_\delta - \beta_0\right)\left(\hat{\beta}_\delta - \beta_0\right)'\right]$. MSE equals the sum of squared bias and variance:

$$MSE\left(\hat{\beta}_\delta\right) = Var\left(\hat{\beta}_\delta\right) + \left[E\left(\hat{\beta}_\delta\right) - \beta_0\right]\left[E\left(\hat{\beta}_\delta\right) - \beta_0\right]'.$$

The estimator bias-variance tradeoff extends to out-of-sample prediction. A scalar Y_f is predicted by $X_f\hat{\beta}_\delta$, and the prediction MSE is defined as $MSE\left(X_f\hat{\beta}_\delta\right) = E\left[\left(Y_f - X_f\hat{\beta}_\delta\right)^2\right]$. We have

$$MSE\left(X_f\hat{\beta}_\delta\right) = \sigma^2 + \sigma^2 X_f\left(X'X + \delta I\right)^{-1}\left(X'X + \delta^2 SNR\right)\left(X'X + \delta I\right)^{-1} X_f',$$

where $SNR = \beta_0\beta_0'/\sigma^2$ denotes the signal-to-noise ratio.

Whether ridge regression outperforms standard linear regression in out-of-sample prediction depends on whether $MSE\left(X_f\hat{\beta}_\delta\right)$ is minimized at $\delta = 0$. In cases of extremely noisy data such that $SNR \to 0$, regularization is most effective, reducing the prediction MSE as δ grows, making a null model preferable. In more realistic scenarios, the prediction MSE likely forms a U-shaped function of δ. As $\delta \to 0$, the MSE tends to $\sigma^2 + \sigma^2 X_f\left(X'X\right)^{-1} X_f'$, namely the MSE of the OLS regression. As $\delta \to \infty$, the MSE tends to $\sigma^2 + X_f\beta_0\beta_0' X_f'$, namely the MSE of a null model. Moreover, it is possible to find a sufficiently small δ such that $I - \delta \cdot SNR$ is positive definite, so that $X'X + \delta^2 SNR$ is smaller than $X'X + \delta^2 I$, and the ridge estimator outperforms the OLS. The intuition is that a sufficiently small regularization parameter induces little bias but reduces variance, resulting in the decrease of the prediction MSE.

3.2.4 High-dimensional Regressions

Big data can refer to scenarios characterized by a large number of observations n or a large number of predictors d. In the case of tall data where $n > d$, ridge regression is employed to address multicollinearity and improve out-of-sample prediction. When the sample size is enormous, big data ridge regression can be efficiently implemented using the NIG summation operator.

In some fields, researchers encounter high-dimensional regression problems where $d > n$. This over-parameterized regime transforms regression into an interpolation problem, as the model perfectly fits all data points. The traditional motivation for ridge regression in this context is to prevent overfitting, i.e., fitting the noise.

Although Equation (3.1) remains valid for high-dimensional ridge regressions, inverting a $d \times d$ matrix can be computationally inefficient or even infeasible. An alternative approach defines the ridge estimator as the posterior mean of β in a Bayesian framework. The Gaussian model $Y = X\beta + \sigma\varepsilon$ and the prior $\left(\beta, \sigma^2\right) \sim NIG\left(0, \delta I, a, b\right)$ indicate that the joint distribution of β and Y is multivariate normal, conditional on σ^2:

$$\begin{pmatrix} \beta \\ Y \end{pmatrix} \middle| \sigma^2 \sim N\left[\begin{pmatrix} 0 \\ 0 \end{pmatrix}, \sigma^2\delta^{-1}\begin{pmatrix} I & X' \\ X & XX' + \delta I \end{pmatrix}\right].$$

The posterior conditional mean is

$$E\left(\beta \left| Y, \sigma^2\right.\right) = X'\left(XX' + \delta I\right)^{-1} Y.$$

Since the conditional mean does not depend on σ^2, we have $E\left(\beta \left| Y\right.\right) = E\left(\beta \left| Y, \sigma^2\right.\right)$, which provides an alternative formula of the ridge estimator

$$\hat{\beta}_\delta = X'\left(XX' + \delta I\right)^{-1} Y. \tag{3.2}$$

Equations (3.1) and (3.2) are algebraically identical (as shown in Exercise 3.4), but they differ in computational complexity. Suppose that inverting a $n \times n$ matrix requires $2n^3$ floating-point operators (flops), and the matrix multiplication of $n \times d$ and $d \times m$ matrices requires $2ndm$ flops. The computing cost of (3.1) is approximately $2d^2n + 2d^3$ flops, while that of (3.2) is $2dn^2 + 2n^3$. If $n \gg d$, the former is more computationally efficient, as complexities of two methods are $O(n)$ and $O(n^3)$ respectively. If $d \gg n$, the latter is more computationally efficient, as complexities are $O(d^3)$ and $O(d)$ respectively.

Equation (3.2) indicates the limiting behavior of the ridge estimator in high-dimensional regressions. When $d > n$, Equation (3.1) is not well-defined at $\delta = 0$ due to the non-invertibility of $X'X$, but Equation (3.2) remains applicable:

$$\lim_{\delta \to 0} X'(XX' + \delta I)^{-1} Y = X'(XX')^{-1} Y.$$

This corresponds to the minimum-norm OLS estimator, equivalent to the SVD-based pseudo-inverse of X times Y. Essentially, the ridge estimator approaches the minimum-norm OLS estimator as the regularization parameter tends to zero.

Conventional wisdom holds that regularization is crucial for high-dimensional regressions. However, insights from deep learning, where numerous parameters are trained with limited data, challenge this notion and urge researchers to rethink generalization (Zhang et al., 2021). Surprises in high-dimensional ridge-less regressions are documented (Hastie et al., 2022). Bartlett et al. (2020) argue that over-parameterization is vital for benign over-fitting, with the minimum-norm interpolation offering near-optimal prediction accuracy. Kobak et al. (2020) show that the optimal regularization parameter for high-dimensional ridge regressions can be negative, suggesting that the ridge estimator is asymptotically equivalent to the minimum-norm OLS estimator in an augmented regression with randomly generated low-variance artificial predictors.

Specifically, consider the linear regression $Y = X\beta + \sigma\varepsilon$. We simulate q artificial predictors X_q by randomly sampling from $N(0, q^{-1}\delta)$ and compute the minimum-norm OLS estimator with the concatenated predictors $\tilde{X} = \begin{pmatrix} X & X_q \end{pmatrix}$. This yields $\tilde{X}'\left(\tilde{X}\tilde{X}'\right)^{-1} Y$, where the first d elements are given by $X'(XX' + X_qX_q')^{-1} Y$. As $q \to \infty$, we have $X_qX_q' \to \delta I$, so the minimum-norm OLS estimator converges to the ridge estimator $X'(XX')^{-1} Y$. This implies that a large pool of low-variance predictors inherently provides shrinkage, while the response variable is predicted by high-variance directions, embodying the concept of implicit ridge regularization.

3.3 Lasso Regression

Shrinkage under ridge regression often proves inadequate for two primary reasons. First, while coefficients are reduced, none are exactly zero, meaning no predictors are strictly excluded from the model. Second, the necessity to specify an ad hoc shrinkage parameter can complicate the implementation of ridge regression.

The lasso regression, introduced by Tibshirani (1996), is a widely used technique for regularization and variable selection. It addresses the limitations of ridge regression through adaptive shrinkage, allowing the data to determine an appropriate level of shrinkage for each variable. In some cases, the shrinkage can be substantial enough to reduce coefficients to zero, thereby achieving sparsity and enabling automatic variable selection. The Bayesian lasso was developed by Park and Casella (2008) and Hans (2009), and extended to the elastic net (a convex combination of ridge and lasso penalties) by Li and Lin (2010). Lykou and Ntzoufras (2013) incorporated binary variable indicators into lasso-type shrinkage. Hierarchical shrinkage can also be used in time-varying-parameter models to achieve parsimony (Belmonte et al., 2014).

3.3.1 Prior and Posterior

From a Bayesian perspective, shrinkage results from an informative prior centered at the origin. Ridge regression uses the prior $NIG(0, \delta I, a, b)$, fixing the shrinkage parameter δ for all variables. The concept of adaptive shrinkage involves randomizing and customizing regularization for each predictor, allowing the data to determine the extent of shrinkage through a hierarchical prior structure: $NIG(0, \Lambda(\psi), a, b)$, where latent variables have the prior distribution $p(\psi)$. The posterior distribution $p(\psi | Y)$ reflects the estimated strength of shrinkage and the associated uncertainty.

Bayesian lasso corresponds to a double exponential prior, which can be represented as a normal-exponential scale mixture, wherein latent variables are assigned an exponential prior. This specification is advantageous because: 1) the posterior conditional distributions of latent variables are tractable, 2) marginalization over latent variables results in a recognizable prior on the predictor coefficients, 3) the posterior mode corresponds to the solution of a penalized least squares problem under 1-norm regularization, with a tractable optimizer available along each coordinate, and 4) some variables have their posterior mode at zero, promoting sparsity which enhances model interpretability.

We now delve into the technical aspects of the scale mixture representation of the double exponential prior distribution. Readers may choose to bypass some details without losing the overarching understanding of lasso.

Definition 3.2 (Double exponential distribution) *If a random variable x has the density function*

$$f(x; \mu, \sigma) = \frac{1}{2\sigma} e^{-\frac{|x-\mu|}{\sigma}},$$

it follows a double exponential distribution (also known as Laplace distribution) with the location parameter μ and the scale parameter σ.

Proposition 3.2 (Normal-exponential scale mixture) *Let ψ be a latent variable following an exponential distribution: $p(\psi) = \frac{1}{2} e^{-\frac{\psi}{2}}$. Conditional on ψ, the random variable x follows a normal distribution with mean μ and variance $\psi\sigma^2$:*

$$p(x | \psi) = (2\pi)^{-\frac{1}{2}} \psi^{-\frac{1}{2}} \sigma^{-1} e^{-\frac{(x-\mu)^2}{2\psi\sigma^2}}.$$

Marginalization by $p(x) = \int p(\psi) p(x | \psi) d\psi$ yields a double exponential distribution with parameters μ and σ.

Proposition 3.3 (Lasso prior) *Consider two specifications of hierarchical priors for the lasso regression $Y = X\beta + \sigma\varepsilon$:*

1. *Exponential-NIG specification:*

$$p(\beta, \sigma^2, \psi) = p(\psi) p(\beta, \sigma^2 | \psi),$$

$$p(\psi) \propto \prod_{j=1}^{d} e^{-\frac{1}{2}\delta^2 \psi_j},$$

$$p\left(\beta,\sigma^2\left|\psi\right.\right)\propto\left(\sigma^2\right)^{-\left(a+\frac{d}{2}+1\right)}e^{-\sigma^{-2}\left[b+\frac{1}{2}\beta'\Lambda(\psi)\beta\right]},$$

where $\Lambda(\psi)=diag\left(\psi_1^{-1},\ldots,\psi_d^{-1}\right)$*. Then, marginalization over* ψ *yields a lasso prior:*

$$p\left(\beta,\sigma^2\right)\propto\left(\sigma^2\right)^{-\left(a+\frac{d}{2}+1\right)}e^{-b\sigma^{-2}-\delta\sigma^{-1}\sum_{j=1}^{d}|\beta_j|}.$$

2. *Inverse-gamma-exponential-normal specification:*

$$p\left(\beta,\sigma^2,\psi\right)=p\left(\sigma^2\right)p\left(\psi\left|\sigma^2\right.\right)p\left(\beta\left|\sigma^2,\psi\right.\right),$$

$$p\left(\sigma^2\right)\propto\left(\sigma^2\right)^{-(a+1)}e^{-b\sigma^{-2}},$$

$$p\left(\psi\left|\sigma^2\right.\right)\propto\prod_{j=1}^{d}e^{-\frac{1}{2}\delta^2\sigma^{-2}\psi_j},$$

$$p\left(\beta\left|\sigma^2,\psi\right.\right)\propto\left(\sigma^2\right)^{-\frac{d}{2}}e^{-\frac{1}{2}\sigma^{-2}\beta'\Lambda(\psi)\beta}.$$

Then, marginalization over ψ *yields a lasso prior:*

$$p\left(\beta,\sigma^2\right)\propto\left(\sigma^2\right)^{-(a+d+1)}e^{-b\sigma^{-2}-\delta\sigma^{-2}\sum_{j=1}^{d}|\beta_j|}.$$

The likelihood function of lasso is the same as that of the Gaussian linear regression:

$$p\left(Y\left|\beta,\sigma^2\right.\right)\propto\left(\sigma^2\right)^{-\frac{n}{2}}e^{-\frac{1}{2}\sigma^{-2}(Y-X\beta)'(Y-X\beta)}.$$

The posterior density is proportional to the prior times the likelihood. Proposition 3.3 provides two prior specifications. Under the first specification, the posterior takes the form

$$p\left(\beta,\sigma^2\left|Y\right.\right)\propto\left(\sigma^2\right)^{-\left(a+\frac{n+d}{2}+1\right)}e^{-b\sigma^{-2}-\frac{1}{2}\sigma^{-2}(Y-X\beta)'(Y-X\beta)-\delta\sigma^{-1}\sum_{j=1}^{d}|\beta_j|}.$$

Under the second specification, we have

$$p\left(\beta,\sigma^2\left|Y\right.\right)\propto\left(\sigma^2\right)^{-\left(a+\frac{n}{2}+d+1\right)}e^{-b\sigma^{-2}-\frac{1}{2}\sigma^{-2}(Y-X\beta)'(Y-X\beta)-\delta\sigma^{-2}\sum_{j=1}^{d}|\beta_j|},$$

which indicates that the MAP estimator (i.e., the posterior mode) is obtain by a penalized least squares problem under the 1-norm penalty.

$$\min_{\beta}\frac{1}{2}\left(Y-X\beta\right)'\left(Y-X\beta\right)+\delta\sum_{j=1}^{d}|\beta_j|.$$

As for the first specification, the MAP estimator is obtained by the same optimization problem, conditional on $\sigma^2=1$. Park and Casella (2008) note that the appeal of the first specification is that the posterior density is concave and unimodal.

3.3.2 Classical Estimation

The lasso optimization problem does not have a closed-form solution when optimizing all variables simultaneously. However, an explicit solution can be derived by optimizing one variable at a time, conditional on the remaining variables. Due to the 1-norm penalty, it is likely that some optimized variables result in a corner solution of being zero. Thinking like an economist, we can derive both internal and corner solutions for the lasso estimator using the well-known marginal analysis in economics.

Algorithm 3.1 Coordinate descent of lasso

Input: data X, Y, and regularization parameter δ.
Output: MAP estimator.
1 Prepare sufficient statistics $X'X$ and $X'Y$.
2 Cycle through univariate optimizations for $j = 1, \ldots, d$:
  2.1 Compute OLS estimator $b_j = \left(X_j'X_j\right)^{-1} X_j'Y_{-j}$.
  2.2 Compute analytic optimizer $\beta_j = sign\left(b_j\right)\left(\left|b_j\right| - \left(X_j'X_j\right)^{-1}\delta\right)^+$.
3 Repeat step 2 until convergence criteria are met.

Let X_j and β_j be the jth predictor data (as a column vector) and the corresponding coefficient. Consider optimization with respect to β_j, while holding the remaining parameters fixed. The objective function to be minimized is

$$\frac{1}{2}\left(Y_{-j} - X_j\beta_j\right)'\left(Y_{-j} - X_j\beta_j\right) + \delta\left|\beta_j\right| + \delta\sum_{i\neq j}\left|\beta_i\right|,$$

where $Y_{-j} = Y - \sum_{i\neq j} X_i\beta_i$. The first term $\left(Y_{-j} - X_j\beta_j\right)'\left(Y_{-j} - X_j\beta_j\right)$ is the sum of squared residuals (SSR), and second term $\delta\left|\beta_j\right|$ corresponds to the lasso penalty.

If $\delta = 0$, the OLS estimator $b_j = \left(X_j'X_j\right)^{-1} X_j'Y_{-j}$ minimizes SSR. If $\delta > 0$, the penalized optimal solution departs from the OLS estimator. We examine two scenarios: $b_j \geq 0$ and $b_j < 0$.

Consider first the scenario where $b_j \geq 0$. The optimal solution must be non-negative; otherwise, a marginal increase in the variable would reduce both the SSR and the penalty. Additionally, the optimal solution must not exceed the OLS estimator; otherwise, a marginal decrease in the variable would reduce both the SSR and the penalty. Within the interval $\beta_j \in [0, b_j]$, the marginal change in SSR increases as β_j decreases, while the marginal penalty remains constant. The interior solution is an equilibrium where the two marginal effects are equal: $X_j'\left(Y_{-j} - X_j\beta_j\right) = \delta$, resulting in $\beta_j = b_j - \left(X_j'X_j\right)^{-1}\delta$. However, a corner solution arises when β_j drops to zero, yet the marginal change in SSR remains lower than the marginal penalty. Consequently, the optimal solution is

$$\beta_j = \max\left[0, b_j - \left(X_j'X_j\right)^{-1}\delta\right].$$

Similarly, the optimal solution in the scenario where $b_j < 0$ is given by

$$\beta_j = \min\left[0, b_j + \left(X_j'X_j\right)^{-1}\delta\right].$$

Combining two cases yields the analytic solution of the univariate optimization problem in a compact form:

$$\beta_j = sign\left(b_j\right)\left(\left|b_j\right| - \left(X_j'X_j\right)^{-1}\delta\right)^+,$$

where $sign(\cdot)$ is the sign function and $(x)^+ = \max(x, 0)$ is the censoring function.

Various methods are available for computing the lasso estimator, including sub-gradient methods (due to non-differentiability of the lasso objective function), least-angle regression (Efron et al., 2004) and coordinate descent (Friedman et al., 2010). Coordinate descent is a multivariate optimization method that iteratively cycles through univariate optimizations for each parameter, holding the remaining parameters fixed at their current values. Algorithm 3.1 outlines the coordinate descent procedure for lasso estimation.

3.3.3 Bayesian Estimation

The appeal of Bayesian analysis lies in its ability to provide uncertainty estimation alongside point estimation. Although the posterior distribution for lasso is not in a readily recognizable form, the posterior conditional distributions are tractable, as detailed in Proposition 3.4. By data augmentation, the Gibbs sampler cycles through the posterior conditional distributions for both model parameters and latent variables.

Proposition 3.4 (Posterior conditionals of lasso) *Consider the lasso regression* $Y = X\beta + \sigma\varepsilon$ *with two prior specifications in Proposition 3.3. Let* $NIG\left(\tilde{\mu}, \tilde{\Lambda}, \tilde{a}, \tilde{b}\right)$ *be the posterior distribution under the non-informative prior (Proposition 1.2). Let* $\gamma_j = \psi_j^{-1}$ *be the reparametrized latent variable. The posterior conditional distributions* $p\left(\beta, \sigma^2 \left| Y, \psi\right.\right)$ *and* $p\left(\psi \left| Y, \beta, \sigma^2\right.\right)$ *are tractable.*

1. *Exponential-NIG specification:*
$p\left(\beta, \sigma^2 \left| Y, \psi\right.\right)$ *follows* $NIG\left(\bar{\mu}, \bar{\Lambda}, \bar{a}, \bar{b}\right)$, *where*

$$NIG\left(\bar{\mu}, \bar{\Lambda}, \bar{a}, \bar{b}\right) = NIG\left(0, \Lambda\left(\psi\right), a, b\right) + NIG\left(\tilde{\mu}, \tilde{\Lambda}, \tilde{a}, \tilde{b}\right).$$

$p\left(\psi \left| Y, \beta, \sigma^2\right.\right) = \prod_{j=1}^{d} p\left(\psi_j \left| \beta_j, \sigma^2\right.\right)$, *where the reparametrized density is*

$$p\left(\gamma_j \left| \beta_j, \sigma^2\right.\right) \propto \gamma_j^{-\frac{3}{2}} \exp\left(-\frac{\beta_j^2}{2\sigma^2}\gamma_j - \frac{1}{2}\delta^2\gamma_j^{-1}\right).$$

which is the inverse Gaussian distribution with parameters $\left|\frac{\delta\sigma}{\beta_j}\right|$ *and* δ^2.

2. *Inverse-gamma-exponential-normal specification:*
$p\left(\beta, \sigma^2 \left| Y, \psi\right.\right)$ *follows* $NIG\left(\bar{\mu}, \bar{\Lambda}, \bar{a}, \bar{b}\right)$, *where*

$$NIG\left(\bar{\mu}, \bar{\Lambda}, \bar{a}, \bar{b}\right) = NIG\left(0, \Lambda\left(\psi\right), a + d, b + \frac{1}{2}\delta^2 \sum_{j=1}^{d} \psi_j\right) + NIG\left(\tilde{\mu}, \tilde{\Lambda}, \tilde{a}, \tilde{b}\right).$$

$p\left(\psi \left| Y, \beta, \sigma^2\right.\right) = \prod_{j=1}^{d} p\left(\psi_j \left| \beta_j, \sigma^2\right.\right)$, *where the reparametrized density is*

$$p\left(\gamma_j \left| \beta_j, \sigma^2\right.\right) \propto \gamma_j^{-\frac{3}{2}} \exp\left(-\frac{\beta_j^2}{2\sigma^2}\gamma_j - \frac{1}{2}\delta^2\sigma^{-2}\gamma_j^{-1}\right).$$

which is the inverse Gaussian distribution with parameters $\left|\frac{\delta}{\beta_j}\right|$ *and* $\delta^2\sigma^{-2}$.

Draws produced by the Gibbs sampler are correlated. It is possible to reduce correlations by using a marginalized MH sampler, with the model parameters integrated out. The posterior density can be decomposed as

$$p\left(\beta, \sigma^2, \psi \left| Y\right.\right) = p\left(\psi \left| Y\right.\right) p\left(\beta, \sigma^2 \left| Y, \psi\right.\right).$$

Proposition 3.4 shows that $p\left(\beta, \sigma^2 \left| Y, \psi\right.\right)$ is tractable, allowing the marginalized MCMC sampler to focus on latent variables with the target distribution

$$p\left(\psi \left| Y\right.\right) \propto p\left(\psi\right) p\left(Y \left| \psi\right.\right) = p\left(\psi\right) \frac{p\left(\beta, \sigma^2 \left| \psi\right.\right) p\left(Y \left| \beta, \sigma^2, \psi\right.\right)}{p\left(\beta, \sigma^2 \left| Y, \psi\right.\right)}.$$

The density $p(\psi|Y)$ can be evaluated via the following components: 1) $p(\psi)$ is the density of an exponential distribution, 2) $p(\beta, \sigma^2 |\psi)$ is the density of $NIG(0, \Lambda(\psi), a, b)$, and 3) $p(\beta, \sigma^2 |Y, \psi)$ is the density of $NIG(\bar{\mu}, \bar{\Lambda}, \bar{a}, \bar{b})$ in Proposition 3.4. The parameters can be evaluated at an arbitrary value, such as $\beta = 0$ and $\sigma^2 = 1$. It is unnecessary to evaluate the likelihood $p(Y|\beta, \sigma^2, \psi)$, since it does not depend on ψ.

Consider an independence MH sampler with the proposal

$$r(\psi^*; \psi) \propto \prod_{j=1}^{d} e^{-\frac{1}{2}\delta^2 \psi_j^*}.$$

The proposal aligns with the prior $p(\psi)$, simplifying the MH acceptance probability

$$\alpha(\psi^*; \psi) = \min\left[1, \frac{p(\beta, \sigma^2 |\psi^*)\, p(\beta, \sigma^2 |Y, \psi)}{p(\beta, \sigma^2 |\psi)\, p(\beta, \sigma^2 |Y, \psi^*)}\right].$$

The marginalized MH sampler updates latent variables without dependency on the model parameters, which are fully adapted to the latent variables and $p(\beta, \sigma^2 |Y, \psi)$ can be sampled from NIG distributions.

The marginalized MH sampler overshadows the Gibbs sampler in terms of generalization to adaptive shrinkage beyond lasso. The key advantage is that marginalization does not require $p(\psi|Y, \beta, \sigma^2)$ to be a recognizable distribution. The marginalized sampler merely evaluates the density $p(\psi|Y)$ pointwise up to a normalizing constant, making it applicable to any hierarchical prior specification $NIG(0, \Lambda(\psi), a, b)$, where the latent variable ψ does not necessarily follow an exponential distribution.

3.3.4 Big Data Bayesian Lasso

When the number of observations is large, the Gibbs sampler can become computationally intensive, as each iteration involves sweeping over the entire dataset. Fortunately, Bayesian lasso differs from standard linear regression only in its prior specification, with the NIG summation operator separating the prior from the classical estimator (as shown in the tilde-form NIG in Proposition 1.5). This separation allows for the implementation of a big data Bayesian lasso with a single pass of the data using the NIG summation operator. The data is scanned once for the tilde-form NIG, after which the Gibbs sampler iteratively combines NIG distributions under different parameters without the need to reload the data, as illustrated in Algorithm 3.2.

Regarding big data algorithms for numerical optimization of the MAP estimator and the marginalized MH sampler, a simple approach is to replace the original data with pseudo data extracted from $NIG(\tilde{\mu}_j, \tilde{\Lambda}_j, \tilde{a}_j, \tilde{b}_j)$ by Proposition 1.3. The original data is scanned once and then substituted with pseudo data, allowing existing algorithms designed for smaller datasets to be applied with minimal adaptation.

3.4 Stochastic Search Variable Selection

Stochastic Search Variable Selection (SSVS), introduced by George and McCulloch (1993); George and Mcculloch (1997), is a widely used Bayesian shrinkage and variable selection method characterized by a hierarchical mixture prior incorporating both small and large

Algorithm 3.2 Big data Bayesian lasso

Input: data X, Y, and prior $p\left(\beta, \sigma^2, \psi\right)$ in Proposition 3.3.
Output: Posterior samples from $p\left(\beta, \sigma^2, \psi \,|Y\right)$.
1 Partition big data into J subsets X_j, Y_j, $j = 1, \ldots, J$.
2 Run subset regressions separately for $NIG\left(\tilde{\mu}_j, \tilde{\Lambda}_j, \tilde{a}_j, \tilde{b}_j\right)$.
3 Sum up subset posterior distributions

$$NIG\left(\tilde{\mu}, \tilde{\Lambda}, \tilde{a}, \tilde{b}\right) = \sum_{j=1}^{J} NIG\left(\tilde{\mu}_j, \tilde{\Lambda}_j, \tilde{a}_j, \tilde{b}_j\right).$$

4 Cycle through posterior conditional distributions:
4.1 Sample β, σ^2 from $NIG\left(\bar{\mu}, \bar{\Lambda}, \bar{a}, \bar{b}\right)$, conditional on ψ, where

$$NIG\left(\bar{\mu}, \bar{\Lambda}, \bar{a}, \bar{b}\right) = NIG\left(0, \Lambda\left(\psi\right), a, b\right) + NIG\left(\tilde{\mu}, \tilde{\Lambda}, \tilde{a}, \tilde{b}\right), or$$

$$NIG\left(\bar{\mu}, \bar{\Lambda}, \bar{a}, \bar{b}\right) = NIG\left(0, \Lambda\left(\psi\right), a + d, b + \frac{1}{2}\delta^2 \sum_{j=1}^{d} \psi_j\right) + NIG\left(\tilde{\mu}, \tilde{\Lambda}, \tilde{a}, \tilde{b}\right).$$

4.2 Sample ψ^{-1} from inverse Gaussian (Proposition 3.4), conditional on β, σ^2.

variance terms. SSVS provides a simulation-based solution to the challenge of computing the closed-form posterior distribution, which, although theoretically possible, is computationally prohibitive. SSVS addresses the issue of calculating posterior probabilities by simulating posterior samples using MCMC methods. SSVS is also useful in searching for parameter restrictions in large Bayesian vector autoregressions (George et al., 2008).

3.4.1 Prior Specification

Let $\gamma_j \in \{0, 1\}$ indicate the inclusion or exclusion of the jth predictor in the regression model. The prior on the latent vector $\gamma = (\gamma_1, \ldots, \gamma_d)$ incorporates non-sample information regarding which variables should be included in the model (see Exercise 3.6 for variable selection examples). A popular choice is an independent prior such that

$$p\left(\gamma\right) = \prod_{j=1}^{d} w_j^{\gamma_j} \left(1 - w_j\right)^{1-\gamma_j}, \tag{3.3}$$

where w_j is the prior probability of variable inclusion. The parameters β and σ^2 are assigned a conditionally conjugate prior

$$\beta, \sigma^2 \,|\gamma \sim NIG\left(0, \Lambda\left(\gamma\right), a, b\right), \tag{3.4}$$

$$\Lambda\left(\gamma\right) = \left[diag\left(V^{\gamma_1} v^{1-\gamma_1}, \ldots, V^{\gamma_d} v^{1-\gamma_d}\right)\right]^{-1},$$

with the hyperparameters $V > v > 0$. Two interpretations of the prior specification are available.

First, SSVS can be interpreted as an adaptive shrinkage method through a hierarchical prior similar to that of Bayesian lasso. In SSVS, the latent variable γ serves as the discrete counterpart to the continuous latent variable ψ in lasso. The data determine the extent of shrinkage for each variable. For the jth predictor. If $\gamma_j = 1$, the prior of β_j has a large variance $\sigma^2 V$, resulting in mild shrinkage. Conversely, if $\gamma_j = 0$, the prior has a small variance $\sigma^2 v$, leading to strong shrinkage. The value of γ_j is not user-specified; instead, the data determine the posterior probability distribution of γ_j.

Second, SSVS can be interpreted as a variable selection method, where the binary vector γ defines 2^d possible models. For example, with $d = 2$, there are four models: $(1,1), (1,0), (0,1), (0,0)$, corresponding to regressions with both predictors, the first predictor only, the second predictor only, and no predictors, respectively. SSVS can be viewed as Bayesian linear regression under a mixture NIG prior with 2^d components:

$$p\left(\beta, \sigma^2\right) = \sum_{\gamma} p\left(\gamma\right) p\left(\beta, \sigma^2 \left| \gamma\right.\right),$$

where $p\left(\beta, \sigma^2 \left|\gamma\right.\right)$ corresponds to the prior NIG under a specific model γ. The weight is given by the prior model probability $p\left(\gamma\right)$.

3.4.2 Posterior Simulation

A mixture prior results in a mixture posterior:

$$p\left(\beta, \sigma^2 \left|Y\right.\right) = \sum_{\gamma} p\left(\gamma \left|Y\right.\right) p\left(\beta, \sigma^2 \left|Y, \gamma\right.\right),$$

where $p\left(\beta, \sigma^2 \left|Y, \gamma\right.\right)$ corresponds to the posterior NIG under a specific model γ. Bayesian averaging combines the model-specific results, weighted by the posterior model probability $p\left(\gamma \left|Y\right.\right)$, using observations to assign higher posterior probabilities to more promising components.

Conceptually, the posterior is analytically tractable as a mixture of NIG distributions. However, enumerating all 2^d components of the mixture distribution for a closed-form solution becomes infeasible with large model dimensionality. The Gibbs sampler in SSVS bypasses the computational burden of calculating posterior probabilities by simulating posterior samples.

Because the hierarchical prior structure in SSVS resembles that in Bayesian lasso, the posterior simulation procedures are also similar. The core of SSVS is a Gibbs sampler that iteratively generates samples from $p\left(\beta, \sigma^2 \left|Y, \gamma\right.\right)$ and $p\left(\gamma \left|Y, \beta, \sigma^2\right.\right)$. On the one hand, conditional on a specific γ, the model reduces to the conjugate Bayesian linear regression. On the other hand, conditional on the parameters β, σ^2, the posterior conditional probability of γ is proportional to the product of the prior probability and the Gaussian density. An efficient big data algorithm is presented in Algorithm 3.3, serving as the discrete counterpart to Algorithm 3.2.

As an alternative to the Gibbs sampler, it is possible to marginalize over parameters and design a marginalized MH sampler with the target distribution $p\left(\gamma \left|Y\right.\right)$, which can be evaluated pointwise by $p\left(\gamma \left|Y\right.\right) \propto p\left(\gamma\right) p\left(Y \left|\gamma\right.\right)$, where

$$p\left(Y \left|\gamma\right.\right) = \frac{p\left(\beta, \sigma^2 \left|\gamma\right.\right) p\left(Y \left|\beta, \sigma^2, \gamma\right.\right)}{p\left(\beta, \sigma^2 \left|Y, \gamma\right.\right)}.$$

Given a model γ, the marginal likelihood $p\left(Y \left|\gamma\right.\right)$ is analytically tractable because $p\left(\beta, \sigma^2 \left|\gamma\right.\right)$ and $p\left(\beta, \sigma^2 \left|Y, \gamma\right.\right)$ are the prior and posterior NIG densities, respectively, and $p\left(Y \left|\beta, \sigma^2, \gamma\right.\right)$ is the Gaussian likelihood. Marginalization is the cure of degenerate distributions when the prior variance is specified as $v = 0$.

Algorithm 3.3 Big data SSVS

Input: data X, Y, and prior specified by Equations (3.3) and (3.4).
Output: posterior samples from $p\left(\beta, \sigma^2, \gamma \,|Y\right)$.
1 Partition big data into J subsets X_j, Y_j, $j = 1, \ldots, J$.
2 Run subset regressions separately for $NIG\left(\tilde{\mu}_j, \tilde{\Lambda}_j, \tilde{a}_j, \tilde{b}_j\right)$.
3 Sum up subset posterior distributions

$$NIG\left(\tilde{\mu}, \tilde{\Lambda}, \tilde{a}, \tilde{b}\right) = \sum_{j=1}^{J} NIG\left(\tilde{\mu}_j, \tilde{\Lambda}_j, \tilde{a}_j, \tilde{b}_j\right).$$

4 Cycle through posterior conditional distributions:
4.1 Sample β, σ^2 from $NIG\left(\bar{\mu}, \bar{\Lambda}, \bar{a}, \bar{b}\right)$, conditional on γ, where

$$NIG\left(\bar{\mu}, \bar{\Lambda}, \bar{a}, \bar{b}\right) = NIG\left(0, \Lambda\left(\gamma\right), a, b\right) + NIG\left(\tilde{\mu}, \tilde{\Lambda}, \tilde{a}, \tilde{b}\right).$$

4.2 Sample binary $\gamma_j, j = 1, \ldots, d$ conditional on β, σ^2, with the probability

$$P\left(\gamma_j = 1 \,|Y, \beta, \sigma^2\right) = \frac{w_j \phi\left(\beta_j; 0, \sigma^2 V\right)}{w_j \phi\left(\beta_j; 0, \sigma^2 V\right) + \left(1 - w_j\right) \phi\left(\beta_j; 0, \sigma^2 v\right)}.$$

3.5 MCMC Model Composition

A notable limitation of the Gibbs sampler for SSVS is its dependency on the prior variance v, a user-specified hyperparameter, which must be small yet positive. If this condition is not met, specifically if $v = 0$, Algorithm 3.3 results in a reducible Markov chain, violating the convergence conditions of MCMC methods. While a small prior variance exerts shrinkage effects on some coefficients, it does not strictly exclude variables from the regression model. The MCMC Model Composition (MC^3) approach, proposed by Madigan et al. (1995) and Raftery et al. (1997), addresses these degenerate cases by integrating out model parameters. Similar marginalization techniques are discussed in the works of Geweke (1996) and Smith and Kohn (1996).

3.5.1 Marginalization

Consider the prior specification in Equations (3.3) and (3.4) with $v = 0$. For a given γ, the vector β is partitioned into two groups: β_γ and β_o, representing variables included in and excluded from the regression, respectively. According to Equation (3.4), the prior of β_γ, σ^2 is $NIG\left(0, V^{-1}I, a, b\right)$, while the prior of β_o is degenerate due to zero variance. Consequently, β_o effectively becomes a vector of zeros, simplifying the linear regression model as

$$Y = X_\gamma \beta_\gamma + \sigma\varepsilon,$$

where X_γ is a submatrix of X with the columns selected by γ.

The Gibbs sampler in Algorithm 3.3 fails under $v = 0$, but this issue can be circumvented by integrating out the parameters β_γ, σ^2 and executing a marginalized MCMC simulation targeting the distribution $p\left(\gamma \,|Y\right)$, which is proportional to the product of the prior probability $p\left(\gamma\right)$ and the marginal likelihood $p\left(Y \,|\gamma\right)$. The marginal likelihood can be evaluated by Bayes' rule:

$$p(Y|\gamma) = \frac{p(\beta_\gamma, \sigma^2|\gamma)\, p(Y|\beta_\gamma, \sigma^2, \gamma)}{p(\beta_\gamma, \sigma^2|Y, \gamma)},$$

where $p(\beta_\gamma, \sigma^2|\gamma)$ and $p(\beta_\gamma, \sigma^2|Y, \gamma)$ are the prior and posterior NIG densities. Note that the identity holds for all β_γ and σ^2 values. By setting $\beta_\gamma = 0$ and choosing an arbitrary σ^2, the likelihood function $p(Y|\beta_\gamma, \sigma^2, \gamma)$ remains constant across γ values, allowing the likelihood function to be omitted when evaluating $p(\gamma|Y)$ up to a normalizing constant.

Marginalized MCMC sampling can be implemented in two ways. The first is a random walk MH sampler. At each iteration, one element of γ is randomly altered as a proposal, $p(\gamma|Y)$ is evaluated, and the proposal is accepted based on the standard MH algorithm. The second method is a marginalized Gibbs sampler that cycles through each element of γ. Bernoulli draws are generated from $p(\gamma_j|Y, \gamma_{-j})$, $j = 1, \ldots, d$.

3.5.2 Big Data Algorithm

For each proposed update of γ in the marginalized sampler, it is necessary to evaluate the posterior NIG density $p(\beta_\gamma, \sigma^2|Y, \gamma)$, which depends on the data X_γ, Y. When the number of observations is large, iterating over the entire dataset at each MCMC step is computationally prohibitive. The goal is to develop an efficient algorithm requiring only a single data pass. The strategy involves scanning the data once and computing the full posterior $NIG\left(\tilde{\mu}, \tilde{\Lambda}, \tilde{a}, \tilde{b}\right)$ by Algorithm 1.1. For a specific γ that selects a subset of variables, the subset posterior $NIG\left(\tilde{\mu}_\gamma, \tilde{\Lambda}_{\gamma\gamma}, \tilde{a}_\gamma, \tilde{b}_\gamma\right)$ is calculated by reducing dimensionality of $NIG\left(\tilde{\mu}, \tilde{\Lambda}, \tilde{a}, \tilde{b}\right)$ without reloading the original data, as described in the following dimension reduction proposition:

Proposition 3.5 (Dimension reduction of NIG) *Let* $(\beta, \sigma^2) \sim NIG\left(\tilde{\mu}, \tilde{\Lambda}, \tilde{a}, \tilde{b}\right)$. *Let* γ *be the selection vector and* $o = 1 - \gamma$ *be the complement selection vector. Partition* $\beta = \begin{pmatrix} \beta_\gamma \\ \beta_o \end{pmatrix}$, $\tilde{\mu} = \begin{pmatrix} \mu_\gamma \\ \mu_o \end{pmatrix}$, $\tilde{\Lambda} = \begin{pmatrix} \Lambda_{\gamma\gamma} & \Lambda_{\gamma o} \\ \Lambda_{o\gamma} & \Lambda_{oo} \end{pmatrix}$, *where* β_γ *and* β_o *are the sub-vectors of* β *selected by* γ, o, *respectively. Assume their lengths are* d_γ *and* d_o. *Other sub-vectors and sub-matrices are defined similarly.*

1. Given $\beta_o = 0$, *the conditional distribution* $(\beta_\gamma, \sigma^2)|\beta_o$ *is* $NIG\left(\tilde{\mu}_\gamma, \tilde{\Lambda}_{\gamma\gamma}, \tilde{a}_\gamma, \tilde{b}_\gamma\right)$, *where*

$$\tilde{\mu}_\gamma = \mu_\gamma + \Lambda_{\gamma\gamma}^{-1}\Lambda_{\gamma o}\mu_o,$$

$$\tilde{\Lambda}_{\gamma\gamma} = \Lambda_{\gamma\gamma},$$

$$\tilde{a}_\gamma = \tilde{a} + \frac{d_o}{2},$$

$$\tilde{b}_\gamma = \tilde{b} + \frac{1}{2}\mu_o'\left(\Lambda_{oo} - \Lambda_{o\gamma}\Lambda_{\gamma\gamma}^{-1}\Lambda_{\gamma o}\right)\mu_o.$$

2. If $NIG\left(\tilde{\mu}, \tilde{\Lambda}, \tilde{a}, \tilde{b}\right)$ *is the full posterior with* X, Y *as in Proposition 1.2, then* $NIG\left(\tilde{\mu}_\gamma, \tilde{\Lambda}_{\gamma\gamma}, \tilde{a}_\gamma, \tilde{b}_\gamma\right)$ *is the subset posterior with* X_γ, Y.

As an application of Proposition 3.5, Algorithm 3.4 illustrates the big data MC^3 by the random walk MH sampler with one pass of data.

The MCMC sampler in Algorithm 3.4 focuses on sampling latent variables from $p(\gamma|Y)$. When parameter draws are needed, two steps for parameter updates can be added to the

Algorithm 3.4 Big data MC^3

Input: data X, Y, and prior specified by Equations (3.3) and (3.4) with $v = 0$.
Output: posterior samples from $p(\gamma\,|Y)$.
1 Partition big data into J subsets X_j, Y_j, $j = 1, \ldots, J$.
2 Run subset regressions separately for $NIG\left(\tilde{\mu}_j, \tilde{\Lambda}_j, \tilde{a}_j, \tilde{b}_j\right)$.
3 Sum up subset posterior distributions

$$NIG\left(\tilde{\mu}, \tilde{\Lambda}, \tilde{a}, \tilde{b}\right) = \sum_{j=1}^{J} NIG\left(\tilde{\mu}_j, \tilde{\Lambda}_j, \tilde{a}_j, \tilde{b}_j\right).$$

4 Iterate the following steps for $t = 1, 2, \ldots$
4.1 Propose γ by randomly changing one element of γ_{t-1}.
4.2 Reduce $NIG\left(\tilde{\mu}, \tilde{\Lambda}, \tilde{a}, \tilde{b}\right)$ to $NIG\left(\tilde{\mu}_\gamma, \tilde{\Lambda}_{\gamma\gamma}, \tilde{a}_\gamma, \tilde{b}_\gamma\right)$ by Proposition 3.5.
4.3 Incorporate prior information by NIG summation

$$NIG\left(\bar{\mu}_\gamma, \bar{\Lambda}_{\gamma\gamma}, \bar{a}_\gamma, \bar{b}_\gamma\right) = NIG\left(0, V^{-1}I, a, b\right) + NIG\left(\tilde{\mu}_\gamma, \tilde{\Lambda}_{\gamma\gamma}, \tilde{a}_\gamma, \tilde{b}_\gamma\right).$$

4.4 Evaluate marginal likelihood using $NIG\left(0, V^{-1}I, a, b\right)$ and $NIG\left(\bar{\mu}_\gamma, \bar{\Lambda}_{\gamma\gamma}, \bar{a}_\gamma, \bar{b}_\gamma\right)$:

$$p(Y\,|\gamma) \propto \frac{p\left(\beta_\gamma, \sigma^2\,|\gamma\right)}{p\left(\beta_\gamma, \sigma^2\,|Y, \gamma\right)}.$$

4.5 Evaluate posterior probability up to a normalizing constant

$$p(\gamma\,|Y) \propto p(\gamma)\, p(Y\,|\gamma).$$

4.6 Compute acceptance probability $\alpha = \min\left[1, \frac{p(\gamma|Y)}{p(\gamma_{t-1}|Y)}\right]$.
4.7 Set $\gamma_t = \begin{cases} \gamma & \textit{with probability } \alpha \\ \gamma_{t-1} & \textit{with probability } 1-\alpha \end{cases}$.

algorithm: 1) generate β_γ, σ^2 from $NIG\left(\bar{\mu}_\gamma, \bar{\Lambda}_{\gamma\gamma}, \bar{a}_\gamma, \bar{b}_\gamma\right)$, conditional on a specific γ, and 2) set $\beta_o = 0$. Parameter updates can be performed at an arbitrary frequency, without perturbing the target distribution of the MCMC sampler.

3.6 Software and Computational Considerations

Implementing Bayesian methods on matrix-oriented computing platforms like MATLAB is advantageous as the code closely resembles mathematical formulae, providing maximum clarity. This style is particularly recommended for prototyping. However, due to the finite precision of numerical values representing real numbers, a literal translation of mathematical formulae into computer programs can occasionally yield inaccurate or incorrect results. To enhance runtime efficiency and numerical stability, it is important to address certain computational issues.

3.6.1 Ridge Estimator

Consider computing the ridge estimator $\hat{\beta}_\delta = (X'X + \delta I)^{-1} X'Y$. Below are alternative implementations:

```
b1 = inv(X'*X+delta*eye(d))*X'*Y;
b2 = (X'*X+delta*eye(d)) \ (X'*Y);
[U,S,V] = svd(X,'econ'); D = diag(S).^2;
b3 = V*diag(1./(D+delta))*V'*(X'*Y);
b4 = V*(V'*(X'*Y)./(D+delta));
```

In the code snippet, `b1` is literal translation of the ridge formula. While it provides clarity, it is neither efficient nor numerically stable due to the direct matrix inversion. `b2` changes the order of matrix multiplication by computing `(X'*Y)` first and solving a linear equation instead of performing direct matrix inversion, enhancing stability. Regarding `b3` and `b4`, these implementations utilize diagonalization, which not only improves numerical conditions but also facilitates multiple ridge regressions with varying regularization parameters. Specifically, `b4` supports a vector-valued δ, say `delta = 2.^(-5:5)`. The eigenvalues and eigenvectors are reused across a collection of ridge estimators, optimizing computational efficiency.

3.6.2 Binary Variables in SSVS

In SSVS, the hyperparameter v is positive for Gibbs sampling, where the binary indicator γ_j is sampled from a Bernoulli distribution:

$$P\left(\gamma_j = 1 \,|Y, \beta, \sigma^2\right) \propto w_j \phi\left(\beta_j; 0, \sigma^2 V\right),$$

$$P\left(\gamma_j = 0 \,|Y, \beta, \sigma^2\right) \propto (1 - w_j)\, \phi\left(\beta_j; 0, \sigma^2 v\right).$$

The following code snippet provides a literal translation of the formulae:

```
P1 = w * normpdf(b,0,sqrt(sigma2*V));
P0 = (1-w) * normpdf(b,0,sqrt(sigma2*v));
P = P1/(P1+P0);
```

The program functions correctly if the hyperparameter v is not too small. However, it is preferable to decrease v toward zero to approximately exclude the jth variable from the regression when $\gamma_j = 0$. To support a smaller v, the program is revised as follows:

```
logP1 = log(w) - 0.5*log(V) - 0.5*b*b/(sigma2*V);
logP0 = log1p(-w) - 0.5*log(v) - 0.5*b*b/(sigma2*v);
constant = max(logP1,logP0);
expLogP1 = exp(logP1 - constant);
expLogP0 = exp(logP0 - constant);
P = expLogP1/(expLogP1+expLogP0);
```

The revised program computes log probabilities and rescales variables, preventing underflow. This treatment enhances numeric stability and accommodates a smaller v during the implementation of the Gibbs sampler in SSVS.

3.7 An Application on Highest Paying Jobs

The Labor Condition Application (LCA) data, disclosed by the Office of Foreign Labor Certification in the U.S. Department of Labor, is a large administrative dataset. It provides valuable insights into the highest-paying jobs and top employers for technical positions. When an employer sponsors an H1B visa, attorneys file the LCA on behalf of employees. This legal document records the job title, prevailing wage, and actual wage offered for the position.

Qian (2018) analyzed Bayesian linear regressions and variable selection methods using 3 million LCA records from 2011 to 2016. As a follow-up, we update this study with the 2021 LCA data, comprising approximately 0.77 million observations. Our objectives are twofold. From a social-economic perspective, the job market has shifted in response to the pandemic and inflation. Our empirical results reveal post-pandemic transitions and highlight promising careers and workplaces. From a statistical perspective, our analysis explores sample size effects on Bayesian variable selection results.

3.7.1 Descriptive Statistics

In the regression, predictors include 200 high-frequency (HF) words extracted from job titles, 100 HF employer names, 9 census division and region indicators, 10 indicators for the most affluent metropolitan areas, employment length and part-time position indicators. The total number of predictors amounts to 322.

Table 3.1 showcases job-related HF words and average wages for subsamples containing these keywords. The HF words cover various categories, such as occupations (e.g., Engineer, Architect), industries (e.g., Software, Finance), experience levels (e.g., Senior, Assistant), positions (e.g., Director, Lead), skills (e.g., Database, Java), and countries (e.g., USA, India). The table indicates that software developers and engineers are the largest group of H1B visa holders, with an average wage of 118K. Some HF words correlate with higher wages, such as Manager (132K) and Senior (120K), while other jobs such as Biologist (76.7K) and Accountant (88.8K) have lower wages.

The response variable in the regression is the annual wage offered for H1B positions. The median wage has increased from 72.4K to 104K (where K stands for thousand dollars) since 2016. H1B wages are typically higher than the average American income, as these technical jobs often require advanced degrees. The 43.7% nominal wage growth surpasses inflation, reflecting productivity growth driven by advancements in information technology and artificial intelligence. However, the income distribution has a standard deviation as large as 43.5K, highlighting unbalanced growth across industries and occupations. Comparing Table 3.2 to Table 3 in Qian (2018), we observe that employees in high-tech industries have seen a 20% - 40% wage increase over five years, while the increase is smaller in many consulting companies.

3.7.2 Linear Regression Results

To benchmark variable selection results, we run Bayesian linear regressions under a non-informative prior, with the posterior mean corresponding to the OLS estimator.

Predictors with the largest positive and negative coefficients are listed in Table 3.3, which is comparable to Table 4 in Qian (2018). Technology giants are generous employers, contributing to large positive coefficients. For example, Google employees earn 35K more than those hired by companies outside the top 100 HF employers. Empirical results suggest that working with Facebook (+51.5K) and Apple (+43.6K) may also be well-paid. However, 2021 data indicates some financial institutions now have large negative coefficients, which are not seen in the data of 2011 - 2016. Also, predictors like Doordash (+57.9) and Epidemiologist (-31.5) are new on the 2021 list, reflecting pandemic-related phenomenon.

From a statistical perspective, we note that the posterior standard deviations of many coefficients in 2021 are about twice as large as those obtained from 2011 - 2016 data. This

TABLE 3.1
Key words of job titles and wage offers.

Word	Frequency	Wage	Word	Frequency	Wage
Software	3.30E+05	118.2	Operation	2.00E+04	117.3
Developer	3.20E+05	118.7	Designer	1.90E+04	108.0
Engineer	3.00E+05	118.7	Intelligence	1.80E+04	108.3
Application	2.60E+05	117.1	Hospital	1.80E+04	116.8
System	1.50E+05	117.6	Database	1.70E+04	103.0
Computer	1.50E+05	110.7	Teacher	1.70E+04	86.7
Analyst	1.50E+05	104.4	Director	1.60E+04	162.8
Senior	1.40E+05	120.1	Assistant	1.60E+04	99.1
Manager	1.10E+05	132.0	Staff	1.60E+04	138.9
Technology	5.80E+04	109.1	Mechanical	1.50E+04	96.9
Architect	5.00E+04	112.3	Accountant	1.40E+04	88.8
Information	4.80E+04	138.8	IT	1.30E+04	110.8
Researcher	4.50E+04	100.0	Statistician	1.30E+04	116.5
Specialist	4.00E+04	97.2	Professor	1.20E+04	109.3
Business	3.60E+04	109.0	Occupational	1.20E+04	91.5
Data	3.60E+04	116.7	Network	1.20E+04	103.0
Scientist	3.50E+04	103.4	Solution	1.10E+04	125.2
Associate	3.40E+04	102.9	Marketing	1.10E+04	120.9
Lead	3.20E+04	112.8	Auditor	1.10E+04	88.4
Project	3.00E+04	111.6	Physician	9.76E+03	212.3
Consultant	2.80E+04	106.0	Physicist	9.32E+03	81.4
Program	2.80E+04	106.4	Industrial	9.31E+03	102.1
Quality	2.80E+04	97.2	Biologist	8.81E+03	76.7
Tester	2.70E+04	97.6	USA	8.81E+03	92.0
Electronic	2.40E+04	116.2	Postdoctoral	8.22E+03	58.9
Assurance	2.40E+04	95.6	Science	8.09E+03	100.5
Principal	2.30E+04	154.4	Chemist	8.00E+03	77.2
Finance	2.30E+04	127.5	Vice	7.99E+03	162.3
Product	2.10E+04	132.2	Java	7.85E+03	100.9
Administrator	2.10E+04	99.0	Security	7.71E+03	122.3

is not surprising, as the sample size of the latter is roughly four times larger than that of the former. Recall that the posterior covariance matrix is proportional to $\left(\sum_{i=1}^{n} X_i'X_i\right)^{-1}$. If $X_1, \ldots, X_n$ are random samples from a population, we expect that the posterior variance of coefficients will decrease proportionally as the sample size increases.

3.7.3 Variable Selection Results

The computer programs for Bayesian lasso, SSVS and MC^3 are the same as those used by Qian (2018), ensuring that variable selection results in Table 3.4 are comparable to Table 5 in that study.

Bayesian lasso regressions are conducted using a sequence of regularization parameters: 500, 1000, 2000, and 5000 for HF words and employers. Geographic dummy variables are excluded from regularization due to their economic significance, warranting their inclusion in the regression model. Adjusting the regularization parameters does not increase the computational burden, as the dataset is scanned once, regardless of the multiple rounds of lasso regressions. This is achieved through NIG summations under different regularization parameters.

TABLE 3.2
Top employers and wage offers.

Employer	Frequency	Wage	Employer	Frequency	Wage
Cognizant	1.90E+04	92.5	Randstad	1.90E+03	116.3
Google	1.60E+04	156.0	Paypal	2.00E+03	149.2
Amazon	2.10E+04	129.2	Kforce	1.80E+03	100.5
Tata	1.20E+04	79.9	Linkedin	1.80E+03	147.3
Microsoft	1.20E+04	143.9	Tesla	1.70E+03	133.6
Ernst Young	1.00E+04	116.1	Tekorg	1.70E+03	80.2
Facebook	9.40E+03	174.1	Oracle	1.90E+03	122.0
Infosys	8.20E+03	86.4	PWC	2.70E+03	130.9
HCL	7.90E+03	99.5	Ebay	1.40E+03	161.4
Apple	6.70E+03	164.6	Uber	1.40E+03	150.8
Accenture	6.30E+03	107.6	CGI	1.50E+03	100.3
Intel	6.10E+03	107.9	Citi Bank	1.30E+03	137.7
Deloitte	9.20E+03	107.4	Intuit	1.30E+03	139.1
Capgemini	4.80E+03	97.1	Adobe	1.20E+03	141.0
Walmart	4.80E+03	130.9	Cummins	1.40E+03	82.9
Wipro	4.20E+03	81.0	Anthem	1.20E+03	99.2
JP Morgan	3.90E+03	129.8	Synechron	1.20E+03	105.5
Qualcomm	4.50E+03	123.8	Mckinsey	1.20E+03	141.6
Compunnel	3.50E+03	100.7	BOA	1.10E+03	133.8
Tech Mahindra	3.50E+03	97.2	Rivian	1.20E+03	141.0
Salesforce	3.10E+03	140.0	General Motors	1.20E+03	110.5
IBM	5.10E+03	104.6	Virtusa	1.60E+03	103.1
Cisco	2.90E+03	140.4	Lttechnology	1.20E+03	84.0
Goldman Sachs	3.60E+03	124.7	Blackrock	1.30E+03	99.4
Mindtree	2.10E+03	94.8	UST Global	1.00E+03	97.6
Vmware	2.10E+03	159.5	Capital One	1.20E+03	100.7
Atos	2.10E+03	95.4	DFS	1.10E+03	109.5
Mphasis	2.00E+03	98.2	Nvidia	1.00E+03	159.3
Fidelity	2.40E+03	102.5	Optum	1.00E+03	110.2
Larsen	2.00E+03	92.9	Comcast	9.82E+02	107.4

The magnitude of lasso shrinkage is inversely related to "t-statistics", defined as the ratio of the point estimator to the standard deviation (error). For instance, Software is a HF word with a large t-statistics 33.59, surviving lasso variable selection under moderate regularization parameters of 500, 1000, and 2000. In contrast with empirical results in Qian (2018), a regularization parameter of 5000 appears excessively large for the 2021 data, where the sample size is only a quarter of that in 2011–2016 data. An overly large regularization parameter significantly reduces all coefficients, resulting in too many zeros that defeat the purpose of variable selection.

To implement SSVS, we specify the large and small variances of the mixture distribution as tuning parameters. The prior does not significantly impact empirical results. We experimented with several pairs, such as 10/0.001, 100/0.001, and 1000/0.001, yielding similar numeric results. SSVS selects approximately 210 out of 322 predictors.

Table 3.4 also presents regression results for the most frequently visited model by MC^3. The empirical results closely resemble those obtained by SSVS. However, computing the marginal likelihood is more computationally intensive compared to the Gibbs sampler used in SSVS, which may explain popularity of SSVS among practitioners.

TABLE 3.3
Bayesian linear regressions for highest paying jobs.

Word	Coeff	Sd	Wage	Word	Coeff	Sd	Wage
Physician	102.74	0.53	212.29	Resident	−124.03	0.74	69.63
Lawyer	76.65	0.58	175.06	Counselors	−51.17	1.09	49.01
Dentist	59.15	0.89	160.17	Fellow	−42.91	0.55	67.81
Doordash	57.90	1.15	185.29	Clinical	−41.20	1.42	58.24
Facebook	51.50	0.33	174.09	Landscape	−32.65	1.52	71.24
Executive	50.63	0.72	190.51	Epidemiologist	−31.45	0.56	79.12
Surgeon	48.58	0.63	210.97	Education	−29.54	0.89	58.19
Director	44.16	0.26	162.80	Scrum	−29.48	1.36	99.50
Apple	43.76	0.39	164.57	Fidelity	−26.84	0.64	102.45
Twitter	42.05	1.13	174.22	Infosys	−25.32	0.94	86.38
BostonConsult.	38.37	1.13	150.86	Tekorg	−23.18	0.82	80.19
Principal	36.12	0.26	154.44	Accountant	−22.96	0.43	88.76
Partner	35.08	0.98	167.74	Assistant	−22.81	0.4	99.11
Google	35.03	0.26	155.97	Junior	−22.58	0.93	80.60
Professor	34.15	0.64	109.32	Mindtree	−21.88	0.69	94.79
Nvidia	32.67	0.98	159.34	Environmental	−21.68	0.91	81.63
Mckinsey	31.33	0.92	141.63	Blackrock	−21.4	0.86	99.42
HP	31.20	1.03	148.39	Wipro	−21.39	0.49	81.02
Ebay	31.11	0.85	161.41	Food	−21.30	1.10	80.56
Oath	27.98	1.05	161.11	Capital One	−21.13	0.92	100.7
Pwc	27.24	0.73	130.89	Worker	−20.73	0.97	64.44
Rivian	24.97	0.91	140.98	Coordinator	−20.08	0.70	70.50
Ernst Young	24.90	0.33	116.06	BOA	−19.67	0.97	133.79
Expert	24.88	0.90	130.46	Social	−19.44	0.88	68.22
Staff	23.53	0.29	138.85	Amer Airlines	−18.57	1.21	91.59
Master	23.48	0.97	113.39	Cognizant	−18.56	0.27	92.53
Hospital	22.92	0.36	116.83	Language	−18.30	0.76	68.42
Citi Bank	22.39	0.87	137.65	Tata	−18.26	0.31	79.89
Vmware	21.36	0.78	159.46	Instructor	−18.14	0.67	72.42
Manager	21.18	0.14	131.98	Qualcomm	−18.10	0.50	123.79

3.8 Bibliographic Notes

In contrast to the century-old 2-norm regularized ridge regression, the 1-norm regularized lasso regression, introduced by Tibshirani (1996), emerged from obscurity and quickly gained popularity in statistics, inspiring numerous extensions. Zou and Hastie (2005) proposed elastic regularization, which combines 1-norm and 2-norm penalties. Bayesian lasso models are explored in works by Park and Casella (2008) and Hans (2009). Li and Lin (2010) employed the Gibbs sampler for Bayesian estimation of the elastic net.

In this chapter, we studied two popular Bayesian model selection methods: SSVS by George and McCulloch (1993); George and Mcculloch (1997), as well as MC^3 by Madigan et al. (1995) and Raftery et al. (1997). Other model selection techniques are available: indicator-in-regression method by Kuo and Mallick (1998), pseudo-prior specifications by Carlin and Chib (1995), reversible jump MCMC by Green (1995), composite space approach by Godsill (2001), Bayesian calibration to frequentist information criteria by George and Foster (2000), among others.

TABLE 3.4
Bayesian variable selection by lasso (with regularization at 500, 1000 and 5000), SSVS and MC3 in highest paying jobs regressions.

Words	Coeff (t-stat)	L500	L1000	L2000	L5000	SSVS	MC3
Software	8.06 (33.59)	9.04	9.00	8.38	5.94	6.45	7.86
USA	8.29 (9.29)	−0.03	−0.06	−0.06	−0.01	8.65	8.37
Finance	15.43 (59.60)	18.00	18.46	17.03	11.56	15.23	15.45
Marketing	4.31 (10.37)	2.92	0.14	0.01	0.00	4.30	4.16
Professor	34.15 (53.06)	29.62	23.75	12.50	0.12	34.44	33.80
IT	−1.47 (−5.14)	−0.66	−0.11	−0.02	0.00	0.00	0.00
Sales	5.66 (14.25)	5.77	4.33	0.21	0.01	5.50	5.48
Resident	−124.03 (−166.61)	−109.77	−97.25	−71.80	−1.38	−123.59	−124.00
Secondary	4.29 (3.43)	0.03	0.00	−0.01	0.00	0.00	0.00
Lawyer	76.65 (132.36)	74.06	68.68	55.59	16.70	76.64	76.55
Artist	−14.85 (−15.34)	−0.76	−0.04	−0.01	0.00	−15.39	−15.15
Family	2.78 (2.43)	0.02	0.01	0.00	0.00	0.00	0.00
Integration	2.50 (4.42)	0.13	0.03	0.01	0.00	0.00	2.49
Cost	−4.85 (−2.45)	−0.24	−0.03	−0.01	0.00	0.00	−4.88
Warehouse	−0.54 (−0.57)	0.01	0.00	0.00	0.00	0.00	−0.44
Estimator	−10.82 (−5.12)	−0.24	−0.03	−0.01	0.00	−15.92	−10.87
IBM	0.27 (0.61)	0.03	0.01	0.00	0.00	0.00	0.00
Intel	−7.81 (−16.10)	−5.84	−3.37	−0.05	−0.01	−6.92	−7.59
Mphasis	−2.43 (−3.35)	−0.03	−0.01	0.00	0.00	0.00	0.00
New England	93.47 (415.90)	89.67	88.61	89.01	91.19	93.65	93.59
New York	11.74 (31.39)	13.12	13.54	13.80	14.89	11.80	11.77

Recent developments in deep learning models, which involve millions of parameters, have sparked renewed interest in ridge and ridge-less regressions. Hastie et al. (2022) document unexpected findings in high-dimensional ridge-less regressions. Bartlett et al. (2020) discuss the phenomena of over-parameterization and benign overfitting. Zou et al. (2024) justify benign overfitting in the context of linear regressions estimated using constant-stepsize stochastic gradient descent algorithms. Kobak et al. (2020) demonstrate that the optimal shrinkage parameter can even be negative due to the implicit ridge regularization in high-dimensional regressions. These advances highlight the evolving landscape of statistical modeling.

3.9 Proof of Propositions

Proposition 3.1 (Multicollinearity and precision matrix)

The definition of eigenvalue and eigenvector indicates that $X'Xv_j = \lambda_j v_j$. *Normalize the eigenvectors such that* $v'v = 1$. *It follows that*

$\|Xv_j\|^2 = v_j'X'Xv_j = v_j'\lambda_j v_j = \lambda_j$.

If $\lambda_j = 0$, *then* $\|Xv_j\|^2 = 0$, *resulting in* $Xv_j = 0$.

If λ_j *is a small positive value below the tolerance level,* $\|Xv_j\| = \sqrt{\lambda_j}$ *leads to inexact multicollinearity, with the linear combination coefficients* v_j.

Proposition 3.2 (Normal-exponential scale mixture)

The marginal distribution is obtained by integrating out the latent variable:

$p(x) = \int p(\psi)\, p(x\,|\psi)\, d\psi = \frac{1}{2}(2\pi)^{-\frac{1}{2}}\sigma^{-1}\int_0^\infty \psi^{-\frac{1}{2}} e^{-\frac{\psi}{2}-\frac{(x-\mu)^2}{2\psi\sigma^2}} d\psi.$

The integrand is the kernel of the reciprocal inverse Gaussian density, so the integration has an explicit solution:

$\int_0^\infty \psi^{-\frac{1}{2}} e^{-\frac{\psi}{2}-\frac{(x-\mu)^2}{2\psi\sigma^2}} d\psi = (2\pi)^{\frac{1}{2}} e^{-\frac{|x-\mu|}{\sigma}}.$

It follows that $p(x) = \frac{1}{2\sigma}e^{-\frac{|x-\mu|}{\sigma}}$, *namely the density of a double exponential distribution with parameters* μ *and* σ.

Proposition 3.3 (lasso prior)

1. By the definition of the NIG distribution, $NIG(0, \Lambda(\psi), a, b)$ *indicates that* σ^2 *follows an inverse gamma distribution* $IG(a, b)$, *while* $\beta\,|\sigma^2, \psi$ *follows a multivariate normal distribution with mean zero and precision matrix* $\sigma^{-2}\Lambda(\psi)$. *The covariance matrix is* $\sigma^2\Lambda^{-1}(\psi)$. *Since the covariance matrix is diagonal, elements in the vector* β *are conditionally independent. That is,* $\beta_j\,|\sigma^2, \psi$ *follows a univariate normal distribution with mean zero and variance* $\psi_j\sigma^2$. *The prior* $p(\psi) \propto \prod_{j=1}^d e^{-\frac{1}{2}\delta^2\psi_j}$ *implies that* $p(\psi_j) \propto e^{-\frac{1}{2}\delta^2\psi_j}$, *which is an exponential distribution with mean* $2\delta^{-2}$. *We reparametrize it by* $\tilde{\psi}_j = \delta^2\psi_j$, *and the exponential distribution is preserved after rescaling:* $p\left(\tilde{\psi}_j\right) \propto \frac{1}{2}e^{-\frac{1}{2}\tilde{\psi}_j}$. *Because* $\beta_j\left|\sigma^2, \tilde{\psi}_j\right.$ *follows a univariate normal distribution with mean zero and variance* $\tilde{\psi}_j\delta^{-2}\sigma^2$. *Proposition 3.2 indicates that* $\beta_j\,|\sigma^2$ *follows a double exponential distribution with parameters* $(0, \delta^{-1}\sigma)$. *We have* $p(\beta_j\,|\sigma^2) = \frac{1}{2\delta^{-1}\sigma}e^{-\delta\sigma^{-1}|\beta_j|}$. *It follows that* $p(\beta, \sigma^2) \propto (\sigma^2)^{-(a+\frac{d}{2}+1)} e^{-b\sigma^{-2}-\delta\sigma^{-1}\sum_{j=1}^d|\beta_j|}$.

2. The prior $p(\psi\,|\sigma^2) \propto \prod_{j=1}^d e^{-\frac{1}{2}\delta^2\sigma^{-2}\psi_j}$ *implies that* $p(\psi_j\,|\sigma^2) \propto e^{-\frac{1}{2}\delta^2\sigma^{-2}\psi_j}$, *which is an exponential distribution with mean* $2\delta^{-2}\sigma^2$. *We reparametrize it by* $\tilde{\psi}_j = \delta^2\sigma^{-2}\psi_j$, *and the exponential distribution is preserved after rescaling:* $p\left(\tilde{\psi}_j\,|\sigma^2\right) \propto \frac{1}{2}e^{-\frac{1}{2}\tilde{\psi}_j}$. *Because* $\beta_j\left|\sigma^2, \tilde{\psi}_j\right.$ *follows a univariate normal distribution with mean zero and variance* $\tilde{\psi}_j\delta^{-2}\sigma^4$. *Proposition 3.2 indicates that* $\beta_j\,|\sigma^2$ *follows a double exponential distribution with parameters* $(0, \delta^{-1}\sigma^2)$. *We have* $p(\beta_j\,|\sigma^2) = \frac{1}{2\delta^{-1}\sigma^2}e^{-\delta\sigma^{-2}|\beta_j|}$. *It follows that* $p(\beta, \sigma^2) \propto (\sigma^2)^{-(a+d+1)} e^{-b\sigma^{-2}-\delta\sigma^{-2}\sum_{j=1}^d|\beta_j|}$.

Proposition 3.4 (Posterior conditionals of lasso)

For the first prior specification, $p(\beta, \sigma^2\,|Y, \psi) \propto p(\beta, \sigma^2\,|\psi)\, p(Y\,|\beta, \sigma^2, \psi)$, *where* $p(\beta, \sigma^2\,|\psi)$ *is* $NIG(0, \Lambda(\psi), a, b)$ *and* $p(Y\,|\beta, \sigma^2, \psi)$ *is the Gaussian likelihood. By Proposition 1.5,* $NIG(\bar{\mu}, \bar{\Lambda}, \bar{a}, \bar{b}) = NIG(0, \Lambda(\psi), a, b) + NIG\left(\tilde{\mu}, \tilde{\Lambda}, \tilde{a}, \tilde{b}\right)$. *In Bayesian lasso regression, the latent variable* ψ_j *follows an exponential distribution with mean* $2\delta^{-2}$. *We have* $p(\psi_j) = \frac{1}{2}\delta^2 e^{-\frac{1}{2}\delta^2\psi_j}$. *By changing the variable* $\gamma_j = \psi_j^{-1}$, *the probability distribution becomes* $P(\gamma_j \le c) = P(\psi_j \ge c^{-1}) = e^{-\frac{1}{2}\delta^2 c^{-1}}$. *Taking derivatives yields the reparametrized density* $p(\gamma_j) = \frac{1}{2}\delta^2\gamma_j^{-2}e^{-\frac{1}{2}\delta^2\gamma_j^{-1}}$. *It follows that*

$$
\begin{aligned}
& p(\gamma_j\,|\beta_j, \sigma^2) \\
& \propto p(\gamma_j)\, p(\beta_j\,|\gamma_j, \sigma^2) \\
& \propto \gamma_j^{-2}\exp\left[-\tfrac{1}{2}\delta^2\gamma_j^{-1}\right]\gamma_j^{1/2}\exp\left[-\tfrac{\beta_j^2}{2\sigma^2}\gamma_j\right] \\
& \propto \gamma_j^{-\frac{3}{2}}\exp\left[-\tfrac{1}{2}\tfrac{\beta_j^2}{\sigma^2}\gamma_j - \tfrac{1}{2}\delta^2\gamma_j^{-1}\right] \\
& \propto \gamma_j^{-\frac{3}{2}}\exp\left[-\tfrac{\lambda(\gamma_j-\mu)^2}{2\mu^2\gamma_j}\right]
\end{aligned}
$$

where $\mu = \left|\frac{\delta\sigma}{\beta_j}\right|$ *and* $\lambda = \delta^2$. *This is an inverse Gaussian density with parameters* μ *and* λ.

For the second prior specification, note that $p(\sigma^2 | \psi)$ *is different from* $p(\sigma^2)$, *but remains an inverse gamma distribution:*

$$\begin{aligned}
&p(\sigma^2 | \psi) \\
&\propto p(\sigma^2) p(\psi | \sigma^2) \\
&\propto (\sigma^2)^{-(a+1)} e^{-b\sigma^{-2}} \prod_{j=1}^d \sigma^{-2} e^{-\frac{1}{2}\delta^2\sigma^{-2}\psi_j} \\
&\propto (\sigma^2)^{-(a+d+1)} e^{-\sigma^{-2}\left(b+\frac{1}{2}\delta^2\sum_{j=1}^d \psi_j\right)}
\end{aligned}$$

which is recognized as $IG\left(a+d, b+\frac{1}{2}\delta^2\sum_{j=1}^d \psi_j\right)$. *Additionally, the prior specification indicates that* $p(\beta | \sigma^2, \psi)$ *follows a normal distribution with mean zero and covariance matrix* $\sigma^2\Lambda^{-1}(\psi)$. *By the definition of NIG distribution,* $p(\beta, \sigma^2 | \psi)$ *follows* $NIG\left(0, \Lambda(\psi), a+d, b+\frac{1}{2}\delta^2\sum_{j=1}^d \psi_j\right)$. *By Proposition 1.5,*

$$NIG(\bar{\mu}, \bar{\Lambda}, \bar{a}, \bar{b}) = NIG\left(0, \Lambda(\psi), a+d, b+\tfrac{1}{2}\delta^2\textstyle\sum_{j=1}^d \psi_j\right) + NIG\left(\tilde{\mu}, \tilde{\Lambda}, \tilde{a}, \tilde{b}\right).$$

The prior of the latent variable is specified as an exponential distribution $p(\psi | \sigma^2) \propto \prod_{j=1}^d e^{-\frac{1}{2}\delta^2\sigma^{-2}\psi_j}$. *By changing the variable* $\gamma_j = \psi_j^{-1}$, *the probability distribution becomes* $P(\gamma_j \le c) = P(\psi_j \ge c^{-1}) = e^{-\frac{1}{2}\delta^2\sigma^{-2}c^{-1}}$. *Taking derivatives yields the reparametrized density* $p(\gamma_j) = \frac{1}{2}\delta^2\sigma^{-2}\gamma_j^{-2}e^{-\frac{1}{2}\delta^2\sigma^{-2}\gamma_j^{-1}}$. *It follows that*

$$\begin{aligned}
&p(\gamma_j | \beta_j, \sigma^2) \\
&\propto p(\gamma_j) p(\beta_j | \gamma_j, \sigma^2) \\
&\propto \gamma_j^{-2} \exp\left[-\tfrac{1}{2}\delta^2\sigma^{-2}\gamma_j^{-1}\right] \gamma_j^{1/2} \exp\left[-\tfrac{\beta_j^2}{2\sigma^2}\gamma_j\right] \\
&\propto \gamma_j^{-\frac{3}{2}} \exp\left[-\tfrac{1}{2}\tfrac{\beta_j^2}{\sigma^2}\gamma_j - \tfrac{1}{2}\delta^2\sigma^{-2}\gamma_j^{-1}\right] \\
&\propto \gamma_j^{-\frac{3}{2}} \exp\left[-\tfrac{\lambda(\gamma_j-\mu)^2}{2\mu^2\gamma_j}\right]
\end{aligned}$$

where $\mu = \left|\frac{\delta}{\beta_j}\right|$ *and* $\lambda = \delta^2\sigma^{-2}$. *This is an inverse Gaussian density with parameters* μ *and* λ.

Proposition 3.5 (Dimension reduction of NIG)

For the first part of the proposition, we note that $(\beta, \sigma^2) \sim NIG\left(\tilde{\mu}, \tilde{\Lambda}, \tilde{a}, \tilde{b}\right)$ *refers to* $\beta | \sigma^2 \sim N\left(\tilde{\mu}, \sigma^2\tilde{\Lambda}^{-1}\right)$ *and* $\sigma^2 \sim IG\left(\tilde{a}, \tilde{b}\right)$. *Denote* $V = \tilde{\Lambda}^{-1}$ *and partition it as* $V = \begin{pmatrix} V_{\gamma\gamma} & V_{\gamma o} \\ V_{o\gamma} & V_{oo} \end{pmatrix}$. *The matrix inversion rule requires*

$$\begin{aligned}
V_{\gamma\gamma} &= \Lambda_{\gamma\gamma}^{-1} + \Lambda_{\gamma\gamma}^{-1}\Lambda_{\gamma o}G^{-1}\Lambda_{o\gamma}\Lambda_{\gamma\gamma}^{-1}, \\
V_{\gamma o} &= -\Lambda_{\gamma\gamma}^{-1}\Lambda_{\gamma o}G^{-1}, \\
V_{o\gamma} &= -G^{-1}\Lambda_{o\gamma}\Lambda_{\gamma\gamma}^{-1}, \\
V_{oo} &= G^{-1}, \\
G &= \Lambda_{oo} - \Lambda_{o\gamma}\Lambda_{\gamma\gamma}^{-1}\Lambda_{\gamma o}.
\end{aligned}$$

Since $\beta | \sigma^2$ *is multivariate normal, its conditional distribution (given* $\beta_o = 0$*) preserves normality:* $\beta_\gamma | \beta_o, \sigma^2 \sim N\left(\tilde{\mu}_\gamma, \sigma^2\tilde{\Lambda}_{\gamma\gamma}^{-1}\right)$, *where*

$$\tilde{\mu}_\gamma = \mu_\gamma + V_{\gamma o}V_{oo}^{-1}(0 - \mu_o) = \mu_\gamma + \Lambda_{\gamma\gamma}^{-1}\Lambda_{\gamma o}\mu_o,$$

$$\tilde{\Lambda}_{\gamma\gamma} = \left(V_{\gamma\gamma} - V_{\gamma o} V_{oo}^{-1} V_{o\gamma}\right)^{-1} = \Lambda_{\gamma\gamma}.$$

Additionally, the conditional distribution $\sigma^2 \,|\beta_o$ is given by

$$p\left(\sigma^2 \,|\beta_o\right) \propto \left(\sigma^2\right)^{-\left(\tilde{a}+\frac{d_o}{2}+1\right)} e^{-\sigma^{-2}\left[\tilde{b}+\frac{1}{2}\mu_o'\left(\Lambda_{oo}-\Lambda_{o\gamma}\Lambda_{\gamma\gamma}^{-1}\Lambda_{\gamma o}\right)\mu_o\right]},$$

which is the density of $IG\left(\tilde{a}_\gamma, \tilde{b}_\gamma\right)$, where

$$\tilde{a}_\gamma = \tilde{a} + \frac{d_o}{2},$$

$$\tilde{b}_\gamma = \tilde{b} + \frac{1}{2}\mu_o' \left(\Lambda_{oo} - \Lambda_{o\gamma}\Lambda_{\gamma\gamma}^{-1}\Lambda_{\gamma o}\right)\mu_o.$$

For the second part of the proposition, we partition the predictors as $X = \begin{pmatrix} X_\gamma & X_o \end{pmatrix}$. Proposition 1.2 indicates that

$$\tilde{\Lambda} = X'X = \begin{pmatrix} \Lambda_{\gamma\gamma} & \Lambda_{\gamma o} \\ \Lambda_{o\gamma} & \Lambda_{oo} \end{pmatrix},$$

$$\tilde{\mu} = \left(X'X\right)^{-1} X'Y = \begin{pmatrix} \mu_\gamma \\ \mu_o \end{pmatrix},$$

$$\tilde{a} = \frac{n-d}{2},$$

$$2\tilde{b} = Y'Y - Y'X\left(X'X\right)^{-1} X'Y = Y'Y - Z_\gamma'\mu_\gamma - Z_o'\mu_o,$$

where

$$\Lambda_{\gamma\gamma} = X_\gamma'X_\gamma, \Lambda_{\gamma o} = X_\gamma'X_o, \Lambda_{o\gamma} = X_o'X_\gamma, \Lambda_{oo} = X_o'X_o,$$

$$Z_\gamma = X_\gamma'Y, Z_o = X_o'Y,$$

$$\mu_\gamma = \left(\Lambda_{\gamma\gamma}^{-1} + \Lambda_{\gamma\gamma}^{-1}\Lambda_{\gamma o}G^{-1}\Lambda_{o\gamma}\Lambda_{\gamma\gamma}^{-1}\right) Z_\gamma - \Lambda_{\gamma\gamma}^{-1}\Lambda_{\gamma o}G^{-1}Z_o,$$

$$\mu_o = -G^{-1}\Lambda_{o\gamma}\Lambda_{\gamma\gamma}^{-1}Z_\gamma + G^{-1}Z_o.$$

It follows that

$$\tilde{\mu}_\gamma = \mu_\gamma + \Lambda_{\gamma\gamma}^{-1}\Lambda_{\gamma o}\mu_o = \Lambda_{\gamma\gamma}^{-1}Z_\gamma = \left(X_\gamma'X_\gamma\right)^{-1} X_\gamma'Y,$$

$$\tilde{\Lambda}_{\gamma\gamma} = \Lambda_{\gamma\gamma} = X_\gamma'X_\gamma,$$

$$\tilde{a}_\gamma = \tilde{a} + \frac{d_o}{2} = \frac{n-d}{2} + \frac{d_o}{2} = \frac{n-d_\gamma}{2},$$

$$\tilde{b}_\gamma = \tilde{b} + \frac{1}{2}\mu_o' \left(\Lambda_{oo} - \Lambda_{o\gamma}\Lambda_{\gamma\gamma}^{-1}\Lambda_{\gamma o}\right)\mu_o = \frac{1}{2}Y'Y - \frac{1}{2}Y'X_\gamma\left(X_\gamma'X_\gamma\right)^{-1} X_\gamma'Y.$$

We note that $NIG\left(\tilde{\mu}_\gamma, \tilde{\Lambda}_{\gamma\gamma}, \tilde{a}_\gamma, \tilde{b}_\gamma\right)$ is the posterior obtained by Proposition 1.2, with X_γ being the predictors.

3.10 Exercises

Exercise 3.1 (Multicollinearity and principal components) *Suppose that the predictor matrix X is centered (i.e., the column-wise mean is zero), and let $\frac{1}{n}X'X$ be the sample covariance matrix. Throughout this exercise, we calculate the sample variance and covariance by dividing n instead of $n-1$.*

1. Show that $X'X$ and $\frac{1}{n}X'X$ have the same eigenvectors $v_1, \ldots, v_d$, while eigenvalues of the latter are rescaled as $\frac{1}{n}\lambda_1, \ldots, \frac{1}{n}\lambda_d$.

2. We arrange eigenvalues in the decreasing order. The jth principal component is defined as $Xv_j, j = 1, \ldots, d$. Show that the sample variance of Xv_j is $\frac{1}{n}\lambda_j$, and the sample covariance between Xv_j and Xv_k, $j \neq k$ is zero.

3. Rephrase Proposition 3.1 in terms of principal components. How to detect multicollinearity by inspecting the principal components?

4. How do the above results change if X is not centered?

Exercise 3.2 (Multicollinearity and precision matrix) *Consider a linear regression with two predictors (a and b). Let the data matrix be*

$$X = \begin{pmatrix} X_a & X_b \end{pmatrix} = \begin{pmatrix} X_{a1} & X_{b1} \\ \cdots & \cdots \\ X_{an} & X_{bn} \end{pmatrix}.$$

1. Suppose that predictors are orthogonal and the precision matrix is diagonal:

$$X'X = \begin{pmatrix} \sum_{i=1}^{n} X_{ai}^2 & 0 \\ 0 & \sum_{i=1}^{n} X_{bi}^2 \end{pmatrix}.$$

As we can see, the precision matrix is determined by the variability of the two predictors. We run Bayesian linear regression under the non-informative prior. How does the predictor variability affect the posterior variance?

2. Suppose that predictors are scaled such that the precision matrix takes the form:

$$X'X = \begin{pmatrix} 1 & \rho \\ \rho & 1 \end{pmatrix},$$

where ρ measures the correlation between two variables. Compute eigenvalues and eigenvectors of the precision matrix, and then express the posterior variance as a function of eigenvalues. How does the correlation parameter affect the posterior variance and covariance?

Exercise 3.3 (Posterior multicollinearity) *Suppose that data are subject to perfect multicollinearity: $Xc = 0$. Under the proper prior $NIG(\mu, \Lambda, a, b)$, the posterior $NIG(\bar{\mu}, \bar{\Lambda}, \bar{a}, \bar{b})$ is generally well-defined (See Proposition 1.1). Identification is achieved by the prior, from which the extracted pseudo-data supplement the collinear observations. However, some linear combination of the coefficients may fail to update in the posterior. Consider a specific combination: $\gamma = c'\Lambda\beta$.*

1. Show that the prior distribution is $\gamma \,|\sigma^2 \sim N(c'\Lambda\mu, \sigma^2 c'\Lambda c)$.

2. Show that the posterior distribution is $\gamma \,|Y, \sigma^2 \sim N(c'\Lambda\mu, \sigma^2 c'\Lambda c)$.

Exercise 3.4 (Ridge estimator) *Show that the two forms of the ridge estimator are numerically equivalent:*

1. $(X'X + \delta I_{d\times d})^{-1} X'Y$.
2. $X'(XX' + \delta I_{n\times n})^{-1} Y$.

Exercise 3.5 (Posterior latent variable in lasso) *In Bayesian lasso regression, the latent variable ψ_j follows the exponential distribution with the mean $2\delta^2$:*

$$p(\psi_j) = \frac{1}{2}\delta^2 e^{-\frac{1}{2}\delta^2\psi_j}.$$

Conditional on ψ_j and σ^2, the regression coefficient β_j follows the univariate normal distribution with the mean zero and the variance $\psi_j\sigma^2$.

$$p(\beta_j\,|\psi_j, \sigma^2) = (2\pi)^{-\frac{1}{2}}\sigma^{-1}\psi_j^{-\frac{1}{2}}e^{-\frac{\beta_j^2}{2\psi_j\sigma^2}}.$$

1. *Let $\gamma_j = \psi_j^{-1}$. Show that its density function is given by*

$$p(\gamma_j) = \frac{1}{2}\delta^2\gamma_j^{-2}e^{-\frac{1}{2}\delta^2\gamma_j^{-1}}.$$

2. *Consider the posterior conditional distribution $p(\gamma_j\,|\beta_j, \sigma^2) \propto p(\gamma_j)\,p(\beta_j\,|\psi_j, \sigma^2)$. Show that*

$$p(\gamma_j\,|\beta_j, \sigma^2) \propto \gamma_j^{-\frac{3}{2}}\exp\left[-\frac{1}{2}\frac{\beta_j^2}{\sigma^2}\gamma_j - \frac{1}{2}\delta^2\gamma_j^{-1}\right].$$

3. *Look up the inverse Gaussian distribution, and show that the result corresponds to the inverse Gaussian density with the location parameter $\left|\frac{\delta\sigma}{\beta_j}\right|$ and the scale parameter δ^2.*

Exercise 3.6 (Prior on the models) *Consider Bayesian model selection with d variables. Let $\gamma_j \in \{0, 1\}$ be the binary indicator for the jth predictor. The prior on the latent vector $\gamma = (\gamma_1, \ldots, \gamma_d)$ reflects our prior belief on the model selection.*

1. *If we have no preference on any model, how to specify the prior?*
2. *If we adhere to the principle of parsimony, how to specify the prior?*
3. *If we must include the jth predictor in the model, how to specify the prior?*
4. *If we must include the jth or $(j+1)^{th}$ predictor in the model, how to specify the prior?*
5. *If at most one of the jth or $(j+1)^{th}$ predictors can be in the model, how to specify the prior?*

Exercise 3.7 (Mixture priors and posteriors) *Let θ be the model parameters and y be the data. Let $p(\theta)$, $p(y\,|\theta)$, $p(y)$, $p(\theta\,|y)$ be the prior, likelihood, marginal likelihood and posterior, respectively. We specify a mixture prior with k classes:*

$$p(\theta) = \sum_{i=1}^{k} w_i p_i(\theta),$$

where w_i is the prior weight of the ith class, and $p_i(\theta)$ is the class-specific prior. Show that

1. the marginal likelihood is a mixture of class-specific marginal likelihoods, with the same weight

$$p(y) = \sum_{i=1}^{k} w_i p_i(y),$$

where $p_i(y) = \int p_i(\theta) p(y|\theta) d\theta$ is the class-specific marginal likelihood.

2. the posterior is a mixture of class-specific posteriors, with the updated weight

$$p(\theta|y) = \sum_{i=1}^{k} \bar{w}_i p_i(\theta|y),$$

where $p_i(\theta|y) = \frac{p_i(\theta)p(y|\theta)}{p_i(y)}$ is the class-specific posterior, and $\bar{w}_i = \frac{w_i p_i(y)}{p(y)}$ is the updated weight.

Exercise 3.8 (Marginal likelihood under the g prior) *Professor Zeller's g prior is widely used in Bayesian model selection. It is a special case of the conjugate prior, where the prior precision is proportional to the data precision. Conditional on the disturbance variance σ^2, the coefficient β follows multivariate normal distribution with the mean zero and the variance $\sigma^2 g (X'X)^{-1}$, where g is a positive scalar.*

1. Show that the posterior distribution (conditional on σ^2) is multivariate normal with the mean $\frac{g}{1+g}\hat{\beta}$ and the variance $\sigma^2 \frac{g}{1+g}(X'X)^{-1}$, where $\hat{\beta} = (X'X)^{-1} X'Y$. Does the result indicate shrinkage? How are the results related to $\bar{\mu}$ and $\bar{\Lambda}$ in Proposition 1.1?

2. By Bayes' rule, the marginal likelihood can be evaluated as

$$p(Y|\sigma^2) = \frac{p(\beta|\sigma^2) p(Y|\beta,\sigma^2)}{p(\beta|Y,\sigma^2)}.$$

As the identity holds for all β, we simply evaluate the marginal likelihood at $\beta = 0$. Show that

$$p(Y|\sigma^2) = (2\pi\sigma^2)^{-\frac{n}{2}} (1+g)^{-\frac{d}{2}} e^{-\frac{1}{2}\sigma^{-2}(Y'Y - \bar{\mu}'\bar{\Lambda}\bar{\mu})},$$

where $\bar{\mu} = \frac{g}{1+g}(X'X)^{-1} X'Y$ and $\bar{\Lambda} = \frac{1+g}{g} X'X$.

3. For the purpose of variable selection, we may change the number of predictors d and the predictor matrix X, but the response variable Y is fixed. All fixed terms can be treated as the proportionality constant, which simplifies the marginal likelihood evaluation. Show that

$$p(Y|\sigma^2) \propto (1+g)^{-\frac{d}{2}} e^{-\frac{1}{2}\sigma^{-2}\frac{g}{1+g}(Y-X\hat{\beta})'(Y-X\hat{\beta})}.$$

4. The marginal likelihood can also be obtained by the weighted average of the likelihood function, with the weights given by the prior. That is,

$$p(Y|\sigma^2) = \int p(\beta|\sigma^2) p(Y|\beta,\sigma^2) d\beta.$$

Compute the above integral. Is the result the same?

4

Correlation, Heteroscedasticity and Non-Gaussian Regressions

In linear regressions, the assumptions of independent and identically distributed Gaussian errors are often adopted for convenience. This is primarily because the Gaussian likelihood, when paired with a NIG conjugate prior, allows the posterior distribution to be analytically tractable. However, real-world data frequently exhibit regression residuals characterized by correlation, heteroscedasticity, skewness, and kurtosis. These empirical anomalies necessitate the exploration of more flexible model specifications that can accommodate non-diagonal covariance matrices, stochastic volatilities, skewness, fat tails, and multi-modal distributions. Although such models typically lack closed-form posterior distributions, conditionally conjugate structures can be exploited to design efficient MCMC samplers for approximating these posteriors.

4.1 General Error Covariance Matrix

Consider a linear regression model expressed in matrix form:

$$Y = X\beta + \varepsilon,$$

where the error term ε follows a multivariate normal distribution, with the covariance matrix $\Sigma = E(\varepsilon\varepsilon')$ assuming a form more general than $\sigma^2 I$. When the diagonal elements of the covariance matrix are not uniform, the situation is termed heteroscedasticity. Moreover, non-zero off-diagonal elements indicate correlation among the errors.

Heteroscedasticity and correlations can be addressed by data transformation: $\tilde{Y} = L^{-1}Y$ and $\tilde{X} = L^{-1}X$, where L is the "square root" of the covariance matrix $\Sigma = LL'$. Two prominent decomposition methods facilitate this transformation. The first is the Cholesky decomposition, which yields L as a lower triangular matrix. The second method involves diagonalization, represented as $\Sigma = UDU'$, where D is a diagonal matrix containing eigenvalues, and U is an orthonormal matrix of eigenvectors. Consequently, $L = UD^{1/2}$. Alternatively, a symmetric matrix square root can be constructed as $L = UD^{1/2}U'$.

Each method has its advantages and limitations: while the eigenvalue-based approach is applicable to positive semi-definite matrices, the Cholesky decomposition is restricted to positive definite matrices. However, eigenvalue algorithms are generally more computationally intensive than performing the Cholesky decomposition. Beyond these general-purpose methods, specific structural properties of the covariance matrix can be leveraged for efficient decomposition. For instance, in panel data regressions with random effects (refer to Section 8.6), the quasi-within transformation is used to manage correlations arising from unobserved heterogeneity.

DOI: 10.1201/9781003564027-4

The transformation of data restores the standard Gaussian linear regression:

$$\tilde{Y} = \tilde{X}\beta + L^{-1}\varepsilon,$$

where the transformed noise term $L^{-1}\varepsilon$ satisfies $E\left(L^{-1}\varepsilon\varepsilon' L^{-1\prime}\right) = I$.

The Gaussian likelihood function

$$p\left(Y\,|\beta,\Sigma\right) \propto e^{-\frac{1}{2}\left(\tilde{Y}-\tilde{X}\beta\right)'\left(\tilde{Y}-\tilde{X}\beta\right)} = e^{-\frac{1}{2}\left(Y-X\beta\right)'\Sigma^{-1}\left(Y-X\beta\right)}, \tag{4.1}$$

can be paired with a conditionally conjugate prior on β:

$$p\left(\beta\right) \propto e^{-\frac{1}{2}\left(\beta-\mu\right)'V^{-1}\left(\beta-\mu\right)}, \tag{4.2}$$

with μ and V as the prior mean and variance, respectively. Normality is preserved in the posterior conditional distribution shown in Proposition 4.1, which can be viewed as a corollary of Proposition 1.1 under transformed data conditional on $\sigma^2 = 1$. The posterior precision

$$\bar{\Lambda} = V^{-1} + \tilde{X}'\tilde{X} = V^{-1} + X'\Sigma^{-1}X$$

represents the inverse covariance matrix $\bar{V}$ in Proposition 4.1. Similarly, the posterior mean is $\left(V^{-1} + \tilde{X}'\tilde{X}\right)^{-1}\left(V^{-1}\mu + \tilde{X}'\tilde{Y}\right)$, which equals $\bar{\mu}$ in the proposition.

Proposition 4.1 (Posterior under general covariance matrix) *Under the prior (4.2) and the likelihood (4.1), $\beta\,|Y,\Sigma$ follows a multivariate normal distribution with mean $\bar{\mu}$ and covariance matrix $\bar{V}$, where*

$$\bar{\mu} = \left(V^{-1} + X'\Sigma^{-1}X\right)^{-1}\left(V^{-1}\mu + X'\Sigma^{-1}Y\right),$$

$$\bar{V} = \left(V^{-1} + X'\Sigma^{-1}X\right)^{-1}.$$

Proposition 4.1 addresses the posterior conditional distribution of the coefficients β. The subsequent discussion in this section focuses on the posterior conditional distribution of the covariance matrix Σ. Estimating an unconstrained covariance matrix is often impractical, as it typically possesses a specific structure governed by a limited number of structural parameters.

4.1.1 Seemingly Unrelated Regression

Seemingly Unrelated Regression (SUR), introduced by Zeller (1962), is a multivariate system wherein the predictors may differ across equations, yet the disturbances are correlated. Consider a system comprising m equations and n observations. Let Y_{ij} represent the response variable for the ith observation in the jth equation, and X_{ij} denote the corresponding predictors, expressed as a row vector that may vary in length across equations. The SUR model is formulated as

$$Y_{ij} = X_{ij}\beta_j + \varepsilon_{ij},$$

where $i = 1, \ldots, n$ and $j = 1, \ldots, m$. The disturbances $\varepsilon_{i1}, \ldots, \varepsilon_{im}$ follow a multivariate normal distribution with a non-diagonal covariance matrix Σ_m.

Although the data have two dimensions (m and n), SUR can be conceptualized as a linear regression by stacking mn data points in two steps. The first step is

$$Y_i = X_i\beta + \varepsilon_i,$$

where $Y_i = \begin{pmatrix} Y_{i1} \\ \cdots \\ Y_{im} \end{pmatrix}$, $X_i = diag\,(X_{i1}, \ldots, X_{im})$, $\beta = \begin{pmatrix} \beta_1 \\ \cdots \\ \beta_m \end{pmatrix}$ and $\varepsilon_i = \begin{pmatrix} \varepsilon_{i1} \\ \cdots \\ \varepsilon_{im} \end{pmatrix}$. The second step is

$$Y = X\beta + \varepsilon,$$

where $Y = \begin{pmatrix} Y_1 \\ \cdots \\ Y_n \end{pmatrix}$, $X = \begin{pmatrix} X_1 \\ \cdots \\ X_n \end{pmatrix}$ and $\varepsilon = \begin{pmatrix} \varepsilon_1 \\ \cdots \\ \varepsilon_n \end{pmatrix}$.

SUR exemplifies regressions under a general error covariance matrix, as the $mn \times mn$ positive definite matrix takes the form

$$\Sigma = E\left(\varepsilon\varepsilon'\right) = I_n \otimes \Sigma_m.$$

Bayesian model specification is completed by incorporating priors for β and Σ_m. When all equations share an identical set of predictors, the conjugate NIW prior is applicable, and as Proposition 1.9 elucidates, the posterior distribution retains the NIW form. Conversely, when predictors differ across equations, conditional conjugacy can be achieved by specifying a normal prior for β and an inverse Wishart prior for Σ_m such that

$$p\left(\beta, \Sigma_m\right) = p\left(\beta\right) p\left(\Sigma_m\right),$$

where $p\left(\beta\right)$ is given by Equation (4.2), and

$$p\left(\Sigma_m\right) \propto \left|\Sigma_m\right|^{-\frac{v+m+1}{2}} e^{-\frac{1}{2}tr\left(\Omega\Sigma_m^{-1}\right)}.$$

By conditional conjugacy, we mean the posterior conditional distributions maintain the normal and inverse Wishart distributions, thus facilitating the application of the Gibbs sampler for posterior simulation (Percy, 1992). Proposition 4.1 indicates that $\beta\,|Y, \Sigma_m$ follows a multivariate normal distribution with mean $\bar{\mu}$ and covariance matrix $\bar{V}$, where

$$\bar{V} = \left[V^{-1} + X'\left(I_n \otimes \Sigma_m\right)^{-1} X\right]^{-1},$$

$$\bar{\mu} = \bar{V}\left[V^{-1}\mu + X'\left(I_n \otimes \Sigma_m\right)^{-1} Y\right].$$

The posterior conditional distribution of the covariance matrix can be derived by $p\left(\Sigma_m\,|Y, \beta\right) \propto p\left(\Sigma_m\right) p\left(Y\,|\beta, \Sigma_m\right)$, where the likelihood function is

$$p\left(Y\,|\beta, \Sigma_m\right) \propto \left|\Sigma_m\right|^{-\frac{n}{2}} e^{-\frac{1}{2}tr\left[\sum_{i=1}^{n}(Y_i - X_i\beta)(Y_i - X_i\beta)'\Sigma_m^{-1}\right]}.$$

It follows that

$$p\left(\Sigma_m\,|Y, \beta\right) \propto \left|\Sigma_m\right|^{-\frac{\bar{v}+m+1}{2}} e^{-\frac{1}{2}tr\left(\bar{\Omega}\Sigma_m^{-1}\right)},$$

where

$$\bar{\Omega} = \Omega + \sum_{i=1}^{n}\left(Y_i - X_i\beta\right)\left(Y_i - X_i\beta\right)',$$

$$\bar{v} = v + n.$$

The posterior conditional distribution remains an inverse Wishart distribution with updated parameters $\bar{\Omega}$ and $\bar{v}$.

The Gibbs sampler cycles through the posterior conditional distributions. In the context of big data applications, the computational demand of scanning the entire dataset at each iteration of the Gibbs sampler can be substantial. Consider a marginalized MCMC sampler targeting $p(\beta|Y) \propto p(\beta)p(Y|\beta)$, with the partially marginalized likelihood function

$$p(Y|\beta) = \frac{p(\Sigma_m|\beta)p(Y|\beta,\Sigma_m)}{p(\Sigma_m|Y,\beta)},$$

where $p(\Sigma_m|\beta)$ and $p(\Sigma_m|Y,\beta)$ are inverse Wishart densities with parameters Ω, v and $\bar{\Omega}, \bar{v}$, respectively, and $p(Y|\beta,\Sigma_m)$ is the Gaussian likelihood. Consequently, the partially marginalized likelihood function is analytically tractable, taking the form

$$p(Y|\beta) \propto |\bar{\Omega}|^{-\frac{\bar{v}}{2}}.$$

Since the marginalized target density $p(\beta|Y)$ can be evaluated pointwise up to a normalizing constant, the MH sampler can be used for posterior inference. This is a big data algorithm that requires only one pass of data, since $\bar{\Omega}$ is determined by sufficient statistics (e.g., $\sum_i X_i'X_i$). Specifically, the (j,k) element of $\bar{\Omega}$ is given by

$\Omega_{jk} + \sum_{i=1}^{n}(Y_{ij} - X_{ij}\beta_j)(Y_{ik} - X_{ik}\beta_k)$

$= \Omega_{jk} + \sum_i Y_{ij}Y_{ik} - \sum_i Y_{ik}X_{ij}\beta_j - \sum_i Y_{ij}X_{ik}\beta_k + \beta_j'\sum_i X_{ij}'X_{ik}\beta_k.$

4.1.2 Autoregression

In contrast to cross-sectional data, a defining characteristic of time series data is the presence of autocorrelations. A parsimonious approach to modeling these autocorrelations is through autoregressive models or regressions with autoregressive disturbances. Consider a linear regression model with correlated disturbances:

$$Y_t = X_t\beta + \varepsilon_t,$$

$$\varepsilon_t = \phi\varepsilon_{t-1} + \sigma u_t,$$

where u_t follows the standard normal distribution. Assume that $|\rho| < 1$ and the initial value ε_0 follows the stationary distribution $N\left(0, \frac{\sigma^2}{1-\rho^2}\right)$.

Data matrices for time series can be structured analogously to cross-sectional data by stacking observations $t = 1, \ldots, n$. The matrix form of the model is

$$Y = X\beta + \varepsilon,$$

where $Y = \begin{pmatrix} Y_1 \\ \cdots \\ Y_n \end{pmatrix}$, $X = \begin{pmatrix} X_1 \\ \cdots \\ X_n \end{pmatrix}$ and $\varepsilon = \begin{pmatrix} \varepsilon_1 \\ \cdots \\ \varepsilon_n \end{pmatrix}$.

This scenario exemplifies regressions with a general error covariance matrix, where $\Sigma = E(\varepsilon\varepsilon')$ is non-diagonal (see Exercise 4.1):

$$\Sigma = \frac{\sigma^2}{1-\rho^2}\begin{pmatrix} 1 & \rho & \cdots & \rho^{n-1} \\ \rho & 1 & \cdots & \rho^{n-2} \\ \cdots & \cdots & \ddots & \cdots \\ \rho^{n-1} & \rho^{n-2} & \cdots & 1 \end{pmatrix}.$$

Regarding Bayesian estimation, the Gibbs sampler cycles through the posterior conditional distributions $p(\beta|Y,\rho,\sigma^2)$ and $p(\rho,\sigma^2|Y,\beta)$. Assume that β is assigned a normal prior with mean μ and variance V. Proposition 4.1 indicates that $\beta|Y,\rho,\sigma^2$ follows a multivariate normal distribution with mean $\bar{\mu}$ and covariance matrix $\bar{V}$, where

$$\bar{V} = \left(V^{-1} + X'\Sigma^{-1}X\right)^{-1},$$

$$\bar{\mu} = \bar{V}\left(V^{-1}\mu + X'\Sigma^{-1}Y\right).$$

Computing $X'\Sigma^{-1}X$ through direct matrix multiplication and inversion can be computationally demanding, especially with a large number of observations. A more efficient approach involves Cholesky decomposition of the covariance matrix, which is analytically tractable in the autoregressive model:

$$X'\Sigma^{-1}X = \left(L^{-1}X\right)\left(L^{-1}X\right),$$

where

$$L^{-1} = \sigma^{-1}\begin{pmatrix} \sqrt{1-\rho^2} & 0 & 0 & \cdots & 0 & 0 \\ -\rho & 1 & 0 & \cdots & 0 & 0 \\ 0 & -\rho & 1 & \cdots & 0 & 0 \\ \cdots & \cdots & \cdots & \ddots & \cdots & \cdots \\ 0 & 0 & 0 & \cdots & -\rho & 1 \end{pmatrix}.$$

The matrix decomposition process can be interpreted as data transformation. Let $\tilde{X} = L^{-1}X$ be the transformed predictors and $\tilde{Y} = L^{-1}Y$ be the transformed response variable. The first observation is rescaled, while the subsequent observations are quasi-differenced: $\tilde{X}_t = \sigma^{-1}\left(X_t - \rho X_{t-1}\right)$ and $\tilde{Y}_t = \sigma^{-1}\left(Y_t - \rho Y_{t-1}\right)$. The transformed data are then utilized in a standard regression

$$\tilde{Y}_t = \tilde{X}_t\beta + u_t,$$

where u_t follows the standard normal distribution.

Although the exact distribution of $p\left(\rho, \sigma^2 \left| Y, \beta\right.\right)$ is not analytically tractable, it assumes a simplified form when conditioned on a presample value $\varepsilon_0 = 0$. Given Y and β, we infer $\varepsilon_t = Y_t - X_t\beta$, which serves as the data for the regression $\varepsilon_t = \phi\varepsilon_{t-1} + \sigma u_t$. Assume that ρ, σ^2 have the prior $NIG\left(\mu_\rho, \Lambda_\rho, a, b\right)$. Proposition 1.1 provides the posterior $NIG\left(\bar{\mu}_\rho, \bar{\Lambda}_\rho, \bar{a}, \bar{b}\right)$, where

$$\bar{\mu}_\rho = \left(\Lambda_\rho + \sum_{t=1}^{n}\varepsilon_{t-1}^2\right)^{-1}\left(\Lambda_\rho\mu_\rho + \sum_{t=1}^{n}\varepsilon_t\varepsilon_{t-1}\right),$$

$$\bar{\Lambda}_\rho = \Lambda_\rho + \sum_{t=1}^{n}\varepsilon_{t-1}^2,$$

$$\bar{a} = a + \frac{n}{2},$$

$$\bar{b} = b + \frac{1}{2}\sum_{t=1}^{n}\varepsilon_t^2 + \frac{1}{2}\mu_\rho'\Lambda_\rho\mu_\rho - \frac{1}{2}\bar{\mu}_\rho'\bar{\Lambda}_\rho\bar{\mu}_\rho.$$

To impose a stationarity constraint $|\rho| < 1$, the NIG prior can be truncated. Consequently, the posterior NIG distribution is also truncated. Rejection sampling can be employed by generating unconstrained NIG proposals until the constraint is satisfied.

While the Gibbs sampler is the preferred method for estimating this model, alternative approaches exist. For instance, sampling from the joint posterior distribution $p\left(\beta, \rho, \sigma^2 \left| Y\right.\right)$ by the MH sampler, since the likelihood function can be computed as

$$p\left(Y \left| \beta, \rho, \sigma^2\right.\right) = \prod_{t=1}^{n} p\left(Y_t \left| Y_{t-1}, \beta, \rho, \sigma^2\right.\right),$$

where $p\left(Y_t \left| Y_{t-1}, \beta, \rho, \sigma^2\right.\right)$ is a normal density with mean $X_t\beta + \phi\left(Y_{t-1} - X_{t-1}\beta\right)$ and variance σ^2.

From the perspective of time series analysis, the current model is a specific instance of the regression models with $ARMA\,(p,q)$ errors. The seminal paper by Chib and Greenberg (1994) discusses the Gibbs and MH sampling using the state space representation of ARMA models (see Section 8.1).

4.1.3 Heteroscedasticity

To relax the homoscedasticity assumption in linear regression, a multiplicative form of heterogeneity can be introduced to the regression (Aitkin, 1987; Cepeda and Gamerman, 2001):

$$Y_i = X_i\beta + e^{\frac{1}{2}Z_i}\varepsilon_i,$$
$$Z_i = W_i\alpha,$$

where ε_i follows the standard normal distribution. The regression error, defined as $Y_i - X_i\beta$, has a variance of e^{Z_i}, which is a deterministic function of the predictors W_i. This assumption will be relaxed in the subsequent section. The model specification is completed by incorporating priors. Let the prior of β, α be normal with means μ_β, μ_α and precision matrices $\Lambda_\beta, \Lambda_\alpha$ (the covariance matrices are $\Lambda_\beta^{-1}, \Lambda_\alpha^{-1}$).

Li and Tobias (2009) propose a Metropolis-within-Gibbs sampler that iteratively draws from the conditional distributions $p(\beta\,|Y,\alpha)$ and $p(\alpha\,|Y,\beta)$. The former follows a conjugate normal distribution, while the latter does not conform to a known distribution form, necessitating the inclusion of a Metropolis component.

Specifically, the error covariance matrix assumes a diagonal form:

$$\Sigma = diag\left(e^{Z_1}, \ldots, e^{Z_n}\right).$$

By Proposition 4.1, the distribution $p(\beta\,|Y,\alpha)$ are multivariate normal with mean $\bar{\mu}_\beta$ and precision $\bar{\Lambda}_\beta$, where

$$\bar{\Lambda}_\beta = \Lambda_\beta + X'\Sigma^{-1}X,$$
$$\bar{\mu}_\beta = \bar{\Lambda}_\beta^{-1}\left(\Lambda_\beta\mu_\beta + X'\Sigma^{-1}Y\right).$$

The diagonal matrix Σ simplifies data transformation: $X'\Sigma^{-1}X = \sum_{i=1}^{n} e^{-Z_i}X_i'X_i$ and $X'\Sigma^{-1}Y = \sum_{i=1}^{n} e^{-Z_i}X_i'Y_i$. The observations can be homogenized through dividing X_i, Y_i by $e^{\frac{1}{2}Z_i}$, thereby eliminating heteroscedasticity and reverting to the standard Bayesian linear regression framework.

The parameter vector α governs the conditional variance of Gaussian observations. Conditional on the parameters α and β, the observation Y_i follows a normal distribution with mean $X_i\beta$ and variance e^{Z_i}. The likelihood function is given by

$$p(Y\,|\alpha,\beta) \propto e^{-\frac{1}{2}\sum_{i=1}^{n}\left[Z_i + e^{-Z_i}(Y_i - X_i\beta)^2\right]}.$$

It follows that the posterior conditional distribution of α is

$$p(\alpha\,|Y,\beta) \propto e^{-\frac{1}{2}\left\{(\alpha-\mu_\alpha)\Lambda_\alpha(\alpha-\mu_\alpha) + \sum_{i=1}^{n}\left[Z_i + e^{-Z_i}(Y_i - X_i\beta)^2\right]\right\}}.$$

Although the density does not possess a recognizable form, it can be evaluated pointwise up to a normalizing constant, prompting the use of the MH algorithm to sample from α, hence the Metropolis-within-Gibbs sampler.

Alternatively, the parameter vector β can be integrated out for a marginalized MH sampler with the target distribution $p(\alpha\,|Y) \propto p(\alpha)\,p(Y\,|\alpha)$, where

$$p(Y\,|\alpha) = \frac{p(\beta\,|\alpha)\,p(Y\,|\alpha,\beta)}{p(\beta\,|Y,\alpha)}.$$

Note that $p(\beta\,|\alpha)$ and $p(\beta\,|Y,\alpha)$ are normal densities with parameters μ_β, Λ_β and $\bar{\mu}_\beta, \bar{\Lambda}_\beta$, respectively. It follows that the partially marginalized likelihood can be evaluated up to a normalizing constant by

$$p(Y\,|\alpha) \propto \left|\bar{\Lambda}_\beta\right|^{-\frac{1}{2}} e^{-\frac{1}{2}\sum_{i=1}^{n} Z_i - \frac{1}{2}\sum_{i=1}^{n} e^{-Z_i} Y_i^2 + \frac{1}{2}\bar{\mu}_\beta' \bar{\Lambda}_\beta \bar{\mu}_\beta}.$$

Upon further inspection at the equation $Y_i = X_i\beta + e^{\frac{1}{2}Z_i}\varepsilon_i$, we note that $p(Y\,|\alpha)$is a multivariate normal density, with mean $X\mu_\beta$ and covariance matrix $X\Lambda_\beta^{-1}X' + \Sigma$. While it may be tempting to evaluate a n dimensional normal density directly, the computational complexity amounts to $O\left(n^3\right)$, rendering it inefficient when the number of observations is large.

4.2 Stochastic Volatility

In the previous section, the conditional variance is assumed to be a deterministic function of predictors. A natural extension involves randomizing the conditional variance, leading to the stochastic volatility, a concept widely embraced by researchers and practitioners in economics and finance.

Following our previous notations, we specify the stochastic volatility model as

$$Y_i = X_i\beta + e^{\frac{1}{2}Z_i}\varepsilon_i,$$

$$Z_i = W_i\alpha + \sigma u_i,$$

where ε_i and u_i follow the standard normal distributions. The observations exhibit conditional heteroskedasticity in the form of e^{Z_i}, yet Z_i functions as a latent variable. In cross-sectional regressions, where the index i represents a microeconomic unit, the conditional variance is determined by observed factors W_i and unit-specific unobserved heterogeneity u_i. In time series regressions, where the index i represents a point in time, the model accommodates autoregressive conditional heteroscedasticity if the predictors W_i include lagged variables such as Z_{i-1}, Y_{i-1}^2, etc.

Stochastic volatility diverges from Gaussian linear regression and poses significant estimation challenges due to the lack of a closed-form likelihood function. This complexity prompts the use of various approximation methods, primarily within the nonlinear state space framework, which will be explored in Chapter 7. Here, we present a novel solution that bypasses the state space model and simulated likelihood functions, offering an exact and scalable MCMC approach suitable for big data applications.

The stochastic volatility model features conditionally linear sub-structures: conditioning on the latent variables simplifies the model as two linear regressions, allowing for the specification of conditionally conjugate priors. Let the prior of β be multivariate normal with mean μ_β and precision Λ_β (the covariance matrix is Λ_β^{-1}). Let the prior of α, σ^2 be $NIG\,(\mu_\alpha, \Lambda_\alpha, a, b)$.

Given the latent variables, heteroscedasticity can be eliminated by redefining $e^{-\frac{1}{2}Z_i}Y_i$ as the response variable and $e^{-\frac{1}{2}Z_i}X_i$ as predictors, resulting in a standard Bayesian linear regression. The posterior conditional distribution $p(\beta\,|Y,Z)$ is multivariate normal with mean $\bar{\mu}_\beta$ and precision $\bar{\Lambda}_\beta$, where

$$\bar{\Lambda}_\beta = \Lambda_\beta + \sum_{i=1}^{n} e^{-Z_i} X_i' X_i,$$

$$\bar{\mu}_\beta = \bar{\Lambda}_\beta^{-1}\left(\Lambda_\beta \mu_\beta + \sum_{i=1}^{n} e^{-Z_i} X_i' Y_i\right).$$

Given the latent variables, the equation $Z_i = W_i\alpha + \sigma u_i$ can be viewed as another linear regression, even if W_i contains autoregressive terms like Z_{i-1}. By Proposition 1.1, the posterior conditional distribution $p(\alpha, \sigma^2 | Y, Z)$ follows $NIG(\bar{\mu}_\alpha, \bar{\Lambda}_\alpha, \bar{a}, \bar{b})$, where

$$\bar{\mu}_\alpha = \bar{\Lambda}_\alpha^{-1}\left(\Lambda_\alpha \mu_\alpha + \sum_{i=1}^{n} W_i' Z_i\right),$$

$$\bar{\Lambda}_\alpha = \Lambda_\alpha + \sum_{i=1}^{n} W_i' W_i,$$

$$\bar{a} = a + \frac{n}{2},$$

$$\bar{b} = b + \frac{1}{2}\sum_{i=1}^{n} Z_i^2 + \frac{1}{2}\mu_\alpha' \Lambda_\alpha \mu_\alpha - \frac{1}{2}\bar{\mu}_\alpha' \bar{\Lambda}_\alpha \bar{\mu}_\alpha.$$

Whenever the posterior density is expressible in a closed form, the marginal likelihood can be evaluated exactly using Bayes' rule. The double marginalized subsampling is well-suited for Bayesian estimation of the stochastic volatility model.

First, decompose the joint posterior distribution as

$$p(\beta, \alpha, \sigma^2, Z | Y) = p(Z | Y) p(\beta | Y, Z) p(\alpha, \sigma^2 | Z).$$

Second, further decompose $p(Z | Y)$ into the product of two marginal likelihood terms by Bayes' rule:

$$p(Z | Y) = p(Z) p(Y | Z),$$

where

$$p(Z) = \frac{p(\alpha, \sigma^2) p(Z | \alpha, \sigma^2)}{p(\alpha, \sigma^2 | Z)},$$

$$p(Y | Z) = \frac{p(\beta | Z) p(Y | \beta, Z)}{p(\beta | Y, Z)}.$$

On the one hand, $p(\alpha, \sigma^2)$ and $p(\alpha, \sigma^2 | Z)$ are densities of $NIG(\mu_\alpha, \Lambda_\alpha, a, b)$ and $NIG(\bar{\mu}_\alpha, \bar{\Lambda}_\alpha, \bar{a}, \bar{b})$, respectively. Consequently, the term $p(Z)$ can be evaluated as

$$p(Z) \propto (\bar{b})^{-\bar{a}} |\bar{\Lambda}_\alpha|^{-\frac{1}{2}}.$$

On the other hand, $p(\beta | Z)$ and $p(\beta | Y, Z)$ are densities of multivariate normal distributions with parameters μ_β, Λ_β and $\bar{\mu}_\beta, \bar{\Lambda}_\beta$, respectively. Consequently, the term $p(Y | Z)$ can be evaluated as

$$p(Y | Z) \propto |\bar{\Lambda}_\beta|^{-\frac{1}{2}} e^{-\frac{1}{2}\sum_{i=1}^{n} Z_i - \frac{1}{2}\sum_{i=1}^{n} e^{-Z_i} Y_i^2 + \frac{1}{2}\bar{\mu}_\beta' \bar{\Lambda}_\beta \bar{\mu}_\beta}.$$

The target distribution $p(Z | Y)$ can be evaluated up to a normalizing constant, enabling a marginalized MCMC method that focuses on sampling the latent variables. Conditional on the posterior draws of Z, posterior simulation for β can be performed using the multivariate normal distribution, as well as simulating α, σ^2 from $NIG(\bar{\mu}_\alpha, \bar{\Lambda}_\alpha, \bar{a}, \bar{b})$.

Notably, the density $p(Z\,|Y)$ is fully determined by a set of cumulative statistics, akin to sufficient statistics but dependent on Z:

$$S_1(Z) = \sum_{i=1}^{n} e^{-Z_i} X_i' X_i,$$

$$S_2(Z) = \sum_{i=1}^{n} e^{-Z_i} X_i' Y_i,$$

$$S_3(Z) = \sum_{i=1}^{n} e^{-Z_i} Y_i^2,$$

$$S_4(Z) = \sum_{i=1}^{n} Z_i,$$

$$S_5(Z) = \sum_{i=1}^{n} W_i' W_i,$$

$$S_6(Z) = \sum_{i=1}^{n} W_i' Z_i,$$

$$S_7(Z) = \sum_{i=1}^{n} Z_i^2.$$

The MH sampler tracks these cumulative statistics under the current draw of Z. Consider a proposal Z^* that differs from Z by one element (e.g., the ith element in the vector). Instead of recalculating $S_1(Z^*)$ and $S_2(Z^*)$ by sweeping over the entire dataset, the ith data point is processed for a rank-one update of cumulative statistics:

$$S_1(Z^*) = S_1(Z) - e^{-Z_i} X_i' X_i + e^{-Z_i^*} X_i' X_i,$$
$$S_2(Z^*) = S_2(Z) - e^{-Z_i} X_i' Y_i + e^{-Z_i^*} X_i' Y_i,$$

and similarly for the remaining cumulative statistics.

The random-walk MH sampler for double marginalized subsampling is detailed in Algorithm 4.1. This algorithm is designed for big data applications, allowing parameters β, α, σ^2 to be updated with each individual observation or at an arbitrary frequency, in contrast with traditional MCMC methods that update parameters after processing the entire dataset. Algorithm 4.1 is both computationally and statistically efficient, as it employs a blocking (grouping) strategy, jointly updating parameters and latent variables, thereby reducing correlations of MCMC draws and enhancing the mixing of the Markov chain

4.3 Fat Tail and Scale Mixture Distribution

A notable limitation of Gaussian regressions is that the tails of normal distributions are too thin to adequately capture the low-probability large shocks in macroeconomic data (Fagiolo et al., 2008; Ascari et al., 2015). Similarly, financial data suggest that the probabilities of extreme profits or losses can be significantly higher than those predicted by normal distributions. Consequently, risk management necessitates research on the tails of distributions, in addition to their central tendencies.

Algorithm 4.1 Double marginalized subsampling for stochastic volatility

Input: data Y, X, W, initial value Z, and tuning parameter λ.
Output: posterior samples from $p\left(\beta, \alpha, \sigma^2, Z \,|Y\right)$.
1 Compute cumulative statistics $S_j\left(Z\right)$, $j = 1, \ldots, 7$ by one sweep over data.
2 Iterate the following steps:
  2.1 Draw a random index $i \in \{1, \ldots, n\}$.
  2.2 Propose Z^* such that $Z^*_{-i} = Z_{-i}$, $Z^*_i = Z_i + \lambda e_i$, $e_i \sim N\left(0, 1\right)$.
  2.3 Access i^{th} data point only.
  2.4 Compute cumulative statistics $S_j\left(Z^*\right)$ by rank-one update of $S_j\left(Z\right)$.
  2.5 Evaluate $p\left(Z^* \,|Y\right)$ by cumulative statistics.
  2.6 Accept proposal (set $Z = Z^*$) with probability $\min\left[1, \frac{p(Z^* |Y)}{p(Z|Y)}\right]$.
  2.7 Sample β from $p\left(\beta \,|Y, Z\right)$ by cumulative statistics.
  2.8 Sample α, σ^2 from $p\left(\alpha, \sigma^2 \,|Z\right)$ by cumulative statistics.

Many distributions possess heavier tails than normal distributions. However, deviating from normality increases Bayesian estimation difficulties due to the absence of conjugate distributions. This section focuses on scale mixtures of normal distributions, which strike a balance between flexibility and tractability. Mixture distributions accommodate excess kurtosis and fat tails compared to normal distributions. Data augmentation and conditionally conjugate structures facilitate MCMC sampling from the posterior conditional distributions.

Consider a scale mixture regression model

$$Y_i = X_i\beta + e_i, \tag{4.3}$$

$$e_i = \sigma\sqrt{Z_i}\varepsilon_i, \tag{4.4}$$

where ε_i represents a standard normal noise term, implying that $e_i \,|Z_i$ follows a normal distribution with the conditional variance $Z_i\sigma^2$. However, the variance is randomized by a latent variable Z_i, akin to stochastic volatility discussed in Section 4.2. Here, we relax the normality assumption, allowing the latent variable to follow log-normal, exponential, inverse gamma, or any positive-valued distribution.

Integrating out the latent variable yields the marginal distribution

$$p\left(e_i\right) = \int p\left(Z_i\right) p\left(e_i \,|Z_i\right) dZ_i.$$

Generally, this integral is not analytically tractable, implying that the likelihood function $p\left(Y_i \,|\beta, \sigma^2\right)$ cannot be evaluated in closed form. This complexity contributes to the difficulty of estimating the stochastic volatility and the scale mixture models.

The reminder of this section explores two special cases before delving into general solutions for scale mixture models. The special cases involve inverse gamma and t distributions, as well as exponential and double exponential distributions. For both cases, the Gibbs sampler cycles through the posterior conditional distributions for posterior inference. In more general scenarios where Z_i is a positive random variable without tractable posterior conditionals, a marginalized subsampling method can be employed to handle the scale mixture distribution, provided that the density function $p\left(Z_i\right)$ can be evaluated pointwise. Even in scenarios where the density function cannot be evaluated, pseudo marginal MCMC methods serve as a last resort, assuming $p\left(Z_i\right)$ can be approximated by an unbiased estimator via Monte Carlo simulations.

4.3.1 Kurtosis

Kurtosis is a statistical measure that describes the tails of a distribution. Higher kurtosis indicates a greater propensity to produce outliers, which are extreme values that deviate significantly from the center of the distribution.

Definition 4.1 (Kurtosis) *Let μ and σ be the mean and standard deviation of a random variable x. Kurtosis is defined as the scaled fourth moment:*

$$\kappa = E\left[\left(\frac{x-\mu}{\sigma}\right)^4\right].$$

The kurtosis of normal distributions is always equal to 3. A distribution with kurtosis larger than 3 is termed leptokurtic. Proposition 4.2demonstrates that scale mixtures of normal distributions must be leptokurtic, regardless of the distribution of latent variables. Additionally, kurtosis increases with the variance of latent variables.

Proposition 4.2 (Leptokurtic scale mixture) *Consider a scale mixture of normal variates $e_i = \sigma\sqrt{Z_i}\varepsilon_i$, where the latent variable Z_i has mean μ_z and variance σ_z^2. Independently, ε_i follows the standard normal distribution. The kurtosis of e_i is given by*

$$\kappa = 3 + 3\frac{\sigma_z^2}{\mu_z^2}.$$

4.3.2 Scale Mixture t Distribution

Proposition 4.2 indicates that scale mixture distributions can accommodate extreme values and fat tails. However, if a fat-tail model is not analytically tractable, it presents a higher estimation risk, necessitating numerical methods to approximate an intractable integral. The t distribution strikes a balance between flexibility and tractability, as its scale mixture representation restores conjugacy when conditioned on latent variables.

Definition 4.2 (t distribution) *If a random variable x has the density*

$$f\left(x;\mu,\sigma^2,v\right) \propto \left[1+\frac{1}{v}\left(\frac{x-\mu}{\sigma}\right)^2\right]^{-\frac{v+1}{2}},$$

it follows the t distribution $t\left(\mu,\sigma^2,v\right)$, where μ,σ^2,v are the location, scale and degrees of freedom parameters, respectively.

Exercise 4.2 explores characteristics of t distributions. Regarding tail behavior, a smaller degrees of freedom results in a fatter tail. When $v = 1$, the t distribution reduces to the Cauchy distribution, whose tail is so thick that even the expectation diverges. As $v \to \infty$, the t distribution approaches the normal distribution with a thin tail.

A key statistical property of the t distribution is its representation as a scale mixture of normal distributions, where the latent variable follows an inverse gamma distribution. For classical inference, t statistics are constructed via this scale mixture (see Exercise 4.2). For Bayesian inference, the scale mixture facilitates the Gibbs sampler through data augmentation (Geweke, 1993).

Proposition 4.3 (Scale mixture t distribution) *If $\tau \sim IG(a,b)$ and $x|\tau \sim N\left(\mu,\tau\sigma^2\right)$, then $x \sim t\left(\mu,\frac{b}{a}\sigma^2,2a\right)$.*

The mean and variance of the inverse gamma distribution $IG(a,b)$ are $\frac{b}{a-1}$ and $\frac{b^2}{(a-1)^2(a-2)}$, respectively. By Proposition 4.2, the kurtosis of the t distribution equals $3+\frac{6}{v-4}$. The leptokurtic property implies that the t distribution is more tolerant of extreme values, compared to the normal distribution. The t distribution is frequently employed in robust regression to detect and accommodate outliers. It may be tempting to fix the degrees of freedom at $v=5$ to ensure a fat tail with a finite fourth moment. Alternatively, it is valid to specify smaller degrees of freedom or treat it as an unknown parameter, allowing the data to estimate it.

Consider Bayesian estimation of the t-distributed linear regression. Equations (4.3) and (4.4) are reproduced below for convenience.

$$Y_i = X_i\beta + e_i,$$

$$e_i = \sigma\sqrt{Z_i}\varepsilon_i,$$

where ε_i follows the standard normal distribution, and the latent variable Z_i follows $IG\left(\frac{v}{2},\frac{v}{2}\right)$. By Proposition 4.3, the marginal distribution of e_i is $t\left(0,\sigma^2,v\right)$.

The essence of the scale mixture is that the model simplifies to a standard Bayesian linear regression when conditioned on Z_i. Redefining the response variable and predictors as $\tilde{Y}_i = Z_i^{-\frac{1}{2}}Y_i$ and $\tilde{X}_i = Z_i^{-\frac{1}{2}}X_i$ leads to a Gaussian linear regression:

$$\tilde{Y}_i = \tilde{X}_i\beta + \sigma\varepsilon_i.$$

Under the conjugate prior $NIG(\mu,\Lambda,a,b)$, the posterior $\beta,\sigma^2\,|Y,Z$ follows $NIG(\bar{\mu},\bar{\Lambda},\bar{a},\bar{b})$, where

$$\bar{\mu} = \bar{\Lambda}^{-1}\left(\Lambda\mu + \sum_{i=1}^{n} Z_i^{-1}X_i'Y_i\right),$$

$$\bar{\Lambda} = \Lambda + \sum_{i=1}^{n} Z_i^{-1}X_i'X_i,$$

$$\bar{a} = a + \frac{n}{2},$$

$$\bar{b} = b + \frac{1}{2}\sum_{i=1}^{n} Z_i^{-1}Y_i^2 + \frac{1}{2}\mu'\Lambda\mu - \frac{1}{2}\bar{\mu}'\bar{\Lambda}\bar{\mu}.$$

The scale mixture representation also implies that the model reduces to a trivial regression conditional on β,σ^2. The equation $Y_i = X_i\beta+\sigma\sqrt{Z_i}\varepsilon_i$ can be interpreted as a Bayesian linear regression with 1) a response variable $(Y_i - X_i\beta)/\sigma$, 2) no predictors (as an empty matrix), 3) only one observation (i.e., the ith data point), and 4) a latent variable Z_i with the prior $IG\left(\frac{v}{2},\frac{v}{2}\right)$. By Proposition 1.1, the posterior $Z_i\,\left|Y,\beta,\sigma^2\right.$ follows $IG(a_z,b_z)$, where

$$a_z = \frac{v+1}{2},$$

$$b_z = \frac{v}{2} + \frac{1}{2}\left(\frac{Y_i - X_i\beta}{\sigma}\right)^2.$$

If the degrees of freedom parameter v is fixed at a user-specified value, the Gibbs sampler cycles through the full posterior conditionals $p\left(\beta,\sigma^2\,|Y,Z\right)$ and $p\left(Z_i\,|Y,\beta,\sigma^2\right)$, $i=1,\ldots,n$. Alternatively, uncertainty in v be addressed by assigning it a prior, such as a uniform distribution in the interval $[\underline{v},\bar{v}]$. The Gibbs sampler also updates v from $p(v\,|Y,Z)$. Although

Algorithm 4.2 Metropolis-within-Gibbs sampler of t-distributed regression

Input: data X, Y, initial value Z, and tuning parameter λ.
Output: posterior samples from $p(\beta, \sigma^2, Z, v \mid Y)$.
Iterate the following steps:
1 Sample β, σ^2 from $NIG(\bar{\mu}, \bar{\Lambda}, \bar{a}, \bar{b})$, given the current value Z, v.
2 Sample $Z_i, i = 1, \ldots, n$ from $IG(a_z, b_z)$, given the current value β, σ^2, v.
3 Propose $v^* = v + \lambda u$, $u \sim N(0, 1)$.
If $v^* \in [\underline{v}, \bar{v}]$, accept proposal with probability $\min\left[1, \frac{p(Z|v^*)}{p(Z|v)}\right]$.

this is not a common distribution, the posterior conditional density can be evaluated using the density of $IG\left(\frac{v}{2}, \frac{v}{2}\right)$:

$$p(v \mid Y, Z) \propto p(Z \mid v) I(\underline{v} \leq v \leq \bar{v}),$$

$$p(Z \mid v) = \exp\left[-n \ln \Gamma\left(\frac{v}{2}\right) + \frac{vn}{2} \ln \frac{v}{2} - \left(\frac{v}{2} + 1\right) \sum_{i=1}^{n} \ln Z_i - \frac{v}{2} \sum_{i=1}^{n} Z_i^{-1}\right].$$

The Metropolis-within-Gibbs sampler is detailed in Algorithm 4.2.

4.3.3 Double Exponential Distribution

Inspired by Bayesian lasso (Park and Casella, 2008) and t-distributed regressions (Geweke, 1993), we consider Equations (4.3) and (4.4), in which the latent variable Z_i follows an exponential distribution:

$$p(Z_i) = \frac{1}{2} e^{-\frac{1}{2} Z_i}.$$

By marginalization over the latent variable, Proposition 3.2 indicates that the regression error $e_i = Y_i - X_i \beta$ follows a double exponential distribution:

$$p(e_i) = \int p(Z_i) p(e_i \mid Z_i) \, dZ_i = \frac{1}{2\sigma} e^{-\frac{|e_i|}{\sigma}}.$$

The tail of the double exponential distribution is thicker than that of the normal distribution, since e^{-x^2} converges to zero faster than $e^{-|x|}$ as $x \to \infty$. Moreover, the double exponential distribution exhibits excess kurtosis. The mean and variance of the latent variable are 2 and 4, respectively. By Proposition 4.2, the kurtosis of the double exponential distribution equals 6, which is twice that of the normal distribution.

Bayesian parameter estimation of the double exponential model can be implemented by the Gibbs sampler that cycles through the full posterior conditionals $p(\beta, \sigma^2 \mid Y, Z)$ and $p(Z_i \mid Y, \beta, \sigma^2)$. The former follows $NIG(\bar{\mu}, \bar{\Lambda}, \bar{a}, \bar{b})$ under the prior $NIG(\mu, \Lambda, a, b)$, mirroring the t-distributed regression. The latter is also analytically tractable. Similar to Bayesian lasso regression, the posterior conditional distribution of Z_i^{-1} is inverse Gaussian. Proposition 3.4 indicates that

$$p(Z_i^{-1} \mid Y, \beta, \sigma^2) \propto Z_i^{\frac{3}{2}} \exp\left[-\frac{1}{2}\left(\frac{Y_i - X_i \beta}{\sigma}\right)^2 Z_i^{-1} - \frac{1}{2} Z_i\right],$$

corresponding to an inverse Gaussian density with parameters $\left|\frac{\sigma}{Y_i - X_i \beta}\right|$ and 1.

4.3.4 Marginalized Subsampling and Big Data Algorithm

A closed-form scalar mixture is an exception rather than the rule. Apart from some special cases like inverse gamma and exponential mixtures, marginal distributions and posterior conditional distributions generally do not have a recognizable form, and the convenience of Gibbs sampling is lost.

Consider the scalar mixture model in its general form. Equations (4.3) and (4.4) are reproduced here:

$$Y_i = X_i\beta + e_i,$$
$$e_i = \sigma\sqrt{Z_i}\varepsilon_i,$$

where the latent variable Z_i is a positive random variable with an arbitrary distribution. The only requirement is the ability to evaluate the density function $p(Z_i)$ pointwise up to a normalizing constant.

Decompose the joint distribution of model parameters and latent variables as

$$p\left(\beta,\sigma^2,Z\,|Y\right) = p(Z\,|Y)\,p\left(\beta,\sigma^2\,|Y,Z\right).$$

Under a conjugate prior $NIG(\mu,\Lambda,a,b)$, the posterior $p\left(\beta,\sigma^2\,|Y,Z\right)$ is $NIG\left(\bar{\mu},\bar{\Lambda},\bar{a},\bar{b}\right)$, where

$$\bar{\mu} = \bar{\Lambda}^{-1}\left(\Lambda\mu + \sum_{i=1}^{n} Z_i^{-1}X_i'Y_i\right),$$
$$\bar{\Lambda} = \Lambda + \sum_{i=1}^{n} Z_i^{-1}X_i'X_i,$$
$$\bar{a} = a + \frac{n}{2},$$
$$\bar{b} = b + \frac{1}{2}\sum_{i=1}^{n} Z_i^{-1}Y_i^2 + \frac{1}{2}\mu'\Lambda\mu - \frac{1}{2}\bar{\mu}'\bar{\Lambda}\bar{\mu}.$$

A marginalized MH sampler integrates out parameters β,σ^2 and evaluates $p(Z\,|Y) \propto p(Z)\,p(Y\,|Z)$, in which the marginal likelihood (conditional on Z) is given by Bayes' rule

$$p(Y\,|Z) = \frac{p\left(\beta,\sigma^2\,|Z\right)p\left(Y\,|\beta,\sigma^2,Z\right)}{p\left(\beta,\sigma^2\,|Y,Z\right)},$$

where $p\left(\beta,\sigma^2\,|Z\right)$ and $p\left(\beta,\sigma^2\,|Y,Z\right)$ are densities of $NIG(\mu,\Lambda,a,b)$ and $NIG\left(\bar{\mu},\bar{\Lambda},\bar{a},\bar{b}\right)$, respectively, and $p\left(Y\,|\beta,\sigma^2,Z\right)$ is the Gaussian likelihood. Evaluating densities at an arbitrary point such as $\beta=0,\sigma^2=1$ yields

$$p(Y\,|Z) \propto \left(\bar{b}\right)^{-\bar{a}}\left|\bar{\Lambda}\right|^{-\frac{1}{2}}e^{-\frac{1}{2}\sum_{i=1}^{n}\ln Z_i}.$$

Note that the marginal likelihood is fully determined by a set of cumulative statistics (i.e., a summable function of X_i,Y_i,Z_i):

$$S_1(Z) = \sum_{i=1}^{n} Z_i^{-1}X_i'X_i,$$
$$S_2(Z) = \sum_{i=1}^{n} Z_i^{-1}X_i'Y_i,$$

Algorithm 4.3 Marginalized subsampling of scale mixture regression

Input: data X, Y, initial value Z, and tuning parameter λ.
Output: posterior samples from $p(\beta, \sigma^2, Z \,|Y)$.
1 Compute cumulative statistics $S_j(Z)$, $j = 1, \ldots, 4$ by one sweep over data.
2 Iterate the following steps:
  2.1 Draw a random index $i \in \{1, \ldots, n\}$.
  2.2 Propose Z^* such that $Z^*_{-i} = Z_{-i}$, $Z^*_i = Z_i + \lambda u_i$, $u_i \sim N(0,1)$.
  2.3 Read i^{th} data point only.
  2.4 Compute cumulative statistics $S_j(Z^*)$ by rank-one update of $S_j(Z)$.
  2.5 Evaluate $p(Z^* \,|Y) \propto p(Z^*)\, p(Y \,|Z^*)$.
  2.6 Accept proposal (set $Z = Z^*$) with probability $\min\left[1, \frac{p(Z^* |Y)}{p(Z|Y)}\right]$.
  2.7 Sample β, σ^2 from $NIG(\bar{\mu}, \bar{\Lambda}, \bar{a}, \bar{b})$.

$$S_3(Z) = \sum_{i=1}^{n} Z_i^{-1} Y_i^2,$$

$$S_4(Z) = \sum_{i=1}^{n} \ln Z_i.$$

As a big data algorithm, subsampling updates one element in the vector Z by reading only one data point, avoiding the need to sweep over the entire dataset. Subsampling is achieved by rank-one update of cumulative statistics. Algorithm 4.3 summarizes the marginalized subsampling method for the scale mixture regression.

4.3.5 Pseudo Marginal Method

Algorithm 4.3 is applicable to mixture regressions in which the distributions of latent variables are arbitrary. However, it requires evaluating the density $p(Z)$ pointwise up to a normalizing constant. Sometimes, it is desirable to specify a hierarchical structure to allow the data to determine the exact distribution of the latent variable. Taking the t-distributed regression as an example, we specify $Z_i \,|v \sim IG\left(\frac{v}{2}, \frac{v}{2}\right)$ with a prior distribution on v (e.g., uniform in an interval). The density

$$p(Z) = \int p(v) \prod_{i=1}^{n} p(Z_i \,|v)\, dv$$

cannot be evaluated exactly, regardless of how the prior on v is specified. As an alternative to Algorithm 4.2, we consider the pseudo marginal version of Algorithm 4.3.The former is limited to t-distributed regression, but the latter provides a general solution to scale mixture regressions.

The pseudo marginal method constructs a non-negative and unbiased estimator of $p(Z)$ such that

$$\hat{p}(Z) = \frac{1}{M} \sum_{m=1}^{M} p(Z \,|v_m),$$

where v_m, $m = 1, \ldots, M$ are independent draws from the prior. The estimator is unbiased because

$$E[\hat{p}(Z)] = \frac{1}{M} \sum_{m=1}^{M} \int p(Z \,|v_m)\, p(v_m)\, dv_m = p(Z).$$

Algorithm 4.4 Pseudo marginal subsampling of scale mixture regression

Input: data X, Y, initial value Z, and tuning parameter λ.
Output: posterior samples from $p\left(\beta, \sigma^2, Z \mid Y\right)$.
1 Compute cumulative statistics $S_j(Z)$ by one sweep over data.
2 Iterate the following steps:
 2.1 Draw a random index $i \in \{1, \ldots, n\}$.
 2.2 Propose Z^* such that $Z_{-i}^* = Z_{-i}$, $Z_i^* = Z_i + \lambda u_i$, $u_i \sim N(0,1)$.
 2.3 Read i^{th} data point only.
 2.4 Compute cumulative statistics $S_j(Z^*)$ by rank-one update of $S_j(Z)$.
 2.5 Calculate $\hat{p}(Z^*)$ as an unbiased estimator of $p(Z^*)$.
 2.6 Evaluate $\hat{p}(Z^* \mid Y) \propto \hat{p}(Z^*)\, p(Y \mid Z^*)$.
 2.7 Accept proposal (set $Z = Z^*$) with probability $\min\left[1, \frac{\hat{p}(Z^* \mid Y)}{\hat{p}(Z \mid Y)}\right]$.
 2.8 Sample β, σ^2 from $NIG\left(\bar{\mu}, \bar{\Lambda}, \bar{a}, \bar{b}\right)$.

In the pseudo marginal MCMC sampling framework (Andrieu and Roberts, 2009), we consider a random-walk MH sampler in which the intractable $p(Z)$ is replaced by its unbiased estimator $\hat{p}(Z)$. We evaluate the acceptance ratio $\frac{\hat{p}(Z^*)p(Y \mid Z^*)}{\hat{p}(Z)p(Y \mid Z)}$, which equals to a verbose expression

$$\frac{\hat{p}(Z^*)\, p(Y \mid Z^*) \prod_{m=1}^{M} p(v_m^*)}{\hat{p}(Z)\, p(Y \mid Z) \prod_{m=1}^{M} p(v_m)} \cdot \frac{q(Z \mid Z^*) \prod_{m=1}^{M} p(v_m)}{q(Z^* \mid Z) \prod_{m=1}^{M} p(v_m^*)},$$

where $q(\cdot)$ is a random-walk transition kernel that satisfies $q(Z \mid Z^*) = q(Z^* \mid Z)$. Also, v_m, $m = 1, \ldots, M$ are random draws previously used for calculating $\hat{p}(Z)$, while v_m^* are random draws currently used for calculating $\hat{p}(Z^*)$.

Denote $\pi(Z, \vec{v}) = \hat{p}(Z)\, p(Y \mid Z) \prod_{m=1}^{M} p(v_m)$, where $\vec{v} = (v_1, \ldots, v_m)$. We have the integral

$$\int \pi(Z, \vec{v})\, d\vec{v} = p(Z)\, p(Y \mid Z).$$

Therefore, the adapted MH sampler with $p(Z)$ approximated by $\hat{p}(Z)$ can be viewed as an extended MH sampler targeting the joint distribution $\pi(Z, \vec{v})$, whose marginal distribution is the exact posterior distribution. For that reason, pseudo marginalization provides an exact approximation.

The pseudo marginal MH sampler remains a big data algorithm, as subsampling updates one element in the vector Z by reading only one data point. This is achieved by rank-one update of cumulative statistics. Algorithm 4.4 summarizes the pseudo marginal extension to Algorithm 4.3.

4.4 Asymmetry and Skewness

Normal distributions are symmetric. The density function of a symmetric distribution satisfies $f(\mu - x) = f(\mu + x)$, which implies that two tails are symmetric about the mean. However, empirical data often exhibit asymmetric tails. For example, extremely large losses are more likely than extremely large profits in some asset returns, resulting in a left-skewed distribution.

4.4.1 Skewness

Skewness is a measure of the asymmetry of a distribution about its mean. If the left tail is longer, the distribution is left-skewed. If the right tail is more stretched, the distribution is right-skewed.

Definition 4.3 (Skewness) *Let μ and σ be the mean and standard deviation of a random variable x. Skewness is defined as the scaled third moment:*

$$\varsigma = E\left[\left(\frac{x-\mu}{\sigma}\right)^3\right].$$

Skewness of a symmetric distribution must be zero (Exercise 4.3). Most positive-valued distributions, such as exponential, lognormal, gamma distributions, are right-skewed. Our aim is to construct a composite distribution on the support $(-\infty, \infty)$ with a tuning parameter that controls the sign and magnitude of skewness. Proposition 4.4 fulfills our goal by taking a linear combination of two random variables.

Proposition 4.4 (Combined skewness) *Consider a linear combination of two random variables $e_i = \delta Z_i + \sigma\varepsilon_i$, where the latent variable Z_i has mean μ_z, variance σ_z^2 and skewness ς_z. Independently, ε_i follows the standard normal distribution. Then, the mean, variance and skewness of e_i are*

$$\mu_e = \delta\mu_z,$$
$$\sigma_e^2 = \sigma^2 + \delta^2\sigma_z^2,$$
$$\varsigma_e = \frac{\sigma_z^3}{\sigma_e^3}\delta^3\varsigma_z.$$

4.4.2 Skew Normal Distribution

Proposition 4.4 is applicable to latent variables with different distributions. For example, if Z_i follows the exponential distribution with mean μ_z, variance μ_z^2 and skewness 2, then the combined variable e_i has the skewness

$$\varsigma_e = \frac{2\mu_z^3}{(\sigma^2 + \delta^2\mu_z^2)^{3/2}}\delta^3.$$

Another example involves specifying Z_i as a half-normal latent variable:

$$p(Z_i) \propto e^{-\frac{1}{2}Z_i^2} I(Z_i > 0),$$

which is a special case of truncated normal distributions. The half-normal distribution has mean $\sqrt{\frac{2}{\pi}}$, variance $1 - \frac{2}{\pi}$ and skewness $\frac{\sqrt{2}(4-\pi)}{(\pi-2)^{3/2}}$. Exercise 4.4 shows that the combined variable e_i follows a skewed normal distribution (Azzalini and Dalla Valle, 1996; Chen et al., 1999). Proposition 4.4 provides its skewness

$$\varsigma_e = \frac{\left(1-\frac{2}{\pi}\right)^{3/2}}{\left[\sigma^2 + \delta^2\left(1-\frac{2}{\pi}\right)\right]^{3/2}} \frac{\sqrt{2}\,(4-\pi)}{(\pi-2)^{3/2}}\delta^3.$$

The parameter δ determines the sign and magnitude of skewness. When $\delta > 0$, the right tail is longer, and the distribution is right-skewed. When $\delta < 0$, the left tail is stretched, and the distribution is left-skewed.

Algorithm 4.5 Gibbs sampler of skew normal regression

Input: data X, Y, and initial value Z.
Output: posterior samples from $p\left(\beta, \sigma^2, Z \,|Y\right)$.
Iterate the following steps:
1 Sample β, σ^2 from $NIG\left(\bar{\mu}, \bar{\Lambda}, \bar{a}, \bar{b}\right)$, given Z.
2 Sample $Z_i, i = 1, \ldots, n$ from $N\left[\frac{\delta(Y_i - \tilde{X}_i\tilde{\beta})}{\sigma^2+\delta^2}, \frac{\sigma^2}{\sigma^2+\delta^2}\right] I\left(Z_i > 0\right)$, given β, σ^2.

The skew normal regression model can be specified as

$$Y_i = \tilde{X}_i\tilde{\beta} + e_i,$$
$$e_i = \delta Z_i + \sigma\varepsilon_i,$$

where ε_i follows the standard normal distribution and Z_i is a half-normal latent variable. The appeal of the half-normal specification is a conditionally conjugate structure that facilitates Bayesian inference by Gibbs sampling.

Conditioning on latent variables results in a standard linear regression $Y_i = X_i\beta + \sigma\varepsilon_i$, where $X_i = \left(\tilde{X}_i, Z_i\right)$ and $\beta = \left(\tilde{\beta}', \delta\right)'$. Under the prior $NIG\left(\mu, \Lambda, a, b\right)$ for β, σ^2, the posterior conditional distribution maintains NIG with updated parameters: $NIG\left(\bar{\mu}, \bar{\Lambda}, \bar{a}, \bar{b}\right)$, where

$$\bar{\mu} = \left(\Lambda + \sum_{i=1}^{n} X_i'X_i\right)^{-1}\left(\Lambda\mu + \sum_{i=1}^{n} X_i'Y_i\right),$$
$$\bar{\Lambda} = \Lambda + \sum_{i=1}^{n} X_i'X_i,$$
$$\bar{a} = a + \frac{n}{2},$$
$$\bar{b} = b + \frac{1}{2}\sum_{i=1}^{n} Y_i^2 + \frac{1}{2}\mu'\Lambda\mu - \frac{1}{2}\bar{\mu}'\bar{\Lambda}\bar{\mu}.$$

Bayesian estimation of latent variable plays a central role in the data augmentation scheme. Conditional on β, σ^2, the equation $Y_i = \tilde{X}_i\tilde{\beta} + \delta Z_i + \sigma\varepsilon_i$ can be interpreted as a linear regression, in which the response variable is $Y_i - \tilde{X}_i\tilde{\beta}$, the predictor data is δ (i.e., only one observation), and Z_i serves as the coefficient with a half-normal prior. For Bayesian linear regressions under the truncated normal prior (see Exercise 4.5), the posterior follows a truncated normal distribution:

$$Z_i \,|Y, \beta, \sigma^2 \sim N\left[\frac{\delta\left(Y_i - \tilde{X}_i\tilde{\beta}\right)}{\sigma^2 + \delta^2}, \frac{\sigma^2}{\sigma^2 + \delta^2}\right] I\left(Z_i > 0\right), i = 1, \ldots, n.$$

Algorithm 4.5 summarizes the Gibbs sampler that cycles through the full posterior conditionals $p\left(\beta, \sigma^2 \,|Y, Z\right)$ and $p\left(Z \,|Y, \beta, \sigma^2\right)$.

For big data applications, it is of interest to integrate out parameters β, σ^2 and subsample the latent vector Z. Marginalization can be achieved by Bayes' rule:

$$p\left(Y \,|Z\right) = \frac{p\left(\beta, \sigma^2 \,|Z\right) p\left(Y \,|\beta, \sigma^2, Z\right)}{p\left(\beta, \sigma^2 \,|Y, Z\right)},$$

Algorithm 4.6 Marginalized subsampling of skew normal regression

Input: data X, Y, initial value Z, and tuning parameter λ.
Output: posterior samples from $p(\beta, \sigma^2, Z \mid Y)$.
1 Compute cumulative statistics $S_j(Z)$, $j = 1, \ldots, 4$ by one sweep over data.
2 Iterate the following steps:
  2.1 Draw a random index $i \in \{1, \ldots, n\}$.
  2.2 Propose Z^* such that $Z^*_{-i} = Z_{-i}$, $Z^*_i = Z_i + \lambda u_i$, $u_i \sim N(0, 1)$.
  2.3 Read i^{th} data point only.
  2.4 Compute cumulative statistics $S_j(Z^*)$ by rank-one update of $S_j(Z)$.
  2.5 Evaluate $p(Z^* \mid Y)$ by cumulative statistics.
  2.6 Accept proposal (set $Z = Z^*$) with probability $\min\left[1, \frac{p(Z^* \mid Y)}{p(Z \mid Y)}\right]$.
  2.7 Sample β, σ^2 from $NIG(\bar{\mu}, \bar{\Lambda}, \bar{a}, \bar{b})$.

where $p(\beta, \sigma^2 \mid Z)$ and $p(\beta, \sigma^2 \mid Y, Z)$ are densities of $NIG(\mu, \Lambda, a, b)$ and $NIG(\bar{\mu}, \bar{\Lambda}, \bar{a}, \bar{b})$, respectively, and $p(Y \mid \beta, \sigma^2, Z)$ is the Gaussian likelihood. Evaluating expressions at an arbitrary point such as $\beta = 0$ and $\sigma^2 = 1$ yields

$$p(Y \mid Z) \propto (\bar{b})^{-\bar{a}} \left|\bar{\Lambda}\right|^{-\frac{1}{2}}.$$

A marginalized MH sampler has the target distribution$p(Z \mid Y) \propto p(Z)\, p(Y \mid Z)$, where

$$p(Z) \propto e^{-\frac{1}{2}\sum_{i=1}^{n} Z_i^2} \prod_{i=1}^{n} I(Z_i > 0).$$

The marginalized posterior density is fully determined by cumulative statistics

$$S_1(Z) = \sum_{i=1}^{n} X_i' X_i,$$

$$S_2(Z) = \sum_{i=1}^{n} X_i' Y_i,$$

$$S_3(Z) = \sum_{i=1}^{n} Y_i^2,$$

$$S_4(Z) = \sum_{i=1}^{n} Z_i^2.$$

Algorithm 4.6 summarizes the marginalized subsampling MH sampler.

4.4.3 Inverse Scale Model

Fernandez and Steel (1998) propose an inverse scale model for regressions with skewed errors. The idea is to split a symmetric density $f(\varepsilon)$ in two halves: the positive part is scaled by $f(\gamma^{-1}\varepsilon)$, while the negative part is scaled by $f(\gamma\varepsilon)$. The parameter γ controls the degree of skewness of the distribution. The modified density still has a single mode at zero, but the densities in the positive and negative orthants are asymmetric.

Consider the inverse-scale linear regression $Y_i = X_i\beta + \sigma\varepsilon_i$, where

$$p(\varepsilon_i \mid \gamma) = \frac{2}{\gamma + \gamma^{-1}} \left[f(\gamma^{-1}\varepsilon_i)\, I(\varepsilon_i \geq 0) + f(\gamma\varepsilon_i)\, I(\varepsilon_i < 0)\right].$$

A symmetric density satisfies $f(\varepsilon) = f(-\varepsilon)$, which implies the integrals

$$\int_0^\infty f\left(\gamma^{-1}\varepsilon\right) d\varepsilon = \frac{1}{2}\gamma,$$

$$\int_0^\infty f(\gamma\varepsilon)\, d\varepsilon = \frac{1}{2}\gamma^{-1}.$$

It follows that $\frac{P(\varepsilon_i \geq 0|\gamma)}{P(\varepsilon_i < 0|\gamma)} = \gamma^2$. The positive part of the distribution is thicker than the negative part if $\gamma > 1$. The skewness of the distribution increases with γ.

Another interpretation of $p(\varepsilon_i|\gamma)$ is a two-component mixture of truncated distributions:

$$p(\varepsilon_i|\gamma) = wp_+(\varepsilon_i) + (1-w)\,p_-(\varepsilon_i),$$

where the weight is $w = \frac{\gamma^2}{\gamma^2+1}$. The positively truncated component $p_+(\varepsilon_i)$ and the negatively truncated component $p_-(\varepsilon_i)$ are proper densities, taking the form:

$$p_+(\varepsilon_i) = 2\gamma^{-1} f\left(\gamma^{-1}\varepsilon_i\right) I(\varepsilon_i \geq 0),$$

$$p_-(\varepsilon_i) = 2\gamma f(\gamma\varepsilon_i)\, I(\varepsilon_i < 0).$$

Fernandez and Steel (1998) specify $f(\cdot)$ as a t distribution, allowing the model to accommodate both skewness and fat tails. For pedagogical purposes, we specify $f(\cdot)$ as the standard normal distribution and focus on skewness. As is shown in Section 4.3, the t distribution is a scale mixture of normal distributions. The extension from normal to t distributions involves the addition of sampling inverse-gamma latent variables.

When $f(\cdot)$ is the standard normal density, the likelihood function takes the form:

$$p\left(Y|\beta,\sigma^2,\gamma\right) = \prod_{i=1}^{n}\left[wp_+(Y_i) + (1-w)\,p_-(Y_i)\right],$$

where $p_+(Y_i)$ and $p_-(Y_i)$ are densities of truncated normal distributions:

$$p_+(Y_i) = \frac{2}{\sqrt{2\pi}\sigma\gamma} e^{-\frac{(Y_i - X_i\beta)^2}{2\sigma^2\gamma^2}} I(Y_i \geq X_i\beta),$$

$$p_-(Y_i) = \frac{2}{\sqrt{2\pi}\sigma\gamma^{-1}} e^{-\frac{(Y_i - X_i\beta)^2}{2\sigma^2\gamma^{-2}}} I(Y_i < X_i\beta).$$

Regardless of the prior specification for β, σ^2, γ, the posterior does not belong to a well-known distribution. We resort to MCMC sampling that evaluates the posterior density $p\left(\beta,\sigma^2,\gamma|Y\right) \propto p\left(\beta,\sigma^2,\gamma\right) p\left(Y|\beta,\sigma^2,\gamma\right)$ up to a normalizing constant. If the sample size is small, we may consider Gibbs sampling. Fernandez and Steel (1998) propose a Gibbs sampler that updates one element of the vector β at a time. The approach sorts observations and samples from truncated normal distributions. See Exercise 4.6 for an illustration.

4.5 Finite Mixture Distribution

A finite mixture distribution is well-suited to a population composed of multiple subpopulations, each described by its own probability distribution. Formally, it is expressed as a weighted average of k distributions:

$$p(Y) = \sum_{j=1}^{k} w_j p_j(Y),$$

where $p_j(Y)$ denotes the density of the jth component (i.e., class or regime) and w_j is the corresponding weight. Here, $p_j(Y)$ is required to be a proper density function that integrates to one over its domain, and the weights must sum to one, ensuring that the mixture is a valid probability distribution:

$$\int p(Y)\, dY = \sum_{j=1}^{k} w_j \int p_j(Y)\, dY = 1.$$

Mixture distributions serve two principal purposes in statistical modeling. First, distributional approximation. Mixture models can approximate complex distributions that exhibit characteristics such as multimodality and skewness, offering a flexible framework for capturing intricate data structures. For instance, while a single Gaussian distribution is unimodal and symmetric, a mixture of Gaussians can model data with multiple peaks and asymmetry. Second, modeling distributional uncertainty. In real-world scenarios, observations may arise from one of several underlying processes, but the specific process generating each observation is unknown. Mixture models address this uncertainty by introducing a latent variable that indicates the class label for each observation. This hierarchical representation involves two steps: selecting a class based on the weights w_j, and then sampling from the corresponding distribution $p_j(Y)$. The latent variable approach allows the marginal distribution to be expressed as a mixture, with posterior distributions providing insights into class membership. Titterington et al. (1985), McLachlan and Peel (2000) and Fruhwirth-Schnatter (2006) provide book-length treatments of finite mixture models. For Bayesian modelling, see Diebolt and Robert (1994) and Richardson and Green (1997).

In Bayesian analysis, the posterior is proportional to the prior times the likelihood. Mixture distributions can be specified in either the prior or the likelihood. This section focuses on mixture distributions for the likelihood function, and we have already studied mixture priors in Chapter 3. For instance, SSVS in Section 3.4 and MC^3 in Section 3.5 employ mixture NIG priors for variable selection in linear regressions. Regardless of mixing in the prior or the likelihood, the latent class labels serve as model selection indicators. Marginalization over the model parameters yields the marginal likelihood of a specific model. The posterior model probability is proportional to the prior model probability times the marginal likelihood. Analysis of mixture distributions can be interpreted as a process of Bayesian model selection and averaging.

4.5.1 Distributional Approximation

Mixture distributions can approximate multimodal and skewed distributions, which may stem from theoretic models or empirical distributions that describe real-world data. The stochastic volatility model outlined in Section 4.2 exemplifies a model-implied distribution:

$$Y_i = X_i\beta + e^{\frac{1}{2}Z_i}\varepsilon_i,$$

where the data Y_i is a nonlinear function of the conditional volatility Z_i. Linearizing the model through data transformation $\tilde{Y}_i = \ln(Y_i - X_i\beta)^2$ yields

$$\tilde{Y}_i = Z_i + \ln \varepsilon_i^2.$$

The linearization method is widely used for stochastic volatility estimation. See Harvey et al. (1994) and Kim et al. (1998). However, linearization results in a loss of normality. If ε_i is a standard normal variate, then $\ln \varepsilon_i^2$ follows $\ln \chi^2(1)$ distribution, characterized by skewness and a thicker left tail, with mean -1.2704 and variance $\frac{\pi^2}{2}$.

The first three rows of Table 4.1 replicate the Gaussian mixture approximation proposed by Kim et al. (1998), alongside an alternative mixture proposal tabulated in the last three rows. As shown in Figure 4.1, both mixtures provide good approximation of $\ln \chi^2(1)$, but parameters are vastly different. This highlights a fundamental property of mixture models: flexibility in approximating complex distributions can induce parameter identifiability issues, as different parameter sets may yield similar likelihoods, complicating parameter estimation due to multiple local maxima.

TABLE 4.1
Gaussian mixture approximation for $\ln \chi^2(1)$ distribution.

class	1	2	3	4	5	6	7
mean	−11.400	−5.243	−9.837	1.507	−0.651	0.525	−2.359
std	2.407	1.617	2.276	0.409	0.800	0.583	1.124
weight	0.007	0.106	0.000	0.044	0.340	0.246	0.258
mean	−9.320	−5.315	−3.415	−1.710	−0.453	0.398	1.193
std	3.279	2.457	1.887	1.312	0.884	0.590	0.499
weight	0.009	0.054	0.134	0.276	0.292	0.149	0.085

4.5.2 Data Augmentation and Gibbs Sampling

Mixture distributions can be interpreted as either an averaging process or a randomization mechanism. The former views the mixture density as a weighted average of component densities, while the latter refers to a two-step data-generating process involving random selection of a component followed by sampling from it. This dual interpretation is crucial for understanding the inherent uncertainty in data generation.

For Gaussian mixtures, assume the observations Y_i, $i = 1, \ldots, n$ are generated by

$$p(Y_i | \theta) = \sum_{j=1}^{k} w_j \phi\left(Y_i, X_i\beta_j, \sigma_j^2\right),$$

where $\theta = \left(w, \beta, \sigma^2\right)$ collects model parameters, and $\phi\left(Y_i, X_i\beta_j, \sigma_j^2\right)$ is the class-specific Gaussian density with mean $X_i\beta_j$ and variance σ_j^2, implying the Gaussian linear regression:

$$Y_i = X_i\beta_j + \sigma_j\varepsilon_{ij}, j = 1, \ldots, k.$$

For the ith observation, there is uncertainty regarding which class it belongs to. Let $Z_i \in \{1, \ldots, k\}$ be an auxiliary variable that indicates the latent class, following a categorical distribution

$$p(Z_i = j | \theta) = w_j.$$

Data are generated by a two-step process. The jth class is selected with the probability w_j. Condition on the selected class, an observation is generated by

$$p(Y_i | Z_i = j, \theta) = \phi\left(Y_i, X_i\beta_j, \sigma_j^2\right).$$

Marginalization over Z_i yields the mixture distribution:

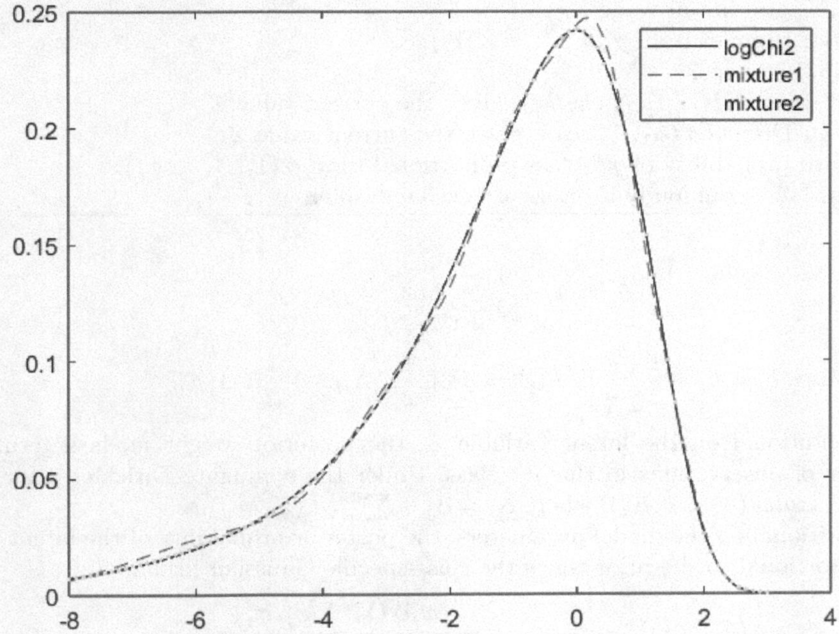

FIGURE 4.1
Gaussian mixture approximation for $\ln \chi^2(1)$ distribution. The solid line plots the target density function. The dashed and dotted lines show two approximated densities.

$$p(Y_i|\theta) = \int p(Z_i|\theta)\, p(Y_i|Z_i,\theta)\, dZ_i = \sum_{j=1}^{k} p(Z_i = j|\theta)\, p(Y_i|Z_i = j,\theta).$$

Bayesian model specification is completed by incorporating priors. Conditionally conjugate distributions are available by specifying $NIG(\mu_j, \Lambda_j, a_j, b_j)$ prior for β_j, σ_j^2, and $Dirichlet(\alpha_1, \ldots, \alpha_k)$ prior for $w = (w_1, \ldots, w_k)$ such that

$$p(w) \propto \prod_{j=1}^{k} w_j^{\alpha_j - 1}.$$

Posterior conditional distributions $p(\beta, \sigma^2|Y, Z, w)$, $p(w|Y, Z, \beta, \sigma^2)$ and $p(Z|Y, w, \beta, \sigma^2)$ are analytically tractable.

First, conditional on the latent variable Z, the posterior distributions of β_j, σ_j^2 are determined by observations in the jth class. By Proposition 1.1, the posterior takes the form $NIG(\bar{\mu}_j, \bar{\Lambda}_j, \bar{a}_j, \bar{b}_j)$, where

$$\bar{\mu}_j = \bar{\Lambda}_j^{-1}\left[\Lambda_j \mu_j + \sum_{i=1}^{n} X_i' Y_i I(Z_i = j)\right],$$

$$\bar{\Lambda}_j = \Lambda_j + \sum_{i=1}^{n} X_i' X_i I(Z_i = j),$$

Algorithm 4.7 Gibbs sampler of Gaussian mixtures

Input: data X, Y, and initial value Z.
Output: posterior samples from $p\left(\beta, \sigma^2, w, Z \,|Y\right)$.
Iterate the following steps:
1 Sample β_j, σ_j^2 from $NIG\left(\bar{\mu}_j, \bar{\Lambda}_j, \bar{a}_j, \bar{b}_j\right)$, given the current value Z.
2 Sample w from $Dirichlet\left(\bar{\alpha}_1, \ldots, \bar{\alpha}_k\right)$, given the current value Z.
3 Sample Z_i with probability of jth class proportional to $w_j \phi\left(Y_i, X_i\beta_j, \sigma_j^2\right)$.
4 Permute class labels randomly if priors are exchangeable.

$$\bar{a}_j = a_j + \frac{1}{2}\sum_{i=1}^{n} I\left(Z_i = j\right),$$

$$\bar{b}_j = b_j + \frac{1}{2}\sum_{i=1}^{n} Y_i^2 I\left(Z_i = j\right) + \frac{1}{2}\mu_j' \Lambda_j \mu_j - \frac{1}{2}\bar{\mu}_j' \bar{\Lambda}_j \bar{\mu}_j.$$

Second, conditional on the latent variable Z, the posterior weight w_j is determined by the number of observations in the jth class. Under the conjugate Dirichlet prior, the posterior is $Dirichlet\left(\bar{\alpha}_1, \ldots, \bar{\alpha}_k\right)$, where $\bar{\alpha}_j = \alpha_j + \sum_{i=1}^{n} I\left(Z_i = j\right)$.

Third, conditional on the model parameters, the posterior probability of the latent variable Z_i is proportional to the prior times the class-specific Gaussian likelihood:

$$p\left(Z_i = j \,|Y, w, \beta, \sigma^2\right) = \frac{w_j \phi\left(Y_i, X_i\beta_j, \sigma_j^2\right)}{\sum_{h=1}^{k} w_h \phi\left(Y_i, X_i\beta_h, \sigma_h^2\right)}.$$

Algorithm 4.7 outlines the Gibbs sampler for Gaussian mixtures, which iteratively updates parameters β, σ^2, w, Z by sampling from NIG, Dirichlet and categorical distributions. Additionally, the algorithm includes a step for random permutation of class labels, as suggested by Fruhwirth-Schnatter (2001). A key challenge in mixture modeling is label switching, where the likelihood remains invariant under permutations of class labels. For instance, switching parameters of any two classes leaves the likelihood unchanged. If the priors are exchangeable, the posterior is invariant to class labels. Exercise 4.7 shows that random permutation corresponds to a Metropolis proposal being accepted with probability one.

4.5.3 Double Marginalized Subsampling

Practical mixture models typically involve multitudinous parameters. In high-dimensional settings, the Gibbs sampler may explore the parameter space slowly, necessitating an enormous number of iterations to escape from a trapping state due to label switching (Marin et al., 2005). This problem can be addressed by double marginalization, which integrates out high-dimensional parameters and focuses on the posterior marginal distribution for latent variables, based on the marginal-conditional decomposition:

$$p\left(Z, w, \beta, \sigma^2 \,|Y\right) = p\left(Z \,|Y\right) p\left(w \,|Y, Z\right) \prod_{j=1}^{k} p\left(\beta_j, \sigma_j^2 \,|Y, Z, w\right),$$

where $p\left(w \,|Y, Z\right)$ and $p\left(\beta_j, \sigma_j^2 \,|Y, Z, w\right)$ are the densities $Dirichlet\left(\bar{\alpha}_1, \ldots, \bar{\alpha}_k\right)$ and $NIG\left(\bar{\mu}_j, \bar{\Lambda}_j, \bar{a}_j, \bar{b}_j\right)$, respectively. The posterior marginal distribution $p\left(Z \,|Y\right)$ can be further decomposed as

$$p\left(Z \,|Y\right) \propto p\left(Z\right) p\left(Y \,|Z\right) = \frac{p\left(w\right) p\left(Z \,|w\right)}{p\left(w \,|Z\right)} \frac{p\left(\beta, \sigma^2 \,|Z\right) p\left(Y \,|Z, \beta, \sigma^2\right)}{p\left(\beta, \sigma^2 \,|Y, Z\right)}.$$

Each of those six terms can be evaluated in a closed form.

For the first three terms, $p(w)$ and $p(w\,|Z)$ are the densities of $Dirichlet\,(\alpha_1,\ldots,\alpha_k)$ and $Dirichlet\,(\bar{\alpha}_1,\ldots,\bar{\alpha}_k)$ respectively, and $p(Z\,|w) = \prod_{i=1}^{n}\sum_{j=1}^{k} w_j I\,(Z_i = j)$. It follows that

$$p(Z) \propto \prod_{j=1}^{k} \Gamma\,(\bar{\alpha}_j)\,.$$

For the remaining three terms, $p\left(\beta,\sigma^2\,|Z\right)$ and $p\left(\beta,\sigma^2\,|Y,Z\right)$ are the product densities of $\prod_{j=1}^{k} NIG\,(\mu_j,\Lambda_j,a_j,b_j)$ and $\prod_{j=1}^{k} NIG\,(\bar{\mu}_j,\bar{\Lambda}_j,\bar{a}_j,\bar{b}_j)$, respectively, and $p\left(Y\,|Z,\beta,\sigma^2\right)$ is the Gaussian likelihood function. It follows that

$$p(Y\,|Z) \propto \prod_{j=1}^{k} \Gamma\,(\bar{a}_j)\,(\bar{b}_j)^{-\bar{a}_j}\,|\bar{\Lambda}_j|^{-\frac{1}{2}}\,.$$

In summary, for a given value of Z, it is feasible to evaluate the posterior marginal density up to a normalizing constant by

$$p(Z\,|Y) \propto \prod_{j=1}^{k} \Gamma\,(\bar{\alpha}_j)\,\Gamma\,(\bar{a}_j)\,(\bar{b}_j)^{-\bar{a}_j}\,|\bar{\Lambda}_j|^{-\frac{1}{2}}\,,$$

which is fully determined by a set of cumulative statistics: for $j = 1,\ldots,k$,

$$S_{1j}\,(Z) = \sum_{i=1}^{n} X_i' X_i I\,(Z_i = j)\,,$$

$$S_{2j}\,(Z) = \sum_{i=1}^{n} X_i' Y_i I\,(Z_i = j)\,,$$

$$S_{3j}\,(Z) = \sum_{i=1}^{n} Y_i^2 I\,(Z_i = j)\,,$$

$$S_{4j}\,(Z) = \sum_{i=1}^{n} I\,(Z_i = j)\,.$$

If a new allocation Z^* differs the current value Z by one element, rank-one update of cumulative statistics can be performed by reading a single observation, rather than sweeping over the full dataset.

Two implementation methods for double marginalized subsampling are available. The first is a general solution by a random-walk MH sampler. A new Z^* is proposed on the basis of the current value Z by randomly sampling an observation index $i \in \{1,\ldots,n\}$ and a class index $j \in \{1,\ldots,k\}$, followed by setting $Z_i^* = j$ and $Z_{-i}^* = Z_{-i}$. The proposal is accepted with probability $\min\left[1, \frac{p(Z^*|Y)}{p(Z|Y)}\right]$. The second method is the single-move marginalized Gibbs sampler specific to discrete-valued latent variables. As a latent class involves k possible values, it is feasible to enumerate them as $p\,(Z_i = j\,|Y, Z_{-i})$, $\forall j = 1,\ldots,k$, where $p\,(Z_i\,|Y, Z_{-i}) \propto p\,(Z\,|Y)$.

Algorithm 4.8 describes a single-move Gibbs sampler for double marginalized subsampling, offering a computationally efficient solution for big data in terms of both large sample size and high-dimensional parameter spaces.

By treating the latent variable Z as a model selection indicator with the prior model probability $p\,(Z)$, the posterior model probability is $p\,(Z\,|Y)$, with all parameters integrated out. The number of models equals k^n, which is too large to be processed analytically. When

Algorithm 4.8 Double marginalized subsampling of Gaussian mixtures

Input: data X, Y, and initial value Z.
Output: posterior samples from $p\left(\beta, \sigma^2, w, Z \,|Y\right)$.
1 Compute cumulative statistics by one sweep over data.
2 Iterate the following steps:
 2.1 Draw a random index $i \in \{1, \ldots, n\}$.
 2.2 Read i^{th} data point only.
 2.3 Compute $p\left(Z_i = j \,|Y, Z_{-i}\right)$, $\forall j = 1, \ldots, k$ by cumulative statistics.
 2.4 Normalize probabilities and sample Z_i from the categorical distribution.
 2.5 Sample β_j, σ_j^2 from $NIG\left(\bar{\mu}_j, \bar{\Lambda}_j, \bar{a}_j, \bar{b}_j\right)$, $j = 1, \ldots, k$.
 2.6 Sample w from $Dirichlet\left(\bar{\alpha}_1, \ldots, \bar{\alpha}_k\right)$.

omitting steps 2.5 and 2.6, Algorithm 4.8 reduces to an MCMC sampler that has $p\left(Z \,|Y\right)$ as its stationary distribution. Therefore, Algorithm 4.8 can be interpreted as MCMC model composition.

4.6 Dirichlet Process Mixture Models

In the parametric mixture models discussed in Section 4.5, the number of classes k is predetermined and finite. This poses a challenge since the true number of classes in a real-world dataset is often unknown and may not be well-represented by a fixed k. To address this limitation, Bayesian nonparametric models offer a flexible alternative where the number of parameters is not fixed a priori and can grow with the sample size, allowing the model complexity to adapt to the data volume. Nonparametric Bayesian analysis was pioneered by Ferguson (1973), and influential studies on mixture Dirichlet processes include Antoniak (1974), Escobar and West (1995) and Neal (2000), among others.

The Dirichlet Process Mixture Model (DPMM) extends the finite mixture model by allowing the number of classes to approach infinity ($k \to \infty$). This nonparametric approach enables the model to infer the number of classes from the data. In contrast to parametric models, where parameters are typically fixed, Bayesian nonparametric models potentially accommodate an infinite number of parameters. However, for a specific dataset, only a finite set of parameters are used, with unused parameters integrated out.

As $k \to \infty$, the Gibbs sampler, as described in Algorithm 4.7, encounters difficulties with DPMM because it is not feasible to sample from a Dirichlet distribution with an infinite number of classes. However, Algorithm 4.8, which employs double marginalized subsampling, can be adapted for DPMM by integrating out the Dirichlet variates. The core of double marginalized subsampling is marginal-conditional decomposition:

$$p\left(Z \,|Y\right) \propto p\left(Z\right) p\left(Y \,|Z\right),$$

where the class indicators $Z = \left(Z_1, \ldots, Z_n\right)'$ are exchangeable, following the marginal distribution

$$p\left(Z\right) \propto \prod_{j=1}^{k} \Gamma\left(\bar{\alpha}_j\right),$$

$$\bar{\alpha}_j = \alpha_j + \sum_{i=1}^{n} I\left(Z_i = j\right).$$

Regarding the hyperparameter specification in DPMM, the Dirichlet prior is specified with a concentration parameter α such that $\alpha_j = \frac{\alpha}{k}$, $\forall j$, as suggested by Ishwaran and Zarepour (2002). The concentration parameter controls modeling flexibility: higher values of α encourage the formation of more classes, reflecting greater uncertainty about the number of underlying processes.

DPMMs can be estimated using a single-move Gibbs sampler. A critical step involves computing the leave-one-out probability for the ith latent variable $p\left(Z_i \,|Z_{-i}\right)$, which is still proportional to the product of gamma functions $\prod_{j=1}^{k} \Gamma\left(\bar{\alpha}_j\right)$.

Let $n_{ij} = \sum_{h \neq i}^{n} I\left(Z_h = j\right)$ represents the count of observations in the jth class, excluding the ith latent variable. It follows that $\bar{\alpha}_j = \frac{\alpha}{k} + n_{ij} + I\left(Z_i = j\right)$. The recursive gamma function $\Gamma\left(n+1\right) = n\Gamma\left(n\right)$ implies that

$$p\left(Z_i = j \,|Z_{-i}\right) \propto \frac{\alpha}{k} + n_{ij}.$$

In the above equation, the term $\prod_{j=1}^{k} \Gamma\left(\frac{\alpha}{k} + n_{ij}\right)$ has been absorbed by the proportionality operator $\propto$. The proportionality constant can be retrieved by summing up probabilities to one, using the equation $\sum_{j=1}^{k} n_{ij} = n - 1$. Specifically, the leave-one-out probability can be articulated as

$$p\left(Z_i = j \,|Z_{-i}\right) = \frac{\frac{\alpha}{k} + n_{ij}}{\alpha + n - 1},$$

for all k in a finite mixture model. DPMM corresponds to the limiting case of $k \to \infty$. In that case, the probability assigned to the jth existing class (i.e., a class occupied by other latent variables Z_{-i}) is proportional to n_{ij}, while the probability of creating a new class is proportional to α. A higher concentration level α leads to more classes. The leave-one-out probability reflects the Matthew effect inherent in DPMMs, where existing classes with more members are more likely to attract new members.

The Chinese restaurant process metaphor provides intuition behind DPMM. In a restaurant with infinitely many food stations, each station serves a line of customers. Each customer takes a "rich get richer" strategy by joining an existing station with probability proportional to the line length, or by exploring a new station with probability proportional to the concentration parameter. Due to exchangeability, the probability of class assignment is invariant of the order of data points.

In DPMM, the number of classes is dynamically adjusted to the data complexity. In a finite sample of the size n, the number of classes is finite (i.e., at most n) The model parameters are integrated out, as we evaluate $p\left(Z_i \,|Z_{-i}\right)$ and $p\left(Y \,|Z\right)$, which aligns with double marginalized subsampling methods.

4.7 Quantile Regression

Classical linear regression assumes that the conditional mean of the response variable is a linear function of predictors, focusing primarily on the central tendency rather than the spread of a distribution. Quantile regression, introduced by Koenker and Bassett (1978), offers a robust alternative by modeling the conditional quantiles of the response variable, providing a more complete picture of covariate effects across the entire distribution and is particularly useful for characterizing the tails of distributions. For instance, Buchinsky (1994) examines changes in the returns to schooling at various points in the wage distributions by quantile regression that describes the entire conditional distributions. An excellent book on the subject is Koenker (2005), covering both theory and applications across multiple fields.

4.7.1 Population and Sample Quantiles

Let $\alpha \in [0,1]$ denote a specified probability. The αth quantile of a distribution is defined as

$$q_\alpha = \inf\{y : F(y) \geq \alpha\},$$

where $F(\cdot)$ is the right-continuous cumulative distribution function (CDF). In cases where the CDF is discontinuous at y, the equation $F(y) = \alpha$ may not have a solution. In such instances, the quantile is the smallest y that satisfies $F(y) \geq \alpha$. If the CDF curve levels off at the point α, resulting in multiple solutions to $F(y) = \alpha$, the quantile is the smallest y among these solutions. For continuous distributions where the CDF is strictly increasing, the quantile is given by the inverse CDF:

$$q_\alpha = F^{-1}(\alpha).$$

In that case, we also have

$$F(q_\alpha) = P(Y \leq q_\alpha) = \alpha.$$

Quantiles serve as essential summary statistics, including the median $q_{1/2}$, the upper quantile $q_{3/4}$, the lower quantile $q_{1/4}$, and tail metrics such as $q_{0.05}$ and $q_{0.95}$. The entire collection of quantiles for all $\alpha \in [0,1]$ provides a complete picture of the distribution. The expectation $E(Y) = \int y dF(y)$ can be conceptualized as an average of quantile functions, using the change of variables $\alpha = F(y)$ for integration:

$$E(Y) = \int_0^1 q_\alpha d\alpha.$$

Given random samples Y_i, $i = 1, \ldots, n$ from the distribution, sample quantiles are derived from the empirical CDF:

$$F_n(y) = \frac{1}{n}\sum_{i=1}^{n} I(Y_i \leq y),$$

$$\hat{q}_\alpha = \min\{y : F_n(y) \geq \alpha\}.$$

The indicator function $I(Y_i \leq y)$ equals one with probability $F(y)$, and zero otherwise. The mean and variance of the indicator variable are $F(y)$ and $F(y)[1 - F(y)]$, respectively. By the Law of Large Numbers, the empirical CDF converges to the true CDF: $F_n(y) \xrightarrow{a.s.} F(y)$ and the sample quantile converges to the true quantile: $\hat{q}_\alpha \xrightarrow{a.s.} q_\alpha$. Moreover, the Central Limit Theorem and the Delta Method provide asymptotical distributions for quantile estimates:

$$\sqrt{n}[F_n(y) - F(y)] \xrightarrow{d} N\{0, F(y)[1 - F(y)]\},$$

$$\sqrt{n}[\hat{q}_\alpha - q_\alpha] \xrightarrow{d} N\left\{0, \frac{\alpha(1-\alpha)}{F'(q_\alpha)^2}\right\}.$$

4.7.2 Quantile under Asymmetric Loss Function

Traditionally associated with data ordering, quantiles can also be viewed through the lens of loss minimization. When predicting Y_i by a constant c, the prediction error $\varepsilon_i = Y_i - c$ can manifest either as a positive value (indicating underestimation) or a negative value (indicating overestimation). In certain situations, the cost of underestimation might exceed that of overestimation, and vice versa, leading to an asymmetric loss function, commonly referred to as the check function:

$$f_\alpha(\varepsilon) = \varepsilon\left[\alpha - I(\varepsilon < 0)\right] = \begin{cases} \alpha\varepsilon & \varepsilon \geq 0 \\ (1-\alpha)(-\varepsilon) & \varepsilon < 0 \end{cases}.$$

A notable instance occurs when $\alpha = \frac{1}{2}$, which aligns with the 1-norm penalty $f_{1/2}(\varepsilon) = \frac{1}{2}|\varepsilon|$. For other values of α, the loss function assigns different weights to positive and negative errors, reflecting the relative costs of underestimation and overestimation. When α approaches zero, a substantial penalty is incurred for $c > Y_i$, prompting a reduction in c to mitigate overestimation. Conversely, when α is close to one, underestimation incurs a significant penalty for $c < Y_i$, suggesting an increase in c to minimize underestimation risk. Proposition 4.5 corroborates our intuition, demonstrating that the solution to this loss minimization problem is the quantile q_α, ensuring that minimization occurs when $\alpha \times 100$ percent of errors are negative.

Proposition 4.5 (Quantile and asymmetric loss function) *Let q_α be the αth quantile and $\hat{q}_\alpha$ be the sample counterpart for the random samples Y_i, $i = 1, \ldots, n$. Loss minimization under the check function $f_\alpha(\varepsilon) = \varepsilon[\alpha - I(\varepsilon < 0)]$ yields*

$$q_\alpha = \arg\min_c E\left[f_\alpha(Y_i - c)\right],$$

$$\hat{q}_\alpha = \arg\min_c \frac{1}{n}\sum_{i=1}^{n} f_\alpha(Y_i - c).$$

4.7.3 Conditional Quantile Functions

Understanding the effect of regressors X on the conditional distribution of Y is of significant interest. While the conditional mean $E(Y|X)$ offers valuable insights, it provides an incomplete depiction of the conditional distribution, motivating the analysis of conditional quantiles.

Let $\alpha \in [0,1]$ denote a specified probability and $q_\alpha(X)$ represent the αth quantile of the conditional distribution of the response variable Y given covariates X. The definition of the conditional quantile mirrors that of the unconditional quantile, with the distinction that the CDF $F(y)$ is replaced by the conditional CDF $F_{Y|X}(y;X)$ such that

$$q_\alpha(X) = \inf\left\{y : F_{Y|X}(y;X) \geq \alpha\right\}.$$

Our analysis is confined to continuous distributions, where quantiles are uniquely determined by the inverse CDF. In this context, the conditional quantile is the unique solution to the equation

$$F_{Y|X}\left[q_\alpha(X);X\right] = P\left[Y \leq q_\alpha(X)|X\right] = \alpha.$$

Conditional on X, the full set of conditional quantiles for all $\alpha \in [0,1]$ offers a complete representation of the conditional distribution, with the conditional mean being the average of conditional quantile functions:

$$E(Y|X) = \int_0^1 q_\alpha(X)\, d\alpha.$$

Unlike the conditional mean, the conditional quantile function is invariant to monotonic transformations: if $\tilde{Y} = g(Y)$ is a monotone mapping, then the conditional quantiles of $\tilde{Y}$ and Y are related through the same mapping $\tilde{q}_\alpha(X) = g[q_\alpha(X)]$. The underlying intuition is that quantiles traverse the monotonic function g, as the transformation does not alter variable rankings.

The conditional quantile function $q_\alpha(X)$ involves two input arguments: α and X. It is invariably an increasing function with respect to α, yet it may assume any form with respect to X. To gain insights into the shape of conditional quantile functions, consider the normal linear regression model, where $Y = X\beta + \sigma\varepsilon$ and ε follows the standard normal distribution. Its inverse CDF, $\Phi^{-1}(\alpha)$, serves as the quantile function of ε. Under monotonic transformation, the conditional quantile function of Y is given by

$$q_\alpha(X) = X\beta + \sigma\Phi^{-1}(\alpha).$$

The functional form of $q_\alpha(X)$ indicates that 1) the conditional quantile function is increasing with α for a given X, 2) the conditional quantile function is linear with respect to X for a given α, assuming homoscedastic regression errors (i.e., constant σ), and 3) the curves $q_\alpha(X)$ and $q_{\tilde{\alpha}}(X)$ are parallel, sharing the same slope β but differing in intercepts. These curves do not intersect.

In heteroscedastic models where σ varies with X, the conditional quantile curves may not be parallel under different α values and may not remain linear with respect to X. Additionally, in nonlinear models such as the log-normal regression $Y = e^{X\beta+\sigma\varepsilon}$, by monotone transformation, the conditional quantile of Y is nonlinear function of X:

$$q_\alpha(X) = e^{X\beta+\sigma\Phi^{-1}(\alpha)}.$$

4.7.4 Classical Quantile Regression

Quantile regression is a powerful tool for describing and predicting specific conditional quantiles of interest. For instance, the concept of Value at Risk typically refers to the 1% or 5% quantile of financial losses. By performing a series of quantile regressions across a range of α values from 0 to 1, researchers can gain a comprehensive understanding of heterogeneous conditional distributions, supplementing the insights provided by conditional means in the classical linear regressions.

While segmenting X to analyze heterogeneous conditional distributions through subsamples is feasible for discrete, low-dimensional covariates, it becomes impractical with numerous regressors and moderate sample sizes. Instead of segmenting data, quantile regression posits a linear form for the αth quantile of the conditional distribution:

$$q_\alpha(X) = X\beta_\alpha,$$

leveraging the entire dataset for estimation. Nonlinear functions can be approximated by polynomial terms, rendering the quantiles approximately linear with respect to β_α, which may vary with α. In practice, incorporating quadratic terms of X into the quantile regression is common.

Define $\varepsilon = Y - X\beta_\alpha$ for a familiar regression form

$$Y = X\beta_\alpha + \varepsilon.$$

Quantile regression does not impose distributional assumptions on the error term ε, which is merely a monotonic transformation of Y. The sole assumption regarding the error term is that the αth conditional quantile of ε given X is zero. It is important to note that the mean of ε is generally not zero and may deviate significantly from zero when α approaches zero or one, necessitating cautious interpretation of the intercept term.

Given the coefficients β_α for all $\alpha \in [0,1]$, the conditional distribution of Y is determined. Conditional on X, random numbers of Y can be generated using the inverse CDF method: $X\beta_U$ where U follows the standard uniform distribution. This represents a random coefficient interpretation of quantile regression (see Exercise 4.10).

In practice, the coefficient β_α is an unknown parameter vector for each α. Classical statistics provides estimators for these parameters and demonstrates their consistency. Before observing data, the response variable Y is predicted using a linear combination of predictors $X\beta$, minimizing an expected loss function with an asymmetric loss function:

$$\min_\beta E\left[f_\alpha\left(Y - X\beta\right)\right],$$

where $f_\alpha(\varepsilon) = \varepsilon\left[\alpha - I\left(\varepsilon < 0\right)\right]$. With some abuse of notation, the density function of the conditional distribution is denoted by $p\left(y\,|X\right)$. The objective function is the sum of two integrals:

$$\min_\beta \int_{-\infty}^{X\beta} (1-\alpha)\left(X\beta - y\right) p\left(y\,|X\right) dy + \int_{X\beta}^{\infty} \alpha\left(y - X\beta\right) p\left(y\,|X\right) dy.$$

Differentiating with respect to β yields the optimal condition:

$$X(1-\alpha)\int_{-\infty}^{X\beta} p\left(y\,|X\right) dy - X\alpha\int_{X\beta}^{\infty} p\left(y\,|X\right) dy = 0.$$

Under the assumption of linear quantiles $q_\alpha(X) = X\beta_\alpha$, the minimizer of the expected asymmetric loss function is $\beta = \beta_\alpha$.

Empirically, with regression data Y_i, X_i, $i = 1, \ldots, n$, quantile regression minimizes the sample average of the asymmetric loss function:

$$\hat{\beta}_\alpha = \arg\min_\beta \frac{1}{n}\sum_{i=1}^{n} f_\alpha\left(Y_i - X_i\beta\right),$$

with a series of quantile regressions performed across a grid of α values.

The true value β_α minimizes the expected loss function in the population, while the estimator $\hat{\beta}_\alpha$ minimizes the sample average of the loss function. Under suitable regularity conditions, M-estimation theory suggests that $\hat{\beta}_\alpha$ is a consistent estimator of β_α.

The quantile regression estimator lacks a closed-form solution, necessitating numerical optimization methods. While mathematical programming techniques (e.g., simplex methods) are feasible, complexity escalates with sample size. Alternatively, smoothing the non-differentiable objective function with a quadratic curve facilitates gradient-based optimization, such as quasi-Newton methods. In the subsequent subsection, quantile regression is reframed as a Bayesian model, enabling efficient parameter estimation via MCMC methods.

4.7.5 Bayesian Quantile Regression

Quantile regression provides a robust framework for modeling the conditional quantiles of a response variable, offering insights beyond those available from classical mean regressions. A key feature of quantile regression lies in its equivalence between minimizing an asymmetric loss function and maximizing the likelihood function under the asymmetric Laplace distribution (ALD). This equivalence, as demonstrated by Yu and Moyeed (2001) and Kozumi and Kobayashi (2011), facilitates Bayesian analysis of quantile regressions using the Gibbs sampler.

Consider a non-Gaussian linear regression:

$$Y_i = X_i\beta + \sigma\varepsilon_i,$$

where the error term ε_i follows the ALD with the density function

$$p(\varepsilon_i) = \alpha(1-\alpha)e^{-f_\alpha(\varepsilon_i)}.$$

Again, the asymmetric loss function is $f_\alpha(\varepsilon) = \varepsilon[\alpha - I(\varepsilon < 0)]$.

The shape of the ALD density function resembles a back-to-back concatenation of an exponential density, proportional to $e^{-\alpha\varepsilon_i}I(\varepsilon_i > 0)$ and a flipped exponential density proportional to $e^{(1-\alpha)\varepsilon_i}I(\varepsilon_i < 0)$. The parameter α controls the skewness of the ALD: as α increases from 0 to 1, the left tail stretches and the right tail contracts, resulting in a distribution more skewed to the left. Unlike the zero-mean noise in standard linear regression, the mean of ε_i in the ALD regression is positive if $\alpha > \frac{1}{2}$ and negative if $\alpha < \frac{1}{2}$. The standard Laplace distribution corresponds to the ALD under $\alpha = \frac{1}{2}$, with a density proportional to $e^{-\frac{1}{2}|\varepsilon_i|}$, which aligns with median regression (see also Section 4.3).

The regression error ε_i has the αth quantile equal to zero: $\int_{-\infty}^{0} p(\varepsilon_i) = \alpha$. Thus, the αth quantile of the conditional distribution of the response variable Y_i is linear with respect to predictors: $q_\alpha(X_i) = X_i\beta$. In contrast with classical quantile regression, the conditional distribution of Y_i in the Bayesian model is fully specified, enabling the formulation of the log-likelihood function:

$$\ln p(Y|\beta,\sigma) = n\ln\alpha(1-\alpha) - n\ln\sigma - \sigma^{-1}\sum_{i=1}^{n} f_\alpha(Y_i - X_i\beta).$$

The maximum likelihood estimator of β aligns with the quantile regression estimator that minimizes the asymmetric loss function:

$$\min_{\beta}\sum_{i=1}^{n} f_\alpha(Y_i - X_i\beta).$$

When a flat prior is specified for β, the maximizer of the posterior density coincides with the maximum likelihood estimator. The appeal of Bayesian quantile regressions lies in the ability to obtain posterior samples via the Gibbs sampler with data augmentation. Proposition 4.6 demonstrates that the ALD can be represented as a mixture of normal distributions, where an exponentially distributed latent variable randomizes the mean and variance of the normal distributions. Additionally, the posterior distribution of the latent variable is analytically tractable, known as the generalized inverse Gaussian distribution for Z if Z^{-1} follows the two-parameter inverse Gaussian distribution.

Proposition 4.6 (Asymmetric Laplace distribution mixture) *Suppose that* $W \sim N(0,1)$ *and* $Z \sim Exp(1)$. *Consider a normal mixture*

$$\varepsilon = \mu Z + \sqrt{sZ}W,$$

where $\mu = \frac{1-2\alpha}{\alpha(1-\alpha)}$ *and* $s = \frac{2}{\alpha(1-\alpha)}$, *for* $\alpha \in (0,1)$.

1. *The marginal distribution of* ε *is the asymmetric Laplace distribution:*

$$p(\varepsilon) = \alpha(1-\alpha)e^{-\varepsilon[\alpha - I(\varepsilon<0)]}.$$

2. The conditional distributions of Z and its reciprocal $\zeta = Z^{-1}$ are

$$p(Z\,|\varepsilon) \propto Z^{-\frac{1}{2}} e^{-\frac{1}{4\alpha(1-\alpha)}Z - \frac{\alpha(1-\alpha)}{4}\varepsilon^2 Z^{-1}},$$

$$p(\zeta\,|\varepsilon) \propto \zeta^{-\frac{3}{2}} e^{-\frac{1}{4\alpha(1-\alpha)}\zeta^{-1} - \frac{\alpha(1-\alpha)}{4}\varepsilon^2 \zeta},$$

as an inverse Gaussian distribution with parameters $\frac{1}{\alpha(1-\alpha)|\varepsilon|}$ *and* $\frac{1}{2\alpha(1-\alpha)}$.

Proposition 4.6 provides a convenient framework for computing summary statistics of the ALD distribution. Since the mean, variance and skewness of Z are 1, 1, 2, respectively, we have

$$E(\varepsilon) = E[E(\varepsilon\,|Z)] = \mu,$$

$$Var(\varepsilon) = Var[E(\varepsilon\,|Z)] + E[Var(\varepsilon\,|Z)] = \mu^2 + s,$$

$$Skew(\varepsilon) = E\left[\frac{\mu Z + \sqrt{sZ}W - \mu}{(\mu^2+s)^{3/2}}\right] = \frac{2\mu^3 + 3\mu s}{(\mu^2+s)^{3/2}}.$$

As μ and s are functions of α, the mean and variance can be expressed as

$$E(\varepsilon) = \frac{1-2\alpha}{\alpha(1-\alpha)},$$

$$Var(\varepsilon) = \frac{1-2\alpha+2\alpha^2}{\alpha^2(1-\alpha)^2}.$$

Proposition 4.6 facilitates Bayesian estimation of quantile regressions by the Gibbs sampler with data augmentation. The linear regression with latent variables takes the form:

$$Y_i = X_i\beta + \mu\sigma Z_i + \sqrt{s\sigma^2 Z_i}W_i,$$

where the latent variables follow the exponential distribution $Z_i \sim Exp(1)$ and W_i follows the standard normal distribution. Bayesian model specification is completed by incorporating prior distributions such as normal and inverse gamma priors: $\beta \sim N(\underline{\mu}, \underline{V})$ and $\sigma \sim IG(\underline{a}, \underline{b})$. To sample from the posterior $p(\beta, Z, \sigma\,|Y)$, the Gibbs sampler cycles through the posterior conditional distributions.

First, $p(\beta\,|Y, Z, \sigma)$ is handled by a linear regression with rescaled data:

$$\frac{Y_i - \mu\sigma Z_i}{\sqrt{s\sigma^2 Z_i}} = \frac{X_i}{\sqrt{s\sigma^2 Z_i}}\beta + W_i.$$

The posterior conditional distribution is $N(\bar{\mu}, \bar{V})$, where

$$\bar{V} = \left(\underline{V}^{-1} + \sum_{i=1}^{n} \frac{X_i' X_i}{s\sigma^2 Z_i}\right)^{-1},$$

$$\bar{\mu} = \bar{V}\left[\underline{V}^{-1}\underline{\mu} + \sum_{i=1}^{n} \frac{X_i'(Y_i - \mu\sigma Z_i)}{s\sigma^2 Z_i}\right].$$

Second, $p(Z\,|Y, \beta, \sigma) = \prod_{i=1}^{n} p(Z_i\,|Y_i, \beta, \sigma)$ is handled by Proposition 4.6. First generate a draw for ζ_i from the inverse Gaussian distribution with the parameters $\frac{1}{\alpha(1-\alpha)}\left|\frac{Y_i - X_i\beta}{\sigma}\right|^{-1}$ and $\frac{1}{2\alpha(1-\alpha)}$, and then use its reciprocal $Z_i = \zeta_i^{-1}$.

Third, $p(\sigma | Y, \beta, Z)$ does not have a tractable form, but several workarounds exist. Regardless of the variance term, the posterior mode of β reproduces the classical quantile regression results. A simple workaround is to fix the variance, for example, $\sigma = 1$. Alternatively, a Metropolis step can be added to the Gibbs sampler, or the variable can be discretized for categorical sampling. Since σ is a scalar-valued random variable, it is feasible to generate a posterior draw directly. Kozumi and Kobayashi (2011) suggest reparametrizing the latent variable to restore the conjugate distribution. By rescaling the latent variable as $\tilde{Z}_i = \sigma Z_i$, the reparametrized regression reads

$$Y_i = X_i\beta + \mu\tilde{Z}_i + \sqrt{s\sigma\tilde{Z}_i}W_i.$$

Since the rescaled latent variable $\tilde{Z}_i$ follows an exponential distribution $Exp(\sigma)$, its posterior conditional distribution remains tractable. Similar to Proposition 4.6 (see Exercise 4.9), $\tilde{Z}_i^{-1}$ follows the inverse Gaussian distribution with the parameters $\frac{1}{\alpha(1-\alpha)|Y_i - X_i\beta|}$ and $\frac{1}{2\alpha(1-\alpha)\sigma}$. The intuition is that $\tilde{Z}_i^{-1} = \sigma^{-1}Z_i^{-1}$, so the original inverse Gaussian parameters is rescaled by σ. In the reparametrized model, σ serves as a variance term, and its posterior conditional distribution has the conjugate inverse gamma form:

$$p\left(\sigma \middle| Y, \beta, \tilde{Z}\right) \propto p(\sigma) \prod_{i=1}^{n} p\left(\tilde{Z}_i | \sigma\right) p\left(Y_i \middle| \beta, \tilde{Z}_i\right),$$

which indicates that $\sigma \Big| Y, \beta, \tilde{Z} \sim IG\left(\bar{a}, \bar{b}\right)$, where

$$\bar{a} = \underline{a} + \frac{3}{2}n,$$

$$\bar{b} = \underline{b} + \sum_{i=1}^{n} \tilde{Z}_i + \sum_{i=1}^{n} \frac{\left(Y_i - X_i\beta - \mu\tilde{Z}_i\right)^2}{2s\tilde{Z}_i}.$$

4.8 Model Assessment and Comparison

This chapter delves into the flexible specifications of regression models designed to accommodate disturbances that are correlated, heterogeneous, skewed, leptokurtic, and multi-modal. During the model-building process, specification error, or misspecification, arises when the proposed distribution or functional form inadequately represents key aspects of the data-generating process. Specification tests refer to assessing the validity of the probabilistic assumptions of the model by identifying deviations from the underlying hypotheses. In applied Bayesian analysis, it is customary to assess the adequacy of model fit and compare it with alternative model specifications.

4.8.1 Posterior Predictive Assessment

Before adopting a flexible model, it is crucial to determine whether a standard regression model is inadequate for the data at hand. Once a flexible model is adopted, evaluating its goodness of fit becomes essential. Gelman et al. (1996) advocate for routine goodness-of-fit assessments using posterior predictive discrepancies between observed and replicated data. They introduce a Bayesian analogue to the classical p-value, a central metric in hypothesis testing. A notable advantage of the Bayesian p-value is its immunity to the complexities associated with nuisance parameters, as all parameters are averaged over their posterior distributions.

Let Y denote the observed data and Y^* represent the replicated data that could have been observed if the data-generating process were repeated. The likelihoods $p(Y|\theta)$ and $p(Y^*|\theta)$ share the same functional form. When conditioned on parameters, observed and replicated data are independent: $p(Y^*|\theta, Y) = p(Y^*|\theta)$, allowing the joint distribution to be factorized as

$$p(Y^*, \theta|Y) = p(\theta|Y)\, p(Y^*|\theta).$$

Marginalizing over parameters yields the posterior predictive distribution of the replicated data:

$$p(Y^*|Y) = \int p(\theta|Y)\, p(Y^*|\theta)\, d\theta.$$

A test statistic $\tau(\cdot)$ as a function of the data. In classical inference, the observed value $\tau(Y)$ is compared to its sampling distribution under the null hypothesis. The p-value serves as a convenient metric indicating the position of $\tau(Y)$ within the tail of the sampling distribution:

$$P[\tau(Y^*) > \tau(Y)|\theta].$$

In the Bayesian framework, the concept of test statistics extends to a function of both data and parameters. The Bayesian p-value measures the position of $\tau(Y, \theta)$ within the tail of the posterior predictive distribution $P[\tau(Y^*, \theta) > \tau(Y, \theta)|Y]$ and is computed as an integral:

$$\int\int I[\tau(Y^*, \theta) > \tau(Y, \theta)]\, p(\theta|Y)\, p(Y^*|\theta)\, dY^* d\theta.$$

The intuition behind the Bayesian p-value is that if the model is appropriate, the observed data should appear plausible within the posterior predictive distribution. Observations located in the extreme tails of this distribution suggest a lack of model fit.

Computing the Bayesian p-value is feasible using Monte Carlo methods. First, sample a parameter θ_t from the posterior distribution $p(\theta_t|Y)$. In practice, the sample is readily available as MCMC output. Second, generate replicated data Y_t^* using the likelihood function $p(Y_t^*|\theta_t)$. Third, compute and compare $\tau(Y_t^*, \theta_t)$ and $\tau(Y, \theta_t)$. Repeating these steps for $t = 1, \ldots, T$ yields the sample analogue of the Bayesian p-value:

$$\frac{1}{T}\sum_{t=1}^{T} I[\tau(Y_t^*, \theta_t) > \tau(Y, \theta_t)].$$

Extreme values close to zero (e.g., < 0.01) or one (e.g., > 0.99) indicate that observations lie in the tail of the posterior predictive distribution, suggesting that the current model is unlikely to have generated the observed data, thus necessitating a model revision.

For instance, consider $\tau(Y, \theta) = y_i$, namely the ith observation. For each MCMC sample θ_t, simulate a replicated data point $\tau(Y_t^*, \theta_t) = y_{i,t}^*$ by the model. The draws $\{y_{i,1}^*, \ldots, y_{i,T}^*\}$ characterizes the posterior predictive distribution $p(y_i^*|Y)$. If the actual observation y_i is among the smallest or largest members of the posterior predictive samples, it provides evidence against the model. In case of discomfort about using the same data for parameter estimation and prediction, the cross-validation method can be employed. Replicated data are simulated from the leaving-one-out posterior predictive distribution $p(y_i^*|Y_{-i})$, where Y_{-i} excludes the ith observation in parameter estimation, and y_i is reserved solely for validation. If the observed y_i is an outlier relative to the replicated data, it implies a poorly fitted model.

4.8.2 Marginal Likelihood for Model Comparison

The marginal likelihood is central to Bayesian model comparison and averaging. Except in certain special cases, such as conjugate Bayesian linear regressions, the marginal likelihood is not analytically tractable. However, it can be approximated using Monte Carlo simulations (Gelfand and Dey, 1994; Chib, 1995; Chib and Jeliazkov, 2001).

A direct Monte Carlo method adheres to the definition of the marginal likelihood: the average of the likelihood weighted by the prior distribution

$$p(Y) = \int p(\theta)\, p(Y\,|\theta)\, d\theta.$$

Let $\theta_t, t = 1, \ldots, T$ be samples drawn from the prior. The likelihood function $p(Y\,|\theta_t)$ is evaluated for each θ_t, resulting in the sample analogue $\frac{1}{T}\sum_{t=1}^{T} p(Y\,|\theta_t)$ as an unbiased and consistent estimator of the marginal likelihood. However, this estimator is often crude due to excessively large Monte Carlo variance, unless the prior distribution is extremely tight.

Gelfand and Dey (1994) propose a more reliable method for computing the marginal likelihood using posterior samples. This approach reinterprets the marginal likelihood as the mathematical expectation with respect to the posterior distribution, which exhibits smaller volatility than samples from the prior distribution. Posterior samples are readily available as the output of MCMC simulations, making it feasible to estimate the marginal likelihood for most econometric models. Proposition 4.7 constructs the posterior expectation for the marginal likelihood.

Proposition 4.7 (Gelfand-Dey marginal likelihood) *Let $p(\theta)$, $p(Y\,|\theta)$ and $p(\theta\,|Y)$ be the prior, likelihood and posterior, respectively. Let $f(\theta)$ be a proper density function with support contained within the support of $p(\theta)$. The marginal likelihood takes the form:*

$$p(Y) = \left\{ E\left[\frac{f(\theta)}{p(\theta)\, p(Y\,|\theta)} \,|Y \right] \right\}^{-1}.$$

Suppose that $\theta_t, t = 1, \ldots, T$ are MCMC samples from the posterior $p(\theta\,|Y)$. By Proposition 4.7, the marginal likelihood can be approximated by the sample analogue:

$$\hat{p}(Y) = \left\{ \frac{1}{T} \sum_{t=1}^{T} \frac{f(\theta_t)}{p(\theta_t)\, p(Y\,|\theta_t)} \right\}^{-1}.$$

The performance of the marginal likelihood estimator depends on the specification of $f(\theta)$. Ideally, $f(\theta)$ should closely resemble the posterior density $p(\theta\,|Y)$. If the numerator and denominator in $\frac{f(\theta)}{p(\theta)p(Y|\theta)}$ are nearly proportional to the posterior density, the ratio becomes nearly constant, minimizing Monte Carlo variation. Additionally, it is desirable to suppress the tails of $f(\theta)$ to ensure $\frac{f(\theta)}{p(\theta)p(Y|\theta)} < \infty, \forall\theta$. Geweke (1999) suggests using a truncated normal density calibrated to the posterior sample mean and variance:

$$f(\theta) = \frac{1}{c}\phi\left(\theta, \hat{\mu}, \hat{\Sigma}\right) I_A(\theta),$$

where $\phi(\cdot)$ denotes a multivariate normal density, with mean $\hat{\mu} = \frac{1}{T}\sum_{t=1}^{T}\theta_t$ and variance $\hat{\Sigma} = \frac{1}{T}\sum_{t=1}^{T}(\theta_t - \hat{\mu})(\theta_t - \hat{\mu})'$. Proposition 4.7 requires $f(\theta)$ as a proper density function, necessitating the normalizing constant

$$c = \int \phi\left(x, \hat{\mu}, \hat{\Sigma}\right) I_A(x)\, dx.$$

It is possible to specify A as a rectangular region to truncate tails, but computing a high-dimensional multivariate normal probability is challenging. Alternatively, a spherical region can be specified as

$$A = \left\{ \theta \in \mathbb{R}^d : (\theta - \hat{\mu})' \hat{\Sigma}^{-1} (\theta - \hat{\mu}) < \chi^2_\alpha (d) \right\},$$

where $\chi^2_\alpha (d)$ denotes the α quantile of the Chi-square distribution with d degrees of freedom (i.e., the dimension of θ). For example, $\chi^2_{0.99} (4) = 13.27$. Consequently, the normalizing constant takes the form:

$$c = \int \phi \left(x, \hat{\mu}, \hat{\Sigma} \right) I_A (x)\, dx = E\left[I_A (X) \right] = P\left[(X - \hat{\mu})' \hat{\Sigma}^{-1} (X - \hat{\mu}) < \chi^2_\alpha (d) \right].$$

We intentionally replaced the symbol θ by X, which follows a multivariate normal distribution $N\left(\hat{\mu}, \hat{\Sigma}\right)$ (distinct from the posterior distribution $p(\theta | Y)$). The standardized variable $\tilde{X} = \hat{\Sigma}^{-1/2} (X - \hat{\mu})$ follows the standard normal distribution. Therefore, $\tilde{X}'\tilde{X} = (X - \hat{\mu})' \hat{\Sigma}^{-1} (X - \hat{\mu})$ follows the Chi-square distribution with d degrees of freedom. The definition of the quantile indicates the normalizing constant

$$c = P\left[\tilde{X}'\tilde{X} < \chi^2_\alpha (d) \right] = \alpha.$$

The proper density to be used by Proposition 4.7 is

$$f(\theta) = \frac{1}{\alpha} \phi \left(\theta, \hat{\mu}, \hat{\Sigma} \right) I_A (\theta),$$

where $\alpha \in (0, 1)$ is a tuning parameter. It is feasible to experiment with a sequence of α values, with minimal additional computational cost.

4.9 Software and Computational Considerations

In this section, we explore implementation issues related to Bayesian SUR and the marginal likelihood evaluation.

4.9.1 Efficient Implementation of Bayesian SUR

Many multivariate econometric models can be effectively reduced to SUR models when conditioned on appropriate latent variables. The Gibbs sampling technique for SUR facilitates Bayesian inference for these models. According to Proposition 4.1, the posterior conditional distribution of the coefficient vector β is normal with mean $\bar{\mu}$ and covariance matrix $\bar{V}$, where

$$\bar{V} = \left[V^{-1} + X' (I_n \otimes \Sigma_m)^{-1} X \right]^{-1},$$

$$\bar{\mu} = \bar{V} \left[V^{-1}\mu + X' (I_n \otimes \Sigma_m)^{-1} Y \right].$$

Directly coding the expression $X' (I_n \otimes \Sigma_m)^{-1} X$ as `X'*inv(kron(eye(n),Sigma))*X` is computationally inefficient and may lead to out-of-memory errors due to the inversion of an $mn \times mn$ matrix. Although the Kronecker product is mathematically elegant, it does not warrant computational efficiency. It is possible to simplify the expression:

$$X'\left(I_n \otimes \Sigma_m\right)^{-1} X = X'\left(I_n \otimes \Sigma_m^{-1}\right) X = \sum_{i=1}^{n} X_i' \Sigma_m^{-1} X_i.$$

Computing $\sum_{i=1}^{n} X_i' \Sigma_m^{-1} X_i$ within a loop may be slow on some computing platform. A more efficient approach is to decompose the covariance matrix as $\Sigma_m = LL'$ using the Cholesky or eigenvalue-based algorithms, resulting in

$$\sum_{i=1}^{n} X_i' \Sigma_m^{-1} X_i = \sum_{i=1}^{n} \left(L^{-1} X_i\right)' \left(L^{-1} X_i\right).$$

Since $X_i = diag\left(X_{i1}, \ldots, X_{im}\right)$ contains many zero elements, computational efficiency improves by eliminating these zeros. Denote

$$L^{-1} = \begin{pmatrix} c_{11} & \cdots & c_{1m} \\ \cdots & \ddots & \cdots \\ c_{m1} & \cdots & c_{mm} \end{pmatrix}.$$

By applying the Cholesky decomposition, the matrix can be further simplified into a lower-triangular form, as a dense matrix

$$L^{-1} X_i = \begin{pmatrix} c_{11} X_{i1} & \cdots & c_{1m} X_{im} \\ \cdots & \ddots & \cdots \\ c_{m1} X_{i1} & \cdots & c_{mm} X_{im} \end{pmatrix},$$

Given that disturbances between equations are correlated, these correlations are removed by transforming $Y_i = X_i \beta + \varepsilon_i$ as $\tilde{Y}_i = \tilde{X}_i \beta + L^{-1} \varepsilon_i$, where $\tilde{Y}_i = L^{-1} Y_i$ and $\tilde{X}_i = L^{-1} X_i$ are the transformed response variables and predictors. Proposition 1.1 is then applicable under the transformed data.

4.9.2 Marginal Likelihood Evaluation

Integrating out parameters yields the marginal likelihood conditional on latent variables Z:

$$p\left(Y \mid Z\right) = \int p\left(\theta \mid Z\right) p\left(Y \mid \theta, Z\right) d\theta.$$

The marginal likelihood is crucial in Bayesian analysis, as it facilitates both parameter estimation and model comparison.

First, the closed-form marginal likelihood facilitates parameter estimation through marginalized MCMC sampling. Data augmentation is a well-established technique in Bayesian econometrics. Even when the posterior $p\left(\theta \mid Y\right)$ is not of a known form, the posterior conditional distributions $p\left(\theta \mid Y, Z\right)$ and $p\left(Z \mid Y, \theta\right)$ may be tractable. In such cases, the Gibbs sampler is often the method of choice for researchers. An often-overlooked fact is that closed-form posterior conditional distributions also enable a marginalized MCMC targeting $p\left(Z \mid Y\right) \propto p\left(Z\right) p\left(Y \mid Z\right)$, where the marginal likelihood can be evaluated by Bayes' rule:

$$p\left(Y \mid Z\right) = \frac{p\left(\theta \mid Z\right) p\left(Y \mid \theta, Z\right)}{p\left(\theta \mid Y, Z\right)}.$$

Marginalized MCMC finds broad applicability since many econometric models exhibit conjugate sub-structures when conditioned on certain latent variables. Concerns about algebraic complexity and computational challenges in evaluating the marginal likelihood are common. However, $p(Y|Z)$ often takes a surprisingly simple form in many models when evaluated up to a normalizing constant.

Regarding debugging marginalized MCMC programs, the first step should be literal translation of Bayes' rule for calculating the marginal likelihood $p(Y|Z)$. The full expressions of $p(\theta|Z)$, $p(Y|\theta, Z)$ and $p(\theta|Y, Z)$ can be evaluated at multiple (including extreme) values of θ. If the program is free of bugs, results remain unchanged with respect to θ. It serves as a safe mode for computing the marginal likelihood, despite slow code. Premature code optimization should be avoided. Once the marginalized MCMC program yields correct results, code optimization can be performed by simplifying expressions that are not functions of Z to enhance runtime performance.

Second, a simulated marginal likelihood facilitates Bayesian model comparison and averaging. When the marginal likelihood is not analytically tractable, the method of Gelfand and Dey (1994) is applicable to a wide range of statistical models, provided that posterior MCMC samples are available and the exact likelihood function can be evaluated. According to Proposition 4.7, the marginal likelihood is approximated by

$$\hat{p}(Y) = \left\{ \frac{1}{T} \sum_{t=1}^{T} \frac{f(\theta_t)}{p(\theta_t)\, p(Y|\theta_t)} \right\}^{-1},$$

where $\theta_t, t = 1, \ldots, T$ are MCMC samples from the posterior $p(\theta|Y)$.

An implementation challenge arises when the marginal likelihood value is small, leading to potential underflow. It is advisable to work with the logarithm of the marginal likelihood. A numerically stable formula is

$$\ln \hat{p}(Y) = -g^* - \ln \left[\frac{1}{T} \sum_{t=1}^{T} e^{g(\theta_t) - g^*} \right],$$

where

$$g(\theta_t) = \ln f(\theta_t) - \ln p(\theta_t) - \ln p(Y|\theta_t),$$
$$g^* = \max_t g(\theta_t).$$

A practical choice of the density function $f(\cdot)$ is the truncated normal density, calibrated to the posterior sample mean $\hat{\mu}$ and variance $\hat{\Sigma}$ such that

$$f(\theta_t) = \frac{1}{\alpha} \phi\left(\theta_t, \hat{\mu}, \hat{\Sigma}\right) I_A(\theta_t),$$
$$A = \left\{ \theta_t \in \mathbb{R}^d : (\theta_t - \hat{\mu})' \hat{\Sigma}^{-1} (\theta_t - \hat{\mu}) < \chi^2_\alpha(d) \right\}.$$

Posterior samples outside the region (i.e., $\theta_t \notin A$) in the average $\frac{1}{T}\sum_{t=1}^{T} \ldots$ are not discarded. Instead, the outsiders should participate in the averaging with $f(\theta_t) = 0$. The tuning parameter α determines which θ_t is an outsider. A good practice is to test a grid of α values. They should yield comparable results, and their average or median value can serve as the marginal likelihood estimator.

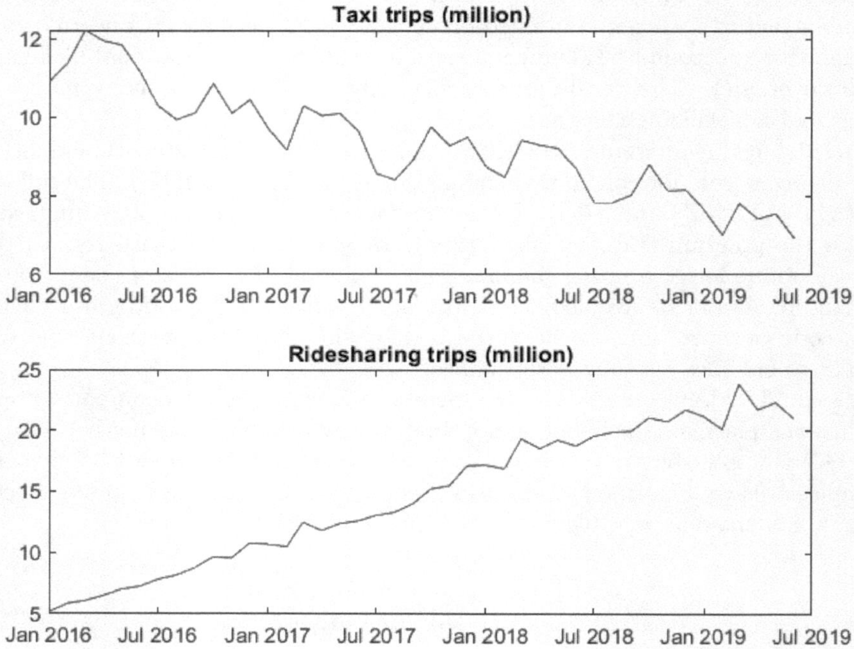

FIGURE 4.2
Monthly time series plot of New York taxi and ridesharing trips.

4.10 An Application of Ridesharing Effect by Non-Gaussian Regressions

Digital ridesharing platforms, such as Uber and Lyft, have transformed the taxi industry through their convenience and competitive pricing. Since 2017, ridesharing trips have surpassed yellow-taxi trips in New York City. The New York City Taxi and Limousine Commission collects and publishes massive data of trip records for both taxis and ridesharing for-hire vehicles. The trip-level data includes pick-up/drop-off dates, times, and locations. From 2016 to 2019, over 390 million trips were recorded for yellow taxis and 615 million for high-volume ridesharing vehicles. Figure 4.2 illustrates monthly time series data, showing a decline in taxi trips from 12 million in early 2016 to 7 million in 2019, while ridesharing trips increased from 5 to 24 million during the same period.

4.10.1 Skew Normal Regressions

We conduct regressions using high-frequency data sampled at 15-minute intervals. The response variable is the number of taxi trips, with predictors including the number of ridesharing trips and hourly dummy variables to capture intra-day seasonality.

Starting with a Gaussian Bayesian linear regression with a non-informative prior, we obtain the ordinary least squares estimator. The regression residuals, calculated by

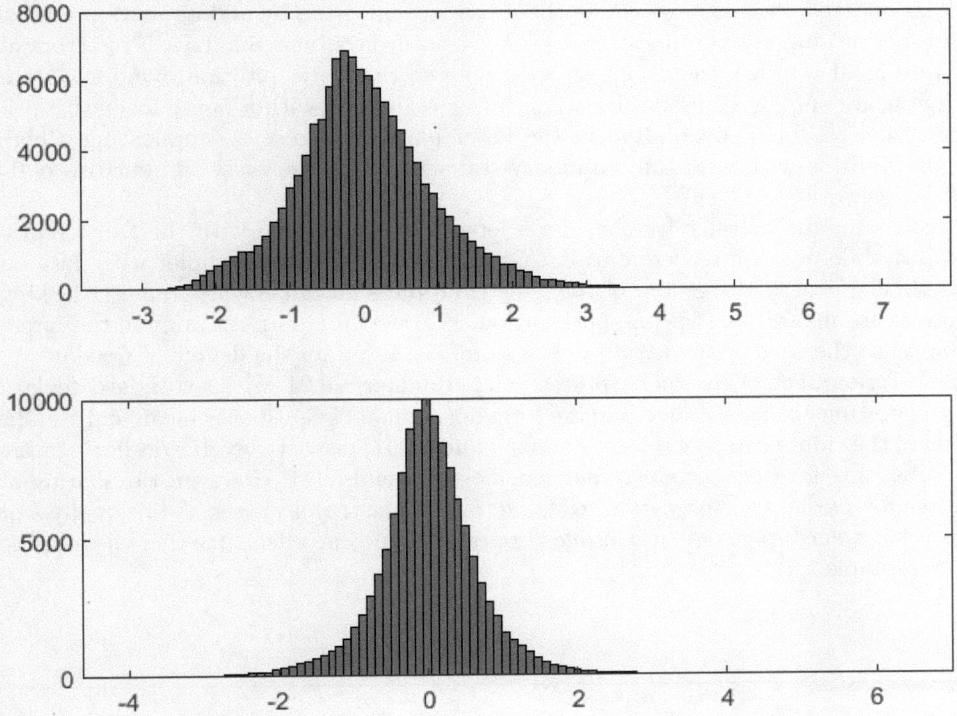

FIGURE 4.3
Residuals of Gaussian regressions under smaller (upper panel) and larger (lower panel) models. Predictors in the smaller model are ridesharing trips and hourly dummy variables. The larger model also has hourly dummy variables of weekdays, weekends, and weather conditions.

subtracting fitted values from observations, are plotted in the upper panel of Figure 4.3. The distribution of residuals is positively skewed, with a thicker right tail than left. Summary statistics show skewness and kurtosis of 0.34 and 3.89, respectively. Classical Jarque-Bera and Kolmogorov-Smirnov tests decisively reject normality.

A parsimonious specification of tail asymmetry is the skew normal distribution, which involves a half-normal latent variable to accommodate asymmetric tails. Bayesian estimation of skew normal regression can be implemented via the Gibbs sampler or marginalized subsampling, as discussed in Section 4.4. Both methods yield similar empirical results, with the estimated coefficient of the latent variable δ being 1.22, which is large compared to the estimated disturbance variance ($\sigma^2 = 0.34$).

An undesirable outcome of the current model specification is the ridesharing trips coefficient of 0.003, which does not reflect the substitution effect seen in Figure 4.2. A plausible reason is the omission of variables in the regression.

4.10.2 Fat Tail Regressions

To address omitted variables, we revise the model specification by adding more predictors, including hourly dummy variables for weekdays, Saturdays, and Sundays. Predictors also incorporate local weather conditions such as temperature, precipitation, wind speed, and snow. Again, we employ Gaussian Bayesian linear regression with a non-informative prior. The regression residuals are plotted in the lower panel of Figure 4.3, indicating a highly peaked distribution with a fat tail. Summary statistics reveal skewness and kurtosis of 0.02 and 6.3, respectively.

To accommodate fat tails, we consider a leptokurtic regression with the t distribution, which has a scale mixture representation that facilitates the Gibbs sampler with data augmentation. The degree of freedom of the t distribution is an unknown parameter lacking a conjugate form, regardless of prior specification. We specify a uniform prior in the interval $(2, 10)$ and use the Metropolis-within-Gibbs sampler to update the degree of freedom.

The posterior mean of the degree of freedom parameter is 3.21, with a standard deviation of 0.04, suggesting a substantial departure from normality. The posterior mean and standard deviation of the ridesharing coefficient are -0.13 and 0.001, respectively. If ridesharing were a perfect substitute for taxis, the regression coefficient would be -1. However, ridesharing also substitutes for public transportation, walking, and other travel means. Additionally, competitive pricing encourages greater usage through the income effect, making the empirical results reasonable.

4.11 Bibliographic Notes

This chapter explores departures from normality and flexible specifications in regression models. While several Bayesian textbooks, such as Koop (2003), Lancaster (2004), Geweke (2005), Koop et al. (2007), Greenberg (2013), cover similar topics, our presentation offers two novel contributions to the field of non-Gaussian regressions.

First, we introduce a new perspective on the stochastic volatility model, extending the heteroscedasticity model of Li and Tobias (2009). Traditionally, the stochastic volatility model is analyzed within the state space framework, as extensively documented in the literature. Notable approaches include the local neighbor sampler by Jacquier et al. (1994, 2004), Gaussian mixture approximation by Kim et al. (1998), and refined approximation by Omori et al. (2007). In contrast, our estimation of the stochastic volatility model eschews reliance on the state space model, the Kalman filter (Kalman, 1960), or the simulation smoother (Carter and Kohn, 1994; Durbin and Koopman, 2002).

Second, we elaborate on a novel MCMC method termed double marginalized subsampling (Qian, 2022), a big data algorithm applicable to various econometric models. This method facilitates Bayesian inference with both tall data (i.e., massive observations) and wide data (i.e., high-dimensional parameters). The Gaussian mixture model exemplifies the scalability of this approach in both dimensions. Celeux et al. (2019) remark on the challenges of conducting Bayesian inference on mixtures of moderate to large dimensions, noting that Gibbs sampling is the only algorithm that scales to high dimensions. However, the limitations of the Gibbs sampler are well-documented, particularly concerning label switching and convergence issues (see Celeux et al. (2000), Jasra et al. (2005), Stephens (2000), Geweke (2007) and Yao and Lindsay (2009) for detailed discussions). Double marginalized subsampling addresses the high-dimensional problem by integrating out most parameters, leaving a vector of latent variables for subsampling.

4.12 Proof of Propositions

Proposition 4.1 (Posterior under a non-diagonal covariance matrix)
Recall the formula for completing the square:
$x'Ax + x'b = \left(x + \frac{1}{2}A^{-1}b\right)' A \left(x + \frac{1}{2}A^{-1}b\right) - \frac{1}{4}b'A^{-1}b.$
The posterior conditional density takes the form
$p(\beta \,|Y, \Sigma) \propto e^{-\frac{1}{2}(\beta-\mu)'V^{-1}(\beta-\mu)-\frac{1}{2}(Y-X\beta)'\Sigma^{-1}(Y-X\beta)}.$
Completing the square for the quadratic function of β *yields*
$(\beta-\mu)' V^{-1} (\beta-\mu) + (Y - X\beta)' \Sigma^{-1} (Y - X\beta)$
$= \beta' \left(V^{-1} + X'\Sigma^{-1}X\right) \beta - 2\beta' \left(V^{-1}\mu + X'\Sigma^{-1}Y\right) + \mu'V^{-1}\mu + Y'\Sigma^{-1}Y$
$= (\beta - \bar{\mu})' \bar{V}^{-1} (\beta - \bar{\mu}) - \bar{\mu}'\bar{V}^{-1}\bar{\mu} + \mu'V^{-1}\mu + Y'\Sigma^{-1}Y$
where $\bar{\mu} = \left(V^{-1} + X'\Sigma^{-1}X\right)^{-1} \left(V^{-1}\mu + X'\Sigma^{-1}Y\right)$ *and* $\bar{V} = \left(V^{-1} + X'\Sigma^{-1}X\right)^{-1}.$
It follows that $p(\beta \,|Y, \Sigma) \propto e^{-\frac{1}{2}(\beta-\bar{\mu})'\bar{V}^{-1}(\beta-\bar{\mu})},$
which is multivariate normal density with the mean $\bar{\mu}$ *and the covariance matrix* $\bar{V}$.

Proposition 4.2 (Leptokurtic scale mixture)
The moments of e_i *can be computed by the law of iterated expectations.*
$E(e_i) = E\left[E\left(\sigma\sqrt{Z_i}\varepsilon_i \,|Z_i\right)\right] = 0.$
$E(e_i^2) = E\left[E\left(\sigma^2 Z_i \varepsilon_i^2 \,|Z_i\right)\right] = E\left(\sigma^2 Z_i\right) = \sigma^2\mu_z.$
$E(e_i^4) = E\left[E\left(\sigma^4 Z_i^2 \varepsilon_i^4 \,|Z_i\right)\right] = E\left(3\sigma^4 Z_i^2\right) = 3\sigma^4\left(\mu_z^2 + \sigma_z^2\right).$
It follows that
$\kappa = \frac{E(e_i^4)}{\left[E(e_i^2)\right]^2} = 3 + 3\frac{\sigma_z^2}{\mu_z^2}.$

Proposition 4.3 (Scale mixture t **distribution)**
The marginal distribution is obtained by integrating out the scale variable:
$p(x) = \int p(\tau)\, p(x \,|\tau)\, d\tau \propto \int \tau^{-(a+1)-\frac{1}{2}} e^{-b\tau^{-1}-\frac{1}{2}\tau^{-1}\left(\frac{x-\mu}{\sigma}\right)^2} d\tau.$
The integrand is proportional to the density of $IG\left[a + \frac{1}{2}, b + \frac{1}{2}\left(\frac{x-\mu}{\sigma}\right)^2\right].$
The integral corresponds to the normalizing constant of that distribution.
$p(x) \propto \left[b + \frac{1}{2}\left(\frac{x-\mu}{\sigma}\right)^2\right]^{-a-\frac{1}{2}} \propto \left[1 + \frac{1}{2b}\left(\frac{x-\mu}{\sigma}\right)^2\right]^{-\frac{2a+1}{2}}.$
We recognize that it is the density of $t\left(\mu, \frac{b}{a}\sigma^2, 2a\right).$

Proposition 4.4 (Combined skewness)
Since Z_i *and* ε_i *are independent, the mean and variance of the linear combination are*
$\mu_e = \delta\mu_z,$
$\sigma_e^2 = \sigma^2 + \delta^2\sigma_z^2.$
Define $\tilde{Z}_i = Z_i - \mu_z$. *By the definition of skewness,*
$\varsigma_e = \sigma_e^{-3} E\left[(e_i - \mu_e)^3\right] = \sigma_e^{-3} E\left[\left(\delta\tilde{Z}_i + \sigma\varepsilon_i\right)^3\right].$
The cubic term can be expanded as

$$
\begin{aligned}
&E\left[\left(\delta\tilde{Z}_i + \sigma\varepsilon_i\right)^3\right] \\
&= \delta^3 E\left(\tilde{Z}_i^3\right) + 3\delta^2\sigma E\left(\tilde{Z}_i^2\right) E(\varepsilon_i) + 3\delta\sigma^2 E\left(\tilde{Z}_i\right) E\left(\varepsilon_i^2\right) + \sigma^3 E\left(\varepsilon_i^3\right) \\
&= \delta^3 E\left(\tilde{Z}_i^3\right)
\end{aligned}
$$

$= \delta^3 \sigma_z^3 \varsigma_z$

It follows that $\varsigma_e = \frac{\sigma_z^3}{\sigma_e^3} \delta^3 \varsigma_z$.

Proposition 4.5 (Quantile and asymmetric loss function)

For the ease of presentation, we assume an absolutely continuous distribution with the density function $p(y)$, *leading to the expected loss function as the sum of two integrals:*

$\min_c \int_{-\infty}^{c} (1-\alpha)(c-y) p(y) dy + \int_{c}^{\infty} \alpha (y-c) p(y) dy$.

Differentiating with respect to c *(it is unnecessary to consider* c *in the upper and lower bounds of the integral due to* $c-c=0$ *in the integrand) yields the first-order condition:*

$\int_{-\infty}^{c} (1-\alpha) p(y) dy + \int_{c}^{\infty} \alpha (-1) p(y) dy = 0$.

The optimal condition is $\alpha = F(y)$, *which indicates that* q_α *is the minimizer of the expected loss function.*

As for the sample analogue, the objective function is the sum of two categories:

$\sum_{i=1}^{n} f_\alpha (Y_i - c) = (1-\alpha) \sum_{Y_i<c} (c - Y_i) + \alpha \sum_{Y_i \geq c} (Y_i - c)$.

Suppose that the optimal c *equals one of the data points* Y_i. *A marginal perturbation of* c *causes the cutoff* Y_i *to change the category. Although the penalty per unit of deviation changes between* α *and* $1-\alpha$, *the distance of deviation is zero:* $c-c=0$. *Consequently, the effect of category change can be ignored, simplifying the optimization problem as*

$\min_c (1-\alpha) \sum_{Y_i<\bar{c}} (c - Y_i) + \alpha \sum_{Y_i \geq \bar{c}} (Y_i - c)$, *where* $\bar{c} = c$.

Taking derivatives with respect to c *yields*

$(1-\alpha) \sum_{Y_i<\bar{c}} 1 - \alpha \sum_{Y_i \geq \bar{c}} 1 = 0$.

It follows that $\frac{1}{n} \sum_{Y_i<c} 1 = \alpha$, *indicating that* $\hat{q}_\alpha$ *is the minimizer under the asymmetric loss function.*

Proposition 4.6 (Asymmetric Laplace distribution mixture)

Conditional on the latent variable Z, *the noise term* ε *follows the normal distribution:* $\varepsilon | Z \sim N(\mu Z, sZ)$. *The marginal distribution of* ε *is obtained by integrating out the exponentially distributed latent variable:*

$p(\varepsilon) = \int p(Z) p(\varepsilon | Z) dZ = \frac{1}{\sqrt{2\pi}} s^{-\frac{1}{2}} e^{\frac{\mu\varepsilon}{s}} \int_0^\infty Z^{-\frac{1}{2}} e^{-\left(1+\frac{\mu^2}{2s}\right) Z - \frac{\varepsilon^2}{2s} Z^{-1}} dZ$.

The integrand is the kernel of the reciprocal inverse Gaussian distribution, yielding an explicit solution:

$\int_0^\infty Z^{-\frac{1}{2}} e^{-bZ - cZ^{-1}} dZ = \sqrt{\pi} e^{-2\sqrt{bc}} b^{-\frac{1}{2}}$.

Here, the parameters b *and* c *are given by*

$b = 1 + \frac{\mu^2}{2s} = \frac{1}{4\alpha(1-\alpha)}$,

$c = \frac{\varepsilon^2}{2s} = \frac{\alpha(1-\alpha)}{4} \varepsilon^2$.

Note that the absolute-value function $|\varepsilon|$ *in the Laplace distribution stems from* $\sqrt{bc} = \frac{1}{4} |\varepsilon|$.

The integral can be simplified as

$\int_0^\infty Z^{-\frac{1}{2}} e^{-bZ - cZ^{-1}} dZ = 2\sqrt{\pi} e^{-\frac{1}{2}|\varepsilon|} \sqrt{\alpha(1-\alpha)}$.

It follows that $p(\varepsilon) = \alpha(1-\alpha) e^{-\frac{1}{2}|\varepsilon| + \frac{1-2\alpha}{2} \varepsilon}$.

If $\varepsilon < 0$, *then* $p(\varepsilon) = \alpha(1-\alpha) e^{(1-\alpha)\varepsilon}$.

If $\varepsilon \geq 0$, *then* $p(\varepsilon) = \alpha(1-\alpha) e^{-\alpha\varepsilon}$.

Combining the two cases results in the marginal density in terms of the asymmetric loss function:

$p(\varepsilon) = \alpha(1-\alpha) e^{-\varepsilon[\alpha - I(\varepsilon<0)]}$.

As for the conditional distribution, it has been shown that

$p(Z | \varepsilon) \propto p(Z) p(\varepsilon | Z) \propto Z^{-\frac{1}{2}} e^{-bZ - cZ^{-1}}$.

This is the reciprocal of the inverse Gaussian distribution (also known as the generalized inverse Gaussian distribution).

By reciprocal, we mean $\zeta = Z^{-1}$. *By the change of variable*

$p(\zeta\,|\varepsilon) \propto \zeta^{-\frac{3}{2}} e^{-c\zeta - b\zeta^{-1}} \propto \zeta^{-\frac{3}{2}} e^{-\frac{2b\left(\zeta-\sqrt{b/c}\right)^2}{2bc^{-1}\zeta}}$,

which is the inverse Gaussian distribution with the parameters $\left(\sqrt{b/c}, 2b\right)$:

$\sqrt{b/c} = \frac{1}{\alpha(1-\alpha)|\varepsilon|}$,

$2b = \frac{1}{2\alpha(1-\alpha)}$.

In summary, the posterior distribution of the latent variable Z^{-1} *follows the inverse Gaussian distribution with the mean parameter* $\frac{1}{\alpha(1-\alpha)|\varepsilon|}$ *and the scale parameter* $\frac{1}{2\alpha(1-\alpha)}$.

Proposition 4.7 (Gelfand-Dey marginal likelihood)
The expectation is taken with respect to the posterior distribution $p(\theta\,|Y)$:

$E\left[\frac{f(\theta)}{p(\theta)p(Y|\theta)}\,|Y\right] = \int_\Theta \frac{f(\theta)p(\theta|Y)}{p(\theta)p(Y|\theta)} d\theta = \frac{\int_\Theta f(\theta)d\theta}{p(Y)} = \frac{1}{p(Y)}$.

4.13 Exercises

Exercise 4.1 (Joint distribution of Autoregression) *Consider the autoregression*

$$\varepsilon_t = \phi\varepsilon_{t-1} + \sigma u_t,$$

where u_t *follows the standard normal distribution. Assume that* $|\rho| < 1$ *and the initial value* ε_0 *follows the stationary distribution* $N\left(0, \frac{\sigma^2}{1-\rho^2}\right)$.

1. *Show that* ε_t *follows* $N\left(0, \frac{\sigma^2}{1-\rho^2}\right)$ *for all* $t \geq 0$, *hence stationarity.*

2. *Show that the joint distribution of* $\varepsilon_1, \ldots, \varepsilon_n$ *is multivariate normal with the covariance matrix*

$$\Sigma = \frac{\sigma^2}{1-\rho^2}\begin{pmatrix} 1 & \rho & \cdots & \rho^{n-1} \\ \rho & 1 & \cdots & \rho^{n-2} \\ \cdots & \cdots & \ddots & \cdots \\ \rho^{n-1} & \rho^{n-2} & \cdots & 1 \end{pmatrix}.$$

Exercise 4.2 (Characteristics of t distribution) *A random variable* x *follows the* $t\left(\mu, \sigma^2, v\right)$ *distribution if its density function is*

$$f\left(x;\mu,\sigma^2,v\right) \propto \left[1 + \frac{1}{v}\left(\frac{x-\mu}{\sigma}\right)^2\right]^{-\frac{v+1}{2}}.$$

1. *Show that the proportionality constant equals*

$$\frac{\Gamma\left(\frac{v+1}{2}\right)}{\Gamma\left(\frac{v}{2}\right)(\pi v)^{\frac{1}{2}}\sigma}.$$

2. Show that the mean and variance are given by

$$E(x) = \mu,$$

$$Var(x) = \frac{v}{v-2}\sigma^2.$$

3. Show the limits of the t distribution:
As $v = 1$, the t distribution reduces to the Cauchy distribution:

$$f(x) \propto \left[1 + \left(\frac{x-\mu}{\sigma}\right)^2\right]^{-1}.$$

As $v \to \infty$, the t distribution approaches the normal distribution

$$f(x) \propto e^{-\frac{(x-\mu)^2}{2\sigma^2}}.$$

4. For frequentist inference, the t statistics is constructed by the normal-chi-square form of the t distribution:

$$t(0, \sigma^2, v) \iff \frac{N(0, \sigma^2)}{\sqrt{\chi^2(v)/v}}.$$

If $z \sim N(0, \sigma^2)$ and $y \sim \chi^2(v)$, then $x = \frac{z}{\sqrt{y/v}}$ follows $t(0, \sigma^2, v)$. Are the results related to the scale mixture representation (Proposition 4.1)? Are the results related to the limits of the t distributions?

5. Show that the tail of $t(\mu, \sigma^2, v)$ is fatter than that of $N(\mu, \sigma^2)$.

6. Consider two t distributions: $t(\mu, \sigma^2, v_1)$ and $t(\mu, \sigma^2, v_2)$, where $v_1 < v_2$. Show that the tail of the former is fatter than that of the latter (that is, a smaller degree of freedom translates to a fatter tail).

Exercise 4.3 (Skewness of a symmetric distribution) *Consider a symmetric distribution whose density function satisfies $f(\mu - x) = f(\mu + x)$, $\forall x \in \mathbb{R}$. Skewness is defined as $\frac{E(X-\mu)^3}{\sigma^3}$, where μ and σ are the mean and standard deviation of the random variable. Show that the skewness of a symmetric distribution equals zero.*

Exercise 4.4 (Characteristics of skewed normal distribution) *Consider $e_i = \delta Z_i + \sigma\varepsilon_i$, where the independently distributed Z_i and ε_i are half normal and standard normal random variables, respectively.*

1. Show that the CDF of e_i evaluated at a point $k \in \mathbb{R}$ takes the form

$$P(e_i \leq k) = \int_0^{\infty} 2\phi(Z_i)\,\Phi\left(\frac{k - \delta Z_i}{\sigma}\right) dZ_i.$$

2. Show that the density of e_i takes the form

$$p(e_i) \propto \phi\left(\frac{e_i}{\sqrt{\sigma^2+\delta^2}}\right)\Phi\left(\frac{\delta e_i}{\sigma\sqrt{\sigma^2+\delta^2}}\right),$$

or in a full form:

$$p(e_i) = 2\phi\left(e_i; 0, \sigma^2+\delta^2\right)\Phi\left(\frac{\delta e_i}{\sigma}; 0, \sigma^2+\delta^2\right).$$

3. Show that the mean, variance and skewness of e_i take the form

$$E(e_i) = \delta\sqrt{2/\pi},$$

$$Var(e_i) = \delta^2\left(1 - \frac{2}{\pi}\right) + \sigma^2,$$

$$Skew(e_i) = \sqrt{2}(4-\pi)\left[\sigma^2\pi + \delta^2(\pi - 2)\right]^{-3/2}\delta^3.$$

Exercise 4.5 (Regression with a truncated normal prior) *Consider Bayesian linear regression $Y_i = X_i\beta + \sigma\varepsilon_i$.*

1. Suppose that σ is known and β has a normal prior $N(\mu, \Lambda^{-1})$, where Λ is the precision matrix and Λ^{-1} is the covariance matrix. Show that the posterior is $N(\bar{\mu}, \bar{\Lambda}^{-1})$, where

$$\bar{\mu} = \left(\Lambda + \sigma^{-2}\sum_{i=1}^{n} X_i'X_i\right)^{-1}\left(\Lambda\mu + \sigma^{-2}\sum_{i=1}^{n} X_i'Y_i\right),$$

$$\bar{\Lambda} = \Lambda + \sigma^{-2}\sum_{i=1}^{n} X_i'X_i.$$

2. Suppose that σ is known and β has a truncated normal prior $N(\mu, \Lambda^{-1}) I(\beta \in B)$, where $B \subset \mathbb{R}^d$. Show that the posterior is $N(\bar{\mu}, \bar{\Lambda}^{-1}) I(\beta \in B)$.

3. What if σ is an unknown parameter? How would you specify the prior $p(\beta, \sigma^2)$ and conduct Bayesian inference?

4. Consider the regression with skewed errors: $Y_i = \tilde{X}_i\tilde{\beta} + \delta Z_i + \sigma\varepsilon_i$. Given the parameters $\tilde{\beta}, \delta, \sigma$, we treat $Y_i - \tilde{X}_i\tilde{\beta}$ as the response variable, δ as the predictor data, and Z_i as the regression coefficient with the half normal prior. Show that

$$Z_i \left| Y, \beta, \sigma^2 \right. \sim N\left[\frac{\delta\left(Y_i - \tilde{X}_i\tilde{\beta}\right)}{\sigma^2 + \delta^2}, \frac{\sigma^2}{\sigma^2 + \delta^2}\right] I(Z_i > 0).$$

5. Take another look at Exercise 4.4. We treat $e_i = \delta Z_i + \sigma\varepsilon_i$ as a linear regression, in which the response variable is e_i, the predictor is δ and the regression coefficient is Z_i. Given the half normal prior $p(Z_i)$, the Gaussian likelihood $p(e_i | Z_i)$, we have the truncated norm posterior $p(Z_i | e_i)$.

$$Z_i | e_i \sim N\left(\frac{\delta e_i}{\sigma^2 + \delta^2}, \frac{\sigma^2}{\sigma^2 + \delta^2}\right) I(Z_i > 0).$$

Apply Bayes' rule and derive the marginal likelihood $p(e_i)$. Is the result same as that in Exercise 4.4?

Exercise 4.6 (Skewed distribution by inverse scale) *Consider a bivariate regression with inverse scale skewed errors:*

$$Y_i = X_{1i}\beta_1 + X_{2i}\beta_2 + \sigma\varepsilon_i,$$

where

$$p(\varepsilon_i) = wp_+(\varepsilon_i) + (1 - w)p_-(\varepsilon_i),$$

$$p_+(\varepsilon_i) = 2\gamma^{-1} f(\gamma^{-1}\varepsilon_i) I(\varepsilon_i \geq 0),$$
$$p_-(\varepsilon_i) = 2\gamma f(\gamma\varepsilon_i) I(\varepsilon_i < 0).$$

To make the matters as simple as possible, assume that γ is known and $\sigma = 1$ is fixed. We have only three observations Y_1, Y_2, Y_3 and all predictors are positive. Also assume that $f(\cdot)$ is the standard normal density. The unknown parameters are β_1 and β_2, which are assigned a non-informative prior $p(\beta_1, \beta_2) \propto 1$. We want to design MCMC samplers that generate posterior draws from $p(\beta_1, \beta_2 | Y)$.

1. To implement a MH sampler, we need to evaluate $p(\beta_1, \beta_2 | Y) \propto p(\beta_1, \beta_2) p(Y | \beta_1, \beta_2)$. In this case, we just evaluate the likelihood function. Show that the likelihood function is given by

$$p(Y | \beta_1, \beta_2) = \prod_{i=1}^{3} [w p_+(Y_i) + (1-w) p_-(Y_i)],$$

where $p_+(Y_i)$ and $p_-(Y_i)$ are truncated normal densities

$$p_+(Y_i) = \frac{2}{\sqrt{2\pi}\gamma} e^{-\frac{(Y_i - X_i\beta)^2}{2\gamma^2}} I(Y_i \geq X_i\beta),$$

$$p_-(Y_i) = \frac{2}{\sqrt{2\pi}\gamma^{-1}} e^{-\frac{(Y_i - X_i\beta)^2}{2\gamma^{-2}}} I(Y_i < X_i\beta).$$

2. A Gibbs sampler cycles through the posterior conditional distributions $p(\beta_1 | Y, \beta_2)$ and $p(\beta_2 | Y, \beta_1)$. Here is some hint to derive $p(\beta_1 | Y, \beta_2)$, whose density is proportional to the likelihood function. Conditioning on β_2, we redefine the response variable as $\tilde{Y}_i = \frac{Y_i - X_{2i}\beta_2}{X_{1i}}$. The indicator functions impose constrains on β_1

$$I(Y_i \geq X_i\beta) \iff I\left(\beta_1 \leq \tilde{Y}_i\right), i = 1, 2, 3.$$

We sort $\tilde{Y}_i$ in ascending order. For brevity, we simply write $\tilde{Y}_1 \leq \tilde{Y}_2 \leq \tilde{Y}_3$. Those are three thresholds that divide $\mathbb{R}$ into four intervals: $\left(-\infty, \tilde{Y}_1\right)$, $\left(\tilde{Y}_1, \tilde{Y}_2\right)$, $\left(\tilde{Y}_2, \tilde{Y}_3\right)$, $\left(\tilde{Y}_3, \infty\right)$.

If $\beta_1 \in \left(-\infty, \tilde{Y}_1\right)$, then $p(Y | \beta_1, \beta_2) = w p_+(Y_1) w p_+(Y_2) w p_+(Y_3)$.

If $\beta_1 \in \left(\tilde{Y}_1, \tilde{Y}_2\right)$, then $p(Y | \beta_1, \beta_2) = (1-w) p_-(Y_1) w p_+(Y_2) w p_+(Y_3)$.

If $\beta_1 \in \left(\tilde{Y}_2, \tilde{Y}_3\right)$, then $p(Y | \beta_1, \beta_2) = (1-w) p_-(Y_1)(1-w) p_-(Y_2) w p_+(Y_3)$.

If $\beta_1 \in \left(\tilde{Y}_3, \infty\right)$, then $p(Y | \beta_1, \beta_2) = (1-w) p_-(Y_1)(1-w) p_-(Y_2)(1-w) p_-(Y_3)$.

Does the partitioned likelihood function indicate the functional form of the posterior conditional distribution $p(\beta_1 | Y, \beta_2)$?

Exercise 4.7 (Random permutation of class labels) *A feature of the finite mixture model is that the likelihood is invariant to permutation of class labels. Taking a two-class mixture as an example. Suppose that Class 1 corresponds to $N(\mu_1, \sigma_1^2)$ with the probability w_1, and Class 2 corresponds to $N(\mu_2, \sigma_2^2)$ with the probability w_2. We switch the labels by assigning Class 1 with $N(\mu_2, \sigma_2^2)$ and w_2, and assigning Class 2 with $N(\mu_1, \sigma_1^2)$ and w_1. The likelihood functions are identical under two assignments.*

Fruhwirth-Schnatter (2001) proposes a random permutation strategy at each iteration of the Gibbs sampler. We permute the indices $\{1, \ldots, k\}$ as $\{\sigma_1, \ldots, \sigma_k\}$. Suppose that the current value is (θ, Z), where θ includes the weights and class-specific parameters. We propose *a new allocation* (θ^*, Z^*)*: if* $Z_i = j$*, then we assign* $Z_i^* = \sigma_j$, $\forall i = 1, \ldots, n$, $j = 1, \ldots, k$. *Also, we assign* $\theta_j^* = \theta_{\sigma_j}$, $\forall j = 1, \ldots, k$.

We interpret random permutation as a MH proposal. What is the MH acceptance probability of the proposal?

Exercise 4.8 (Laplace distribution) *Proposition 4.6 is closely related to Proposition 3.2, as the standard Laplace distribution is a special case of the asymmetric Laplace distribution. Consider Proposition 3.2 under the specification* $\mu = 0$ *and* $\sigma = 2$*. According to Proposition 3.2, if* $p(\psi) = \frac{1}{2}e^{-\frac{\psi}{2}}$ *and* $p(x\,|\psi) = (2\pi)^{-\frac{1}{2}}\psi^{-\frac{1}{2}}\frac{1}{2}e^{-\frac{x^2}{8\psi}}$*, then we have the Laplace marginal distribution* $p(x) = \frac{1}{4}e^{-\frac{|x|}{2}}$*, which corresponds to the asymmetric Laplace distribution under* $\alpha = \frac{1}{2}$*. Note that the latent variable* ψ *in Proposition 3.2 follows* $Exp(2)$*, while the latent variable* Z *in Proposition 4.6 follows* $Exp(1)$*. Is it possible to validate Proposition 3.2 by Proposition 4.6 (i.e., to show the former is a corollary of the latter)?*

Exercise 4.9 (Inverse Gaussian Posterior) *Consider the quantile regression in the mixture form:*

$$Y_i = X_i\beta + \mu\tilde{Z}_i + \sqrt{s\sigma\tilde{Z}_i}W_i,$$

where $W_i \sim N(0,1)$ *and* $\tilde{Z}_i \sim Exp(\sigma)$*. The constants* μ *and* s *are the same as those in Proposition 4.6:* $\mu = \frac{1-2\alpha}{\alpha(1-\alpha)}$ *and* $s = \frac{2}{\alpha(1-\alpha)}$*, for* $\alpha \in (0,1)$*. Show that the posterior conditional distribution* $p\left(\tilde{Z}_i^{-1}\,|Y_i, \beta, \sigma\right)$ *is the inverse Gaussian distribution with the parameters* $\frac{1}{\alpha(1-\alpha)|Y_i - X_i\beta|}$ *and* $\frac{1}{2\alpha(1-\alpha)\sigma}$.

Exercise 4.10 (Generate quantile regression data) *The quantile regression model assumes that the* α^{th} *quantile of* Y *takes a linear form:*

$$q_\alpha(X) = X\beta_\alpha.$$

We interpret it as a structural model where the coefficient β_α *is fully known (as a function of* α*). The full set of quantile regressions for all* $\alpha \in [0,1]$ *provide a complete picture of the conditional distribution of* Y*, because we have fully specified the inverse CDF function. Is it possible to generate data* Y *conditional on* X *by the structural model?*

5

Limited Dependent Variable Models

The dependent variable in standard linear regression is continuous, spanning the entire set of real numbers $\mathbb{R}$. However, a limited dependent variable is confined to a specific subset of $\mathbb{R}$. Understanding these constraints is crucial because they influence the choice of statistical methods for analysis. Real-world economic data often face various constraints:

- *Discrete data*: For instance, a loan officer's decision to approve or deny a mortgage application results in binary data. In machine learning, this is treated as a classification problem, where each discrete value represents a distinct class. In micro-econometric studies, it's viewed as a random-utility discrete choice problem.

- *Truncation or censoring*: Economic data can be subject to inequality constraints. Traditionally, interest rates are non-negative, but recent instances of negative rates maybe lower bounded by cash storage costs, leading to truncation or censoring issues.

- *Incidental truncation*: Not all missing data occur randomly. Sometimes, data are missing due to the truncation of a correlated variable, a phenomenon known as incidental truncation, which affects sample selection and validity of inference.

- *Count data*: Integer-valued data with many zeros, such as the number of doctor visits in a health economics study, are better analyzed as count data rather than through discrete choice or classification frameworks.

- *Duration data*: Nonnegative variates representing the lifespan of an entity or process, such as the time until a machine part fails, are essential for survival analysis and reliability studies.

- *Data in the unit interval*: Variables like proportions, weights, and Gini indices are confined to the range [0,1]. Properly analyzing these data requires methods that account for their bounded nature.

To ease readers into this chapter, we provide a high-level view of what we did in previous chapters and will do in this chapter. The response variable Y_i and the predictors X_i are random samples from a joint distribution $p(X_i, Y_i \,|\theta\,)$, where θ is a set of parameters with a prior distribution $p(\theta)$. For Gaussian linear regressions in Chapter 1, the likelihood function takes the form

$$p(Y_i \,|\theta, X_i\,) = \frac{1}{\sqrt{2\pi}\sigma} e^{-\frac{(Y_i - X_i\beta)^2}{2\sigma^2}},$$

with the NIG prior for $\theta = (\beta, \sigma^2)$. Chapter 3 follows the Gaussian likelihood, but specifies different priors $p(\theta)$ to achieve shrinkage and variable selection. Chapter 4 focuses on departure from the Gaussian likelihood, with flexible forms of $p(Y_i \,|\theta, X_i\,)$ such as the scale or finite mixture distributions to accommodate skewness and fat tails. In this chapter, we consider appropriate models that capture the conditional distributions $p(Y_i \,|\theta, X_i\,)$ for binary, multinomial, truncated, censored, incidentally truncated, count and duration data.

DOI: 10.1201/9781003564027-5

5.1 Probit Model

Consider a binary response variable $Y_i \in \{0,1\}$ and a set of predictors X_i. It may be tempting to specify linear conditional probabilities:

$$p(Y_i = 1 | X_i, \beta) = X_i\beta.$$

The linear probability model has pro and cons. It serves as a natural extension of linear regression, maintaining linearity for the conditional mean $E(Y_i | X_i, \beta) = X_i\beta$, allowing a regression representation: $Y_i = X_i\beta + \varepsilon_i$, where ε_i still satisfies $E(\varepsilon_i | X_i) = 0$, but it cannot be normally distributed. However, a significant flaw of the model is its failure to restrict probabilities within the unit interval.

A well-specified binary response model employs a link function that maps $X_i\beta$ into the unit interval, with any cumulative distribution function (CDF) fulfilling this role. The Probit model, for instance, utilizes the standard normal CDF $\Phi(\cdot)$ such that

$$p(Y_i = 1 | X_i, \beta) = \Phi(X_i\beta),$$

$$p(Y_i = 0 | X_i, \beta) = \Phi(-X_i\beta).$$

It might seem unnecessary to specify $p(Y_i = 0 | X_i, \beta)$, which must equal $1 - p(Y_i = 1 | X_i, \beta)$. A cravat is that $\Phi(-X_i\beta) = 1 - \Phi(X_i\beta)$ acts as an identity in mathematics but not on a computer. The tail of the normal CDF is so thin that $1 - \Phi(X_i\beta)$ can easily underflow. Surprisingly, $1 - \Phi(9)$ results in a numerical zero, if the expression is computed by double-precision floating-point data.

In linear regression, the coefficient β quantifies the marginal effects of regressors on the conditional mean of the response variable. However, in the Probit model, the marginal effects vary with the point of evaluation (Anderson and Newell, 2003):

$$\frac{\partial E(Y_i | X_i, \beta)}{\partial X_i} = \phi(X_i\beta)\beta,$$

where $\phi(\cdot)$ denotes the standard normal density. Empirical studies typically report either average marginal effects $\frac{1}{n}\sum_{i=1}^{n}\phi(X_i\beta)\beta$ or marginal effects for an average individual $\phi(\bar{X}_i\beta)\beta$, where $\bar{X}_i = \frac{1}{n}\sum_{i=1}^{n}X_i$. These two measures of marginal effects are not numerically equivalent.

5.1.1 Data Augmentation and Gibbs Sampling

The likelihood function of the Probit model is formulated by the product of conditional probabilities, $\Phi(X_i\beta)$ or $\Phi(-X_i\beta)$, depending on the value of Y_i:

$$p(Y | \beta) = \prod_{i=1}^{n} [\Phi(X_i\beta)]^{Y_i} [\Phi(-X_i\beta)]^{1-Y_i} = \prod_{i=1}^{n} \Phi[(2Y_i - 1) X_i\beta],$$

without conditioning on the covariates X_i explicitly denoted hereafter.

The posterior is proportional to the product of the prior and the likelihood: $p(\beta | Y) \propto p(\beta)p(Y | \beta)$. Since the likelihood function involves the normal CDF, a recognizable posterior is not available regardless of the prior specification. Nonetheless, data augmentation through a latent variable Z_i leads to a conditionally conjugate linear regression:

$$Z_i = X_i\beta + \varepsilon_i,$$

$$Y_i = I(Z_i > 0),$$

or in a vector/matrix form $Z = X\beta + \varepsilon$ and $Y = I(Z > 0)$. The noise term ε_i follows the standard normal distribution, resulting in

$$p(Y_i = 1 \,|\beta) = p(\varepsilon_i > -X_i\beta \,|\beta) = \Phi(X_i\beta).$$

The variance of ε_i is normalized to one for identification purposes. If ε_i had a variance σ^2, the conditional probability would become $\Phi\left(X_i \frac{\beta}{\sigma}\right)$, implying that β and σ are identified only up to the ratio $\frac{\beta}{\sigma}$, but not separately.

Data augmentation replaces an intractable posterior $p(\beta \,|Y)$ by two tractable posterior conditionals: $p(\beta \,|Y, Z)$ and $p(Z \,|Y, \beta)$, facilitating the Gibbs sampler that cycles through the full posterior conditionals (Albert and Chib, 1993).

On the one hand, given the latent variables Z, the parameters β are managed within the Bayesian linear regression framework. Let the prior be $\beta \sim N\left(\mu, \Lambda^{-1}\right)$, where Λ is the precision matrix (with the covariance matrix being Λ^{-1}). The posterior conditional distribution is $N\left(\bar{\mu}, \bar{\Lambda}^{-1}\right)$, where

$$\bar{\mu} = (\Lambda + X'X)^{-1}(\Lambda\mu + X'Z),$$
$$\bar{\Lambda} = \Lambda + X'X.$$

On the other hand, given the parameters β, the latent variables Z are updated using truncated normal distributions. Normality arises from the data-generating process $Z = X\beta + \varepsilon$, while truncation is induced by the sign restriction imposed by binary observations. If $Y_i = 1$, the posterior distribution is $N(X_i\beta, 1)$ truncated to the interval $(0, \infty)$. If $Y_i = 0$, the posterior is $N(X_i\beta, 1)$ truncated to the interval $(-\infty, 0]$. The posterior conditional distribution can be expressed as

$$p(Z \,|Y, \beta) = \prod_{i=1}^{n} p(Z_i \,|\beta)\left[I(Y_i = 1)\, I(Z_i > 0) + I(Y_i = 0)\, I(Z_i \le 0)\right].$$

The Gibbs sampler for the Probit model is outlined in Algorithm 5.1, which involves generating samples from truncated normal distributions. Several sampling methods are available. The simplest method is rejection sampling: an unrestricted normal proposal is accepted if it falls within the required interval; otherwise, another proposal is made until it fits the interval. This method becomes impractical when the acceptance probability is close to zero. An alternative sampler employs the inverse CDF: a uniform random number U for the transformation

$$X_i\beta + \Phi^{-1}\left[\Phi(-X_i\beta) + \Phi(X_i\beta)\, U\right]$$

yields a truncated normal draw from $N(X_i\beta, 1)\, I_{(0,\infty)}$. Exercise 5.1 provides rationale for the inverse CDF method, which performs satisfactorily even under extreme values such as $X_i\beta \to \infty$. However, the computational burden of calculating the normal CDF and its inverse raises a concern on numerical efficiency. Geweke (1991) develops an efficient sampling algorithm that switches between rejection sampling and inverse CDF contingent on the evaluation points.

5.1.2 Marginalization and Big Data Algorithm

Holmes and Held (2006) highlight the strong posterior correlation between β and Z in the standard Gibbs sampler. They suggest updating these variables jointly to reduce autocorrelation and improve mixing of the Markov chain. Their approach leverages the marginal-conditional decomposition

$$p(\beta, Z \,|Y) = p(Z \,|Y)\, p(\beta \,|Y, Z).$$

Algorithm 5.1 Gibbs sampler of Probit model

Input: data X, Y, and initial value Z.
Output: posterior samples from $p(\beta, Z \mid Y)$.
Iterate the following steps:
1 Sample β from $N\left(\bar{\mu}, \bar{\Lambda}^{-1}\right)$, given the current value Z.
2 Sample Z_i from $\begin{cases} N(X_i\beta, 1)\, I_{(0,\infty)} & Y_i = 1 \\ N(X_i\beta, 1)\, I_{(-\infty,0)} & Y_i = 0 \end{cases}$, given the current value β.

The core idea is to integrate out β and concentrate on the posterior sampling of Z. From a high-level perspective, $p(Z \mid Y)$ must be a truncated multivariate normal distribution, as the normal regression $Z = X\beta + \varepsilon$ paired with the normal prior $N\left(\mu, \Lambda^{-1}\right)$ results in

$$Z \sim N\left(X\mu, I_n + X\Lambda^{-1}X'\right).$$

Conditional on the binary data Y, the random vector Z is restricted to a specific region: either $Y_i = 1, Z_i > 0$ or $Y_i = 0, Z_i \leq 0$, hence the truncation.

When the number of observations is large, directly sampling from a high-dimensional truncated normal distribution for $p(Z \mid Y)$ becomes impractical. However, it is feasible to evaluate the density pointwise up to a normalizing constant: $p(Z \mid Y) \propto p(Z)\, p(Y \mid Z)$, where

$$p(Y \mid Z) = \prod_{i=1}^{n} \left[I(Y_i = 1)\, I(Z_i > 0) + I(Y_i = 0)\, I(Z_i \leq 0)\right],$$

$$p(Z) = \frac{p(\beta)\, p(Z \mid \beta)}{p(\beta \mid Z)}.$$

Since $p(\beta)$, $p(\beta \mid Z)$ and $p(Z \mid \beta)$ are densities of $N\left(\mu, \Lambda^{-1}\right)$, $N\left(\bar{\mu}, \bar{\Lambda}^{-1}\right)$ and $N(X\beta, I_n)$, respectively, evaluating densities at $\beta = 0$ yields

$$p(Z) \propto e^{-\frac{1}{2}\left(Z'Z - \bar{\mu}'\bar{\Lambda}\bar{\mu}\right)},$$

which is algebraically equivalent to the density of $N\left(X\mu, I_n + X\Lambda^{-1}X'\right)$, but it is more numerically stable, avoiding the computation of the determinant and inversion of $n \times n$ matrices. See Exercise 5.2 for details. The density is fully determined by a set of cumulative statistics, which are functions of latent variables in a summable form:

$$S_1(Z) = X'X = \sum_{i=1}^{n} X_i'X_i,$$

$$S_2(Z) = X'Z = \sum_{i=1}^{n} X_i'Z_i,$$

$$S_3(Z) = Z'Z = \sum_{i=1}^{n} Z_i^2.$$

As a big data algorithm, a marginalized MCMC sampler targeting $p(Z \mid Y)$ updates one element in the vector Z at a time by processing a single data point instead of the entire dataset. This is achieved through a rank-one update of cumulative statistics. This technique has been applied to various models in Chapter 4. While a generic MH sampler is applicable to models where $p(Z \mid Y)$ is not of a recognizable form, a marginalized Gibbs sampler is well-suited for the Probit model, which features an analytically tractable $p(Z_i \mid Z_{-i}, Y)$,

Algorithm 5.2 Marginalized MCMC of Probit model

Input: data X, Y, and initial value Z.
Output: posterior samples from $p(\beta, Z \mid Y)$.
Iterate the following steps:
1 Initialize cumulative statistics $\sum_{i=1}^{n} X_i' X_i$ and $\sum_{i=1}^{n} X_i' Z_i$.
2 Compute posterior conditional mean $\bar{\mu}$ and precision $\bar{\Lambda}$.
3 Cache $\tilde{\Lambda}_i = 1 - h_i$, where h_i is i^{th} diagonal element of $X\bar{\Lambda}^{-1}X'$.
4 Iterate the following steps:
  4.1 Draw a random index $i \in \{1, \ldots, n\}$.
  4.2 Read i^{th} data point X_i, Y_i, Z_i.
  4.3 Compute $\tilde{\mu}_i$, given the current Z_i.
  4.4 Sample Z_i from $\begin{cases} N\left(\tilde{\mu}_i, \tilde{\Lambda}_i^{-1}\right) I_{(0,\infty)} & Y_i = 1 \\ N\left(\tilde{\mu}_i, \tilde{\Lambda}_i^{-1}\right) I_{(-\infty,0)} & Y_i = 0 \end{cases}.$
  4.5 Update cumulative statistics and $\bar{\mu}$,given the new Z_i.
  4.6 Sample β from $N\left(\bar{\mu}, \bar{\Lambda}^{-1}\right)$.

as the conditional distribution of a multivariate truncated normal distribution remains a truncated normal distribution. Specifically,

$$p(Z_i \mid Z_{-i}, Y) \propto e^{-\frac{1}{2}(Z'Z - \bar{\mu}'\bar{\Lambda}\bar{\mu})} p(Y_i \mid Z_i),$$

where $p(Y_i \mid Z_i) = I(Y_i = 1) I(Z_i > 0) + I(Y_i = 0) I(Z_i \leq 0)$, which is a binary indicator for truncation. Since $Z'Z - \bar{\mu}'\bar{\Lambda}\bar{\mu}$ is a quadratic function of Z_i, the density $p(Z_i \mid Z_{-i}, Y)$ has a Gaussian kernel. For simplicity, assume that $\mu = 0$ (as in a typical prior setting). Completing the square (see Exercise 5.2) yields the functional form of the truncated normal distribution:

$$Z_i \mid Z_{-i}, Y \sim \begin{cases} N\left(\tilde{\mu}_i, \tilde{\Lambda}_i^{-1}\right) I_{(0,\infty)} & Y_i = 1 \\ N\left(\tilde{\mu}_i, \tilde{\Lambda}_i^{-1}\right) I_{(-\infty,0)} & Y_i = 0 \end{cases},$$

$$\tilde{\mu}_i = \frac{X_i\bar{\mu} - h_i Z_i}{1 - h_i},$$

$$\tilde{\Lambda}_i = 1 - h_i,$$

and h_i is the ith diagonal element of the matrix $X\bar{\Lambda}^{-1}X'$. As the variance term $\tilde{\Lambda}_i^{-1}$ does not depend on Z_i, it can be computed and cached prior to MCMC sampling. Algorithm 5.2 summarizes the marginalized single-move Gibbs sampler for the Probit model.

5.1.3 Extensions to Multiple Categories

The binary Probit model can be extended to an ordered Probit model with multiple categories. For instance, a bond may be classified into 9 categories, ranging from AAA to CCC in descending order. Consider a latent variable regression

$$Z_i = X_i\beta + \varepsilon_i,$$

and the observation is $Y_i = j$, if $Z_i \in (a_{j-1}, a_j]$, $j = 1, \ldots, J$, with the boundaries $a_0 = -\infty$ and $a_J = \infty$. The thresholds $a = (a_1, \ldots, a_{J-1})$ are unknown parameters. The noise term ε_i follows the standard normal distribution, leading to

$$p(Y_i = j \mid a, \beta) = p(a_{j-1} < Z_i \leq a_j \mid a, \beta) = \Phi(a_j - X_i\beta) - \Phi(a_{j-1} - X_i\beta).$$

Parameter identification necessitates normalizing the variance of ε_i to one, as in the binary Probit model. An additional identification constraint is to normalize $a_1 = 0$ or exclude the intercept term from predictors, avoiding the intercept from arbitrarily shifting the thresholds. We consider the regression without an intercept.

Data augmentation facilitates the Gibbs sampler that cycles through the posterior conditional distributions: $p(\beta \,|Y, Z, a)$, $p(Z \,|Y, a, \beta)$ and $p(a \,|Y, Z, \beta)$.

First, conditional on latent variables and thresholds, conditionally conjugate results are available, as in the standard Bayesian linear regression. Let the prior be $\beta \sim N(\mu, \Lambda^{-1})$, where Λ is the precision matrix (the covariance matrix is Λ^{-1}). The posterior conditional distribution is $N(\bar{\mu}, \bar{\Lambda}^{-1})$, where

$$\bar{\mu} = (\Lambda + X'X)^{-1} (\Lambda\mu + X'Z),$$

$$\bar{\Lambda} = \Lambda + X'X.$$

Second, conditional on coefficients and thresholds, the latent variable follows the truncated normal distribution:

$$p(Z_i \,|Y_i, a, \beta) \propto p(Z_i \,|a, \beta)\, p(Y_i \,|Z_i, a, \beta),$$

where $p(Z_i \,|a, \beta)$ follows $N(X_i\beta, 1)$, and $p(Y_i \,|Z_i, a, \beta)$ indicates an appropriate truncation interval: observing $Y_i = j$ implies $Z_i \in (a_{j-1}, a_j]$. Consequently, the latent variable follows a truncated normal distribution $N(X_i\beta, 1)\, I_{(a_{j-1}, a_j]}$, if $Y_i = j$.

Third, conditional on coefficients and latent variables, the thresholds can be updated. One strategy is to update one element (e.g., a_j) at a time, conditional on the remaining thresholds. Let the prior be $p(a_j)$. The posterior should be proportional to the truncated prior. Truncation arises from two sources. First, thresholds are ordered, thus $a_j \in (a_{j-1}, a_{j+1})$. Second, the latent variables impose constraints on the threshold location. Since $Z_i \in (a_{j-1}, a_j]$ under $Y_i = j$, we infer that 1) a_j must be larger than all latent variables in the jth category, and 2) a_j must be smaller than all latent variables in the $(j+1)^{th}$ category. Therefore, the posterior conditional distribution is

$$p(a_j \,|Y, Z, \beta, a_{-j}) \propto p(a_j)\, I(a_{j-1} < a_j a_{j+1})\, d_j,$$

$$d_j = \prod_{i=1}^{n} \left[I(Y_i = j)\, I(a_j \geq Z_i) + I(Y_i = j+1)\, I(a_j < Z_i) \right].$$

To accelerate MCMC convergence, Cowles (1996) proposes a Metropolis-within-Gibbs sampler that updates Z and a jointly, instead of individually from their full posterior conditional distributions. Marginalizing over the latent variables Z necessitates a Metropolis step for updating the thresholds.

It is possible to extend the ordered Probit to a multinomial Probit model with unordered categories, if identification constraints are properly addressed (Keane, 1992). Evaluating or simulating the CDF of multivariate normal distributions can be computationally intensive. Geweke et al. (1994) compare alternative estimation approaches for the multinomial Probit model, including the Gibbs sampler with data augmentation. Chib and Greenberg (1998) show MCMC and marginal likelihood in multinomial Probit models. A more economical alternative is the multinomial logistic regression, which will be discussed in the subsequent section.

5.2 Logistic Regression

A statistical model for classification is typically specified through the conditional probability distribution of the response variable. Two prominent models in this domain are the Probit and Logit, which are distinguished by their use of different CDFs to define these probabilities. The Probit model employs the standard normal CDF, while the Logit model, also known as logistic regression, utilizes the logistic CDF to specify conditional probabilities:

$$p(Y_i = 1 \,|X_i, \beta) = \Lambda (X_i\beta),$$
$$p(Y_i = 0 \,|X_i, \beta) = \Lambda (-X_i\beta),$$

where $\Lambda(x) = (1 + e^{-x})^{-1}$ represents the logistic CDF, characterized by its sigmoid shape. In some contexts, the sigmoid function is synonymous with the logistic CDF. Although $\Lambda(-X_i\beta)$ is algebraically equivalent to $1 - \Lambda(X_i\beta)$, the latter expression should be avoided in numerical computations to prevent cancellation errors.

5.2.1 Comparison between Logit and Probit

Logistic regression is widely favored for classification tasks due to its simplicity, interpretability, and robustness. One of its key advantages is the closed form of the logistic CDF, in contrast to the normal CDF, which is expressed as an integral. Although the normal CDF can be accurately evaluated using numerical methods, it is computationally more intensive than the logistic CDF. In practice, the Logit and Probit models often produce similar classification results, but the Logit model typically executes much faster.

The interpretability of logistic regression coefficients is another significant advantage. The coefficient vector β represents semi-elasticity, measuring the marginal effects of predictors on the log odds ratio:

$$\ln \frac{p(Y_i = 1 \,|X_i, \beta)}{p(Y_i = 0 \,|X_i, \beta)} = X_i\beta.$$

This interpretation provides a foundation for extending logistic regression to multinomial contexts.

Furthermore, data augmentation offers the Logit model a robust regression form with a latent variable Z_i:

$$Z_i = X_i\beta + \varepsilon_i,$$
$$Y_i = I(Z_i > 0),$$

where the error term ε_i follows the logistic distribution, with the density function

$$\Lambda'(x) = \Lambda(x)\,\Lambda(-x) = \frac{e^{-x}}{(1 + e^{-x})^2}.$$

Figure 5.1 demonstrates that the logistic density closely resembles that of a normal distribution $N\left(0, \frac{\pi^2}{3}\right)$, explaining the empirical rule that the Logit estimator is approximately proportional to the Probit estimator, with the former being about $\frac{\pi}{\sqrt{3}}$ times larger. However, the tails of the two distributions differ. As $x \to \infty$, the tail of the logistic density approaches e^{-x}, while the tail of the corresponding normal density is proportional to $e^{-\frac{3}{2\pi^2}x^2}$, converging to zero faster.

Additionally, the kurtosis characterizes the tail behavior. The logistic distribution is a Kolmogorov scale mixture of normal distributions (Andrews and Mallows, 1974). According

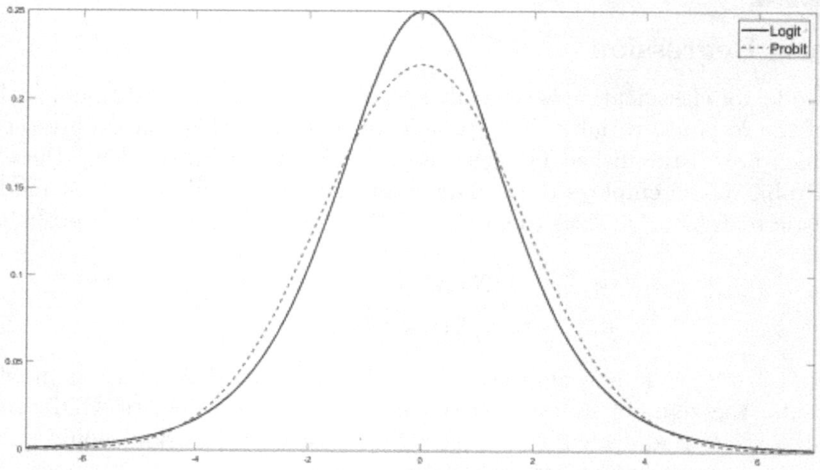

FIGURE 5.1
Density functions of logistic and normal distributions.

to Proposition 4.2, the logistic distribution must be leptokurtic. The fat-tailed nature of the logistic distribution makes the regression more robust to extreme values. In terms of numerical stability, the Probit model suffers from underflow at $1 - \Phi(9)$ with double floating-point numbers, whereas the Logit model does not encounter this issue until $1 - \Lambda(37)$.

5.2.2 MH Sampling

Before MCMC methods were widely used by statisticians, Bayesian logistic regression was typically estimated using importance sampling based on multivariate normal or t proposals (Zellner and Peter, 1984; Dempster et al., 1983). Gelfand and Smith (1990) heralded the arrival of MCMC methods in statistics. The MH and Gibbs samplers have since been indispensable to Bayesian logistic regression. As the logistic likelihood cannot be paired with a natural conjugate prior from common distributions, various MCMC sampling methods have been proposed, including Zeger and Karim (1991), Dellaportas and Smith (1993), Gamerman (1997), Holmes and Held (2006), Fruhwirth-Schnatter and Fruhwirth (2007), Polson et al. (2013), among others.

In the absence of conjugate structures, Bayesian logistic regression is typically handled by the MH sampler with the target distribution $p(\beta | Y) \propto p(\beta) p(Y | \beta)$, where a normal prior is often used:

$$p(\beta) \propto e^{-\frac{1}{2}(\beta-\mu)' V^{-1} (\beta-\mu)},$$

and the likelihood function is given by

$$p(Y | \beta) = \prod_{i=1}^{n} [\Lambda(X_i\beta)]^{Y_i} [\Lambda(-X_i\beta)]^{1-Y_i} = \prod_{i=1}^{n} \Lambda[(2Y_i - 1) X_i\beta].$$

The performance of the MH sampler is heavily influenced by the proposal distribution. A common strategy involves several steps.

First, perform numeric optimization on the posterior log density (see Section 5.8 for implementation details). This step identifies the posterior mode, also known as the maximum-a-posteriori estimator

$$\beta^* = \arg\max_{\beta} \ln p(\beta \,|Y).$$

The objective function for logistic regression is globally concave, since both the normal prior and the logistic likelihood function are concave with respect to β, The Newton-Raphson algorithm is well-suited to numerical optimization under concave functions, usually converging in a few iterations.

Second, compute the Hessian matrix evaluated at the posterior mode:

$$H(\beta) = \frac{\partial^2 \ln p(\beta \,|Y)}{\partial \beta \partial \beta'}.$$

An attraction of logistic regression is the simplicity of the analytical Hessian:

$$H(\beta) = -V^{-1} - \sum_{i=1}^{n} \Lambda(X_i \beta) \Lambda(-X_i \beta) X_i' X_i.$$

Third, approximate $p(\beta \,|Y)$ by a multivariate normal distribution with mean β^* and covariance matrix $-H^{-1}(\beta^*)$, which is justified by the second-order Taylor approximation of the log posterior:

$$p(\beta \,|Y) \approx e^{\ln p(\beta^* |Y) + \frac{1}{2}(\beta - \beta^*)' H(\beta^*)(\beta - \beta^*)},$$

where the first-order derivatives are zero because β^* maximizes the posterior density.

Fourth, implement a random-walk or independence MH sampler with a multivariate normal or t proposal distribution, in which the covariance matrix is proportional to the inverse Hessian: $-cH^{-1}(\beta^*)$, where c is a tuning parameter. A smaller c usually increases the acceptance rate of the proposal draws. It is advisable to tune the proposal variance so that the sampling acceptance rate is roughly 1/4 (Gelman et al., 1997).

From a programming perspective, Bayesian logistic regression can serve as a sandbox for prototyping custom MCMC samplers for more complex models (e.g., Dynamic Stochastic General Equilibrium models in Section 8.8), where high-dimensional numerical optimization on non-concave functions poses significant computational challenges, and the Hessian evaluation often encounters severe numerical issues.

5.2.3 Separation and Bayesian Inference

For many practitioners, the choice between classical and Bayesian inference is largely driven by computational considerations. In the case of logistic regression, maximum likelihood estimation is a tempting choice due to its relative simplicity compared to MCMC methods. This inclination arises because tuning the MH sampler necessitates numerical optimization of the posterior density, which may not be simpler than maximizing the likelihood.

Surprisingly, maximizing the likelihood can be more challenging than optimizing the posterior density in certain datasets, particularly when the sample size is small and includes rarely occurring dummy variables. Empirical researchers may encounter anomalies caused by separation, such as: 1) numerical optimization fails to converge; 2) one or more parameters diverge to infinity, resulting in huge standard errors; 3) the likelihood function may converge to a saturation point; 4) propensity scores of many, if not all, observations tend to extremes (i.e., zero or one), and 5) the Hessian matrix exhibits poor conditioning numbers. Refer to Albert and Anderson (1984), Firth (1993), Heinze and Schemper (2002) and Zorn (2005) for the problems and solutions to separation in the logistic regression.

Definition 5.1 (Separation) *Let $Y_i, i = 1, \ldots, n$ be the binary response variable and S_i be a subset of predictors X_i.*

If there exist a vector b such that 1) $S_i b > 0 \Rightarrow Y_i = 1$, and 2) $S_i b \leq 0 \Rightarrow Y_i = 0$, for all $i = 1, \ldots, n$, then the dataset is completely separated by $S_i b$ (complete separation).

If there exist a vector b such that 1) $Y_i = 1 \Rightarrow S_i b \geq 0$, and at least one observation has $S_i b = 0$, and 2) $Y_i = 0 \Rightarrow S_i b \leq 0$, and at least one observation has $S_i b = 0$, then the dataset is quasi-completely separated by $S_i b$ (quasi-complete separation).

Under complete separation, the predictor combination $S_i b$ perfectly predicts all Y_i. Conversely, Y_i also fully determines the sign of $S_i b$, because $S_i b > 0 \Rightarrow Y_i = 1$ implies that $Y_i = 0 \Rightarrow S_i b \leq 0$. Similarly, $S_i b \leq 0 \Rightarrow Y_i = 0$ implies that $Y_i = 1 \Rightarrow S_i b > 0$.

Under quasi-complete separation, $S_i b$ exactly predicts some, but not all, Y_i. Positive (negative) $S_i b$ always corresponds to $Y_i = 1$ ($Y_i = 0$). However, there is a middle ground where $S_i b = 0$ cannot determine Y_i.

Quasi-complete separation is often observed in small datasets with rarely occurring dummy variables. For instance, if the treated group $D_i = 1$ always has $Y_i = 1$, the separation issue arises, leading to an unidentified treatment effect.

Because the subset of predictors can include an intercept term, Definition 5.1 specifies the separating threshold at zero without loss of generality. In practice, we may encounter a variable Z_i such that $Z_i > c \Rightarrow Y_i = 1$, and $Z_i \leq c \Rightarrow Y_i = 0$. Definition 5.1 remains applicable if we specify $S_i = (Z_i, 1)$ and $b = (1, -c)'$.

To analyze the effects of separation on maximum likelihood estimation, variables are partitioned as $X_i = (S_i, G_i)$, $\beta = (\beta_b; \beta_g)$, and the log-likelihood of the logistic regression is expressed as the sum of three terms:

$$\begin{aligned}\mathcal{L}(\beta_b, \beta_g) = & \sum_{S_i b > 0} \ln \Lambda(S_i \beta_b + G_i \beta_g) + \sum_{S_i b < 0} \ln \Lambda(-S_i \beta_b - G_i \beta_g) \\ & + \sum_{S_i b = 0} \ln \Lambda[(2Y_i - 1)(S_i \beta_b + G_i \beta_g)].\end{aligned}$$

The objective is to maximize the log-likelihood with respect to β_b and β_g. The log-likelihood of the logistic regression must be non-positive. Saturation refers to $\ln \Lambda(\cdot) \to 0$. To derive the maximum likelihood estimator, we let β_b grow in proportional to b and consider the estimator $\hat{\beta}_b = \lim_{k \to \infty} kb$, which pushes the likelihood to saturation: $\sum_{S_i b > 0} \ln \Lambda\left(S_i \hat{\beta}_b + G_i \beta_g\right) \to 0$, and $\sum_{S_i b < 0} \ln \Lambda\left(-S_i \hat{\beta}_b - G_i \beta_g\right) \to 0$, for all β_g. Under complete separation, the middle-ground likelihood reduces to $\sum_{S_i b = 0} \ln \Lambda(-G_i \beta_g)$, which also reaches saturation as β_g tends to negative infinity. In the case of quasi-complete separation, an internal solution for β_g could be obtained, akin to running a logistic regression with the subset data restricted to $S_i b = 0$.

Complete separation has the following consequences: 1) the maximum likelihood estimator of β_b diverges to infinity, proportional to b; 2) the maximum likelihood estimator of β_g either diverges to infinity or remains undetermined if no observation satisfies $S_i b = 0$; 3) the log-likelihood reaches saturation; 4) the Hessian of the log-likelihood is zero, resulting in infinite standard errors for the estimators; and 5) propensity scores are either zero or one, depending on the sign of $S_i b$.

Quasi-complete separation has the following consequences: 1) the maximum likelihood estimator of β_b diverges to infinity, proportional to b; 2) the maximum likelihood estimator of β_g is obtained by the logistic regression using the subset data restricted to $S_i b = 0$; 3) the log-likelihood converges to a negative value, which is the same as the logistic regression result under the subset data ; 4) the estimator standard error of β_b explodes, but that of

β_g may not; and 5) propensity scores of some observations tend to zero or one, depending on the sign of $S_i b$.

Although separation complicates numerical optimization, it does not imply that the separating variables should be discarded. On the contrary, they are influential predictors that transform the regression into a decision tree: specific values of the separating variables fully determine classification, without reference to other predictors. Under complete separation, all data can be perfectly classified by the decision tree. Under quasi-complete separation, some data are perfectly classified by the tree, and the remaining data are handled by the regression.

Bayesian methods resolve the separation issue by incorporating an informative prior, which supplements observations with subjective data. This approach results in a posterior density that is not a monotone function of the parameters, enabling numerical optimization programs to identify a well-defined posterior mode and Hessian matrix, thus facilitating the tuning of MCMC samplers.

Separation is fundamentally a data-specific problem rather than a flaw in the statistical model or optimization method. While the separation problem affects Probit and other classification models, it is especially pertinent to logistic regression, providing a compelling argument for Bayesian inference.

5.2.4 Gibbs sampling

The Probit model is well-suited for Bayesian inference using the Gibbs sampler with data augmentation. This is primarily because the posterior conditional distributions in the Probit model are analytically tractable. However, a natural question arises: is Gibbs sampling applicable to the Logit model as well?

The Logit model, a staple in statistical analysis for binary outcomes, can be adapted for Gibbs sampling. Andrews and Mallows (1974) and Holmes and Held (2006) demonstrate that the logistic distribution can be represented as a scale mixture of normal distributions, using data augmentation with Kolmogorov-distributed latent variables. A technical challenge arises: the Kolmogorov density function manifests as an infinite series

$$f(x) = 8x \sum_{k=1}^{\infty} (-1)^{k-1} k^2 e^{-2k^2 x^2}.$$

The latent variable representation of the logistic regression is

$$Z_i = X_i \beta + \varepsilon_i,$$

$$Y_i = I(Z_i > 0),$$

$$\varepsilon_i | \lambda_i \sim N(0, \lambda_i),$$

where $\lambda_i = 4\psi_i^2$ and ψ_i follows the Kolmogorov distribution. The marginal distribution of ε_i is the logistic distribution. The Gibbs sampler cycles through the posterior conditionals, including $p(\beta | Y, Z, \lambda)$, $p(Z | Y, \beta, \lambda)$ and $p(\lambda | Y, Z, \beta)$.

First, conditional on Y, Z, λ, the model simplifies to linear regression with known heteroscedasticity. Let the prior be $\beta \sim N(\mu, V)$. By standardizing data as $\lambda_i^{-1/2} Z_i$ and $\lambda_i^{-1/2} X_i$, the posterior conditional distribution is $N(\bar{\mu}, \bar{V})$, where

$$\bar{\mu} = \bar{V} \left(V^{-1}\mu + \sum_{i=1}^{n} \lambda_i^{-1} X_i' Z_i \right),$$

$$\bar{V}^{-1} = V^{-1} + \sum_{i=1}^{n} \lambda_i^{-1} X_i' X_i.$$

Second, conditional on Y, β, λ, the latent variable Z_i follows a truncated normal distribution. If $Y_i = 1$, then Z_i follows $N(X_i\beta, \lambda_i)$ truncated to the interval $(0, \infty)$. If $Y_i = 0$, the posterior is $N(X_i\beta, \lambda_i)$ truncated to the interval $(-\infty, 0]$.

Third, conditional on Y, Z, β, the regression error is known as $\varepsilon_i = Z_i - X_i\beta$. By a change of variable, the density of λ_i inherits the Kolmogorov series: $\sum_{k=1}^{\infty} (-1)^{k-1} k^2 e^{-\frac{1}{2}k^2\lambda_i}$. The posterior conditional distribution of λ_i does not have a recognizable form, but the density can be expressed as an infinite series:

$$p(\lambda_i | Y, Z, \beta) \propto \lambda_i^{-\frac{1}{2}} \sum_{k=1}^{\infty} (-1)^{k-1} k^2 e^{-\frac{1}{2}(k^2\lambda_i + \varepsilon_i^2 \lambda_i^{-1})},$$

which is reminiscent of the inverse Gaussian density, allowing for rejection sampling with an inverse Gaussian proposal.

Fruhwirth-Schnatter and Fruhwirth (2007) propose an alternative Gibbs sampling scheme for the Logit model, where the logistic distribution is represented as the difference between two extreme value distributions, which is approximated by finite Gaussian mixture distributions, akin to the mixture approximation in Kim et al. (1998) for the stochastic volatility model (see Section 4.5). The advantages of this approach are simplicity and scalability. First, all steps in the Gibbs sampler involves sampling from standard distributions (e.g., normal, exponential and categorical distributions). Second, the extreme value distribution representation facilitates multivariate extension of the logistic regression in the McFadden random utility framework (Train, 2003).

Polson et al. (2013) develop a novel data augmentation strategy using Polya-Gamma latent variables for the logistic regression and related models with binomial likelihoods, enabling the Gibbs sampler to cycle through multivariate normal draws for the regression coefficients and Polya-Gamma draws for the latent variables. They demonstrate the excellent performance of the Polya-Gamma sampler, compared to the methods of Holmes and Held (2006) and Fruhwirth-Schnatter and Fruhwirth (2007).

Additionally, the logistic log-likelihood function is globally concave, allowing for a single-move Gibbs sampler that updates each element in β using the adaptive rejection sampling (Gilks and Wild, 1992). This strategy is adopted by Dellaportas and Smith (1993) and applicable to other models such as negative binomial regression (see Section 5.5).

In practice, the MH sampler often proves more convenient and efficient for estimating plain-vanilla logistic regression, compared to the Gibbs samplers. However, the Gibbs samplers offer advantages when extending logistic regression to more complex scenarios. For instance, consider mixed-effects logistic regression with longitudinal data (Zeger and Karim, 1991):

$$p(Y_{it} = 1 | X_i, \beta, \alpha) = \Lambda(X_{it}\beta + \alpha_i),$$

where the normally distributed random effects α_i capture individual-specific unobserved heterogeneity. With the unobserved α_i integrated out, the likelihood function lacks an explicit form, complicating the MH sampler implementation. In contrast, the Gibbs sampler with data augmentation remains applicable, with an extra step of sampling α_i from conditionally conjugate normal distributions.

5.2.5 Multinomial Extension

Binary logistic regression can be extended to a multinomial scenario where the response variable has $J \geq 2$ unordered categories: $Y_i \in \{1, \ldots, J\}$. Multinomial logistic regression is widely deployed in classification applications across various fields and is also known as the soft-max, maximum entropy classifier, and random utility discrete choice model, among other names. Refer to Bishop (2006) for soft-max functions in pattern recognition, Berger et al. (1996) for the maximum entropy approach to natural language processing, Train (2003) for random utility discrete choice methods.

The multinomial logistic regression is specified by the conditional probability

$$p\,(Y_i = j\,|X_i, \beta) \propto e^{X_i\beta_j},$$

with the following interpretations.

First, the binary logistic regression is characterized by the log ratio:

$$\ln \frac{p\,(Y_i = 1\,|X_i, \beta)}{p\,(Y_i = 0\,|X_i, \beta)} = X_i\beta.$$

The multinomial extension maintains linearity in the log ratio:

$$\ln \frac{p\,(Y_i = j\,|X_i, \beta)}{p\,(Y_i = J\,|X_i, \beta)} = X_i\beta_j,$$

for $j = 1, \ldots, J-1$. Since probabilities must sum to one, the conditional probability for the Jth category satisfies

$$p\,(Y_i = J\,|X_i, \beta) = \frac{1}{1 + \sum_{j=1}^{J-1} e^{X_i\beta_j}}.$$

Additionally, normalizing $\beta_J = 0$ ensures that conditional probabilities appear symmetric across categories:

$$p\,(Y_i = j\,|X_i, \beta) = \frac{e^{X_i\beta_j}}{\sum_{k=1}^{J} e^{X_i\beta_k}}, j = 1, \ldots, J.$$

Second, $X_i\beta_j$ can be interpreted as the relative score of the jth category. While it might be tempting to select the category with the highest score, this hard-max approach is not differentiable. Therefore, a soft-max specification is preferred for probabilistic selection:

$$p\,(Y_i = j\,|X_i, \beta) = \frac{e^{X_i\beta_j}}{\sum_{k=1}^{J} e^{X_i\beta_k}},$$

where the category with the highest exponential score is selected with probability close to one. When the predictor X_i does not vary across categories, the parameters $\beta_1, \ldots, \beta_J$ are not individually identified, due to

$$\frac{e^{X_i\beta_j}}{\sum_{k=1}^{J} e^{X_i\beta_k}} = \frac{e^{X_i(\beta_j - \beta_J)}}{\sum_{k=1}^{J} e^{X_i(\beta_k - \beta_J)}},$$

prompting the normalization $\beta_J = 0$.

Third, consider a random utility discrete choice model

$$Y_{ij}^* = X_i\beta_j + \varepsilon_{ij},$$

where ε_{ij} is a random variable: to an outside observer, behaviors of a decision maker who chooses the category of the highest utility may appear random. Assume that ε_{ij} follows an extreme value distribution:

$$p\,(\varepsilon_{ij}) = e^{-\varepsilon_{ij} - \exp(-\varepsilon_{ij})}.$$

The main reason for using the extreme value distribution is that it provides a closed-form solution for the probability of selecting the jth category (see Exercise 5.3):

$$p(Y_i = j \mid X_i, \beta) = p\left(Y_{ij}^* \geq Y_{ik}^*, \forall k\right) = \frac{e^{X_i\beta_j}}{\sum_{k=1}^{J} e^{X_i\beta_k}}.$$

It serves as an extension to the binary logistic regression, since the difference of two independent extreme value random variables follows the logistic distribution. The difference of random utilities determines the discrete choice.

Bayesian inference of the multinomial logistic regression can be implemented by the MH or HMC sampler, as the model does not have a conjugate structure. The posterior is proportional to the prior times the likelihood. Observations can be decomposed as $y_{ij} = I(Y_i = j)$, $j = 1, \ldots, J$, and the log-likelihood function takes the form

$$\ln p(Y \mid X, \beta) = \sum_{i=1}^{n} \left[\left(\sum_{j=1}^{J} y_{ij} X_i \beta_j \right) - \ln \sum_{j=1}^{J} e^{X_i\beta_j} \right],$$

with the gradient

$$\frac{\partial \ln p(Y \mid X, \beta)}{\partial \beta_j} = \sum_{i=1}^{n} \left(y_{ij} - \frac{e^{X_i\beta_j}}{\sum_{k=1}^{J} e^{X_i\beta_k}} \right) X_i',$$

where the term $y_{ij} - \frac{e^{X_i\beta_j}}{\sum_{k=1}^{J} e^{X_i\beta_k}}$ is reminiscent of the residual in regression.

When implementing multinomial logistic regression on a computer, it is advisable to subtract $e^{X_i\beta_j}$ by a constant c to prevent overflow caused by exponentiation. This adjustment does not affect gradient computation but is crucial for accurately computing the log-likelihood function as

$$\ln p(Y \mid X, \beta) = \sum_{i=1}^{n} \left[\left(\sum_{j=1}^{J} y_{ij} X_i \beta_j \right) - \ln \sum_{j=1}^{J} e^{X_i\beta_j - c} - c \right].$$

5.2.6 Propensity Score and Treatment Effect

A prominent application of logistic regression is the estimation of the propensity score, which plays a central role in treatment effect identification (Rosenbaum and Rubin, 1983). In the potential outcome framework proposed by Rubin (1974, 1978), a causal statement presumes that an individual exposed to a treatment could alternatively have been exposed to an alternative treatment. Let D_i be a discrete-valued treatment for individual i and Y_{id} be the potential outcome under treatment $D_i = d$. A researcher observes only one of the potential outcomes for each person, which corresponds to the actual treatment received:

$$Y_i = \sum_{d} Y_{id} I(D_i = d).$$

Causal inference relies on two key assumptions: conditional independence and overlap. First, suppose we have covariates X_i that are common causes of both the treatment and potential outcomes, known as confounding variables. Conditional independence refers to the assumption that D_i and Y_{id} are independent conditional on X_i. The assumption is also referred to as unconfoundedness, conditional ignorability, and selection on observables. Second, overlap refers to the positive probability of receiving each treatment level for each possible X_i, resulting in $0 < p(D_i = d|X_i) < 1$. The probability $p(D_i = d|X_i)$ is termed the propensity score.

Under the assumptions of conditional independence and overlap, the potential outcome means (POM) $E(Y_{id})$ can be identified. Contrasts of POM provide the average treatment effect. Various methods exist for treatment effect identification and estimation, including regression adjustment using the outcome model (Imbens and Wooldridge, 2009), inverse probability weighting using the treatment model (Horvitz and Thompson, 1952; Imbens, 2000; Hirano et al., 2003), doubly robust estimation using both outcome and treatment models (Bang and Robins, 2005; Tan, 2010), and propensity score matching (Rosenbaum and Rubin, 1983), among others. We will briefly discuss some industry-standard methods available in mainstream statistical software.

First, with a correctly specified outcome model, the POM is identified as

$$E(Y_{id}) = E[E(Y_{id}|X_i)] = E[E(Y_i|X_i, D_i = d)],$$

where the last equality follows the conditional independence. Overlap is implicit in the formulation, as $E(Y_i|X_i, D_i = d)$ is not well-defined if $p(D_i = d|X_i) = 0$. The outer expectation integrates over the covariate distribution $p(X_i)$, while the inner expectation integrates over the conditional distribution of the outcome $p(Y_i|X_i, D_i)$. Because these distributions involve only observed variables, the POM is identified and can be estimated by regression adjustment. If the outcome variable is continuous, a natural choice for the outcome model is

$$E(Y_i|X_i, D_i = d) = X_i\beta_d.$$

Using the subset data selected by $D_i = d$, we estimate the parameters by

$$\hat{\beta}_d = \left[\sum_{i=1}^{n} X_i'X_i I(D_i = d)\right]^{-1} \left[\sum_{i=1}^{n} X_i'Y_i I(D_i = d)\right].$$

The counterfactual outcome for person i is estimated by $\hat{Y}_{id} = X_i\hat{\beta}_d$. Note that the estimator $\hat{\beta}_d$ is obtained by the regression with the subset data restricted to $D_i = d$, while the predicted value $\hat{Y}_{id}$ covers the full-sample data: $i = 1, \ldots, n$.

POM requires integration over the covariate distribution, which is not explicitly modeled but random samples from the distribution are readily available, namely the covariate data. A moment estimator of the POM is $\frac{1}{n}\sum_{i=1}^{n} \hat{Y}_{id}$.

Second, if an outcome model is unspecified but a treatment model can be reliably estimated for the propensity score, an alternative method to identify the POM is inverse probability weighting. By conditional independence, we have

$$E[Y_{id}I(D_i = d)|X_i] = E(Y_{id}|X_i)\, p(D_i = d|X_i).$$

Because $Y_{id}I(D_i = d) = Y_i I(D_i = d)$, we have

$$E(Y_{id}|X_i) = E\left[\frac{Y_i I(D_i = d)}{p(D_i = d|X_i)}|X_i\right].$$

By the law of iterated expectations, we identify the POM as

$$E(Y_{id}) = E\left[\frac{Y_i I(D_i = d)}{p(D_i = d|X_i)}\right].$$

The outcomes under the treatment $D_i = d$ are weighted by the inverse of the propensity score, hence the name inverse probability weighting. The overlap assumption is crucial for treatment effect identification; otherwise, the denominator is zero under some treatment levels. Exercise 5.13 provides an alternative derivation of inverse probability weighting.

In practice, logistic regression is employed for estimating the propensity score $\hat{p}(D_i = d | X_i)$, resulting in a moment estimator of the POM: $\frac{1}{n}\sum_{i=1}^{n}\tilde{Y}_{id}$, where

$$\tilde{Y}_{id} = \frac{Y_i I(D_i = d)}{\hat{p}(D_i = d | X_i)}.$$

Third, it is possible to combine regression adjustment and inverse probability weighting for a doubly robust estimator:

$$\frac{1}{n}\sum_{i=1}^{n}\left[\tilde{Y}_{id} - \frac{I(D_i = d) - \hat{p}(D_i = d | X_i)}{\hat{p}(D_i = d | X_i)}\hat{Y}_{id}\right].$$

This estimator is consistent if either the outcome or treatment model is correctly specified, hence the name double robustness.

Fourth, if the treatment is binary, conditional independence given high-dimensional covariates is equivalent to conditional independence given the propensity score. This remarkable dimension-reduction property underpins the propensity score matching method developed by Rosenbaum and Rubin (1983). For each treated (control) unit, a match is made with a unit in the control (treatment) group with the same propensity score. The outcome difference provides an intuitive measure of the treatment effect. In practice, the propensity score is estimated using logistic regression, and exact matches by the propensity score are rare. A workaround involves matching by nearest neighbors, within a specified radius (caliper), through stratification, or using a weighted average with a kernel density, among other methods.

5.3 Truncation, Censoring and Tobit Model

In previous sections, we explored specific instances of truncated distributions, including the half-normal component within the skew normal distribution, the posterior distribution of latent variables in the Probit model, and the stationary autoregressive process under a truncated prior specification. In this section, we aim to provide a comprehensive solution to the issues of truncation and censoring. These two concepts are intricately linked: a censored distribution can be conceptualized as a mixture distribution, one component of which is the truncated distribution. We will also elucidate the properties of truncated and censored normal distributions. The Tobit model is a prominent application of censored normal distributions (Tobin, 1958).

5.3.1 Truncation

An absolutely continuous distribution is characterized by a density function. The density of a truncated distribution is proportional to the original density, and the normalizing constant is derived through integration.

Definition 5.2 (Truncated density) *Let f be the density function of a proper distribution that satisfies $\int_{\mathbb{R}} f(y)\,dy = 1$. The function g is called the truncated density of f in the set $A \subset \mathbb{R}$ if*

$$g(y) = \frac{f(y)}{\int_A f(z)\,dz} I_A(y),$$

where the indicator function $I_A(y) = 1$, if $y \in A$, and zero otherwise.

Definition 5.2 indicates that the density function of a truncated distribution is proportional to the original density in the set A. Density increases proportionally to maintain a proper distribution.

From a Bayesian perspective, truncation can be interpreted as g being the posterior distribution conditional on the observation in the set A. To illustrate, consider an interval $A = (a, b)$. Let Y be a random variable with density f and $F(y) = \int_{-\infty}^{y} f(z)\,dz$. Denote an indicator variable

$$D = I_{(a,b)}(Y) = \begin{cases} 1 & Y \in (a,b) \\ 0 & Y \notin (a,b) \end{cases}.$$

By Bayes' rule, the conditional probability is defined as

$$P(Y \le y \mid D = 1) = \frac{P(Y \le y, D = 1)}{P(D = 1)} = \frac{F(y) - F(a)}{F(b) - F(a)}, a < y < b.$$

Taking derivatives with respect to y, we obtain the truncated density:

$$\frac{dP(Y \le y \mid D = 1)}{dy} = \frac{f(y)}{F(b) - F(a)} I_{(a,b)}(y).$$

Truncation may arise due to missing data. Consider latent variables Y_i^*, $i = 1, \ldots, n$ generated from the distribution f. If we can only observe variables $Y_i^* \in A$, the variables $Y_i^* \notin A$ are missing values. The observations are thus sampled from the truncated distribution. This scenario is a specific instance of rejection sampling, where a proposed sample from f is accepted if it falls within the set A; otherwise, it is rejected. Among the accepted samples, the probability that an accepted sample is less than or equal to a point of evaluation y is given by

$$P(Y_i^* \le y \mid Y_i^* \in A) = \frac{P(Y_i^* \le y, Y_i^* \in A)}{P(Y_i^* \in A)} = \frac{\int_{-\infty}^{y} f(z) I_A(z)\,dz}{\int_A f(z)\,dz}.$$

By differentiating with respect to y, we obtain the truncated density:

$$\frac{dP(Y_i^* \le y \mid Y_i^* \in A)}{dy} = \frac{f(y)}{\int_A f(z)\,dz} I_A(y).$$

The truncated normal distribution plays an important role in econometric models. Let $\phi(y; \mu, \sigma^2)$ be the normal density with mean μ and variance σ^2. Assume the truncation set is an interval $A = (a, b)$, where a and b are extended real numbers (including infinity). The truncated normal distribution, denoted by $N(\mu, \sigma^2) I_{(a,b)}$, has the density

$$g(y) = \frac{\phi(y; \mu, \sigma^2)}{\Phi(b; \mu, \sigma^2) - \Phi(a; \mu, \sigma^2)} I_{(a,b)}(y),$$

where $\Phi(b; \mu, \sigma^2)$ is the CDF of $N(\mu, \sigma^2)$ evaluated at b. Proposition 5.1 captures the characteristics of the truncated normal distribution, which will be extensively utilized in the subsequent sections.

Proposition 5.1 (Truncated normal distribution) *Suppose that Y follows $N(\mu, \sigma^2)I_{(a,b)}$. The mean and variance are given by*

$$E(Y) = \mu + \sigma(\lambda_a + \lambda_b),$$

$$Var(Y) = \sigma^2 \left[1 - (\lambda_a + \lambda_b)^2 + \lambda_a \frac{a-\mu}{\sigma} + \lambda_b \frac{b-\mu}{\sigma}\right],$$

where the inverse Mills ratios λ_a and λ_b are defined as

$$\lambda_a = \frac{\phi\left(\frac{a-\mu}{\sigma}\right)}{\Phi\left(\frac{b-\mu}{\sigma}\right) - \Phi\left(\frac{a-\mu}{\sigma}\right)},$$

$$\lambda_b = -\frac{\phi\left(\frac{b-\mu}{\sigma}\right)}{\Phi\left(\frac{b-\mu}{\sigma}\right) - \Phi\left(\frac{a-\mu}{\sigma}\right)}.$$

For the lower-bounded $N(\mu, \sigma^2) I_{(a,\infty)}$, the inverse Mills ratios satisfy $\lambda_a > 0$ and $\lambda_b = 0$, indicating that the mean of the truncated distribution exceeds μ, while the variance is reduced compared to σ^2. Similarly, for the upper-bounded $N(\mu, \sigma^2) I_{(-\infty,b)}$, the mean is lower than μ, and the variance is smaller than σ^2.

5.3.2 Censoring

The distinction between truncation and censoring is best understood in terms of latent variables generated by a distribution. If data are missing when latent variables fall outside a specified interval, the resulting observations follow a truncated distribution. Conversely, if data are recorded at the lower or upper bound rather than the actual realization when a latent variable falls outside the interval, observations follow the censored distribution. In reviewing a dataset, censored data are characterized by a clustering of observations at the sample minimum or maximum, whereas truncated data do not exhibit such clustering, assuming the original distribution is absolutely continuous.

In contrast to truncation, a censored distribution is not continuous and does not have a density function. Nonetheless, it can be defined by its CDF.

Definition 5.3 (Censored distribution) *Let F be the CDF of a distribution. If*

$$G(y) = \begin{cases} 0 & y < a \\ F(y) & y \geq a \end{cases},$$

then G is the CDF of the censored distribution with the lower bound a. If

$$H(y) = \begin{cases} F(y) & y < b \\ 1 & y \geq b \end{cases},$$

then H is the CDF of the censored distribution with the upper bound b.

The function G in Definition 5.3 is a well-defined CDF due to its right-continuity and $G(-\infty) = 0$ and $G(\infty) = 1$. However, the CDF experiences a jump at the censoring point, where this discontinuity introduces a positive probability. Regarding the shape of the censored CDF, the curves of the original and censored CDFs overlap in the non-censored region.

Proposition 5.2 offers a latent variable interpretation of censoring. When a latent variable falls below (or rises above) a threshold, the observed value is the lower (or upper) bound instead of the actual realization.

Proposition 5.2 (Censored variable) *Let Z be a latent variable with the CDF F.*

1. *If we observe $Y = \max(Z, a)$, then the CDF of Y is given by*

$$G(y) = \begin{cases} 0 & y < a \\ F(y) & y \geq a \end{cases}.$$

2. *If we observe $Y = \min(Z, b)$, then the CDF of Y is given by*

$$H(y) = \begin{cases} F(y) & y < b \\ 1 & y \geq b \end{cases}.$$

Although a censored distribution does not have a density, Proposition 5.3 provides a proxy for the density function using the Dirac delta function δ_a such that $\delta_a(y) = 0$, $\forall y \neq a$, and $\int_A \delta_a(y)\,dy = I_A(a)$. We may conceptualize δ_a as the limit of a density function, such as the normal density with mean a and an infinitesimal variance.

Proposition 5.3 (Censored density) *Let f and F be the density and CDF of a distribution.*

1. *Let G be the CDF of the censored distribution*

$$G(y) = \begin{cases} 0 & y < a \\ F(y) & y \geq a \end{cases}.$$

We construct a proxy of the density function such that

$$g(y) = F(a)\,\delta_a(y) + [1 - F(a)]\frac{f(y)}{1 - F(a)} I_{(a,\infty)}(y).$$

Then we have $G(y) = \int_{-\infty}^{y} g(z)\,dz$, $\forall y \in \mathbb{R}$.

2. *Let H be the CDF of the censored distribution*

$$H(y) = \begin{cases} F(y) & y < b \\ 1 & y \geq b \end{cases}.$$

We construct a proxy of the density function such that

$$h(y) = F(b)\frac{f(y)}{F(b)} I_{(-\infty,b)}(y) + [1 - F(b)]\,\delta_b(y).$$

Then we have $H(y) = \int_{-\infty}^{y} h(z)\,dz$, $\forall y \in \mathbb{R}$.

Proposition 5.3 elucidates the relationship between censoring and truncation. The censored distribution G is a two-component mixture: a degenerate component at the censoring point and a truncated distribution over the interval (a, ∞), with weights $F(a)$ and $1 - F(a)$, respectively. Similarly, the censored distribution H is a mixture comprising a truncated distribution over $(-\infty, b)$ and the Dirac delta at the censoring point b, with weights $F(b)$ and $1 - F(b)$, respectively.

Proposition 5.3 provides a convenient method for computing the mean of the censored distribution. By the law of iterated expectations, the mean of a mixture distribution is the weighted average of the component-specific means. For instance, consider a censored normal distribution $N(\mu, \sigma^2)$ at the lower bound a. The mean of the degenerate component is a, and the mean of the truncated normal $N(\mu, \sigma^2)\,I_{(a,\infty)}$ is $\mu + \sigma\lambda_a$, where the inverse Mills ratio λ_a is defined in Proposition 5.1. The weights are $\Phi\left(\frac{a-\mu}{\sigma}\right)$ and $\Phi\left(-\frac{a-\mu}{\sigma}\right)$, respectively. Consequently, the mean of the censored normal distribution is given by

$$E(Y) = \Phi\left(\frac{a - \mu}{\sigma}\right) a + \Phi\left(-\frac{a - \mu}{\sigma}\right)(\mu + \sigma\lambda_a).$$

5.3.3 Tobit Model

In the Tobit model, the response variable is censored due to external constraints such as imperfect observations or internal constraints such as corner solutions. The Tobit model features a latent variable representation:

$$Z_i = X_i\beta + \sigma\varepsilon_i,$$

$$Y_i = \max(Z_i, c),$$

where ε_i follows the standard normal distribution. We observe the censored variable Y_i, but not the latent variable Z_i. With a sufficiently large sample size, some data points will cluster at the sample minimum, identifying the censoring point, which is assumed to be known. Exercise 5.4 extends the Tobit model to accommodate an unknown censoring point.

In economic applications, it is pertinent to study the marginal effect of X_i on Z_i or Y_i. The mean of the censored distribution is given by

$$E\left(Y_i \left| X_i, \beta, \sigma^2\right.\right) = \Phi(\tilde{c}_i)\, c + \Phi(-\tilde{c}_i)\, X_i\beta + \sigma\phi(\tilde{c}_i),$$

where $\tilde{c}_i = \frac{c - X_i\beta}{\sigma}$. It follows that the marginal effect is

$$\frac{\partial E\left(Y_i \left| X_i, \beta, \sigma^2\right.\right)}{\partial X_i} = \Phi(-\tilde{c}_i)\,\beta.$$

In contrast to standard linear regression, where the marginal effect equals β, the marginal effect on the censored variable is adjusted by $\Phi(-\tilde{c}_i)$. The intuition is that $\Phi(\tilde{c}_i)$ is the probability that the observed Y_i remains unchanged due to censoring. With probability $\Phi(-\tilde{c}_i)$, the continuous variable changes by β as the explanatory variable increases marginally. On average, the marginal effect is $\Phi(-\tilde{c}_i)\,\beta$.

Consider Bayesian estimation of the Tobit model. If the latent variables were known, the model would simplify to a standard linear regression. We specify the conjugate prior $NIG(\mu, \Lambda, a, b)$ for β and σ^2, so that the posterior conditional distribution $p\left(\beta, \sigma^2 \left| Y, Z\right.\right)$ is $NIG\left(\bar{\mu}, \bar{\Lambda}, \bar{a}, \bar{b}\right)$, where

$$\bar{\mu} = \left(\Lambda + \sum_{i=1}^{n} X_i' X_i\right)^{-1} \left(\Lambda\mu + \sum_{i=1}^{n} X_i' Z_i\right),$$

$$\bar{\Lambda} = \Lambda + \sum_{i=1}^{n} X_i' X_i,$$

$$\bar{a} = a + \frac{n}{2},$$

$$\bar{b} = b + \frac{1}{2}\sum_{i=1}^{n} Z_i^2 + \frac{1}{2}\mu'\Lambda\mu - \frac{1}{2}\bar{\mu}'\bar{\Lambda}\bar{\mu}.$$

The posterior conditional distribution of the latent variable takes the form

$$p\left(Z \left| Y, \beta, \sigma^2\right.\right) = \prod_{i=1}^{n} p\left(Z_i \left| \beta, \sigma^2\right.\right) p\left(Y_i \left| Z_i\right.\right),$$

which suggests that knowledge of the latent variable is derived from two sources. First, the latent variable is generated from a Gaussian linear regression, as $Z_i \left| \beta, \sigma^2\right.$ follows $N\left(X_i\beta, \sigma^2\right)$. Second, by observing Y_i we can infer whether Z_i is censored or not. If $Y_i > c$, then $Z_i = Y_i$. If $Y_i = c$, then $Z_i \leq c$ (though the exact value of Z_i is unknown). This inference is possible because $p(Y_i | Z_i)$ serves as an indicator

$$\begin{aligned} p(Y_i | Z_i) &= I(Z_i > c, Y_i = Z_i) + I(Z_i \leq c, Y_i = c) \\ &= I(Y_i > c, Z_i = Y_i) + I(Y_i = c, Z_i \leq c). \end{aligned}$$

Note that the event $\{Z_i > c, Y_i = Z_i\}$ is equivalent to $\{Y_i > c, Z_i = Y_i\}$. The former describes the data generating process (if a latent variable is not censored, the observation equals the latent variable), while the latter provides posterior inference (if an observation exceeds the censoring point, we infer that the latent variable equals the observation). Similarly, the event $\{Z_i \leq c, Y_i = c\}$ is equivalent to $\{Y_i = c, Z_i \leq c\}$. The former describes the censored observation at the lower bound, while the latter provides posterior inference: observing the data at the lower bound, we infer that the latent variable cannot exceed the censoring point.

In summary, the posterior conditional distribution of the latent variable can be described as follows: if $Y_i > c$, then $Z_i = Y_i$; if $Y_i = c$, then Z_i follows the truncated normal distribution $N(X_i\beta, \sigma^2) I_{(-\infty,c)}$.

As the full posterior conditional distributions are tractable, the Gibbs sampler can be employed for posterior simulation. The Gibbs sampler alternately generates parameters from NIG distributions and latent variables from truncated normal distributions.

When the sample size is large, it becomes computationally intensive to sweep over the full dataset at each iteration of the Gibbs sampler. Our strategy is to marginalize β, σ^2 and subsample the vector Z. Consider decomposing the posterior density as

$$p(Z, \beta, \sigma^2 | Y) = p(Z | Y) p(\beta, \sigma^2 | Y, Z),$$

where $p(\beta, \sigma^2 | Y, Z)$ is the density of $NIG(\bar{\mu}, \bar{\Lambda}, \bar{a}, \bar{b})$, while $p(Z | Y)$ can be further decomposed as $p(Z | Y) \propto p(Z) p(Y | Z)$, which can be evaluated pointwise by

$$p(Z) \propto (\bar{b})^{\bar{a}},$$

$$p(Y | Z) = \prod_{i=1}^{n} [I(Y_i > c, Z_i = Y_i) + I(Y_i = c, Z_i \leq c)].$$

Because the posterior marginal density $p(Z | Y)$ can be evaluated up to a normalizing constant, MCMC sampling is applicable. Algorithm 5.3 demonstrates a random-walk MH sampler. At each iteration, we subsample an element in the vector Z by reading a single data point for rank-one update of cumulative statistics, which are functions of latent variables in the summable form:

$$S_1(Z) = \sum_{i=1}^{n} X_i' X_i,$$

$$S_2(Z) = \sum_{i=1}^{n} X_i' Z_i,$$

$$S_3(Z) = \sum_{i=1}^{n} Z_i^2.$$

When the MH sampler is initialized, the starting value of Z must satisfy $Z_i = Y_i$ if $Y_i > c$, and $Z_i \leq c$ if $Y_i = c$. During the MCMC sampling, we first subsample an index i. If the ith observation is not censored, we can skip proposing new values, because the proposal will be rejected and $Z_i = Y_i$. If the corresponding observation is censored, we propose a new

Algorithm 5.3 Marginalized subsampling of Tobit model

Input: data X, Y, initial value Z, and tuning parameter λ.
Output: posterior samples from $p\left(\beta, \sigma^2, Z \,|Y\right)$.
1 Compute cumulative statistics $S_j(Z)$, $j = 1, \ldots, 3$ by one sweep over data.
2 Iterate the following steps:
  2.1 Randomly draw an index $i \in \{1, \ldots, n\}$. If $Y_i > c$, jump to step 2.7.
  2.2 Propose Z^* such that $Z^*_{-i} = Z_{-i}$, $Z^*_i = Z_i + \lambda u_i$, $u_i \sim N(0,1)$. If $Z^*_i > c$, jump to step 2.7.
  2.3 Read i^{th} data point only.
  2.4 Compute cumulative statistics $S_j(Z^*)$ by rank-one update of $S_j(Z)$.
  2.5 Evaluate $p(Z^* \,|Y)$ by cumulative statistics.
  2.6 Accept the proposal (set $Z = Z^*$) with the probability $\min\left[1, \frac{p(Z^*|Y)}{p(Z|Y)}\right]$.
  2.7 Sample β, σ^2 from $NIG\left(\bar{\mu}, \bar{\Lambda}, \bar{a}, \bar{b}\right)$.

value Z^*_i. If $Z^*_i > c$, we reject the proposal. Otherwise, we calculate $p(Z^* \,|Y)$ by rank-one update of cumulative statistics and accept the proposal with probability $\min\left[1, \frac{p(Z^*|Y)}{p(Z|Y)}\right]$. Although $p(Y\,|Z)$ is binary by construction, we maintain $p(Y\,|Z) = 1$ throughout the Monte Carlo simulation.

5.4 Incidental Truncation and Sample Selection

Truncation and censoring can be extended to multivariate distributions. If some variables are truncated while others are not, we encounter a problem of incidental truncation. The following definition provides a multivariate extension to Definition 5.2.

Definition 5.4 (Multivariate truncated distribution) *Let f be a multivariate density function. Consider a set $A \subset \mathbb{R}^d$ such that $F(A) = \int_A f(y)\,dy > 0$. The function g is called the truncated density of f in the set A if*

$$g(y) = \frac{f(y)}{F(A)} I_A(y).$$

The truncated density is proportional to the original density in the set A, and zero otherwise. The normalizing constant $F(A)$ involves high dimensional integration that may not be analytically tractable.

5.4.1 Incidental Truncation

In practice, we are likely to encounter one or more truncated variables, while other correlated variables remain unrestricted. What are the marginal and conditional distributions of the truncated and unrestricted variables? Proposition 5.4 provides an answer in the context of multivariate normal (MVN) distributions. Interestingly, the marginal distributions of unrestricted variables are distorted by the truncated variables, which is the essence of incidental truncation.

Proposition 5.4 (Normal truncation) *Consider truncated MVN density*

$$p(y_1, y_2) \propto \phi\left[\begin{pmatrix} y_1 \\ y_2 \end{pmatrix}; \begin{pmatrix} \mu_1 \\ \mu_2 \end{pmatrix}, \begin{pmatrix} \Sigma_{11} & \Sigma_{12} \\ \Sigma_{21} & \Sigma_{22} \end{pmatrix}\right] I_{A_1}(y_1),$$

where $A_1 \subset \mathbb{R}^{d_1}$ indicates that y_1 is truncated, but y_2 is not.

1. The marginal density of y_1 is

$$p(y_1) = \frac{\phi(y_1; \mu_1, \Sigma_{11})}{F(A_1; \mu_1, \Sigma_{11})} I_{A_1}(y_1),$$

where

$$F(A_1; \mu_1, \Sigma_{11}) = \int_{A_1} \phi(y_1; \mu_1, \Sigma_{11})\, dy_1.$$

2. The marginal density of y_2 is

$$p(y_2) = \phi(y_2; \mu_2, \Sigma_{22}) \frac{F(A_1; \bar{\mu}_1, \bar{\Sigma}_{11})}{F(A_1; \mu_1, \Sigma_{11})},$$

where

$$\bar{\mu}_1 = \mu_1 + \Sigma_{12}\Sigma_{22}^{-1}(y_2 - \mu_2),$$
$$\bar{\Sigma}_{11} = \Sigma_{11} - \Sigma_{12}\Sigma_{22}^{-1}\Sigma_{21}.$$

3. The conditional density of y_1 is

$$p(y_1 | y_2) = \frac{\phi(y_1; \bar{\mu}_1, \bar{\Sigma}_{11})}{F(A_1; \bar{\mu}_1, \bar{\Sigma}_{11})} I_{A_1}(y_1).$$

4. The conditional density of y_2 is

$$p(y_2 | y_1) = \phi(y_2; \bar{\mu}_2, \bar{\Sigma}_{22}),$$

where

$$\bar{\mu}_2 = \mu_2 + \Sigma_{21}\Sigma_{11}^{-1}(y_1 - \mu_1),$$
$$\bar{\Sigma}_{22} = \Sigma_{22} - \Sigma_{21}\Sigma_{11}^{-1}\Sigma_{12}.$$

Some remarks on Proposition 5.4 are noteworthy.

First, an unrestricted variable can be viewed as a special case of the truncated variable. Like the truncation indicator $I_{A_1}(y_1)$, we could make a similar indicator $I_{A_2}(y_2)$, where $A_2 = \mathbb{R}^{d_2}$.

Second, Proposition 5.4 indicates that the conditional distributions remain truncated normal, allowing samples from the multivariate truncated distribution to be obtained via a single-move Gibbs sampler that cycles through univariate truncated distributions.

Third, Proposition 5.4 indicates that the marginal distribution of the truncated vector remains truncated MVN, after integrating out the unrestricted variables. The vector y_1 follows $N(\mu_1, \Sigma_{11}) I_{A_1}$, but it does not imply that each element in y_1 follows a univariate truncated normal distribution (see Exercise 5.6). This property differs from that of the conditional distributions, where both vector-wise and element-wise conditional distributions are truncated normal.

Fourth, Proposition 5.4 indicates that the marginal distribution of the unrestricted vector is no longer MVN. Although the vector y_2 is not truncated, its marginal distribution departs from normality due to truncation of other variables. This phenomenon is known as incidental truncation: a truncated variable alters the distribution of correlated variables. Correlation is key to incidental truncation; in the absence of correlation (i.e., $\Sigma_{12} = 0$), incidental truncation disappears.

Proposition 5.5 illustrates incidental truncation in the context of a bivariate normal distribution with a truncated and unrestricted variable. The proposition summarizes the means and variances of the two variables.

Proposition 5.5 (Incidental truncation) *Consider a truncated bivariate normal distribution, in which $y_1 \in (a,b)$ and $y_2 \in \mathbb{R}$:*

$$\begin{pmatrix} y_1 \\ y_2 \end{pmatrix} \sim N\left[\begin{pmatrix} \mu_1 \\ \mu_2 \end{pmatrix}, \begin{pmatrix} \sigma_1^2 & \rho\sigma_1\sigma_2 \\ \rho\sigma_1\sigma_2 & \sigma_2^2 \end{pmatrix}\right] I_{(a,b)\times\mathbb{R}}.$$

1. *The marginal distribution of y_1 is truncated normal, with the mean and variance*

$$E(y_1) = \mu_1 + \sigma_1(\lambda_a + \lambda_b),$$

$$Var(y_1) = \sigma_1^2\left[1 - (\lambda_a + \lambda_b)^2 + \lambda_a\frac{a-\mu_1}{\sigma_1} + \lambda_b\frac{b-\mu_1}{\sigma_1}\right].$$

2. *The marginal distribution of y_2 is skewed, with the mean and variance*

$$E(y_2) = \mu_2 + \rho\sigma_2(\lambda_a + \lambda_b),$$

$$Var(y_2) = \sigma_2^2\left[1 - \rho^2(\lambda_a + \lambda_b)^2 + \rho^2\lambda_a\frac{a-\mu_1}{\sigma_1} + \rho^2\lambda_b\frac{b-\mu_1}{\sigma_1}\right].$$

where

$$\lambda_a = \frac{\phi\left(\frac{a-\mu_1}{\sigma_1}\right)}{\Phi\left(\frac{b-\mu_1}{\sigma_1}\right) - \Phi\left(\frac{a-\mu_1}{\sigma_1}\right)},$$

$$\lambda_b = -\frac{\phi\left(\frac{b-\mu_1}{\sigma_1}\right)}{\Phi\left(\frac{b-\mu_1}{\sigma_1}\right) - \Phi\left(\frac{a-\mu_1}{\sigma_1}\right)}.$$

5.4.2 Sample Selection Model

In many empirical studies, the observations available to researchers are not random samples from the entire population but are instead filtered through specific rules or mechanisms. Truncated data occur when observations are excluded based on their own values. Censored data are observed only within a certain range; values outside this range are not recorded in their entirety. Incidentally truncated data refer to cases where observations are filtered by other variables, not by the values themselves.

When discussing sample selection in econometric analyses, it is crucial to articulate the population of interest. A classic example is the determination of mortgage rates for loan applicants. If an application is rejected, the mortgage rate is unobservable. Researchers only observe mortgage rates of approved applications. If the population of interest is the mortgage rate distribution of approved loans, there is no issue of sample selection. However, if the population of interest is the mortgage rate distribution of all loan applications, the observed data are incidentally truncated by the loan approval decision.

The seminal paper of Heckman (1979) raises the sample selection issue and proposes a two-step estimator for correcting omitted variable bias. Chib et al. (2009) and van Hasselt (2011) develop Bayesian inference in endogeneity and sample selection models.

5.4.2.1 Model Specification

Consider a sample selection model where observations are incidentally truncated, modeled using a Probit regression framework. This involves two equations with latent variables: one for participation choice and the other for the activity level:

$$Z_{1i} = X_{1i}\beta_1 + \varepsilon_{1i},$$

$$Z_{2i} = X_{2i}\beta_2 + \varepsilon_{2i}.$$

The observed outcome is denoted as

$$Y_i = Z_{2i} I\left(Z_{1i} > 0\right).$$

The activity level data Z_{2i} are filtered such that we only observe $Y_i = Z_{2i}$ if $Z_{1i} > 0$. Otherwise, $Y_i = 0$, which serves as a convenient coding for missing values (alternative symbols like NaN or 999 could also be used). Since the continuous random variable Z_{2i} is almost surely non-zero, observing $Y_i = 0$ implies $Z_{1i} \leq 0$. The observed variables can be written in a verbose form:

$$D_i = I\left(Z_{1i} > 0\right),$$

$$Y_i = \begin{cases} Z_{2i} & D_i = 1 \\ 0 & D_i = 0 \end{cases}.$$

The disturbances $\varepsilon_{1i}, \varepsilon_{2i}$ are assumed to follow a bivariate normal distribution

$$N\left[\begin{pmatrix} 0 \\ 0 \end{pmatrix}, \begin{pmatrix} 1 & \rho\sigma \\ \rho\sigma & \sigma^2 \end{pmatrix}\right].$$

For identification purposes, the variance of ε_{1i} is normalized to one, akin to the identification scheme in the Probit model.

The observed variable Y_i follows a two-component mixture distribution (also see Exercise 5.7 for an alternative representation by the CDF):

$$p\left(Y_i\right) = p\left(D_i = 0\right) p\left(Y_i \,|D_i = 0\right) + p\left(D_i = 1\right) p\left(Y_i \,|D_i = 1\right),$$

where 1) $p\left(D_i = 0\right) = \Phi\left(-X_{1i}\beta_1\right)$, which is the probability of $\{Z_{1i} \leq 0\}$ in the Probit model; 2) $p\left(Y_i \,|D_i = 0\right) = \delta_0\left(Y_i\right)$, which is the Dirac Delta function that indicates $Y_i = 0$; 3) $p\left(D_i = 1\right) = \Phi\left(X_{1i}\beta_1\right)$, which is the probability of the event $\{Z_{1i} > 0\}$; and 4)

$$p\left(Y_i \,|D_i = 1\right) = \frac{\phi\left(Y_i; X_{2i}\beta_2, \sigma^2\right) \Phi\left(\bar{\mu}_i / \bar{\sigma}\right)}{\Phi\left(X_{1i}\beta_1\right)},$$

which follows Proposition 5.4, and $\bar{\mu}_i$ and $\bar{\sigma}^2$ represent the conditional mean and variance of the bivariate normal distribution:

$$\bar{\mu}_i = X_{1i}\beta_1 + \rho\sigma^{-1}\left(Y_i - X_{2i}\beta_2\right),$$

$$\bar{\sigma}^2 = 1 - \rho^2.$$

The correlation ρ is a key parameter in understanding incidental truncation. In the subsample of non-missing data ($Y_i \neq 0$), truncation in the participation equation distorts the distribution of the activity level, provided that $\rho \neq 0$. If the correlation were zero, $p\left(Y_i \,|D_i = 1\right)$ would be $\phi\left(Y_i; X_{2i}\beta_2, \sigma^2\right)$. Furthermore, Proposition 5.5 indicates that the conditional mean of non-missing data is

$$E\left(Y_i \,|D_i = 1\right) = X_{2i}\beta_2 + \rho\sigma \frac{\phi\left(X_{1i}\beta_1\right)}{\Phi\left(X_{1i}\beta_1\right)},$$

leading to the Heckman two-step estimator. First, estimate the Probit model with the full-sample data. Second, augment regressors by the inverse Mills ratio for subsample regression with non-missing data.

5.4.2.2 Bayesian Estimation

Bayesian inference of the sample selection model can be achieved by data augmentation and the Gibbs sampling, cycling through the full posterior conditional distributions of parameters $\beta_1, \beta_2, \rho, \sigma$ and latent variables Z_{1i}, Z_{2i}.

First, consider the posterior conditional distribution of Z_{1i}, given the remaining variables. Similar to the Probit model, the sign of the latent variable can be inferred from observations: $Y_i \neq 0$ indicates $Z_{1i} > 0$, and $Y_i = 0$ indicates $Z_{1i} \leq 0$. Due to correlations between two equations, the conditional mean and variance (without considering truncation) of Z_{1i} are

$$\bar{\mu}_{1i} = X_{1i}\beta_1 + \rho\sigma^{-1}\left(Z_{2i} - X_{2i}\beta_2\right),$$
$$\bar{\sigma}_1^2 = 1 - \rho^2.$$

Consequently, Z_{1i} follows a truncated normal distribution:

$$\begin{cases} N\left(\bar{\mu}_{1i}, \bar{\sigma}_1^2\right) I_{(0,\infty)} & Y_i \neq 0 \\ N\left(\bar{\mu}_{1i}, \bar{\sigma}_1^2\right) I_{(-\infty,0)} & Y_i = 0 \end{cases}.$$

Second, consider the posterior conditional distribution of Z_{2i}, given the remaining variables. Observing non-missing data $Y_i \neq 0$ leads to $Z_{2i} = Y_i$, while $Y_i = 0$ provides no information about the sign or value of Z_{2i} (conditional on Z_{1i}). Additionally, due to cross-equation correlations, the conditional mean and variance of Z_{2i} are

$$\bar{\mu}_{2i} = X_{2i}\beta_2 + \rho\sigma\left(Z_{1i} - X_{1i}\beta_1\right),$$
$$\bar{\sigma}_2^2 = \left(1 - \rho^2\right)\sigma^2.$$

Consequently, the posterior conditional distribution is $Z_{2i} = Y_i$ if $Y_i \neq 0$, and $Z_{2i} \sim N\left(\bar{\mu}_{2i}, \bar{\sigma}_2^2\right)$ if $Y_i = 0$.

Third, the posterior conditional distribution of $\beta_1, \beta_2, \rho, \sigma$ depends on the prior specification and the choice of predictors. If the same set of predictors is used in both regressions, it is convenient to specify a conjugate NIW prior. The two equations can be combined as

$$Z_i = X_i B + \varepsilon_i,$$

where $Z_i = (Z_{1i}, Z_{2i})$, $X_i = X_{1i} = X_{2i}$, $B = (\beta_1, \beta_2)$, and $\varepsilon_i = (\varepsilon_{1i}, \varepsilon_{2i})$ with the covariance matrix $\Sigma = \begin{pmatrix} 1 & \rho\sigma \\ \rho\sigma & \sigma^2 \end{pmatrix}$. The corresponding matrix form is:

$$Z = XB + \varepsilon,$$

where $Z = \begin{pmatrix} Z_1 \\ \cdots \\ Z_n \end{pmatrix}$, $X = \begin{pmatrix} X_1 \\ \cdots \\ X_n \end{pmatrix}$ and $\varepsilon = \begin{pmatrix} \varepsilon_1 \\ \cdots \\ \varepsilon_n \end{pmatrix}$.

Under the prior $NIW\left(U, \Lambda, \Omega, v\right)$ for the parameter pair (B, Σ), Proposition 1.8 indicates that the posterior is $NIW\left(\bar{U}, \bar{\Lambda}, \bar{\Omega}, \bar{v}\right)$,where

$$\bar{U} = \left(\Lambda + X'X\right)^{-1}\left(\Lambda U + X'Z\right),$$
$$\bar{\Lambda} = \Lambda + X'X,$$
$$\bar{\Omega} = \Omega + Z'Z + U'\Lambda U - \bar{U}'\bar{\Lambda}\bar{U},$$
$$\bar{v} = n + v.$$

Bypassing the normalization constraint on the covariance matrix Σ necessitates post-processing MCMC samples to focus on identifiable quantities of interest. McCulloch and Rossi (1994) argue that traversing through a non-identified parameter space simplifies the posterior computations and improves mixing of the Markov chain.

It is possible to impose the normalization constraint by Cholesky decomposition of the covariance matrix $\Sigma = LL'$, where

$$L = \begin{pmatrix} 1 & 0 \\ \rho\sigma & \sqrt{1-\rho^2}\sigma \end{pmatrix}.$$

Correlations between error terms in two equations are removed by constructing the uncorrelated variables

$$\begin{pmatrix} v_{1i} \\ v_{2i} \end{pmatrix} = L^{-1} \begin{pmatrix} \varepsilon_{1i} \\ \varepsilon_{2i} \end{pmatrix},$$

resulting in

$$Z_{1i} = X_{1i}\beta_1 + v_{1i},$$

$$Z_{2i} = X_{2i}\beta_2 + \rho\sigma v_{1i} + \sqrt{1-\rho^2}\sigma v_{2i}.$$

The two regressions can be processed sequentially. In the first regression, the posterior conditional distribution of β_1 is normal if its prior is normal. In the second regression, the variable $v_{1i} = Z_{1i} - X_{1i}\beta_1$ serves as an additional predictor, and $\left(1-\rho^2\right)\sigma^2$ acts as the reparametrized variance. Given a draw of $\rho\sigma$ and $\left(1-\rho^2\right)\sigma^2$, the original parameter ρ, σ can be recovered. See Exercise 5.8 for more discussions on sequential estimation of multiple equations.

An alternative solution is proposed by Nobile (2000), who develops a method of sampling the Wishart distribution in which the first diagonal element is constrained to be one. Under the normal prior for β_1, β_2 and a constrained inverse Wishart prior for Σ, the Gibbs sampler cycles through the multivariate normal $p(\beta_1, \beta_2 \,|\Sigma, Y, Z)$ and the constrained inverse Wishart $p(\Sigma \,|\beta_1, \beta_2, Y, Z)$.

5.4.2.3 Big Data Algorithm

When the sample size is large, computational efficiency becomes a concern. If the same predictors are used in both equations and a conjugate NIW prior is adopted, model parameters B, Σ can be integrated out, allowing for subsampling of latent variables from the posterior density $p(Z_1, Z_2 \,|Y) \propto p(Z_1, Z_2)\, p(Y \,|Z_1, Z_2)$, where

$$p(Z_1, Z_2) \propto \left|\bar{\Omega}\right|^{-\bar{v}/2},$$

$$p(Y \,|Z_1, Z_2) \propto \prod_{i=1}^{n} \left[I(Y_i \neq 0, Z_{1i} > 0, Z_{2i} = Y) + I(Y_i = 0, Z_{1i} \leq 0)\right],$$

where $\bar{\Omega}$ and $\bar{v}$ are parameters in the posterior distribution $NIW\left(\bar{U}, \bar{\Lambda}, \bar{\Omega}, \bar{v}\right)$.

A random-walk MH sampler accesses a single data point and proposes a new value for the latent variable pair. With rank-one update of cumulative statistics $S_1(Z) = \sum_{i=1}^{n} X_i' X_i$, $S_2(Z) = \sum_{i=1}^{n} X_i' Z_i'$ and $S_3(Z) = \sum_{i=1}^{n} Z_i Z_i'$, the target density $p(Z_1, Z_2 \,|Y)$ can be evaluated up to a normalizing constant.

In big data applications, where computing costs are prohibitively high, a Bayesian analogue of the Heckman two-step approach can be considered. First, Bayesian Probit regression is conducted, followed by Bayesian linear regression with the inverse Mills ratio as an additional predictor. Big data algorithms are readily available for both steps.

5.4.3 Endogenous Treatment Variable Regression

A fundamental challenge in econometrics is to estimate causal effects in the presence of endogenous variables. The causal inference methods discussed in Section 5.2 rely on the conditional independence assumption: the treatment variable is independent of potential outcomes conditional on observed confounding variables (as common causes of the treatment and outcomes). This assumption is often violated in empirical studies due to unobserved confounders, rendering traditional methods inadequate.

Consider the classic example in labor economics: estimating the return to schooling. Here, the response variable is earnings, and one of the regressors is a dummy variable indicating college education. This regressor is endogenous because of unobserved factors, such as individual ability, which influence both the pursuit of higher education and workplace performance.

Incidental truncation techniques facilitate regression analysis of the treatment effects. Endogeneity of the treatment variable is modeled by a two-equation system in which the disturbances are correlated.

1. Treatment selection: $Z_i = X_{1i}\beta_1 + \varepsilon_{1i}$ and $D_i = I(Z_i > 0)$.
2. Treatment effect: $Y_i = X_{2i}\beta_2 + D_i\alpha + \varepsilon_{2i}$.

In the context of the return to education regression, the observed variables are the outcome Y_i (earning), the treatment D_i (education) and exogenous predictors X_{1i}, X_{2i} (social-economic background observed by researchers), but Z_i is a latent variable that determines Probit-type selection of treatment.

The disturbances $\varepsilon_{1i}, \varepsilon_{2i}$ are correlated due to the unobserved confounding η_i (such as the latent ability). Assume that $\varepsilon_{1i} = \eta_i + \tilde{\varepsilon}_{1i}$ and $\varepsilon_{2i} = \eta_i + \tilde{\varepsilon}_{2i}$, where $\eta_i, \tilde{\varepsilon}_{1i}, \tilde{\varepsilon}_{2i}$ are independent normal errors. Marginalization over the unobserved confounding yields $\varepsilon_{1i}, \varepsilon_{2i}$ as correlated variates, following a bivariate normal distribution

$$N\left[\begin{pmatrix} 0 \\ 0 \end{pmatrix}, \begin{pmatrix} 1 & \rho\sigma \\ \rho\sigma & \sigma^2 \end{pmatrix}\right].$$

The equation $Y_i = X_{2i}\beta_2 + D_i\alpha + \varepsilon_{2i}$ alone is not intended for the standard regression. Regressing Y_i on X_{2i}, D_i does not produce a consistent estimator of β_2, α, because of endogeneity: $p(\varepsilon_{2i} \mid D_i) \neq p(\varepsilon_{2i})$.

Conditional on $D_i = 1$ (among the college graduates), the pair (Y_i, Z_i) follows a truncated bivariate normal distribution. By Proposition 5.4, Y_i is subject to incidental truncation, and its density is given by

$$p(Y_i \mid D_i = 1) = \phi\left(Y_i; X_{2i}\beta_2 + \alpha, \sigma^2\right) \frac{\Phi(\bar{\mu}_i/\bar{\sigma})}{\Phi(X_{1i}\beta_1)},$$

where $\bar{\mu}_i$ and $\bar{\sigma}$ are the conditional mean and variance of the bivariate normal distribution:

$$\bar{\mu}_i = X_{1i}\beta_1 + \rho\sigma^{-1}(Y_i - X_{2i}\beta_2 - D_i\alpha),$$

$$\bar{\sigma}^2 = 1 - \rho^2.$$

The correlation ρ is a key parameter for incidental truncation. If the correlation were zero, $p(Y_i \mid D_i = 1)$ would be $\phi\left(Y_i; X_{2i}\beta_2 + \alpha, \sigma^2\right)$, as in a standard linear regression. Proposition 5.5 indicates that the conditional mean is

$$E(Y_i \mid D_i = 1) = X_{2i}\beta_2 + \alpha + \rho\sigma\frac{\phi(X_{1i}\beta_1)}{\Phi(X_{1i}\beta_1)}.$$

Similarly, conditional on $D_i = 0$, the distribution of Y_i is incidentally truncated due to $Z_i \leq 0$, with the density

$$p(Y_i | D_i = 0) = \phi(Y_i; X_{2i}\beta_2, \sigma^2) \frac{\Phi(-\bar{\mu}_i/\bar{\sigma})}{\Phi(-X_{1i}\beta_1)},$$

and the conditional mean is given by

$$E(Y_i | D_i = 0) = X_{2i}\beta_2 - \rho\sigma \frac{\phi(-X_{1i}\beta_1)}{\Phi(-X_{1i}\beta_1)}.$$

Define the treatment effect of college education as the earning of a college graduate minus the counterfactual earning if he did not attend college. In this regression analysis, the return to schooling is captured by the coefficient α. However, endogeneity of the treatment variable prevents consistent estimation of α by a single-equation regression of Y_i on D_i.Consider a simplified regression in which education is the only regressor (in the absence of $X_{2i}\beta_2$):

$$E(Y_i | D_i = 1) - E(Y_i | D_i = 0) = \alpha + \rho\sigma \frac{\phi(X_{1i}\beta_1)}{\Phi(X_{1i}\beta_1)\Phi(-X_{1i}\beta_1)}.$$

However, it is not equal to α if $\rho \neq 0$. The discrepancy comes from the ability premium. College graduates are talented people. The high earning observed in that group is attributed to both education and endowment. Therefore, the observed earning difference among the two groups over-estimates the treatment effect of college education.

The conditional means $E(Y_i | D_i = 1)$ and $E(Y_i | D_i = 0)$ can be represented in regression forms:

1. In the subsample $\{D_i = 1\}$: $Y_i = X_{2i}\beta_2 + \alpha + \rho\sigma \frac{\phi(X_{1i}\beta_1)}{\Phi(X_{1i}\beta_1)} + noise.$
2. In the subsample $\{D_i = 0\}$: $Y_i = X_{2i}\beta_2 - \rho\sigma \frac{\phi(-X_{1i}\beta_1)}{\Phi(-X_{1i}\beta_1)} + noise.$

The two regressions suggest that the treatment effect can be consistently estimated by the Heckman two-step approach, where the auxiliary variables $\frac{\phi(X_{1i}\beta_1)}{\Phi(X_{1i}\beta_1)}$ and $\frac{\phi(-X_{1i}\beta_1)}{\Phi(-X_{1i}\beta_1)}$ constructed from the Probit model serve as additional regressors to correct the bias induced by endogenous variables.

From the perspective of instrumental regression, the treatment effect is identified because X_{1i} functions as an instrumental variable, which is correlated with D_i but uncorrelated with ε_{2i}. A linear model could be fitted by two-stage least squares, but our model is nonlinear, as the binary variable D_i is specified as a nonlinear function of X_{1i} under the Probit model. The distributional assumption (incidentally truncated normal distribution) is utilized to correct the endogeneity bias.

Bayesian inference of the endogenous treatment model can be implemented by data augmentation and the Gibbs sampler that cycles through the posterior conditional distributions of latent variables Z and parameters $\beta_1, \beta_2, \alpha, \rho, \sigma$.

Consider the posterior conditional distribution $p(Z | Y, D, \beta_1, \beta_2, \alpha, \rho, \sigma)$. On the one hand, the correlation between Y_i and Z_i implies an updated distribution of Z_i with the conditional mean $\bar{\mu}_i$ and variance $\bar{\sigma}^2$. On the other hand, the observed D_i indicates the sign of Z_i. It follows that the posterior conditional distribution of Z_i is a truncated normal distribution:

$$\begin{cases} N(\bar{\mu}_i, \bar{\sigma}^2) I_{(0,\infty)} & if D_i = 1 \\ N(\bar{\mu}_i, \bar{\sigma}^2) I_{(-\infty,0)} & if D_i = 0 \end{cases}.$$

The posterior conditional distribution $p(\beta_1, \beta_2, \alpha, \rho, \sigma | Y, D, Z)$ depends on the prior specification of parameters. The regressors in two equations are asymmetric, as D_i must be in the second, but not the first, equation. Asymmetry disables the conjugate NIW prior. However, a conditionally conjugate normal prior can be specified for β_1, β_2, α, and the posterior conditional distribution preserves normality. Specifically,

$$\beta = \begin{pmatrix} \beta_1 \\ \beta_2 \\ \alpha \end{pmatrix} \sim N\left(\mu_\beta, \Lambda_\beta^{-1}\right),$$

and denote the covariance matrix by $\Sigma = \begin{pmatrix} 1 & \rho\sigma \\ \rho\sigma & \sigma^2 \end{pmatrix}$. The two equations consist of a seemingly unrelated regression

$$\tilde{Z}_i = X_i\beta + \varepsilon_i,$$

where $\tilde{Z}_i = \begin{pmatrix} Z_i \\ Y_i \end{pmatrix}$ and $X_i = \begin{pmatrix} X_{1i} & 0 & 0 \\ 0 & X_{2i} & D_i \end{pmatrix}$. The posterior conditional distribution of β follows Bayesian linear regression results, namely $N\left(\bar{\mu}_\beta, \bar{\Lambda}_\beta^{-1}\right)$, where

$$\bar{\mu}_\beta = \bar{\Lambda}_\beta^{-1}\left(\Lambda_\beta\mu_\beta + \sum_{i=1}^{n} X_i'\Sigma^{-1}\tilde{Z}_i\right),$$

$$\bar{\Lambda}_\beta = \Lambda_\beta + \sum_{i=1}^{n} X_i'\Sigma^{-1}X_i.$$

Regarding the posterior conditional distribution $p(\Sigma\,|Y, D, Z, \beta)$, several sampling methods are available. First, bypass the constraint on the covariance matrix and post-process draws by focusing on identifiable regions. Under the inverse Wishart prior $IW(\Omega, v)$, the posterior conditional distribution of Σ preserves inverse Wishart: $IW(\bar{\Omega}, \bar{v})$, where

$$\bar{\Omega} = \Omega + \sum_{i=1}^{n}\left(\tilde{Z}_i - X_i\beta\right)\left(\tilde{Z}_i - X_i\beta\right)',$$

$$\bar{v} = v + n.$$

Second, use the constrained Wishart sampler proposed by Nobile (2000). Third, replace correlated variables by linear combinations of uncorrelated variables, using the Cholesky factor of the covariance matrix: $\Sigma = LL'$,where $L = \begin{pmatrix} 1 & 0 \\ \rho\sigma & \sqrt{1-\rho^2}\sigma \end{pmatrix}$. It follows that

$$Z_i = X_{1i}\beta_1 + v_{1i},$$

$$Y_i = X_{2i}\beta_2 + D_i\alpha + \rho\sigma v_{1i} + \sqrt{1-\rho^2}\sigma v_{2i}.$$

Conditioning on β_1 allows $v_{1i} = Z_i - X_{1i}\beta_1$ serving as a predictor in the second regression, in which the response variable is $Y_i - X_{2i}\beta_2 - D_i\alpha$, and $\rho\sigma$ is the coefficient of the regressor v_{1i}. Under a normal prior for $\rho\sigma$ and an inverse gamma prior for $\left(1-\rho^2\right)\sigma^2$, Bayesian linear regression provides samples for $\rho\sigma$ and $\left(1-\rho^2\right)\sigma^2$, from which the original parameters ρ, σ can be recovered.

5.5 Count Data and Poisson Regression

In econometrics and statistical modeling, response variables that are nonnegative integers often arise in various contexts. For datasets with a limited range (e.g., binary outcomes such as 0 and 1), Probit or Logit models are often appropriate. Conversely, when the data exhibit a wide range, it is reasonable to approximate them as continuous with a lower bound at zero, leading to the application of truncated or censored regression models (Tobin, 1958). Count data models are particularly suitable for nonnegative integer data characterized by three main features: 1) predominantly small values, 2) a substantial number of zero values, and 3) a minority of large values without a discernible upper limit. For instance, consider a health economist analyzing the annual number of doctor visits. Typically, most individuals visit doctors infrequently, many do not visit at all, while a few patients require frequent medical attention. A pivotal challenge in count data analysis is effectively managing the significant presence of zeros (Cameron and Trivedi, 2013; Winkelmann, 2008).

5.5.1 Poisson Regression

The Poisson distribution is a foundational tool for modeling rare events, where the probability of occurrence π is low, and the number of trials T is large. Assuming the mean $\lambda = \pi T$ is known, the number of occurrences Y_i follows a binomial distribution:

$$P(Y_i = y) = \frac{T!}{y!\,(T-y)!}\left(\frac{\lambda}{T}\right)^y\left(1-\frac{\lambda}{T}\right)^{T-y}, y = 0, 1, \ldots, T.$$

As the number of trials increases and the probability decreases ($T \to \infty$, $\pi \to 0$ and $\lambda = \pi T$ is fixed), the Poisson distribution emerges as the limiting case of the binomial distribution:

$$\lim_{T\to\infty} P(Y_i = y) = \frac{\lambda^y e^{-\lambda}}{y!}.$$

In Poisson regression, each observation Y_i follows a Poisson distribution with an individual-specific mean λ_i, determined by predictors X_i. To ensure a positive Poisson mean, a common specification is $\lambda_i = e^{X_i\beta}$, where the coefficient β indicates semi-elasticity: the percentage change in the conditional mean due to a unit change in the predictor. The log-likelihood of the Poisson regression is expressed as

$$\ln p(Y\,|\beta) = \sum_{i=1}^{n}\left[-e^{X_i\beta} + Y_i X_i\beta - \ln(Y_i!)\right].$$

Bayesian inference for Poisson regression parallels that of logistic regression in two key respects. First, there is no conjugate structure in the likelihood function, rendering the posterior distribution of β analytically intractable, regardless of prior specification. Second, the log-likelihood function is globally concave with respect to β, facilitating the MH sampling with a tailored proposal distribution. A common approach involves numerical optimization (refer to Section 5.8 for implementation details), evaluating the Hessian matrix at the posterior mode, and utilizing the negative inverse Hessian as the proposal covariance matrix for the MH sampler.

Under a normal prior $p(\beta) \propto e^{-\frac{1}{2}(\beta-\mu)'V^{-1}(\beta-\mu)}$, the second-order Taylor approximation of the log posterior is

$$\ln p(\beta\,|Y) \approx \ln p(\beta^*\,|Y) + G(\beta^*)(\beta-\beta^*) + \frac{1}{2}(\beta-\beta^*)'H(\beta^*)(\beta-\beta^*),$$

where the gradient and Hessian are

$$G(\beta) = -V^{-1}(\beta-\mu) + \sum_{i=1}^{n}\left(Y_i - e^{X_i\beta}\right)X_i',$$

$$H(\beta) = -V^{-1} - \sum_{i=1}^{n} e^{X_i\beta} X_i' X_i.$$

Since the Hessian is negative definite for all β, the log posterior is globally concave, allowing a numerical optimization program to efficiently identify the posterior mode β^* such that $G(\beta^*) = 0$. Consequently, $p(\beta | Y)$ is approximately proportional to $e^{\frac{1}{2}(\beta-\beta^*)' H(\beta^*)(\beta-\beta^*)}$, providing a normal approximation with mean β^* and covariance matrix $-H^{-1}(\beta^*)$. This can be employed as the proposal distribution for the MH sampler in Bayesian estimation of Poisson regression. In practice, the proposal covariance matrix may be rescaled as $-cH^{-1}(\beta^*)$, where c is a tuning parameter influencing the acceptance rate of proposal draws. It is advisable to maintain the MH acceptance rate around 25% (Gelman et al., 1997).

5.5.2 Negative Binomial Regression

A limitation of the Poisson distribution is the inherent equality of the mean and variance (i.e., λ_i). It is advantageous to specify an over-dispersion model that maintains the mean while increasing the variance. A mean-preserving spread can be achieved by randomizing the conditional mean of the Poisson distribution:

$$Y_i \left| \tilde{\lambda}_i \sim Poisson\left(\tilde{\lambda}_i\right), \right.$$

$$\tilde{\lambda}_i = e^{X_i\beta + \varepsilon_i},$$

where the random variable ε_i satisfies $E(e^{\varepsilon_i}) = 1$. The law of iterated expectations indicates that

$$E(Y_i | \beta) = E[E(Y_i | \beta, \varepsilon_i)] = e^{X_i\beta},$$

$$Var(Y_i | \beta) = E[Var(Y_i | \beta, \varepsilon_i)] + Var[E(Y_i | \beta, \varepsilon_i)] = e^{X_i\beta} + e^{2X_i\beta} Var(e^{\varepsilon_i}).$$

A robust statistical model balances flexibility and tractability. An over-dispersion model offers greater flexibility than standard Poisson regression, as the latter is a special case of the former under $\varepsilon_i = 0$. Any distributional specification of ε_i results in over-dispersion, as $Var(e^{\varepsilon_i}) \geq 0$. However, the likelihood function generally lacks analytical tractability due to the complexity of the integral

$$p(Y_i | \beta) = \int p(\varepsilon_i) p(Y_i | \beta, \varepsilon_i) d\varepsilon_i.$$

The integral assumes an explicit form in the Poisson-gamma conjugate scenario (Cameron and Trivedi, 1986): if e^{ε_i} has a gamma prior, then the posterior distribution remains gamma, and the likelihood marginalized over the random noise is negative binomial, as demonstrated in Proposition 5.6. Hilbe (2011) provides in-depth analysis on the negative binomial regression for over-dispersed count data.

Proposition 5.6 (Poisson-gamma conjugacy) *Consider the Poisson-gamma hierarchical specification:*

$$Y_i \left| \tilde{\lambda}_i \sim Poisson\left(\tilde{\lambda}_i\right), \right.$$

$$\tilde{\lambda}_i = e^{X_i\beta + \varepsilon_i},$$

where $u_i = e^{\varepsilon_i}$ follows a gamma distribution $Gamma(a, a)$ such that

$$p(u_i | a) \propto u_i^{a-1} e^{-au_i}.$$

1. *The posterior distribution of u_i is Gamma $\left(a+Y_i, a+e^{X_i\beta}\right)$:*

$$p\left(u_i\,|Y_i, a, \beta\right) \propto u_i^{a+Y_i-1} e^{-\left(a+e^{X_i\beta}\right)u_i}.$$

2. *The likelihood marginalized over u_i is negative binomial:*

$$p\left(Y_i\,|a, \beta\right) = \frac{\Gamma\left(a+Y_i\right)}{\Gamma\left(a\right)\Gamma\left(Y_i+1\right)} r_i^{Y_i}\left(1-r_i\right)^a,$$

$$r_i = \frac{1}{1+ae^{-X_i\beta}}.$$

5.5.3 Adaptive Rejection Sampling

In the context of Bayesian estimation of the negative binomial regression, one viable approach is the MH sampling, targeting the posterior distribution

$$p\left(a, \beta\,|Y\right) \propto p\left(a\right) p\left(\beta\right) \prod_{i=1}^{n} p\left(Y_i\,|a, \beta\right),$$

where $p\left(Y_i\,|a, \beta\right)$ represents the negative binomial probability as elucidated in Proposition 5.6. However, formulating an effective proposal distribution for the MH sampler poses challenges, given that $p\left(Y_i\,|a, \beta\right)$ is a complex function of both a and β.

An alternative MCMC strategy involves data augmentation, resulting in an extended target distribution $p\left(u, a, \beta\,|Y\right)$. This approach employs a single-move Gibbs sampler that iteratively cycles through the full posterior conditional distributions, updating one variable at a time. First, $p\left(u_i\,|Y_i, a, \beta\right)$ follows the gamma distribution as specified in Proposition 5.6. Second, $p\left(\beta\,|Y, u, a\right)$ corresponds to the posterior of the Poisson regression. Although this does not conform to a recognizable distribution, the log posterior density is globally concave, implying that the log posterior is concave with respect to each element in β, conditional on the remaining elements. Third, $p\left(a\,|Y, u, \beta\right) \propto p\left(a\right) \prod_{i=1}^{n} u_i^{a-1} e^{-au_i}$ does not align with a common distribution, yet the log posterior density remains concave, provided the log prior of a is concave. Here, the key concept is log concavity.

Adaptive rejection sampling, as introduced by Gilks and Wild (1992), offers an elegant solution for automatically generating samples from a univariate distribution with a concave log density function. The essence of rejection sampling lies in generating proposals from a standard distribution and evaluating the target-to-proposal density ratio, normalized between 0 and 1. A high ratio indicates that the proposal closely resembles a draw from the target distribution, leading to acceptance with the probability determined by the normalized ratio. Proposition 5.7 provides a theoretical foundation for rejection sampling, by studying the acceptance rate and the distribution of accepted draws. A notable challenge in implementing standard rejection sampling is determining the upper bound of the target-to-proposal density ratio. Adaptive rejection sampling addresses this issue by constructing a proposal from the upper envelope of the log-concave target density, ensuring the upper bound inherently equals one. The piecewise linear envelope is automatically generated and refined with each rejected sample, thereby enhancing sampling efficiency.

Proposition 5.7 (Rejection sampling) *Let $f\left(\cdot\right)$ and $s\left(\cdot\right)$ denote unnormalized target and proposal densities, with an upper bound of the density ratio: $\frac{f(x)}{s(x)} \leq M$, $\forall x \in \mathbb{R}$. A proposal θ from $s\left(\cdot\right)$ is accepted by the probability $\frac{f(\theta)}{s(\theta)M}$. Let normalizing constants be $\int f\left(x\right) dx = c_f$ and $\int s\left(x\right) d = c_s$. Let U be a uniform random number independent to θ. Rejection sampling has the following properties:*

1. *acceptance probability:*

$$P\left[U \leq \frac{f(\theta)}{s(\theta)M}\right] = \frac{c_f}{c_s M};$$

2. *CDF of accepted draws:*

$$P\left[\theta \leq c \middle| U \leq \frac{f(\theta)}{s(\theta)M}\right] = \int_{-\infty}^{c} \frac{1}{c_f} f(x)\,dx, \forall c \in \mathbb{R},$$

indicating that accepted proposals are samples from $f(\cdot)$.

Assume the target density $f(x)$ is log-concave. Adaptive rejection sampling commences with three initial points $x_1 < x_2 < x_3$ such that $\frac{d\ln f(x_1)}{dx_1} > 0$ and $\frac{d\ln f(x_3)}{dx_3} < 0$. The tangent lines at these points form an upper envelope of $\ln f(x)$, serving as an unnormalized proposal density function:

$$\ln s(x) = \begin{cases} a_1 x + b_1 & x \leq k_1 \\ a_2 x + b_2 & k_1 < x \leq k_2 \\ a_3 x + b_3 & x > k_2 \end{cases},$$

where $a_i = \frac{d\ln f(x_i)}{dx_i}$, $b_i = \ln f(x_i) - a_i x_i$ and $k_i = -\frac{b_{i+1}-b_i}{a_{i+1}-a_i}$, $i = 1, 2, 3$. Since $s(x)$ is a piecewise exponential density function, it is feasible to compute the CDF of the proposal distribution and generate proposal draws using the inverse CDF method.

The envelope provides the upper bound: $\frac{f(x)}{s(x)} \leq 1$, $\forall x \in \mathbb{R}$. A proposal θ is accepted with probability $\frac{f(\theta)}{s(\theta)}$. According to Proposition 5.7, accepted proposals are valid samples from the target distribution. Moreover, a rejected proposal still contributes to refinement of the upper envelope, thereby increasing the acceptance probability of subsequent proposals.

Adaptive rejection sampling presents both advantages and limitations. In contrast with the MH sampler, which may perform sub-optimally without meticulous tuning, adaptive rejection sampling is fully automatic and free from tuning parameters. While the MH sampler yields asymptotic draws from the target distribution, adaptive rejection sampling is exact and does not require burn-in draws. However, adaptive rejection sampling necessitates a log-concave density function to define a proper envelope. Regarding runtime performance, the adaptive rejection sampler evaluates the target function at multiple points before generating a draw, whereas the standard MH sampler requires only one functional call per draw. Additionally, the former necessitates gradient information to construct tangent lines, whereas the latter does not. Consequently, the sampling speed of the adaptive rejection algorithm may not match that of the MH sampler.

5.6 Duration Data and Survival Analysis

Duration data, a fundamental concept in statistical analysis, pertain to nonnegative random variables that characterize the lifetime of an object or process. Duration data are widely studied across various disciplines, including economics, technometrics, biometrics and medical sciences, where they are often referred to under different terminologies such as event history analysis, life table analysis, reliability and failure time analysis, and survival analysis. Ibrahim et al. (2001), Kalbfleisch and Prentice (2002) and Collett (2015) provide comprehensive overview of Bayesian and classical methods in survival analysis.

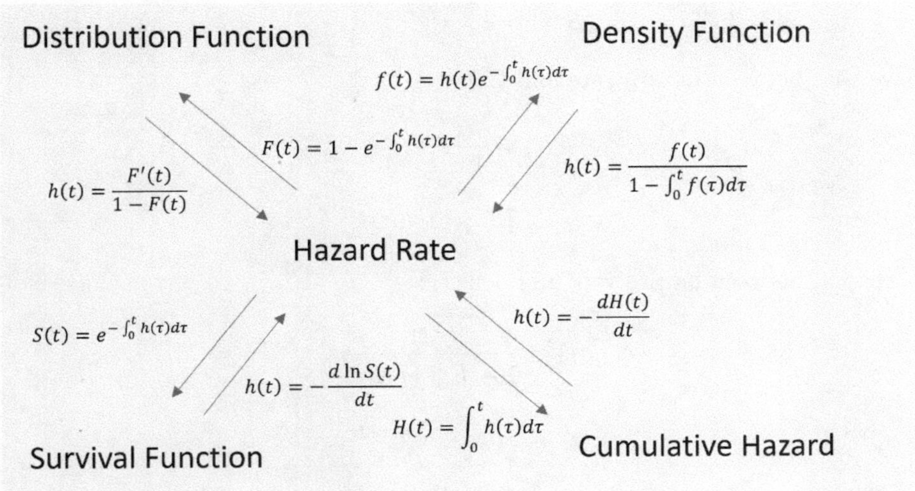

FIGURE 5.2
Hazard rate and probability distribution.

5.6.1 Hazard Rate

The primary variable of interest in duration data analysis is the stochastic lifetime Y_i, modeled as a nonnegative and continuous random variable with a density function $f(t)$ and a (lower) CDF $F(t) = \int_0^t f(\tau)\, d\tau$. The survival function is defined as the complementary (i.e., upper) CDF:

$$S(t) = \int_t^{\infty} f(\tau)\, d\tau,$$

representing the probability that the lifetime exceeds a particular time t.

A salient feature of duration data analysis is hazard rate modeling, which examines the effects of exogenous factors on the risk of instantaneous failure or mortality. The hazard rate, defined as

$$h(t) = \frac{f(t)}{S(t)},$$

quantifies the instantaneous mortality risk for individuals who have survived up to time t. This interpretation of hazard rate is justified by the following equation:

$$\lim_{\Delta t \to 0} \frac{P(t \le Y_i \le t + \Delta t \,|Y_i \ge t)}{\Delta t} = \lim_{\Delta t \to 0} \frac{[F(t + \Delta t) - F(t)]/\Delta t}{S(t)} = \frac{f(t)}{S(t)},$$

where $P(t \le Y_i \le t + \Delta t \,|Y_i \ge t)$ represents the probability of event occurrence immediately after a specific time t, conditional on the lifetime at least t.

For continuous random variables, the hazard rate, survival function, density, and CDF are equivalent representations of a probability distribution. Specifying one allows for the derivation of the others, as depicted in Figure 5.2. The following conversion list will be used extensively throughout our analysis.

1. Conversion between hazard rate and survival function:

$$h(t) = -\frac{d \ln S(t)}{dt},$$

$$S(t) = e^{-\int_0^t h(\tau)d\tau}.$$

2. Conversion between hazard rate and CDF:

$$h(t) = \frac{F'(t)}{1 - F(t)},$$

$$F(t) = 1 - e^{-\int_0^t h(\tau)d\tau}.$$

3. Conversion between hazard rate and density:

$$h(t) = \frac{f(t)}{1 - \int_0^t f(\tau)\, d\tau},$$

$$f(t) = h(t)\, e^{-\int_0^t h(\tau)d\tau}.$$

In empirical studies, an entry point of survival analysis is often the specification of a hazard rate function rather than a density function. Hazard rate functions can exhibit various shapes, reflecting different real-world phenomena. Examples include:

1. Decreasing hazard rates: common in job search durations, where the likelihood of finding a job decreases over time.
2. Increasing hazard rates: observed in the lifespan of durable goods, which are more likely to fail as they age.
3. Hump-shaped hazard rates: typical in the growth of startups, where the risk initially increases and then decreases.
4. Bathtub-shaped hazard rates: seen in population life tables, indicating high initial infant mortality, a long period of low risk during adulthood, and increased risk in old age.

A nonnegative function that integrates to infinity over the interval $(0, \infty)$ constitutes a well-defined hazard function. Below are some popular specifications.

1. Hazard rate as a constant over time:

$$h(t) = b,$$

corresponding to the exponential distribution with the density

$$f(t) = be^{-bt}.$$

2. Log hazard rate as a linear function of time:

$$\ln h(t) = b + at,$$

corresponding to the Gompertz distribution with the density

$$f(t) = \exp\left[b + at - a^{-1}e^{b}\left(e^{at} - 1\right)\right].$$

3. Log hazard rate as a linear function of log time:

$$\ln h(t) = b + a \ln t,$$

corresponding to the Weibull distribution with the density

$$f(t) = e^b t^a \exp\left[-e^b (a+1)^{-1} t^{a+1}\right].$$

A more familiar form of the Weibull density can be obtained by parametrization: $k = a+1$ and $\lambda^{-k} = e^b (a+1)^{-1}$. The reparametrized density function is

$$f(t) = \frac{k}{\lambda}\left(\frac{t}{\lambda}\right)^{k-1} \exp\left[-\left(\frac{t}{\lambda}\right)^k\right].$$

4. Hazard rate as a step function:

$$h(t) = \sum_{j=1}^{J} c_j I_{(a_j, a_{j+1})}(t),$$

where the steps $c_j > 0$. Thresholds have boundaries $a_1 = 0$ and $a_{J+1} = \infty$. The corresponding density function is

$$f(t) = h(t)\, e^{-\int_0^t h(\tau)d\tau},$$

$$\int_0^t h(\tau)\, d\tau = \sum_{j=1}^{J} c_j L_j(t),$$

$$L_j(t) = \begin{cases} a_{j+1} - a_j & t \geq a_{j+1} \\ t - a_j & a_j \leq t < a_{j+1} \\ 0 & t < a_j \end{cases}.$$

Additionally, Exercise 5.9 explores non-monotonic hazard rates using the log-logistic distribution, demonstrating the versatility of hazard rate modeling.

5.6.2 Likelihood and Posterior

When the functional form of the hazard rate is fully specified, the likelihood function is determined by the hazard rate. Bayesian model specification is completed by incorporating a prior on the model parameters, allowing the posterior density to be expressed as the product of the prior and the likelihood up to a normalizing constant.

A notable characteristic of real-world duration data is censoring, which occurs when events are not fully observed due to time or resource constraints. Let $\delta_i = 0$ be a censoring indicator, signifying that the latent lifetime is at least Y_i, while $\delta_i = 1$ indicates that the actual lifetime equals Y_i. Censoring is assumed to be independent and non-informative regarding model parameters. Akin to the Tobit model, the likelihood is determined by the survival function for censored observations and the density function for uncensored observations.

Let θ represent the model parameters, and $f(t;\theta)$, $S(t;\theta)$ denote the density and survival functions conditional on θ. The likelihood function is structured as

$$p(Y|\theta) = \prod_{i=1}^{n} \left[f(Y_i;\theta)\right]^{\delta_i} \left[S(Y_i;\theta)\right]^{1-\delta_i}.$$

Converting the density and survival functions to the hazard rate yields an alternative representation of the likelihood function:

$$p(Y|\theta) = \prod_{i=1}^{n} [h(Y_i;\theta)]^{\delta_i} e^{-\int_0^{Y_i} h(t;\theta)dt}.$$

For instance, the exponential distribution features a constant hazard rate $h(t;\theta) = \theta$, resulting in the likelihood function:

$$p(Y|\theta) = \prod_{i=1}^{n} \theta^{\delta_i} e^{-\theta Y_i} = \theta^{\sum_{i=1}^{n}\delta_i} e^{-\theta \sum_{i=1}^{n} Y_i}.$$

The exponential likelihood has the gamma distribution as a natural conjugate prior: $p(\theta) \propto \theta^{a-1}e^{-b\theta}$, where a and b are hyperparameters. The posterior distribution retains the gamma form: $p(\theta|Y) \propto \theta^{\bar{a}-1}e^{-\bar{b}\theta}$, where

$$\bar{a} = a + \sum_{i=1}^{n} \delta_i,$$

$$\bar{b} = b + \sum_{i=1}^{n} Y_i.$$

The posterior predictive distribution is crucial in survival analysis, providing insights into future observations by integrating out parameters. The posterior predictive density $f_Y(t)$ is obtained by marginalizing the conditional density $f(t;\theta)$ over the posterior distribution of parameters:

$$f_Y(t) = \int f(t;\theta)\, p(\theta|Y)\, d\theta.$$

Similarly, the posterior predictive survival function $S_Y(t)$ is obtained by marginalizing the conditional survival function $S(t;\theta)$ over the posterior distribution of parameters (as explored in Exercise 5.10).

$$S_Y(t) = \int S(t;\theta)\, p(\theta|Y)\, d\theta.$$

In the exponential-gamma conjugate model, the posterior predictive survival function is analytically tractable:

$$S_Y(t) = \int e^{-\theta t} \frac{\bar{b}^{\bar{a}}}{\Gamma(\bar{a})} \theta^{\bar{a}-1} e^{-\bar{b}\theta} d\theta = \left(\frac{\bar{b}}{\bar{b}+t}\right)^{\bar{a}}.$$

Another example of a tractable posterior is the log-normal model, wherein $\ln Y_i$ follows a normal distribution, denoted as $N(\mu, \sigma^2)$. This specification aligns with linear regression, in which $\ln Y_i$ serves as the response variable:

$$\ln Y_i = \mu + \sigma\varepsilon_i,$$

where ε_i follows the standard normal distribution. Without censored data, analytic results are available under the NIG prior, as discussed in Chapter 1. In the presence of censoring, the model resembles a Tobit model, with tractable posterior conditional distributions through data augmentation, allowing posterior inference via the Gibbs sampler.

Analytical tractability, however, is more the exception than the norm. Generally, the posterior distribution lacks a recognizable form. As the posterior density $p(\theta|Y)$ can be evaluated pointwise up to a normalizing constant, the MH sampler may be employed for posterior simulation.

5.6.3 Proportional Hazards Model

Duration data analysis often aims to evaluate the effect of exogenous predictors X_i on the lifetime of an entity. The proportional hazards (PH) model is a widely used approach, incorporating predictors into the hazard rate function by specifying

$$h_i(t) = e^{X_i\beta} h_0(t),$$

where $h_0(t)$ is the baseline hazard rate common to all observations $(i = 1, \ldots, n)$. The hazard rate ratio between two individuals remains time invariant:

$$\frac{h_i(t)}{h_j(t)} = e^{(X_i - X_j)\beta},$$

hence the name proportional hazards.

The PH model is compatible with exponential, Gompertz, and Weibull distributions under fully specified baseline hazard rate functions:

1. Exponential: $h_0(t) = 1$ and $\ln h_i(t) = X_i\beta$.
2. Gompertz: $h_0(t) = e^{at}$ and $\ln h_i(t) = X_i\beta + at$.
3. Weibull: $h_0(t) = t^a$ and $\ln h_i(t) = X_i\beta + a\ln t$.

Consider the marginal effect of an exogenous factor on the hazard rate. In the exponential and Gompertz distributions, the coefficient β represents semi-elasticity, indicating the marginal percentage change in the hazard rate following a unit change in the predictor. For the Weibull distribution, the coefficient β represents elasticity, reflecting the percentage change in the hazard rate in response to a percentage change in the predictor.

Dellaportas and Smith (1993) show that many PH models have log-concave posterior density, allowing for the Gibbs sampler via adaptive rejection sampling (Gilks and Wild, 1992).

The popularity of the PH model is largely due to the partial likelihood method introduced by Cox (1972), allowing consistent estimation of parameters without specifying a baseline hazard function. In Bayesian analysis, a flexible model for the baseline hazard rate is specified, often as a step function:

$$h_0(t) = \sum_{j=1}^{J} c_j I_{(a_j, a_{j+1})}(t),$$

where the steps $c_1, \ldots, c_J$ are unknown parameters, while the thresholds $a_1, \ldots, a_{J+1}$ are user-specified, with the boundaries $a_1 = 0$ and $a_{J+1} = \infty$. It is important to note that thresholds cannot be identified by data if there are no observations in the final step or in two consecutive steps (See Exercise 5.11). A recommended strategy is to begin with a parsimonious model featuring fewer steps and refine the model incrementally.

Let Y_i be the observed lifetime with the censoring indicator δ_i. The likelihood can be expressed in terms of the hazard rate:

$$p(Y|\theta) = \prod_{i=1}^{n} \left[e^{X_i\beta} h_0(Y_i)\right]^{\delta_i} e^{-\int_0^{Y_i} e^{X_i\beta} h_0(t)dt},$$

where the integral of the step function equals the sum of rectangular areas:

$$\int_0^{Y_i} h_0(t)\,dt = \sum_{j=1}^{J} c_j L_{ij},$$

$$L_{ij} = \begin{cases} a_{j+1} - a_j & Y_i \geq a_{j+1} \\ Y_i - a_j & a_j \leq Y_i < a_{j+1} \\ 0 & Y_i < a_j \end{cases}.$$

A normal prior for β and and a gamma prior for c_j can be specified. Although the posterior density does not conform to a standard distribution, it can be evaluated point-wise up to a normalizing constant, making the MH sampler a suitable choice for posterior sampling. For instance, Li (2007) applies Bayesian proportional hazards to timing of high school dropout decisions.

5.6.4 Accelerated Failure Time Model

Compared to the PH model, the accelerated failure time (AFT) model offers an alternative framework for analyzing the effects of predictors on the lifetime of an entity. In the AFT model, predictors determine the lifetime scale parameter that acts multiplicatively on time, adjusting the speed of progression to an event.

Let X_i and Y_i denote the predictors and the stochastic lifetime in the form

$$Y_i = e^{-X_i\beta}\tilde{Y}_i,$$

where the random variable $\tilde{Y}_i$ follows a baseline distribution for all $i = 1, \ldots, n$ independently, with hazard rate $\tilde{h}(t)$, density $\tilde{f}(t)$, CDF $\tilde{F}(t)$, and survival function $\tilde{S}(t)$. Since Y_i is merely a rescaled version of $\tilde{Y}_i$, the hazard rate, density, CDF and survival functions are rescaled as follows:

$$\begin{aligned} h_i(t) &= e^{X_i\beta}\tilde{h}\left(e^{X_i\beta}t\right), \\ f_i(t) &= e^{X_i\beta}\tilde{f}\left(e^{X_i\beta}t\right), \\ F_i(t) &= \tilde{F}\left(e^{X_i\beta}t\right), \\ S_i(t) &= \tilde{S}\left(e^{X_i\beta}t\right). \end{aligned}$$

The AFT model is compatible with log-normal, log-logistic, gamma, and Weibull distributions:

1. Log-normal: $\ln \tilde{Y}_i \sim N\left(0, \sigma^2\right)$ and $\ln Y_i \sim N\left(-X_i\beta, \sigma^2\right)$.
2. Log-logistic: $\ln \tilde{Y}_i \sim logistic\left(0, \sigma^2\right)$ and $\ln Y_i \sim logistic\left(-X_i\beta, \sigma^2\right)$.
3. Gamma: $\tilde{Y}_i \sim Gamma\,(a, b)$ and $Y_i \sim Gamma\left(a, be^{-X_i\beta}\right)$.
4. Weibull: $\tilde{Y}_i \sim Weibull\,(\lambda, k)$ and $Y_i \sim Weibull\left(\lambda e^{-X_i\beta}, k\right)$.

The AFT model can be represented as a regression, by taking logarithm on both sides of $Y_i = e^{-X_i\beta}\tilde{Y}_i$:

$$\ln Y_i = -X_i\beta + \varepsilon_i,$$

where $\varepsilon_i = \ln \tilde{Y}_i$.

If ε_i follows a normal distribution $N\left(0, \sigma^2\right)$, the AFT model reduces to a standard linear regression, with $\ln Y_i$ serving as the response variable. Analytical results from Bayesian linear regression, as discussed in Chapter 1, are applicable when observations are not censored. Otherwise, the Gibbs sampler for the Tobit model is appropriate.

If ε_i follows a logistic distribution, as a Kolmogorov scale mixture of normal distributions (Andrews and Mallows, 1974), the AFT model simplifies to linear regression with logistic errors. The Gibbs sampler with data augmentation, akin to the method by Holmes and Held (2006) for logistic regression, can be employed.

If ε_i follows an extreme value distribution, then Y_i follows a Weibull distribution. However, conjugate results are not available. For Weibull and other AFT distributional specifications, the MH sampler is applicable for posterior inference once the likelihood function is fully specified.

More discussions on Bayesian AFT models and applications can be found in Bedrick et al. (2000) and Zhang and Lawson (2011), among others.

5.6.5 Unobserved Heterogeneity

In a linear regression model $Y_i = X_i\beta + \varepsilon_i$, observed heterogeneity is captured by the predictor term $X_i\beta$, while unobserved heterogeneity is absorbed in the error term ε_i. In the context of duration data and survival analysis, the random effects caused by unobserved covariates are known as frailty (Duchateau and Janssen, 2008; Wienke, 2011). Unobserved heterogeneity may be confounded with the baseline hazard rate, necessitating careful interpretation about the shape of the hazard function. Even if the individual-specific hazard rate remains constant over time (albeit varying among individuals), the average or aggregated hazard rate can exhibit time-varying characteristics.

5.6.5.1 Finite Mixture

To illustrate, consider a PH model with a constant baseline hazard rate and an individual-specific latent variable (frailty):

$$h_i\left(t\,|u_i\right) = u_i,$$

where $h_i\left(t\,|u_i\right)$ refers to the conditional hazard rate function of the ith individual, given frailty u_i. The corresponding conditional density is

$$f_i\left(t\,|u_i\right) = u_i e^{-u_i t}.$$

The lifetime of an individual follows an exponential distribution. The expected lifetime varies from person to person, depending on realizations of u_i.

Frailty, being unobserved by the researcher, is treated as a random variable. In this example, we assume a discrete $u_i \in \{\theta_1, \dots, \theta_m\}$ with probabilities $\{w_1, \dots, w_m\}$. Marginalizing over this distribution leads to the average or aggregated density, as a finite mixture of exponential distributions:

$$f_i\left(t\right) = \sum_{j=1}^{m} w_j \theta_j e^{-\theta_j t}.$$

Converting densities into hazard rates yields the marginalized (average or aggregate) hazard rate function:

$$h_i\left(t\right) = \frac{\sum_{j=1}^{m} w_j \theta_j e^{-\theta_j t}}{\sum_{j=1}^{m} w_j e^{-\theta_j t}}.$$

Interestingly, the marginalized hazard rate curve becomes time-varying, despite the constant individual hazard rate curves. Consider a numeric exercise. Let $m = 2$, $w_1 = w_2 = 1/2$, $\theta_1 = 1/2$, $\theta_2 = 1/4$. We have

$$h_i\left(t\right) = \frac{1}{4}\frac{2 + \exp\left(t/4\right)}{1 + \exp\left(t/4\right)},$$

which is a decreasing function of time. At $t = 0$, the hazard rate ranges between $1/4$ and $1/2$. As $t \to \infty$, the hazard rate converges to $1/4$. The intuition is that frail individuals, with a hazard rate of 1/2, do not persist in the long run, leaving survivors with a hazard rate of 1/4.

A notable advantage of the finite mixture model is the tractability of the marginalized hazard rate for any specification of the conditional hazard rate function $h_i(t\,|u_i)$, achieved through a three-step procedure.

First, converting hazard rate to density, we obtain the conditional density

$$f_i(t\,|u_i) = h_i(t\,|u_i)\, e^{-\int_0^t h_i(\tau|u_i)d\tau}.$$

Second, marginalization over discrete distributions is always tractable, achieved by taking the weighted average of conditional densities. Assume that the latent variable u_i follows a categorical distribution on $\{\theta_1, \ldots, \theta_m\}$ with probabilities $\{w_1, \ldots, w_m\}$. The marginalized density takes the form

$$f_i(t) = \sum_{j=1}^{m} w_j f_i(t\,|\theta_j).$$

Third, converting density to hazard rate, we obtain the marginalized hazard rate

$$h_i(t) = \frac{f_i(t)}{1 - \int_0^t f_i(\tau)\, d\tau}.$$

5.6.5.2 Weibull-gamma Mixture

A mixed PH model assumes a multiplicative form of unobserved heterogeneity:

$$h_i(t\,|u_i) = e^{X_i\beta+\varepsilon_i} h_0(t) = u_i b_i h_0(t),$$

where $b_i = e^{X_i\beta}$ and $u_i = e^{\varepsilon_i}$. A tractable specification is Weibull-gamma mixture: $h_0(t) = t^a$ and $u_i \sim Gamma(\gamma, \gamma)$. The Weibull-gamma mixture is a hierarchical model such that

$$h_i(t\,|u_i) = u_i b_i t^a,$$

$$p(u_i) = \frac{\gamma^\gamma}{\Gamma(\gamma)} u_i^{\gamma-1} e^{-\gamma u_i}.$$

Regarding the latent variable (frailty) u_i, we have $E(u_i) = 1$ and $Var(u_i) = \gamma^{-1}$. As $\gamma \to \infty$, unobserved heterogeneity degenerates to unity, the model simplifies to the standard PH model.

Converting hazard rate to density, conditional on u_i, we have the conditional density function:

$$f_i(t\,|u_i) = u_i b_i t^a \exp\left[-u_i b_i (a+1)^{-1} t^{a+1}\right].$$

The marginalized (average or aggregate) density is obtained by integrating out unobserved heterogeneity. This integral has a closed form due to the gamma density kernel (as the posterior u_i follows the gamma distribution).

$$f_i(t) = \int f_i(t\,|u_i)\, p(u_i)\, du_i = \frac{b_i t^a}{\left(1 + c_i\gamma^{-1}t^{a+1}\right)^{\gamma+1}},$$

where $c_i = b_i(a+1)^{-1}$. Converting density to hazard rate and survival function, we have

$$S_i(t) = \left(1 + c_i\gamma^{-1}t^{a+1}\right)^{-\gamma},$$

$$h_i(t) = \frac{b_i t^a}{1 + c_i\gamma^{-1}t^{a+1}}.$$

Comparing the conditional and marginalized hazard rate curves reveals that the former is monotonic (increasing if $a > 0$; decreasing if $a < 0$), while the shape of the latter is more flexible.

The exponential-gamma mixture is a special case of the Weibull-gamma mixture, obtained by setting $a = 0$. In this scenario, the conditional hazard rate is time-invariant: $h_i(t|u_i) = u_i b_i$, yet the marginalized hazard rate is time-varying: $h_i(t) = b_i\left(1 + c_i\gamma^{-1}t\right)^{-1}$.

The Weibull PH model is the limiting case of Weibull-gamma mixture, obtained by setting $\gamma \to \infty$. In that case, the limiting hazard rate is

$$\lim_{\gamma\to\infty} h_i(t) = b_i t^a,$$

that is,

$$\lim_{\gamma\to\infty} \ln h_i(t) = X_i\beta + a\ln t.$$

It is possible to approximate the Weibull-gamma mixture by a finite mixture. By creating a grid $\theta_j, j = 1, \ldots, m$ covering the support of $Gamma(\gamma, \gamma)$, and evaluating the weight $w_j \propto \theta_j^{\gamma-1} e^{-\gamma\theta_j}$, the marginalized hazard rate can be approximated by

$$h_i(t) \approx \frac{\sum_{j=1}^m w_j f_i(t|\theta_j)}{\sum_{j=1}^m w_j S_i(t|\theta_j)},$$

where

$$f_i(t|\theta_j) = \theta_j b_i t^a \exp\left[-\theta_j b_i (a+1)^{-1} t^{a+1}\right],$$
$$S_i(t|\theta_j) = \exp\left[-\theta_j b_i (a+1)^{-1} t^{a+1}\right].$$

5.7 Beta Regression

In certain regression models, the response variable is confined to the unit interval: $Y_i \in (0, 1)$. This is typical for data such as rates, proportions, weights, and Gini indices. While a transformation like $\ln\frac{Y_i}{1-Y_i}$ extends the range to the entire real line, it renders the regression coefficients less interpretable. To address this, Ferrari and Cribari-Neto (2004) propose a regression model where the response variable follows a Beta distribution, with a heterogeneous mean and precision determined by predictors. A key advantage of Beta regression is the flexibility of the likelihood function, as Beta densities can assume various shapes by adjusting the parameters.

5.7.1 Reparametrized Beta Distribution

Conventionally, the Beta density function is parameterized by the number of successes a and failures b such that

$$f(y; a, b) = \frac{\Gamma(a+b)}{\Gamma(a)\Gamma(b)} y^{a-1}(1-y)^{b-1} I_{(0,1)}(y).$$

The mean and variance are $\frac{a}{a+b}$ and $\frac{ab}{(a+b)^2(a+b+1)}$, respectively. For improved interpretability, we reparametrize a, b as μ, ϕ, maintaining a one-to-one correspondence:

$$\mu = \frac{a}{a+b},$$

$$\phi = a + b,$$

where μ is the mean of the Beta distribution, and ϕ is a precision parameter inversely related to the variance. In terms of successes and failures, μ represents the success rate and ϕ corresponds to the total number of trails. Specifically, the number of successes is given by $a = \mu\phi$, and the number of failures is $b = (1 - \mu)\,\phi$.

The reparametrized Beta density is

$$f(y; \mu, \phi) = \frac{\Gamma(\phi)}{\Gamma(\mu\phi)\,\Gamma[(1-\mu)\,\phi]} y^{\mu\phi - 1} (1-y)^{(1-\mu)\phi - 1} I_{(0,1)}(y).$$

5.7.2 Model Specification and Parameter Estimation

In Beta regression, we assume each observation Y_i follows a Beta distribution with an individual-specific mean μ_i and precision ϕ_i, which are determined by regression-like effects $X_i\beta$ and $W_i\gamma$.

On the one hand, the mean must lie within the unit interval, hence we employ an (inverse) link function by the logistic CDF:

$$\mu_i = \Lambda(X_i\beta) = \frac{1}{1 + e^{-X_i\beta}}.$$

On the other hand, the precision must be positive, so we apply the exponential transformation:

$$\phi_i = e^{W_i\gamma}.$$

With the data-generating process specified, the log-likelihood of Beta regression is formulated accordingly:

$$\ln p(Y\,|\beta, \gamma) = \sum_{i=1}^{n} \ln p(Y_i\,|\beta, \gamma),$$

where

$$\begin{aligned}\ln p(Y_i\,|\beta, \gamma) &= \ln\Gamma(\phi_i) - \ln\Gamma(\mu_i\phi_i) - \ln\Gamma[(1-\mu_i)\,\phi_i] \\ &\quad + (\mu_i\phi_i - 1)\ln Y_i + [(1-\mu_i)\,\phi_i - 1]\ln(1 - Y_i).\end{aligned}$$

Due to the complexity of the likelihood function, the posterior distribution $p(\beta, \gamma\,|Y)$ is not analytically tractable, but the MH sampler provides a general solution to the problem of posterior simulation. Ferrari and Cribari-Neto (2004) provide the gradient and Hessian of the log-likelihood function, facilitating quasi-Newton numerical optimization for the posterior mode and the Hessian evaluated at the mode. These are key steps of specifying an appropriate proposal distribution for the MH sampler.

5.8 Software and Computational Considerations

Limited dependent variable models such as Logit, Probit, and Tobit are supported by most statistical software packages. These models are typically estimated using maximum likelihood methods through numerical optimization, which are useful for both classical and Bayesian estimation. In the absence of conjugate structures, Bayesian models can be estimated using the MH sampler, where the proposal is generally derived from numerical optimization of the posterior density. In this section, we review gradient-based numerical optimization techniques and discuss computational issues.

5.8.1 Gradient Descent/Ascent

A local optimization strategy involves following the steepest direction of the objective function by evaluating the gradient at the current point. Let $f(\theta): \mathbb{R}^d \to \mathbb{R}$ be a differentiable function and θ_k be the current point. For minimization problems, the gradient descent algorithm updates variables by

$$\theta_{k+1} = \theta_k - \lambda \frac{\partial f}{\partial \theta_k},$$

where λ is the step size (also known as the learning rate). Maximization problems are addressed by negating the objective function, and the gradient sign is reversed: $\frac{\partial(-f)}{\partial \theta_k} = -\frac{\partial f}{\partial \theta_k}$. The gradient ascent recursive formula is

$$\theta_{k+1} = \theta_k + \lambda \frac{\partial f}{\partial \theta_k}.$$

The rationale behind the gradient ascent algorithm is the mean value theorem: there exists an intermediate point $\bar{\theta}$ between θ_k and θ_{k+1} such that

$$f(\theta_{k+1}) = f(\theta_k) + \frac{\partial f}{\partial \bar{\theta}'}(\theta_{k+1} - \theta_k) = f(\theta_k) + \lambda \frac{\partial f}{\partial \bar{\theta}'} \frac{\partial f}{\partial \theta_k}.$$

As $\lambda \to 0^+$, the intermediate point approaches the current point, resulting in

$$f(\theta_{k+1}) \to f(\theta_k) + \lambda \left| \frac{\partial f}{\partial \theta_k} \right|^2,$$

where $|x|^2 = x'x$ denotes the squared Euclid norm. It indicates that the objective function increases with each iteration, if the step size is sufficiently small. Proposition 5.8 provides formal results on monotonicity and convergence of the gradient ascent algorithm.

Proposition 5.8 (Gradient ascent) *Let $f: \mathbb{R}^d \to \mathbb{R}$ be the objective function for maximization. Assume that the gradient is Lipschitz continuous: there exists $L > 0$ such that $\left| \frac{\partial f}{\partial y} - \frac{\partial f}{\partial x} \right| \leq L|y - x|$. Consider the gradient ascent iteration $\theta_{k+1} = \theta_k + \lambda \frac{\partial f}{\partial \theta_k}$, where $\lambda < \frac{1}{2L}$.*

1. *Monotonicity:*

$$f(\theta_{k+1}) \geq f(\theta_k) + \frac{1}{2}\lambda \left| \frac{\partial f}{\partial \theta_k} \right|^2.$$

2. *Convergence: if f is globally concave with the maximum at θ^*, then*

$$\lim_{k \to \infty} f(\theta_k) = f(\theta^*),$$

$$f(\theta^*) - f(\theta_k) \leq \frac{1}{2\lambda k} |\theta^* - \theta_0|^2.$$

The gradient ascent algorithm overshadows alternative optimization methods by ensuring a monotonic increase of the objective function at each iteration, provided that the step size is adequately small. For non-concave functions, other numerical optimization methods may underperform, but gradient ascent boasts continuous improvements. Another advantage is its low computational cost and minimal memory usage. It only requires evaluating the gradient vector and storing the parameter vector in memory, making the algorithm ideal for high-dimensional optimization problems. For instance, linear regressions with millions of predictors may fail to be solved analytically due to memory constraints, whereas gradient ascent can find a satisfactory (if not optimal) solution, albeit requiring many iterations to converge.

5.8.2 Stochastic Gradient Descent/Ascent

In many models, the objective function can be expressed in a summable form $f(\theta) = \frac{1}{n}\sum_{i=1}^{n} f_i(\theta)$. Consider the full-sample gradient ascent formula

$$\theta_{k+1} = \theta_k + \lambda\left(\frac{1}{n}\sum_{i=1}^{n}\frac{\partial f_i}{\partial \theta_k}\right),$$

which becomes computationally expensive if the number of observations n is large. The stochastic gradient descent/ascent (Robbins and Monro, 1951) is a big data optimization algorithm that updates parameters based on the gradient of a single observation (or a mini batch of data). For maximization problems, the recursive formula is

$$\theta_{k+1} = \theta_k + \lambda_k \frac{\partial f_i}{\partial \theta_k},$$

where the index i is randomly sampled from $\{1, \ldots, n\}$ at each iteration. It may be tempting to read data sequentially to accelerate computation, but the stochastic gradient $\frac{\partial f_i}{\partial \theta_k}$ is a random variable. Denote the conditional expectation as $E_k(\cdot) = E(\cdot|\theta_1, \ldots, \theta_k)$. The stochastic gradient provides an unbiased estimator of the full-sample gradient:

$$E_k\left(\frac{\partial f_i}{\partial \theta_k}\right) = \frac{1}{n}\sum_{i=1}^{n}\frac{\partial f_i}{\partial \theta_k} = \frac{\partial f}{\partial \theta_k}.$$

It is common to use a mini batch of data for the stochastic gradient, as the gradient estimator remains unbiased while its variance decreases. It is also possible to incorporate a momentum term into the stochastic gradient, corresponding to exponential smoothing of previously used gradients.

The stochastic gradient is an imperfect measure of the true gradient. A noise-ridden gradient does not necessarily undermine numerical optimization. For complex problems, noise in the gradient may help an optimization program escape saddle points. From a programming perspective, randomness is not a bug but a feature.

Convergence of the stochastic gradient algorithm critically depends on diminishing step sizes $\{\lambda_k\}$, which satisfy $\sum_{i=1}^{\infty}\lambda_k = \infty$ and $\sum_{i=1}^{\infty}\lambda_k^2 < \infty$. A common specification is $\lambda_k = k^{-\alpha}$, $\frac{1}{2} < \alpha \leq 1$. Due to noise in the gradient, $\{\theta_k\}$ becomes a random sequence, whose convergence mode depends on regularity conditions. We assume strong concavity of the objective function in Proposition 5.9 to simplify the proof of mean-square convergence of the random sequence.

Proposition 5.9 (Stochastic gradient) *Let $f : \mathbb{R}^d \to \mathbb{R}$ be a strongly concave function: there exist $\delta > 0$ such that*

$$f(y) \leq f(x) + \frac{\partial f}{\partial x'}(y - x) - \frac{\delta}{2}|y - x|^2.$$

Consider the stochastic gradient ascent iteration $\theta_{k+1} = \theta_k + \lambda_k \frac{\partial f_i}{\partial \theta_k}$, where $\lambda_k = \delta^{-1}k^{-1}$, and the stochastic gradient is an unbiased estimator of the gradient with a finite variance: $E_k\left(\frac{\partial f_i}{\partial \theta_k}\right) = \frac{\partial f}{\partial \theta_k}$ and $E_k\left|\frac{\partial f_i}{\partial \theta_k}\right|^2 < M$.

Then, the random sequence $\{\theta_k\}$ converges in mean square to the maximizer of the objective function θ^*:

$$E|\theta_k - \theta^*|^2 \leq k^{-1}C,$$

where $C = \max\left(\delta^{-2}M, |\theta_1 - \theta^*|^2\right)$.

5.8.3 Newton-Raphson Method

A nonlinear objective function may not have a closed-form optimum, but a quadratic function does have. The Newton-Raphson optimization method iteratively approximates the objective function using a sequence of quadratic functions constructed from the gradient and Hessian. Let f be a continuously differentiable function and θ_k be the current point. Consider the second-order Taylor approximation by a quadratic function $\tilde{f}$ such that

$$\tilde{f}(\theta) = f(\theta_k) + \frac{\partial f}{\partial \theta_k'}(\theta - \theta_k) + \frac{1}{2}(\theta - \theta_k)' \frac{\partial^2 f}{\partial \theta_k \partial \theta_k'}(\theta - \theta_k).$$

The quadratic function has a closed-form optimizer obtained by solving the first-order condition

$$\frac{\partial f}{\partial \theta_k} + \frac{\partial^2 f}{\partial \theta_k \partial \theta_k'}(\theta - \theta_k) = 0,$$

which provides the solution $\tilde{\theta}^* = \theta_k - \left(\frac{\partial^2 f}{\partial \theta_k \partial \theta_k'}\right)^{-1} \frac{\partial f}{\partial \theta_k}$. Since $\tilde{\theta}^*$ is not the same as the optimum of the original objective function, the Newton-Raphson algorithm sets $\theta_{k+1} = \tilde{\theta}^*$ for the next iteration by quadratic approximation around $\tilde{\theta}^*$, hence the recursive formula:

$$\theta_{k+1} = \theta_k - \left(\frac{\partial^2 f}{\partial \theta_k \partial \theta_k'}\right)^{-1} \frac{\partial f}{\partial \theta_k}.$$

Unlike the gradient descent/ascent method, the Newton-Raphson recursive formulas are identical for both maximization and minimization problems, as $\min f(\theta)$ is equivalent to $\max -f(\theta)$, and

$$\left(\frac{\partial^2 (-f)}{\partial \theta_k \partial \theta_k'}\right)^{-1} \frac{\partial (-f)}{\partial \theta_k} = \left(\frac{\partial^2 f}{\partial \theta_k \partial \theta_k'}\right)^{-1} \frac{\partial f}{\partial \theta_k}.$$

If the objective function is globally concave and the Hessian matrix takes a simple form, the Newton-Raphson algorithm is computationally efficient and usually converges in a few iterations. In econometrics, the likelihood functions of Probit, Logit and reparametrized Tobit models are concave. In statistical software, the Newton-Raphson can be the default algorithm for logistic regressions, as the Hessian is easy to evaluate.

Unfortunately, many objective functions encountered in practice are so complex that Newton-Raphson iterations are likely to get stuck at a saddle point. The root cause is that the solution of the first-order condition of a non-concave quadratic function is not necessarily a maximum. If the sliced (profiled) objective function is convex for a variable, the Newton-Raphson formula may push that variable in the wrong direction, causing the objective function value to decrease. This is a fatal problem.

The risk of Newton-Raphson algorithm can be illustrated by a toy example:

$$\max_{-1 \leq x, y \leq 2} x^2 - y^2.$$

The optimal solution is $\begin{pmatrix} x \\ y \end{pmatrix} = \begin{pmatrix} 2 \\ 0 \end{pmatrix}$, while $\begin{pmatrix} 0 \\ 0 \end{pmatrix}$ is a saddle point. Assume that the initial point is $\begin{pmatrix} 1 \\ 1 \end{pmatrix}$. The Newton-Raphson formula

$$\begin{pmatrix} x_{k+1} \\ y_{k+1} \end{pmatrix} = \begin{pmatrix} x_k \\ y_k \end{pmatrix} - \begin{pmatrix} 2 & 0 \\ 0 & -2 \end{pmatrix}^{-1} \begin{pmatrix} 2x_k \\ -2y_k \end{pmatrix}$$

indicates that the algorithm converges to the saddle point in one iteration. Although the variable y moves in the correct direction, the variable x moves in the wrong direction because the objective function is convex with respect to x. In contrast, the gradient ascent formula is

$$\begin{pmatrix} x_{k+1} \\ y_{k+1} \end{pmatrix} = \begin{pmatrix} x_k \\ y_k \end{pmatrix} + \lambda \begin{pmatrix} 2x_k \\ -2y_k \end{pmatrix},$$

which pushes both x and y toward the optimal point, provided that the step size is sufficiently small.

For high-dimensional maximization problems, the gradient ascent method is slow but safe, while the Newton-Raphson algorithm is fast but risky. A recommended strategy is the coordinate ascent optimization that applies the Newton-Raphson method to one variable at a time, conditional on other variables. This coordinate optimization strategy is reminiscent of the Gibbs sampler that cycles through the full posterior conditionals. After all, there is no saddle point for a univariate optimization problem. The sign of the scalar-valued Hessian is informative. We apply a Newton-Raphson iteration to a variable only if the Hessian evaluated at the current point is negative. Proposition 5.10 discusses the convergence of the Newton-Raphson algorithm for univariate optimization problems. This is an application of the fixed-point theorem, which is widely used in other areas of economics.

Proposition 5.10 (Newton-Raphson) *Let $f : [a, b] \to \mathbb{R}$ be a continuously differentiable and concave function, with the maximum θ^* such that $f'(\theta^*) = 0$. Consider Newton-Raphson iteration with the initial value θ_0:*

$$\theta_{k+1} = \theta_k - \frac{f'(\theta_k)}{f''(\theta_k)} \equiv T(\theta_k).$$

1. *There exists $\delta > 0$ such that $\{\theta_k\}$ converges to θ^*, if $\theta_0 \in [\theta^* - \delta, \theta^* + \delta]$.*
2. $\lim_{k\to\infty} \frac{|\theta_{k+1}-\theta^*|}{|\theta_k-\theta^*|^2} = \frac{1}{2}T''(\theta^*)$.

5.8.4 BFGS and L-BFGS

The Newton-Raphson algorithm faces three computational challenges in high-dimensional numerical optimization. First, it can be trapped at a saddle point if the Hessian is not negative definite. Second, evaluating the Hessian matrix is computationally intensive and often inaccurate. Computer programs frequently issue warning messages about ill-conditioned numeric Hessians. Third, storing a large Hessian matrix in computer memory can be costly, potentially leading to out-of-memory errors.

The Broyden-Fletcher-Goldfarb-Shanno (BFGS) algorithm addresses these issues by recursively approximating the Hessian matrix. First, the approximated Hessian is constructed using negative definite matrices. Second, it uses gradient information to approximate the Hessian instead of evaluating it directly. Third, the limited-memory version of the BFGS algorithm only requires vector-valued data, avoiding the need to store matrices.

As a gradient-based numerical optimization algorithm, the transition from θ_{k-1} to θ_k is accompanied by a gradient transition from $\frac{\partial f}{\partial \theta_{k-1}}$ to $\frac{\partial f}{\partial \theta_k}$. For univariate optimization, it is possible to recover the numeric Hessian by finite differencing:

$$\frac{\partial^2 f}{\partial^2 \theta_k} \approx \frac{\frac{\partial f}{\partial \theta_k} - \frac{\partial f}{\partial \theta_{k-1}}}{\theta_k - \theta_{k-1}}.$$

However, in multivariate optimization problems, obtaining partial derivatives when all variables change simultaneously is not feasible. Despite this limitation, the gradient transition provides information about the curvature of the Hessian matrix. The BFGS algorithm recursively learns the Hessian matrix by collecting gradient transition information at each iteration.

Consider the first-order Taylor approximation of the gradient:

$$\frac{\partial f}{\partial \theta_{k-1}} \approx \frac{\partial f}{\partial \theta_k} + \frac{\partial^2 f}{\partial \theta_k \partial \theta_k'} (\theta_{k-1} - \theta_k),$$

where the high order terms are ignored. Denote $\Delta\theta_k = \theta_k - \theta_{k-1}$ and $\Delta g_k = \frac{\partial f}{\partial \theta_k} - \frac{\partial f}{\partial \theta_{k-1}}$, which are connected by the secant equations:

$$\Delta\theta_k = -A_k \Delta g_k,$$

where A_k ideally represents the negative inverse Hessian $-\left(\frac{\partial^2 f}{\partial \theta_k \partial \theta_k'}\right)^{-1}$. The $\frac{d(d+1)}{2}$ unknowns in A_k are not uniquely determined by d secant equations, providing an opportunity to smooth the sequence $\{A_k\}$ and ensure a positive definite matrix. By smoothing, we mean that the change from A_{k-1} to A_k should be small. We minimize the distance between A_{k-1} and A_k, subject to the constraint that secant equations must be satisfied:

$$A_k = \min_A |A - A_{k-1}|,$$

$$s.t. \; \Delta\theta_k = -A_k \Delta g_k, \; A = A'.$$

The solution of the optimization problem is

$$A_k = \left(I - \frac{\Delta g_k \Delta \theta_k'}{\Delta g_k' \Delta \theta_k}\right)' A_{k-1} \left(I - \frac{\Delta g_k \Delta \theta_k'}{\Delta g_k' \Delta \theta_k}\right) - \frac{\Delta \theta_k \Delta \theta_k'}{\Delta g_k' \Delta \theta_k}.$$

The BFGS algorithm uses A_k as an approximation of the negative inverse Hessian in the quasi-Newton recursive formula:

$$\theta_{k+1} = \theta_k + \lambda_k A_k \frac{\partial f}{\partial \theta_k},$$

where the step size λ_k is a tuning parameter for stabilizing the algorithm. A line search may be performed to find a reasonable step size that increases the objective function value. A small step size can be specified in the first few iterations since A_k has not yet accumulated sufficient information from previous gradients. As iterations progress, it gathers more information about the Hessian curvature, and $-A_k^{-1}$ usually provides a good approximation of the true Hessian.

For Bayesian inference, the BFGS algorithm is valuable for the MH sampler tuning. A popular strategy is to maximize the posterior density and set the proposal covariance matrix proportional to the negative inverse Hessian evaluated at the posterior mode. The intermediate variable A_k used in BFGS algorithm can be recycled for the MH sampler. The true Hessian is too expensive to evaluate, but A_k is readily available at no additional computing cost. Additionally, the numeric Hessian is often ill-conditioned and difficult to invert, but A_k is positive definite by construction. Without costly matrix inversion, a properly rescaled A_k can serve as a good proposal covariance matrix for the MH sampler.

For high-dimensional optimization problems, it is costly to compute and store a $d \times d$ dense matrix A_k, which is constructed by inner and outer products of $d \times 1$ vectors $\Delta\theta_i$, Δg_i, $i = 1, \ldots, k$. It is desirable to compress data by storing vectors instead of matrices.

The limited memory BFGS (L-BFGS) algorithm maintains a fixed-length history of $\Delta\theta_i$, Δg_i, $i = k, k-1, \ldots, k-h$, and replaces matrix-matrix multiplications with vector-vector multiplications in the BFGS recursive formula. "Limited memory" has a dual nature. First, gradient vectors in the remote past are forgotten, and only recent values are utilized. Second, the computer memory consumption of L-BFGS is substantially smaller than that of BFGS, as the latter involves recursion of covariance matrices, while the former only stores and manipulates vector-valued variables in memory.

Denote $b_k = \frac{\partial f}{\partial \theta_k}$. The BFGS recursive formula takes the form

$$\theta_{k+1} = \theta_k + \lambda_k \left(I - \frac{\Delta g_k \Delta \theta_k'}{\Delta g_k' \Delta \theta_k} \right)' A_{k-1} \left(b_k - \frac{\Delta g_k \Delta \theta_k' b_k}{\Delta g_k' \Delta \theta_k} \right) - \lambda_k \frac{\Delta \theta_k \Delta \theta_k' b_k}{\Delta g_k' \Delta \theta_k}.$$

Reading the above equation from right to left, we note that many operations have been vectorized. However, we still need to simplify the matrix-vector multiplication $A_{k-1} b_{k-1}$, where we define a backward recursion

$$b_{k-i-1} = b_{k-i} - \frac{\Delta g_{k-i} \Delta \theta_{k-i}' b_{k-i}}{\Delta g_{k-i}' \Delta \theta_{k-i}},$$

for $i = 0, 1, \ldots, h$. We also have a forward recursion

$$A_{k-i} b_{k-i} = \left(I - \frac{\Delta g_{k-i} \Delta \theta_{k-i}'}{\Delta g_{k-i}' \Delta \theta_{k-i}} \right)' A_{k-i-1} b_{k-i-1} - \frac{\Delta \theta_{k-i} \Delta \theta_{k-i}' b_{k-i}}{\Delta g_{k-i}' \Delta \theta_{k-i}}.$$

The h-step L-BFGS algorithm proceeds as follows. First, compute $b_k, b_{k-1}, \ldots, b_{k-h}$ by backward recursion. Second, initialize $A_{k-h} = I$, and compute $A_{k-h} b_{k-h}, \ldots, A_{k-1} b_{k-1}$ by forward recursion. Third, update θ_k by

$$\theta_{k+1} = \theta_k + \lambda_k \left(A_{k-1} b_{k-1} - \frac{\Delta \theta_k \Delta g_k' A_{k-1} b_{k-1}}{\Delta g_k' \Delta \theta_k} \right) - \lambda_k \frac{\Delta \theta_k \Delta \theta_k' b_k}{\Delta g_k' \Delta \theta_k}.$$

Both backward and forward recursions involve vector-vector multiplications, without storing or computing any square matrices explicitly. If we specify $h < k$, L-BFGS approximates the original algorithm. If $h = k$, L-BFGS reproduces BFGS results.

5.9 An Application of Loan Actions and Mortgage Rates

In Chapter 1, we examined mortgage rates for applicants who successfully obtained loans using data from the Home Mortgage Disclosure Act (HMDA). However, not all loan applications were approved. Typically, an applicant is offered a mortgage rate only after approval. If an application is denied, the applicant does not know the interest rate. Heuristically, mortgage rates and loan actions are correlated; applicants with a high probability of loan approval are likely to receive lower mortgage rates. This negative correlation suggests incidental truncation of mortgage rates, as described by the sample selection model discussed in Section 5.4.

5.9.1 Logit and Probit Models for Loan Actions

To address the question "to loan or not to loan", we employ Logit and Probit models for the binary response variable: loan originated or denied. The predictors include the debt-to-income ratio (DIR), loan-to-value ratio (LVR), loan size, popular loan terms (15 or 30 years), fixed-rate indicator, subordinate lien, applicant age, gender, joint filing status, and regional income. Lacking prior information on predictor coefficients, we use a diffuse prior. Our goal is to study factors influencing loan actions for home buyers, using 2021 HMDA loan records for single-family home purchases. The sample size is approximately 5.6 million.

We begin with maximum likelihood estimation of the Logit model and then use the inverse Hessian evaluated at the maximum likelihood estimator as the covariance matrix for the proposal distribution in the MH sampler. This approach works well for logistic regressions as the likelihood function is globally concave, allowing the Newton-Raphson method to converge quickly to the maximum. As shown in the first two columns of Table 5.1, classical and Bayesian estimators are similar for all coefficients. The posterior means (maximum likelihood estimators) have the expected signs, and the posterior standard deviations (standard errors) are small. For example, the coefficients for DIR and LVR are significantly negative, as these factors are critical for loan decisions. Empirical results suggest strategies for home buyers to enhance their chances of loan approval: opt for a fixed-rate mortgage of 15 or 30 years, consider a jumbo loan, avoid subordinate liens, apply for a mortgage earlier in life, and find a co-applicant for joint filing.

The third column of Table 5.1 presents Bayesian estimators of the Probit regression using the Gibbs sampler. Although Logit and Probit estimators have the same signs for all variables, their magnitudes differ. A comparison of the second and third columns reveals that coefficients are nearly proportional, with Logit estimators being about 1.8 times larger than Probit estimators.

5.9.2 Sample Selection Models for Mortgage Rates

If mortgage rates for all applicants were available, we could perform a linear regression. However, 11% of loan applications have missing data, and these are not missing at random: none of the denied applications report an interest rate, while 98% of the 5.1 million originated loans do, indicating mortgage rate sample selection based on loan actions. We observe mortgage rates only if loans are originated. Given the potential negative correlation between loan approval probabilities and mortgage rates, we employ a sample selection model to address incidental truncation of mortgage rates.

Loan action: $Z_{1i} = X_i\beta_1 + \varepsilon_{1i}$,

Mortgage rate: $Z_{2i} = X_i\beta_2 + \varepsilon_{2i}$,

Observation: $Y_i = Z_{2i} I\,(Z_{1i} > 0)$.

Bivariate normal disturbances ε_{1i} and ε_{2i} have the covariance matrix Σ.

The two-step procedure (Heckman, 1979) addresses incidental truncation at a low computational cost. Using the Probit estimator, we construct the inverse Mills ratio as an auxiliary predictor to correct bias in the mortgage rate regression. The results are shown in the fourth column of Table 5.1, suggesting that predictors increasing the probability of loan approval decrease the mortgage rate, as the coefficient signs in the fourth column are opposite to those in the third column (Probit estimator). The empirical results confirm that home buyers with low DIR and LVR receive significantly better mortgage rates. Table 5.1 also highlights mortgage market characteristics in 2021, such as fixed rates being lower than adjustable rates and jumbo loan interest rates being more favorable.

The sample selection model involves two correlated equations, and an efficient estimator addresses the joint distribution $p\,(\beta_1, \beta_2, \Sigma, Z_1, Z_2\,|Y\,)$, which can be simulated using the marginalized subsampling technique discussed in Section 5.4. The posterior draws are post-processed to focus on identifiable quantities of interest. The last column of Table 5.1 reports empirical results, indicating that DIR and LVR remain the most important factors for mortgage rates.

TABLE 5.1
Logit and Probit modeling for loan actions, estimated by maximum likelihood, MH, and Gibbs samplers, with two-step and joint estimation for mortgage rates.

	Logit MLE	Logit MH	Probit Gibbs	Two-step	Joint
Intercept	−2.973	−2.975	−1.569	6.198	5.735
	(0.091)	(0.090)	(0.045)	(0.016)	(0.013)
Debt/Income	−6.372	−6.372	−2.873	0.521	0.209
	(0.013)	(0.013)	(0.006)	(0.007)	(0.010)
Loan/Value	−0.461	−0.460	−0.260	0.134	0.104
	(0.013)	(0.013)	(0.006)	(0.002)	(0.002)
LoanSize	3.005	3.006	1.419	−1.462	−1.302
	(0.030)	(0.030)	(0.015)	(0.005)	(0.005)
PopularTerm	0.374	0.375	0.218	−0.230	−0.200
	(0.009)	(0.009)	(0.005)	(0.002)	(0.001)
FixedRate	0.597	0.598	0.316	−0.100	−0.059
	(0.008)	(0.008)	(0.004)	(0.002)	(0.002)
SubordinateLien	−1.007	−1.007	−0.578	1.214	1.113
	(0.014)	(0.014)	(0.008)	(0.004)	(0.004)
Age	−0.550	−0.550	−0.314	0.043	0.003
	(0.013)	(0.013)	(0.006)	(0.002)	(0.002)
Female	0.124	0.124	0.068	−0.001	0.007
	(0.004)	(0.004)	(0.002)	(0.001)	(0.001)
JointFilers	0.307	0.306	0.163	−0.013	0.005
	(0.004)	(0.004)	(0.002)	(0.001)	(0.001)
RegionIncome	3.409	3.409	1.884	−1.047	−0.815
	(0.085)	(0.083)	(0.041)	(0.014)	(0.011)

5.10 Bibliographic Notes

Limited dependent variable models are extensively utilized in micro-econometrics. It is worthwhile to consult relevant chapters in the textbooks by Greene (2008), Wooldridge (2010), Cameron and Trivedi (2005), Train (2003), Manski and McFadden (1981), among others. These authors are leading researchers in the field of econometrics.

For the Probit model, the Gibbs sampler with data augmentation is introduced by Albert and Chib (1993). Holmes and Held (2006) recommend jointly updating model parameters and latent variables.

Logistic regression is a common example for demonstrating novel MCMC methods, though the simple MH sampler may perform exceptionally well. Andrews and Mallows (1974) demonstrate that the logistic distribution is a scale mixture of normal distributions. Building on this mixture representation, Holmes and Held (2006) discuss Gibbs sampling with data augmentation. The Gibbs samplers for logistic regression have various forms, such as the single-move rejection sampling by Dellaportas and Smith (1993), finite Gaussian mixture sampler proposed by Fruhwirth-Schnatter and Fruhwirth (2007), and Polya-Gamma sampler developed by Polson et al. (2013).

The Tobit model originates from the study of Tobin (1958) on censored household expenditure on durable goods. Traditionally, it is estimated using numerical maximum likelihood. The reparametrized likelihood function (Olsen, 1978) facilitates numerical optimization by ensuring the Hessian is negative definite. The textbook by Amemiya (1985) dedicates an entire chapter to the Tobit model and its generalizations, which encompass five types of Tobit models, including incidental truncation and sample selection. Besides maximum likelihood estimation, Amemiya (1985) also explores alternative estimation methods like the E-M algorithm and Heckman two-step regression. Chib (1992) develops Bayesian inference for the Tobit model using the Gibbs sampler with data augmentation. The Gibbs sampler is also applicable to the sample selection model, where the covariance matrix is constrained due to parameter identification. There are differing perspectives on MCMC sampling with non-identified parameters; see McCulloch and Rossi (1994) and Li and Tobias (2009).

Count and duration data models are employed in econometrics and other statistical fields. Numerous Bayesian textbooks in general statistics feature chapters on Poisson and survival analysis, including works by Bolstad (2010), Christensen et al. (2011), Congdon (2014), among others.

Logistic and Poisson regressions are instances of generalized linear models (Nelder and Wedderburn, 1972; McCullagh and Nelder, 1989), which are extensively studied in Bayesian statistics literature such as Zeger and Karim (1991), Dellaportas and Smith (1993), Gelfand and Sahu (1999). Most of these Bayesian methods utilize the Gibbs sampler.

5.11 Proof of Propositions

Proposition 5.1 (Truncated normal distribution)

The mean and variance of truncated normal distributions can be obtained by the following integrals:

$$\int_a^b e^{-\frac{y^2}{2}} dy = \sqrt{2\pi}\left[\Phi(b) - \Phi(a)\right],$$

$$\int_a^b y e^{-\frac{y^2}{2}} dy = \sqrt{2\pi}\left[\phi(a) - \phi(b)\right],$$

$$\int_a^b y^2 e^{-\frac{y^2}{2}} dy = \sqrt{2\pi}\left[a\phi(a) - b\phi(b) + \Phi(b) - \Phi(a)\right].$$

The first integral is the Gaussian kernel. The second integral can be obtained by the change of variable by $z = -\frac{y^2}{2}$ *. The third integral can be obtained by integration by parts.*

Proposition 5.2 (Censored variable)

1. Let $y \in \mathbb{R}$ *be a point of evaluation.*

$$G(y) = P(Y \le y) = P(\max(Z, a) \le y) = P(Z \le y \, and \, a \le y).$$

It follows that $G(y) = 0$*, if* $y < a$*, and* $G(y) = F(y)$*, if* $y \ge a$*.*

2. Similarly,

$$H(y) = P(Y \le y) = P(\min(Z, b) \le y) = P(Z \le y \, or \, b \le y).$$

It follows that $H(y) = F(y)$*, if* $y < b$*, and* $H(y) = 1$*, if* $y \ge b$*.*

Proposition 5.3 (Censored density)

The integrals can be obtained as follows:

If $y < a$*, then* $\int_{-\infty}^{y} \delta_a(z)\, dz = 0$*,* $\int_{-\infty}^{y} \frac{f(z)}{1-F(a)} I_{(a,\infty)}(z)\, dz = 0$*, and* $G(y) = 0$*.*

If $y \geq a$, *then* $\int_{-\infty}^{y} \delta_a(z)\,dz = 1$, $\int_{-\infty}^{y} \frac{f(z)}{1-F(a)} I_{(a,\infty)}(z)\,dz = \frac{F(y)-F(a)}{1-F(a)}$, *and* $G(y) = F(y)$.

If $y < b$, *then* $\int_{-\infty}^{y} \frac{f(z)}{F(b)} I_{(-\infty,b)}(z)\,dz = \frac{F(y)}{F(b)}$, $\int_{-\infty}^{y} \delta_b(z)\,dz = 0$, *and* $G(y) = F(y)$.

If $y \geq b$, *then* $\int_{-\infty}^{y} \frac{f(z)}{F(b)} I_{(-\infty,b)}(z)\,dz = 1$, $\int_{-\infty}^{y} \delta_b(z)\,dz = 1$, *and* $G(y) = 1$.

Proposition 5.4 (Truncated multivariate normal distribution)

For the unrestricted MVN, the joint, marginal and conditional distributions are all MVNs. The joint density equals the product of marginal and conditional densities. We have two forms:

Form 1: $\phi(y,\mu,\Sigma) = \phi(y_1;\mu_1,\Sigma_{11})\,\phi\left(y_2;\bar{\mu}_2,\bar{\Sigma}_{22}\right)$.

Form 2: $\phi(y,\mu,\Sigma) = \phi(y_2;\mu_2,\Sigma_{22})\,\phi\left(y_1;\bar{\mu}_1,\bar{\Sigma}_{11}\right)$.

The conditional means and variances are given by

$\bar{\mu}_1 = \mu_1 + \Sigma_{12}\Sigma_{22}^{-1}(y_2-\mu_2)$, $\bar{\Sigma}_{11} = \Sigma_{11} - \Sigma_{12}\Sigma_{22}^{-1}\Sigma_{21}$.

$\bar{\mu}_2 = \mu_2 + \Sigma_{21}\Sigma_{11}^{-1}(y_1-\mu_1)$, $\bar{\Sigma}_{22} = \Sigma_{22} - \Sigma_{21}\Sigma_{11}^{-1}\Sigma_{12}$.

For the truncated MVN where $y_1 \in A_1 \subset \mathbb{R}^{d_1}$ *and* $y_2 \in \mathbb{R}^{d_2}$, *we denote* $A = A_1 \times \mathbb{R}^{d_2}$. *By Definition 5.4, the joint density takes the form:*

$$p(y) = \frac{\phi(y,\mu,\Sigma)}{\int_A \phi(y,\mu,\Sigma)\,dy} I_A(y) = \frac{\phi(y,\mu,\Sigma)}{F(A_1;\mu_1,\Sigma_{11})} I_{A_1}(y_1),$$

where $F(A_1;\mu_1,\Sigma_{11}) = \int_{A_1} \phi(y_1;\mu_1,\Sigma_{11})\,dy_1$.

To show the marginal density of y_1, *we resort to Form 1:*

$$p(y_1) = \int \frac{\phi(y_1;\mu_1,\Sigma_{11})\,\phi\left(y_2;\bar{\mu}_2,\bar{\Sigma}_{22}\right)}{F(A_1;\mu_1,\Sigma_{11})} I_{A_1}(y_1)\,dy_2.$$

Because $\int \phi\left(y_2;\bar{\mu}_2,\bar{\Sigma}_{22}\right)dy_2 = 1$, *we have*

$$p(y_1) = \frac{\phi(y_1;\mu_1,\Sigma_{11})}{F(A_1;\mu_1,\Sigma_{11})} I_{A_1}(y_1).$$

To show the marginal density of y_2, *we resort to Form 2:*

$$p(y_2) = \int \frac{\phi(y_2;\mu_2,\Sigma_{22})\,\phi\left(y_1;\bar{\mu}_1,\bar{\Sigma}_{11}\right)}{F(A_1;\mu_1,\Sigma_{11})} I_{A_1}(y_1)\,dy_1.$$

Because $\int \phi\left(y_1;\bar{\mu}_1,\bar{\Sigma}_{11}\right) I_{A_1}(y_1)\,dy_1 = F\left(A_1;\bar{\mu}_1,\bar{\Sigma}_{11}\right)$, *we have*

$$p(y_2) = \phi(y_2;\mu_2,\Sigma_{22})\,\frac{F\left(A_1;\bar{\mu}_1,\bar{\Sigma}_{11}\right)}{F(A_1;\mu_1,\Sigma_{11})}.$$

To show the conditional density of y_1, *given a specific* y_2, *we resort to Form 2:*

$$p(y_1\,|y_2) = \frac{p(y)}{p(y_2)},$$

where

$$p(y) = \frac{\phi(y_2;\mu_2,\Sigma_{22})\,\phi\left(y_1;\bar{\mu}_1,\bar{\Sigma}_{11}\right)}{F(A_1;\mu_1,\Sigma_{11})} I_{A_1}(y_1).$$

It follows that

$$p(y_1\,|y_2) = \frac{\phi\left(y_1;\bar{\mu}_1,\bar{\Sigma}_{11}\right)}{F\left(A_1;\bar{\mu}_1,\bar{\Sigma}_{11}\right)} I_{A_1}(y_1).$$

To show the conditional density of y_2, given a specific y_1, we resort to Form 1:

$$p(y_2 | y_1) = \frac{p(y)}{p(y_1)},$$

where

$$p(y) = \frac{\phi(y_1; \mu_1, \Sigma_{11}) \phi(y_2; \bar{\mu}_2, \bar{\Sigma}_{22})}{F(A_1; \mu_1, \Sigma_{11})} I_{A_1}(y_1).$$

It follows that

$$p(y_2 | y_1) = \phi(y_2; \bar{\mu}_2, \bar{\Sigma}_{22}).$$

Proposition 5.5 (Incidental truncation)

As is shown in Proposition 5.4, y_1 follows the truncated normal distribution, and its mean and variance are given by Proposition 5.1.

$$E(y_1) = \mu_1 + \sigma_1(\lambda_a + \lambda_b),$$

$$Var(y_1) = \sigma_1^2 \left[1 - (\lambda_a + \lambda_b)^2 + \lambda_a \frac{a - \mu_1}{\sigma_1} + \lambda_b \frac{b - \mu_1}{\sigma_1}\right].$$

Although it is feasible to calculate the integral $E(y_2) = \int y_2 p(y_2) dy_2$, the law of iterated expectations provides more convenient calculation.

$$E(y_2) = E[E(y_2 | y_1)].$$

By Proposition 5.4,

$$E(y_2 | y_1) = \bar{\mu}_2 = \mu_2 + \rho\sigma_2\sigma_1^{-1}(y_1 - \mu_1).$$

So we have

$$E(y_2) = \mu_2 + \rho\sigma_2\sigma_1^{-1}[E(y_1) - \mu_1] = \mu_2 + \rho\sigma_2(\lambda_a + \lambda_b).$$

The variance of y_2 equals the mean of the conditional variance plus the variance of the conditional mean:

$$Var(y_2) = E[Var(y_2 | y_1)] + Var[E(y_2 | y_1)].$$

By Proposition 5.4, the conditional variance is

$$Var(y_2 | y_1) = \sigma_2^2 - \rho^2\sigma_2^2.$$

The variance of the conditional mean is

$$Var[E(y_2 | y_1)] = \rho^2\sigma_2^2\sigma_1^{-2} Var(y_1).$$

So we have

$$Var(y_2) = \sigma_2^2 \left[1 - \rho^2(\lambda_a + \lambda_b)^2 + \rho^2\lambda_a \frac{a - \mu_1}{\sigma_1} + \rho^2\lambda_b \frac{b - \mu_1}{\sigma_1}\right].$$

Proposition 5.6 (Poisson-gamma conjugacy)

For brevity, we drop the subscript i in all variables, without conditioning on a, β explicitly denoted.

$p(u|Y)$
$\propto p(u)\,p(Y|u)$
$\propto u^{a-1}e^{-au}\left(e^{X\beta}u\right)^{Y}e^{-e^{X\beta}u}$
$\propto u^{a+Y-1}e^{-\left(a+e^{X\beta}\right)u}$

We recognize that it is the density of Gamma $\left(a+Y, a+e^{X\beta}\right)$.

By Bayes' rule

$p(Y)$
$=\frac{\Gamma^{-1}(a)a^{a}u^{a-1}e^{-au}\left(e^{X\beta}u\right)^{Y}e^{-e^{X\beta}u}(Y!)^{-1}}{\Gamma^{-1}(a+Y)(a+e^{X\beta})^{a+Y}u^{a+Y-1}e^{-\left(a+e^{X\beta}\right)u}}$
$=\frac{\Gamma(a+Y)a^{a}\left(e^{X\beta}\right)^{Y}}{\Gamma(a)\Gamma(Y+1)(a+e^{X\beta})^{a+Y}}$
$=\frac{\Gamma(a+Y)}{\Gamma(a)\Gamma(Y+1)}r^{Y}(1-r)^{a}$,

where $r=\frac{e^{X\beta}}{a+e^{X\beta}}$.

This is the negative binomial probability distribution.

Proposition 5.7 (Rejection sampling)

First, consider a simple scenario that both $f(\cdot)$ *and* $s(\cdot)$ *are proper density functions (that integrate to one). The joint density of* θ *and* U *is* $s(\theta)\,I_{(0,1)}(U)$, *so the acceptance probability is*

$$P\left[U\leq\frac{f(\theta)}{s(\theta)\,M}\right]=\int_{-\infty}^{\infty}\int_{0}^{\frac{f(\theta)}{s(\theta)M}}s(\theta)\,dU\,d\theta=\int_{-\infty}^{\infty}\frac{f(\theta)}{M}d\theta=\frac{1}{M}.$$

In the above derivation, M must be an upper bound to ensure $\frac{f(\theta)}{s(\theta)M}\leq 1$.

The CDF of accepted draws is the conditional probability

$$P\left[\theta\leq c\middle|U\leq\frac{f(\theta)}{s(\theta)\,M}\right]=\frac{P\left[\theta\leq c, U\leq\frac{f(\theta)}{s(\theta)M}\right]}{P\left[U\leq\frac{f(\theta)}{s(\theta)M}\right]},$$

where

$$P\left[\theta\leq c, U\leq\frac{f(\theta)}{s(\theta)\,M}\right]=\int_{-\infty}^{c}\int_{0}^{\frac{f(\theta)}{s(\theta)M}}s(\theta)\,dU\,d\theta=\frac{1}{M}\int_{-\infty}^{c}f(\theta)\,d\theta.$$

It follows that

$$P\left[\theta\leq c\middle|U\leq\frac{f(\theta)}{s(\theta)\,M}\right]=\int_{-\infty}^{c}f(\theta)\,d\theta.$$

Finally, consider unnormalized densities such that $\int_{-\infty}^{\infty}f(x)\,dx=c_f$ *and* $\int_{-\infty}^{\infty}s(x)\,dx=c_s$. *The integrals are calculated in the similar way, and we have*

$$P\left[U\leq\frac{f(\theta)}{s(\theta)\,M}\right]=\frac{c_f}{c_s M},$$

$$P\left[\theta\leq c, U\leq\frac{f(\theta)}{s(\theta)\,M}\right]=\frac{1}{c_s M}\int_{-\infty}^{c}f(\theta)\,d\theta,$$

$$P\left[\theta\leq c\middle|U\leq\frac{f(\theta)}{s(\theta)\,M}\right]=\int_{-\infty}^{c}\frac{1}{c_f}f(\theta)\,d\theta.$$

Proposition 5.8 (Gradient ascent)

By the mean value theorem, there exists an intermediate point z between θ_k and θ_{k+1} such that

$$\begin{aligned}
&f(\theta_k)\\
&= f(\theta_{k+1}) + \tfrac{\partial f}{\partial z'}(\theta_k - \theta_{k+1})\\
&= f(\theta_{k+1}) + \tfrac{\partial f}{\partial \theta_k'}(\theta_k - \theta_{k+1}) + \left(\tfrac{\partial f}{\partial z'} - \tfrac{\partial f}{\partial \theta_k'}\right)(\theta_k - \theta_{k+1})\\
&\le f(\theta_{k+1}) + \tfrac{\partial f}{\partial \theta_k'}(\theta_k - \theta_{k+1}) + \left|\tfrac{\partial f}{\partial z'} - \tfrac{\partial f}{\partial \theta_k'}\right| |\theta_k - \theta_{k+1}|\\
&\le f(\theta_{k+1}) + \tfrac{\partial f}{\partial \theta_k'}(\theta_k - \theta_{k+1}) + L|z - \theta_k| |\theta_k - \theta_{k+1}|\\
&\le f(\theta_{k+1}) + \tfrac{\partial f}{\partial \theta_k'}(\theta_k - \theta_{k+1}) + L|\theta_k - \theta_{k+1}|^2\\
&= f(\theta_{k+1}) + \tfrac{\partial f}{\partial \theta_k'}\left(-\lambda\tfrac{\partial f}{\partial \theta_k}\right) + L\left|-\lambda\tfrac{\partial f}{\partial \theta_k}\right|^2\\
&= f(\theta_{k+1}) - \lambda(1 - L\lambda)\left|\tfrac{\partial f}{\partial \theta_k}\right|^2\\
&\le f(\theta_{k+1}) - \tfrac{1}{2}\lambda\left|\tfrac{\partial f}{\partial \theta_k}\right|^2.
\end{aligned}$$

In the above derivation, the four inequalities correspond to Cauchy-Schwartz inequality, Lipschitz continuity, triangular inequality and $\lambda < \frac{1}{2L}$, respectively. It follows that

$$f(\theta_{k+1}) \ge f(\theta_k) + \frac{1}{2}\lambda\left|\frac{\partial f}{\partial \theta_k}\right|^2.$$

Note that monotonicity does not depend on concavity of the objective function. The objective function increases at each iteration, provided that 1) the objective function is smooth enough, and 2) the step size is small enough.

To show convergence of the gradient ascent algorithm, we assume the objective function is globally concave, so that

$$\begin{aligned}
&f(\theta^*)\\
&\le f(\theta_k) + \tfrac{\partial f}{\partial \theta_k'}(\theta^* - \theta_k)\\
&\le f(\theta_{k+1}) - \tfrac{1}{2}\lambda\left|\tfrac{\partial f}{\partial \theta_k}\right|^2 + \tfrac{\partial f}{\partial \theta_k'}(\theta^* - \theta_k)\\
&= f(\theta_{k+1}) - \tfrac{1}{2\lambda}\left[\lambda^2\left|\tfrac{\partial f}{\partial \theta_k}\right|^2 + 2\lambda\tfrac{\partial f}{\partial \theta_k'}(\theta_k - \theta^*) + |\theta_k - \theta^*|^2 - |\theta_k - \theta^*|^2\right]\\
&= f(\theta_{k+1}) - \tfrac{1}{2\lambda}\left[\left|\lambda\tfrac{\partial f}{\partial \theta_k'} + \theta_k - \theta^*\right|^2 - |\theta_k - \theta^*|^2\right]\\
&= f(\theta_{k+1}) - \tfrac{1}{2\lambda}\left[|\theta_{k+1} - \theta^*|^2 - |\theta_k - \theta^*|^2\right].
\end{aligned}$$

It follows that

$$f(\theta^*) - f(\theta_{k+1}) \le \frac{1}{2\lambda}\left[|\theta_k - \theta^*|^2 - |\theta_{k+1} - \theta^*|^2\right].$$

This is a telescope series. Summing up the series for $k = 0, 1, 2, \ldots$, we have

$$\frac{1}{k}\sum_{i=0}^{k-1}\left[f(\theta^*) - f(\theta_{i+1})\right] \le \frac{1}{2\lambda k}|\theta_0 - \theta^*|^2.$$

Because the series $\{f(\theta_{i+1})\}$ is monotone increasing, we have the inequality

$$f(\theta^*) - f(\theta_k) \le \frac{1}{2\lambda k}|\theta_0 - \theta^*|^2.$$

As $k \to \infty$, the term $\frac{1}{2\lambda k}\left|\theta_0-\theta^*\right|^2$ tends to zero, hence convergence:

$$\lim_{k\to\infty} f\left(\theta_k\right)=f\left(\theta^*\right).$$

Proposition 5.9 (Stochastic gradient)

By strong concavity, we have inequalities for all realizations of θ_k:

$$f\left(\theta^*\right)\le f\left(\theta_k\right)+\frac{\partial f}{\partial\theta_k'}\left(\theta^*-\theta_k\right)-\frac{\delta}{2}\left|\theta^*-\theta_k\right|^2,$$

$$f\left(\theta_k\right)\le f\left(\theta^*\right)+\frac{\partial f}{\partial\theta^{*\prime}}\left(\theta_k-\theta^*\right)-\frac{\delta}{2}\left|\theta_k-\theta^*\right|^2.$$

Because θ^* is the maximizer, we have $\frac{\partial f}{\partial\theta^{*\prime}}=0$. It follows that

$$\frac{\partial f}{\partial\theta_k'}\left(\theta_k-\theta^*\right)\le-\delta\left|\theta_k-\theta^*\right|^2.$$

Consider the conditional expectation:

$$\begin{aligned}
&E_k\left|\theta_{k+1}-\theta^*\right|^2\\
&=E_k\left|\theta_k+\lambda_k\tfrac{\partial f_i}{\partial\theta_k}-\theta^*\right|^2\\
&=\left|\theta_k-\theta^*\right|^2+\lambda_k^2E_k\left|\tfrac{\partial f_i}{\partial\theta_k}\right|^2+2\lambda_kE_k\left[\tfrac{\partial f_i}{\partial\theta_k'}\left(\theta_k-\theta^*\right)\right]\\
&=\left|\theta_k-\theta^*\right|^2+\lambda_k^2E_k\left|\tfrac{\partial f_i}{\partial\theta_k}\right|^2+2\lambda_k\tfrac{\partial f}{\partial\theta_k'}\left(\theta_k-\theta^*\right)\\
&\le\left|\theta_k-\theta^*\right|^2+\lambda_k^2E_k\left|\tfrac{\partial f_i}{\partial\theta_k}\right|^2-2\lambda_k\delta\left|\theta_k-\theta^*\right|^2\\
&\le\left(1-\tfrac{2}{k}\right)\left|\theta_k-\theta^*\right|^2+\tfrac{1}{k^2}C.
\end{aligned}$$

By construction, $E\left|\theta_1-\theta^*\right|^2\le C$. By induction, if $E\left|\theta_k-\theta^*\right|^2\le k^{-1}C$, then we apply the law of iterated expectations,

$$\begin{aligned}
&E\left|\theta_{k+1}-\theta^*\right|^2\\
&=E\left\{E_k\left|\theta_{k+1}-\theta^*\right|^2\right\}\\
&\le E\left\{\left(1-\tfrac{2}{k}\right)\left|\theta_k-\theta^*\right|^2+\tfrac{1}{k^2}C\right\}\\
&=\left(1-\tfrac{2}{k}\right)E\left|\theta_k-\theta^*\right|^2+\tfrac{1}{k^2}C\\
&\le\left(1-\tfrac{2}{k}\right)\tfrac{1}{k}C+\tfrac{1}{k^2}C\\
&\le\tfrac{1}{k+1}C.
\end{aligned}$$

So we conclude that $E\left|\theta_k-\theta^*\right|^2\le k^{-1}C$, $\forall k\ge1$.

Taking the limit, we have the mean-square convergence:

$$\lim_{k\to\infty}E\left|\theta_k-\theta^*\right|^2=0.$$

Proposition 5.10 (Newton-Raphson)

Let $T(\theta)=\theta-\frac{f'(\theta)}{f''(\theta)}$. Because $f'(\theta^*)=0$, we have $T(\theta^*)=\theta^*$. That is, θ^* is a fixed point of the mapping T.

As f is differentiable, so is T, whose derivative is

$$T'(\theta)=\frac{f'(\theta)f'''(\theta)}{\left[f''(\theta)\right]^2}.$$

Note that $T'(\theta^*) = 0$. *Because of continuity, for an arbitrary discount factor* $\beta \in (0,1)$, *there exist* $\delta > 0$ *such that* $|T'(\theta)| < \beta$, $\forall \theta \in [\theta^* - \delta, \theta^* + \delta]$. *By the mean value theorem, there exists an intermediate point* ξ *between* $x, y \in [\theta^* - \delta, \theta^* + \delta]$ *such that*

$$|T(x) - T(y)| = |T'(\xi)(x-y)| < \beta |x-y|,$$

which indicates that T *is a contraction mapping in the interval* $\theta_0 \in [\theta^* - \delta, \theta^* + \delta]$. *By the fixed point theorem, the Newton-Raphson iterations* $\theta_{k+1} = T(\theta_k)$ *converges to the fixed point* θ^*.

Moreover, consider the second-order Taylor approximation:

$$T(\theta) = T(\theta^*) + T'(\theta^*)(\theta - \theta^*) + \frac{1}{2}T''(\xi)(\theta - \theta^*)^2,$$

where the intermediate point ξ *is between* θ *and* θ^*. *As* $T(\theta^*) = \theta^*$ *and* $T'(\theta^*) = 0$, *we have*

$$T(\theta) = \theta^* + \frac{1}{2}T''(\xi)(\theta - \theta^*)^2.$$

Evaluating the function at $\theta = \theta_k$, *we have*

$$\theta_{k+1} = \theta^* + \frac{1}{2}T''(\xi)(\theta_k - \theta^*)^2.$$

It follows that

$$\lim_{k\to\infty} \frac{|\theta_{k+1} - \theta^*|}{|\theta_k - \theta^*|^2} = \lim_{k\to\infty} \frac{1}{2}T''(\xi) = \frac{1}{2}T''(\theta^*).$$

5.12 Exercises

Exercise 5.1 (Truncated normal sampling) *Truncated normal distribution* $N(\mu, \sigma^2) I_{(a,b)}$ *is defined by its density function*

$$f(x) = \frac{\phi(x;\mu,\sigma^2)}{\Phi(b;\mu,\sigma^2) - \Phi(a;\mu,\sigma^2)} I_{(a,b)}(x),$$

where ϕ *is the normal density and* Φ *is the normal CDF.*

1. Show that the truncated normal CDF and inverse CDF are given by

$$F(x) = \frac{\Phi(x;\mu,\sigma^2) - \Phi(a;\mu,\sigma^2)}{\Phi(b;\mu,\sigma^2) - \Phi(a;\mu,\sigma^2)},$$

$$F^{-1}(U) = \mu + \sigma\Phi^{-1}\left\{\Phi(a;\mu,\sigma^2) + U\left[\Phi(b;\mu,\sigma^2) - \Phi(a;\mu,\sigma^2)\right]\right\},$$

where Φ^{-1} *is the standard normal inverse CDF.*

2. Suppose that we generate a standard uniform random draw U, *show that* $F^{-1}(U)$ *is a truncated normal draw from* $N(\mu, \sigma^2) I_{(a,b)}$. *This is called the inverse CDF sampling.*

3. Suppose that we generate a proposal draw from $N(\mu, \sigma^2)$. *If it falls into the interval* (a, b), *then we accept it. Otherwise, we keep on generating proposals until a draw falls into the interval. This is a version of rejection sampling. Show that an accepted draw follows the truncated normal distribution.*

Exercise 5.2 (Posterior of Probit model) *The latent variable representation of the Probit model indicates that* $Z = X\beta + \varepsilon$, *where* $\varepsilon \sim N(0, I_n)$. *The prior is specified as* $\beta \sim N(\mu, \Lambda^{-1})$, *where* Λ *is the precision matrix (the covariance matrix is* Λ^{-1}*).*

1. Show that $Z \sim N(X\mu, I_n + X\Lambda^{-1}X')$.

2. Show that the density of $N(X\mu, I_n + X\Lambda^{-1}X')$ *is proportional to* $e^{-\frac{1}{2}(Z'Z-\bar{\mu}'\bar{\Lambda}\bar{\mu})}$, *where* $\bar{\mu} = (\Lambda + X'X)^{-1}(\Lambda\mu + X'Z)$ *and* $\bar{\Lambda} = \Lambda + X'X$.

3. Consider an alternative derivation of $p(Z)$ *by Bayes' rule:*

$$p(Z) = \frac{p(\beta)\, p(Z\,|\beta)}{p(\beta\,|Z)}.$$

Show that the result is the same as that in 2.

4. Show that $p(Z\,|Y)$ *follows the multivariate truncated normal distribution.*

5. Assume that $\mu = 0$. *Show that* $p(Z_i\,|Z_{-i}, Y)$ *follows the univariate truncated normal distribution*

$$Z_i\,|Z_{-i}, Y \sim \begin{cases} N\left(\tilde{\mu}_i, \tilde{\Lambda}_i^{-1}\right) I_{(0,\infty)} & Y_i = 1 \\ N\left(\tilde{\mu}_i, \tilde{\Lambda}_i^{-1}\right) I_{(-\infty,0)} & Y_i = 0 \end{cases},$$

$$\tilde{\mu}_i = \frac{X_i\bar{\mu} - h_i Z_i}{1 - h_i},$$

$$\tilde{\Lambda}_i = 1 - h_i,$$

and h_i *is the* i^{th} *diagonal element of the matrix* $X\bar{\Lambda}^{-1}X'$.

6. The functional form of $\tilde{\mu}_i = \frac{X_i\bar{\mu} - h_i Z_i}{1-h_i}$ *seems to suggest that it is a function of* Z_i. *Is it a problem?*

Exercise 5.3 (Random utility discrete choice) *Let the utility of person* i *selecting the* j^{th} *option be*

$$Y_{ij}^* = X_i\beta_j + \varepsilon_{ij},$$

where ε_{ij} *follows the extreme value distribution with the density function*

$$f(x) = e^{-x-\exp(-x)}.$$

1. Show that the CDF of the extreme value distribution is

$$F(x) = e^{-\exp(-x)}.$$

2. The decision maker selects the category of the highest utility. Show that

$$p(Y_i = 1\,|X_i, \beta) = p(Y_{i1}^* \geq Y_{ik}^*, \forall k\,|X_i, \beta).$$

3. Use the PDF and CDF to calculate the probability

$$p(Y_i = 1\,|X_i, \beta) = \int f(\varepsilon_{i1})\, p(Y_i = 1\,|X_i, \beta, \varepsilon_{i1})\, d\varepsilon_{i1},$$

where

$$p(Y_i = 1\,|X_i, \beta, \varepsilon_{i1}) = \prod_{j=2}^{J} F\left[\varepsilon_{i1} + X_i(\beta_1 - \beta_j)\right].$$

4. Calculate the integral and show

$$p(Y_i = 1\,|X_i, \beta) = \frac{e^{X_i\beta_1}}{\sum_{j=1}^{J} e^{X_i\beta_j}}.$$

Exercise 5.4 (Tobit with an unknown censoring point) *Consider the Tobit model, where the imperfectly observed response variable is censored at the lower bound c.*

1. Is it possible that all observations are strictly larger than c? In that event, can we identify the censoring point from the dataset alone? What do we know about the censoring point?

2. We consider Bayesian inference of the Tobit model when the censoring point is unknown. For simplicity, we assume a uniform prior in the interval $(\underline{c}, \bar{c})$*. The remaining model specifications are the same as those presented in the text, and our aim is parameter estimation by Gibbs sampling. What are the full posterior conditional distributions* $p(\beta, \sigma^2 | Y, Z, c)$, $p(Z | Y, \beta, \sigma^2, c)$ *and* $p(c | Y, Z, \beta, \sigma^2)$*?*

Exercise 5.5 (CDF and Expectation) *It is possible to obtain the expectation of a nonnegative random variable directly from CDF. The Tobit model provides an example.*

$$Z_i = X_i\beta + \sigma\varepsilon_i,$$

$$Y_i = \max(Z_i, c),$$

where ε_i *follows the standard normal distribution. By Proposition 5.2, the CDF of* Y_i *in the Tobit model takes the form*

$$F_{Y_i}(y) = \Phi\left(\frac{y - X_i\beta}{\sigma}\right) I_{[c,\infty)}(y).$$

1. To construct a nonnegative random variable, we shift Y_i *by defining* $W_i = Y_i - c$*. What is the CDF of* W_i*?*

2. The expectation of a nonnegative random variable can be calculated by the formula

$$E(W_i) = \int_0^\infty [1 - F_{W_i}(w)]\, dw,$$

where $F_{W_i}(w)$ *is the CDF of the nonnegative random variable. Show that*

$$E(W_i) = \sigma\phi\left(\frac{c - X_i\beta}{\sigma}\right) - (c - X_i\beta)\,\Phi\left(-\frac{c - X_i\beta}{\sigma}\right).$$

3. Show that

$$E(Y_i) = \Phi\left(\frac{c - X_i\beta}{\sigma}\right) c + \Phi\left(-\frac{c - X_i\beta}{\sigma}\right) X_i\beta + \sigma\phi\left(\frac{c - X_i\beta}{\sigma}\right).$$

Exercise 5.6 (Truncated normal) *Suppose that the joint distribution of* $(y_1, \ldots, y_d)$ *is a truncated MVN. Consider the* i^{th} *variable* y_i *and the remaining variables* $y_{-i} = (y_1, \ldots, y_{i-1}, y_{i+1}, \ldots, y_d)$*. Show that*

1. The conditional distribution $p(y_i | y_{-i})$ *is univariate truncated normal.*

2. The marginal distribution $p(y_i)$ *is not truncated normal in general, but there are two exceptions:* y_{-i} *is unrestricted, or no correlation between* y_i *and* y_{-i}*.*

Exercise 5.7 (CDF in the sample selection model) *Consider a sample selection model*

$$Z_{1i} = X_{1i}\beta_1 + \varepsilon_{1i},$$

$$Z_{2i} = X_{2i}\beta_2 + \varepsilon_{2i},$$

$$Y_i = Z_{2i} I(Z_{1i} > 0),$$

where the disturbances follow $N\left[\begin{pmatrix} 0 \\ 0 \end{pmatrix}, \begin{pmatrix} 1 & \rho\sigma \\ \rho\sigma & \sigma^2 \end{pmatrix}\right]$. *The observation* Y_i *is neither discrete nor absolutely continuous. Instead, it follows a mixture distribution. The CDF of any distribution is always well defined, so we consider* $P(Y_i \le c)$, *where* c *is a point of evaluation.*

1. *We decompose the CDF as the sum of two terms:*

$$P(Y_i \le c) = P(Y_i \le c, Z_{1i} > 0) + P(Y_i \le c, Z_{1i} \le 0).$$

Show that

$$P(Y_i \le c, Z_{1i} > 0) = \Phi\left[\begin{pmatrix} 0 \\ c \end{pmatrix}; \begin{pmatrix} -X_{1i}\beta_1 \\ X_{2i}\beta_2 \end{pmatrix}, \begin{pmatrix} 1 & -\rho\sigma \\ -\rho\sigma & \sigma^2 \end{pmatrix}\right],$$

$$P(Y_i \le c, Z_{1i} \le 0) = \Phi(-X_{1i}\beta_1) I(c \ge 0).$$

2. *CDF must be right-continuous, but not necessarily continuous. Show that it has a discontinuity at the origin, hence a positive probability mass:*

$$P(Y_i = 0) = \Phi(-X_{1i}\beta_1).$$

3. *Although the distribution is not absolutely continuous and does not have a density function, we construct a special function as a proxy of the density:*

$$p(Y_i) = p(D_i = 0)\, p(Y_i | D_i = 0) + p(D_i = 1)\, p(Y_i | D_i = 1),$$

where 1) $p(D_i = 0) = \Phi(-X_{1i}\beta_1)$, *2)* $p(Y_i | D_i = 0) = \delta_0(Y_i)$, *3)* $p(D_i = 1) = \Phi(X_{1i}\beta_1)$, *and 4)*

$$p(Y_i | D_i = 1) = \frac{\phi(Y_i; X_{2i}\beta_2, \sigma^2)\, \Phi(\bar{\mu}_i / \bar{\sigma})}{\Phi(X_{1i}\beta_1)},$$

$$\bar{\mu}_i = X_{1i}\beta_1 + \rho\sigma^{-1}(Y_i - X_{2i}\beta_2),$$

$$\bar{\sigma}^2 = 1 - \rho^2.$$

We treat $p(Y_i)$ *as a function of* Y_i. *With some abuse of notation, we still denote the function as* $p(\cdot)$. *Show that it satisfies the standard properties of a density function:*

$$p(c) \ge 0, \forall c \in \mathbb{R},$$

$$\int_{-\infty}^{c} p(y)\, dy = P(Y_i \le c).$$

4. *Show that the expectation is given by*

$$E(Y_i) = \Phi(X_{1i}\beta_1)\left[X_{2i}\beta_2 + \rho\sigma \frac{\phi(X_{1i}\beta_1)}{\Phi(X_{1i}\beta_1)}\right].$$

5. *Show that the marginal effect with respect to* X_{2i} *is given by*

$$\frac{\partial E(Y_i)}{\partial X_{2i}} = \Phi(X_{1i}\beta_1)\, \beta_2.$$

Exercise 5.8 (Sequential estimation of multiple equations) *Consider the two-equation system in the sample selection model:*

$$Z_{1i} = X_{1i}\beta_1 + v_{1i},$$

$$Z_{2i} = X_{2i}\beta_2 + \rho\sigma v_{1i} + \sqrt{1-\rho^2}\sigma v_{2i},$$

where v_{1i} and v_{2i} are independent standard normal disturbances. Our aim is to sample from the posterior distribution $p(\beta_1, \beta_2, \rho, \sigma \,|Z_1, Z_2)$ under carefully specified priors.

1. If we adopt a normal prior for β_1, say $N(\mu_1, \Lambda_1^{-1})$, and only use the observations $Z_1 = (Z_{11}, \ldots, Z_{1n})$, what is the posterior distribution $p(\beta_1 \,|Z_1)$?

2. The observations include both Z_1 and Z_2. What is the posterior distribution $p(\beta_1 \,|Z_1, Z_2)$? Could we specify a prior for β_2, ρ, σ such that $p(\beta_1 \,|Z_1) = p(\beta_1 \,|Z_1, Z_2)$, or at least the equality holds approximately?

3. Consider the marginal-conditional decomposition:

$$p(\beta_1, \beta_2, \rho, \sigma \,|Z_1, Z_2) = p(\beta_1 \,|Z_1, Z_2)\, p(\beta_2, \rho, \sigma \,|Z_1, Z_2, \beta_1).$$

We handle two equations sequentially. We first handle the first equation by sampling β_1 from a marginalized posterior, and then address the second equation by sampling β_2, ρ, σ from the conditional posterior. How to sample from $p(\beta_2, \rho, \sigma \,|Z_1, Z_2, \beta_1)$?

4. Design a Gibbs sampler that cycles through the posterior conditional distributions $p(\beta_1, \beta_2 \,|Z_1, Z_2, \rho, \sigma)$ and $p(\rho, \sigma \,|Z_1, Z_2, \beta_1, \beta_2)$.

Exercise 5.9 (Hazard rate of log-logistic distribution) *Suppose that the log lifetime $(\ln Y_i)$ follows a logistic distribution with the location parameter μ and scale parameter σ. That is, $\sigma^{-1}(\ln Y_i - \mu)$ has the standard logistic CDF*

$$\Lambda(x) = \frac{1}{1+e^{-x}}.$$

Show that the hazard rate of Y_i takes the form

$$h(t) = \Lambda\left[\sigma^{-1}(\ln t - \mu)\right]\sigma^{-1}t^{-1},$$

which is hump-shaped (invert bathtub) if $\sigma < 1$.

Exercise 5.10 (Posterior predictive survival function) *The posterior predictive density $f_Y(t)$ is obtained by marginalizing the conditional density $f(t;\theta)$ over the posterior distribution of parameters:*

$$f_Y(t) = \int f(t;\theta)\, p(\theta \,|Y)\, d\theta.$$

1. Is the posterior predictive survival function $S_Y(t)$ obtained by marginalizing the conditional survival function $S(t;\theta)$ over the posterior distribution of parameters?

2. Is the posterior predictive hazard rate $h_Y(t)$ obtained by marginalizing the conditional hazard rate $h(t;\theta)$ over the posterior distribution of parameters?

Exercise 5.11 (Identification of step hazard rate) *Suppose that the hazard rate is a step function:*

$$h(t;\theta) = \sum_{j=1}^{J} c_j I_{(a_j, a_{j+1})}(t),$$

where the steps $\theta = (c_1, \ldots, c_J)$ are unknown parameters, while the thresholds $a_1, \ldots, a_{J+1}$ are specified by users, with the boundaries $a_1 = 0$ and $a_{J+1} = \infty$. Let Y_i be the observed lifetime with the censoring indicator δ_i, $i = 1, \ldots, n$. The likelihood function is given by

$$p(Y \,|\theta) = \prod_{i=1}^{n} \left[h(Y_i;\theta)\right]^{\delta_i} e^{-\int_0^{Y_i} h(t;\theta)dt}$$

1. Show that c_J is unidentified by data if $a_J > Y_i$, $\forall i = 1, \ldots, n$.

2. Show that c_j and c_{j+1} are not individually identified by data if no observations fall in two consecutive intervals (a_j, a_{j+1}) and (a_{j+1}, a_{j+2}).

Exercise 5.12 (Weibull distribution) *The Weibull distribution belongs to both PH and AFT models, but they provide different interpretations on how an exogenous factor affects the lifetime. For example, a patient consults doctors about smoking. Denote the smoking status by $X_i \in \{0, 1\}$.*

Based on a Weibull PH model

$$\ln h_i(t) = c + X_i\beta + a \ln t,$$

the first doctor suggests that smoking increases the hazard rate by β percent.

Based on a Weibull AFT model

$$f_i(t) = \frac{k}{\lambda}\left(\frac{t}{\lambda}\right)^{k-1} \exp\left[-\left(\frac{t}{\lambda}\right)^k\right],$$

where the scale parameter takes the form $\lambda = e^{-\tilde{c} - X_i\tilde{\beta}}$, the second doctor suggests that smoking reduces the expected lifetime by $\tilde{\beta}$ percent.

Could the suggestions of both doctors be correct? Suppose that parameters β, c, a in the PH model and parameters $\tilde{\beta}, \tilde{c}, k$ in the AFT model are estimated by maximum likelihood with the same dataset. Is it possible to convert the estimators of β, c, a to $\tilde{\beta}, \tilde{c}, k$, and vice versa?

Exercise 5.13 (Inverse probability weighting) *Under the assumption of conditional independence and overlap, we can identify the potential outcome means and the average treatment effect by either the outcome model or the treatment model (or both). In the text, we showed regression adjustment by the outcome model with the identity*

$$E(Y_{id}) = E[E(Y_i | X_i, D_i = d)],$$

where the two expectation operators are essentially double integrals:

$$E[E(Y_i | X_i, D_i = d)] = \int p(X_i) \int Y_i p(Y_i | X_i, D_i = d) \, dY_i dX_i.$$

Because the treatment variable is discrete, we have a trivial integral

$$\int I(D_i = d) \, dD_i = \sum_d I(D_i = d) = 1.$$

Combine three integrals and show the identity

$$E(Y_{id}) = \int\int\int \frac{Y_i I(D_i = d)}{p(D_i = d | X_i)} p(X_i, D_i, Y_i) \, dY_i dX_i dD_i = E\left[\frac{Y_i I(D_i = d)}{p(D_i = d | X_i)}\right].$$

6

Linear State Space Models

When an econometrician meets an engineer, they often find common ground in discussing state space models. These models, as highlighted in the seminal paper of Kalman (1960), provide a recursive algorithm for linear filtering in control systems and have seen a resurgence in time series econometrics. The state space framework is remarkably versatile, accommodating a wide array of econometric models, such as autoregressive moving average (ARMA), vector autoregressions, dynamic factor models, panel data and time-varying-parameter regressions, stochastic volatility models, and structural time series models with trend, seasonal and regression components.

In contrast to the traditional ARMA model that describes observed time series, the state space model incorporates latent variables, which can be 1) unobserved components of times series such as trends and seasonality (Harvey and Durbin, 1986), 2) missing values in the dataset (Rubin, 1976), 3) out-of-sample variables in the conditional forecast (Waggoner and Zha, 1999), 4) disturbances in regression (Koopman, 1993), 5) auxiliary variables constructed for data augmentation (Tanner and Wong, 1987), 6) disaggregated or high-frequency data hidden in aggregated low-frequency observations (Schorfheide and Song, 2015), and 7) unknown parameters in the model (Chopin, 2002). Any latent variables can be put in the state vector. A time series model often has multiple state space representations by judicious choice of states. Different representations can be observationally equivalent, as the likelihood functions are the same.

The Kalman filter is central to linear state space models. While not inherently Bayesian, studying the Kalman filter from a Bayesian perspective offers the most insight. The Kalman filter updates the posterior state distribution with each new observation, a process known as forward filtering. This sequential learning algorithm is particularly well-suited for big data, as it can process one data point at a time, making it scalable and efficient.

Once all data are processed, we can apply smoothing techniques to refine our estimates of past states. The smoother can either provide a summary of the posterior state distribution or simulate a state path conditional on the observed data. While state estimation aligns with the Bayesian paradigm, parameter estimation can be approached from either classical or Bayesian perspectives. The Kalman filter provides the likelihood function as a by-product of state filtering, which facilitates likelihood-based inference. Also, the simulation smoother samples from the posterior conditional distribution of states, which greatly aids in the Gibbs sampling of state space models.

Given the complexity of state space theory and the breadth of its applications, we dedicate three chapters to this topic. Chapters 6 and 7 delve into theoretical analysis of linear and nonlinear models, while Chapter 8 explores practical applications of state space models across various econometric contexts.

DOI: 10.1201/9781003564027-6

6.1 Model Specification

State space models encompass two primary types of time series: latent states S_t and observable measurements Y_t. State space models are defined by three fundamental distributions: the states evolve by a Markov transition process $p(S_t | S_{t-1})$, starting with an initial state distribution $p(S_0)$. Observations provide an imperfect measurement of states, characterized by an observation distribution $p(Y_t | S_t)$.

In particular, Gaussian linear state space models take the form

$$p(S_t | S_{t-1}) = \phi(S_t; A_t S_{t-1}, B_t B_t'),$$
$$p(Y_t | S_t) = \phi(Y_t; C_t S_t, D_t D_t'),$$
$$p(S_0) = \phi(S_0; \mu_0, V_0),$$

where $\phi(\cdot)$ denotes the multivariate normal density. Alternatively, Gaussian linear state space models can be articulated through transition and observation equations, as in dynamic linear models (Petris et al., 2009):

$$S_t = A_t S_{t-1} + B_t u_t, \tag{6.1}$$
$$Y_t = C_t S_t + D_t \varepsilon_t, \tag{6.2}$$

where u_t and ε_t are independent and identically distributed standard normal disturbances. The state space coefficient matrices are determined by underlying parameters θ such as $A_t(\theta)$, $B_t(\theta)$, $C_t(\theta)$, $D_t(\theta)$. Specifying the parameter mapping is the process of casting a time series model in the state space form, which is inherently non-unique. A structural time series model often has multiple state space representations differing by the specification of the state vector.

6.1.1 An Introductory Example

To illustrate the non-uniqueness of state space forms in a time series model, consider a stationary autoregressive process: $y_t = c + \rho y_{t-1} + \sigma e_t$, where $|\rho| < 1$ and e_t follows the standard normal distribution.

Form 1 (intercept as a static state):

$$S_t = \begin{pmatrix} y_t \\ c \end{pmatrix}, Y_t = y_t,$$

$$A_t = \begin{pmatrix} \rho & 1 \\ 0 & 1 \end{pmatrix}, B_t = \begin{pmatrix} \sigma \\ 0 \end{pmatrix}, C_t = \begin{pmatrix} 1 & 0 \end{pmatrix}, D_t = 0.$$

Form 2 (constant one as a static state):

$$S_t = \begin{pmatrix} y_t \\ 1 \end{pmatrix}, Y_t = y_t,$$

$$A_t = \begin{pmatrix} \rho & c \\ 0 & 1 \end{pmatrix}, B_t = \begin{pmatrix} \sigma \\ 0 \end{pmatrix}, C_t = \begin{pmatrix} 1 & 0 \end{pmatrix}, D_t = 0.$$

Form 3 (centered states):

$$S_t = y_t - \frac{c}{1-\rho}, Y_t = y_t - \frac{c}{1-\rho},$$

$$A_t = \rho, B_t = \sigma, C_t = 1, D_t = 0.$$

As demonstrated in the example, the selection of states is pivotal to model specification. Form 1 treats the intercept c as a static state, while Form 2 and 3 view it as a parameter. State estimation is performed analytically via the Kalman filter, whereas parameter estimation necessitates numerical optimization methods. The Kalman filter applied to state space Form 1 yields the concentrated (profiled) likelihood. Moreover, the shorter state vector in Form 3 results in faster computation. Although Form 3 may seem trivial, as the state is directly observed, it remains a valid specification. Casting a time series model in state space form is both a science and an art. Any unobserved dynamic or static variable can be incorporated into the state vector, and even an observed variable may act as a perfectly measured state.

The initial state distribution is indispensable to model specification. Most users prefer initialization by the stationary distribution. In this example, the stationary mean and variance of $\{y_t\}$ are $\frac{c}{1-\rho}$ and $\frac{\sigma^2}{1-\rho^2}$, respectively. For the auxiliary static state c or 1, the variance of the constant one must be zero, and the variance of an unknown constant can be infinite. Although diffuse initialization is widely used, workarounds using proper priors are available. We will address initialization issues in Section 6.6.

6.1.2 Time-varying Coefficients and Dimensions

The flexibility of the state space model allows for time-varying transition and observation equations, with the possibility of changing dimensions of states and observations over time (Jungbacker et al., 2011; Chan et al., 2012; Qian, 2014). This flexibility does not complicate state and parameter estimation, as the Kalman filter processes one data point at a time.

Missing data and mixed frequency sampling can be conveniently handled by time-varying models (Ansley and Kohn, 1983). For example, consider an autoregressive process $y_t = c + \rho y_{t-1} + \sigma e_t$, where y_t is observed only on odd dates ($t = 1, 3, \ldots$), with data missing on even dates ($t = 2, 4, \ldots$). Multiple state space representations are possible. We examine three forms sharing the same transition equation

$$\begin{pmatrix} y_t \\ 1 \end{pmatrix} = \begin{pmatrix} \rho & c \\ 0 & 1 \end{pmatrix} \begin{pmatrix} y_{t-1} \\ 1 \end{pmatrix} + \begin{pmatrix} \sigma \\ 0 \end{pmatrix} e_t,$$

but the observation equations differ.

1. (time-varying dimension of observations):
   In odd dates, $Y_t = y_t$, $C_t = \begin{pmatrix} 1 & 0 \end{pmatrix}$, $D_t = 0_{1\times 0}$.
   In even dates, $Y_t = 0_{0\times 1}$, $C_t = 0_{0\times 2}$, $D_t = 0_{0\times 0}$.
2. (re-coding missing data as zero):
   In odd dates, $C_t = \begin{pmatrix} 1 & 0 \end{pmatrix}$, $D_t = 0$.
   In even dates, $C_t = \begin{pmatrix} 0 & 0 \end{pmatrix}$, $D_t = 0$.
3. (noisy observation by an arbitrary value):
   In odd dates, $C_t = \begin{pmatrix} 1 & 0 \end{pmatrix}$, $D_t = 0$.
   In even dates, $C_t = \begin{pmatrix} 0 & 0 \end{pmatrix}$, $D_t = 10^7$.

In Form 1, a missing value is treated as 0×1 empty matrix, with coefficients C_t and D_t being 0×2 and 0×0 empty matrices, respectively. In Form 2, missing values are recoded as zeros, simplifying the observation equation to a trivial identity: $0 = 0$. Form 3 allows missing data to be represented by arbitrary real numbers, with an inflated noise variance ensuring that missing data remain uninformative for state estimation.

Another application of time-varying coefficients is the univariate treatment of multivariate series. Consider $n \times 1$ vector Y_t, $t = 1, \ldots, T$. Instead of batch-processing n data points in T batches, the nT individual data points can be processed one at a time. The idea is to divide a period into n sub-periods. In the first sub-period, states evolve according to the Markov transition, while in subsequent sub-periods, states remain unchanged. The approach is well-suited for observations that are uncorrelated conditional on states (i.e. $D_t D_t'$ is a diagonal matrix). Even with correlated observations, univariate treatment remains feasible through state expansion:

$$\begin{pmatrix} S_t \\ \varepsilon_t \end{pmatrix} = \begin{pmatrix} A_t & 0 \\ 0 & 0 \end{pmatrix} \begin{pmatrix} S_{t-1} \\ \varepsilon_{t-1} \end{pmatrix} + \begin{pmatrix} B_t & 0 \\ 0 & I \end{pmatrix} \begin{pmatrix} u_t \\ \varepsilon_t \end{pmatrix},$$

$$Y_t = \begin{pmatrix} C_t & D_t \end{pmatrix} \begin{pmatrix} S_t \\ \varepsilon_t \end{pmatrix}.$$

6.1.3 Lagged Observations

In the baseline state space model, state evolution does not depend on observations. The data generating process can be separated as two stages. First, the state series $S_0, S_1, \ldots, S_T$ are generated by transition equations without reference to observation equations. Second, observations Y_t are generated conditional on S_t in each period $t = 1, \ldots, T$, without feedback to states.

Extending the model to support observational feedback enriches model dynamics without increasing the complexity of the Kalman filter. Denote sequences as $S_{0:T} = (S_0, S_1, \ldots, S_T)$ and $Y_{1:T} = (Y_1, \ldots, Y_T)$. An abstract state space model is defined by the joint distribution of states and observations:

$$p(S_{0:T}, Y_{1:T}) = p(S_0) \prod_{t=1}^{T} p(S_t | S_{0:t-1}, Y_{1:t-1}) \, p(Y_t | S_{0:t}, Y_{1:t-1}),$$

which consist of three components: 1) initial distribution $p(S_0)$, 2) Markov transition: S_t depends on S_{t-1}, not on $S_{t-2}, S_{t-3}, \ldots$, with potential path dependence on past observations:

$$p(S_t | S_{0:t-1}, Y_{1:t-1}) = p(S_t | S_{t-1}, Y_{1:t-1}),$$

and 3) instantaneous observations: Y_t depends on S_t, not on $S_{t-1}, S_{t-2}, \ldots$, with potential path dependence on past observations:

$$p(Y_t | S_{0:t}, Y_{1:t-1}) = p(Y_t | S_t, Y_{1:t-1}).$$

In terms of transition and observation Equations (6.1) and (6.2), the matrices A_t, B_t, C_t, D_t can be functions of past observations $Y_{1:t-1}$. For example, consider a time-varying-parameter AR(1) model:

$$y_t = \rho_t y_{t-1} + \sigma e_t,$$

$$\rho_t = \rho_{t-1} + \sigma_\rho u_t.$$

A state space form can be

$$S_t = \rho_t, Y_t = y_t,$$

$$A_t = 1, B_t = \sigma_\rho, C_t = y_{t-1}, D_t = \sigma,$$

with lagged observations in the matrix C_t.

6.2 Kalman Filter

The primary objective of state estimation is to compute the posterior distribution of states, conditional on observations and parameters. There are two principal types of state estimates: filtering and smoothing. The state filtering distribution refers to $p(S_t|Y_{1:t},\theta)$, while the state smoothing distribution refers to $p(S_t|Y_{1:T},\theta)$, $t=1,\ldots,T$.

The Kalman filter serves as a sequential learning algorithm, providing the optimal state estimator for linear state space models. Utilizing the previous state estimation, the Kalman filter predicts forthcoming observations, compares these predictions with actual data, and subsequently updates the state distribution based on prediction errors. The Kalman filter is inherently a big data algorithm, as it does not require loading the entire dataset into the computer memory for batch processing. Instead, the forward recursion sequentially processes one data point at a time.

Consider the Gaussian linear state space model specified by Equations (6.1) and (6.2), where the transition and observation equations can be expressed in matrix form:

$$\begin{pmatrix} S_t \\ Y_t \end{pmatrix} = \begin{pmatrix} A_t & B_t & 0 \\ C_tA_t & C_tB_t & D_t \end{pmatrix} \begin{pmatrix} S_{t-1} \\ u_t \\ \varepsilon_t \end{pmatrix}.$$

Since S_t, Y_t are linear functions of normally distributed random variables $S_{t-1}, u_t, \varepsilon_t$, the predictive distribution $p(S_t, Y_t|Y_{1:t-1},\theta)$ is multivariate normal (see Exercise 6.1). Means and variances suffice for summarizing normal distributions. We define

$$\begin{aligned} S_{t|r} &= E(S_t|Y_{1:r},\theta), \\ P_{t|r} &= Var(S_t|Y_{1:r},\theta), \\ Y_{t|r} &= E(Y_t|Y_{1:r},\theta), \\ O_{t|r} &= Var(Y_t|Y_{1:r},\theta). \end{aligned}$$

The Kalman filter is initialized with $S_{0|0}=\mu_0$ and $P_{0|0}=V_0$. In the forward recursion for period $t=1,\ldots,T$, based on the previous state estimator $N\left(S_{t-1|t-1}, P_{t-1|t-1}\right)$, the predictive distribution $p(S_t, Y_t|Y_{1:t-1},\theta)$ is given by

$$N\left[\begin{pmatrix} S_{t|t-1} \\ Y_{t|t-1} \end{pmatrix}, \begin{pmatrix} P_{t|t-1} & L_{t|t-1} \\ L'_{t|t-1} & O_{t|t-1} \end{pmatrix}\right],$$

where

$$\begin{aligned} S_{t|t-1} &= A_tS_{t-1|t-1}, \\ Y_{t|t-1} &= C_tS_{t|t-1}, \\ P_{t|t-1} &= A_tP_{t-1|t-1}A_t' + B_tB_t', \\ O_{t|t-1} &= C_tP_{t|t-1}C_t' + D_tD_t', \\ L_{t|t-1} &= P_{t|t-1}C_t'. \end{aligned}$$

The key step of the Kalman filter is updating the state distribution using observations. The update rule adheres to the multivariate normal theory: conditional distributions maintain normality. The filtering distribution $p(S_t|Y_{1:t},\theta)$ is given by $N\left(S_{t|t}, P_{t|t}\right)$, where

$$S_{t|t} = S_{t|t-1} + L_{t|t-1}O_{t|t-1}^{-1}\left(Y_t - Y_{t|t-1}\right),$$

Algorithm 6.1 Kalman filter

Input: data Y and initial state distribution $\left(S_{0|0}, P_{0|0}\right)$.
Output: state filtering distribution $\left(S_{t|t}, P_{t|t}\right)$.
Iterate the following distributions for $t = 1, \ldots, T$:
1 State predictive distribution $\left(S_{t|t-1}, P_{t|t-1}\right)$

$$S_{t|t-1} = A_t S_{t-1|t-1},$$

$$P_{t|t-1} = A_t P_{t-1|t-1} A_t' + B_t B_t'.$$

2 Data predictive distribution $\left(Y_{t|t-1}, O_{t|t-1}\right)$

$$Y_{t|t-1} = C_t S_{t|t-1},$$

$$O_{t|t-1} = C_t P_{t|t-1} C_t' + D_t D_t'.$$

3 State filtering distribution $\left(S_{t|t}, P_{t|t}\right)$

$$S_{t|t} = S_{t|t-1} + P_{t|t-1} C_t' O_{t|t-1}^{-1} \left(Y_t - Y_{t|t-1}\right),$$

$$P_{t|t} = P_{t|t-1} - P_{t|t-1} C_t' O_{t|t-1}^{-1} C_t P_{t|t-1}.$$

$$P_{t|t} = P_{t|t-1} - L_{t|t-1} O_{t|t-1}^{-1} L_{t|t-1}'.$$

The intuition behind the Kalman filter is that states are learned from prediction errors (i.e., $Y_t - Y_{t|t-1}$), and the learning rate depends on data volatility (i.e., $O_{t|t-1}$) and correlations between states and observations (i.e., $L_{t|t-1}$).

Algorithm 6.1 summarizes the Kalman filter with prediction and update steps in the forward recursion. We intentionally simplify the notation $N\left(S_{t|t}, P_{t|t}\right)$ as $\left(S_{t|t}, P_{t|t}\right)$. The reason is that the Kalman filer can be derived without dependency on normality, with the linear projection theory justifying the optimality of the state estimator. Nevertheless, the distributional assumption supports Bayesian interpretations of the Kalman filter, where the filtering distribution is viewed as the posterior state distribution conditional on observations and parameters. The Kalman filter formula follows directly from well-known properties of multivariate normal distributions.

Further insights into the Kalman filter are obtained by examining the predictive distribution. First, the assumption that u_t and ε_t are independent can be relaxed. In the presence of correlations, $S_{t-1}, u_t, \varepsilon_t$ still follow a multivariate normal distribution, and the predictive distribution $p\left(S_t \,|Y_{1:t}, \theta\right)$ preserves normality (see Exercises 6.1 and 6.2). Second, the assumption of instantaneous observations can be relaxed such that $Y_t = C_t S_t + D_t \varepsilon_t + E_t S_{t-1}$. Since S_t, Y_t remain linear functions of $S_{t-1}, u_t, \varepsilon_t$, the Kalman filter proceeds with minimal changes (see Exercise 6.3). Third, it is feasible to share u_t between transition and observation equations, preserving a multivariate normal predictive distribution, thereby maintaining the applicability of the Kalman filter (see Exercise 6.4).

6.3 More on Kalman Filter

The Kalman filter derives state estimates from two sources: the model and data. The transition equation proposes a model-implied state estimator based on previous estimations. The observation equation updates the state estimator using data. It is possible to separate the roles of the model and data. The model solely determines the state variance recursion, with no dependency on observations. Additionally, the model determines the Kalman gain, which quantifies the learning rate from prediction errors.

In this section, we explore alternative forms of the Kalman filter. Although algebraically equivalent, certain forms facilitate theoretical analysis of convergence of the Kalman filter, while others enhance numerical stability during computer implementation.

6.3.1 Riccati Equation

The forward recursion of the Kalman filter involves six key variables:

$$S_{t|t-1}, P_{t|t-1}, Y_{t|t-1}, O_{t|t-1}, S_{t|t}, P_{t|t},$$

which can be categorized into two groups: the conditional means $S_{t|t-1}, Y_{t|t-1}, S_{t|t}$ and the conditional variances $P_{t|t-1}, O_{t|t-1}, P_{t|t}$. The variance recursion formulae can be consolidated as

$$\begin{aligned} P_{t+1|t} &= A_{t+1}P_{t|t-1}A'_{t+1} + B_{t+1}B'_{t+1} \\ &\quad - A_{t+1}P_{t|t-1}C'_t\left(C_tP_{t|t-1}C'_t + D_tD'_t\right)^{-1}C_tP_{t|t-1}A'_{t+1}, \end{aligned}$$

which is known as the discrete-time algebraic Riccati equation (DARE) for the variance sequence $\{P_{t+1|t}\}$, $t = 0, 1, 2, \ldots$. This recursion is governed by model coefficients A_t, B_t, C_t, D_t with no reference to actual data, implying that we can determine the uncertainty of state estimation even before observing the data.

In most state space models, the variance sequence converges as the Kalman filter progresses. Proposition 6.1 provides sufficient conditions under which the DARE variance sequence converges to a steady state. Even if the eigenvalue conditions are not satisfied, convergence results may be established by alternative proofs. See Section 6.8 for an example, in which the transition equation involves a unit root (i.e., an eigenvalue equals one).

Proposition 6.1 (Kalman Filter Convergence) *Consider a linear state space model with time-invariant coefficients A, B, C, D. If all eigenvalues of A lie within the unit circle, and either $Q = BB'$ or $R = DD'$ is positive definite, then the variance sequence $\{P_{t+1|t}\}$ converges to a steady P^* such that*

$$P^* = AP^*A' - AP^*C'\left(CP^*C' + R\right)^{-1}CP^*A' + Q.$$

6.3.2 Kalman Gain

The recursive formulae of the conditional mean group (i.e., $S_{t|t-1}, Y_{t|t-1}, S_{t|t}$) can be combined as

$$S_{t+1|t} = A_{t+1}S_{t|t-1} + K_tv_t,$$

where $v_t = Y_t - C_tS_{t|t-1}$ is the prediction error, and

$$K_t = A_{t+1}P_{t|t-1}C'_t\left(C_tP_{t|t-1}C'_t + D_tD'_t\right)^{-1}$$

represents the Kalman gain, determined by the model coefficients without dependency on observations. As we iterate through the variance sequence $\{P_{t+1|t}\}$, we can compute the Kalman gain as well.

The Kalman gain weights in prediction errors for updating the model-implied state estimator. The Kalman gain increases with correlations between states and observations, and it decreases with data volatility. For instance, if observations are excessively noisy, extracting signals from imprecise data becomes challenging. In such cases, the Kalman gain is small, and the model-implied state estimator prevails.

It may be tempting to view the Kalman gain as a mathematical construct derived from model coefficients A_t, B_t, C_t, D_t. Although the Kalman gain varies over time, it converges asymptotically, along with the convergence of the Riccati equation, as indicated by Proposition 6.1. With a steady Kalman gain, we envision a subjective state space model:

$$S_{t+1|t} = A_{t+1} S_{t|t-1} + K_t v_t,$$

$$Y_t = C_t S_{t|t-1} + v_t.$$

The model is subjective in that the states represent our beliefs about the true states. Updating beliefs are reminiscent of exponential smoothing: a weighted sum of the momentum (i.e., $A_{t+1} S_{t|t-1}$) and the prediction error (i.e., v_t), with the weight determined by the Kalman gain.

6.3.3 Numerically Stable Filter

Consider the state and data prediction errors:

$$\eta_t = S_t - S_{t|t-1},$$

$$v_t = Y_t - Y_{t|t-1}.$$

By the law of iterated expectations, the random variables η_t and v_t have means 0 and variances $P_{t|t-1}$ and $O_{t|t-1}$ respectively:

$$E(\eta_t) = E[E(\eta_t | Y_{1:t-1})] = 0,$$

$$Var(\eta_t) = E\left\{E\left[(S_t - S_{t|t-1})(S_t - S_{t|t-1})' | Y_{1:t-1}\right]\right\} = P_{t|t-1},$$

$$E(v_t) = E[E(v_t | Y_{1:t-1})] = 0,$$

$$Var(v_t) = E\left\{E\left[(Y_t - Y_{t|t-1})(Y_t - Y_{t|t-1})' | Y_{1:t-1}\right]\right\} = O_{t|t-1}.$$

Moreover, the data prediction errors follow a white noise process, as $E(v_t v_{t-j}) = 0$, $\forall j \geq 1$, but the state prediction errors exhibit serial correlations. Interestingly, the state and data prediction errors provide a companion state space system:

$$\eta_{t+1} = J_t \eta_t + B_{t+1} u_{t+1} - K_t D_t \varepsilon_t,$$

$$v_t = C_t \eta_t + D_t \varepsilon_t,$$

where $J_t = A_{t+1} - K_t C_t$.

By examining the transition equation for η_{t+1} and taking the expectation of the squared η_{t+1}, we obtain the Joseph stable form of the Riccati equation:

$$P_{t+1|t} = J_t P_{t|t-1} J_t' + B_{t+1} B_{t+1}' + K_t D_t D_t' K_t'.$$

This recursive formula is numerically stable because $P_{t+1|t}$, as the sum of three positive semi-definite matrices, maintains symmetry and positive semi-definiteness throughout the forward recursion. The stable form is well-suited for the Kalman filter initialized with a large variance, whereas the regular Kalman filter might suffer catastrophic cancellation errors in numerical computation.

The stable form is reminiscent of the Lyapunov equation, and the vectorized sequence evolves by

$$vec\left(P_{t+1|t}\right) = \left(J_t \otimes J_t\right) vec\left(P_{t|t-1}\right) + vec\left(B_{t+1}B'_{t+1} + K_t D_t D'_t K'_t\right).$$

In the proof of Proposition 6.1, we demonstrated that all eigenvalues of the steady state $J^* = A - K^*C$ are inside the unit circle, as are the eigenvalues of $J^* \otimes J^*$. This justifies the convergence of the variance sequence $\{P_{t+1|t}\}$.

6.3.4 Square Root Filter

When the Kalman filter was first implemented on computers several decades ago, engineers were surprised by performance degradation. After a few iterations, the state covariance matrices could lose symmetry and positive definiteness: a theoretical impossibility was observed on computers. Researchers found that this anomaly could be avoided, or at least mitigated, by tracking the matrix square root, leading to the development of the square root Kalman filter.

Using Cholesky or eigenvalue decomposition, we can find the square root of a covariance matrix. Let $P_{t|t-1} = \tilde{P}_{t|t-1}\tilde{P}'_{t|t-1}$. The square root filter focuses on the sequence $\left\{\tilde{P}_{t|t-1}\right\}$, instead of $\{P_{t|t-1}\}$, in the forward recursion.

To recursively compute $\tilde{P}_{t+1|t}$ for a given $\tilde{P}_{t|t-1}$, we construct a utility matrix:

$$U_t = \begin{pmatrix} C_t\tilde{P}_{t|t-1} & D_t & 0 \\ A_{t+1}\tilde{P}_{t|t-1} & 0 & B_{t+1} \end{pmatrix}.$$

The idea is that $U_tU'_t$ contains all components of the Riccati equation:

$$U_tU'_t = \begin{pmatrix} C_tP_{t|t-1}C'_t + D_tD'_t & C_tP_{t|t-1}A'_{t+1} \\ A_{t+1}P_{t|t-1}C'_t & A_{t+1}P_{t|t-1}A'_{t+1} + B_{t+1}B'_{t+1} \end{pmatrix}.$$

The matrix $\tilde{P}_{t+1|t}$ can be computed from QR decomposition: $U'_t = q_tr_t$, where q_t is an orthonormal matrix and r_t takes the form:

$$r'_t = \begin{pmatrix} U^*_{1t} & 0 & 0 \\ U^*_{2t} & U^*_{3t} & 0 \end{pmatrix}.$$

Since $U_tU'_t = r_tr'_t$, the matrices U^*_{1t}, U^*_{2t} and U^*_{3t} can be obtained by comparing coefficients of the following equation:

$$\begin{pmatrix} U^*_{1t}U^{*\prime}_{1t} & U^*_{1t}U^{*\prime}_{2t} \\ U^*_{2t}U^{*\prime}_{1t} & U^*_{2t}U^{*\prime}_{2t} + U^*_{3t}U^{*\prime}_{3t} \end{pmatrix} = \begin{pmatrix} C_tP_{t|t-1}C'_t + D_tD'_t & C_tP_{t|t-1}A'_{t+1} \\ A_{t+1}P_{t|t-1}C'_t & A_{t+1}P_{t|t-1}A'_{t+1} + B_{t+1}B'_{t+1} \end{pmatrix},$$

resulting in

$$U^*_{1t} = \tilde{O}_{t|t-1},$$
$$U^{*\prime}_{2t} = A_{t+1}P_{t|t-1}C'_t\tilde{O}^{-1}_{t|t-1},$$
$$U^*_{3t} = \tilde{P}_{t+1|t},$$

where $O_{t|t-1} = \tilde{O}_{t|t-1}\tilde{O}'_{t|t-1}$.

In summary, the square root filter iterates the Riccati equation by tracking the square root sequence $\left\{\tilde{P}_{t|t-1}\right\}$. For a given $\tilde{P}_{t|t-1}$, we construct a utility matrix and perform QR decomposition: $U'_t = q_tr_t$, which reveals $\tilde{P}_{t+1|t}$ by examining the matrix element in r_t.

6.3.5 Chandrasekhar Recursion

In the forward recursion of the state variance sequence $\{P_{t+1|t}\}$, the most computationally intensive operation is matrix multiplication $AP_{t|t-1}A'$, especially when the state dimension is high. The Chandrasekhar recursion provides an alternative filtering algorithm for a special class of state space models characterized by three conditions: 1) coefficients A, B, C, D are time-invariant, 2) states are initialized by the stationary distribution, and 3) observations contain no missing values. The Chandrasekhar recursion focuses on the first-differenced variance sequence $\{P_{t+1|t} - P_{t|t-1}\}$, which is of reduced rank and computationally efficient, particularly when the number of states significantly exceeds the number of observed series. Herbst (2015) demonstrates that this method is ideal for Dynamic Stochastic General Equilibrium (DSGE) models.

When the transition equation $S_t = AS_{t-1} + Bu_t$ admits a stationary distribution, the stationary variance satisfies the Lyapunov equation

$$P_{0|0} = AP_{0|0}A' + BB',$$

which also implies that $P_{1|0} = P_{0|0}$. Let $\Delta P_{t+1|t} = P_{t+1|t} - P_{t|t-1}$ be the first-differenced state variance, which possesses several useful features.

First, the matrix $\Delta P_{t+1|t}$ is negative semi-definite due to stationarity (i.e., the state distribution remains unchanged by shifting the time index)

$$P_{t+1|t} = Var\left(S_{t+1} \,|Y_{1:t}\right) \leq Var\left(S_{t+1} \,|Y_{2:t}\right) = Var\left(S_t \,|Y_{1:t-1}\right) = P_{t|t-1},$$

where the inequality follows from the decreasing posterior variance conditional on a larger information set.

Second, the sequence $\{\Delta P_{t+1|t}\}$ converges to zero, as the monotone decreasing sequence $\{P_{t+1|t}\}$ is bounded below by zero.

Third, the matrix $\Delta P_{t+1|t}$ is of low rank, which is typically equal to the number of observed series (i.e., the dimension of Y_t), whereas the rank of the covariance matrix $P_{t+1|t}$ can be as large as the number of states (i.e., the dimension of S_t). This rank reduction is the root cause of the computational efficiency of the Chandrasekhar recursion.

Consider the variance recursion by the Kalman filter in the first two periods:

$$P_{2|1} = AP_{1|0}A' - AP_{1|0}C'\left(CP_{1|0}C' + DD'\right)^{-1}CP_{1|0}A' + BB'.$$

Stationarity requires $P_{1|0} = AP_{1|0}A' + BB'$, which indicates that the change of state variances is non-positive:

$$\Delta P_{2|1} = -AP_{1|0}C'\left(CP_{1|0}C' + DD'\right)^{-1}CP_{1|0}A'.$$

The rank of $\Delta P_{2|1}$ is no larger than the rank of the covariance matrix $Var\left(Y_1\right) = CP_{1|0}C' + DD'$, and the rank is usually equal to the number of observed series. For univariate observations, the change in state variances is a rank-one matrix.

The Chandrasekhar recursion is initialized by decomposing $\Delta P_{2|1}$ as the product of low-rank matrices:

$$\Delta P_{2|1} = -W_1 M_1 W_1',$$

where $W_1 = AP_{1|0}C'$ and $M_1 = \left(CP_{1|0}C' + DD'\right)^{-1}$.

If the state space model coefficients A, B, C, D are time-invariant, the variance recursion of the Kalman filter can be formulated in terms of first-differenced variances:

$$\Delta P_{t+1|t} = J_{t-1}\left(\Delta P_{t|t-1} - \Delta P_{t|t-1}C'O_{t|t-1}^{-1}C\Delta P_{t|t-1}\right)J_{t-1}',$$

where

$$J_t = A - K_t C,$$
$$K_t = AP_{t|t-1}C'O_{t|t-1}^{-1},$$
$$O_{t|t-1} = CP_{t|t-1}C' + DD'.$$

By induction, the current-period decomposition $\Delta P_{t|t-1} = -W_{t-1}M_{t-1}W_{t-1}'$ leads to the next-period decomposition

$$\Delta P_{t+1|t} = -W_t M_t W_t',$$

where

$$W_t = J_{t-1}W_{t-1},$$
$$M_t = M_{t-1} + M_{t-1}W_{t-1}'C'O_{t|t-1}^{-1}CW_{t-1}M_{t-1}.$$

Since M_t is always a symmetric and positive definite matrix, the change in state variances $\Delta P_{t+1|t}$ is always a symmetric and negative semi-definite matrix (with some eigenvalues being zero). Symmetry and definiteness prevent numerical issues when implementing the Kalman filter on computers, making the Chandrasekhar recursion a numerically stable algorithm for state filtering and likelihood evaluation. If the number of states exceeds the number of observed series, the rank of $\Delta P_{t+1|t}$ is determined by the rank M_t, which is further determined by the rank of $O_{t|t-1}$. Typically, this rank equals the number of observed series, making the Chandrasekhar recursion computationally efficient compared to the standard Kalman filter.

When the eigenvalues of A are inside the unit circle, the sequence $\{J_t\}$ tends to a steady state J^*, whose eigenvalues are also inside the unit circle. Consequently, $W_t = J_{t-1}W_{t-1}$ should decrease to zero as the filter moves forward, with the sequence $\{\Delta P_{t+1|t}\}$ converging to zero as well.

Algorithm 6.2 summarizes the Chandrasekhar filtering process, where the core steps are 2.1 and 2.4, while other steps reproduce the standard Kalman filter output variables. The primary difference between the Chandrasekhar method and the standard Kalman filter lies in updating the state variance $P_{t+1|t}$. The standard Kalman filter requires matrix multiplication $AP_{t|t-1}A'$, whereas the Chandrasekhar recursion uses matrix addition $P_{t+1|t} = P_{t|t-1} + \Delta P_{t+1|t}$, with $\Delta P_{t+1|t}$ updated by low-rank matrix multiplication of the form $-W_t M_t W_t'$. The computational complexity of $\Delta P_{t+1|t}$ is $O\left(m^2 n\right)$, where m and n are dimensions of S_t and Y_t, respectively. In contrast, the computing cost of matrix multiplication $AP_{t|t-1}A'$ amounts to $O\left(m^3\right)$. Therefore, the Chandrasekhar filtering algorithm is more computationally efficient than the standard Kalman filter, if the number of states is larger than the number of observed series.

In Algorithm 6.2, we separate the conditional variance updates (step 2) and the conditional mean updates (step 3), but it is possible to incorporate step 3 into step 2, as soon as $\Delta P_{t+1|t}$ is updated in each period. The separation indicates that the variance update (either by the level or first-differenced series) only depends on the model coefficients A, B, C, D, but not on the data Y.

Additionally, the time indices of some variables in Algorithm 6.2 are shifted for consistency with the standard Kalman filter formulae. For example, $\Delta P_{t+1|t}$, $t = 1, \ldots, T-1$ is written in an equivalent form: $\Delta P_{t|t-1}, t = 2, \ldots, T$.

Algorithm 6.2 Chandrasekhar recursion for Kalman filter

Input: time-invariant coefficients A, B, C, D and data Y.
Output: Kalman filter results $(S_{t|t}, P_{t|t})$, $(S_{t|t-1}, P_{t|t-1})$ and $(Y_{t|t-1}, O_{t|t-1})$.
1 Initialize state variance by stationary distribution:

$$P_{1|0} = AP_{1|0}A' + BB',$$

and initialize low-rank matrices

$$W_1 = AP_{1|0}C',$$

$$M_1 = \left(CP_{1|0}C' + DD'\right)^{-1}.$$

2 Iterate the following steps of variance update for $t = 2, \ldots, T$:
2.1 Compute differenced state variances by low-rank multiplication:

$$\Delta P_{t|t-1} = -W_{t-1}M_{t-1}W'_{t-1}.$$

2.2 Recover predictive variances for states and observations:

$$P_{t|t-1} = P_{t-1|t-2} + \Delta P_{t|t-1},$$

$$O_{t|t-1} = CP_{t|t-1}C' + DD'.$$

2.3 Compute Kalman gain (to be used in the next iteration):

$$K_t = AP_{t|t-1}C'O^{-1}_{t|t-1},$$

$$J_t = A - K_tC.$$

2.4 Update low-rank matrices (to be used in the next iteration):

$$W_t = J_{t-1}W_{t-1},$$

$$M_t = M_{t-1} + M_{t-1}W'_{t-1}C'O^{-1}_{t|t-1}CW_{t-1}M_{t-1}.$$

2.5 Recover filtering variance (for compatibility with Kalman filter)

$$P_{t|t} = P_{t|t-1} - P_{t|t-1}C'O^{-1}_{t|t-1}CP_{t|t-1}.$$

3 Recover conditional mean terms by variance terms

$$S_{t|t-1} = AS_{t-1|t-1},$$

$$Y_{t|t-1} = CS_{t|t-1},$$

$$S_{t|t} = S_{t|t-1} + P_{t|t-1}C'O^{-1}_{t|t-1}\left(Y_t - Y_{t|t-1}\right).$$

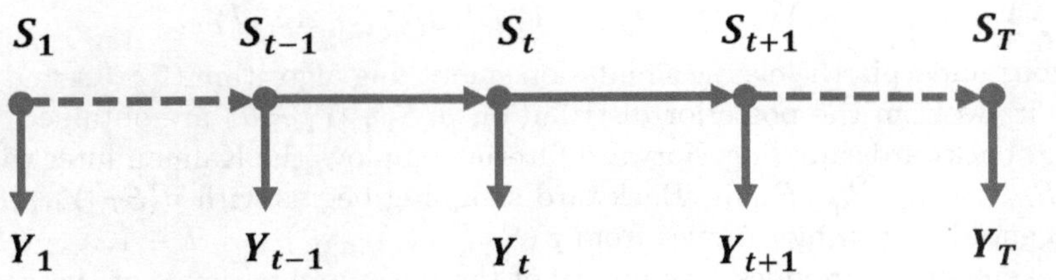

FIGURE 6.1
Directed acyclic graph of state space models.

6.4 Kalman Smoother and Simulation Smoothing

The forward recursion of the Kalman filter is an online learning algorithm, where data are accessed sequentially to produce state filtering distributions $p(S_t | Y_{1:t}, \theta)$, $t = 1, \ldots, T$. Once the full sample data are available, it becomes pertinent to perform backward recursion for smoothing distributions $p(S_t | Y_{1:T}, \theta)$, $t = T, T-1, \ldots, 1$. In the context of Bayesian inference, it is also important to generate a sequence of draws from the posterior distribution $p(S_{1:T} | Y_{1:T}, \theta)$ by forward filtering and backward sampling, which is typically employed in the Gibbs sampler that cycles thorough $p(\theta | Y_{1:T}, S_{1:T})$ and $p(S_{1:T} | Y_{1:T}, \theta)$.

The classic smoothing algorithm is derived from the marginal-conditional decomposition of the joint posterior distribution

$$p(S_{1:T} | Y_{1:T}, \theta) = \prod_{t=1}^{T} p(S_t | S_{t+1:T}, Y_{1:T}, \theta) = \prod_{t=1}^{T} p(S_t | S_{t+1}, Y_{1:t}, \theta).$$

The underlying intuition is that the state S_t is learned by three channels: S_{t+1}, $Y_{1:t-1}$ and Y_t, which represent future, past and present variables, respectively. Given these channels, other conditioning variables, such as $S_{t+2:T}$ and $Y_{t+1:T}$ become redundant. See Exercise 6.6 for a derivation.

The conditional independence structure in the state space model is depicted Figure 6.1 as a directed acyclic graph (DAG). Since the node S_{t+1} blocks the chain from S_t to S_T in the DAG, the nodes $S_{t+2:T}$ must be independent to S_t conditional on S_{t+1}. Similarly, the node S_{t+1} also blocks the chains from S_t to Y_{t+j}, $\forall j \geq 1$, thus the nodes $Y_{t+1:T}$ are independent to S_t conditional on S_{t+1}. Additionally, the DAG in Figure 6.1 demonstrates the Markov property in the state space model: the node S_{t-1} blocks the fork between Y_{t-1} and S_t, as well as the chain from S_1 to S_t. Therefore, the future evolution of S_t depends solely on S_{t-1} without path dependency.

In Gaussian linear state space models, the joint distribution $p(S_t, S_{t+1} | Y_{1:t}, \theta)$ is multivariate normal:

$$N\left[\begin{pmatrix} S_{t|t} \\ S_{t+1|t} \end{pmatrix}, \begin{pmatrix} P_{t|t} & P_{t|t} A'_{t+1} \\ A_{t+1} P_{t|t} & P_{t+1|t} \end{pmatrix}\right].$$

By Bayes' rule, the conditional distribution $p(S_t | S_{t+1}, Y_{1:t}, \theta)$ is normal with mean and variance

$$E(S_t | S_{t+1}, Y_{1:t}, \theta) = S_{t|t} + P_{t|t} A'_{t+1} P^{-1}_{t+1|t} (S_{t+1} - S_{t+1|t}),$$

$$Var\left(S_t \left| S_{t+1}, Y_{1:t}, \theta\right.\right) = P_{t|t} - P_{t|t} A'_{t+1} P_{t+1|t}^{-1} A_{t+1} P_{t|t}.$$

These equations underpin the classic simulation smoothing algorithm (Carter and Kohn, 1994), in which draws from the posterior distribution $p\left(S_{1:T} \left| Y_{1:T}, \theta\right.\right)$ are obtained by forward filtering and backward sampling. Forward filtering employs the Kalman filter to recursively compute $S_{t|t}, S_{t+1|t}, P_{t|t}, P_{t+1|t}$. Backward sampling begins with $p\left(S_T \left| Y_{1:T}, \theta\right.\right)$ and proceeds by sequentially drawing samples from $p\left(S_t \left| S_{t+1}, Y_{1:t}, \theta\right.\right)$, $t = T-1, \ldots, 1$.

The pointwise smoother provides a summary of the mean and variance of the posterior state distribution at each time point. By the law of iterated expectations, we have

$$S_{t|T} = S_{t|t} + P_{t|t} A'_{t+1} P_{t+1|t}^{-1} \left(S_{t+1|T} - S_{t+1|t}\right).$$

This is a recursive formula. First, we run the Kalman filter and save $S_{t|t}, S_{t+1|t}, P_{t|t}, P_{t+1|t}$. Second, we start from $S_{T|T}$ and recursively compute $S_{T-1|T}, \ldots, S_{1|T}$. Similarly, the smoothing variances also follow a recursive form (See Exercise 6.7):

$$P_{t|T} = P_{t|t} - G_t \left(P_{t+1|t} - P_{t+1|T}\right) G'_t,$$

where $G_t = P_{t|t} A'_{t+1} P_{t+1|t}^{-1}$. With outputs from the Kalman filter, backward recursion commences from $P_{T|T}$ and proceeds with $P_{T-1|T}, \ldots, P_{1|T}$. A larger set of conditioning variables reduces the uncertainty of state estimation, implying that $P_{t+1|t} - P_{t+1|T}$ and $P_{t|t} - P_{t|T}$ are positive semi-definite matrices.

The classic smoothing algorithm has limitations. First, it is computationally intensive to invert the matrix $P_{t+1|t}$ if the state dimension is high. Second, the state covariance matrix can be positive semi-definite (with some eigenvalues being zero), leading to non-invertibility of the state covariance matrix. Third, the smoothing formulae appear counter-intuitive. Compared to the filtering distribution, the smoothing distribution incorporates additional observations in the set of conditioning variables. However, the backward recursion uses the state precision $P_{t+1|t}^{-1}$ rather than the observation precision $O_{t|t-1}^{-1}$ in the backward recursion. Utilizing the latter would be more computationally efficient, as the observation dimension is typically lower than the state dimension.

De Jong (1989) introduces an improved smoothing algorithm that addresses the invertibility issue inherent in the classic smoothing approach. The revised formula reads

$$r_t = C'_t O_{t|t-1}^{-1} v_t + J'_t r_{t+1},$$

where $r_t = P_{t|t-1}^{-1} \left(S_{t|T} - S_{t|t-1}\right)$, $v_t = Y_t - Y_{t|t-1}$, $J_t = A_{t+1} - K_t C_t$ and $K_t = A_{t+1} P_{t|t-1} C'_t O_{t|t-1}^{-1}$. The revised formula is equivalent to the classic smoothing formula through an equation $P_{t|t} A'_{t+1} = P_{t|t-1} J'_t$. Nevertheless, the revised formula is more computationally efficient, as it avoids inverting the state covariance matrix. The improved smoothing algorithm commences from $r_{T+1} = 0$ and recursively computes $r_T, r_{T-1}, \ldots, r_1$. The state smoother can be recovered by

$$S_{t|T} = S_{t|t-1} + P_{t|t-1} r_t.$$

See Exercise 6.8 for technical details on the improved smoothing algorithm.

Last but not least, Durbin and Koopman (2002) develop a clever method of simulation smoothing, exploiting a unique property of the normal distribution: the conditional variance is not a function of the conditioning variable. The algorithm involves three steps: 1) simulate states and artificial data by sampling from the joint distribution $p\left(S_{0:T}^*, Y_{1:T}^*\right)$, 2) apply pointwise smoothing to both artificial and actual data: $S_{t|T} = E\left(S_t \left| Y_{1:T}\right.\right)$ and $S_{t|T}^* = E\left(S_t^* \left| Y_{1:T}^*\right.\right)$, and 3) adjust the mean of the simulated states: $S_t = S_t^* - S_{t|T}^* + S_{t|T}$. Consequently, $S_{0:T}$ is sampled from the smoothing distribution $p\left(S_{0:T} \left| Y_{1:T}\right.\right)$. Durbin and Koopman (2002) describe it as "a simple and efficient simulation smoother", and Exercise 6.9 explores the rationale behind the algorithm.

6.5 Parameter Estimation

Unknown variables in state space models include time-varying states and time-invariant parameters. The distinction between states and parameters is crucial in state space modeling. State estimation is tackled using the Kalman filter and smoother, which recursively compute the posterior state distributions conditional on observations and parameters. In contrast, parameter estimation is typically addressed through numerical maximum likelihood methods or Bayesian MCMC simulation, wherein states are integrated out of the likelihood function.

6.5.1 Likelihood function

For state space models, the likelihood function is expressed in terms of observational densities conditional on parameters, with states integrated out:

$$\begin{aligned} p(Y_{1:T}\,|\theta) &= \int p(Y_{1:T}, S_{0:T}\,|\theta)\, dS_{0:T} \\ &= \int p(S_0\,|\theta) \prod_{t=1}^{T} p(S_t\,|S_{t-1},\theta)\, p(Y_t\,|S_t,\theta)\, dS_{0:T}. \end{aligned}$$

The high-dimensional integral does not pose computational burden in linear state space models, as the closed-form likelihood function is readily available as a by-product of the Kalman filter. Algorithm 6.1 indicates that the predictive distribution $p(Y_t\,|Y_{1:t-1},\theta)$ is given by the normal density $\phi\left(Y_t; Y_{t|t-1}, O_{t|t-1}\right)$. Upon completing the forward recursion of the Kalman filter, the likelihood function is already obtained in the prediction error decomposition form:

$$p(Y_{1:T}\,|\theta) = \prod_{t=1}^{T} \phi\left(Y_t; Y_{t|t-1}, O_{t|t-1}\right).$$

An intriguing question is whether $p(Y_{1:T}\,|\theta)$ is a multivariate normal distribution. The answer is affirmative if the coefficients A_t, B_t, C_t, D_t are deterministic. In that case, closed-form filtering and smoothing results can be derived without the Kalman filter (see Exercise 6.5). However, if the coefficients are functions of past observations, the joint distribution $p(Y_{1:T}\,|\theta)$ is not necessarily normal, even if $p(Y_t\,|Y_{1:t-1},\theta)$ follows a normal distribution at each point in time. An example is the time-varying-parameter regression where $C_t = Y_{t-1}$. Regardless of distributional forms, the Kalman filter provides an exact and efficient algorithm for evaluating the density function $p(Y_{1:T}\,|\theta)$ for each value of θ.

6.5.2 Likelihood-based Inference

With the ability to evaluate the likelihood function pointwise, numerical optimization methods can be applied to the log-likelihood $\ln p(Y_{1:T}\,|\theta)$ as the objective function. Maximizing this objective function with respect to θ yields the maximum likelihood estimator, which generally provides a consistent estimator of unknown parameters.

If the objective function is differentiable with respect to parameters, it is desirable to employ quasi-Newton numerical optimization methods (refer to Section 5.8), which utilize gradient information to update parameters. Given that the likelihood function derived from the Kalman filter takes an iterative form, gradient computation is not straightforward. Koopman and Shephard (1992) discuss the exact gradient of the log-likelihood. They conclude that the gradient associated with system matrices A_t and C_t can be tedious. In practice, numerical differentiation via finite differencing or automatic differentiation in the forward mode may be used.

The likelihood function is fundamental to Bayesian analysis, as it allows the decomposition of the posterior distribution $p(\theta, S_{1:T} | Y_{1:T})$ as

$$p(\theta, S_{1:T} | Y_{1:T}) = p(\theta | Y_{1:T}) p(S_{1:T} | Y_{1:T}, \theta),$$
$$p(\theta | Y_{1:T}) \propto p(\theta) p(Y_{1:T} | \theta).$$

A marginalized MH sampler employs the Kalman filter to evaluate the likelihood function, updating parameters with the MH sampler, and then invokes the simulation smoother for state estimation (see Section 6.4). This method is extensively used in the macro-econometric analysis of Dynamic Stochastic General Equilibrium models. For pioneering work, see Smets and Wouters (2007).

An alternative posterior simulation method is the Gibbs sampler that cycles through the full posterior conditionals $p(\theta | Y_{1:T}, S_{1:T})$ and $p(S_{1:T} | Y_{1:T}, \theta)$. The approach is applicable to many time series models where $p(\theta | Y_{1:T}, S_{1:T})$ assumes a conjugate form. For instance, the dynamic factor model in Diebold et al. (2006) involves latent factors (as state variables) following a vector autoregressive (VAR) process. Given the factors sampled from the simulation smoother of the state space model, the transition equation simplifies to a conjugate Bayesian VAR model, where coefficients and the covariance matrix can be sampled from multivariate normal and inverse Wishart conjugate distributions, as discussed in Chapter 1.

6.5.3 Regression Effect Estimation

Regressions play a central role in econometrics, and it is often desirable to incorporate regression effects into the state and/or observation equations. Equations (6.1) and (6.2) can be expanded by exogenous predictors X_{1t} and X_{2t} such as

$$S_t = A_t S_{t-1} + X_{1t}\beta_1 + B_t u_t,$$

$$Y_t = C_t S_t + X_{2t}\beta_2 + D_t \varepsilon_t,$$

where the matrices A_t, B_t, C_t, D_t are determined by underlying parameters θ. Let $\beta = [\beta_1; \beta_2]$ represent the vertically concatenated coefficients.

The inclusion of regression effects does not complicate the Kalman filter, as these effects are treated as deterministic terms similar to intercepts. Consequently, the variance terms of the Kalman filter $P_{t|t-1}, O_{t|t-1}, P_{t|t}$ remain unchanged, while the mean terms are adjusted by

$$S_{t|t-1} = A_t S_{t-1|t-1} + X_{1t}\beta_1,$$
$$Y_{t|t-1} = C_t S_{t|t-1} + X_{2t}\beta_2.$$

The forward recursion of the Kalman filter yields the likelihood function in the prediction error decomposition form:

$$p(Y | \beta, \theta) = \prod_{t=1}^{T} \phi(Y_t; Y_{t|t-1}, O_{t|t-1}).$$

Ideally, the unknown parameters β and θ can be fitted by the maximum likelihood estimation (MLE):

$$\left(\hat{\beta}, \hat{\theta}\right) = \arg\max_{\beta,\theta} p(Y | \beta, \theta),$$

which does not have a closed-form solution. While it is feasible to use quasi-Newton methods to optimize β and θ simultaneously, high-dimensional numerical optimization is computationally intensive and can be numerically unstable, with the optimization process potentially becoming trapped at local maxima or saddle points. We regard the theoretic MLE results $\hat{\beta}, \hat{\theta}$ as benchmarks for comparison with alternative estimators. Our objective is to replicate the high-dimensional MLE for β and θ using a closed-form maximizer for β and a low dimensional numerical maximizer for θ.

A promising approach to handling the regression effect is to incorporate the regression coefficients into the state vector such that $\beta_t = \beta_{t-1} = \beta$, $\forall t$. This leads to an augmented state space model

$$\begin{pmatrix} S_t \\ \beta_1 \\ \beta_2 \end{pmatrix} = \begin{pmatrix} A_t & X_{1t} & 0 \\ 0 & I & 0 \\ 0 & 0 & I \end{pmatrix} \begin{pmatrix} S_{t-1} \\ \beta_1 \\ \beta_2 \end{pmatrix} + \begin{pmatrix} B_t \\ 0 \\ 0 \end{pmatrix} u_t,$$

$$Y_t = \begin{pmatrix} C_t & 0 & X_{2t} \end{pmatrix} \begin{pmatrix} S_t \\ \beta_1 \\ \beta_2 \end{pmatrix} + D_t \varepsilon_t.$$

Given that the coefficients β have been incorporated in the state vector, it is tempting to estimate the remaining parameters θ by solving a smaller-scale optimization problem:

$$\tilde{\theta} = \arg\max_{\theta} p\left(Y \mid \theta\right),$$

where $p\left(Y \mid \theta\right)$ is the likelihood function of the augmented state space model. Note that $p\left(Y \mid \theta\right)$ is obtained by marginalization over β such that

$$p\left(Y \mid \theta\right) = \int p\left(\beta \mid \theta\right) p\left(Y \mid \beta, \theta\right) d\beta,$$

where a normal prior $p\left(\beta \mid \theta\right)$ is specified for initializing the Kalman filter on the augmented state space model. We consider a proper normal prior $N\left(\underline{\mu}, \underline{\Omega}\right)$, where the variances in $\underline{\Omega}$ do not need to be large. Under a proper prior, it is safe to apply the ordinary Kalman filter.

The ordinary Kalman filter on the augmented model completes forward recursion by returning at least three variables: the posterior mean of the filtering distribution $\mu\left(\theta\right) = E\left(\beta \mid Y, \theta\right)$, the posterior variance of the filtering distribution $\Omega\left(\theta\right) = Var\left(\beta \mid Y, \theta\right)$ and the likelihood $p\left(Y \mid \theta\right)$.

Given the estimated parameters $\tilde{\theta}$, it is tempting to take the posterior mean as the state estimator: $\tilde{\beta} = \mu\left(\tilde{\theta}\right)$. Unfortunately, the state estimator $\tilde{\beta}$ differs from the direct MLE $\hat{\beta}$, and $\tilde{\theta} \neq \hat{\theta}$ as well, regardless of the prior specification for $p\left(\beta \mid \theta\right)$.

To identify the source of discrepancy in parameter estimation, we apply Bayes' rule

$$p\left(\beta \mid Y, \theta\right) = \frac{p\left(\beta \mid \theta\right) p\left(Y \mid \beta, \theta\right)}{p\left(Y \mid \theta\right)}$$

to rearrange the objective function for the MLE:

$$\left(\hat{\beta}, \hat{\theta}\right) = \arg\max_{\beta, \theta} \frac{p\left(\beta \mid Y, \theta\right) p\left(Y \mid \theta\right)}{p\left(\beta \mid \theta\right)}.$$

Optimization can be handled by a two-step approach. First, maximizing β for a given θ:

$$\beta^*\left(\theta\right) = \arg\max_{\beta} \frac{p\left(\beta \mid Y, \theta\right) p\left(Y \mid \theta\right)}{p\left(\beta \mid \theta\right)}.$$

Second, maximizing θ over the concentrated likelihood function:

$$\theta^* = \arg\max_{\theta} \frac{p\left(\beta^*\left(\theta\right) | Y, \theta\right) p\left(Y | \theta\right)}{p\left(\beta^*\left(\theta\right) | \theta\right)}.$$

The optimizer θ^* is identical to $\hat{\theta}$, and $\beta^*\left(\theta^*\right)$ equals to $\hat{\beta}$.

Because the prior $p\left(\beta | \theta\right)$ and the posterior $p\left(\beta | Y, \theta\right)$ follow normal distributions $N\left(\underline{\mu}, \underline{\Omega}\right)$ and $N\left[\mu\left(\theta\right), \Omega\left(\theta\right)\right]$, respectively, the objective function in the first-step optimization problem takes a quadratic form due to the normal densities:

$$\beta^*\left(\theta\right) = \arg\max_{\beta} -\left[\beta - \mu\left(\theta\right)\right]' \Omega^{-1}\left(\theta\right)\left[\beta - \mu\left(\theta\right)\right] + \left(\beta - \underline{\mu}\right)' \underline{\Omega}^{-1}\left(\beta - \underline{\mu}\right),$$

which has a unique maximizer

$$\beta^*\left(\theta\right) = \left[\Omega^{-1}\left(\theta\right) - \underline{\Omega}^{-1}\right]^{-1}\left[\Omega^{-1}\left(\theta\right)\mu\left(\theta\right) - \underline{\Omega}^{-1}\underline{\mu}\right].$$

The closed-form solution indicates that $\beta^*\left(\theta\right)$ is a weighted average of the posterior mean $\mu\left(\theta\right)$ obtained from the Kalman filter and the prior mean $\underline{\mu}$ specified by users. This is an application of the Bayesian principle that the posterior is a compromise between the prior and the data. Here, we solve an inverse problem to reproduce MLE results from the user-specified prior and the posterior obtained via the Kalman filter.

By substituting $\beta^*\left(\theta\right)$ back into the objective function, we obtain the maximizer of the concentrated likelihood with respect to θ, thereby reproducing MLE results:

$$\theta^* = \arg\max_{\theta} \frac{\phi\left[\beta^*\left(\theta\right); \mu\left(\theta\right), \Omega\left(\theta\right)\right] p\left(Y | \theta\right)}{\phi\left[\beta^*\left(\theta\right); \underline{\mu}, \underline{\Omega}\right]}.$$

The appeal of likelihood concentration is to reduce dimensionality for numerical optimization. The computational saving is substantial if the length of β is much higher than that of θ.

Under a proper prior $p\left(\beta | \theta\right)$, the posterior $p\left(\beta | Y, \theta\right)$ is also proper, provided that the sample size exceeds the number of unknown parameters, ensuring a finite (marginal) likelihood $p\left(Y | \theta\right)$. This approach avoids technical difficulties associated with improper priors and divergent objective functions.

It is of interest to consider the limiting case that the prior variance tends to infinity (i.e., the precision matrix approaches zero): $\underline{\Omega}^{-1} \to 0$, which collapses to a diffuse prior $p\left(\beta | \theta\right) \propto 1$. In that case, the exact MLE takes a surprisingly simple form:

$$\beta^*\left(\theta\right) = \mu\left(\theta\right),$$

$$\theta^* = \arg\max_{\theta} \left|\Omega\left(\theta\right)\right|^{-1/2} p\left(Y | \theta\right).$$

Intuitively, the maximizer of the likelihood function is also the maximizer of the posterior density under a diffuse prior. For a normal distribution, the maximizer of the posterior density coincides with the posterior mean, thus $\beta^*\left(\theta\right) = \mu\left(\theta\right)$. To evaluate the normal density at the posterior mean/mode, we have

$$\phi\left[\beta^*\left(\theta\right); \mu\left(\theta\right), \Omega\left(\theta\right)\right] \propto \left|\Omega\left(\theta\right)\right|^{-1/2}.$$

If we omit the term $\left|\Omega\left(\theta\right)\right|^{-1/2}$ in the objective function and maximize $p\left(Y | \theta\right)$ alone, we cannot reproduce MLE results.

A concern on the diffuse prior is the finiteness of the objective function. Bayesian paradigm indicates that the marginal likelihood is infinite under a diffuse prior, implying that $p(Y|\theta)$ diverges as $\underline{\Omega}^{-1} \to 0$. However, the ratio $\frac{p(Y|\theta)}{p(\beta|\theta)}$ remains finite due to the identity:

$$\frac{p(Y|\theta)}{p(\beta|\theta)} = \frac{p(Y|\beta,\theta)}{p(\beta|Y,\theta)},$$

where the right-hand-side of the equation is finite, provided that the posterior is proper. The posterior should be proper in almost all practical models unless the number of observations is smaller than the number of regressors. If we really want to ensure finiteness of the objective function under the diffuse prior, we could write the optimization problem as

$$\theta^* = \arg\max_{\theta} |\Omega(\theta)|^{-1/2} \frac{p(Y|\theta)}{\phi\left[\beta^*(\theta);\underline{\mu},\underline{\Omega}\right]},$$

in which the diffuse prior density $\phi\left[\beta^*(\theta);\underline{\mu},\underline{\Omega}\right]$ does not depend on θ when $\underline{\Omega}^{-1} \to 0$, so it has no effect on numerical optimization with respect to θ. However, the term $|\Omega(\theta)|^{-1/2}$ is crucial for reproducing the exact MLE when we augment the state vector and rely on the Kalman filter to provide a state estimator for β .

In summary, the MLE results can be reproduced using the ordinary Kalman filter on the augmented state space model, regardless of whether large or small variances are specified for the initial states. Initializing with a large variance has its pros and cons. It is intuitive and computationally efficient, as we continue to use the posterior mean as the state estimator $\beta^*(\theta) = \mu(\theta)$, and maximize a simple objective function $|\Omega(\theta)|^{-1/2} p(Y|\theta)$ by numerical methods for the MLE. However, we should be cautious of potential numerical issues induced by initializing the Kalman filter with a large variance. In the next section, we will demonstrate a numerically stable implementation of the Kalman filter that can avoid catastrophic cancellation errors.

6.5.4 Likelihood Concentration on Scale Parameter

The MLE is obtained by maximizing the likelihood function with respect to the unknown parameters in the model. Likelihood concentration is applicable when the maximizer for a subset of parameters has a closed-form solution, conditional on the remaining parameters. It is advantageous to utilize the analytical solution whenever available and to apply numerical methods to a smaller set of parameters.

We illustrate the likelihood concentration approach for a state space model specified by Equations (6.1) and (6.2), incorporating a scale parameter σ for the observations:

$$S_t = A_t S_{t-1} + B_t u_t,$$

$$Y_t = \sigma\left(C_t S_t + D_t \varepsilon_t\right),$$

where the coefficients A_t, B_t, C_t, D_t are determined by underlying parameters θ. Our objective is to derive the closed-form MLE of σ conditional on θ. This is feasible because Y_t/σ represents the rescaled observation, but the variance recursion of the Kalman filter does not depend on observations (refer to the Riccati equation in Section 6.3).

According to Algorithm 6.1, the Kalman filter involves the forward recursion of the following variables:

$$P_{t|t-1} = A_t P_{t-1|t-1} A_t' + B_t B_t',$$

$$P_{t|t} = P_{t|t-1} - P_{t|t-1} C_t' \left(C_t P_{t|t-1} C_t' + D_t D_t'\right)^{-1} C_t P_{t|t-1},$$

$$S_{t|t} = A_t S_{t-1|t-1} + \sigma^{-1} P_{t|t-1} C_t' \left(C_t P_{t|t-1} C_t' + D_t D_t'\right)^{-1} \left(Y_t - \sigma C_t S_{t|t-1}\right).$$

The scale parameter σ has no impact on the variances $P_{t|t-1}$ and $P_{t|t}$, $\forall t$, because σC_t and σD_t scale proportionally with σ. Although the state estimator $S_{t|t}$ depends on σ, it is possible to eliminate dependence by rescaling states. Define $\tilde{S}_{t|t} = \sigma S_{t|t}$ and assume that $S_{0|0} = 0$. We note that the rescaled states evolve without dependency on σ:

$$\tilde{S}_{t|t} = A_t \tilde{S}_{t-1|t-1} + P_{t|t-1} C_t' \left(C_t P_{t|t-1} C_t' + D_t D_t'\right)^{-1} \left(Y_t - C_t \tilde{S}_{t-1|t-1}\right).$$

The recursive formulae for $\tilde{S}_{t|t}, P_{t|t-1}, P_{t|t}$ are identical to those for the original state space model without σ.

The log-likelihood function takes the form

$$\ln L\left(\sigma^2, \theta\right) = -\frac{T}{2} \ln 2\pi - \frac{T}{2} \ln \sigma^2 - \frac{1}{2} \sum_{t=1}^{T} \ln \left(C_t P_{t|t-1} C_t' + D_t D_t'\right),$$
$$-\frac{1}{2} \sum_{t=1}^{T} \frac{\left(Y_t - C_t \tilde{S}_{t-1|t-1}\right)^2}{\sigma^2 \left(C_t P_{t|t-1} C_t' + D_t D_t'\right)}.$$

Taking the derivative with respect to σ^2 conditional on θ, we obtain the maximizer of the likelihood function:

$$\hat{\sigma}^2(\theta) = \frac{1}{T} \sum_{t=1}^{T} \frac{\left(Y_t - C_t \tilde{S}_{t-1|t-1}\right)^2}{C_t P_{t|t-1} C_t' + D_t D_t'}.$$

The analytical result is intuitive, as the variance estimator corresponds to the mean squared errors of prediction. We employ numerical methods to optimize the concentrated likelihood function $\ln L\left(\sigma^2, \theta\right)$ with respect to θ.

The local level model serves as an example of the likelihood concentration approach.

$$\mu_t = \mu_{t-1} + \sigma_u \eta_t,$$
$$Y_t = \mu_t + \sigma \varepsilon_t,$$

Reparametrizing $\sigma_u = q\sigma$ and $\ddot{\mu}_t = \sigma^{-1} \mu_t$, we have

$$\ddot{\mu}_t = \ddot{\mu}_{t-1} + q\eta_t,$$
$$Y_t = \sigma\left(\ddot{\mu}_t + \varepsilon_t\right).$$

The likelihood concentration approach reduces a two-dimensional numerical optimization problem to a univariate optimization problem by leveraging the analytical MLE of σ conditional on q:

$$\hat{\sigma}^2(q) = \frac{1}{T} \sum_{t=1}^{T} \frac{\left(Y_t - \tilde{S}_{t-1|t-1}\right)^2}{P_{t|t-1} + 1},$$

where the Kalman filter output variables $P_{t|t-1}$ and $\tilde{S}_{t-1|t-1}$ depend on q but not on σ:

$$P_{t+1|t} = \frac{P_{t|t-1}}{P_{t|t-1} + 1} + q^2,$$

$$\tilde{S}_{t|t} = \tilde{S}_{t-1|t-1} + \frac{P_{t|t-1}}{P_{t|t-1} + 1} \left(Y_t - \tilde{S}_{t-1|t-1}\right).$$

6.6 Initialization of State Space Models

The initial state distribution is an indispensable component of the state space model specification, in addition to the transition and observation equations. The initial distribution has an impact on the state filtering distributions, although its effect typically diminishes as the Kalman filter progresses. Here, we explore several initialization strategies.

6.6.1 Initialization by Stationary Distribution

Ideally, states can be initialized using the stationary distribution, if it exists. Consider the state equation with an intercept term (refer to Exercise 6.10 for the rationale behind adding the intercept):

$$S_t = c + AS_{t-1} + Bu_t.$$

If all eigenvalues of the matrix A lie within the unit circle, the state series is stationary with mean $\bar{\mu}$ and variance $\bar{V}$ obtained by solving the equations

$$\bar{\mu} = c + A\bar{\mu},$$

$$\bar{V} = A\bar{V}A' + BB',$$

resulting in the closed-form solutions:

$$\bar{\mu} = (I - A)^{-1} c,$$

$$vec\left(\bar{V}\right) = (I - A \otimes A)^{-1} vec\left(BB'\right).$$

However, this strategy can be computationally demanding in high-dimensional models, as deriving an explicit solution for $\bar{V}$ involves a complexity of $O\left(m^6\right)$, where m is the state dimension. For large m, it is more efficient to solve the Lyapunov equation using Schur decomposition. In practice, computing the initial distribution might take longer than executing the Kalman filter. However, the effects of the initial distribution typically fade quickly as the Kalman filter advances. With a sufficiently large sample size, the final filtering results may be similar across alternative initial distributions.

6.6.2 Initial States as Regression Coefficients

For random walk states, the stationary distribution is unavailable. An alternative approach is to treat the initial states as fixed but unknown parameters, which can be estimated along with the other model parameters. This approach is appealing because estimating the initial states does not require numerical optimization. It is possible to convert the initial states to regression coefficients, for which the closed-form maximum likelihood estimator is available, conditional on the remaining model parameters.

Consider the transition and observation equations specified in Equation (6.1) and (6.2):

$$S_t = A_t S_{t-1} + B_t u_t,$$

$$Y_t = C_t S_t + D_t \varepsilon_t.$$

To isolate the initial state S_0 from the state series, we define $\tilde{S}_t = S_t - S_0$ for $t \geq 0$. The definition is valid if the state vector does not contain lagged variables. The reparametrized transition and observation equations take the form

$$\tilde{S}_t = A_t \tilde{S}_{t-1} + (A_t - I)\, S_0 + B_t u_t,$$

$$Y_t = C_t\tilde{S}_t + C_tS_0 + D_t\varepsilon_t,$$

where the reparametrized state series is initialized at the origin: $\tilde{S}_0 = 0$. The initial state S_0 is converted to the regression coefficients in the transition equation as $(A_t - I)\, S_0$ and in the observation equation as C_tS_0. The method discussed in Section 6.5 supports the same regression coefficients in both transition and observation equations. The closed-form maximum likelihood estimator of S_0 is available conditional on other model parameters.

This approach can be illustrated by a local level model:

$$\mu_t = \mu_{t-1} + \sigma_u\eta_t,$$

$$Y_t = \mu_t + \sigma_\varepsilon\varepsilon_t,$$

where ε_t and η_t are independent standard normal noise. If $\sigma_u = 0$, then μ_t reduces to a time-invariant intercept. If $\sigma_u > 0$, the local level follows a random walk process that does not have a stationary distribution. By reparametrizing the state as $\tilde{\mu}_t = \mu_t - \mu_0$, we have

$$\tilde{\mu}_t = \tilde{\mu}_{t-1} + \sigma_u\eta_t,$$

$$Y_t = \tilde{\mu}_t + \mu_0 + \sigma_\varepsilon\varepsilon_t.$$

The initial state μ_0 is converted to an intercept term in the observation equation, and the reparametrized state sequence $\{\tilde{\mu}_t\}$ is initialized at zero by construction: $\tilde{\mu}_0 = 0$.

Alternatively, it is of interest to normalize the state as $\ddot{\mu}_t = \sigma_u^{-1}(\mu_t - \mu_0)$, facilitating a multivariate extension of the local level model such that

$$\ddot{\mu}_t = \ddot{\mu}_{t-1} + \eta_t,$$

$$Y_t = \sigma_u\ddot{\mu}_t + \mu_0 + \sigma_\varepsilon\varepsilon_t,$$

where Y_t can be vector-valued if $\sigma_u, \mu_0, \sigma_\varepsilon$ represent vectors. In this context, the scalar random variable $\ddot{\mu}_t$ serves as an unobserved factor for the vector series Y_t. The normalized state sequence $\{\ddot{\mu}_t\}$ is initialized at zero and subject to unit-variance shocks subsequently. The standard deviation σ_u is converted to the state loading in the observation equation.

The local trend model provides another example with initial states converted to regression coefficients. Let τ_t be the local trend, of which the first difference $\Delta\tau_t = c - \tau_{t-1}$ follows a random walk process:

$$\Delta\tau_t = \Delta\tau_{t-1} + \sigma_u u_t,$$

where the initial slope $\Delta\tau_0$ and and the initial intercept τ_0 are unknown parameters. The observations provide noisy measurement of the local trend:

$$Y_t = \tau_t + \sigma_\varepsilon\varepsilon_t.$$

The state-space representation of the local trend model is the following:

$$\begin{pmatrix} \tau_t \\ \tau_{t-1} \end{pmatrix} = \begin{pmatrix} 2 & -1 \\ 1 & 0 \end{pmatrix} \begin{pmatrix} \tau_{t-1} \\ \tau_{t-2} \end{pmatrix} + \begin{pmatrix} \sigma_u \\ 0 \end{pmatrix} u_t,$$

$$Y_t = \begin{pmatrix} 1 & 0 \end{pmatrix} \begin{pmatrix} \tau_t \\ \tau_{t-1} \end{pmatrix} + \sigma_\varepsilon\varepsilon_t,$$

with the initial state vector determined by the initial intercept and slope:

$$\begin{pmatrix} \tau_0 \\ \tau_{-1} \end{pmatrix} = \begin{pmatrix} \tau_0 \\ \tau_0 - \Delta\tau_0 \end{pmatrix}.$$

Note that lagged variables are in the state vector, which invalidates reparameterization like

$$\begin{pmatrix} \tilde{\tau}_t \\ \tilde{\tau}_{t-1} \end{pmatrix} = \begin{pmatrix} \tau_t \\ \tau_{t-1} \end{pmatrix} - \begin{pmatrix} \tau_0 \\ \tau_{-1} \end{pmatrix}.$$

However, it is valid to normalize the state variable as

$$\ddot{\tau}_t = \sigma_u^{-1} \left(\tau_t - \tau_0 - \Delta\tau_0 t \right).$$

It follows that the reparametrized state space model is given by

$$\begin{pmatrix} \ddot{\tau}_t \\ \ddot{\tau}_{t-1} \end{pmatrix} = \begin{pmatrix} 2 & -1 \\ 1 & 0 \end{pmatrix} \begin{pmatrix} \ddot{\tau}_{t-1} \\ \ddot{\tau}_{t-2} \end{pmatrix} + \begin{pmatrix} 1 \\ 0 \end{pmatrix} u_t,$$

$$Y_t = \tau_0 + \Delta\tau_0 t + \begin{pmatrix} \sigma_u & 0 \end{pmatrix} \begin{pmatrix} \ddot{\tau}_t \\ \ddot{\tau}_{t-1} \end{pmatrix} + \sigma_\varepsilon \varepsilon_t,$$

where the initial states are fixed at zero by construction: $\ddot{\tau}_0 = 0$ and $\ddot{\tau}_{-1} = 0$.

The reparametrized model clarifies that the local trend model is a stochastic extension to the deterministic linear trend $\tau_0 + \Delta\tau_0 t$. The normalized local trend is initialized at the origin and driven by unit-variance shocks in subsequent periods. Once the initial intercept and slope are converted to regression effects in the observation equation, we apply the likelihood concentration approach to obtain the closed-form maximum likelihood estimator using the Kalman filter.

6.6.3 Approximate Diffuse Initialization

Diffuse initialization refers to setting the initial state variance to infinity. This approach is supported by two main arguments. First, if a state follows a random walk process, its variance would be infinite if it originated in the infinitely remote past. Second, from a Bayesian perspective, a random variable with infinite variance corresponds to a non-informative prior, reflecting lack of knowledge about the variable.

The ordinary Kalman filter requires a finite variance to initialize the forward recursion. It might be tempting to approximate diffuse initialization using a normal distribution with a large variance, such as $N(0, \kappa I)$, where the variance κ could be 10^7. However, large variance initialization is disregarded in the literature due to concerns about significant rounding errors. It is important to identify the source of rounding errors in numerical computing and assess the impact of κ on the performance of the Kalman filter.

Computers typically represent real numbers as double-precision floating-point numbers. The discrete floating-point numbers cannot exactly represent all real numbers, leading to round-off errors. Double-type data have a precision of 16 decimal digits and an upper bound of 1.8×10^{308}. Regardless of whether κ is set to a larger value like 10^7 or a smaller one like 10^3, the number is represented with the same precision and is unlikely to overflow during Kalman filtering.

Rounding errors are inherent in floating-point arithmetic, but operations like addition, multiplication, and division of two large positive numbers do not typically result in significant accuracy loss or erroneous outcomes. For instance, the computation of $ae^b \times ce^d$ involves multiplying significands ac and adding exponents $b+d$ with extra digits, followed by rounding to double precision. However, catastrophic cancellation errors can occur when subtracting two nearly equal large numbers, as the resulting difference might lose significance. In extreme cases, all significant digits of precision can be lost.

The ordinary Kalman filter iterates state variances using two equivalent forms (i.e., the Riccati equation and its stable form):

Form 1: $P_{t+1|t} = A_{t+1}\left(P_{t|t-1} - P_{t|t-1}C_t'O_{t|t-1}^{-1}C_tP_{t|t-1}\right)A_{t+1}' + B_{t+1}B_{t+1}'$.
Form 2: $P_{t+1|t} = J_tP_{t|t-1}J_t' + B_{t+1}B_{t+1}' + K_tD_tD_t'K_t'$.
The Kalman gain is defined as

$$K_t = A_{t+1}P_{t|t-1}C_t'O_{t|t-1}^{-1},$$

$$O_{t|t-1} = C_tP_{t|t-1}C_t' + D_tD_t',$$

$$J_t = A_{t+1} - K_tC_t.$$

The two forms are algebraically identical. In an ideal scenario with exact real number representation, they would yield identical results. However, finite-precision floating-point arithmetic may not align with mathematical expectations. Form 2 is numerically stable, computing predictive variance as the sum of three positive semi-definite matrices, whereas Form 1 involves subtraction. Under large variance initialization, the variance sequence commences from a large value and decreases significantly in the subsequent periods. In Form 1, variance reduction is achieved by subtracting two large positive numbers, potentially leading to catastrophic cancellation errors. In contrast, Form 2 reduces variance by multiplying a large number by nearly zero, avoiding cancellation or absorption issues.

We illustrate large variance initialization using the local level model, where $A_t = 1$, $B_t = \sigma_u$, $C_t = 1$, $D_t = \sigma_\varepsilon$. Suppose that the Kalman filter is initialized with $P_{1|0} = \kappa$. Form 1 updates the variance as

$$P_{2|1} = \kappa - \frac{\kappa^2}{\kappa + \sigma_\varepsilon^2} + \sigma_u^2.$$

As $\kappa \to \infty$, catastrophic cancellation errors loom larger. The difference between κ and $\frac{\kappa^2}{\kappa+\sigma_\varepsilon^2}$ should be a small but positive, converging to σ_ε^2, yet the floating-point subtraction could yield zero if κ is excessively large. Consequently, the theoretic value of $P_{2|1}$ converges to $\sigma_\varepsilon^2 + \sigma_u^2$, but the actual value returned by the computer could be σ_u^2.

Form 2 remains well-behaved as $\kappa \to \infty$. The theoretic value of the Kalman gain is

$$K_1 = \frac{\kappa}{\kappa + \sigma_\varepsilon^2} \to 1,$$

$$J_1 = \frac{\sigma_\varepsilon^2}{\kappa + \sigma_\varepsilon^2} \to 0.$$

The actual value of K_1 and J_1 obtained from the computer should approximate one and zero, respectively. A striking feature of Form 2 is that the impact of large κ is absorbed by the multiplication operation: $J_1P_{1|0}J_1' \to 0$, resulting in $P_{2|1}$ being close to the true value $\sigma_\varepsilon^2 + \sigma_u^2$.

In summary, large variance initialization in the ordinary Kalman filter does not necessarily lead to significant rounding errors. Forms 1 and 2 provide different ways to offset the large κ impact on the Kalman filter. Form 1 involves subtracting two nearly equal large numbers (in the form of $\kappa - \tilde{\kappa}$, where $\tilde{\kappa} \approx \kappa$), while Form 2 multiplies a large number by nearly zero (in the form of $\kappa \times o$, where $o \approx 0$). In floating-point arithmetic, subtraction can cause significant rounding errors, whereas multiplication does not. Form 2 is the numerically stable choice for the Kalman filter initiated with a large variance, because it avoids cancellation errors, maintains positive semi-definiteness, and preserves symmetry.

6.6.4 Exact Initialization

The ordinary Kalman filter does not accommodate an initial state with infinite variance, but modifications to the algorithm allow filtering under such conditions. Exact initialization methods are proposed in the literature by Koopman (1997) and De Jong (1991). The method of Koopman (1997) employs matrix inverse expansion with big-O analysis, while the idea of De Jong (1991) is to separate initial states from the Kalman filter based on an intriguing property: the conditional mean variables in Kalman filtering are linear with respect to the initial states.

The Kalman filter initialized with fixed (i.e., zero variance) initial states satisfies a linear equation (refer to Exercise 6.11 for a derivation):

$$S_{t|t}(S_0, 0) = S_{t|t}(0, 0) + S_{t|t}(I, 0)\, S_0,$$

where $S_{t|t}(S_0, 0)$ indicates the filtered states $S_{t|t}$ under the initial distribution $S_{0|0} = S_0$ and $P_{0|0} = 0$. Using the same transition and observation equations and identical data, we run the Kalman filter again with a different initial distribution $S_{0|0} = 0$ and $P_{0|0} = 0$, and the filtering output is denoted by $S_{t|t}(0, 0)$. Additionally, initializing with $S_{0|0} = I$ and $P_{0|0} = 0$ and applying the Kalman filter formulae with all observations set to zero yields the matrix-valued $S_{t|t}(I, 0)$.

Note that $S_{t|t}(S_0, 0)$ represents the filtering results conditional on the initial states S_0. Without conditioning on the parameters θ explicitly denoted, the conditional expectation is denoted as

$$S_{t|t}(S_0, 0) = E\left(S_t \,|Y_{1:t}, S_0\right).$$

However, S_0 is a random variable with the distribution $N(\mu_0, V_0)$, where the variance V_0 can be infinite. By the law of iterated expectations, the conditional expectation takes the form

$$E\left(S_t \,|Y_{1:t}\right) = E\left[E\left(S_t \,|Y_{1:t}, S_0\right)\right],$$

resulting in the equation

$$S_{t|t}(\mu_0, V_0) = S_{t|t}(0, 0) + S_{t|t}(I, 0)\, E\left(S_0 \,|Y_{1:t}\right),$$

where $E\left(S_0 \,|Y_{1:t}\right)$ is the smoother of the initial states. By induction, it can be shown (see Exercise 6.11) that other intermediate variables in the Kalman filter are also linear with respect to the initial states. In particular, the data prediction errors satisfy

$$v_t(S_0, 0) = v_t(0, 0) + v_t(I, 0)\, S_0.$$

By Bayes' rule,

$$\begin{aligned} p\left(S_0 \,|Y_{1:t}\right) &\propto p(S_0)\, p\left(Y_{1:t} \,|S_0\right) \\ &= p(S_0) \prod_{\tau=1}^{t} \phi\left[v_\tau(S_0, 0); 0, O_{t|t-1}\right] \\ &= p(S_0) \prod_{\tau=1}^{t} \phi\left[v_\tau(0, 0); -v_\tau(I, 0)\, S_0, O_{t|t-1}\right]. \end{aligned}$$

This scenario can be interpreted as Bayesian linear regression for S_0, with the prior $p(S_0)$. In each period $\tau = 1, \ldots, T$, the response variable is $v_t(0, 0)$, the predictor is

$-v_t(I,0)$, and the disturbance variance is $O_{t|t-1}$. Standard Bayesian linear regression results suggest that $p(S_0|Y_{1:t})$ follows $N(\bar{\mu}_0, \bar{V}_0)$, where

$$\bar{V}_0 = Var(S_0|Y_{1:t}) = \left[V_0^{-1} + \sum_{\tau=1}^{t} v_\tau(I,0)' O_{t|t-1}^{-1} v_\tau(I,0)\right]^{-1},$$

$$\bar{\mu}_0 = E(S_0|Y_{1:t}) = \bar{V}_0 \left[V_0^{-1}\mu_0 - \sum_{\tau=1}^{t} v_\tau(I,0)' O_{t|t-1}^{-1} v_\tau(0,0)\right].$$

The problem of diffuse initial distribution is resolved, as an infinite variance V_0 corresponds to zero precision $V_0^{-1} = 0$. Bayesian linear regression with a diffuse prior is well-defined. The posterior (the smoothing distribution of initial states) remains proper if the sample size is sufficiently large (typically $t > m$). In summary, the filtering formula under exact initialization is

$$S_{t|t}(\mu_0, V_0) = S_{t|t}(0,0) + S_{t|t}(I,0)\bar{\mu}_0.$$

6.6.5 Comparison of Fixed and Diffuse Initialization

When the stationary distribution is unavailable, two competing initialization methods exist. First, treat initial states as fixed but unknown parameters to be estimated by MLE. Second, consider initial states as diffuse random variables with infinite variances for state estimation using the diffuse Kalman filter. The question arises whether these methods yield equivalent results.

Consider a simplified model without unknown parameters. If the initial states are the sole unknowns, the MLE takes the form

$$\hat{S}_0 = \arg\max_{S_0} p(Y|S_0),$$

where the likelihood function $p(Y|S_0)$ is computed by the ordinary Kalman filter initialized with a constant (i.e. $S_{0|0} = S_0$ and $P_{0|0} = 0$).

Bayes' rule indicates that

$$p(S_0|Y) = \frac{p(S_0)\,p(Y|S_0)}{p(Y)},$$

so that the objective function for optimization can be written as

$$\hat{S}_0 = \arg\max_{S_0} \frac{p(S_0|Y)\,p(Y)}{p(S_0)}.$$

Under a diffuse prior $p(S_0) \propto 1$ or a normal approximation $p(S_0) = \phi(S_0; 0, \kappa I)$ with $\kappa \to 0$, the objective function for optimization is simplified as

$$\hat{S}_0 = \arg\max_{S_0} p(S_0|Y),$$

where the posterior $p(S_0|Y)$ can be obtained by the diffuse Kalman filter or the ordinary Kalman filter initialized by a large variance. Since $p(S_0|Y)$ is normally distributed, the maximizer of the posterior density equals the posterior mean:

$$\hat{S}_0 = E(S_0|Y).$$

This completes the proof that the MLE $\hat{S}_0$ matches the state estimator $E(S_0 | Y)$ obtained by the Kalman filter under diffuse initialization in a model without other unknown parameters.

If the model includes unknown parameters θ, they can be estimated alongside the initial states:

$$\left(\hat{S}_0, \hat{\theta}\right) = \arg\max_{S_0,\theta} p(Y | S_0, \theta).$$

Alternatively, simultaneous optimization over S_0, θ can be replaced by the likelihood profiling: optimizing S_0 conditional on θ, followed by optimizing the concentrated likelihood function with respect to θ:

$$S_0^*(\theta) = \arg\max_{S_0} p(Y | S_0, \theta),$$

$$\theta^* = \arg\max_{\theta} p(Y | S_0^*(\theta), \theta).$$

The likelihood concentration approach reproduces the MLE:

$$\hat{\theta} = \theta^*,$$

$$\hat{S}_0 = S_0^*(\theta^*).$$

By Bayes' rule, the objective function can be expressed as

$$S_0^*(\theta) = \arg\max_{S_0} \frac{p(S_0 | Y, \theta) p(Y | \theta)}{p(S_0 | \theta)}.$$

Specifying a diffuse prior $p(S_0 | \theta) \propto 1$ (or a normal approximation with a large variance) simplifies the optimization objective function as $p(S_0 | Y, \theta)$, with the closed-form optimizer

$$S_0^*(\theta) = E(S_0 | Y, \theta),$$

as the maximizer of the posterior density equals the posterior mean under the normal distribution. The posterior mean can be obtained via the diffuse Kalman filter or the ordinary Kalman filter with a large variance initialization.

For numerical optimization of the concentrated likelihood function, two options exist:

$$\theta^* = \arg\max_{\theta} p(Y | S_0^*(\theta), \theta),$$

$$\theta^* = \arg\max_{\theta} |Var(S_0 | Y, \theta)|^{-1/2} \frac{p(Y | \theta)}{p(S_0^*(\theta) | \theta)},$$

where the determinant term $|Var(S_0 | Y, \theta)|^{-1/2}$ arises from the multivariate normal posterior density $p(S_0 | Y, \theta)$ evaluated at the maximizer $S_0^*(\theta)$. The term $Var(S_0 | Y, \theta)$ represents the smoothed variance of initial states, which is an intermediate variable used in the diffuse Kalman filter (or it can be obtained by the ordinary Kalman smoother initialized with a large variance).

Note that the ratio $\frac{p(Y|\theta)}{p(S_0^*(\theta)|\theta)}$ is finite, despite divergence of $p(Y | \theta)$ under a diffuse prior. When employing the ordinary Kalman filter with a large variance approximation, the likelihood function $p(Y | \theta)$ becomes readily available upon completing the forward filtering process. Alternatively, using the diffuse Kalman filter, the ratio aligns with the diffuse likelihood as described by De Jong (1988).

In summary, initial states can be approached as either unknown parameters or diffuse random variables. The MLE can be reproduced by the state estimator derived from the diffuse Kalman filter, provided that the determinant term is incorporated into the objective function for optimizing unknown parameters in the model.

6.7 Software and Computational Considerations

When the Kalman filter was first implemented on computers, researchers were surprised by performance degradation during forward filtering. After several iterations, the covariance matrix of filtered states sometimes lost positive definiteness, a theoretical impossibility. This issue led to the development of square root filtering algorithms, which significantly improved the numerical stability of the Kalman filter. However, this does not imply that the covariance filter is inapplicable. A straightforward method to prevent numerical issues in the covariance filter is to enforce symmetry in the covariance matrix of filtered states at each iteration. Specifically, we can replace $P_{t|t}$ by $\frac{1}{2}\left(P_{t|t} + P_{t|t}'\right)$ in Algorithm 6.1.

A computational consideration for the Kalman filter is initialization using the stationary distribution. When the number of states is large, calculating the stationary distribution by solving the Lyapunov equation can be time-consuming, potentially taking longer than the Kalman filtering process itself. However, in most applications, the effect of initialization on the filter diminishes rapidly after a few iterations. A suggested approach is to approximate the diffuse initialization by specifying a large variance for the initial states before applying the numerically stable version of the Kalman filter.

Additionally, univariate treatment of multivariate series is highly recommended for implementing the Kalman filter, whenever the covariance matrix $D_t D_t'$ is diagonal. This approach can substantially reduce the number of floating-point operations, compared to batch processing of multivariate data. Even if the matrix $D_t D_t'$ is not diagonal, univariate treatment is still feasible by expanding the state vector.

6.8 An Application on Exchange Rates

We consider an application of the local level model applied to daily U.S. dollar to euro spot exchange rates from 1999 to 2023, comprising a sample size of 6,450 observations.

The local level model is specified as

$$\mu_t = \mu_{t-1} + \sigma_u \eta_t, \tag{6.3}$$

$$Y_t = \mu_t + \sigma_\varepsilon \varepsilon_t. \tag{6.4}$$

The state variable is normalized by defining $\ddot{\mu}_t = \sigma_u^{-1}(\mu_t - \mu_0)$, resulting in

$$\ddot{\mu}_t = \ddot{\mu}_{t-1} + \eta_t, \tag{6.5}$$

$$Y_t = \sigma_u \ddot{\mu}_t + \mu_0 + \sigma_\varepsilon \varepsilon_t, \tag{6.6}$$

and the normalized state is initialized at the origin by construction: $\ddot{\mu}_0 = 0$.

6.8.1 Maximum Likelihood Estimation

The unknown parameters can be estimated by maximum likelihood. We compare the following estimation methods:

1. Run the Kalman filter based on Equations (6.3) and (6.4) for the likelihood function $p(Y\,|\sigma_u, \sigma_\varepsilon, \mu_0)$, and maximize the likelihood function with respect to $\sigma_u, \sigma_\varepsilon, \mu_0$.

2. Run the Kalman filter based on Equations (6.5) and (6.6) for the likelihood function $p(Y|\sigma_u, \sigma_\varepsilon, \mu_0)$, and maximize the likelihood function with respect to $\sigma_u, \sigma_\varepsilon, \mu_0$.
3. Concentrate the regression effect (i.e., intercept) μ_0 in Equation (6.6) and maximize the concentrated likelihood function with respect to $\sigma_u, \sigma_\varepsilon$.
4. Concentrate the scale parameter σ_ε in Equation (6.6) and maximize the concentrated likelihood function with respect to σ_u, μ_0.
5. Treat the initial state μ_0 in Equation (6.3) as a diffuse random variable and view $\sigma_u, \sigma_\varepsilon$ as unknown parameters for maximum likelihood estimation.
6. Reserve the first observation as the presample data for initializing the state as $N(Y_1, \sigma_\varepsilon^2)$, and the remaining data are used for parameter estimation of $\sigma_u, \sigma_\varepsilon$.

TABLE 6.1
Full-sample (upper panel) and sub-sample (lower panel) parameter estimation of local level model.

	1	2	3	4	5	6
σ_u	0.0067	0.0067	0.0067	0.0067	0.0068	0.0068
σ_ε	0.2298	0.2298	0.2298	0.2298	0.2298	0.2298
μ_0	1.0837	1.0837	1.0837	1.0837	1.0838	1.0838

	1	2	3	4	5	6
σ_u	0.2481	0.2481	0.2482	0.2481	0.2531	0.2529
σ_ε	0.3581	0.3581	0.3581	0.3581	0.3560	0.3561
μ_0	2.3949	2.3949	2.3949	2.3949	2.3949	2.3949

Table 6.1 presents parameter estimation results using these six methods on full-sample and sub-sample data (first 200 observations). The first four methods yield identical results, differing from those obtained by the last two methods. This discrepancy reflects differing perspectives on the initial state as an unknown parameter versus a diffuse random variable, with the impact diminishing as sample size increases. A few thousand observations can nearly eliminate initialization effects.

6.8.2 Initialization of the Kalman Filter

Given the estimated parameters σ_u and σ_ε in Equations (6.3) and (6.4), we compare the Kalman filter under different initial state distributions: $\mu_0 \sim N(0, \kappa)$, where $\kappa = \{10^2, 10^7, 10^{14}, 10^{16}\}$. We compute the state variance sequence $\{P_{t+1|t}\}$ in both the regular form (Form 1) and the stable form (Form 2) discussed in Section 6.3.

Inspecting Table 6.2 row by row, we note that the Kalman filter in the stable form always provides consistent results, regardless of the magnitude of κ. Performance of the regular form varies with κ. For moderate variances like 10^2 and 10^7, the results align with the stable form. However, results deteriorate slightly at $\kappa = 10^{14}$, with significant errors at $\kappa = 10^{16}$ due to the limitations of double-precision floating-point numbers, leading to catastrophic cancellation errors. Specifically, double-precision floating numbers have at most 16 digits of precision, and all significant digits can be lost in subtracting two large numbers, hence an erroneous result of $P_{2|1}$ close to σ_u^2, while the true value should be around $\sigma_u^2 + \sigma_\varepsilon^2$.

Inspecting Table 6.2 column by column, we note that the variance sequence $\{P_{t+1|t}\}$ drops significantly in initial periods and converges to a steady state within 50 observations. Both regular and stable forms of the Kalman filter provide the same asymptotic variance.

TABLE 6.2
Kalman filter in regular and stable forms under different initial state variances $\kappa = \{10^2, 10^7, 10^{14}, 10^{16}\}$.

t	KF10^2	Stable10^2	KF10^7	Stable10^7	KF10^{14}	Stable10^{14}	KF10^{16}	Stable10^{16}
1	100	100	1.00E7	1.00E7	1.0E14	1.0E14	1.0E16	1.0E16
2	0.0528	0.0528	0.0528	0.0528	0.0469	0.0528	0.0000	0.0528
3	0.0265	0.0265	0.0265	0.0265	0.0249	0.0265	0.0001	0.0265
4	0.0177	0.0177	0.0177	0.0177	0.017	0.0177	0.0001	0.0177
5	0.0133	0.0133	0.0133	0.0133	0.0129	0.0133	0.0002	0.0133
10	0.0060	0.0060	0.0060	0.0060	0.0059	0.0060	0.0004	0.0060
20	0.0031	0.0031	0.0031	0.0031	0.0031	0.0031	0.0008	0.0031
30	0.0023	0.0023	0.0023	0.0023	0.0023	0.0023	0.0011	0.0023
40	0.0019	0.0019	0.0019	0.0019	0.0019	0.0019	0.0013	0.0019
50	0.0018	0.0018	0.0018	0.0018	0.0018	0.0018	0.0014	0.0018
1000	0.0016	0.0016	0.0016	0.0016	0.0016	0.0016	0.0016	0.0016
2000	0.0016	0.0016	0.0016	0.0016	0.0016	0.0016	0.0016	0.0016
3000	0.0016	0.0016	0.0016	0.0016	0.0016	0.0016	0.0016	0.0016
4000	0.0016	0.0016	0.0016	0.0016	0.0016	0.0016	0.0016	0.0016
5000	0.0016	0.0016	0.0016	0.0016	0.0016	0.0016	0.0016	0.0016

Even in the worst-case scenario with $\kappa = 10^{16}$, the Kalman filter tolerates initial numeric errors, eventually converging to the desired steady state.

Empirical results indicate the existence and attainability of the steady state of the variance sequence via Riccati equation iterations. Proposition 6.1 provides a sufficient condition for convergence when all eigenvalues of the transition matrix lie within the unit circle, but it is not a necessary condition. Despite states following a unit root process, the variance sequence converges, as demonstrated by an alternative proof using the fixed-point theorem.

The recursive formula for the variance sequence

$$P_{t+1|t} = \frac{P_{t|t-1}\sigma_\varepsilon^2}{P_{t|t-1} + \sigma_\varepsilon^2} + \sigma_u^2$$

can be interpreted as a mapping $f(x) = \frac{x\sigma_\varepsilon^2}{x+\sigma_\varepsilon^2} + \sigma_u^2$ with $x > \sigma_u^2$, which satisfies an inequality:

$$|f(x_1) - f(x_2)| = \frac{\sigma_\varepsilon^4}{(x_1 + \sigma_\varepsilon^2)(x_2 + \sigma_\varepsilon^2)} |x_1 - x_2| < \beta |x_1 - x_2|,$$

where the discount factor $\beta = \frac{\sigma_\varepsilon^4}{(\sigma_u^2+\sigma_\varepsilon^2)^2} < 1$. This is a contraction mapping, under which iterations converge to a unique fixed point such that

$$P^* = \frac{P^*\sigma_\varepsilon^2}{P^* + \sigma_\varepsilon^2} + \sigma_u^2.$$

The explicit solution of the quadratic equation is

$$P^* = \frac{\sigma_u^2 + \sqrt{4\sigma_u^2\sigma_\varepsilon^2 + \sigma_u^4}}{2}.$$

The fixed-point theorem guarantees convergence from an arbitrary initial value, corroborated by empirical results in Table 6.2, showing $P^* = 0.0016$ across all initial state distributions, even in the worst-case scenario that the initial value is corrupted by catastrophic cancellation errors. The Kalman filter illustrates a phenomenon in Bayesian analysis: the prior can be washed out as observations accumulate.

6.9 Bibliographic Notes

The Kalman filter stands as one of the most influential algorithms of the 20th century. In his groundbreaking paper, Kalman (1960) introduced a recursive algorithm designed for control system problems, specifically for online filtering and prediction. Statisticians later adapted the Kalman filter for time series analysis, as the state space framework enables the exact evaluation of the likelihood function for models such as ARMA and vector autoregressions, accommodating missing data, temporal aggregation and mixed frequency regressions. See Jones (1980), Ansley and Kohn (1983), Harvey and Pierse (1984), Zadrozny (1988), Mariano and Murasawa (2003, 2010), among others.

Beyond forward filtering and likelihood evaluation, smoothing and simulation smoothing are essential for Bayesian statistical inference. The classic smoothing algorithm was introduced by Carter and Kohn (1994) and subsequently enhanced by De Jong (1989), Durbin and Koopman (2002), and Briers et al. (2010).

To address non-stationary time series within the state space framework, diffuse initialization has garnered interest among researchers. Exact initialization techniques were developed by Koopman (1997) and De Jong (1991) to accommodate these challenges.

6.10 Proof of Propositions

Proposition 6.1 (Kalman Filter Convergence)

Let the initial state variance be V_0*, which has an impact on the sequence* $\{P_{t+1|t}\}$, $t = 0, 1, 2, \ldots$*. However, we will show that the consequence converges to the same* P^* *regardless of the initial variance specification.*

First, consider initialization by the stationary distribution: $V_0 = AV_0A' + Q$*. Let the corresponding sequence be* $\left\{P^{**}_{t|t-1}\right\}$*. Because of strictly stationarity, the joint distribution* $p\left(S_{0:t-1}, Y_{1:t-1}\right)$ *is identical to* $p\left(S_{1:t}, Y_{2:t}\right)$*, without conditioning on the model parameters explicitly denoted. It follows that*

$$P^{**}_{t+1|t} = Var\left(S_{t+1}\left|Y_{1:t}\right.\right) \leq Var\left(S_{t+1}\left|Y_{2:t}\right.\right) = Var\left(S_t\left|Y_{1:t-1}\right.\right) = P^{**}_{t|t-1}.$$

The monotone decreasing sequence $\left\{P^{**}_{t|t-1}\right\}$ *is lower bounded by zero, and therefore converges to* P^{**} *such that*

$$P^{**} = AP^{**}A' - AP^{**}C'\left(CP^{**}C' + R\right)^{-1}CP^{**}A + Q.$$

Second, consider initialization by a fixed state S_0 *(so that the state filtering distributions are conditional on* S_0*). In that case,* $V_0 = 0$*. Let the corresponding sequence be* $\left\{\underline{P}_{t|t-1}\right\}$*. Because of temporal homogeneity,* $p\left(S_t\left|Y_{1:t-1}, S_0\right.\right)$ *is identical to* $p\left(S_{t+1}\left|Y_{2:t}, S_1\right.\right)$*. It follows that*

$$Var\left(S_t\left|Y_{1:t-1}, S_0\right.\right) = Var\left(S_{t+1}\left|Y_{2:t}, S_1\right.\right) = Var\left(S_{t+1}\left|Y_{1:t}, S_1, S_0\right.\right) \leq Var\left(S_{t+1}\left|Y_{1:t}, S_0\right.\right),$$

that is, $\underline{P}_{t|t-1} \leq \underline{P}_{t+1|t}$*. The monotone increasing sequence* $\left\{\underline{P}_{t|t-1}\right\}$ *is upper bounded by* $P^{**}_{1|0}$*, and therefore converges to* P^* *such that*

$$P^* = AP^*A' - AP^*C'\left(CP^*C' + R\right)^{-1}CP^*A + Q.$$

Third, consider initialization by an arbitrary V_0, but we run a suboptimal filter, which induces a larger variance of state estimation compared to the optimal filter in the Joseph stable form:

$$P_{t+1|t} = J_t P_{t|t-1} J_t' + Q + K_t R K_t',$$

where

$$K_t = A P_{t|t-1} C' \left(C P_{t|t-1} C' + R\right)^{-1},$$
$$J_t = A - K_t C.$$

The suboptimal filter replaces K_t, J_t by K^*, J^*, where

$$K^* = A P^* C' \left(C P^* C' + R\right)^{-1},$$
$$J^* = A - K^* C.$$

Let the corresponding sequence be $\left\{\bar{P}_{t|t-1}\right\}$ such that

$$\bar{P}_{t+1|t} = J^* \bar{P}_{t|t-1} J^{*\prime} + Q + K^* R K^{*\prime}.$$

Let λ^*, z^* be the eigenvalue and eigenvector of $J^{*\prime}$. Left-multiply $z^{*\prime}$ and right-multiply z^* to both sides of $P^* = J^* P^* J^{*\prime} + Q + K^* R K^{*\prime}$, we have

$$\left(1 - \lambda^{*2}\right) z^{*\prime} P^* z^* = z^{*\prime} Q z^* + z^{*\prime} K^* R K^{*\prime} z^*.$$

Because the eigenvalues of A are in the unit circle, as well as Q or R (or both) is positive definite, we must have $|\lambda^*| < 1$. It follows that all eigenvalues of $J^* \otimes J^*$ are in the unit circle, so that the vectorized sequence

$$vec\left(\bar{P}_{t+1|t}\right) = \left(J^* \otimes J^*\right) vec\left(\bar{P}_{t|t-1}\right) + vec\left(Q + K^* R K^{*\prime}\right)$$

must converge to the same point under an arbitrary V_0. If we put $V_0 = P^*$, we see that the sequence $\left\{\bar{P}_{t|t-1}\right\}$ converges to P^*.

Fourth, consider the standard Kalman filter under an arbitrary initial variance V_0. Let the corresponding sequence be $\left\{P_{t|t-1}\right\}$, which is squeezed by the lower bound $\left\{\underline{P}_{t|t-1}\right\}$ and the upper bound $\left\{\bar{P}_{t|t-1}\right\}$, both of which converge to the same point P^*. Therefore, the sequence $\left\{P_{t|t-1}\right\}$ also converges to P^*.

6.11 Exercises

Exercise 6.1 (Normality of filtering distribution) *Consider the Gaussian linear state space model*

$$S_t = A_t S_{t-1} + B_t u_t,$$
$$Y_t = C_t S_t + D_t \varepsilon_t,$$

where $\{u_t\}$ and $\{\varepsilon_t\}$ are independent standard normal noise series, and independent to the normally distributed initial state S_0. The coefficients A_t, B_t, C_t, D_t are either deterministic or functions of $Y_{1:t-1}$.

1. Show that $p(S_1, Y_1)$ follows a multivariate normal distribution.

2. Show by induction that $p(S_t, Y_t | Y_{1:t-1}), t \geq 1$ follows a multivariate normal distribution.

3. Normality of $p(S_t, Y_t | Y_{1:t-1})$ implies that the incremental likelihood function $p(Y_t | Y_{1:t-1})$ takes the form of a normal density. Is the full-sample likelihood function $p(Y_{1:T}) = \prod_{t=1}^{T} p(Y_t | Y_{1:t-1})$ a multivariate normal density as well?

Exercise 6.2 (Kalman filter with correlated disturbances) *In the text, we derived the Kalman filter under the assumption of* $E(u_t\varepsilon_t') = 0$. *We consider relaxing that assumption.*

1. Suppose that $E(u_t\varepsilon_t') = \rho$. *Is the Kalman filter still applicable? If yes, how to adapt the recursive formulae?*

2. Suppose that your favorite econometric software only supports Kalman filtering with $E(u_t\varepsilon_t') = 0$, *but you must work with a state space model with* $E(u_t\varepsilon_t') = \rho$. *Is there a workaround?*

Exercise 6.3 (Extension of instantaneous observation) *Consider an extension to the state space model, in which* Y_t *is a function of* S_t *and* S_{t-1}:

$$S_t = A_t S_{t-1} + B_t u_t,$$

$$Y_t = C_t S_t + D_t \varepsilon_t + E_t S_{t-1}.$$

1. Is the Kalman filter still applicable? If yes, how to adapt the recursive formulae?

2. Suppose that your favorite econometric software only supports the standard Kalman filter. How to cast the model in the standard state space form?

Exercise 6.4 (Kalman filter with common disturbances) *Consider a state space model in which the disturbances are shared by transition and observation equations:*

$$S_{t+1} = AS_t + \bar{K}v_t,$$

$$Y_t = CS_t + v_t,$$

where $v_t \sim N\left(0, \bar{O}\right)$. *Note that the timing convention here slightly differs from that in a standard state space model.*

1. Is the Kalman filter still applicable? If yes, how to adapt the recursive formulae?

2. Suppose that your favorite econometric software only supports the standard Kalman filter. How to cast the model in the standard state space form?

Exercise 6.5 (Explicit solutions of filtering and smoothing) *The Kalman filter offers a computationally efficient method for state filtering. However, it is possible to obtain an explicit smoothing distribution without resorting to the Kalman filter. Consider the Gaussian linear state space model*

$$S_t = A_t S_{t-1} + B_t u_t,$$

$$Y_t = C_t S_t + D_t \varepsilon_t,$$

where $\{u_t\}$ *and* $\{\varepsilon_t\}$ *are independent standard normal noise series, and independent to the normally distributed initial state* $S_0 \sim N(\mu_0, V_0)$. *The coefficients* A_t, B_t, C_t, D_t *are deterministic (no dependency on* $Y_{1:t-1}$*). To obtain the joint distribution of states* $p(S_{0:T})$, *we write the transition equation in the matrix form* $\Lambda S_{0:T} = \mu$, *where*

$$\Lambda = \begin{pmatrix} I & 0 & 0 & 0 & 0 \\ -A_1 & I & 0 & 0 & 0 \\ \cdots & \cdots & \ddots & \cdots & \cdots \\ 0 & 0 & 0 & -A_T & I \end{pmatrix},$$

$$\mu = \begin{pmatrix} S_0 \\ B_1 u_1 \\ \cdots \\ B_T u_T \end{pmatrix}.$$

Note that the matrix Λ *is of full rank, with the inverse matrix*

$$\Lambda^{-1} = \begin{pmatrix} I & 0 & 0 & \cdots & 0 & 0 \\ A_1 & I & 0 & \cdots & 0 & 0 \\ A_2 A_1 & A_2 & I & \cdots & 0 & 0 \\ \cdots & \cdots & \cdots & \ddots & \cdots & \cdots \\ \prod_{t=1}^{T} A_t & \prod_{t=2}^{T} A_t & \prod_{t=3}^{T} A_t & \cdots & A_T & I \end{pmatrix}.$$

The transition equation can be inverted as $S_{0:T} = \Lambda^{-1}\mu$. *Similarly, we can write the observation equation in the matrix form*

$$Y_{1:T} = \Gamma S_{0:T} + \Delta \varepsilon_{1:T},$$

where $\Gamma = [0, diag\,(C_1, \ldots, C_T)]$ *and* $\Delta = diag\,(D_1, \ldots, D_T)$.

1. Is the state distribution $p\,(S_{0:T})$ *Gaussian? If yes, what is the mean and covariance matrix?*

2. Given the state sequence $S_{0:T}$, *what is the conditional likelihood function, namely* $p\,(Y_{1:T}\,|S_{0:T})$*? What is the likelihood function* $p\,(Y_{1:T})$*?*

3. Conditional on the observations, what is the state filtering/smoothing distribution, namely $p\,(S_{0:T}\,|Y_{1:T})$*?*

4. If an explicit solution is available, why do we use the Kalman filter and smoother?

Exercise 6.6 (Smoothing distribution) *The state space model has two fundamental properties: Markov transition and instantaneous observation.*

$$p\,(S_t\,|S_{0:t-1}, Y_{1:t-1}) = p\,(S_t\,|S_{t-1}, Y_{1:t-1}),$$

$$p\,(Y_t\,|S_{0:t}, Y_{1:t-1}) = p\,(Y_t\,|S_t, Y_{1:t-1}).$$

Given those properties, show that the smoothing distribution satisfies

$$p\,(S_t\,|S_{t+1:T}, Y_{1:T}) = p\,(S_t\,|S_{t+1}, Y_{1:t}).$$

Exercise 6.7 (Smoothing uncertainty) *The smoothing distribution* $p\,(S_t\,|Y_{1:T})$ *is summarized by the mean and variance:*

$$S_{t|T} = E\,(S_t\,|Y_{1:T}),$$

$$P_{t|T} = Var\,(S_t\,|Y_{1:T}).$$

In the text, we derived the backward recursion of the mean sequence $\{S_{t|T}\}$, $t = T, T-1, \ldots, 1$. *The variance sequence* $\{P_{t|T}\}$ *can be computed similarly by backward recursion. The idea is that the unconditional variance equals the mean of the conditional variance plus the variance of the conditional mean.*

$$P_{t|T} = E\,[Var\,(S_t\,|S_{t+1}, Y_{1:T})] + Var\,[E\,(S_t\,|S_{t+1}, Y_{1:T})].$$

1. *Show that*

$$E\left[Var\left(S_t\left|S_{t+1},Y_{1:T}\right.\right)\right]=Var\left(S_t\left|S_{t+1},Y_{1:T}\right.\right),$$

where

$$Var\left(S_t\left|S_{t+1},Y_{1:T}\right.\right)=P_{t|t}-P_{t|t}A'_{t+1}P^{-1}_{t+1|t}A_{t+1}P_{t|t}.$$

2. *Show that*

$$Var\left[E\left(S_t\left|S_{t+1},Y_{1:T}\right.\right)\right]=P_{t|t}A'_{t+1}P^{-1}_{t+1|t}P_{t+1|T}P^{-1}_{t+1|t}A_{t+1}P_{t|t}.$$

3. *Show that backward recursion takes the form*

$$P_{t|T}=P_{t|t}-G_t\left(P_{t+1|t}-P_{t+1|T}\right)G'_t,$$

where $G_t=P_{t|t}A'_{t+1}P^{-1}_{t+1|t}$.

Exercise 6.8 (Improved smoothing algorithm) *In Section 6.3, we defined a few intermediate variables:*

$$v_t=Y_t-Y_{t|t-1},$$
$$J_t=A_{t+1}-K_tC_t,$$
$$K_t=A_{t+1}P_{t|t-1}C'_tO^{-1}_{t|t-1}.$$

1. *Show the identity*

$$P_{t|t}A'_{t+1}=P_{t|t-1}J'_t.$$

2. *Show that the recursion formula*

$$S_{t|T}=S_{t|t}+P_{t|t}A'_{t+1}P^{-1}_{t+1|t}\left(S_{t+1|T}-S_{t+1|t}\right)$$

can be written as

$$r_t=C'_tO^{-1}_{t|t-1}v_t+J'_tr_{t+1},$$

where $r_t=P^{-1}_{t|t-1}\left(S_{t|T}-S_{t|t}\right)$.

3. *Backward recursion of the sequence* $\left\{P_{t|T}\right\}$ *can be adapted in a similar way. Show that the recursive formula*

$$P_{t|T}=P_{t|t}-G_t\left(P_{t+1|t}-P_{t+1|T}\right)G'_t$$

can be rewritten as

$$N_t=C'_tO^{-1}_{t|t-1}C_t+J'_tN_{t+1}J_t,$$

where $N_t=P^{-1}_{t|t-1}\left(P_{t|t-1}-P_{t|T}\right)P^{-1}_{t|t-1}$.

Exercise 6.9 (An alternative simulation smoother) *Durbin and Koopman (2002) develop a clever method of simulation smoothing that leverages a special feature of the Gaussian distribution: the conditional variance is not a function of the conditioning variable. Suppose that*

$$\begin{pmatrix}X\\Y\end{pmatrix}\sim N\left[\begin{pmatrix}\mu_1\\\mu_2\end{pmatrix},\begin{pmatrix}\Omega_{11}&\Omega_{12}\\\Omega_{21}&\Omega_{22}\end{pmatrix}\right].$$

The conditional distribution takes the form:

$$X\,|Y \sim N\left[f(Y), \Psi\right],$$

where the conditional mean is a linear function of Y*, but the conditional variance is not.*

$$f(Y) = \mu_1 + \Omega_{12}\Omega_{22}^{-1}(Y - \mu_2),$$

$$\Psi = \Omega_{11} - \Omega_{12}\Omega_{22}^{-1}\Omega_{21}.$$

1. Let the realization (i.e., the observed data) of Y *be* y*. We also generate some artificial data* X^* *and* Y^* *from the joint distribution* $N\left[\begin{pmatrix}\mu_1\\ \mu_2\end{pmatrix}, \begin{pmatrix}\Omega_{11} & \Omega_{12}\\ \Omega_{21} & \Omega_{22}\end{pmatrix}\right]$*, and then compute*

$$Z = X^* - f(Y^*) + f(y).$$

Show that Z *is sampled from the conditional distribution* $N\left[f(y), \Psi\right]$.
2. Consider the simulation smoothing algorithm:
First, generate $p(S_{0:T}^*, Y_{1:T}^*)$.
Second, smooth $S_{t|T} = E(S_t\,|Y_{1:T})$.
Third, smooth $S_{t|T}^* = E(S_t^*\,|Y_{1:T}^*)$.
Fourth, compute $S_t = S_t^* - S_{t|T}^* + S_{t|T}$.
Show that $S_{0:T}$ *is sampled from the smoothing distribution* $p(S_{0:T}\,|Y_{1:T})$.

Exercise 6.10 (Stationary distribution initialization) *When we discuss initialization by the stationary distribution, we explicitly put an intercept term in the state equation:*

$$S_t = c + AS_{t-1} + Bu_t.$$

If all eigenvalues of A *are in the unit circle, we have the stationary distribution with the mean* $\bar{\mu} = (I - A)^{-1}c$*. It is tempting to incorporate an intercept in the state vector such that*

$$\begin{pmatrix} S_t \\ c \end{pmatrix} = \tilde{A}\begin{pmatrix} S_{t-1} \\ c \end{pmatrix} + \begin{pmatrix} B \\ 0 \end{pmatrix} u_t,$$

where $\tilde{A} = \begin{pmatrix} A & I \\ 0 & 1 \end{pmatrix}$*. The aim of the exercise is to show that the enlarged state space form is not suitable for computing the stationary distribution.*

1. Show that the enlarged transition matrix $\tilde{A}$ *contains a unit root (eigenvalue), even if all eigenvalues of* A *are in the unit circle.*

2. Show that $I - \tilde{A}$ *is not invertible. The stationary mean cannot be computed by solving an equation like* $\tilde{\mu} = \tilde{A}\tilde{\mu}$*, although a solution* $\tilde{\mu} = \begin{pmatrix} \bar{\mu} \\ c \end{pmatrix}$ *satisfies the equation.*

Exercise 6.11 (Diffuse Kalman filter) *Consider three Kalman filters, all of which use the same transition and observation equations. The first is initialized by* $S_{0|0} = S_0$ *and* $P_{0|0} = 0$*. The second is initialized by* $S_{0|0} = 0$ *and* $P_{0|0} = 0$*. The third is initialized by* $S_{0|0} = I$ *and* $P_{0|0} = 0$*. Note that even though* $S_{0|0}$ *becomes matrix valued, the Kalman filtering formulae remain unchanged. In the first two Kalman filters, we use actual data* Y_t*, while all observations are zeroed for the third one. We want to show that the Kalman filter outputs are linear with respect to the initial states.*

State prediction:

$$S_{t|t-1}(S_0,0) = S_{t|t-1}(0,0) + S_{t|t-1}(I,0)\,S_0.$$

Data prediction:

$$Y_{t|t-1}(S_0,0) = Y_{t|t-1}(0,0) + Y_{t|t-1}(I,0)\,S_0.$$

Prediction error:

$$v_t(S_0,0) = v_t(0,0) + v_t(I,0)\,S_0.$$

State filtering:

$$S_{t|t}(S_0,0) = S_{t|t}(0,0) + S_{t|t}(I,0)\,S_0.$$

1. Show that the variance terms, namely $P_{t|t-1}, O_{t|t-1}, L_{t|t-1}, P_{t|t}$ *are the same for three Kalman filters.*

2. Show that linearity holds in the first period:

$$S_{1|1}(S_0,0) = S_{1|1}(0,0) + S_{1|1}(I,0)\,S_0.$$

3. Show by induction that if linearity holds in the period $t-1$,

$$S_{t-1|t-1}(S_0,0) = S_{t-1|t-1}(0,0) + S_{t-1|t-1}(I,0)\,S_0,$$

then linearity also holds in the period t:

$$S_{t|t}(S_0,0) = S_{t|t}(0,0) + S_{t|t}(I,0)\,S_0.$$

Exercise 6.12 (Decay of initial value in Kalman filtering) *The initial distribution has an impact on the Kalman filter, but in most applications the impact decays quickly as the Kalman filter moves forward. In this exercise, we investigate asymptotic effects of initialization on the Kalman filter.*

1. Consider two Kalman filters, both of which use the same transition and observation equation, as well as the same data. The only difference is initialization: one specifies $S_0 \sim N(0,V_0)$, *and the other specifies* $S_0 \sim N(\mu_0,V_0)$. *Show that the Kalman filter outputs are linear with respect to initial states:*

$$S_{t|t-1}(\mu_0,V_0) = S_{t|t-1}(0,V_0) + Z_t\mu_0,$$

where $S_{t|t-1}(\mu_0,V_0)$ *and* $S_{t|t-1}(0,V_0)$ *refer to* $S_{t|t-1}$ *under the initialization of* $N(\mu_0,V_0)$ *and* $N(0,V_0)$, *respectively. The sequence* $\{Z_t\}$ *is computed recursively by*

$$Z_t = J_{t-1}Z_{t-1},$$

with the starting value $Z_1 = A_1$ *and* $J_t = A_{t+1} - K_tC_t$.

2. Proposition 6.1 indicates that under the regularity conditions: a) coefficients A, B, C, D *are time-invariant; b) all eigenvalues of* A *are in the unit circle; and c) either* $Q = BB'$ *or* $R = DD'$ *is positive definite, the variance sequence* $\{P_{t+1|t}\}$ *converges to* P^*. *Meanwhile, the sequence* $\{J_t\}$ *also converges to* J^* *such that*

$$P^* = J^*P^*J^{*\prime} + Q + K^*RK^{*\prime}.$$

Show that the sequence $\{Z_t\}$ *decays to zero, so that the impact of* μ_0 *decays as well.*

3. Third, consider two Kalman filters initialized by different initial state means and/or variances, say $N(\mu_0,V_0)$ *and* $N\left(\tilde{\mu}_0,\tilde{V}_0\right)$. *Under the regularity conditions specified above, show that the discrepancy of the Kalman filter output converges to zero:*

$$\lim_{t\to\infty}\left|S_{t|t}(\mu_0,V_0) - S_{t|t}\left(\tilde{\mu}_0,\tilde{V}_0\right)\right| = 0.$$

Exercise 6.13 (Smooth initial states) *Consider a linear state space model with time-invariant coefficients:*

$$S_t = AS_{t-1} + Bu_t,$$

$$Y_t = CS_t + D\varepsilon_t,$$

where the eigenvalues of A are inside the unit circle, so it is possible to initialize the states S_0 by the stationary distribution.

Suppose that your favorite Kalman filter and smoother software can produce state estimates $p(S_t | Y_{1:T})$ for periods $t = 1, \ldots, T$. However, the software does not report the initial state smoothing results, namely $p(S_0 | Y_{1:T})$.

1. If the initial state distribution is specified as the stationary distribution, how to get the initial state smoothing results from the software?

2. If the initial state distribution is different from the stationary distribution, how to get the initial state smoothing results from the software?

Exercise 6.14 (Chandrasekhar recursion with missing data) *Suppose that the coefficients A, B, C, D in the state space model are time-invariant, and the states are initialized by the stationary distribution. However, observations in some periods are missing at random.*

1. Does the standard Kalman filter (Algorithm 6.1) work in the presence of missing data? If yes, explain how the filtering formula should be adapted. If no, explain why the algorithm fails.

2. Does the Chandrasekhar recursion (Algorithm 6.2) work in the presence of missing data? If yes, explain how the filtering formula should be adapted. If no, explain why the algorithm fails.

7

Nonlinear State Space Models

The Kalman filter and smoother provide optimal state estimation for Gaussian linear state space models. However, when dealing with nonlinear and non-Gaussian state space models, analytical tractability is compromised. Our objective is to approximate filtering and smoothing distributions while maintaining reasonable computational costs. This chapter delves into a range of model specifications, with varying degrees of nonlinearity.

First, the transition and observation equations remain linear, but the error terms become non-Gaussian:

$$S_t = A_t S_{t-1} + B_t u_t, \tag{7.1}$$

$$Y_t = C_t S_t + D_t \varepsilon_t. \tag{7.2}$$

If the innovations follow special distributions such as Student t, double exponential or skew normal distributions, an effective estimation approach is data augmentation, which exploits Gaussian linear structures conditional on latent variables. The Metropolis-within-Gibbs samplers for these non-Gaussian models are elaborated in Section 7.1.

Second, the transition and observation equations become nonlinear:

$$S_t = A_t \left(S_{t-1}\right) + B_t u_t, \tag{7.3}$$

$$Y_t = C_t \left(S_t\right) + D_t \varepsilon_t. \tag{7.4}$$

If $A_t(\cdot)$ and $C_t(\cdot)$ are differentiable functions, the extended Kalman filter provides a solution by linearizing the model through first-order approximation prior to applying the Kalman filter. See Section 7.2. If $A_t(\cdot)$ and $C_t(\cdot)$ are not differentiable, the unscented filter can be employed. This method discretizes distributions using carefully selected sigma points to achieve mean-variance matching, as discussed in Section 7.3.

Third, the states become discrete and characterized by a transition matrix, while the observations may follow any distribution (not necessarily expressed in explicit equation forms). Here, a discrete version of the Kalman filter offers exact state filtering and smoothing distributions, forming the basis of the Hidden Markov model (also known as the Markov switching model), which is explored in Section 7.4.

Fourth, a more general form of the nonlinear and non-Gaussian state space model is defined by the transition distribution $p(S_t | S_{t-1})$ and the observation distribution $p(Y_t | S_t)$. Sequential Monte Carlo (SMC) methods offer powerful yet computationally intensive solutions for these general state space models. SMC becomes even more powerful when it is combined with MCMC. On the one hand, SMC can be integrated into MCMC for approximating an intractable target density by an unbiased estimator (Andrieu and Roberts, 2009; Andrieu et al., 2010). On the other hand, MCMC can be integrated into SMC to promote particle diversity (Gilks and Berzuini, 2001; Chopin, 2002; Chopin et al., 2013). A significant portion of this chapter is dedicated to SMC methods for particle filtering, smoothing, parameter estimation, and parameter learning.

DOI: 10.1201/9781003564027-7

7.1 Special Non-Gaussian State Space Models

While closed-form filters and smoothers are generally unavailable for non-Gaussian and nonlinear state space models, efficient MCMC methods exist for certain specialized non-Gaussian models. These models feature innovations that follow distributions such as Student t, finite Gaussian mixture, double exponential, and skew normal distributions. A common characteristic of these distributions is their representation as a mixture of Gaussian distributions. Data augmentation simplifies the Gibbs sampler by reducing the model to a Gaussian linear form conditional on latent variables.

Consider the transition and observation equations in Equations (7.1) and (7.2), where the coefficient matrices A_t, B_t, C_t and D_t are implicitly determined by underlying parameters θ. Non-Gaussian distributions may be specified in either the state or observation equations. For simplicity, we assume that u_t in the transition equation follows the standard normal distribution, while ε_t in the observation equation does not. The marginal distribution $p(\varepsilon_t)$ can be Student t, finite Gaussian mixture, double exponential or skew normal, but the conditional distribution $p(\varepsilon_t | Z_t)$ is normal, given the latent variable Z_t. MCMC methods are employed to generate samples from the posterior distribution $p(\theta, S, Z | Y)$.

7.1.1 Scale Mixture t Distribution

Bayesian inference in t-distributed linear regression was developed by Geweke (1993) and applied to state space models in Chib and Ramamurthy (2014) and Curdia et al. (2014). Consider the innovation process as a scale mixture such that

$$\varepsilon_t = \sqrt{Z_t}\tilde{\varepsilon}_t,$$

where the $n \times 1$ vector $\tilde{\varepsilon}_t$ follows the multivariate normal distribution with an identity covariance matrix (i.e., a vector of independent standard normal variates) and the scalar-valued latent variable Z_t follows an inverse gamma distribution $IG\left(\frac{v}{2}, \frac{v}{2}\right)$. Specifically,

$$p(Z_t) \propto Z_t^{-\left(\frac{v}{2}+1\right)} e^{-\frac{v}{2}Z_t^{-1}},$$

$$p(\varepsilon_t | Z_t) \propto Z_t^{-\frac{n}{2}} e^{-\frac{1}{2}Z_t^{-1}\varepsilon_t'\varepsilon_t}.$$

The marginal distribution can be derived by either integrating out the latent variable

$$p(\varepsilon_t) = \int p(Z_t)\, p(\varepsilon_t | Z_t)\, dZ_t,$$

or applying Bayes' rule

$$p(\varepsilon_t) = \frac{p(Z_t)\, p(\varepsilon_t | Z_t)}{p(Z_t | \varepsilon_t)},$$

where $p(Z_t | \varepsilon_t)$ remains an inverse gamma distribution $IG\left(\frac{v+n}{2}, \frac{v+\varepsilon_t'\varepsilon_t}{2}\right)$. Both methods yield an explicit marginal distribution

$$p(\varepsilon_t) \propto \left(1 + \frac{1}{v}\varepsilon_t'\varepsilon_t\right)^{-\frac{v+n}{2}},$$

which is recognized as a multivariate t distribution $MVT(0, I, v)$, as a multivariate extension of Proposition 4.3.

The appeal of the scale mixture representation of the t distribution is that the Gaussian linear state space model is restored conditional on the latent variables. Two methods are available for sampling from the posterior $p(\theta, S, Z | Y)$. The first is the Gibbs sampler, which cycles through the conditional distributions $p(\theta | Y, S, Z)$, $p(S | Y, \theta, Z)$ and $p(Z | Y, \theta, S)$. The second strategy involves the marginalized Metropolis-within-Gibbs sampler, which jointly updates parameters and states based on the marginal-conditional decomposition:

$$p(\theta, S | Y, Z) = p(\theta | Y, Z) \, p(S | Y, \theta, Z),$$
$$p(\theta | Y, Z) \propto p(\theta | Z) \, p(Y | \theta, Z).$$

Marginalization over states is accomplished via the Kalman filter, applicable to the Gaussian linear model conditional on the latent variables. As the likelihood function $p(Y | \theta, Z)$ is evaluated by the Kalman filter, the posterior density $p(\theta | Y, Z)$ is evaluated up to a normalizing constant. Parameter updates are implemented through the MH sampling, while state updates are handled by the simulation smoother of the state space model.

The marginalized sampler cycles through $p(\theta, S | Y, Z)$ and $p(Z | Y, \theta, S)$. The latter takes the form

$$p(Z_t | Y, \theta, S) \propto p(Z_t) \, p(Y_t | Z_t, \theta, S),$$
$$p(Z_t) \propto Z_t^{-\left(\frac{v}{2}+1\right)} e^{-\frac{v}{2} Z_t^{-1}},$$
$$p(Y_t | Z_t, \theta, S) \propto Z_t^{-\frac{n}{2}} e^{-\frac{(Y_t - C_t S_t)' (D_t D_t')^{-1} (Y_t - C_t S_t)}{2} Z_t^{-1}}.$$

Consequently, the posterior conditional distribution of the latent variable Z_t remains inverse gamma

$$IG\left[\frac{v+n}{2}, \frac{v + (Y_t - C_t S_t)' (D_t D_t')^{-1} (Y_t - C_t S_t)}{2}\right].$$

Alternatively, when conditioned on states and parameters, the observation equation can be interpreted as a linear regression model

$$\tilde{Y}_t = C_t S_t + D_t \sqrt{Z_t} \tilde{\varepsilon}_t,$$

where the response variable is $\tilde{Y}_t = D_t^{-1} (Y_t - C_t S_t)$, with the predictor matrix being empty (i.e., no predictors), and the innovation variance (i.e., Z_t) has an inverse gamma prior $IG\left(\frac{v}{2}, \frac{v}{2}\right)$. The number of observations is n (the length of Y_t). As a special case of conjugate Bayesian linear regressions, the posterior conditional distribution of Z_t remain inverse gamma:

$$IG\left(\frac{v+n}{2}, \frac{v + \tilde{Y}_t' \tilde{Y}_t}{2}\right).$$

If the degrees of freedom $v \to \infty$, the t distribution converges to the normal distribution. Meanwhile, $IG(\infty, \infty)$ reduces to a degenerate distribution, as its mean converges to one, and the variance collapses to zero, resulting in degeneracy as a constant $Z_t = 1$.

Regarding the degrees of freedom parameter v, it is acceptable to specify a fixed value, such as $v = 5$ to ensure a fat-tail distribution. Alternatively, the degrees of freedom parameter can be estimated alongside other unknown variables in the model. A Metropolis component may be incorporated into the sampler, as its posterior conditional distribution lacks a closed form, regardless of prior specification. If a uniform prior is adopted within an interval $(\underline{v}, \bar{v})$, the posterior conditional density can be evaluated as

$$p(v | Y, \theta, S, Z) = p(v | Z) \propto \prod_{t=1}^{T} p(Z_t | v) \, I_{(\underline{v}, \bar{v})}(v),$$

where $p(Z_t|v)$ is the density of $IG\left(\frac{v}{2},\frac{v}{2}\right)$:

$$p(v|Z) \propto \exp\left[-T\ln\Gamma\left(\frac{v}{2}\right)+\frac{vT}{2}\ln\frac{v}{2}-\left(\frac{v}{2}+1\right)\sum_{t=1}^{T}\ln Z_t-\frac{v}{2}\sum_{t=1}^{T}Z_t^{-1}\right] I_{(\underline{v},\bar{v})}(v).$$

7.1.2 Finite Gaussian Mixture Distribution

A finite mixture distribution consists of random components from different distributions. The mixture density is the weighted average of these component densities, offering flexibility to accommodate skewed, leptokurtic, and multi-modal distributions. Consider a Gaussian mixture

$$p(\varepsilon_t) = \sum_{j=1}^{k} w_j \phi(\varepsilon_t; \mu_j, \Sigma_j),$$

where $\phi(\cdot)$ denotes the multivariate normal density, and μ_j, Σ_j are component-specific mean and variance. The finite mixture distribution has a hierarchical representation. Let $Z_t \in \{1, \ldots, k\}$ be a latent variable that follows a categorical distribution

$$p(Z_t = j) = w_j.$$

The jth component is selected with the probability w_j. The conditional distribution is Gaussian:

$$p(\varepsilon_t|Z_t = j) = \phi(\varepsilon_t; \mu_j, \Sigma_j).$$

Marginalization over the latent variable yields the marginal distribution

$$p(\varepsilon_t) = \int p(Z_t)\, p(\varepsilon_t|Z_t)\, dZ_t = \sum_{j=1}^{k} w_j \phi(\varepsilon_t; \mu_j, \Sigma_j).$$

In the finite mixture state space model, the unknown variables include static parameters, dynamic states, and latent variables. The Metropolis-within-Gibbs sampler cycles through the posterior conditional distributions. Parameters and states are sampled jointly via the marginal-conditional decomposition

$$p(\theta, S|Y, Z) = p(\theta|Y, Z)\, p(S|Y, \theta, Z).$$

Given the latent variables, the model reduces to a Gaussian linear state space model, where the observation equation is expressed as

$$\tilde{Y}_t = C_t S_t + D_t \Sigma_{Z_t}^{1/2} \tilde{\varepsilon}_t,$$

where $\tilde{Y}_t$ is the deflated observation such that $\tilde{Y}_t = Y_t - D_t \mu_{Z_t}$ and $\Sigma_j^{1/2}$ is the Cholesky factor of Σ_j, and $\tilde{\varepsilon}_t$ follows the standard normal distribution. The Kalman filter computes the likelihood $p(Y|\theta, Z)$, with the states integrated out. Consequently, the MH sampler updates the parameters by evaluating the posterior density $p(\theta|Y, Z) \propto p(\theta|Z)\, p(Y|\theta, Z)$. Additionally, the simulation smoother updates states from $p(S|Y, \theta, Z)$ through forward filtering and backward sampling.

To update the latent variables, we consider the posterior conditional distribution

$$p(Z|Y, \theta, S) \propto \prod_{t=1}^{T} p(Z_t)\, p(Y_t|Z_t, \theta, S).$$

The observation equation of the state space model indicates that

$$p(Y_t | Z_t, \theta, S) = \phi(Y_t; C_t S_t + D_t \mu_{Z_t}, D_t \Sigma_{Z_t} D_t').$$

The discrete Z_t makes the posterior a categorical distribution:

$$\begin{aligned} p(Z_t = j | Y, \theta, S) &\propto w_j \phi(Y_t; C_t S_t + D_t \mu_j, D_t \Sigma_j D_t') \\ &\propto w_j |D_t \Sigma_j D_t'|^{-\frac{1}{2}} e^{-\frac{1}{2}(Y_t - C_t S_t - D_t \mu_j)' (D_t \Sigma_j D_t')^{-1} (Y_t - C_t S_t - D_t \mu_j)}. \end{aligned}$$

Alternatively, the posterior can be derived from the innovation

$$\varepsilon_t = D_t^{-1}(Y_t - C_t S_t),$$

if D_t is an invertible matrix. The latent variable has the posterior

$$p(Z_t = j | Y, \theta, S) \propto w_j \phi(\varepsilon_t; \mu_j, \Sigma_j) \propto w_j |\Sigma_j|^{-\frac{1}{2}} e^{-\frac{1}{2}(\varepsilon_t - \mu_j)' \Sigma_j^{-1} (\varepsilon_t - \mu_j)}.$$

Are the parameters μ_j, Σ_j estimable? The answer is both yes and no. On the one hand, they can be sampled from the posterior conditional distribution $p(\mu_j, \Sigma_j | Y, \theta, S, Z)$, which follows a normal-inverse-Wishart conjugate distribution, using the subsample selected by $Z_t = j$. On the other hand, the Gaussian mixture is so flexible that different parameters may yield similar densities, making reliable estimation of these parameters challenging.

7.1.3 Double Exponential Distribution

Assume that the observation Y_t is scalar valued. If not, we may apply the technique of univariate treatment of multivariate series by dividing a period into multiple sub-periods, each observing a single data point. Similar to the t-distributed state space model, we specify the innovation process as a scale mixture such that

$$\varepsilon_t = \sqrt{Z_t} \tilde{\varepsilon}_t,$$

where $\tilde{\varepsilon}_t$ follows the standard normal distribution, and the latent variable Z_t follows an exponential distribution

$$p(Z_t) = \frac{1}{2} e^{-\frac{1}{2} Z_t}.$$

By marginalization over the latent variable, we have the marginal distribution

$$p(\varepsilon_t) = \int p(Z_t) p(\varepsilon_t | Z_t) dZ_t = \frac{1}{2}(2\pi)^{-\frac{1}{2}} \int_0^\infty z^{-\frac{1}{2}} e^{-\frac{1}{2}(z + z^{-1}\varepsilon_t^2)} dz,$$

where the integrand takes the form of reciprocal inverse Gaussian density, allowing the integral to have an explicit solution

$$\int_0^\infty z^{-\frac{1}{2}} e^{-\frac{1}{2}(z + z^{-1}\varepsilon_t^2)} dz = (2\pi)^{\frac{1}{2}} e^{-|\varepsilon_t|}.$$

It follows that

$$p(\varepsilon_t) = \frac{1}{2} e^{-|\varepsilon_t|}.$$

We recognize that the marginal distribution is double exponential, with the location parameter 0 and the scale parameter 1. Without loss of generality, a non-zero location can be accommodated by deflating observations, and a scale parameter can be adjusted by the coefficient D_t.

The posterior distribution of parameters, states and latent variables can be simulated by the Metropolis-within-Gibbs sampler that cycles through the posterior conditional distributions. First, parameters and states are sampled jointly via marginal-conditional decomposition, following the same procedure as the t-distributed state space model. Second, latent variables are sampled from their posterior conditional distribution

$$p(Z_t|Y,\theta,S) \propto p(Z_t)\,p(\varepsilon_t|Z_t) \propto Z_t^{-\frac{1}{2}} e^{-\frac{1}{2}\left(Z_t+Z_t^{-1}\varepsilon_t^2\right)}.$$

Although $p(Z_t|Y,\theta,S)$ does not appear to be a common distribution, its reciprocal does. By reparametrizing the latent variable as $\gamma_t = Z_t^{-1}$, and the posterior conditional density is given by

$$p(\gamma_t|Y,\theta,S) \propto \gamma_t^{-\frac{3}{2}} e^{-\frac{1}{2}\left(\gamma_t^{-1}+\gamma_t\varepsilon_t^2\right)},$$

which is the inverse Gaussian distribution with the mean parameter $\left|\varepsilon_t^{-1}\right|$ and the scale parameter 1.

7.1.4 Skew Normal Distribution

Assume that the observation Y_t is scalar valued, and the innovation process takes the form

$$\varepsilon_t = \delta Z_t + v_t,$$

where v_t follows the standard normal distribution, while the latent variable Z_t follows a half normal distribution, which is the standard normal distribution truncated to the positive region. Alternatively, this can be interpreted as the distribution for the absolute value of a standard normal variate. The marginal density is given by the integral

$$p(\varepsilon_t) = \int p(Z_t)\,p(\varepsilon_t|Z_t)\,dZ_t = \int_0^\infty \frac{1}{\pi} e^{-\frac{1}{2}\left[z^2+(\varepsilon_t-\delta z)^2\right]} dz.$$

By completing the square, we derive

$$z^2 + (\varepsilon_t - \delta z)^2 = \left(\delta^2+1\right)\left(z - \frac{\delta\varepsilon_t}{\delta^2+1}\right)^2 + \frac{\varepsilon_t^2}{\delta^2+1}.$$

It follows that

$$p(\varepsilon_t) = 2\phi\left(\varepsilon_t; 0, \delta^2+1\right)\Phi\left(\delta\varepsilon_t; 0, \delta^2+1\right),$$

which is recognized as the density of the skew normal $SN\left(0, \delta^2+1, \delta\right)$. Compared to the skew normal regression model specified in Section 4.4, it does not contain a scale parameter σ, as in $SN\left(0, \delta^2+1, \delta/\sigma\right)$. The reason is that the state space model coefficient D_t incorporates the scale parameter. For the purpose of identification, we normalize $\sigma = 1$.

The skew normal distribution can be represented as the weighted sum of a half normal latent variable and a standard normal variate. The former is always positive, so the weight determines the skewness. If $\delta < 0$, the distribution is skewed to the left, with a fat tail in the negative region of the density function; if $\delta > 0$, the distribution is skewed to the right, with a fat tail in the positive portion of the density function; if $\delta = 0$, it reduces to the standard normal distribution.

To estimate the state space model with the skew normal distribution, we apply the Metropolis-within-Gibbs sampler with data augmentation.

First, conditional on the latent variable, we have a Gaussian linear state space model, in which we treat $Y_t - \delta Z_t$ as the deflated observation. We may simultaneously update parameters and states by the Kalman filter and the simulation smoother, following the same procedure as the t-distributed state space model.

Second, conditional on the states and parameters (including δ), the latent variable Z_t is handled by a Bayesian linear regression with a single data point: the transformed response variable is $D_t^{-1}(Y_t - C_t S_t)$, the predictor is δ, and the noise variance equals one. Additionally, the latent variable is subject to truncation due to the half-normal prior, resulting in a truncated normal posterior distribution $N(\bar{\mu}_t, \bar{\sigma}^2)\, I(Z_t > 0)$, where

$$\bar{\mu}_t = \left(\delta^2 + 1\right)^{-1} \delta D_t^{-1} \left(Y_t - C_t S_t\right),$$

$$\bar{\sigma}^2 = \left(\delta^2 + 1\right)^{-1}.$$

Third, conditional on the states, parameters (excluding δ), and the latent variable, the skewness parameter δ is handled by a conjugate Bayesian linear regression (with T data points), in which the transformed response variable is $D_t^{-1}(Y_t - C_t S_t)$, the predictor is Z_t and the noise variance equals one. Assuming δ has a normal prior $N(\mu_\delta, \sigma_\delta^2)$, the posterior follows a normal distribution $N(\bar{\mu}_\delta, \bar{\sigma}_\delta^2)$, where

$$\bar{\mu}_\delta = \bar{\sigma}_\delta^2 \left[\sigma_\delta^{-2}\mu_\delta + \sum_{t=1}^{T} D_t^{-1}\left(Y_t - C_t S_t\right) Z_t\right],$$

$$\bar{\sigma}_\delta^2 = \left(\sigma_\delta^{-2} + \sum_{t=1}^{T} Z_t^2\right)^{-1}.$$

The third step may be combined with the first step, if the observation equation $Y_t = C_t S_t + D_t \delta Z_t + D_t v_t$ takes the form of a linear regression. In that case, it is possible to update δ along with other model parameters to improve the convergence of the MCMC chain.

7.2 Extended Kalman Filter

The Kalman filter provides optimal state estimation for linear state space models. In the case of nonlinear state space models, a natural solution is to linearize the model and apply the Kalman filter. The extended Kalman filter requires differentiability of nonlinear functions $A_t(\cdot)$ and $C_t(\cdot)$ in Equations (7.3) and (7.4). The core idea of the extended Kalman filter is to linearize the transition and observation equations using a first-order Taylor approximation around the current state estimator. In a locally linear model, the state prediction and update rules align with the standard Kalman filter, which keeps track of a set of intermediate variables $S_{t|t-1}, P_{t|t-1}, Y_{t|t-1}, O_{t|t-1}, S_{t|t}, P_{t|t}$ in the forward recursion.

Consider forward filtering in period $t = 1, \ldots, T$. Given the previous state estimator $S_{t-1|t-1}$ and $P_{t-1|t-1}$, the transition equation is linearized around $S_{t|t-1}$:

$$S_t \approx A_t\left(S_{t-1|t-1}\right) + \mathbb{A}_t\left(S_{t-1} - S_{t-1|t-1}\right) + B_t u_t,$$

where the gradient $\mathbb{A}_t = \frac{\partial A_t(S_{t-1|t-1})}{\partial S_{t-1|t-1}}$ is evaluated at $S_{t-1|t-1}$. The predictive mean and variance of S_t are approximated by

$$S_{t|t-1} = A_t\left(S_{t-1|t-1}\right),$$

$$P_{t|t-1} = \mathbb{A}_t P_{t-1|t-1} \mathbb{A}_t' + B_t B_t'.$$

Similarly, the observation equation is linearized around $S_{t|t-1}$,

$$Y_t \approx C_t\left(S_{t|t-1}\right) + \mathbb{C}_t\left(S_t - S_{t|t-1}\right) + D_t\varepsilon_t.$$

where the gradient $\mathbb{C}_t = \frac{\partial C_t(S_{t|t-1})}{\partial S_{t|t-1}}$ is evaluated at $S_{t|t-1}$. The predictive mean and variance of Y_t are approximated by

$$\begin{aligned} Y_{t|t-1} &= C_t\left(S_{t|t-1}\right), \\ O_{t|t-1} &= \mathbb{C}_t P_{t|t-1}\mathbb{C}_t' + D_t D_t', \\ L_{t|t-1} &= P_{t|t-1}\mathbb{C}_t'. \end{aligned}$$

The update step mirrors that of the standard Kalman filter. Once observations for period t are available, the mean and variance of the filtering distribution of states are updated by

$$S_{t|t} = S_{t|t-1} + L_{t|t-1}O_{t|t-1}^{-1}\left(Y_t - Y_{t|t-1}\right),$$

$$P_{t|t} = P_{t|t-1} - L_{t|t-1}O_{t|t-1}^{-1}L_{t|t-1}'.$$

The extended Kalman filter is widely used in tracking and control of nonlinear state space systems. Although the first-order approximation may seem crude, it typically performs well due to reinforced learning from each observation. An update might not be optimal, but any overshoot or under-reaction is likely corrected in the subsequent period as the filter responds to larger prediction errors. The Kalman filter copes with sub-optimality by incorporating feedback from the data: a good state estimator is rewarded with a smaller prediction error in the next period, resulting in a smaller state update to maintain momentum. Conversely, a poor state estimator leads to a larger prediction error, prompting the filter to apply a larger state update.

The absence of optimality also implies non-unique solution to nonlinear filtering, allowing for algorithmic extensions in various directions. For example, the center point for linearizing the observation equation could be revised to $S_{t|t}$ after updating the states based on an initial linearization around $S_{t|t-1}$. Additionally, consider the predictive mean approximation by $S_{t|t-1} = A_t\left(S_{t-1|t-1}\right)$. The Jensen's inequality suggests that the concavity of the nonlinear function affects the predictive mean, which can be addressed with a second-order approximation incorporating a Hessian term. It implies that both $S_{t-1|t-1}$ and $P_{t-1|t-1}$ contribute to the predictive mean approximation. Naturally, there is a tradeoff between filtering accuracy and computational intensity. The art of nonlinear filtering lies in finding a satisfactory filter with manageable computational costs.

7.3 Unscented Kalman Filter

As an alternative to the extended Kalman filter that approximates nonlinear functions, the unscented Kalman filter (Julier and Uhlmann, 2004) resorts to a discrete approximation of distributions. This is achieved through carefully selected points and weights that match the moments of continuous distributions. Nonlinearity poses no obstacle for discrete random variables, as the nonlinear functions are simply evaluated at these discrete points without changing the weights. A key advantage of the unscented filter is that it does not require differentiability of nonlinear functions, as it bypasses the need for gradient information in distributional approximation.

7.3.1 Unscented Transformation

We consider the unscented transformation in a generic setting. Let X be a $m \times 1$ random vector with mean μ and covariance matrix Ω. We are interest in the distribution of $Y = g(X)$, where g is a nonlinear function that is not necessarily differentiable. The essence of the unscented transformation is to approximate the distribution of X with a categorical distribution of only $2m+1$ discrete points, matching the mean and covariance matrix. The approximation algorithm proceeds as follows:

First, apply the Cholesky decomposition to the covariance matrix: $\Omega = LL'$, where $L = (L_1, \ldots, L_m)$ and $\Omega = \sum_{i=1}^m L_i L_i'$.

Second, specify a tuning parameter k and construct a set of symmetric points x_i, $i = -m, \ldots, -1, 0, 1, \ldots, m$ such that

$$x_i = \begin{cases} \mu + \sqrt{m+k} L_i & i > 0 \\ \mu & i = 0 \\ \mu - \sqrt{m+k} L_{-i} & i < 0 \end{cases}.$$

Third, assign weights to these points: $w_0 = \frac{k}{m+k}$ and $w_i = \frac{1}{2(m+k)}$, $\forall i \neq 0$.

The mean and variance of the categorical distribution $\{x_i, w_i\}$ align with the original distribution:

$$\sum_{i=-m}^{m} w_i x_i = \mu,$$

$$\sum_{i=-m}^{m} w_i (x_i - \mu)(x_i - \mu)' = \sum_{i=1}^{m} L_i L_i' = \Omega.$$

The random variable Y, being a nonlinear function of X, is then approximated by the categorical distribution $\{g(x_i), w_i\}$, with the mean and variance

$$\mu_Y = \sum_{i=-m}^{m} w_i g(x_i),$$

$$V_Y = \sum_{i=-m}^{m} w_i [g(x_i) - \mu_Y][g(x_i) - \mu_Y]'.$$

7.3.2 Unscented Filter

The unscented transformation can be applied to the nonlinear state space model described in Equations (7.3) and (7.4), where the nonlinear functions $A_t(\cdot)$ and $C_t(\cdot)$ are not necessarily differentiable. Similar to the Kalman filter, the following set of intermediate variables will be used in the forward recursion: $S_{t|t-1}, P_{t|t-1}, Y_{t|t-1}, O_{t|t-1}, S_{t|t}, P_{t|t}$. Given the previous state estimator $S_{t-1|t-1}, P_{t-1|t-1}$, we consider forward filtering in period t.

Two approaches can be used to apply unscented transformation. A straightforward way is to discretize S_{t-1} in the transition equation, followed by discretizing S_t again in the observation equation. An alternative method is to expand the states as

$$\tilde{S}_{t-1} = \begin{pmatrix} S_{t-1} \\ u_t \end{pmatrix},$$

with the mean and variance

$$\tilde{S}_{t-1|t-1} = \begin{pmatrix} S_{t-1|t-1} \\ 0 \end{pmatrix},$$

$$\tilde{P}_{t-1|t-1} = \begin{pmatrix} P_{t-1|t-1} & 0 \\ 0 & I \end{pmatrix}.$$

Apply the Cholesky decomposition: $\tilde{P}_{t-1|t-1} = LL'$, where $L = (L_1, \ldots, L_m)$ and m is the length of $\tilde{S}_{t-1}$. The points and weights for unscented transformation are given by

$$x_i = \begin{cases} \tilde{S}_{t-1|t-1} + \sqrt{m+k}L_i & i > 0 \\ \tilde{S}_{t-1|t-1} & i = 0\,, \\ \tilde{S}_{t-1|t-1} - \sqrt{m+k}L_{-i} & i < 0 \end{cases}$$

$$w_i = \begin{cases} \frac{1}{2(m+k)} & i > 0 \\ \frac{k}{m+k} & i = 0\,. \\ \frac{1}{2(m+k)} & i < 0 \end{cases}$$

The transition equation can be rewritten in a compact form:

$$S_t = \tilde{A}_t\left(\tilde{S}_{t-1}\right) = A_t\left(S_{t-1}\right) + B_t u_t.$$

The distribution of S_t is approximated by the categorical distribution $\left\{\tilde{A}_t\left(x_i\right), w_i\right\}$, from which the predictive mean and variance are derived:

$$S_{t|t-1} = \sum_{i=-m}^{m} w_i \tilde{A}_t\left(x_i\right),$$

$$P_{t|t-1} = \sum_{i=-m}^{m} w_i \left[\tilde{A}_t\left(x_i\right) - S_{t|t-1}\right]\left[\tilde{A}_t\left(x_i\right) - S_{t|t-1}\right]'.$$

Meanwhile, Y_t is partially discretized as the sum of a discrete part $C_t\left(S_t\right)$ and a continuous part $D_t\varepsilon_t$. The former is represented by the categorical distribution $\left\{C_t\left[\tilde{A}_t\left(x_i\right)\right], w_i\right\}$, with the predictive mean and variance

$$Y_{t|t-1} = \sum_{i=-m}^{m} w_i C_t\left[\tilde{A}_t\left(x_i\right)\right],$$

$$O_{t|t-1} = D_t D_t' + \sum_{i=-m}^{m} w_i \left\{C_t\left[\tilde{A}_t\left(x_i\right)\right] - Y_{t|t-1}\right\}\left\{C_t\left[\tilde{A}_t\left(x_i\right)\right] - Y_{t|t-1}\right\}',$$

$$L_{t|t-1} = \sum_{i=-m}^{m} w_i \left\{\tilde{A}_t\left(x_i\right) - S_{t|t-1}\right\}\left\{C_t\left[\tilde{A}_t\left(x_i\right)\right] - Y_{t|t-1}\right\}'.$$

The update step aligns with the standard Kalman filter:

$$S_{t|t} = S_{t|t-1} + L_{t|t-1} O_{t|t-1}^{-1}\left(Y_t - Y_{t|t-1}\right),$$

$$P_{t|t} = P_{t|t-1} - L_{t|t-1} O_{t|t-1}^{-1} L_{t|t-1}'.$$

The unscented Kalman filter exhibits several notable features. 1) High accuracy. Julier and Uhlmann (2004) demonstrate that the unscented transformation achieves a second-order approximation, whereas the extended Kalman filter is limited to first-order approximation. 2) Low computational cost. The unscented transformation selects a small number of discrete points to approximate a high-dimensional random vector, while the particle filter typically requires sampling thousands of points for Monte Carlo approximation. 3) Non-differentiability. Unlike the extended Kalman filter that requires differentiable functions, the unscented transformation is applicable to non-differentiable functions. 4) Deterministic output. Given the model and data, the results of the unscented Kalman filter are deterministic, free from Monte Carlo noise. This ensures that the approximated log-likelihood is deterministic, facilitating numerical optimization. 5) Downward compatibility. The unscented filter reduces to the standard Kalman filter if the transition and observation equations are linear (see Exercise 7.1).

7.4 Hidden Markov Model

In this section, we explore a specific case of nonlinear and non-Gaussian state space models, where states are discrete, such as $S_t \in \{1, \ldots, m\}$. The fundamental assumptions of state space models are maintained. First, Markov transition: the transition distribution $p(S_t | S_{t-1})$ is represented by a transition matrix, with the probability of transitioning from state i to j denoted by $q(i,j)$. Second, instantaneous observation: the measurement distribution $p(Y_t | S_t)$ can be any density function if Y_t is continuous, or a probability mass function if Y_t is discrete. This model is known as the Hidden Markov Model (HMM) or the Markov Switching Model (Cappe et al., 2004; Fruhwirth-Schnatter, 2006).

Given the previous filtering results $p(S_{t-1} | Y_{1:t-1})$, forward filtering in state space models involves the following steps:

1. State prediction:
$$p(S_t | Y_{1:t-1}) = \int p(S_{t-1} | Y_{1:t-1}) p(S_t | S_{t-1}) dS_{t-1}.$$

2. Observation prediction:
$$p(Y_t | Y_{1:t-1}) = \int p(S_t | Y_{1:t-1}) p(Y_t | S_t) dS_t.$$

3. Sate update:
$$p(S_t | Y_{1:t}) = \frac{p(S_t | Y_{1:t-1}) p(Y_t | S_t)}{p(Y_t | Y_{1:t-1})}.$$

For general nonlinear and non-Gaussian state space models, these integrals are not analytically tractable. However, there are two notable exceptions. The first is Gaussian linear state space models, where the predictive and filtering distributions are multivariate normal, allowing the integrals in explicit forms. The second is discrete state models, where integration reduces to summation, which is always tractable. The discrete version of the state prediction formula reads

$$p(S_t = s_t | Y_{1:t-1}) = \sum_{s_{t-1}} p(S_{t-1} = s_{t-1} | Y_{1:t-1}) q(s_{t-1}, s_t).$$

Similarly, the discrete version of the observation prediction is

$$p(Y_t | Y_{1:t-1}) = \sum_{s_t} p(S_t = s_t | Y_{1:t-1}) p(Y_t | S_t = s_t).$$

Algorithm 7.1 Kalman filter for HMM

Input: data Y, state transition $q(s_{t-1}, s_t)$, observation distribution $p(Y_t | S_t)$, initial state distribution $p(S_0) = p(S_0 | Y_{1:0})$.
Output: state filtering distribution $p(S_t | Y_{1:t})$.
Iterate the following distributions for $t = 1, \ldots, T$:
1 State predictive distribution:

$$p(S_t = s_t | Y_{1:t-1}) = \sum_{s_{t-1}} p(S_{t-1} = s_{t-1} | Y_{1:t-1}) \, q(s_{t-1}, s_t).$$

2 Data predictive distribution:

$$p(Y_t | Y_{1:t-1}) = \sum_{s_t} p(S_t = s_t | Y_{1:t-1}) \, p(Y_t | S_t = s_t).$$

3 State filtering distribution:

$$p(S_t = s_t | Y_{1:t}) = \frac{p(S_t = s_t | Y_{1:t-1}) \, p(Y_t | S_t = s_t)}{p(Y_t | Y_{1:t-1})}.$$

Algorithm 7.1 summarizes the discrete version of the Kalman filter.

Traditionally, state estimation in HMMs is addressed using the Baum-Welch algorithm (Baum et al., 1970), which tracks forward and backward variables defined as

$$p(S_t = s_t | Y_{1:T}) \propto p(S_t = s_t, Y_{1:t}) \, p(Y_{t+1:T} | S_t = s_t),$$

where $F_t(s_t) = p(S_t = s_t, Y_{1:t})$ is called the forward variable, and $B_t(s_t) = p(Y_{t+1:T} | S_t = s_t)$ is called the backward variable. As the name suggests, the forward variable is obtained by forward recursion:

$$F_t(s_t) = \sum_{s_{t-1}} F_{t-1}(s_{t-1}) \, q(s_{t-1}, s_t) \, p(Y_t | S_t = s_t).$$

Similarly, the backward variable is obtained by backward recursion:

$$B_t(s_t) = \sum_{s_{t+1}} q(s_t, s_{t+1}) \, p(Y_{t+1} | S_{t+1} = s_{t+1}) \, B_{t+1}(s_{t+1}).$$

The forward variable, as it represents the joint probability of an increasing number of variables, tends to decrease over time, potentially leading to numerical underflow on computers. To mitigate the numerical risk, the forward variable can be rescaled to ensure summing to one, effectively converting $p(S_t = s_t, Y_{1:t})$ to $p(S_t = s_t | Y_{1:t})$. This normalization strategy makes the forward variable recursion numerically equivalent to the Kalman filter (Exercise 7.2).

The Baum-Welch algorithm serves multiple purposes:

1. State filtering:

$$p(S_t = s_t | Y_{1:t}) \propto F_t(s_t).$$

2. State smoothing:

$$p(S_t = s_t | Y_{1:T}) \propto F_t(s_t) \, B_t(s_t).$$

3. Likelihood evaluation:

$$p(Y_{1:T}) = \sum_{s_t} F_t(s_t) B_t = \sum_{s_T} F_T(s_T).$$

Additionally, the joint distribution of adjacent states can be evaluated using forward and backward variables:

$$p(S_t = s_t, S_{t+1} = s_{t+1} | Y_{1:T}) \propto F_t(s_t) q(s_t, s_{t+1}) p(Y_{t+1} | S_{t+1} = s_{t+1}) B_{t+1}(s_{t+1}),$$

facilitating estimation of the transition matrix from data.

Chib (1996) develops simulation smoothing for states $S_{0:T}$ jointly, enabling full Bayesian analysis by the Gibbs sampler with data augmentation.

7.5 Particle Filter

We have previously discussed filtering methods tailored for state space models with specific structures. The remainder of this chapter focuses on general-form state space models defined by the transition density $p(S_t | S_{t-1})$ and the observation density $p(Y_t | S_t)$.

Our primary interest lies in the state filtering distributions $p(S_{0:t} | Y_{1:t})$, $t = 1, \ldots, T$, which may not be analytically tractable. However, the density can be evaluated pointwise up to a normalizing constant:

$$p(S_{0:t} | Y_{1:t}) \propto p(S_0) \prod_{\tau=1}^{t} p(S_\tau | S_{\tau-1}) p(Y_\tau | S_\tau), \tag{7.5}$$

in which the unknown normalizing constant is of interest, as it corresponds to the likelihood function $p(Y_{1:t})$, without conditioning on the model parameters explicitly denoted. Computing the likelihood function is essential for parameter estimation using either maximum likelihood or Bayesian MCMC sampling.

The filtering distributions adhere to fundamental recursions:

$$p(S_{0:t} | Y_{1:t}) = p(S_{0:t-1} | Y_{1:t-1}) \frac{p(S_t | S_{t-1}) p(Y_t | S_t)}{p(Y_t | Y_{1:t-1})}. \tag{7.6}$$

The particle filter, introduced by Gordon et al. (1993), is a powerful sequential Monte Carlo method for approximating a series of target distributions, as described in Equation (7.5) or (7.6). Developed within the importance sampling framework, particle filtering involves generating particles from a well-chosen proposal density and assigning them importance weights. The weighted particles represent a categorical distribution that approximates the target distribution, and the sample average of the unnormalized weights approximates the likelihood function.

7.5.1 Importance Sampling

Consider a high-dimensional target distribution from which we wish to draw random samples. The density function is proportional to $g(x)$, but the normalizing constant $k = \int g(x)\,dx$ is not analytically tractable. Although the function $g(x)$ can be evaluated pointwise, direct sampling is not feasible. This scenario is typical in nonlinear state space models, where the filtering distribution can only be evaluated up to an unknown normalizing constant.

For univariate distributions, various deterministic or stochastic methods are available for numerical integration and sampling (e.g., Gauss-Legendre quadrature, rejection sampling). For multivariate distributions, high-dimensional integration and sampling are typically managed within the importance sampling or MCMC frameworks. It is also possible to integrate MCMC into importance sampling (e.g., the resample-move method) and vice versa (e.g., the pseudo marginal MH sampler).

The idea behind importance sampling is to compute the weighted average of samples from a proposal density $f(x)$ that is easy to evaluate and sample from. A critical requirement is that the support of the proposal distribution must include that of the target distribution, meaning $f(x) > 0$ whenever $g(x) > 0$. This is generally not restrictive, as Gaussian proposals, which have positive density everywhere, are often used.

The importance sampling algorithm involves two steps. First, generate random samples (i.e., particles) x_i, $i = 1, \ldots, n$ from the proposal distribution. Second, assign unnormalized weights to these particles:

$$w(x_i) = \frac{g(x_i)}{f(x_i)}.$$

Since the weights of all particles must sum to one, the normalized weights are

$$\tilde{w}(x_i) = \frac{w(x_i)}{\sum_{j=1}^{n} w(x_j)}.$$

The particles and weights form a categorical distribution: discrete points $x_1, \ldots, x_n$ associated with probabilities $\tilde{w}(x_1), \ldots, \tilde{w}(x_n)$. These weighted particles approximate the target distribution. By distributional approximation, we mean that the empirical cumulative distribution function (CDF) converges to the CDF of the target distribution as the number of particles increases, as demonstrated in Proposition 7.1. See also Geweke (1989a) for Monte Carlo integration theorems in Bayesian inference.

Proposition 7.1 (Importance sampling) *Suppose that the target density is proportional to $g(x)$, with an unknown normalizing constant $k = \int g(x)\,dx$. Let $F(c) = \int_{-\infty}^{c} k^{-1} g(x)\,dx$ be the CDF of the target distribution evaluated at a point c. Let x_i, $i = 1, \ldots, n$ be random particles generated from a proposal density $f(x)$, with the unnormalized weights $w(x_i) = \frac{g(x_i)}{f(x_i)}$ and the normalized weights $\tilde{w}(x_i) = \frac{w(x_i)}{\sum_{j=1}^{n} w(x_j)}$. The empirical CDF is defined as*

$$F_n(c) = \sum_{i=1}^{n} \tilde{w}(x_i)\, I(x_i \leq c).$$

1. The mean of unnormalized weights is an unbiased and consistent estimator of the normalizing constant

$$E[w(x_i)] = k,$$
$$\frac{1}{n}\sum_{i=1}^{n} w(x_i) \xrightarrow{a.s.} k.$$

2. The empirical CDF is a consistent estimator of the CDF of the target distribution

$$F_n(c) \xrightarrow{a.s.} F(c).$$

The importance sampling takes a new look from the perspective of consistent estimation of expectations instead of distributional approximation. Exercise 7.3 highlights a subtle difference between approximating expectations and distributions. Exercise 7.4 illustrates the importance sampling as a variance reduction technique. Exercise 7.5 discusses selecting proposals to minimize the variance of importance weights.

7.5.2 Resampling and Reweighting

Resampling and reweighting are central to nonlinear filtering and smoothing algorithms. Resampling discards unpromising particles and mitigates weight degeneracy, while reweighting redirects existing particles to approximate a new target distribution. Resampling and reweighting enable switching between different target distributions, such as the predictive, filtering, and smoothing distributions in nonlinear state space models.

Let x_i and $\tilde{w}(x_i)$, $i = 1, \ldots, n$ represent the weighted particles generated by the importance sampler. Multinomial resampling refers to random sampling with replacement a new set of particles $\hat{x}_j$, $j = 1, \ldots, J$ (usually $J = n$) from the existing particles, drawing from the standard multinomial (categorical) distribution: $P(\hat{x}_j = x_i) = \tilde{w}(x_i)$.

The frequency that x_i appears in the resampled particles should approximately match the probability $\tilde{w}(x_i)$, implying that particles with lower weights are likely to be discarded during resampling. Alternative resampling methods, such as systematic and residual resampling, can speed up simulations and reduce Monte Carlo variation by aligning frequencies with probabilities more efficiently (Kitagawa, 1996; Liu and Chen, 1998; Carpenter et al., 1999).

Resampling can be viewed as a special case of importance sampling. If the proposal distribution matches the target distribution, the importance sampler uses equal-weight particles to approximate the target distribution. Random draws are generated from the categorical distribution, and the equal-weight resampled particles approximate both the categorical and original target distributions.

Through importance sampling and resampling, draws from an intractable target distribution $g(x)$ can be obtained in two steps. First, sample particles from a proposal distribution $f(x)$, and assign them weights proportional to $\frac{g(x)}{f(x)}$. Second, resample with replacement from the weighted particles. The equal-weight resampled particles approximate the target distribution. The two-step procedure is also known as the weighted bootstrap.

Reweighting is a process of weight correction due to changes in target distributions, such as transition from filtering to smoothing distributions. Reweighting recycles existing particles and adjusts importance weights for the new target distribution.

Reweighting adheres to the importance sampling scheme. Let x_i, $i = 1, \ldots, n$ be particles generated from a proposal density $f(x)$. If the target density is proportional to $g(x)$, the unnormalized importance weights are $w(x_i) = \frac{g(x_i)}{f(x_i)}$. If the target density changes to $G(x)$, the new weights are corrected as

$$W(x_i) = \frac{G(x_i)}{f(x_i)} = w(x_i)\frac{G(x_i)}{g(x_i)},$$

which indicates that the existing particles are reweighted to reflect the revision in the target distribution from $g(x)$ to $G(x)$.

In nonlinear state space models, the revised target density is often closely related to the original target density in a multiplicative form, such as $G(x) = g(x)h(x)$. In that case, the weighted particles $\{x_i, w(x_i)\}$ approximate the original target $g(x)$, while the reweighted particles $\{x_i, w(x_i)h(x_i)\}$ approximate the new target $G(x)$. Furthermore, in a common scenario $G(x) \propto g(x)h(x)$, the reweighting scheme still maintains $w(x_i)h(x_i)$ as the unnormalized weights.

7.5.3 Sequential Monte Carlo

Sequential Monte Carlo (SMC) refers to recursive applications of importance sampling and resampling techniques to a sequence of target distributions (Del Moral et al., 2006). In the context of nonlinear state space models, the particle filter employs SMC to simulate a sequence of filtering distributions $p(S_{0:t}|Y_{1:t})$, $t = 1, \ldots, T$, which can be evaluated pointwise up to a normalizing constant. The target density $p(S_{0:t}|Y_{1:t})$ is proportional to

$$g_t(S_{0:t}) = p(S_{0:t}, Y_{1:t}) = p(S_0) \prod_{\tau=1}^{t} p(S_\tau|S_{\tau-1}) p(Y_\tau|S_\tau),$$

and the normalizing constant corresponds to the likelihood function (without conditioning on underlying parameters explicitly denoted):

$$k_t = \int g_t(S_{0:t}) dS_{0:t} = p(Y_{1:t}).$$

To clarify our notation, $g_t(\cdot)$ refers to a specific function, while $p(\cdot)$ is used generically to denote the density of a distribution. The subscript t indicates that the functional form of $g_t(\cdot)$ is time-varying, and the dimension increases over time.

SMC adheres to the importance sampling procedure: generate particles from a proposal function $f_t(S_{0:t})$ and compute the unnormalized weights $w_t(S_{0:t}) = \frac{g_t(S_{0:t})}{f_t(S_{0:t})}$ in every period $t = 1, \ldots, T$. The proposal distributions are carefully specified in a recursive manner, by recycling or resampling from the previous particles. Two types of proposal distributions are available.

First, the recycling-type proposal:

$$f_t(S_{0:t}) = f_{t-1}(S_{0:t-1}) q(S_t; S_{t-1}, Y_t),$$

where $f_{t-1}(S_{0:t-1})$ represents the previously used proposal function and $q(S_t; S_{t-1}, Y_t)$ is an incremental proposal function for generating particles in period t. By recycling, we mean that the previous particles $S_{0:t-1}$ are accepted without any modification. The unnormalized importance weights $\frac{g_t(S_{0:t})}{f_t(S_{0:t})}$ under the recycling-type proposal take a recursive form

$$w_t(S_{0:t}) = \alpha_t w_{t-1}(S_{0:t-1}),$$

where

$$\alpha_t = \frac{p(S_t|S_{t-1}) p(Y_t|S_t)}{q(S_t; S_{t-1}, Y_t)}$$

is known as the incremental importance weight.

Second, the resampling-type proposal:

$$f_t(S_{0:t}) = p(S_{0:t-1}|Y_{1:t-1}) q(S_t; S_{t-1}, Y_t).$$

By resampling, we mean that the previous particles $S_{0:t-1}$ are randomly sampled with replacement, with probabilities proportional to their weights. The resampled particles are approximate draws from $p(S_{0:t-1}|Y_{1:t-1})$. The unnormalized importance weights $\frac{g_t(S_{0:t})}{f_t(S_{0:t})}$ under the resampling-type proposal take the form

$$w_t(S_{0:t}) = \alpha_t p(Y_{1:t-1}).$$

Although the term $p(Y_{1:t-1})$ is not analytically tractable, it is merely a constant that does not change with particles. As we normalize the weights, this constant plays no role. While it may be tempting to omit the constant when computing weights, we recommend approximating $p(Y_{1:t-1})$ by the sample average of the weights in period $t-1$. According to Proposition 7.1, the sample average of the unnormalized weights in period t is an unbiased and consistent estimator of the likelihood function $p(Y_{1:t})$. This is particularly useful for likelihood-based parameter estimation.

In summary, we have the option to choose between the recycling-type or resampling-type proposal distribution, and the selected type can change in different periods. Both types of proposals generate new particles S_t from $q(\cdot)$, but the existing particles $S_{0:t-1}$ are treated differently, which implies different weighting schemes. Recycling-type proposals reuse the existing particles generated by $f_{t-1}(S_{0:t-1})$ and augments them by sampling S_t from $q(\cdot)$. Resampling-type proposals resample the existing particles to obtain approximated draws from $p(S_{0:t-1}|Y_{1:t-1})$, and then simulate the particles S_t from $q(\cdot)$. According to the importance sampling paradigm, the former sets the unnormalized weights as $\alpha_t w_{t-1}(S_{0:t-1})$, while the latter uses the unnormalized weights $\alpha_t p(Y_{1:t-1})$.

Algorithm 7.2 seamlessly integrates two types of proposals and manages the weight recursion using a unified formula. Whenever particles are resampled, the weights are reset accordingly. The normalized weights must be $1/n$ for all resampled particles, but the unnormalized weights are reset as the normalizing constant of the target distribution to maintain the recursive formula $w_t(S_{0:t}) = \alpha_t w_{t-1}(S_{0:t-1})$. The sample average of these unnormalized weights consistently approximates the likelihood function throughout the forward recursion, regardless of the choice of proposals.

The desire for a judicious choice of proposals in each period stems from the pros and cons of resampling. If we exclusively use the recycling-type proposal for all periods (i.e., no resampling), the particle filter will degenerate after a few iterations. Conversely, if we apply the resampling-type proposal in every period, the particle filter remains functional but may not operate efficiently.

The primary purpose of resampling is to prevent degeneracy by aligning the proposal distribution with the target distribution. Even if there is a significant discrepancy between the proposal and target distributions in some periods (especially the initial ones), the particle filter can correct past approximation errors through resampling. This allows for continuous improvement of proposals over a long time series of filtering. Without resampling, the discrepancy between the proposal and target distributions will increase over time, eventually causing the particle filter to degenerate, with a single particle dominating all weights.

We emphasize that resampling aims to prevent future degeneration but does not necessarily provide a better approximation of past and current target distributions. Resampling increases Monte Carlo noise and reduces the number of distinct particles, as heavy-weight particles are likely to be duplicated while light-weight particles may be eliminated. This loss of diversity is known as sample impoverishment. To balance degeneration and impoverishment, particles should be resampled only if the weight inequality is significant. A measure of weight equality is the effective sample size (ESS):

$$ESS = \frac{1}{\sum_{i=1}^{n} \tilde{w}_{t,i}^2},$$

where $\tilde{w}_{t,i}$ is the normalized importance weight of the ith particle. The best-case scenario occurs when all particles are equally weighted, resulting in $\tilde{w}_{t,i} = 1/n$ and $ESS = n$, which corresponds to an ideal case where the proposal and target distributions are identical. The worst-case scenario is when a single particle holds all the weights, leading to $ESS = 1$. A rule of thumb is that $ESS < n/2$ indicates a substantial discrepancy between the proposal and target distributions, prompting resampling to realign the proposal distribution.

On the surface, the particle filter approximates the high-dimensional target distribution $p(S_{0:t}|Y_{1:t})$, but only the marginal distribution $p(S_t|Y_{1:t})$ can be well represented by the weighted particles. Due to the loss of particle diversity caused by resampling, the smoothing distributions $p(S_\tau|Y_{1:t})$, $\tau < t$ are often poorly approximated during the forward recursion, necessitating an additional backward recursion for better approximation of these smoothing distributions. In practice, we only track the current particles S_t instead of the entire

Algorithm 7.2 Particle filter

Input: data $Y_{1:T}$, proposal $q(S_t; S_{t-1}, Y_t)$, initial particles $S_{0,i}$ sampled from $p(S_0)$ with uniform weights $w_{0,i} = 1$, $i = 1, \ldots, n$.
Output: weighted particles for approximating $p(S_t \mid Y_{1:t})$ and $p(Y_{1:t})$.
Iterate the following steps for $t = 1, \ldots, T$:
1 Generate particles $S_{t,i}$, $i = 1, \ldots, n$ from the proposal $q(\cdot)$.
2 Compute incremental, unnormalized and normalized importance weights

$$\alpha_{t,i} = \frac{p(S_{t,i} \mid S_{t-1,i})\, p(Y_t \mid S_{t,i})}{q(S_{t,i}; S_{t-1,i}, Y_t)},$$

$$w_{t,i} = \alpha_{t,i} w_{t-1,i},$$

$$\tilde{w}_{t,i} = \frac{w_{t,i}}{\sum_{j=1}^{n} w_{t,j}}.$$

3 Evaluate likelihood function (if parameter estimation is of interest)

$$p(Y_{1:t}) = k_t = \frac{1}{n} \sum_{i=1}^{n} w_{t,i}.$$

4 Summarize filtering distribution by weighted particles, such as expectations

$$E(S_t \mid Y_{1:t}) = \frac{1}{n} \sum_{i=1}^{n} \tilde{w}_{t,i} S_{t,i}.$$

5 Check effective sample size

$$ESS = \frac{1}{\sum_{i=1}^{n} \tilde{w}_{t,i}^2}.$$

6 Resample particles and reset weights $w_{t,i} = k_t$, $\forall i$, if ESS is low.

sequence, which also reduces the computational memory burden. A typical workflow of the particle filter for the nonlinear state space model is summarized in Algorithm 7.2.

Additionally, Algorithm 7.2 can automatically handle out-of-sample forecasts, which are essential in statistical data analysis. In the state space framework, forecasting can be viewed as a special case of filtering with missing data in the forecast horizon. The particle filter proceeds with $p(Y_t \mid S_t) \propto 1$ whenever Y_t is missing. See Exercise 7.6 for further discussions.

7.5.4 Choice of Transition Kernel

To implement Algorithm 7.2, users must specify a transition kernel $q(S_t; S_{t-1}, Y_t)$, which significantly influences the performance of the particle filter. Here, we discuss some popular specifications.

7.5.4.1 Bootstrap Filter

The bootstrap particle filter (Gordon et al., 1993) makes proposals by the state transition:

$$q(S_t; S_{t-1}, Y_t) = p(S_t \mid S_{t-1}),$$

and the incremental weight is given by the observation density:

$$\alpha_t = p(Y_t \mid S_t).$$

Note that the term "the bootstrap" has a different meaning in statistics (e.g., the bootstrap standard error). In the context of particle filtering, the bootstrap method refers to particle generation using the prior distribution of the state transition, without reliance on observations, as the data Y_t is absent in the transition kernel.

When resampling in every period, the bootstrap particle filter can be interpreted as a Monte Carlo version of the Kalman filter through prediction and update steps:

$$p(S_t | Y_{1:t-1}) = \int p(S_{t-1} | Y_{1:t-1}) \, p(S_t | S_{t-1}) \, dS_{t-1},$$

$$p(S_t | Y_{1:t}) \propto p(S_t | Y_{1:t-1}) \, p(Y_t | S_t).$$

First, resampling particles in period $t-1$ generates approximate draws from $p(S_{t-1} | Y_{1:t-1})$. Under the bootstrap transition kernel $p(S_t | S_{t-1})$, the particles S_t follow the predictive distribution $p(S_t | Y_{1:t-1})$. This is the prediction step executed through the Monte Carlo simulation. Second, reweighting changes the target from the predictive distribution to the filtering distribution. Samples from $p(S_t | Y_{1:t-1})$ weighted by $p(Y_t | S_t)$ approximate the filtering distribution $p(S_t | Y_{1:t})$. This update step is also performed through Monte Carlo simulation. See Exercise 7.7 for more discussions.

The bootstrap particle filter has pros and cons. Simplicity is a key advantage of the bootstrap method. Compared to alternative proposals, the bootstrap filter has minimal computing cost, allowing for the use of more particles to compensate for unbalanced weights. Another strength of the bootstrap proposal is its applicability to complex state space models where the transition density $p(S_t | S_{t-1})$ is difficult to evaluate but easy to sample. In that case, the bootstrap filter is the unique choice, as its importance weights do not depend on the transition density.

However, a notable weakness of the bootstrap proposal is its blind particle generation, which does not incorporate information from Y_t. Practically, this means the bootstrap filter may require frequent resampling. Additionally, the bootstrap filter struggles with perfectly observed data, which corresponds to an observation equation $Y_t = C_t(S_t)$, because the incremental weight $p(Y_t | S_t)$ is not well defined without observation noise. For example, the state space representation of the ARMA model has an observation equation free of noise, rendering the bootstrap filter inapplicable.

Herbst and Schorfheide (2019) note that the larger the measurement error variance, the more accurate the bootstrap particle filter becomes, as the variance of the particle weights decreases. This insight leads to tempered particle filtering, which uses a sequence of bridging distributions with inflated variances.

7.5.4.2 Conditionally Optimal Filter

To incorporate the current observation into the proposal, we consider the marginal-conditional decomposition:

$$p(S_{0:t} | Y_{1:t}) = p(S_{0:t-1} | Y_{1:t}) \, p(S_t | S_{t-1}, Y_t).$$

The smoothing distribution $p(S_{0:t-1} | Y_{1:t})$ can be approximated by the filtering distribution $p(S_{0:t-1} | Y_{1:t-1})$ using the resampled particles. If we specify a proposal

$$q(S_t; S_{t-1}, Y_t) = p(S_t | S_{t-1}, Y_t),$$

the weights have minimum variance conditional on $S_{0:t-1}$. For that reason, it is known as the conditionally optimal importance function (Doucet et al., 2000). Whenever $p(S_t | S_{t-1}, Y_t)$ is tractable, the incremental importance weights can be evaluated in a closed form:

$$\alpha_t = \frac{p(S_t | S_{t-1}) \, p(Y_t | S_t)}{p(S_t | S_{t-1}, Y_t)} = p(Y_t | S_{t-1}).$$

The conditionally optimal proposal can be interpreted as the one-step filtering distribution conditional on a fixed initial state, and the incremental weights correspond to the one-period likelihood function. Although the optimal proposal is not analytically tractable in most nonlinear state space models, there are a few special cases where it is applicable. First, conditional Gaussian models where the state transition is nonlinear, but the measurement equation is linear: $S_t = A_t(S_{t-1}) + B_t u_t$ and $Y_t = C_t S_t + D_t \varepsilon_t$. Second, nonlinear models with deterministic state transition: $S_t = A_t(S_{t-1})$ and $Y_t = C_t(S_t) + D_t \varepsilon_t$. Third, special models with linear components. Exercise 7.8 illustrates an inequality constrained state space model.

A prominent feature of the optimal proposal is that the incremental weights do not depend on the current particles. This allows for switching the order of importance sampling and resampling to preserve particle diversity. We recalculate the particle weights by multiplying $p(Y_t|S_{t-1})$, resample S_{t-1}, and sample S_t using the proposal $p(S_t|S_{t-1}, Y_t)$. It is legitimate to reverse the order of sampling and resampling, because the reweighted and resampled particles approximate the smoothing distribution $p(S_{0:t-1}|Y_{1:t})$, and S_t can be fully adapted to the resampled particles: $(S_{0:t-1}, S_t)$ jointly follows the filtering distribution $p(S_{0:t}|Y_{1:t})$. See Exercise 7.9 for more discussions. The attraction of the reweight-resample-sample strategy is to incorporate information from Y_t in resampling, ensuring that particles compatible with the current observation have a good chance of surviving, which preserves particle diversity.

7.5.4.3 Unscented Particle Filter

The intractability of the optimal proposal provides opportunities for alternative methods that incorporate current observations into proposal specifications. One such example is the unscented particle filter, which approximates the optimal proposal by an unscented transformation (Julier and Uhlmann, 2004).

Consider the nonlinear state space model specified by Equations (7.3) and (7.4). Although the one-step filtering distribution $p(S_t|S_{t-1}, Y_t)$ is not analytically tractable, the predictive distribution $p(S_t|S_{t-1})$ is multivariate normal with mean $A_t(S_{t-1})$ and covariance matrix $B_t B_t'$. We discretize S_t by an unscented transformation, approximating it by a categorical distribution: $\{x_i, w_i\}$, $i = -m, \ldots, 0, \ldots, m$ such that

$$x_i = \begin{cases} A_t(S_{t-1}) + \sqrt{m+k} L_i & i > 0 \\ A_t(S_{t-1}) & i = 0 \\ A_t(S_{t-1}) - \sqrt{m+k} L_{-i} & i < 0 \end{cases},$$

where $(L_1, \ldots, L_m)$ is the Cholesky factor of $B_t B_t'$. The weights are $w_0 = \frac{k}{m+k}$ and $w_i = \frac{1}{2(m+k)}, \forall i \neq 0$. Discretization facilitates state propagation to the nonlinear observation equation $Y_t = C_t(S_t) + D_t \varepsilon_t$. The predictive mean and variance of Y_t are given by

$$Y_{t|t-1} = \sum_{i=-m}^{m} w_i C_t(x_i),$$

$$O_{t|t-1} = D_t D_t' + \sum_{i=-m}^{m} w_i \left[C_t(x_i) - Y_{t|t-1}\right]\left[C_t(x_i) - Y_{t|t-1}\right]',$$

$$L_{t|t-1} = \sum_{i=-m}^{m} w_i \left[x_i - A_t(S_{t-1})\right]\left[C_t(x_i) - Y_{t|t-1}\right]'.$$

The Kalman filter provides the updating rule for the state filtering:

$$S_{t|t} = A_t(S_{t-1}) + L_{t|t-1} O_{t|t-1}^{-1} (Y_t - Y_{t|t-1}),$$

$$P_{t|t} = B_t B_t' - L_{t|t-1} O_{t|t-1}^{-1} L_{t|t-1}'.$$

The unscented particle filter makes proposals using

$$q(S_t; S_{t-1}, Y_t) = \phi(S_t; S_{t|t}, P_{t|t}),$$

which is an unscented approximation of the optimal proposal$p(S_t | S_{t-1}, Y_t)$, and the corresponding incremental weight is

$$\alpha_t = \frac{p(S_t | S_{t-1}) p(Y_t | S_t)}{\phi(S_t; S_{t|t}, P_{t|t})}.$$

A computational issue arises when we attempt to implement the unscented transformation for each S_{t-1}, which is represented by thousands of particles, resulting in excessively high computational costs. A more economical strategy is to select a representative particle (e.g., the mean of the particles) and apply the unscented transformation just once. Assume that $O_{t|t-1}$, $L_{t|t-1}$ and $P_{t|t}$ computed under the representative particle are applicable to all particles. For each particle, we may approximate $Y_{t|t-1}$ and $S_{t|t}$ by

$$Y_{t|t-1} = C_t(A_t(S_{t-1})),$$

$$S_{t|t} = A_t(S_{t-1}) + L_{t|t-1} O_{t|t-1}^{-1} (Y_t - Y_{t|t-1}).$$

The one-pass unscented filter substantially reduces the computational costs of the unscented particle filter, but it remains a valid SMC approach, provided that the importance weights are computed correctly.

7.5.5 Rao-Blackwellization

Monte Carlo methods are computationally intensive and should be considered a last resort. Many nonlinear state space models contain linear components: conditional on a subset of states, the remaining states evolve within a Gaussian linear sub-model. It is advantageous to apply the Kalman filter to isolate the linear components, allowing the particle filter to focus on the nonlinear components, thereby reducing Monte Carlo noise. The combination of Kalman and particle filters is known as Rao-Blackwellization (Doucet et al., 2001), also referred to as mixture Kalman filters (Chen and Liu, 2000) or marginalized particle filtering (Schon et al., 2005).

Suppose the state variables consist of conditionally linear states L_t and nonlinear states S_t. By conditional linearity, we mean the posterior conditional distribution $p(L_{0:t} | S_{0:t}, Y_{1:t})$ is analytically tractable by the Kalman filter and smoother, as in a Gaussian linear sub-model. Additionally, the Kalman filter provides the exact conditional likelihood $p(Y_{1:t} | S_{0:t})$ as a by-product of the forward recursion.

The state filtering distribution can be decomposed into tractable and intractable parts:

$$p(L_{0:t}, S_{0:t} | Y_{1:t}) = p(L_{0:t} | S_{0:t}, Y_{1:t}) p(S_{0:t} | Y_{1:t}),$$

where $p(S_{0:t} | Y_{1:t})$ is the target distribution for the particle filter. The target density can be evaluated up to a normalizing constant:

$$p(S_{0:t} | Y_{1:t}) \propto p(S_{0:t}) p(Y_{1:t} | S_{0:t}),$$

where the normalizing constant is the likelihood function $p(Y_{1:t})$, while the conditional likelihood function $p(Y_{1:t} | S_{0:t})$ can be evaluated exactly by the Kalman filter on the Gaussian linear sub-model.

Following the importance sampling scheme, we specify a proposal distribution $f_t(S_{0:t})$ and compute the unnormalized weights $w_t(S_{0:t}) = \frac{p(S_{0:t}, Y_{1:t})}{f_t(S_{0:t})}$ in every period $t = 1, \ldots, T$. Like the standard particle filter, we have two types of proposals.

First, the recycling-type proposal:

$$f_t(S_{0:t}) = f_{t-1}(S_{0:t-1})\, q(S_t; S_{t-1}, Y_t),$$

with the unnormalized importance weights

$$w_t(S_{0:t}) = \alpha_t w_{t-1}(S_{0:t-1}),$$

where the incremental importance weights

$$\alpha_t = \frac{p(S_t, Y_t | S_{0:t-1}, Y_{1:t-1})}{q(S_t; S_{t-1}, Y_t)}$$

are evaluated with the aid of the Kalman filter that marginalizes over the linear states.

Second, the resampling-type proposal:

$$f_t(S_{0:t}) = p(S_{0:t-1} | Y_{1:t-1})\, q(S_t; S_{t-1}, Y_t),$$

with the unnormalized importance weights

$$w_t(S_{0:t}) = \alpha_t p(Y_{1:t-1}).$$

We have the option to choose between the recycling-type or resampling-type proposal distribution in each period of the particle filtering. Whenever we resample particles, we also reset weights to ensure that the sample average of the unnormalized weights always approximates the likelihood function.

7.5.6 Auxiliary Particle Filter

Introduced by Pitt and Shephard (1999) and analyzed in Johansen and Doucet (2008) and Douc et al. (2009), the auxiliary particle filter is closely related to the conditionally optimal filter, making it useful to recap the optimal filter that reverses the order of importance sampling and resampling: 1) reweight particles by multiplying $p(Y_t | S_{t-1})$, 2) resample particles from the one-step smoothing distribution $p(S_{0:t-1} | Y_{1:t})$, and 3) sample the fully adapted particles from the one-step filtering distribution $p(S_t | S_{t-1}, Y_t)$. A limitation of the optimal filter is the requirement for the one-step filtering distribution to be analytically tractable.

The auxiliary particle filter extends the reweight-resample-sample strategy to state space models where the one-step filtering distribution is not tractable. The approach involves replacing the intractable distribution with a proposal distribution to facilitate reweighting and resampling. The discrepancy between the target and proposal distributions is corrected using importance weights. Interestingly, the auxiliary particle filter can be interpreted from various perspectives, yet they converge on the filtering algorithm. Our interpretation aligns with the standard importance sampling paradigm, involving the following steps: 1) specify the target distribution, 2) generate particles from a carefully designed proposal distribution, 3) evaluate the importance weights, and 4) handle the normalizing constant in the target and proposal densities for evaluating the likelihood function.

First, the target distributions are specified as $p(S_{0:t}|Y_{1:t})$, $t = 1, \ldots, T$, which adhere to the fundamental recursions:

$$p(S_{0:t}|Y_{1:t}) = \frac{p(S_{0:t-1}|Y_{1:t-1})\, p(S_t|S_{t-1})\, p(Y_t|S_t)}{p(Y_t|Y_{1:t-1})}.$$

Second, the proposal distributions are constructed by a hypothetical state space model identical to the original model up to period $t-1$, enabling reuse of $p(S_{0:t-1}|Y_{1:t-1})$. However, a local change occurs in period t such that data are generated by S_{t-1} without dependency on S_t. Let $\tilde{p}(\cdot)$ be the probability distributions under the hypothetical model. The proposal density takes the form

$$\tilde{p}(S_{0:t}|Y_{1:t}) = \frac{p(S_{0:t-1}|Y_{1:t-1})\, \tilde{p}(Y_t|S_{t-1})\, \tilde{p}(S_t|S_{t-1}, Y_t)}{\tilde{p}(Y_t|Y_{1:t-1})},$$

where $\tilde{p}(Y_t|S_{t-1})$ and $\tilde{p}(S_t|S_{t-1}, Y_t)$ are user-specified. Unlike the intractable normalizing constant $p(Y_t|Y_{1:t-1})$ in the target distribution, it is feasible to compute the constant $\tilde{p}(Y_t|Y_{1:t-1})$ in the proposal distribution by direct Monte Carlo simulation of the integral

$$\tilde{p}(Y_t|Y_{1:t-1}) = \int p(S_{t-1}|Y_{1:t-1})\, \tilde{p}(Y_t|S_{t-1})\, dS_{t-1}.$$

The main distinction between the hypothetical and original models is that $\tilde{p}(Y_t|S_{t-1})$ is tractable but $p(Y_t|S_{t-1})$ may not be. It is optional to simply the proposal by $\tilde{p}(S_t|S_{t-1}, Y_t) = p(S_t|S_{t-1})$, leaving the transition equation unchanged. For example, suppose the original model takes the form

$$S_t = A_t(S_{t-1}) + B_t u_t,$$
$$Y_t = C_t(S_t) + D_t \varepsilon_t.$$

It is tempting to specify a hypothetical model with a local change:

$$S_t = A_t(S_{t-1}) + B_t u_t,$$
$$Y_t = C_t(A_t(S_{t-1})) + \tilde{D}_t \varepsilon_t.$$

In practice, it is desirable to specify $\tilde{D}_t$ larger than D_t, or specify a fat-tail distribution for ε_t in the hypothetical model, rendering $\tilde{p}(Y_t|S_{t-1})$ more diffuse than $p(Y_t|S_{t-1})$, as the dispersion of the proposal distribution determines the performance of importance sampling. If the tail of the proposal distribution is too thin, the variance of the importance weights could be excessively large.

The essence of the auxiliary particle filter is the reweight-resample-sample strategy for generating particles from the proposal distribution, which has the marginal-conditional decomposition:

$$\tilde{p}(S_{0:t}|Y_{1:t}) = \tilde{p}(S_{0:t-1}|Y_{1:t})\, \tilde{p}(S_t|S_{t-1}, Y_t),$$

where

$$\tilde{p}(S_{0:t-1}|Y_{1:t}) \propto p(S_{0:t-1}|Y_{1:t-1})\, \tilde{p}(Y_t|S_{t-1}).$$

The auxiliary particle filter is an iterative algorithm, where the previous iteration produces weighted particles that approximate $p(S_{0:t-1}|Y_{1:t-1})$. By multiplying the previous weights by $\tilde{p}(Y_t|S_{t-1})$, the reweighted particles approximate the one-step smoothing distribution $\tilde{p}(S_{0:t-1}|Y_{1:t})$ under the hypothetical model. Given the reweighted particles, we resample $S_{0:t-1}$ and then sample S_t from $\tilde{p}(S_t|S_{t-1}, Y_t)$. The reweight-resample-sample process is analogous to the conditionally optimal particle filtering. The auxiliary particle filter avoids evaluating the intractable (true) density $p(Y_t|S_{t-1})$ by evaluating the tractable (hypothetical) density $\tilde{p}(Y_t|S_{t-1})$.

Third, the true and hypothetical densities differ. Within the importance sampling framework, samples drawn from the proposal distribution must be weighted appropriately:

$$w(S_{0:t}) = \frac{p(S_{0:t}|Y_{1:t})}{\tilde{p}(S_{0:t}|Y_{1:t})} = \frac{p(Y_t|S_t)\,p(S_t|S_{t-1})}{\tilde{p}(Y_t|S_{t-1})\,\tilde{p}(S_t|S_{t-1},Y_t)}\frac{\tilde{p}(Y_t|Y_{1:t-1})}{p(Y_t|Y_{1:t-1})}.$$

Importance sampling inherently includes a self-normalizing mechanism, allowing importance weights to be evaluated up to a normalizing constant. Specifically, there is a constant $p(Y_t|Y_{1:t-1})$ in the target density and a constant $\tilde{p}(Y_t|Y_{1:t-1})$ in the proposal density. When estimating state filtering distributions, these constant terms can be safely ignored. However, evaluating the likelihood function for parameter estimation necessitates computing $\tilde{p}(Y_t|Y_{1:t-1})$ via the weighted average of particles (refer to step 1 of Algorithm 7.3).

Like the standard particle filter, the auxiliary particle filter has the target distribution $p(S_{0:t}|Y_{1:t})$. However, only the marginal distribution $p(S_t|Y_{1:t})$ can avoid degeneracy. Additionally, tracking the full sequence $S_{0:t}$ in the forward recursion is computationally demanding. As demonstrated in Algorithm 7.3, a practical workflow of the auxiliary filter focuses on S_{t-1} and S_t at each time step of the forward filtering. Exercise 7.10 offers an alternative target distribution specification as $p(S_{t-1},S_t|Y_{1:t})$, while the filtering algorithm remains unchanged.

The auxiliary particle filter accommodates the standard particle filter with resampling in every period as a special case, by specifying $\tilde{p}(Y_t|S_{t-1})$ as an arbitrary density (or simply a nonnegative function) that does not depend on S_{t-1}. We propose a hypothetical model where the observation Y_t is non-informative about the states, implying that the smoothing distribution $\tilde{p}(S_{0:t-1}|Y_{1:t})$ is identical to the filtering distribution $p(S_{0:t-1}|Y_{1:t-1})$. We resample $S_{0:t-1}$ from the filtering distribution, generate new particles by $\tilde{p}(S_t|S_{t-1},Y_t)$, with the importance weights proportional to $\frac{p(Y_t|S_t)p(S_t|S_{t-1})}{\tilde{p}(S_t|S_{t-1},Y_t)}$, akin to the incremental importance weights in the standard particle filter.

For instance, to replicate the bootstrap filter within the auxiliary particle filter framework, we specify $\tilde{p}(Y_t|S_{t-1}) \propto 1$ and $\tilde{p}(S_t|S_{t-1},Y_t) = p(S_t|S_{t-1})$, with the unnormalized importance weights $p(Y_t|S_t)$. Algorithm 7.3 simplifies to the bootstrap filter with resampling in each period.

Similarly, to reproduce the conditionally optimal filter, we specify $\tilde{p}(Y_t|S_{t-1}) \propto 1$ and $\tilde{p}(S_t|S_{t-1},Y_t) = p(S_t|S_{t-1},Y_t)$, with weights proportional to $p(Y_t|S_{t-1})$.

The auxiliary particle filtering method also justifies the conditionally optimal filter that switches the order of sampling and resampling, through the following specification:

$$\tilde{p}(Y_t|S_{t-1}) = p(Y_t|S_{t-1}),$$
$$\tilde{p}(S_t|S_{t-1},Y_t) = p(S_t|S_{t-1},Y_t).$$

Since the hypothetical model mirrors the original model (proposal and target distributions are identical), step 5 of Algorithm 7.3 results in uniform weights $w_{t,i} = p(Y_t|Y_{1:t-1})$ and $\tilde{w}_{t,i} = \frac{1}{n}$, $\forall t,i$. We simply resample from the smoothing distribution $p(S_{t-1}|Y_{1:t})$ and generate fully adapted particles by $p(S_t|S_{t-1},Y_t)$. Thus, Algorithm 7.3 effectively reproduces the conditionally optimal particle filter with reverse-order sampling and resampling.

Algorithm 7.3 Auxiliary particle filter

Input: data $Y_{1:T}$, proposals $\tilde{p}(Y_t | S_{t-1})$ and $\tilde{p}(S_t | S_{t-1}, Y_t)$, initial particles $S_{0,i}$ sampled from $p(S_0)$ with weights $\tilde{w}_{0,i} = \frac{1}{n}$, $i = 1, \ldots, n$.
Output: weighted particles for approximating $p(S_t | Y_{1:t})$ and $p(Y_{1:t})$.
Iterate the following steps for $t = 1, \ldots, T$:
1 Compute normalizing constant by weighted particles $\{S_{t-1,i}, \tilde{w}_{t-1,i}\}$:

$$\tilde{p}(Y_t | Y_{1:t-1}) = \sum_{i=1}^{n} \tilde{w}_{t-1,i} \tilde{p}(Y_t | S_{t-1,i}).$$

2 Reset particle weights:

$$\tilde{w}^*_{t-1,i} \propto \tilde{w}_{t-1,i} \tilde{p}(Y_t | S_{t-1,i}).$$

3 Resample particles $S^*_{t-1,i}$ from reweighted particles $\{S_{t-1,i}, \tilde{w}^*_{t-1,i}\}$.
4 Sample particles $S_{t,i}$ from $\tilde{p}(S_{t,i} | S^*_{t-1,i}, Y_t)$.
5 Evaluate importance weights:

$$w_{t,i} = \frac{p(Y_t | S_{t,i}) \, p(S_{t,i} | S^*_{t-1,i})}{\tilde{p}(Y_t | S^*_{t-1,i}) \, \tilde{p}(S_{t,i} | S^*_{t-1,i}, Y_t)} \tilde{p}(Y_t | Y_{1:t-1}),$$

$$\tilde{w}_{t,i} = \frac{w_{t,i}}{\sum_{j=1}^{n} w_{t,j}}.$$

6 Evaluate likelihood function (if parameter estimation is of interest)

$$p(Y_t | Y_{1:t-1}) = \frac{1}{n} \sum_{i=1}^{n} w_{t,i},$$

$$p(Y_{1:t}) = p(Y_{1:t-1}) \, p(Y_t | Y_{1:t-1}).$$

7 Summarize filtering distribution (e.g., expectations) by weighted particles $\{S_{t,i}, \tilde{w}_{t,i}\}$:

$$E(S_t | Y_{1:t}) = \frac{1}{n} \sum_{i=1}^{n} \tilde{w}_{t,i} S_{t,i}.$$

7.6 Particle Smoothing

Filtering and smoothing are two primary types of state estimation conditional on observations and model parameters. While forward filtering utilizes the current information set at each time step for online updating of state distributions, backward smoothing utilizes the entire dataset to derive the posterior distributions of states. The applications of smoothing extend beyond state estimation, playing a crucial role in parameter estimation through various mechanisms: 1) evaluating gradients for maximum likelihood estimation, 2) computing posterior expectations for the Expectation-Maximization algorithm, and 3) simulating smoothed state sequences from posterior conditional distributions for the Gibbs sampler.

Despite the particle filter having target distributions $p(S_{0:t}|Y_{1:t})$, $t = 1, \ldots, T$, it does not function as a simulation smoother. Although the marginal distribution $p(S_t|Y_{1:t})$ can be well approximated by diverse particles, the smoothing distributions $p(S_\tau|Y_{1:t}), \tau < t$ often face degeneration due to resampling. The main theme of smoothing algorithms is to prevent degeneracy by reweighting particles generated by the particle filter. This reweighting process transforms filtering distributions into smoothing distributions, approximated by existing particles.

Nonlinear smoothing is computationally more demanding than filtering. The complexity of filtering is $O(nT)$, whereas smoothing can reach $O(n^2T)$, where n represents the number of particles and T denotes the sample size. Smoothing complexity can be reduced to $O(nT)$ in some models, such as the stochastic volatility model where the state transition is Gaussian and linear.

7.6.1 Simulation Smoothing

The simulation smoother operates through a forward filtering and backward sampling (FFBS) algorithm. The initial step involves forward recursion via the standard particle filter, which generates weighted particles approximating the filtering distributions $p(S_t|Y_{1:t})$, $t = 1, \ldots, T$. The subsequent step is backward recursion, which simulates $S_T, S_{T-1}, \ldots, S_0$, which jointly follow the smoothing distribution $p(S_{0:T}|Y_{1:T})$. FFBS relies on the marginal-conditional decomposition:

$$p(S_{0:T}|Y_{1:T}) = p(S_T|Y_{1:T}) \prod_{t=0}^{T-1} p(S_t|S_{t+1}, Y_{1:t}).$$

Backward recursion commences from the final state of forward filtering $p(S_T|Y_{1:T})$, from which a particle is resampled as the smoothing sample for S_T. Two methods are available for sampling from $p(S_t|S_{t+1}, Y_{1:t})$, $t = T-1, ..., 0$.

The first method is to reweight and resample. The filtering and smoothing distributions satisfy

$$p(S_t|S_{t+1}, Y_{1:t}) \propto p(S_t|Y_{1:t})\, p(S_{t+1}|S_t).$$

By multiplying the filtering weights with the transition density $p(S_{t+1}|S_t)$, the reweighted particles approximate the one-step smoothing distribution $p(S_t|S_{t+1}, Y_{1:t})$. Resampling from the reweighted particles yields the smoothed sample for S_t.

The reweight-resample method is computationally intensive due to the exhaustive evaluation of smoothing weights for all particles. The simulation smoother has the complexity $O(nT)$ for generating a single path of the posterior sample.

It is of interest to summarize the smoothing distribution $p(S_{0:T}|Y_{1:T})$ by expectations such as $E(S_t|Y_{1:T})$. By applying FFBS to simulate n paths S_{ti}, $i = 1, \ldots, n$, expectations can be approximated via sample analogues $\frac{1}{n}\sum_{i=1}^{n} S_{ti}$, with a complexity of $O\left(n^2 T\right)$. It is more computationally efficient to simulate n paths simultaneously, rather than processing each path separately. The initial step involves simulating S_{Ti}, $i = 1, \ldots, n$ from $p\left(S_T \left| Y_{1:T}\right.\right)$, equivalent to standard resampling. Since some resampled particles are duplicated, the reweighting procedures are the same for the duplicated particles, thereby reducing the computational burden of calculating smoothing weights.

The second method is rejection sampling. Let the target and proposal distributions be $p\left(S_t \left| S_{t+1}, Y_{1:t}\right.\right)$ and $p\left(S_t \left| Y_{1:t}\right.\right)$, respectively. Their density ratio is proportional to the transition density $p\left(S_{t+1} \left| S_t\right.\right)$. In the cases of nonlinear but Gaussian transition equations such that $S_t = A_t\left(S_{t-1}\right) + B_t u_t$, the upper bound of the target-proposal density ratio is analytically tractable, because the Gaussian density is maximized at $S_t = A_t\left(S_{t-1}\right)$. The rejection sampling involves three steps: 1) resampling a particle from the filtering distribution as a proposal, 2) evaluating the transition density at the proposal, and 3) accepting the proposal with probability equal to the transition density divided by the upper bound.

The rejection sampling circumvents exhaustive weight evaluations for all particles, reducing complexity to $O\left(T\right)$ for each path of the posterior sample. However, if the acceptance probability is low during certain periods, the number of iterations for rejection sampling may become prohibitively large. To mitigate this computational bottleneck, an adaptive FFBS that switches to the reweight-resample method when rejection sampling becomes inefficient is recommended.

7.6.2 Forward-backward Smoothing

As an alternative to simulation smoothing, which approximates the joint distribution $p\left(S_{0:T} \left| Y_{1:T}\right.\right)$, it is feasible to compute the marginal distribution $p\left(S_t \left| Y_{1:T}\right.\right)$ via backward recursion:

$$p\left(S_t \left| Y_{1:T}\right.\right) = \int p\left(S_{t+1} \left| Y_{1:T}\right.\right) p\left(S_t \left| S_{t+1}, Y_{1:t}\right.\right) dS_{t+1},$$

where

$$p\left(S_t \left| S_{t+1}, Y_{1:t}\right.\right) = \frac{p\left(S_t \left| Y_{1:t}\right.\right) p\left(S_{t+1} \left| S_t\right.\right)}{p\left(S_{t+1} \left| Y_{1:t}\right.\right)}.$$

Rearrange terms and express the marginal smoothing distribution as

$$p\left(S_t \left| Y_{1:T}\right.\right) = p\left(S_t \left| Y_{1:t}\right.\right) \int p\left(S_{t+1} \left| Y_{1:T}\right.\right) \frac{p\left(S_{t+1} \left| S_t\right.\right)}{p\left(S_{t+1} \left| Y_{1:t}\right.\right)} dS_{t+1}.$$

Let $\{S_{t,i}, w_{t,i}\}$ represent the weighted particles that approximate the filtering distribution $p\left(S_t \left| Y_{1:t}\right.\right)$. The smoothing distribution $p\left(S_t \left| Y_{1:T}\right.\right)$ can be approximated by reweighted particles $\left\{S_{t,i}, w_{t,i}^*\right\}$, where the weights are re-calculated as

$$w_{t,i}^* = w_{t,i} \int p\left(S_{t+1} \left| Y_{1:T}\right.\right) \frac{p\left(S_{t+1} \left| S_{t,i}\right.\right)}{p\left(S_{t+1} \left| Y_{1:t}\right.\right)} dS_{t+1}.$$

The reweighting scheme involves an integral, representing a conditional expectation with respect to the smoothing distribution $p\left(S_{t+1} \left| Y_{1:T}\right.\right)$. Backward recursion has already generated weighted particles $\left\{S_{t+1,i}, w_{t+1,i}^*\right\}$ for smoothing $p\left(S_{t+1} \left| Y_{1:T}\right.\right)$. Therefore, the integral can be approximated by the sample analogue of the expectation:

$$\int p\left(S_{t+1} \left| Y_{1:T}\right.\right) \frac{p\left(S_{t+1} \left| S_{t,i}\right.\right)}{p\left(S_{t+1} \left| Y_{1:t}\right.\right)} dS_{t+1} = \sum_{j=1}^{n} w_{t+1,j}^* \frac{p\left(S_{t+1,j} \left| S_{t,i}\right.\right)}{p\left(S_{t+1,j} \left| Y_{1:t}\right.\right)}.$$

Unlike simulation smoothing, where the normalizing constant $p(S_{t+1}|Y_{1:t})$ can be omitted, here it is necessary to compute the hidden integral (as it depends on the index j):

$$p(S_{t+1,j}|Y_{1:t}) = \int p(S_t|Y_{1:t})\, p(S_{t+1,j}|S_t)\, dS_t.$$

The integral can be handled by weighted particles $\{S_{t,i}, w_{t,i}\}$ for the filtering distribution $p(S_t|Y_{1:t})$, with the sample average approximating the expectation:

$$p(S_{t+1,j}|Y_{1:t}) = \sum_{k=1}^{n} w_{t,k}\, p(S_{t+1,j}|S_{t,k}).$$

In summary, the marginal smoothing distribution involves double integrals, approximated by two sets of weighted particles: one for the forward filtering distribution and the other for the backward smoothing distribution, hence the name forward-backward smoothing. The algorithm is computationally intensive, with a complexity of $O(n^2T)$. The particles used in backward smoothing are identical to those in forward filtering, but the weights are recalculated to incorporate information from $Y_{1:T}$. The combined reweighting formula is

$$w^*_{t,i} = w_{t,i} \sum_{j=1}^{n} \frac{w^*_{t+1,j}\, p(S_{t+1,j}|S_{t,i})}{\sum_{k=1}^{n} w_{t,k}\, p(S_{t+1,j}|S_{t,k})},$$

where backward recursion is initialized by $w^*_{T,i} = w_{T,i}$. Given that the normalizing constant $p(S_{t+1}|Y_{1:T})$ has been explicitly addressed, the weights $w^*_{t,i}$ always sum to one.

7.6.3 Forward Smoothing

Smoothing is typically associated with backward recursion, but it is possible to leverage forward recursion alone to compute the smoothing expectation

$$E[f(S_{0:t})|Y_{1:t}] = \int f(S_{0:t})\, p(S_{0:t}|Y_{1:t})\, dS_{0:t},$$

if the function of interest takes a summation form:

$$f(S_{0:t}) = \sum_{\tau=1}^{t} f_\tau(S_{\tau-1}, S_\tau).$$

Forward smoothing is computationally efficient for processing in a series of expectations for $t = 1, \ldots, T$. An application of forward smoothing is to approximate the gradient of the log-likelihood function at each time point for nonlinear state space models.

The core idea of forward smoothing is to reduce high-dimensional expectations (with respect to $S_{0:t}$) to one-dimensional expectations (with respect to S_t) of the value function. The law of iterated expectations indicates that

$$E[f(S_{0:t})|Y_{1:t}] = E[V_t(S_t)|Y_{1:t}] = \int V_t(S_t)\, p(S_t|Y_{1:t})\, dS_t,$$

where the value function $V_t(S_t)$ is defined as the expectation of $f(S_{0:t})$ for a given S_t:

$$V_t(S_t) = E[f(S_{0:t})|S_t, Y_{1:t}] = \int f(S_{0:t})\, p(S_{0:t-1}|S_t, Y_{1:t-1})\, dS_{0:t-1}.$$

Since the function of interest takes a summation form:

$$f(S_{0:t}) = f_t(S_{t-1}, S_t) + \sum_{\tau=1}^{t-1} f_\tau(S_{\tau-1}, S_\tau),$$

the value function has a recursive representation, allowing $V_t(S_t)$ to be expressed as a function of $V_{t-1}(S_{t-1})$ such that

$$V_t(S_t) = \int [V_{t-1}(S_{t-1}) + f_t(S_{t-1}, S_t)]\, p(S_{t-1} | S_t, Y_{1:t-1})\, dS_{t-1}.$$

The value function iteration relies on the decomposition of the smoothing distribution as

$$p(S_{0:t} | Y_{1:t}) = p(S_t | Y_{1:t}) \prod_{\tau=1}^{t} p(S_{\tau-1} | S_\tau, Y_{1:\tau-1}),$$

which is also at the core of simulation smoothing. The difference is that backward smoothing commences upon completion of the particle filter, whereas forward smoothing proceeds with each iteration of the particle filter.

Forward smoothing is implemented in conjunction with forward filtering, which generates weighted particles $\{S_{t-1,j}, w_{t-1,j}\}$ that approximate the filtering distribution $p(S_{t-1} | Y_{1:t-1})$. By multiplying the weights by the transition density, the reweighted particles approximate the one-step smoothing distribution $p(S_{t-1} | S_t, Y_{1:t-1})$. Given the previous-state value function $V_{t-1}(S_{t-1})$, the current-state value function is computed as

$$V_t(S_{t,i}) = \frac{\sum_{j=1}^{n} w_{t-1,j}\, p(S_{t,i} | S_{t-1,j})\, [V_{t-1}(S_{t-1,j}) + f_t(S_{t-1,j}, S_{t,i})]}{\sum_{j=1}^{n} w_{t-1,j}\, p(S_{t,i} | S_{t-1,j})},$$

with the denominator normalizing the weights (ensuring they sum to one). With the value function, the sample analogue of the expectation is computed:

$$E[f(S_{0:t}) | Y_{1:t}] = \sum_{i=1}^{n} w_{t,i} V_t(S_{t,i}).$$

Since the value function iteration involves particles from two periods: $S_{t-1,j}$ and $S_{t,i}$, $i, j = 1, \ldots, n$, the computational complexity of forward smoothing amounts to $O(n^2 T)$, mirroring that of backward smoothing. However, the appeal of forward smoothing lies in incurring no additional computational costs for obtaining a series of expectations for $t = 1, \ldots, T$.

7.6.4 Two-filter Smoothing

As the name suggests, the two-filter smoothing method involves two independent filters: a standard particle filter that moves forward and an information filter that moves backward. The particles generated by the backward information filter are reweighted to approximate the smoothing distributions.

The idea of two-filtering smoothing is to decompose the smoothing distribution into forward and backward components:

$$p(S_t | Y_{1:T}) \propto p(S_t | Y_{1:t-1})\, p(Y_{t:T} | S_t),$$

where the state predictive distribution $p(S_t | Y_{1:t-1})$ is estimated by a forward particle filter, while the conditional likelihood $p(Y_{t:T} | S_t)$ is managed through backward recursion:

$$p(S_t | Y_{1:t-1}) = \int p(S_{t-1} | Y_{1:t-1})\, p(S_t | S_{t-1})\, dS_{t-1},$$

$$p(Y_{t:T}|S_t) = \int p(Y_t|S_t)\, p(S_{t+1}|S_t)\, p(Y_{t+1:T}|S_{t+1})\, dS_{t+1}.$$

A significant challenge of backward recursion is that $p(Y_{t:T}|S_t)$ is not necessarily a proper distribution with respect to S_t. Transformations are required before approximating it with weighted particles.

It is helpful to consider a special case that the transition equation is Gaussian and linear, such as $S_t = AS_{t=1} + Bu_t$. If the process is stationary, it is convenient to evaluate and sample from the stationary distribution $p(S_t)$, allowing the transformation of the conditional likelihood into the posterior-to-prior ratio:

$$p(Y_{t:T}|S_t) \propto \frac{p(S_t|Y_{t:T})}{p(S_t)},$$

where the posterior $p(S_t|Y_{t:T})$ takes a recursive form:

$$p(S_t|Y_{t:T}) \propto p(S_t)\, p(Y_t|S_t) \int p(S_{t+1}|S_t) \frac{p(S_{t+1}|Y_{t+1:T})}{p(S_{t+1})} dS_{t+1}.$$

The backward information filter is initialized by particles generated from $p(S_T)$ with weights proportional to $p(Y_T|S_T)$. The weighted particles approximate $p(S_T|Y_T)$. The backward recursion can be described as follows: given the weighted particles $\left\{\tilde{S}_{t+1,j}, \tilde{w}_{t+1,j}\right\}$ that approximate $p(S_{t+1}|Y_{t+1:T})$, we sample $\tilde{S}_{t,i}$ from the prior $p(S_t)$ with importance weights

$$\tilde{w}_{t,i} \propto p\left(Y_t \middle| \tilde{S}_{t,i}\right) \sum_{j=1}^{n} \tilde{w}_{t+1,j} \frac{p\left(\tilde{S}_{t+1,j} \middle| \tilde{S}_{t,i}\right)}{p\left(\tilde{S}_{t+1,j}\right)}.$$

The weighted particles $\left\{\tilde{S}_{t,i}, \tilde{w}_{t,i}\right\}$ should approximate the posterior $p(S_t|Y_{1:T})$.

Independent of the backward recursion, a standard particle filter moves forward, generating weighted particles $\{S_{t-1,k}, w_{t-1,k}\}$ that approximate the filtering distribution $p(S_{t-1}|Y_{1:t-1})$. The predictive state distribution can be computed using sample analogues:

$$p(S_t|Y_{1:t-1}) = \sum_{k=1}^{n} w_{t-1,k}\, p(S_t|S_{t-1,k}).$$

The final step is to combine the two sets of particles to approximate the smoothing distribution, which takes the form

$$p(S_t|Y_{1:T}) \propto p(S_t|Y_{1:t-1}) \frac{p(S_t|Y_{t:T})}{p(S_t)}.$$

Given the weighted particles $\left\{\tilde{S}_{t,i}, \tilde{w}_{t,i}\right\}$ that approximate $p(S_t|Y_{t:T})$, we reuse $\tilde{S}_{t,i}$ and reweight them as

$$w^*_{t,i} = \tilde{w}_{t,i} \frac{\sum_{k=1}^{n} w_{t-1,k}\, p\left(\tilde{S}_{t,i} \middle| S_{t-1,k}\right)}{p\left(\tilde{S}_{t,i}\right)}.$$

The reweighted particles $\left\{\tilde{S}_{t,i}, w^*_{t,i}\right\}$ should approximate the smoothing distribution $p(S_t|Y_{1:T})$.

7.6.5 Generalized Two-filter Smoothing

The two-filter smoothing is applicable when $p(S_t)$ is tractable. However, in a general state-space model, evaluating or sampling from it is often infeasible. Briers et al. (2010) propose an artificial distribution $\gamma_t(S_t)$ as a substitute for $p(S_t)$.

A state space model is fully specified by three components: 1) a transition density $p(S_t | S_{t-1})$, 2) an observation density $p(Y_t | S_t)$, and 3) an initial state distribution $p(S_0)$. Although the likelihood function $p(Y_{1:T})$ is determined by all components, the conditional likelihood $p(Y_{1:T} | S_0)$ does not depend on the initial state distribution. Using an artificial distribution $\gamma_0(S_0)$ as a substitute for $p(S_0)$ does not alter the conditional likelihood.

The Markovian state transition permits shifting the time index by t periods. We make a local modification to the model such that $p(S_t)$ is replaced by an artificial distribution $\gamma_t(S_t)$, while the transition and observation densities from period t to T remain unchanged. Such a local change of the initial distribution does not affect the conditional likelihood $p(Y_{t:T} | S_t)$, resulting in

$$p(Y_{t:T} | S_t) \propto \frac{\tilde{p}(S_t | Y_{t:T})}{\gamma_t(S_t)},$$

where $\tilde{p}(S_t | Y_{t:T})$ refers to the artificial posterior under the locally modified model. Local changes are made in each period by specifying a series of artificial priors $\gamma_t(S_t)$, $t = 0, 1, \ldots, T$, resulting in a series of artificial posteriors. The artificial posterior-to-prior ratio is always proportional to the true conditional likelihood under the original model (also see Proposition 1 in Briers et al. (2010) for a technical proof).

Provided that the artificial priors are easy to evaluate and sample, the artificial posteriors can be computed by backward recursion:

$$\tilde{p}(S_t | Y_{t:T}) \propto \gamma_t(S_t)\, p(Y_t | S_t) \int p(S_{t+1} | S_t) \frac{\tilde{p}(S_{t+1} | Y_{t+1:T})}{\gamma_{t+1}(S_{t+1})} dS_{t+1}.$$

Given the weighted particles $\left\{\tilde{S}_{t+1,j}, \tilde{w}_{t+1,j}\right\}$ that approximate $\tilde{p}(S_{t+1} | Y_{t+1:T})$, we sample $\tilde{S}_{t,i}$ from the artificial prior $\gamma_t(S_t)$ with importance weights

$$\tilde{w}_{t,i} \propto p\left(Y_t \middle| \tilde{S}_{t,i}\right) \sum_{j=1}^{n} \tilde{w}_{t+1,j} \frac{p\left(\tilde{S}_{t+1,j} \middle| \tilde{S}_{t,i}\right)}{\gamma_{t+1}\left(\tilde{S}_{t+1,j}\right)}.$$

The weighted particles $\left\{\tilde{S}_{t,i}, \tilde{w}_{t,i}\right\}$ should approximate $\tilde{p}(S_t | Y_{t:T})$. These particles are reweighted to estimate the smoothing distribution:

$$w^*_{t,i} = \tilde{w}_{t,i} \frac{\sum_{k=1}^{n} w_{t-1,k}\, p\left(\tilde{S}_{t,i} | S_{t-1,k}\right)}{\gamma_t\left(\tilde{S}_{t,i}\right)},$$

where the weighted particles $\{S_{t-1,k}, w_{t-1,k}\}$ are obtained from a standard particle filter that approximates the filtering distribution $p(S_{t-1} | Y_{1:t-1})$.

In summary, forward-backward smoothing and FFBS reuse and reweight particles generated by the forward filter, while two-filter smoothing methods reweight particles generated by the backward information filter. The two-filter approach may outperform the forward-backward approach when the support of the smoothing distribution substantially differs from that of the filtering distribution. However, the performance of two-filter smoothing depends on the specification of the artificial prior, highlighting a fundamental problem of importance sampling: weights are unevenly distributed among particles if there is a significant discrepancy between the target and proposal distributions.

7.7 Parameter Estimation

In state space models, the unknown variables include dynamic states $S_{0:T}$ and static parameters θ. It might be tempting to treat parameters as states with a trivial transition $\theta_t = \theta_{t-1}$, allowing state estimation and parameter estimation to be jointly processed by the particle filter in a self-organized manner. However, this approach has two significant drawbacks. First, it may introduce a higher degree of nonlinearity. Second, unlike the Kalman filter, which can successfully handle zero-variance variables, the particle filter tends to degenerate under deterministic state transitions like $p(S_t | S_{t-1}) = I(S_t = S_{t-1})$, as particles are confined to the initial set of samples and cannot be renewed. In this section, we explore alternative approaches for parameter estimation in nonlinear state space models.

7.7.1 Particle MCMC Methods

The posterior distribution $p(S_{0:T}, \theta | Y_{1:T})$ can be decomposed as

$$p(S_{0:T}, \theta | Y_{1:T}) = p(\theta | Y_{1:T}) \, p(S_{0:T} | Y_{1:T}, \theta).$$

On the one hand, $p(S_{0:T} | Y_{1:T}, \theta)$ can be approximated by the particle filter and smoother. On the other hand, $p(\theta | Y_{1:T})$ can be handled by MCMC sampling if an estimator of the likelihood function is available. The challenge lies in combining the particle filter and MCMC sampler in a principled manner, without perturbing the target distribution.

If the focus is solely on parameter estimation, the pseudo-marginal method proposed by Andrieu and Roberts (2009) offers an elegant solution, using the particle filter to evaluate the target density for the MH sampler. The marginal target distribution is $p(\theta | Y_{1:T}) \propto p(\theta) \, p(Y_{1:T} | \theta)$. Although the likelihood function $p(Y_{1:T} | \theta)$ is not analytically tractable, the particle filter provides an unbiased estimator of the likelihood function through the sample mean of the unnormalized importance weights. As demonstrated in Proposition 7.2, at each iteration of the MH sampler, we run the particle filter, compute the importance weights, and evaluate the MH acceptance probability by replacing the intractable likelihood with the average weights. The noisy but unbiased estimator does not perturb the target distribution.

For both state and parameter estimation, the particle marginal Metropolis-Hastings (PMMH) method by Andrieu et al. (2010) offers a comprehensive solution that combines the particle filter and MH sampler. The joint target distribution is $p(S_{0:T}, \theta | Y_{1:T})$. If it were possible to generate an exact state path from $p(S_{0:T} | Y_{1:T}, \theta)$, Proposition 7.2 could be applied, followed by sampling the state path perfectly adapted to the proposed parameter transition. However, it is not feasible to generate an exact draw, but possible to obtain an approximated draw by resampling from the weighted particles. Proposition 7.3 shows that such an approximation is exact in the sense that the constructed transition kernel leaves the target distribution invariant.

A comparison between Proposition 7.3 and 7.2 reveals that the PMMH sampler is closely related to the pseudo-marginal method. The former includes an additional resampling step for an approximated draw from $p(S_{0:T} | Y_{1:T}, \theta)$ using the weighted particles. Additionally, the PMMH sampler resembles the multiple-try MH sampler discussed in Chapter 2. A commonality is that they can be justified by the standard MH sampler on an extended state space with auxiliary variables.

The technique used by the PMMH sampler can be described at a high level. First, we specify an extended target distribution that admits $p(S_{0:T}, \theta | Y_{1:T})$ as the marginal distribution. Second, we specify an extended proposal distribution conformable to the PMMH algorithm. Third, we run the standard MH sampler with the extended target and proposal distributions. The probability of accepting proposals under the extended MH sampler is the same as that under the PMMH algorithm.

Particle MCMC methods also include the particle Gibbs sampler. The standard Gibbs sampler cycles through $p(\theta\,|Y_{1:T}, S_{0:T})$ and $p(S_{0:T}\,|Y_{1:T}, \theta)$. The former is problem-specific, and the particle filter and smoother provide an approximation of the latter. The particle Gibbs sampler by Andrieu et al. (2010) features conditional update of $n-1$ particles, leaving a selected particle fixed. Exercise 7.11 illustrates the full posterior conditionals for the particle Gibbs sampler.

Proposition 7.2 (Pseudo-marginal MH sampler) *Let $p(\theta\,|Y_{1:T}) \propto p(\theta)\,p(Y_{1:T}\,|\theta)$ be the target distribution, with the states integrated out. Let $r(\theta^*, \theta)$ be a proposal transition kernel of parameters. If θ follows the target distribution, then $\tilde{\theta}$, obtained by the following steps, has the same distribution.*

1 Propose θ^* by the transition kernel $r(\theta^*, \theta)$.

2 Apply particle filter and obtain unnormalized weights $w^*_{T,i}$, $i = 1, \ldots, n$.

3 Approximate the likelihood function by

$$\hat{p}(Y_{1:T}\,|\theta) = \frac{1}{n}\sum_{i=1}^{n} w^*_{T,i}.$$

4 Compute the MH-style probability, with $p(Y_{1:T}\,|\theta)$ replaced by $\hat{p}(Y_{1:T}\,|\theta)$:

$$\alpha = \min\left[1, \frac{p(\theta^*)\,\hat{p}(Y_{1:T}\,|\theta^*)\,r(\theta;\theta^*)}{p(\theta)\,\hat{p}(Y_{1:T}\,|\theta)\,r(\theta^*;\theta)}\right].$$

5 Accept the proposal by the MH-style probability:

$$\tilde{\theta} = \begin{cases} \theta^* & \text{with probability } \alpha \\ \theta & \text{with probability } 1-\alpha \end{cases}.$$

Proposition 7.3 (PMMH sampler) *Let $p(S_{0:T}, \theta\,|Y_{1:T})$ be the target distribution. Let $r(\theta^*, \theta)$ be a proposal transition kernel of parameters. If $(S_{0:T}, \theta)$ follows the target distribution, then $\left(\tilde{S}_{0:T}, \tilde{\theta}\right)$, obtained by the following steps, has the same distribution.*

1 Propose θ^* by the transition kernel $r(\theta^*, \theta)$.

2 Apply particle filter and obtain particles $S^*_{0:T,i}$ and weights $w^*_{T,i}$, $i = 1, \ldots, n$.

3 Sample one particle $S^*_{0:T}$ among $\{S^*_{0:T,1}, \ldots, S^*_{0:T,n}\}$ by weights $\left\{\frac{w^*_{T,1}}{\sum w^*_{T,i}}, \ldots, \frac{w^*_{T,n}}{\sum w^*_{T,i}}\right\}$.

4 Approximate the likelihood function by

$$\hat{p}(Y_{1:T}\,|\theta) = \frac{1}{n}\sum_{i=1}^{n} w^*_{T,i}.$$

5 Compute the MH-style probability, with $p(Y_{1:T}\,|\theta)$ replaced by $\hat{p}(Y_{1:T}\,|\theta)$:

$$\alpha = \min\left[1, \frac{p(\theta^*)\,\hat{p}(Y_{1:T}\,|\theta^*)\,r(\theta;\theta^*)}{p(\theta)\,\hat{p}(Y_{1:T}\,|\theta)\,r(\theta^*;\theta)}\right].$$

6 Accept the proposed state-parameter pair by the MH-style probability:

$$\left(\tilde{S}_{0:T}, \tilde{\theta}\right) = \begin{cases} (S^*_{0:T}, \theta^*) & \text{with probability } \alpha \\ (S_{0:T}, \theta) & \text{with probability } 1-\alpha \end{cases}.$$

The pseudo-marginal and particle MCMC samplers provide asymptotically exact sampling from the posterior distributions. However, a naive implementation may result in excessively high variance when evaluating the likelihood. The sampler might become stuck due to a spuriously large likelihood value inflated by the Monte Carlo noise. The root cause is that the likelihood function approximated by the particle filter is not a continuous function of parameters due to resampling. Even a slight change in parameters can lead to a significant jump in the estimated likelihood value.

Deligiannidis et al. (2018) propose correlating random variables using Hilbert sort. See Chapter 2 for details on the correlated pseudo-marginal method. The concept of Hilbert space filling involves mapping multidimensional particles to one-dimensional indices that preserve locality. Sorting particles may alleviate the problem of the non-smooth likelihood function.

7.7.2 Gradient and Hessian

The performance of pseudo-marginal and PMMH methods hinges on the proposal distribution used for parameter updates. In the standard MH sampler, a common tuning strategy involves numerical optimization on the posterior density $p(\theta | Y_{1:T})$ and constructing the proposal covariance matrix using the inverse Hessian evaluated at the posterior mode. However, gradient-based numerical optimization becomes challenging when the objective function is polluted by Monte Carlo noise, as the numerical gradient computed through finite differencing can significantly amplify the noise, resulting in a poor approximation or even erroneous result. This issue is also encountered in classical inference during maximum simulated likelihood estimation (Flury and Shephard, 2011).

The simulation smoother for nonlinear state space models is valuable for computing the gradient and Hessian of the log-likelihood function (as well as the posterior density). The underlying concept is data augmentation: the complete-data log-likelihood $\ln p(S_{0:T}, Y_{1:T} | \theta)$ can be evaluated exactly (without Monte Carlo noise) as the sum of the transition and observation log densities. As demonstrated in Proposition 7.4, the gradient and Hessian of the log-likelihood function can be represented by expected complete-data log-likelihood derivatives. It is important to note that this expectation is taken with respect to the posterior state distribution $p(S_{0:T} | Y_{1:T}, \theta)$. In practice, the simulation smoother or forward smoothing methods can be employed to construct the sample analog of this expectation using posterior samples, and numerical derivates can be accurately computed.

Additionally, the simulation smoother facilitates the Expectation-Maximization (EM) algorithm, which maximizes the expected complete-data log-likelihood with respect to the posterior state distribution. Refer to Exercise 7.12 for an example of the EM algorithm applied to a time-varying-parameter Poisson regression.

Proposition 7.4 (Gradient and Hessian by simulation smoothing) *Let the gradient and Hessian of the log-likelihood function be*

$$g(\theta) = \frac{\partial \ln p(Y_{1:T} | \theta)}{\partial \theta},$$

$$h(\theta) = \frac{\partial^2 \ln p(Y_{1:T} | \theta)}{\partial \theta \partial \theta'}.$$

Let the gradient and Hessian of the complete-data log-likelihood function be

$$G(\theta, S_{0:T}) = \frac{\partial \ln p(S_{0:T}, Y_{1:T} | \theta)}{\partial \theta},$$

$$H(\theta, S_{0:T}) = \frac{\partial^2 \ln p(S_{0:T}, Y_{1:T} | \theta)}{\partial \theta \partial \theta'}.$$

Taking the expectation $E(\cdot|Y_{1:T},\theta)$ *with respect to the posterior state distribution* $p(S_{0:T}|Y_{1:T},\theta)$, *we have*

$$g(\theta) = E\left[G(\theta, S_{0:T})\,|Y_{1:T},\theta\right],$$
$$h(\theta) = E\left[H(\theta, S_{0:T})\,|Y_{1:T},\theta\right] + E\left[G(\theta, S_{0:T})\,G(\theta, S_{0:T})'\,|Y_{1:T},\theta\right] - g(\theta)\,g(\theta)'.$$

7.8 Parameter Learning

Parameter learning refers to sequential parameter estimation using cumulative observations. This process is an online algorithm, as parameters are updated whenever new data become available. After the final parameter update using the last data point, the results are comparable to those obtained from offline parameter estimation using the entire dataset.

Parameter learning is well-suited to big data applications, as it does not require loading the entire dataset for in-memory computing. Unlike conventional MCMC methods that sweep over the entire dataset at each iteration of parameter update, parameter learning can be a one-pass algorithm that sequentially processes each data point.

7.8.1 Sequential Particle Filter Method for Static Models

In Bayesian inference, the posterior distribution of parameters is proportional to the product of the prior and the incremental likelihood terms:

$$p(\theta\,|Y_{1:T}) \propto p(\theta)\prod_{t=1}^{T} p(Y_t\,|Y_{1:t-1},\theta).$$

Chopin (2002) introduces a dynamic structure to the static model, allowing observations to arrive sequentially, either one at a time or in mini-batches, thus creating a sequence of target distributions: $p(\theta\,|Y_{1:t})$, $t = 1,\ldots,T$.

Consider a naive sequential sampler. Initially, particles are sampled from the prior $p(\theta)$ with importance weights proportional to $p(Y_1\,|\theta)$. The weighted particles approximate $p(\theta\,|Y_1)$. Optionally, resampling can be performed. Subsequently, particles are reweighted by multiplying the weights by $p(Y_2\,|Y_1,\theta)$, allowing the reweighted particles to approximate $p(\theta\,|Y_{1:2})$. Similarly, reweighting by $p(Y_t\,|Y_{1:t-1},\theta)$ for previous particles provides an approximation of $p(\theta\,|Y_{1:t})$.

Unfortunately, the naive sampler suffers from degeneration or impoverishment after a few iterations, as particles remain confined to the initially sampled set from the prior distribution. Without resampling, weights concentrate on a single particle; with resampling, all particles become identical.

The degeneracy problem can be mitigated by the resample-move method of Gilks and Berzuini (2001), which rejuvenates particles through a transition kernel that maintains the target distribution as its stationary distribution.

The resample-move method is an extension of standard importance sampling. Suppose the target distribution is proportional to $g(x)$, with an unknown normalizing constant $k = \int g(x)\,dx$. Particles are generated from a proposal distribution $f(x)$, with the unnormalized importance weights $w(x) = \frac{g(x)}{f(x)}$. The resample-move method jitters particles using a transition kernel $q(x^*;x)$ with the target distribution as its stationary distribution. In other words, x^* follows the target distribution if x follows the same distribution:

$$\int g(x)\,q(x^*;x)\,dx = g(x^*).$$

Let the original and moved particles be x_i and x_i^*, $i = 1, \ldots, n$, respectively, both having weights proportional to $w(x_i)$. Standard importance sampling uses $\{x_i, w(x_i)\}$ to approximate the target distribution, while the resample-move method uses $\{x_i^*, w(x_i)\}$ instead. We emphasize that weights for x_i^* remain $w(x_i)$, rather than $w(x_i^*)$. Proposition 7.5 demonstrates that the empirical CDF under the resample-move method converges to the CDF of the target distribution, providing a distributional approximation.

Proposition 7.5 (Resample-move) *Suppose that the target density is proportional to $g(x)$, with an unknown normalizing constant $k = \int g(x)\,dx$. Let $F(c) = \int_{-\infty}^{c} k^{-1} g(x)\,dx$ be the CDF of the target distribution evaluated at a point c. Let x_i, $i = 1, \ldots, n$ be particles generated from a proposal density $f(x)$, with the unnormalized weights $w(x_i) = \frac{g(x_i)}{f(x_i)}$ and the normalized weights $\tilde{w}(x_i) = \frac{w(x_i)}{\sum_{j=1}^{n} w(x_j)}$. Let $q(x_i^*; x_i)$ be a transition kernel that has the target distribution as its stationary distribution:*

$$\int g(x_i)\, q(x_i^*; x_i)\, dx_i = g(x_i^*).$$

As particles are moved from x_i to x_i^* without changing weights, the empirical CDF

$$F_n(c) = \sum_{i=1}^{n} \tilde{w}(x_i)\, I(x_i^* \le c)$$

is a consistent estimator of the CDF of the target distribution:

$$F_n(c) \xrightarrow{a.s.} F(c).$$

As the name suggests, the resample-move method involves two steps: resampling and moving. Proposition 7.5 justifies the latter. To address degeneracy, it is beneficial to resample particles before moving them. In the worst-case scenario, weights concentrate on a single particle, and all resampled particles are identical. However, each resampled particle is then moved by the transition kernel, resulting in distinct jittered particles, so that particle diversity is restored. While technically correct to move particles without resampling, the weights of moved particles remain unchanged, and weight degeneracy is resolved only if resampling precedes moving.

In practice, the transition kernel is typically constructed by an iteration of the Gibbs or MH sampling, ensuring that the transition kernel has the target distribution as its stationary distribution. Chopin (2002) suggests using an independence MH kernel, where the proposal is a multivariate normal distribution with mean and variance matching the sample analogues of the weighted particles.

The resample-move steps in SMC involve a single application of the MH transition kernel to multiple particles, whereas the standard MH sampler involves multiple applications of the transition kernel to a single "particle" (i.e., the initial value of the MH sampler). It implies that the ergodicity of the Markov chain is not critical, as the average of many short chains is used instead of a long chain.

A computational issue arises when the resample-move method is incorporated into sequential parameter learning in a static model. As the transition kernel is applied to each particle separately, it can be computationally intensive. Additionally, applying the transition kernel requires evaluating $p(Y_{1:t} | \theta)$. As t increases in the forward recursion, the resample-move operations become more computationally demanding. Fortunately, the target distributions $p(\theta | Y_{1:t})$ tend to stabilize as t increases, reducing the need for frequent resample-move steps.

7.8.2 SMC-Square

The sequential particle filter method for static models by Chopin (2002) requires computing the incremental likelihood function $p(Y_t|Y_{1:t-1}, \theta)$, but the likelihood function is not analytically tractable in nonlinear state space models. The SMC^2 method proposed by Chopin et al. (2013) addresses intractability by approximating the likelihood function using a nested particle filter for each θ particle. Fulop and Li (2013) developed a similar approach known as the marginalized resample-move method.

Since the importance weights are generated by another importance sampler, these weights are inherently random. The random-weight importance sampling is justified by extending the target and proposal distributions to include auxiliary variables underlying the random weights, as demonstrated in Proposition 7.6. Similar to the pseudo-marginal methods by Andrieu and Roberts (2009) and particle MCMC methods by Andrieu et al. (2010), random-weight importance sampling achieves an exact approximation of the target distribution, provided we have a nonnegative and unbiased estimator of the importance weights.

When comparing Proposition 7.6 to Proposition 7.1, it is evident that random-weight importance sampling adheres to the standard importance sampling framework. In practice, this involves replacing an intractable target density with a nonnegative and unbiased estimator when evaluating the importance weights.

Proposition 7.6 (Random-weight importance sampling) *Suppose that the target density is proportional to $g(x)$, with an unknown normalizing constant $k = \int g(x)\,dx$. Let $F(c) = \int_{-\infty}^{c} k^{-1} g(x)\,dx$ be the CDF of the target distribution evaluated at a point c.*

Suppose that we are unable to evaluate $g(x)$ but have access to a nonnegative and unbiased estimator $\hat{g}(x, u)$, where the random variables u follow the distribution $p(u)$. An unbiased estimator refers to

$$\int \hat{g}(x, u)\, p(u)\, du = g(x).$$

Let $x_1, \ldots, x_n$ be particles generated from a proposal density $f(x)$, with random weights $w(x_i, u_i) = \frac{\hat{g}(x_i, u_i)}{f(x_i)}$, where u_i is sampled from $p(u)$. Consider the empirical CDF of the weighted particles:

$$F_n(c) = \sum_{i=1}^{n} \tilde{w}(x_i, u_i)\, I(x_i \le c),$$

where weights are normalized as $\tilde{w}(x_i, u_i) = \frac{w(x_i, u_i)}{\sum_{j=1}^{n} w(x_j, u_j)}$.

1. *We have an unbiased and consistent estimator of the normalizing constant*

$$E[w(x_i, u_i)] = k,$$

$$\frac{1}{n}\sum_{i=1}^{n} w(x_i, u_i) \xrightarrow{a.s.} k.$$

2. *We have a consistent estimator of the CDF of the target distribution*

$$F_n(c) \xrightarrow{a.s.} F(c).$$

For parameter estimation in nonlinear state space models, SMC^2 utilizes two layers of particle filters. The outer-layer SMC, which is the particle filter for parameters θ, has the target distribution $p(\theta|Y_{1:t})$. Since the importance weights are not analytically tractable, the inner-layer SMC, or the particle filter for states $S_{1:t}$, is employed to obtain an unbiased estimator of $p(Y_t | Y_{1:t-1}, \theta)$ for each θ particle. This step is justified by random-weight importance sampling. Additionally, to prevent degeneracy of the outer-layer particles, the resample-move method is used to rejuvenate particles, typically using a MH kernel with the stationary distribution $p(\theta | Y_{1:t})$, which is evaluated by an unbiased estimator. This step is justified by the pseudo-marginal MH sampler.

7.9 Software and Computational Considerations

Implementing MCMC methods, such as the Gibbs sampler, are prone to programming errors (refer to Cook et al. (2006) and Geweke (2004) for MCMC software validation methods). In contrast, the particle filter is recognized in the literature for its ease of implementation. SMC templates can be readily adapted for use in many nonlinear models with minimal modifications.

The primary step in implementing SMC is computing the importance weights. Similar to other models that involve weights, such as finite mixture models and hidden Markov models, the best practice is to focus on log densities before normalizing weights to ensure numerical stability. Suppose the target density is proportional to $g(x)$ and the proposal density is $f(x)$. Let $X = (x_1, \ldots, x_n)$ represent the concatenated particles generated from the proposal distribution. A template for computing weights can be as follows:

```
logWeights = log(g(X)) - log(f(X));
logWeights(isnan(logWeights)) = -Inf;
logWeights = logWeights - max(logWeights);
unnormalizedWeights = exp(logWeights);
normalizedWeights = unnormalizedWeights ./ sum(unnormalizedWeights);
```

While the particle filter offers a robust solution for nonlinear state space models, its numerical performance can sometimes be unsatisfactory in practice. Users may encounter a learning curve when applying it to real-world problems. Understanding the limitations is as crucial as appreciating the strengths. First, the particle filter may not perform well under diffuse initial states if initial particles are generated from $p(S_0)$. Second, a noise-free observation equation (such as ARMA in state space form) can lead to degeneracy of importance weights, as $p(Y_t | S_t)$ collapses to a single point. Third, deterministic transitions of some state variables (e.g., AR(2) states) invalidate reweighting for particle smoothing, as $p(S_t | S_{t-1})$ collapses. These limitations appear in certain proposal specifications (e.g., the bootstrap kernel), but not in others. In practice, selecting an appropriate SMC proposal is the key to successful implementation of the particle filter.

Additionally, some limitations can be avoided. For instance, it might be tempting to estimate parameters in nonlinear state space models using maximum simulated likelihood. However, the likelihood function estimated by the particle filter is not smooth, and quasi-Newton numerical optimization may become stuck. This limitation arises from poor approximation of gradients through finite differencing. An alternative method for computing gradients using the simulation smoother can address this issue (see Proposition 7.4).

7.10 An Application on Inequality Constrained Time-varying-parameter Regression

The time-varying-parameter regression is a prominent application of state space models, with notable studies by Cogley and Sargent (2005), Primiceri (2005), and Stock and Watson (2007). These studies address parameter uncertainty and instability using random coefficients assumed to follow random walk or autoregressive processes. For instance, an AR(2) model with random-walk coefficients can be specified as

$$\begin{aligned} Y_t &= \phi_0 + \phi_{1t} Y_{t-1} + \phi_{2t} Y_{t-2} + \varepsilon_t, \\ \phi_{1t} &= \phi_{1,t-1} + u_{1t}, \\ \phi_{2t} &= \phi_{2,t-1} + u_{2t}, \end{aligned}$$

where the independent Gaussian disturbances ε_t, u_{1t} and u_{2t} have variances σ_ε^2, σ_1^2 and σ_2^2, respectively.

7.10.1 Inequality Constraints in State Space Models

It is desirable to impose restrictions on ϕ_{1t} and ϕ_{2t} to ensure the time series remains non-explosive at each point in time. The stationarity conditions for an AR(2) process are defined by triangular conditions:

$$\begin{aligned} \phi_{2t} + \phi_{1t} &\leq 1, \\ \phi_{2t} - \phi_{1t} &\leq 1, \\ \phi_{2t} &\geq -1, \end{aligned}$$

under which the eigenvalues of the AR(2) process never fall outside the unit circle.

We consider a state space model where the transition equation is subject to inequality constraints on state variables (Qian, 2019). A similar model is discussed by Koop and Potter (2011). This model is non-Gaussian because the transition density is defined by a truncated normal distribution. It is also nonlinear because the truncation probability depends on the state variables. Equation (7.1) subject to the inequality constraints $S_t \in \Omega_t$ provides the transition density

$$p\left(S_t \left| S_{t-1}\right.\right) = \frac{\phi\left(S_t; A_t S_{t-1}, B_t B_t'\right)}{F\left(A_t S_{t-1}, B_t B_t', \Omega_t\right)} I\left(S_t \in \Omega_t\right),$$

where $\phi(\cdot)$ denotes the normal density, and $F(\cdot)$ denotes the normal probability in a measurable set Ω_t. Note that $F(\cdot)$ in the denominator depends on S_{t-1}, introducing nonlinearity.

Inequality constrained state space models can be handled by the particle filter, in which the conditionally optimal proposal is available as a truncated normal distribution:

$$p\left(S_t \left| S_{t-1}, Y_t\right.\right) = \frac{\phi\left(S_t; \mu_t, \Sigma_t\right)}{F\left(\mu_t, \Sigma_t, \Omega_t\right)} I\left(S_t \in \Omega_t\right),$$

where

$$\begin{aligned} \mu_t &= A_t S_{t-1} + B_t B_t' C_t' \left(C_t B_t B_t' C_t' + D_t D_t'\right)^{-1} \left(Y_t - C_t A_t S_{t-1}\right), \\ \Sigma_t &= B_t B_t' - B_t B_t' C_t' \left(C_t B_t B_t' C_t' + D_t D_t'\right)^{-1} C_t B_t B_t'. \end{aligned}$$

Interestingly, the incremental importance weight takes the form of incidentally truncated normal density. See Greene (2008, p.883) for sample selection econometric models.

$$p\left(Y_t \left| S_{t-1}\right.\right) = \phi\left(Y_t; C_t A_t S_{t-1}, C_t B_t B_t' C_t' + D_t D_t'\right) \frac{F\left(\mu_t, \Sigma_t, \Omega_t\right)}{F\left(A_t S_{t-1}, B_t B_t', \Omega_t\right)}.$$

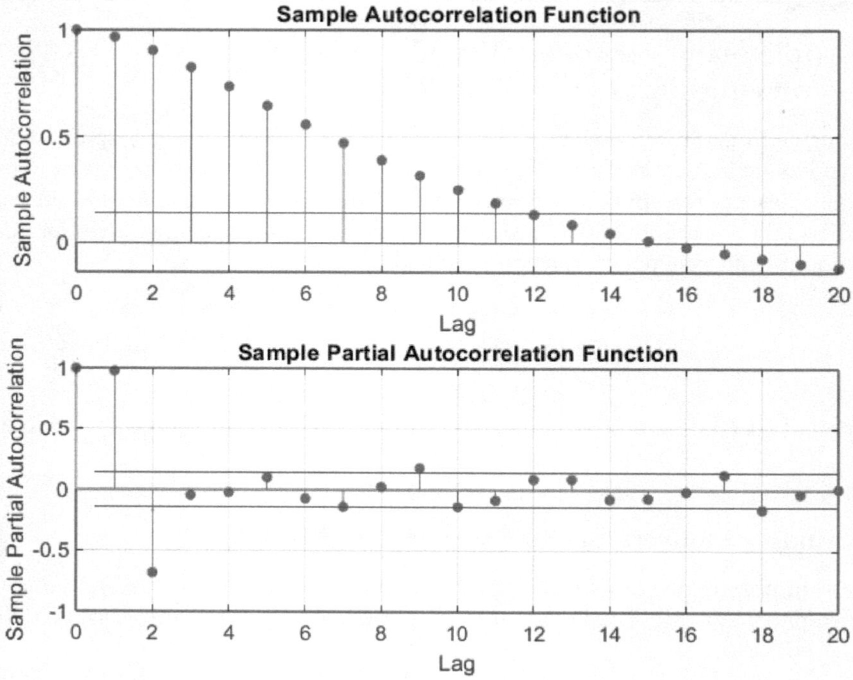

FIGURE 7.1
Autocorrelations and partial correlations in unemployment rates.

7.10.2 Empirical Results

The time series analyzed in this study is quarterly U.S. unemployment rate series from 1969 to 2020. Observations range from 3.4 to 10.7, with a mean of 6.1 and a standard deviation of 1.6. Figure 7.1 indicates that the time series exhibits geometrically decaying autocorrelations and a clear truncation after two lags in the partial autocorrelations. According to the Box and Jenkins (1976) approach, the series is well-suited for an AR(2) model. However, ten-year rolling window AR(2) regressions reveal fluctuations in the coefficients ϕ_{1t} and ϕ_{2t}, with persistency of the time series (measured by $\phi_{2t} + \phi_{1t}$) showing counter-cyclical movements. Persistency increases during economic recessions. Based on the rolling window estimation results, the model parameters are calibrated as $\phi_0 = 0.360$, $\sigma_\varepsilon = 0.225$, $\sigma_1 = 0.056$ and $\sigma_2 = 0.053$.

Figure 7.2 presents state estimation using the (unconstrained) Kalman filter. The filtered series for $\phi_{2t} + \phi_{1t}$ exhibit large spikes exceeding the unity upper bound. However, the other two inequality constraints are unlikely to tightly bind the states, as the estimated $\phi_{2t} - \phi_{1t}$ range from -1.77 to -0.78, and the estimated ϕ_{2t} ranges between -0.43 and 0.11.

Monte Carlo simulations are computationally expensive. It is prudent to impose a constraint only if there is a significant probability that the unrestricted filter violates it. Parsimony helps minimize Monte Carlo variance. Figure 7.2 indicates that the unconstrained Kalman filter results violate the constraint $\phi_{2t} + \phi_{1t} \leq 1$ in 19 quarters (mostly during economic recessions), necessitating the particle filter for enforcing the inequality constraint.

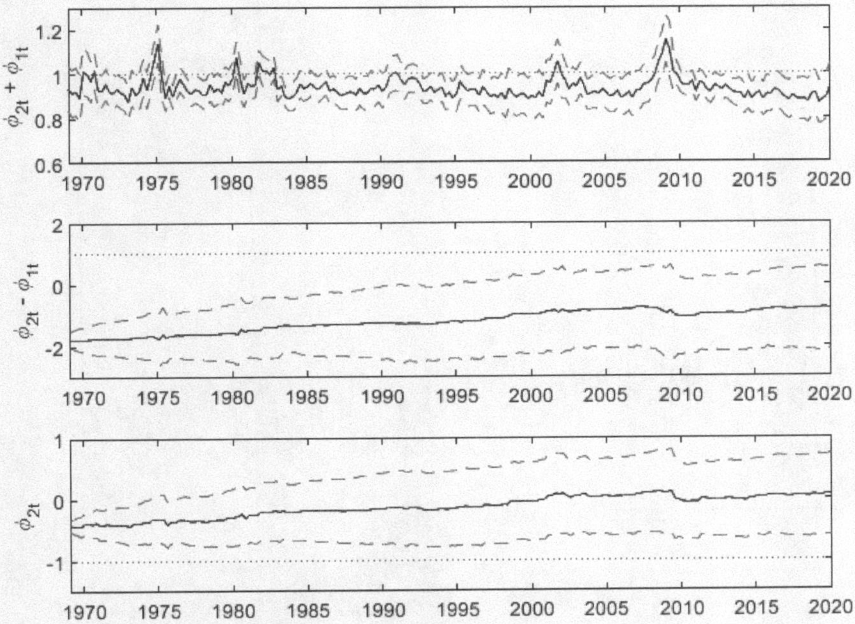

FIGURE 7.2
Time-varying-parameter AR(2) regression by unconstrained Kalman filter.

Figure 7.3 demonstrates that the particle filter effectively suppresses all spikes that violate the constraint $\phi_{2t} + \phi_{1t} \leq 1$. For instance, the unconstrained estimator in the first quarter of 2009 is 1.152, while the constrained estimator is 0.981.

7.11 Bibliographic Notes

The Kalman filter (Kalman, 1960) is optimal for linear filtering. For nonlinear and/or non-Gaussian state space models, the unscented filter by Julier and Uhlmann (2004) offers a deterministic approximation at a low computational cost. For certain non-Gaussian models with Student t, skew normal, and double exponential distributions, it is possible to exploit the conditional linear substructure for Gibbs sampling, as discussed by Chib and Ramamurthy (2014).

Pioneered by Gordon et al. (1993), the particle filter is a powerful tool for nonlinear filtering within the importance sampling framework. For theoretical analysis of sequential importance sampling, see Liu and Chen (1998), Chopin (2004), Doucet and Johansen (2009). The auxiliary particle filter by Pitt and Shephard (1999) is closely related to the particle filter, with Johansen and Doucet (2008) providing an interpretation of the auxiliary particle filter as sequential importance sampling.

Particle smoothing methods include forward-backward smoothing (Doucet et al., 2000), forward smoothing (Del Moral et al., 2010), and two-filter smoothing (Briers et al., 2010). Compared to particle filtering, smoothing is more computationally intensive with a

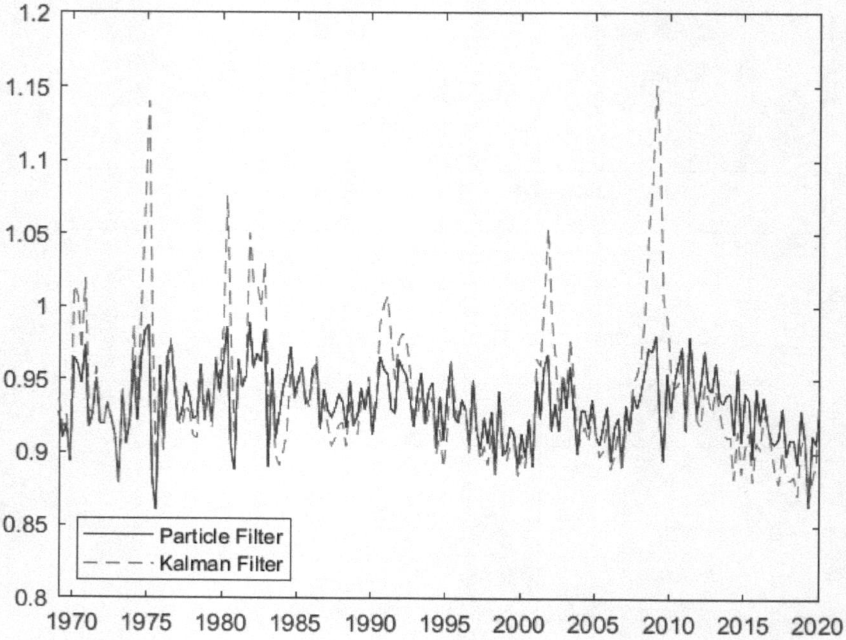

FIGURE 7.3
Time-varying-parameter AR(2) regression by inequality constrained particle filter.

complexity of $O\left(n^2T\right)$, but it is possible to reduce the complexity to $O\left(nT\right)$, as demonstrated by Douc et al. (2011).

Parameters in nonlinear state space models can be estimated either offline or online. For a comprehensive review of particle methods for parameter estimation, see Kantas et al. (2015). The pseudo-marginal MCMC (Andrieu and Roberts, 2009) and particle marginal MCMC (Andrieu et al., 2010) combine SMC and MCMC in a principled manner. The correlated pseudo-marginal method by Deligiannidis et al. (2018) significantly enhances the numerical stability of the sampler. Additionally, SMC^2 by Chopin et al. (2013) and Fulop and Li (2013) provide sequential parameter learning methods that update parameters as new data become available.

7.12 Proof of Propositions

Proposition 7.1 (Importance sampling)

Because $x_1, \ldots, x_n$ are random variables, the importance weights and the empirical CDF are random variables as well. Consider the mean of the unnormalized weight:

$$E\left[w\left(x_i\right)\right] = \int w\left(x_i\right) f\left(x_i\right) dx_i = \int g\left(x_i\right) dx_i = k.$$

By the law of large numbers,

$$\frac{1}{n}\sum_{i=1}^{n} w\left(x_i\right) \xrightarrow{a.s.} k.$$

Similarly, we have

$$E\left[w\left(x_i\right) I\left(x_i \le c\right)\right] = \int_{-\infty}^{c} w\left(x_i\right) f\left(x_i\right) dx_i = kF\left(c\right),$$

$$\frac{1}{n}\sum_{i=1}^{n} w\left(x_i\right) I\left(x_i \le c\right) \xrightarrow{a.s.} kF\left(c\right).$$

The empirical CDF takes the form

$$F_n\left(c\right) = \frac{\frac{1}{n}\sum_{i=1}^{n} w\left(x_i\right) I\left(x_i \le c\right)}{\frac{1}{n}\sum_{i=1}^{n} w\left(x_i\right)}.$$

By the continuous mapping theorem, we have $F_n\left(c\right) \xrightarrow{a.s.} F\left(c\right)$*. We have shown that the empirical CDF converges to the true CDF at each point of evaluation. The pointwise convergence can be enhanced to the uniform convergence by the Glivenko-Cantelli theorem.*

Proposition 7.2 (Pseudo-marginal MH sampler)

Because $\hat{p}\left(Y_{1:T} \left|\theta\right.\right)$ *is a non-negative and unbiased estimator of* $p\left(Y_{1:T} \left|\theta\right.\right)$*, we apply the pseudo-marginal MH sampler. On the extended state space, it is a standard MH sampler, so that* $p\left(\theta \left|Y_{1:T}\right.\right)$ *is the stationary distribution of the Markov chain, and the parameter update leaves the target distribution invariant.*

Proposition 7.3 (PMMH sampler)

The proof is similar to arguments for the independence multiple-try MH Sampling. We will show that the PMMH sampler is a standard MH sampler with an extended target and proposal distributions.

The particle filter follows the importance sampling paradigm. Given the target distribution $p\left(S_{0:T} \left|Y_{1:T}, \theta\right.\right) \propto p\left(S_{0:T}, Y_{1:T} \left|\theta\right.\right)$*, we specify a proposal distribution* $f\left(S_{0:T}\right)$ *and compute the unnormalized importance weights:*

$$w\left(S_{0:T}, \theta\right) = \frac{p\left(S_{0:T}, Y_{1:T} \left|\theta\right.\right)}{f\left(S_{0:T}\right)}.$$

For brevity, hereafter we drop the temporal subscript. For example, we write $S_{0:T}$ *as* S*,* $Y_{1:T}$ *as* Y*. The subscript in state variables refers to particle indices, like* $S_1, \ldots, S_n$*, where* n *is the number of particles.*

The extended state space has: 1) the static parameter θ*, 2) state particles* $S_1, \ldots, S_n$*, 3) an integer index* $\tau \in \{1, \ldots, n\}$ *that indicates selecting one particle, as described in step 3 of the algorithm, and 4) the selected particle* $S \equiv S_\tau$*. Let* $\Theta = (\theta, S, \tau, S_1, \ldots, S_n)$*.*

The extended target distribution can be described as follows: θ *and* S *follows the original target distribution* $p\left(S, \theta \left|Y\right.\right)$*. The index* τ *is generated by a uniform distribution between 1 and* n*. Conditional on* θ, S, τ*, we set* $S_\tau = S$*. The remaining particles* $S_i, i \neq \tau$ *are* $n-1$ *independent draws from* $f\left(\cdot\right)$*. The extended target distribution admits* $p\left(S, \theta \left|Y\right.\right)$ *as its marginal distribution.*

The extended target density takes the form:

$$\tilde{\pi}(\Theta) = \frac{1}{n} p(S, \theta | Y) I(S_\tau = S) \prod_{i \neq \tau} f(S_i).$$

We propose a transition from Θ *to* Θ^*. *The transition process is conformable to the PMMH algorithm. First, we generate* θ^* *by the parameter transition kernel* $r(\theta^*; \theta)$. *Second, in the particle filter we propose* S_i^*, $i = 1, \ldots, n$ *independently from* $f(\cdot)$. *Third, we compute the unnormalized importance weights:*

$$w(S_i^*, \theta^*) = \frac{p(S_i^*, Y | \theta^*)}{f(S_i^*)}.$$

Fourth, to select a particle, we generate the index τ^* *from a discrete categorical distribution with the probability* $\left\{ \frac{w(S_1^*, \theta^*)}{\sum w(S_i^*, \theta^*)}, \ldots, \frac{w(S_n^*, \theta^*)}{\sum w(S_i^*, \theta^*)} \right\}$. *Finally, we set* $S^* = S_{\tau^*}^*$. *That is, we sample* S^* *among* $\{S_1^*, \ldots, S_n^*\}$ *according to the importance weights.*

The extended proposal transition density takes the form:

$$\tilde{r}(\Theta^*; \Theta) = r(\theta^*; \theta) \prod_{i=1}^{n} f(S_i^*) \frac{w(S_{\tau^*}^*, \theta^*)}{\sum w(S_i^*, \theta^*)} I(S_{\tau^*}^* = S^*).$$

Given the extended target and proposal distributions, we consider a standard MH sampler. The MH acceptance probability takes the form $\min(1, \rho)$ *such that*

$$\begin{aligned} \rho &= \frac{\tilde{\pi}(\Theta^*) \tilde{r}(\Theta; \Theta^*)}{\tilde{\pi}(\Theta) \tilde{r}(\Theta^*; \Theta)} \\ &= \frac{r(\theta; \theta^*) \sum w(S_i^*, \theta^*)}{r(\theta^*; \theta) \sum w(S_i, \theta)} \frac{p(S^*, \theta^* | Y) p(S, Y | \theta)}{p(S, \theta | Y) p(S^*, Y | \theta^*)} \\ &= \frac{r(\theta; \theta^*) \hat{p}(Y_{1:T} | \theta^*)}{r(\theta^*; \theta) \hat{p}(Y_{1:T} | \theta)} \frac{p(\theta^*)}{p(\theta)}, \end{aligned}$$

where $\hat{p}(Y_{1:T} | \theta) = \frac{1}{n} \sum w(S_i, \theta)$ *is the likelihood function approximated by the sample average of the weights.*

As all the $f(S_i)$ *and* $f(S_i^*)$ *terms are cancelled in the expression of* ρ, *we recognize that this is the acceptance probability described in the PMMH algorithm. Therefore, the PMMH is a standard MH sampler on the extended state space. The state-parameter update leaves the target distribution invariant.*

Proposition 7.4 (Gradient and Hessian by simulation smoothing)
The derivatives of the log-likelihood function can be written as

$$g(\theta) = \frac{1}{p(Y_{1:T} | \theta)} \frac{\partial p(Y_{1:T} | \theta)}{\partial \theta} = \frac{1}{p(Y_{1:T} | \theta)} \frac{\partial \int p(S_{0:T}, Y_{1:T} | \theta) dS_{0:T}}{\partial \theta}.$$

Under suitable regularity conditions, the order the integration and differentiation can be exchanged:

$$\frac{\partial \int p(S_{0:T}, Y_{1:T} | \theta) dS_{0:T}}{\partial \theta} = \int \frac{\partial p(S_{0:T}, Y_{1:T} | \theta)}{\partial \theta} dS_{0:T}.$$

It follows that

$$g(\theta) = \int \frac{\partial \ln p(S_{0:T}, Y_{1:T} | \theta)}{\partial \theta} \frac{p(S_{0:T}, Y_{1:T} | \theta)}{p(Y_{1:T} | \theta)} dS_{0:T}.$$

We recognize that $\frac{p(S_{0:T}, Y_{1:T}|\theta)}{p(Y_{1:T}|\theta)}$ *is the posterior density* $p(S_{0:T}|Y_{1:T},\theta)$. *That is,* $g(\theta)$ *equals the expectation of* $G(\theta, S_{0:T})$ *with respect to the posterior state distribution.*

The Hessian is obtained by differentiating the gradient. Switching the order of integration and differentiation, we have

$$h(\theta) = \frac{\partial g(\theta)}{\partial \theta'} = \int \frac{\partial G(\theta, S_{0:T})\, p(S_{0:T}|Y_{1:T},\theta)}{\partial \theta'} dS_{0:T}.$$

Note that both $G(\theta, S_{0:T})$ *and* $p(S_{0:T}|Y_{1:T},\theta) = \frac{p(S_{0:T}, Y_{1:T}|\theta)}{p(Y_{1:T}|\theta)}$ *are functions of* θ. *The derivative of* $g(\theta)$ *equals the sum of the following three terms:*

$$\int \frac{\partial G(\theta, S_{0:T})}{\partial \theta'} p(S_{0:T}|Y_{1:T},\theta)\, dS_{0:T},$$

$$\int G(\theta, S_{0:T}) \frac{1}{p(Y_{1:T}|\theta)} \frac{\partial p(S_{0:T}, Y_{1:T}|\theta)}{\partial \theta'} dS_{0:T},$$

$$-\int G(\theta, S_{0:T}) \frac{p(S_{0:T}, Y_{1:T}|\theta)}{p(Y_{1:T}|\theta)} \frac{\partial \ln p(Y_{1:T}|\theta)}{\partial \theta'} dS_{0:T}.$$

The first term is the expected Hessian of the complete-data log-likelihood: $E[H(\theta, S_{0:T})|Y_{1:T},\theta]$. *The second term is the expected outer product of gradient (OPG) of the complete-data log-likelihood* $E\left[G(\theta, S_{0:T})\, G(\theta, S_{0:T})'|Y_{1:T},\theta\right]$. *The third term is the OPG of the standard log-likelihood* $-g(\theta)\, g(\theta)'$. *Summing up the three terms, we have*

$$h(\theta) = E\left[H(\theta, S_{0:T})|Y_{1:T},\theta\right] + E\left[G(\theta, S_{0:T})\, G(\theta, S_{0:T})'|Y_{1:T},\theta\right] - g(\theta)\, g(\theta)'.$$

Proposition 7.5 (Resample-move)

Consider the expectation:

$$E\left[w(x_i)\, I(x_i^* \le c)\right] = \int\int w(x_i)\, I(x_i^* \le c)\, f(x_i)\, q(x_i^*; x_i)\, dx_i dx_i^*.$$

Because the transition kernel has the target distribution as its stationary distribution:

$$\int g(x_i)\, q(x_i^*; x_i)\, dx_i = g(x_i^*),$$

we have

$$E\left[w(x_i)\, I(x_i^* \le c)\right] = \int g(x_i^*)\, I(x_i^* \le c)\, dx_i^* = kF(c).$$

By the law of large numbers, we have

$$\frac{1}{n}\sum_{i=1}^{n} w(x_i)\, I(x_i^* \le c) \xrightarrow{a.s.} kF(c).$$

Similarly, $\frac{1}{n}\sum_{i=1}^{n} w(x_i) \xrightarrow{a.s.} k$. *It follows that* $F_n(c) \xrightarrow{a.s.} F(c)$.

Proposition 7.6 (Random-weight importance sampling)

Consider joint sampling of (x, u) *in the standard importance sampling framework. Let the extended target density be*

$$G(x, u) = \hat{g}(x, u) p(u).$$

Because $\hat{g}(x, u)$ *is a nonnegative and unbiased estimator of* $g(x)$ *such that* $\int \hat{g}(x, u) p(u) du = g(x)$, *we have* $G(x, u) \geq 0$ *and*

$$\int \int G(x, u) dudx = \int g(x) dx = k.$$

That is, $k^{-1}G(x, u)$ *is a proper density function for the joint distribution of* (x, u), *which admits the original target density* $k^{-1}g(x)$ *as the marginal density.*

Let the extended proposal density be

$$F(x, u) = f(x) p(u).$$

By the standard importance sampling scheme, we generate particle pairs (x_i, u_i), $ii = 1, \ldots, n$ *from the extended proposal distribution, and compute the unnormalized importance weights*

$$w(x_i, u_i) = \frac{G(x, u)}{F(x, u)} = \frac{\hat{g}(x_i, u_i)}{f(x_i)}.$$

Consider the expected weights:

$$E[w(x_i, u_i)] = \int \int w(x_i, u_i) F(x_i, u_i) du_i dx_i = \int g(x_i) dx_i = k.$$

By the law of large numbers,

$$\frac{1}{n}\sum_{i=1}^{n} w(x_i, u_i) \xrightarrow{a.s.} k.$$

Similarly, we have

$$\begin{aligned} E[w(x_i, u_i) I(x_i \leq c)] &= \int \int w(x_i, u_i) I(x_i \leq c) F(x_i, u_i) du_i dx_i \\ &= \int g(x_i) I(x_i \leq c) dx_i \\ &= kF(c). \end{aligned}$$

By the law of large numbers,

$$\frac{1}{n}\sum_{i=1}^{n} w(x_i, u_i) I(x_i \leq c) \xrightarrow{a.s.} kF(c).$$

The empirical CDF takes the form

$$F_n(c) = \frac{\frac{1}{n}\sum_{i=1}^{n} w(x_i, u_i) I(x_i \leq c)}{\frac{1}{n}\sum_{i=1}^{n} w(x_i, u_i)}$$

By the continuous mapping theorem, we have $F_n(c) \xrightarrow{a.s.} F(c)$.

7.13 Exercises

Exercise 7.1 (Downward compatibility of unscented filter) *Suppose that we apply the unscented Kalman filter to a linear state space model, in which*

$$S_t = \tilde{A}_t\left(\tilde{S}_{t-1}\right) = \begin{pmatrix} A_t & B_t \end{pmatrix}\begin{pmatrix} S_{t-1} \\ u_t \end{pmatrix} = A_t S_{t-1} + B_t u_t,$$

$$Y_t = C_t\left(S_t\right) + D_t\varepsilon_t = C_t S_t + D_t\varepsilon_t.$$

The points and weights $\{x_i, w_i\}$ are constructed by the mean and variance matching:

$$\sum_{i=-m}^{m} w_i x_i = \bar{x} = \begin{pmatrix} S_{t-1|t-1} \\ 0 \end{pmatrix},$$

$$\sum_{i=-m}^{m} w_i\left(x_i - \bar{x}\right)\left(x_i - \bar{x}\right)' = \begin{pmatrix} P_{t-1|t-1} & 0 \\ 0 & I \end{pmatrix}.$$

The unscented filtering involves forward recursion by the following equations:

$$S_{t|t-1} = \sum_{i=-m}^{m} w_i \tilde{A}_t\left(x_i\right),$$

$$P_{t|t-1} = \sum_{i=-m}^{m} w_i\left[\tilde{A}_t\left(x_i\right) - S_{t|t-1}\right]\left[\tilde{A}_t\left(x_i\right) - S_{t|t-1}\right]',$$

$$Y_{t|t-1} = \sum_{i=-m}^{m} w_i C_t\left[\tilde{A}_t\left(x_i\right)\right],$$

$$O_{t|t-1} = D_t D_t' + \sum_{i=-m}^{m} w_i\left\{C_t\left[\tilde{A}_t\left(x_i\right)\right] - Y_{t|t-1}\right\}\left\{C_t\left[\tilde{A}_t\left(x_i\right)\right] - Y_{t|t-1}\right\}',$$

$$L_{t|t-1} = \sum_{i=-m}^{m} w_i\left\{\tilde{A}_t\left(x_i\right) - S_{t|t-1}\right\}\left\{C_t\left[\tilde{A}_t\left(x_i\right)\right] - Y_{t|t-1}\right\}',$$

$$S_{t|t} = S_{t|t-1} + L_{t|t-1} O_{t|t-1}^{-1}\left(Y_t - Y_{t|t-1}\right),$$

$$P_{t|t} = P_{t|t-1} - L_{t|t-1} O_{t|t-1}^{-1} L_{t|t-1}'.$$

Show that the unscented filter reduces to the standard Kalman filter if the model is linear, and the equations reduce to

$$S_{t|t-1} = A_t S_{t-1|t-1},$$

$$P_{t|t-1} = A_t P_{t-1|t-1} A_t' + B_t B_t',$$

$$Y_{t|t-1} = C_t S_{t|t-1},$$

$$O_{t|t-1} = C_t P_{t|t-1} C_t' + D_t D_t',$$

$$S_{t|t} = S_{t|t-1} + L_{t|t-1} O_{t|t-1}^{-1}\left(Y_t - Y_{t|t-1}\right),$$

$$P_{t|t} = P_{t|t-1} - L_{t|t-1} O_{t|t-1}^{-1} L_{t|t-1}'.$$

Exercise 7.2 (Forward variable recursion) *In the Baum-Welch algorithm, the forward variable is defined as*

$$F_t(s_t) = p(S_t = s_t, Y_{1:t}).$$

In practice, we normalized the forward variable by

$$\tilde{F}_t(s_t) = \frac{F_t(s_t)}{\sum_{i=1}^m F_t(i)},$$

where the normalization constant satisfies $\sum_{i=1}^m F_t(i) = p(Y_{1:t})$. *Show that the forward variable recursion formula is exactly the same as the Kalman filter.*

Exercise 7.3 (Importance sampling for integral approximation) *There is a subtle difference between approximating a distribution and the mean of a distribution. The latter approximates an integral, while the former focuses on the function to be integrated. Let* $g(\cdot)$ *be the target density with the mean* $E(X)$. *We generate particles* $x_1, \ldots, x_n$ *from a proposal density* $f(\cdot)$, *and assign each particle* x_i *the unnormalized weight* $w(x_i) = \frac{g(x_i)}{f(x_i)}$ *and the normalized weight* $\tilde{w}(x_i) = \frac{w(x_i)}{\sum_{j=1}^n w(x_j)}$.

1. The weighted particles represent a categorical distribution that approximates the target distribution. A natural estimator of $E(X)$ *is*

$$\hat{E}(X) = \sum_{i=1}^n x_i \tilde{w}(x_i) = \frac{\sum_{i=1}^n x_i w(x_i)}{\sum_{i=1}^n w(x_i)}.$$

Show that $\hat{E}(X)$ *is a consistent, but not unbiased estimator of* $E(X)$.

2. Suppose that $g(\cdot)$ *is proper density function, so the mean is given by*

$$E(X) = \int x g(x)\, dx = \int x w(x) f(x)\, dx = E[x_i w(x_i)],$$

where the random variable x_i *has the density* $f(\cdot)$, *while the random variable* X *has the density* $g(\cdot)$. *We approximate the integral by the important sampling. Consider the estimator*

$$\tilde{E}(X) = \frac{1}{n}\sum_{i=1}^n x_i w(x_i).$$

Show that $\tilde{E}(X)$ *is a consistent and unbiased estimator of* $E(X)$.

3. Suppose that the target density is known up to a proportionality constant:

$$k = \int g(x)\, dx,$$

which is approximated by the sample average of the unnormalized weights:

$$\hat{k} = \frac{1}{n}\sum_{i=1}^n w(x_i).$$

In that case, the mean is given by

$$E(X) = \int x k^{-1} g(x)\, dx = \int x k^{-1} w(x) f(x)\, dx = k^{-1} E[x_i w(x_i)],$$

where the random variable x_i has the density $f(\cdot)$, while the random variable X has the density $k^{-1}g(\cdot)$. Consider the estimator

$$\dot{E}(X) = \hat{k}^{-1}\frac{1}{n}\sum_{i=1}^{n} x_i w(x_i).$$

Show that $\dot{E}(X)$ is equivalent to $\hat{E}(X)$.

4. Discuss similarities and differences of approximating a distribution and integral (e.g., the mean of a distribution) by the importance sampling.

Exercise 7.4 (Variance reduction by importance sampling) *The importance sampling is a variance reduction technique in Monte Carlo integral approximation. Consider the expectation of an exponential distribution $Exp\left(\frac{1}{2}\right)$:*

$$E(X) = \int_0^{\infty} xg(x)\,dx,$$

where the density is $g(x) = 2e^{-2x}$. A straightforward Monte Carlo strategy is to sample from the distribution and take the average. Monte Carlo simulation provides a noisy estimator of the population expectation. It is possible to reduce the noise by the importance sampling.

Consider a proposal distribution from $Exp\left(\lambda^{-1}\right)$: $f(x) = \lambda e^{-\lambda x}$, where λ is a tuning parameter. We generate random samples (i.e., particles) $X_i, i = 1, \ldots, n$ from the proposal distribution. Because

$$E(X) = \int_0^{\infty} x\frac{g(x)}{f(x)}f(x)\,dx = \int_0^{\infty} 2\lambda^{-1}xe^{(\lambda-2)x}f(x)\,dx,$$

an importance sampling estimator approximates the expectation by the sample analogue:

$$\hat{E}_\lambda = \frac{1}{n}\sum_{i=1}^{n} 2\lambda^{-1}X_i e^{(\lambda-2)X_i}.$$

1. Show that $\hat{E}_\lambda$ is an unbiased estimator of $E(X)$, for all $\lambda > 0$.

2. Show that an efficient estimator, which minimizes the variance of $\hat{E}_\lambda$, is obtained by $\lambda = 1$. Note that $\lambda = 2$ corresponds to the ideal case that the proposal is identical to the target distribution, but it is not the optimal proposal. By selecting $\lambda = 1$, we achieve variance reduction by the importance sampling.

3. Suppose that we want to use the importance sampling to approximate other characteristics of the distribution (e.g., variance, skewness, kurtosis, etc.). Is $\lambda = 1$ still the optimal proposal?

Exercise 7.5 (Variance of importance weight) *The choice of proposal has a great impact on performance of the importance sampling. Let the target distribution be*

$$g(x) = 2e^{-2x}I(x > 0).$$

Our proposal is

$$f(x) = \lambda e^{-\lambda x}I(x > 0),$$

where λ *is a tuning parameter. We generate particles* $x_i, i = 1, \dots, n$ *from the proposal and assign the unnormalized importance weights*

$$w(x_i) = 2\lambda^{-1} e^{(\lambda-2)x_i} I(x_i > 0).$$

1. Compute the mean and variance of the importance weights.

2. Show that the variance is minimized at $\lambda = 2$ *(when the proposal is the same as the target distribution, the variance is minimized). Also show that the variance can be infinite, if* λ *is poorly chosen.*

3. Now we extend the static importance sampling to sequential Monte Carlo. Consider particle filtering with a sequence of target distributions

$$g_t(x_{1:t}) = \prod_{\tau=1}^{t} 2e^{-2x_\tau} I(x_\tau > 0), t = 1, 2, \dots$$

Suppose that our proposal is formulated by

$$f_t(x_{1:t}) = f_{t-1}(x_{1:t-1}) \lambda e^{-\lambda x_t} I(x_t > 0).$$

The unnormalized weight is

$$w_t = \prod_{\tau=1}^{t} 2\lambda^{-1} e^{(\lambda-2)x_\tau} I(x_\tau > 0).$$

Show that the weight variance will explode over time. In the absence of resampling, the particle filter is unlikely to work in practice.

Exercise 7.6 (Forecast by Kalman and particle filters with missing data) *In the state space framework, out-of-sample forecast is a special case of forward filtering, where data are missing in the forecast horizon. Missing data can be thought as artificial observations that are not informative about states, or as an empty vector of data.*

1. Consider a linear state space model where the observation in period t *is not informative about states. We discuss two forms of non-informative data. First, assume that the noise variance tends to infinity. We have* $D_t \to \infty$ *in the observation equation* $Y_t = C_t S_t + D_t \varepsilon_t$. *Show that the filtering distribution* $(S_{t|t}, P_{t|t})$ *reduces to the predictive distribution* $(S_{t|t-1}, P_{t|t-1})$ *by the Kalman filter. Second, assume that* Y_t *is an empty vector. Consequently, the number of rows of the matrices* C_t *and* D_t *equal zero as well. Does the Kalman filter work with the empty vectors and matrices? Are the filtering and predictive distributions same under empty data?*

2. Consider a nonlinear state space model where the observation in period t *is not informative about states such that* $p(Y_t | S_t) \propto 1$. *Show that the filtering distribution reduces to the predictive distribution:* $p(S_t | Y_{1:t}) = p(S_t | Y_{1:t-1})$.

3. Extend the nonlinear filtering results to multi-step forecast. For example, assume that $p(Y_t | S_t) \propto 1$ *and* $p(Y_{t+1} | S_{t+1}) \propto 1$. *Show that* $p(S_{t+1} | Y_{1:t+1}) = p(S_{t+1} | Y_{1:t-1})$.

4. Consider the bootstrap particle filter, where particles are proposed by the state transition $p(S_t | S_{t-1})$ *and the incremental importance weight is the observation density* $p(Y_t | S_t)$. *Suppose that the observation in period* t *is not informative about states such that* $p(Y_t | S_t) \propto 1$. *Does the bootstrap filter still work? If yes, describe the behavior of the filter. If not, explain the cause of failure.*

Exercise 7.7 (Bootstrap particle filter as Monte Carlo version of Kalman filter) *Consider a nonlinear state space model with the transition density $p(S_t|S_{t-1})$ and the observation density $p(Y_t|S_t)$. The state filtering evolves in prediction and update steps.*

Prediction: $p(S_t|Y_{1:t-1}) = \int p(S_t|S_{t-1})\, p(S_{t-1}|Y_{1:t-1})\, dS_{t-1}$.

Update: $p(S_t|Y_{1:t}) = \frac{p(S_t|Y_{1:t-1})p(Y_t|S_t)}{\int p(S_t|Y_{1:t-1})p(Y_t|S_t)dS_t}$.

The prediction and update steps are analytically tractable under the linear state space model, which facilitates the Kalman filter. However, the integrals involved in the prediction and update steps do not have a closed form in a general nonlinear state space model. The bootstrap particle filter can be interpreted as a Monte Carlo version of the Kalman filter, as it resorts to the simulation-based prediction and update steps.

1. Suppose that we implement the bootstrap particle filter with resampling in every time period. Particles in period $t-1$ approximately follow the filtering distribution $p(S_{t-1}|Y_{1:t-1})$. Conditional on S_{t-1}, new particles are generated by the bootstrap proposal $p(S_t|S_{t-1})$. Show that the marginal distribution of S_t is the predictive distribution $p(S_t|Y_{1:t-1})$.

2. Suppose that the predictive particles S_t are weighted by $p(Y_t|S_t)$, as in the bootstrap particle filter with resampling. Show that the empirical distribution of the weighted particles approximates the filtering distribution $p(S_t|Y_{1:t})$.

Exercise 7.8 (Optimal importance function) *Consider an inequality constrained state space model*

$$S_t = A_t S_{t-1} + B_t u_t,$$
$$Y_t = C_t S_t + D_t \varepsilon_t,$$
$$Z_t = I(S_t \geq 0).$$

In addition to the regular observation equation, we also observe the sign of the states, hence a nonlinear state space model. The particle filter handles nonlinear filtering by importance sampling.

1. Show that the optimal transition kernel is given by

$$p(S_t|S_{t-1}, Y_t, Z_t) = \phi(S_t, \mu_t, \Sigma_t)\left[I(S_t \geq 0)\, I(Z_t = 1) + I(S_t < 0)\, I(Z_t = 0)\right],$$

where

$$\mu_t = A_t S_{t-1} + B_t B_t' C_t' \left(C_t B_t B_t' C_t' + D_t D_t'\right)^{-1} (Y_t - C_t A_t S_{t-1}),$$
$$\Sigma_t = B_t B_t' - B_t B_t' C_t' \left(C_t B_t B_t' C_t' + D_t D_t'\right)^{-1} C_t B_t B_t'.$$

2. Show that the incremental weight takes the form

$$p(Y_t, Z_t|S_{t-1}) = \phi\left(Y_t, C_t A_t S_{t-1}, C_t B_t B_t' C_t' + D_t D_t'\right)$$
$$\left[\Phi(0, -\mu_t, \Sigma_t)\, I(Z_t = 1) + \Phi(0, \mu_t, \Sigma_t)\, I(Z_t = 0)\right].$$

Exercise 7.9 (Reverse sampling and resampling) *For nonlinear and non-Gaussian state space models, the particle filter has the target distribution $p(S_{0:t}|Y_{1:t})$. Suppose that the conditionally optimal transition kernel $p(S_t|S_{t-1}, Y_t)$ is tractable. The corresponding incremental importance weight takes the form $p(Y_t|S_{t-1})$. The standard filtering proceeds by the following steps:*

i) sample period-t particles by the transition kernel:

$$f_t(S_{0:t}) = f_{t-1}(S_{0:t-1})\, p(S_t|S_{t-1}, Y_t).$$

ii) calculate importance weights for period-t particles:

$$w_t(S_{0:t}) = w_{t-1}(S_{0:t-1})\,p(Y_t\,|S_{t-1}).$$

iii) resample period-t particles.

Consider reversing the order of sampling and resampling. The reweight-resample-sample strategy involves the following steps:

i) recalculate weights for period-(t − 1) particles $w_{t-1}(S_{0:t-1})\,p(Y_t\,|S_{t-1})$.

ii) resample period-(t − 1) particles.

iii) sample fully adapted period-t particles by the transition kernel: $p(S_t\,|S_{t-1}, Y_t)$.

Show the following:

1. Weighted particles $\{S_{0:t-1}, w_{t-1}(S_{0:t-1})\}$ *approximate the filtering distribution* $p(S_{0:t-1}\,|Y_{1:t-1})$.

2. Reweighted particles $\{S_{0:t-1}, w_{t-1}(S_{0:t-1})\,p(Y_t\,|S_{t-1})\}$ *approximate the smoothing distribution* $p(S_{0:t-1}\,|Y_{1:t})$.

3. The particles $S_{0:t}$ *approximately follow* $p(S_{0:t}\,|Y_{1:t})$.

4. Switching the order of sampling and resampling improves particle diversity.

Exercise 7.10 (Standard and auxiliary particle filters) *Even if we specify the target distribution as* $p(S_{0:t}\,|Y_{1:t})$, $t = 1, \ldots, T$, *we can at best approximate the marginal distribution* $p(S_t\,|Y_{1:t})$, *as the lagged states could be poorly represented by degenerate or impoverished particles. A more realistic specification of the target distribution can be* $p(S_{t-1}, S_t\,|Y_{1:t})$ *for the sequential Monte Carlo, in which we approximate the marginal distribution* $p(S_t\,|Y_{1:t})$ *based on the previous approximation for* $p(S_{t-1}\,|Y_{1:t-1})$. *The target distributions for two consecutive periods satisfy fundamental recursions:*

$$p(S_{t-1}, S_t\,|Y_{1:t}) = \frac{p(S_{t-1}\,|Y_{1:t-1})\,p(S_t\,|S_{t-1})\,p(Y_t\,|S_t)}{p(Y_t\,|Y_{1:t-1})}.$$

1. Is the standard particle filter (Algorithm 7.2) applicable under the new target distribution?

2. Is the auxiliary particle filter (Algorithm 7.3) applicable under the new target distribution?

3. Discuss similarities and differences between the standard and auxiliary particle filters.

Exercise 7.11 (Particle Gibbs sampler) *The particle MCMC methods include the particle marginal MH (PMMH) sampler and the particle Gibbs sampler. The PMMH meets the expectation of researchers, as the PMMH algorithm was implemented in empirical studies (e.g., estimating DSGE models) before the rigorous PMMH theory was developed. However, the particle Gibbs sampler may be surprising to some researchers, as the algorithm provides a constrained particle filter with one of the n particles fixed in simulation.*

For notational simplicity, we drop the time index, writing $S_{0:T}$ *as* S *and* $Y_{1:T}$ *as* Y. *The subscript in state variables refers to particle indices, like* $S_1, \ldots, S_n$, *where n is the number of particles.*

The particle filter follows the importance sampling paradigm. Given the target distribution $p(S\,|Y, \theta) \propto p(S, Y\,|\theta)$, *we specify a proposal distribution* $f(S)$ *and compute the unnormalized importance weights*

$$w(S, \theta) = \frac{p(S, Y\,|\theta)}{f(S)}.$$

We have 1) the static parameter θ, 2) state particles $S_1, \ldots, S_n$, 3) an integer index $\tau \in \{1, \ldots, n\}$ that indicates selecting one particle, as described in step 3 of the algorithm, and 4) the selected particle $S \equiv S_\tau$. The extended target density is specified as

$$\tilde{\pi}(\theta, S, \tau, S_1, \ldots, S_n) = \frac{1}{n} p(S, \theta \,|Y) \, I(S_\tau = S) \prod_{i \neq \tau} f(S_i).$$

1. *Show that the full posterior conditional distributions take the form:*

$$p(\theta \,|\cdot) \propto p(\theta \,|Y, S),$$

$$p(S, \tau \,|\cdot) \propto w(S_\tau, \theta) \, I(S_\tau = S),$$

$$p(S_1, \ldots, S_n \,|\cdot) \propto I(S_\tau = S) \prod_{i \neq \tau} f(S_i).$$

2. *How do particles update in the particle Gibbs sampler?*

Exercise 7.12 (EM algorithm by simulation smoothing) *For models that involve latent variables, maximum likelihood estimation by quasi-Newton numerical optimization could be computationally challenging, as the likelihood function (that integrates out the latent variables) may not be analytically tractable. The EM algorithm addresses the problem by maximizing the posterior expectation of the complete-data log-likelihood function. In the context of nonlinear state space models, an iteration of the parameter update (from θ_k to θ_{k+1}) by the EM algorithm takes the form:*

$$\theta_{k+1} = \arg\max_{\theta} \int \ln p(S_{0:T}, Y_{1:T} \,|\theta) \, p(S_{0:T} \,|Y_{1:T}, \theta_k) \, dS_{0:T},$$

where the complete-data log-likelihood function is the sum of the transition and observation log densities:

$$\ln p(S_{0:T}, Y_{1:T} \,|\theta) = \sum_{t=1}^{T} \ln p(S_t \,|S_{t-1}, \theta) + \ln p(Y_t \,|S_t, \theta).$$

For complex models, the expectation step and/or the maximization step may not have a closed-form solution. The simulation smoother provides a sample analogue of the expectation, and numerical optimization methods can be used in the maximization step.

We illustrate the EM algorithm by a time-varying-parameter Poisson regression:

$$S_t = S_{t-1} + \sigma u_t,$$

$$Y_t \sim Poisson\left(e^{S_t + X_t \beta}\right),$$

where u_t follows the standard normal distribution. Provided that the predictors X_t have an intercept term, the random-walk state variable can be initialized at zero: $S_0 = 0$. In other words, the initial state is treated as an unknown parameter that is reparametrized by the intercept term in the predictors. The parameters to be estimated are $\theta = (\beta, \sigma^2)$.

1. *Show that the complete-data log-likelihood takes the form:*

$$\ln p(S_{0:T}, Y_{1:T} \,|\theta) = -\frac{T}{2} \ln 2\pi - \frac{T}{2} \ln \sigma^2 - \frac{1}{2\sigma^2} \sum_{t=1}^{T} (S_t - S_{t-1})^2$$

$$+\sum_{t=1}^{T}\left[Y_t\left(S_t+X_t\beta\right)-e^{S_t+X_t\beta}-\ln Y_t!\right].$$

2. At each iteration of the EM algorithm, we run the simulation smoother conditional on the current parameter θ_k. *With samples from* $p\left(S_{0:T}\left|Y_{1:T},\theta_k\right.\right)$, *we can approximate the expectation*

$$\int \ln p\left(S_{0:T},Y_{1:T}\left|\theta\right.\right)p\left(S_{0:T}\left|Y_{1:T},\theta_k\right.\right)dS_{0:T}.$$

Which terms in the expected complete-data log-likelihood are irrelevant to the maximization step and can be dropped? Show that the expected complete-data log-likelihood can be simplified as

$$-\frac{T}{2}\ln\sigma^2-\frac{1}{2\sigma^2}C_1+\sum_{t=1}^{T}\left[Y_tX_t\beta-C_{2t}e^{X_t\beta}\right]+C_3,$$

where

$$C_1=\int\sum_{t=1}^{T}\left(S_t-S_{t-1}\right)^2p\left(S_{0:T}\left|Y_{1:T},\theta_k\right.\right)dS_{0:T},$$

$$C_{2t}=\int e^{S_t}p\left(S_{0:T}\left|Y_{1:T},\theta_k\right.\right)dS_{0:T},$$

and C_3 *refers to terms irrelevant to the maximization step. Note that* C_1, C_{2t} *and* C_3 *are not functions of* θ, *although they are determined by* θ_k *when we compute the sample analogue of the posterior expectation by the simulation smoother. Therefore, we simply treat* C_1, C_{2t} *and* C_3 *as "constants" in the maximization step.*

3. We maximize the expected complete-data log-likelihood with respect to β *and* σ^2. *Does the maximization problem have a closed-form solution? If yes, write down the analytical solution. If not, discuss how to obtain a numerical solution. (Hint: consider maximum likelihood estimation of the standard linear regression and the Poisson regression.)*

4. The beauty of the EM is that the likelihood function $p\left(Y_{1:T}\left|\theta\right.\right)$ *should increase with every iteration. Explain how to monitor the change of the likelihood function as we implement the EM algorithm.*

Exercise 7.13 (Tempered particle filtering) *Consider a nonlinear state space model, in which the observations are normally distributed conditional on the states:*

$$Y_t=C\left(S_t\right)+D\varepsilon_t,$$

where ε_t *follows standard normal distributions. The target distribution for the sequential Monte Carlo is specified as*

$$p\left(S_{t-1},S_t\left|Y_{1:t}\right.\right)=\frac{p\left(S_{t-1}\left|Y_{1:t-1}\right.\right)p\left(S_t\left|S_{t-1}\right.\right)p\left(Y_t\left|S_t\right.\right)}{p\left(Y_t\left|Y_{1:t-1}\right.\right)},$$

$$p\left(Y_t\left|S_t\right.\right)=\phi\left[Y_t;C\left(S_t\right),DD'\right],$$

where ϕ *denotes the multivariate normal density. As an iterative algorithm, we have particles of* S_{t-1} *for approximating* $p\left(S_{t-1}\left|Y_{1:t-1}\right.\right)$. *The bootstrap particle filter proposes* S_t *by the transition distribution* $p\left(S_t\left|S_{t-1}\right.\right)$, *and the incremental importance weights are determined by* $\phi\left[Y_t;C\left(S_t\right),DD'\right]$. *As the particles are generated blindly without guidance of the observation* Y_t, *the weights under the bootstrap particle filter can be volatile.*

The tempered particle filtering proposed by Herbst and Schorfheide (2019) features a sequence of bridging target distributions with inflated variances

$$p_n(S_{t-1}, S_t | Y_{1:t}) = \frac{p(S_{t-1} | Y_{1:t-1}) p(S_t | S_{t-1}) p_n(Y_t | S_t)}{p_n(Y_t | Y_{1:t-1})},$$

$$p_n(Y_t | S_t) = \phi[Y_t; C(S_t), DD'/k_n],$$

for $n = 1, \ldots, N$ *and* $k_1 > k_2 > \ldots > k_N = 1$. *We start from the bootstrap particle filter under the inflated target distribution* $p_1(S_{t-1}, S_t | Y_{1:t})$, *which should produce nearly equal-weighted particles because of the large observation variance. Our aim is to mutate the particles for a sequence of less-inflated target distributions* $p_2(S_{t-1}, S_t | Y_{1:t})$, ..., $p_N(S_{t-1}, S_t | Y_{1:t})$, *which eventually restores the original target distribution* $p(S_{t-1}, S_t | Y_{1:t})$. *The tempering iterations cycles through the correction (reweighting), selection (resampling) and mutation (resample-move) steps.*

1.correction: we have weighted particles for approximating $p_{n-1}(S_{t-1}, S_t | Y_{1:t})$. *How do we update the particle weights for approximating* $p_n(S_{t-1}, S_t | Y_{1:t})$ *instead?*

2. selection: why do we select/resample particles for approximating $p_n(S_{t-1}, S_t | Y_{1:t})$*?*

3. mutation: how do we mutate/diversify particles for approximating $p_n(S_{t-1}, S_t | Y_{1:t})$*?*

8

Applications of State Space Models

State space models are highly versatile and find applications across various time series contexts. In univariate settings, autoregressive moving average (ARMA) and unobserved component models are prominent applications. ARMA models focus on the observed time series, providing a concise parameterization for capturing autocorrelations, while unobserved component models emphasize latent factors beyond direct observations, including local level, trend, seasonality, and intervention. The state space framework is also well-suited for multivariate time series models. The vector autoregression (VAR) model serves as a cornerstone for macroeconomic forecasting and policy analysis, and Dynamic Stochastic General Equilibrium (DSGE) models, incorporating economic theory and structural shocks, have gained traction in recent years. Additionally, time-varying parameter and dynamic factor models are widely used in macroeconomics and finance forecasting. In the realm of nonlinear state space models, stochastic volatility stands out as a successful application, particularly in risk and volatility management.

8.1 ARMA Models

A distinguishing feature of time series data, in contrast to cross-sectional data, is the correlation of variables across different time points, a phenomenon known as autocorrelation or serial correlation. It implies that past observations provide valuable information for predicting future values. The ARMA model offers a parsimonious framework for characterizing autocorrelations through two primary mechanisms: the feedback of past observations and the propagation of overlapping disturbances over time.

8.1.1 Introductory Example by AR(1)

To illustrate, consider a zero-mean $AR(1)$ model

$$Y_t = \phi Y_{t-1} + \sigma \varepsilon_t,$$

where the disturbances ε_t are independent and identically distributed as standard normal variates. The model is readily in the state space form, with the coefficients $A = \phi$, $B = \sigma$, $C = 1$, $D = 0$. The state variable coincides with the observation, assuming no missing data.

The $AR(1)$ process is stationary if $|\phi| < 1$. The concept of stationarity can be examined from various perspectives. First, assuming the initial value Y_0 follows $N(0, \gamma_0)$ with $\gamma_0 = \frac{\sigma^2}{1-\phi^2}$, the marginal distributions of Y_t, $\forall t \geq 1$ remain $N(0, \gamma_0)$ over time. Second, the joint distribution of $Y_{1:T}$ is multivariate normal with mean 0 and covariance matrix Ω, in which the (i, j) element takes the form $\Omega_{ij} = \gamma_0 \phi^{|i-j|}$. Consequently, the joint distribution of $Y_{s:t}$ is time-invariant for all $s \leq t$. Third, even without the normality assumption, the linear nature of the model alone suggests that the unconditional means, variances and covariances

DOI: 10.1201/9781003564027-8

are time-invariant. Fourth, regardless of the initial distribution, the means and variances asymptotically converge to the stationary values, namely 0 and γ_0, respectively.

An autoregressive process is driven by its past values, resulting in autocorrelations. Denote the autocovariance by $\gamma_j = E(Y_t Y_{t-j})$, which can be computed by the recursion $\gamma_j = \phi\gamma_{j-1}$, initialized with the variance $\gamma_0 = \frac{\sigma^2}{1-\phi^2}$. Autocorrelations indicate that past observations are informative for prediction. The $AR(1)$ possesses the Markov property, whereby the most recent value encapsulates the entire history, resulting in

$$E(Y_{t+h} | Y_{1:t}) = p(Y_{t+h} | Y_t), \forall h \geq 1.$$

The conditional mean and variance of the predictive distributions are

$$E(Y_{t+h} | Y_t) = \phi^h Y_t,$$
$$Var(Y_{t+h} | Y_t) = (\phi^{h-1} + \cdots + \phi + 1)\sigma^2.$$

As the forecast horizon extends, these conditional values converge to the unconditional mean and variance:

$$\lim_{h\to\infty} E(Y_{t+h} | Y_t) = 0,$$
$$\lim_{h\to\infty} Var(Y_{t+h} | Y_t) = \gamma_0.$$

A notable property of a stationary $AR(1)$ process is reversibility. If the time script of the stationary series is reversed as $\tilde{Y}_t = Y_{T-t}$, $t = 1, \ldots, T$, the resulting series remains an $AR(1)$ process, following the same Markov chain as the original series: $\tilde{Y}_t \left| \tilde{Y}_{t-1} \sim N\left(\phi\tilde{Y}_{t-1}, \sigma^2\right)\right.$, as $Y_t | Y_{t+1} \sim N(\phi Y_{t+1}, \sigma^2)$. The reversibility property can be proved by symmetry in the joint distribution $Y_{1:T} \sim N(0, \Omega)$, where elements in the covariance matrix take the form $\Omega_{ij} = \gamma_0 \phi^{|i-j|}$.

8.1.2 ARMA in State Space Form

The $ARMA(p, q)$ model has p autoregressive lags and q moving average disturbances:

$$Y_t = \sum_{j=1}^{p} \phi_j Y_{t-j} + \sigma\varepsilon_t + \sigma \sum_{j=1}^{q} \psi_j \varepsilon_{t-j}.$$

For simplicity, we omit the intercept and focus on the autocorrelation analysis of the time series model.

A straightforward state space representation of the ARMA model involves current and lagged observations and disturbances in the state vector:

$$S_t = (Y_t, Y_{t-1}, \ldots, Y_{t-p+1}, \varepsilon_t, \varepsilon_{t-1}, \ldots, \varepsilon_{t-q+1})'.$$

The state space coefficients take the form

$$A = \begin{pmatrix} F_{p\times p} & G_{p\times q} \\ 0_{q\times p} & E_{q\times q} \end{pmatrix}, B = \sigma(1, 0_{1\times p-1}, 1, 0_{1\times q-1})', C = (1, 0_{1\times p+q-1}), D = 0.$$

The autoregressive coefficients $\phi_1, \ldots, \phi_p$ are incorporated in the companion matrix

$$F_{p\times p} = \begin{pmatrix} \phi_1 & \phi_2 & \cdots & \phi_{p-1} & \phi_p \\ 1 & 0 & \cdots & 0 & 0 \\ 0 & 1 & \cdots & 0 & 0 \\ \cdots & \cdots & \ddots & \cdots & \cdots \\ 0 & 0 & \cdots & 1 & 0 \end{pmatrix}.$$

Similarly, the moving average coefficients $\psi_1, \ldots, \psi_q$ are captured by

$$G_{p\times q} = \begin{pmatrix} \psi_1 & \psi_2 & \cdots & \psi_q \\ 0 & 0 & \cdots & 0 \\ 0 & 0 & \cdots & 0 \\ \cdots & \cdots & \ddots & \cdots \\ 0 & 0 & \cdots & 0 \end{pmatrix}.$$

The lagged disturbances are handled by the matrix

$$E_{q\times q} = \begin{pmatrix} 0 & 0 & \cdots & 0 & 0 \\ 1 & 0 & \cdots & 0 & 0 \\ 0 & 1 & \cdots & 0 & 0 \\ \cdots & \cdots & \ddots & \cdots & \cdots \\ 0 & 0 & \cdots & 1 & 0 \end{pmatrix}.$$

The state space representation of an ARMA model is not unique. The aforementioned form is particularly interpretable, as it comprises states of lagged observations and disturbances. However, the length of the state vector is $p+q$, and the transition matrix contains numerous zero entries. It is possible to streamline the state vector by leveraging the property that a weighted sum of lagged AR series constitutes an ARMA process (refer to Exercise 8.1):

$$Y_t = Z_t + \sum_{j=1}^{q} \psi_j Z_{t-j},$$

where

$$Z_t = \sum_{j=1}^{p} \phi_j Z_{t-j} + \sigma\varepsilon_t.$$

We specify the state vector $(Z_t, Z_{t-1}, \ldots, Z_{t-r+1})'$, where $r = \max(p, q+1)$. The state space coefficients are $A = F_{r\times r}$, $B = (\sigma, 0_{1\times r-1})'$, $C = (1, \psi_1, \ldots, \psi_{r-1})$, $D = 0$.

Alternatively, an inexact state space formulation retains only the moving average terms as states, treating the autoregressive terms as regression effects. The state space model is specified by $A = E_{q+1\times q+1}, B = (\sigma, 0_{1\times q})', C = (1, \psi_1, \ldots, \psi_q), D = 0$. By reserving the initial p observations as presample data, the autoregressive terms $\sum_{j=1}^{p} \phi_j Y_{t-j}$ serve as the regression effects, which can be concentrated out of the likelihood function to simplify the maximum likelihood estimation. Additionally, Bayesian linear regression techniques are applicable to the autoregressive terms.

8.1.3 Features of State Space Model

Traditional ARMA analysis encompasses a variety of topics, such as 1) solving the Yule-Walker equations for autocorrelation patterns, 2) backcasting disturbances from observations, 3) evaluating the log-likelihood for parameter estimation, 4) performing out-of-sample forecasts, and 5) selecting lag orders using information criteria. Refer to Box and Jenkins (1976), Hamilton (1994) for in-depth discussions. Alternatively, these topics can be addressed within a unified framework by representing the ARMA model in state space form.

First, as an alternative to solving the Yule-Walker equations, the transition equation in the state space model indicates that ARMA autocovariances can be computed recursively as $\Gamma_j = A\Gamma_{j-1}$, where $\Gamma_j = E\left(S_t S'_{t-j}\right)$ and the stationary variance Γ_0 will be addressed in the next subsection.

Second, as an alternative to backcasting disturbances, the Kalman smoother offers optimal state/disturbance estimation via the exact posterior distribution $p(\varepsilon_t | Y_{1:T}, \theta)$.

Third, as an alternative to evaluating the conditional likelihood function $p(Y_{p+1:T} | Y_{1:p}, \theta)$, the Kalman filter can compute the exact likelihood function of the full-sample observations $p(Y_{1:T} | \theta)$, thereby avoiding parameter estimation conditional on the presample data $Y_{1:p}$. Exact Bayes inference in regression models with ARMA errors is developed by Chib and Greenberg (1994) using the Metropolis-within-Gibbs sampling based on the state space representation of ARMA models.

Fourth, as an alternative to imputing missing values by heuristic methods, the Kalman filter handles out-of-sample forecasting in the same way as processing missing data. The Kalman filter, comprising prediction and update steps, skips the update step when encountering missing data, thereby generating the predictive distribution $p(Y_{T+h} | Y_{1:T}, \theta)$, $h \geq 1$.

Fifth, as an alternative to selecting lag orders based on the conditional likelihood, the Kalman filter can compute the exact full-sample log-likelihood, facilitating likelihood-based inference, including lag order selection based on Akaike and Bayesian information criteria (AIC and BIC), which balance model fit (higher likelihood) against parsimony (fewer parameters):

$$AIC = -2 \ln p(Y_{1:T} | \theta) + 2k,$$

$$BIC = -2 \ln p(Y_{1:T} | \theta) + k \ln T,$$

where k denotes the number of parameters in the model.

A complexity arises when lag order selection relies on the conditional likelihood $p(Y_{p+1:T} | Y_{1:p}, \theta)$, as the estimation sample size diminishes with more lags. A caveat is that conditional likelihood values are not directly comparable across models with different lags unless sample size adjustments are made. Otherwise, AIC and BIC are likely to favor models without lags, since the log-likelihood value is typically negative. The empirical anomaly is absent if the exact likelihood is evaluated by the Kalman filter.

A prudent recommendation for lag order specification is to prioritize parsimony, given the notorious difficulty in estimating moving average components in time series models. We suggest setting $q \leq 1$ for practical ARMA modeling.

8.1.4 Stationarity

It is desirable to initialize the state space model by the stationary distribution, if available. Iterating the transition equation $S_t = AS_{t-1} + Bu_t$ yields

$$S_t = A^t S_0 + \sum_{\tau=0}^{t-1} A^\tau B u_{t-\tau}.$$

The transition matrix can be decomposed by eigenvalues and eigenvectors such that $A = V \Lambda V^{-1}$, where Λ is a diagonal matrix of eigenvalues and V comprises the eigenvectors. The decomposition simplifies computing the matrix power $A^t = V \Lambda^t V^{-1}$. If all eigenvalues lie within the unit circle, then $\Lambda^t \to 0$ as $t \to \infty$, indicating that the influence of initial states diminishes over time.

For the state space representation of the ARMA model, the transition matrix is structured as a block upper triangular matrix $A = \begin{pmatrix} F & G \\ 0 & E \end{pmatrix}$, without matrix dimensions explicitly denoted. Its eigenvalues are solutions to the equation

$$|F - \lambda I| \cdot |E - \lambda I| = 0.$$

Note that $E - \lambda I$ is a lower triangular matrix, resulting in $|E - \lambda I| = (-\lambda)^q$, which corresponds to zero eigenvalues. Non-zero eigenvalues are derived solely from $|F - \lambda I| = 0$. The companion matrix F possesses the property (see Exercise 8.2)

$$|F - \lambda I| = (\phi_1 - \lambda)\,\lambda^{p-1} + \phi_2\lambda^{p-2} + \phi_3\lambda^{p-3} + \cdots + \phi_p.$$

It follows that the eigenvalues are the solutions to the characteristic equation

$$\lambda^p - \phi_1\lambda^{p-1} - \phi_2\lambda^{p-2} - \cdots - \phi_p = 0.$$

To compute the stationary variance, we consider asymptotic states:

$$S_t = \sum_{\tau=0}^{\infty} A^\tau B u_{t-\tau}.$$

Denote $\Gamma_0 = E\left(S_t S_t'\right) = \sum_{\tau=0}^{\infty} A^\tau B B' A^{\tau\prime}$, which satisfies

$$\Gamma_0 = A\Gamma_0 A' + BB'.$$

This is known as the Lyapunov equation, which has an explicit solution

$$vec\left(\Gamma_0\right) = \left(I - A \otimes A\right)^{-1} vec\left(BB'\right),$$

provided that all eigenvalues of A (and $A \otimes A$ as well) reside within the unit circle (refer to Exercise 8.3). Stationarity can be interpreted in two ways. First, if S_0 follows $N\left(0, \Gamma_0\right)$, then $S_t, \forall t \geq 1$ follows the same distribution. Second, if S_0 follows an arbitrary distribution, asymptotically S_t approaches $N\left(0, \Gamma_0\right)$.

In summary, the stationarity condition for an ARMA model is entirely dictated by the autoregressive coefficients, regardless of the moving average coefficients. If all eigenvalues of the companion matrix (and the transition matrix) are less than one in absolute value, the effect of initial states vanishes over time. The stationary variance is characterized by the Lyapunov equation, while stationary covariances can be recursively computed using the transition equation.

8.2 Unobserved Component Models

Structural time series analysis (Harvey, 1989) emerges as an alternative to the ARMA framework (Box and Jenkins, 1976). An observed time series can be conceptualized as the sum of various unobserved components, such as the local level, local linear trend, seasonality, interventions, business cycles, and irregular components. For instance, Harvey and Durbin (1986) and Durbin and Koopman (2012) study the effects of the seat belt law on road accidents. The observed casualty series is decomposed into local level, trigonometric seasonal effect, intervention and irregular components. These additive components function as state variables, which can be estimated using the Kalman filter. For clarity of exposition, we focus on one unobserved component at a time. Nonetheless, due to their additive nature, multiple unobserved components can be integrated into a composite model in practice.

8.2.1 Local Level

The local level model is a stochastic extension of the intercept (constant) in a time series. Let μ_t be the local level component of the observed series Y_t such that

$$Y_t = \mu_t + \sigma_\varepsilon \varepsilon_t,$$

$$\mu_t = \mu_{t-1} + \sigma_u u_t,$$

where ε_t and u_t are independent standard normal disturbances. If $\sigma_u = 0$, then μ_t reduces to a time-invariant intercept. If $\sigma_u > 0$, then μ_t can also be interpreted as a random coefficient, a time-varying parameter, or an unobserved random effect.

The initialization of the random walk state deserves careful consideration. It is evident that μ_t lacks a stationary distribution, and it may be tempting to initialize it using a diffuse distribution. The effect of diffuse initialization is similar to setting μ_0 as an unknown parameter to be estimated by data. It implies that the model should not explicitly or implicitly include another intercept term, as this would lead to an identification problem.

Additional insights on the local level component can be gained by reparameterization with a normalized state variable:

$$\ddot{\mu}_t = \sigma_u^{-1} (\mu_t - \mu_0).$$

By construction, the initial state of the normalized variable is fixed at the origin: $\ddot{\mu}_0 = 0$. The normalized state continues to follow a random walk process, but the disturbance variance is normalized to one. The reparametrized local level model

$$Y_t = \sigma_u \ddot{\mu}_t + \mu_0 + \sigma_\varepsilon \varepsilon_t,$$

$$\ddot{\mu}_t = \ddot{\mu}_{t-1} + \eta_t,$$

clarifies that μ_0 serves as an intercept in the observation equation. From a Bayesian perspective, the parameter σ_u can be treated as a regression coefficient with a normal prior, instead of σ_u^2 as a variance term with an inverse gamma prior.

Reparameterization facilitates multivariate extension, where a univariate factor $\ddot{\mu}_t$ drives multivariate observations Y_t, akin to a dynamic factor model (refer to Section 8.4). The factor loading is given by a vector-valued σ_u. Additionally, μ_0 and σ_ε are treated as vectors in the multivariate model.

8.2.2 Local Trend

The local trend model is a stochastic extension of the linear trend in a time series. Let τ_t be the local trend component of the observed series Y_t such that

$$Y_t = \tau_t + \sigma_\varepsilon \varepsilon_t,$$

$$\Delta\tau_t = \Delta\tau_{t-1} + \sigma_u u_t,$$

where Δ is the first difference operator: $\Delta\tau_t = \tau_t - \tau_{t-1}$. If $\sigma_u = 0$, then τ_t reduces to a deterministic linear trend, as $\Delta\tau_t = \Delta\tau_{t-1} = c$ indicates that $\tau_t = ct+\tau_0$. If $\sigma_u > 0$, the trend becomes stochastic. By taking $(\tau_t, \tau_{t-1})'$ as state variables, the state space representation is

$$A = \begin{pmatrix} 2 & -1 \\ 1 & 0 \end{pmatrix}, B = \begin{pmatrix} \sigma_u \\ 0 \end{pmatrix}, C = \begin{pmatrix} 1 & 0 \end{pmatrix}, D = \sigma_\varepsilon.$$

Since all eigenvalues of A equal one, initializing states with the stationary distribution is not feasible. Diffuse initialization is akin to estimating τ_0 and τ_{-1} as if they were unknown parameters. The estimated τ_0 serves as an intercept and $\Delta\tau_0 = \tau_0 - \tau_{-1}$ as the initial trend. The hidden intercept in the local trend may cause an identification problem if another intercept appears elsewhere in the model.

Reparameterization sheds light on the relation between deterministic and stochastic trends. By normalizing the state variable as

$$\ddot{\tau}_t = \sigma_u^{-1} (\tau_t - \tau_0 - \Delta\tau_0 t),$$

the reparametrized state space model reads

$$Y_t = \tau_0 + \Delta\tau_0 t + \begin{pmatrix} \sigma_u & 0 \end{pmatrix} \begin{pmatrix} \ddot{\tau}_t \\ \ddot{\tau}_{t-1} \end{pmatrix} + \sigma_\varepsilon \varepsilon_t,$$

$$\begin{pmatrix} \ddot{\tau}_t \\ \ddot{\tau}_{t-1} \end{pmatrix} = \begin{pmatrix} 2 & -1 \\ 1 & 0 \end{pmatrix} \begin{pmatrix} \ddot{\tau}_{t-1} \\ \ddot{\tau}_{t-2} \end{pmatrix} + \begin{pmatrix} 1 \\ 0 \end{pmatrix} u_t,$$

where the initial states are fixed at zero by construction: $\ddot{\tau}_0 = 0$ and $\ddot{\tau}_{-1} = 0$. In the reparametrized model, it is clear that the local trend model has a deterministic linear trend $\tau_0 + \Delta\tau_0 t$, alongside a stochastic trend component $\ddot{\tau}_t$, which is initialized at the origin and driven by unit-variance shocks in subsequent periods.

Reparameterization of the local trend model enables a multivariate extension, where a common stochastic trend $\ddot{\tau}_t$ drives multiple time series Y_t. The reparametrized model takes the same form, except that τ_0, $\Delta\tau_0$, σ_u and σ_ε are expanded as vectors.

8.2.3 Intervention, Seasonality and Cycle

The intervention component resembles regression effects:

$$Y_t = \gamma d_t + \sigma_\varepsilon \varepsilon_t,$$

where d_t is typically a dummy variable that equals one after the intervention, and the coefficient γ measures the intervention effect.

Seasonality components can be either deterministic or stochastic. Deterministic seasonality can be handled by intervention with seasonal dummy variables. Assuming k seasons, dummy variables d_{it}, $i = 1, \ldots, k$ can be constructed such that $\sum_{i=1}^k d_{it} = 1$, $\forall t$. The deterministic seasonal model functions as a linear regression

$$Y_t = \sum_{i=1}^{k} \gamma_i d_{it} + \sigma_\varepsilon \varepsilon_t.$$

If a model also includes an intercept, either explicitly or implicitly (via the local level or trend), normalization is needed for model identification, either by constraining on coefficients $\sum_{i=1}^k \gamma_i = 0$ or by omitting one dummy variable (see Exercise 8.4). Under the coefficient constraint, the seasonal effects repeat cyclically over time with a periodicity of k:

$$\gamma_1, \ldots, \gamma_k, \gamma_1, \ldots, \gamma_k, \gamma_1, \ldots, \gamma_k, \cdots$$

It is important to note that any k consecutive seasonal effects sum up to zero: $\gamma_t + \sum_{i=1}^{k-1} \gamma_{t-i} = 0$, $\forall t$. This identity enables a stochastic extension of the seasonal effects such that $\gamma_t + \sum_{i=1}^{k-1} \gamma_{t-i} = \sigma_u u_t$, where u_t represents Gaussian noise that randomizes seasonal effects. A stochastic seasonal model takes the form

$$Y_t = \gamma_t + \sigma_\varepsilon \varepsilon_t,$$

$$\gamma_t = -\sum_{i=1}^{k-1} \gamma_{t-i} + \sigma_u u_t.$$

By taking $(\gamma_t, \gamma_{t-1}, \cdots, \gamma_{t-k+1})'$ as state variables, the state space representation is

$$A = \begin{pmatrix} -1 & -1 & \cdots & -1 & -1 \\ 1 & 0 & \cdots & 0 & 0 \\ \cdots & \cdots & \ddots & \cdots & \cdots \\ 0 & 0 & \cdots & 1 & 0 \end{pmatrix}, B = \begin{pmatrix} \sigma_u \\ 0 \\ \cdots \\ 0 \end{pmatrix}, C = \begin{pmatrix} 1 & 0 & \cdots & 0 \end{pmatrix}, D = \sigma_\varepsilon.$$

Business cycles are recurrent (but not exactly periodic) transitions between recessions and expansions. Randomized trigonometric series provides a simple characterization of business cycles. Let ψ_t be the cycle component of the observed series Y_t such that

$$Y_t = \psi_t + \sigma_\varepsilon \varepsilon_t,$$
$$\begin{pmatrix} \psi_t \\ \psi_t^* \end{pmatrix} = A \begin{pmatrix} \psi_{t-1} \\ \psi_{t-1}^* \end{pmatrix} + \begin{pmatrix} \sigma_u u_t \\ \sigma_u u_t^* \end{pmatrix},$$

where $A = \begin{pmatrix} \cos\omega & \sin\omega \\ -\sin\omega & \cos\omega \end{pmatrix}$, and the parameter ω indicates the frequency. The auxiliary variable ψ_t^* accompanies ψ_t, and both exhibit cyclical behaviors, discoverable through diagonalization of the transition matrix: $A = V\Lambda V^{-1}$, where the complex eigenvalues and eigenvectors are

$$\Lambda = \begin{pmatrix} \cos\omega - \sin\omega i & 0 \\ 0 & \cos\omega + \sin\omega i \end{pmatrix},$$
$$V = \frac{\sqrt{2}}{2} \begin{pmatrix} 1 & 1 \\ -i & i \end{pmatrix}.$$

Since two eigenvalues reside exactly on the unit circle, ψ_t and ψ_t^* exhibit random oscillation (neither convergence nor divergence). If $\sigma_u = 0$, then ψ_t simplifies to a deterministic trigonometric series.

In summary, a structural time series consists of multiple unobserved components including 1) local level μ_t, 2) local linear trend τ_t, 3) seasonality γ_t, 4) intervention effect γd_t, 5) business cycle ψ_t, and 6) irregular component ε_t. The observed time series Y_t is modeled as a function (typically the sum) of unobserved components. To express a structural time series model in the state space form, we stack the unobserved components into the state vector.

The unobserved component model is flexible, allowing for any combination of unobserved components, but identification can be a concern. From the perspective of programming, if we encounter difficulties with the inversion of the Hessian matrix during parameter estimation, it is often due to identification issues caused by diffuse initialization of non-stationary components. A useful debugging strategy is to fix the initial values of some states and then estimate the model.

8.2.4 More on Irregular Component and Outliers

In the previous subsections, we assumed that the irregular component follows a standard normal distribution, which often has tails too thin to accommodate the extreme values observed in real-world datasets, particularly in financial time series.

The most widely used fat-tailed specification is the t distribution, providing a straightforward yet effective way of accommodating and detecting outliers. Other distribution families, such as the generalized extreme value, generalized Pareto, and stable distributions, are also used for modeling tail events. However, incorporating these distributions into the state space model results in a loss of analytical tractability, making it extremely costly to approximate the posterior distributions of states and parameters. In contrast, the Gaussian linear structure is maintained when conditioned on the latent variables underlying the scale mixture representation of the t distribution. Consequently, the Kalman filter and the simulation smoother remain applicable.

Consider a state space model with t-distributed error:

$$S_t = A_t S_{t-1} + B_t u_t,$$

$$Y_t = C_t S_t + D_t \varepsilon_t,$$

where $\varepsilon_t \sim MVT(0, I, v)$ and $u_t \sim MVT(0, I, q)$.

By the inverse gamma scale mixture representation of the multivariate t distribution, the disturbance term can be represented as $\varepsilon_t = \sqrt{z_t}\tilde{\varepsilon}_t$, where $\tilde{\varepsilon}_t \sim N(0, I)$ and the latent variable $z_t \sim IG\left(\frac{v}{2}, \frac{v}{2}\right)$. Marginalize over the latent variable yields

$$p(\varepsilon_t) \propto \left(1 + \frac{1}{v}\varepsilon_t' \varepsilon_t\right)^{-\frac{v+n}{2}},$$

which is recognized as the density function of $MVT(0, I, v)$. Similarly, the error term in the state equation can be expressed as $u_t = \sqrt{w_t}\tilde{u}_t$, where $\tilde{u}_t \sim N(0, I)$ and $w_t \sim IG\left(\frac{q}{2}, \frac{q}{2}\right)$.

Assume that the state space coefficients are determined by underlying parameters $A_t(\theta)$, $B_t(\theta)$, $C_t(\theta)$ and $D_t(\theta)$, with a prior $p(\theta)$. In this model, the unknown variables include 1) parameters θ, 2) states $S \equiv \{S_t\}$, and 3) latent variables $z \equiv \{z_t\}$ and $w \equiv \{w_t\}$. The posterior distribution $p(\theta, S, z, w \,|Y)$ can be simulated using the Metropolis-within-Gibbs sampler, which cycles through the posterior conditional distributions.

First, parameters and states can be updated jointly, by the marginal-conditional decomposition:

$$p(\theta, S \,|Y, z, w) \propto p(\theta \,|Y, z, w)\, p(S \,|Y, \theta, z, w).$$

Conditional on the latent variables z, w, we have a Gaussian linear state space model:

$$S_t = A_t S_{t-1} + B_t \sqrt{w_t}\tilde{u}_t,$$

$$Y_t = C_t S_t + D_t \sqrt{z_t}\tilde{\varepsilon}_t.$$

The Kalman filter computes the likelihood $p(Y \,|\theta, z, w)$, with the states integrated out. Consequently, the posterior density $p(\theta \,|Y, z, w) \propto p(\theta)\, p(Y \,|\theta, z, w)$ can be evaluated for updating parameters by the MH sampler. Also, the simulation smoother updates states from $p(S \,|Y, \theta, z, w)$ by forward filtering and backward sampling.

Second, the latent variables can be updated in a block by

$$p(z, w \,|Y, S, \theta) = \prod_{t=1}^{T} p(z_t \,|Y_t, S_t, \theta)\, p(w_t \,|Y_t, S_t, \theta).$$

Conditional on S_t, θ, the observation equation simplifies to a linear regression:

$$Y_t = C_t(\theta) S_t + D_t(\theta) \sqrt{z_t}\tilde{\varepsilon}_t,$$

where $D_t^{-1}(Y_t - C_t S_t)$ serves as the response variable, without predictors (as an empty matrix), while z_t functions as the disturbance variance with an inverse gamma prior $IG\left(\frac{v}{2}, \frac{v}{2}\right)$. Consequently, the posterior distribution $p(z_t \,|Y_t, S_t, \theta)$ remains inverse gamma:

$$IG\left[\frac{v+n}{2}, \frac{v + (Y_t - C_t S_t)' D_t^{-1\prime} D_t^{-1} (Y_t - C_t S_t)}{2}\right].$$

If Y_t is a scalar, the inverse D_t^{-1} is well-defined. However, with multivariate observations, D_t may not be a square matrix. A workaround is to take the pseudo inverse or add a small positive scalar λ (i.e., ridge regularization) to diagonal elements of the covariance matrix $D_t D_t'$, allowing for inversion as $(D_t D_t' + \lambda)^{-1}$.

Similarly, the posterior distribution $p(w_t \,|Y_t, S_t, \theta)$ remains inverse gamma

$$IG\left[\frac{q+m}{2}, \frac{q + (S_t - A_t S_{t-1})' B_t^{-1\prime} B_t^{-1} (S_t - A_t S_{t-1})}{2}\right].$$

If the degrees of freedom $v \to \infty$, the t distribution converges to the normal distribution. Meanwhile, $IG(\infty, \infty)$ reduces to a degenerate distribution, as the mean converges to one and the variance converges to zero, resulting in $z_t = 1$ without randomness.

It is feasible to estimate the degrees of freedom v along with other unknown variables in the model. A Metropolis component can be added to the sampler, as its posterior conditional distribution does not have a closed form, regardless of the prior specification. Under a uniform prior in an interval $(\underline{v}, \bar{v})$, the posterior conditional density can be evaluated as

$$p(v\,|Y, S, \theta, z, w) \propto \prod_{t=1}^{T} p(z_t\,|v)\, I(\underline{v} < v < \bar{v}),$$

where $p(z_t\,|v)$ is the density of $IG\left(\frac{v}{2}, \frac{v}{2}\right)$:

$$p(z\,|v) \propto \exp\left[-T\ln\Gamma\left(\frac{v}{2}\right) + \frac{vT}{2}\ln\frac{v}{2} - \left(\frac{v}{2}+1\right)\sum_{t=1}^{T}\ln z_t - \frac{v}{2}\sum_{t=1}^{T} z_t^{-1}\right].$$

Durbin and Koopman (2012) provide a convincing example of outlier detection using t-distributed state space models. Consider the quarterly time series of UK gas demand from 1960 to 1986. Figure 8.1 indicates that the time series exhibits strong seasonality and an upward trend, prompting us to specify unobserved components: a local trend τ_t, seasonal term γ_t, and an irregular component ε_t such that

$$Y_t = \tau_t + \gamma_t + \sigma_\varepsilon \varepsilon_t,$$

$$\Delta\tau_t = \Delta\tau_{t-1} + \sigma_\tau u_{1t},$$

$$\gamma_t = -\sum_{i=1}^{3} \gamma_{t-i} + \sigma_\gamma u_{2t}.$$

The state-space model takes the form:

$$A = \begin{pmatrix} 2 & -1 & 0 & 0 & 0 \\ 1 & 0 & 0 & 0 & 0 \\ 0 & 0 & -1 & -1 & -1 \\ 0 & 0 & 1 & 0 & 0 \\ 0 & 0 & 0 & 1 & 0 \end{pmatrix}, B = \begin{pmatrix} \sigma_\tau & 0 \\ 0 & 0 \\ 0 & \sigma_\gamma \\ 0 & 0 \\ 0 & 0 \end{pmatrix},$$

$$C = \begin{pmatrix} 1 & 0 & 1 & 0 & 0 \end{pmatrix}, D = \sigma_\varepsilon.$$

Although a gas supply disruption occurred around 1970, visual inspection of the time series plot does not reveal any apparent outliers. Surprisingly, Figure 8.2 demonstrates that the t-distributed state space model successfully detects outliers, whereas the normal model does not. The bottom-right panel plots the smoothed irregular components under the t-distribution, while the bottom-left panel shows the results of the normal model. The large spike around 1970 confirms the anomaly in gas consumption in response to the gas supply disruption.

Durbin and Koopman (2012) handle fat-tailed distributions using an iterative mode approximation method for the smoothed states. In contrast, we leverage the inverse-gamma mixture of the t distribution and employ the Metropolis-within-Gibbs sampler. Although the posterior simulation methods differ, the empirical results align.

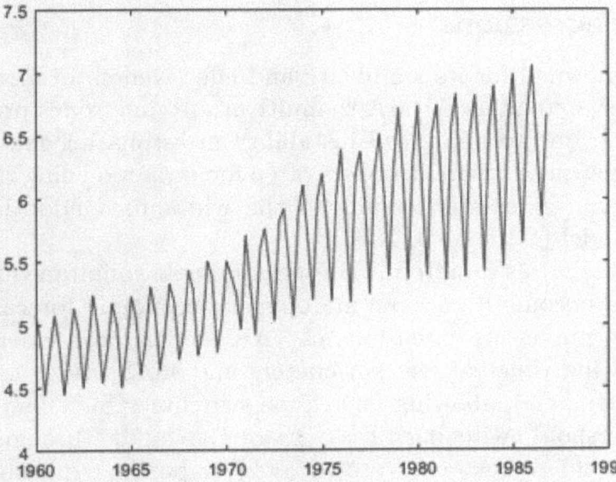

FIGURE 8.1
Quarterly demand for gas from 1960 to 1986.

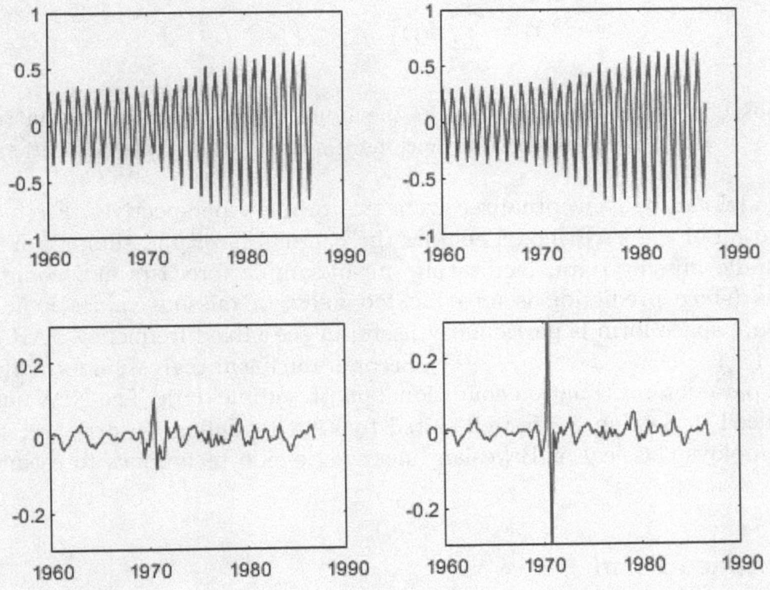

FIGURE 8.2
Smoothed seasonal and irregular components of gas consumption. Top left panel: Gaussian seasonal components. Top right panel: fat-tail seasonal components. Bottom left panel: Gaussian irregular components. Bottom right panel: fat-tail irregular components.

8.3 Vector Autoregressions

The ARMA model is renowned for its simplicity and effectiveness in modeling univariate time series data. However, extending ARMA to multivariate time series presents significant computational challenges, due to numerical instability in estimating multivariate moving average components. Consequently, practitioners often focus on modeling the autoregressive components akin to linear regression, leading to the widespread adoption of the vector autoregression (VAR) model (Lutkepohl, 2005).

Introduced by Sims (1980) as an alternative to large-scale simultaneous equation models, the VAR model has become a cornerstone of macroeconomic forecasting and policy analysis. Over the years, numerous extensions of VAR models have emerged in the literature, such as incorporating time-varying parameters and stochastic volatility (Primiceri, 2005; Cogley and Sargent, 2005), allowing for regime switching (Sims and Zha, 2006; Sims et al., 2008), setting threshold switching (Tsay, 1998; Balke, 2000), augmenting dynamic factors (Bernanke et al., 2005; Stock and Watson, 2002), imposing reduced-rank restrictions for cointegrations (Johansen, 1991; Strachan and Inder, 2004), enforcing shrinkage (Litterman, 1986; Banbura et al., 2010), encapsulating DSGE priors (Del Negro and Schorfheide, 2004; Del Negro et al., 2007), supporting mixed frequency data (Ghysels, 2016; Schorfheide and Song, 2015; Qian, 2016), operating on panel data (Holtz-Eakin et al., 1988; Love and Zicchino, 2006), accommodating large shocks (Lenza and Primiceri, 2022; Carriero et al., 2024). Some of these models will be discussed in subsequent sections.

Let Y_t be $m \times 1$ vector time series. A reduced form VAR model is specified as

$$Y_t = \sum_{j=1}^{p} \Phi_j Y_{t-j} + \varepsilon_t.$$

Contemporaneous effects are incorporated in the non-diagonal covariance matrix of disturbances: $E\left(\varepsilon_t \varepsilon_t'\right) = \Sigma$, allowing for interactions among the variables at the same time point.

The VAR model can be conceptualized from two primary perspectives. First, the state space representation of the VAR model enables the Kalman filter and simulation smoother to efficiently handle missing data, facilitating out-of-sample forecasts and counterfactual scenario analysis (since prediction is an educated guess of missing values in forecasting periods). The state space form is particularly useful in the mixed frequency VAR and real-time forecasting (Schorfheide and Song, 2015). Second, the linear regression form of the VAR model simplifies parameter estimation conditional on presample data. The NIW summation operator introduced in Section 1.8 is well-suited to VAR modeling. In practice, statistical software often employs classical or Bayesian linear regression techniques to estimate VAR models.

8.3.1 State Space Form

By stacking current and lagged observations into the state vector

$$S_t = \left(Y_t', Y_{t-1}', \ldots, Y_{t-p+1}'\right)',$$

the VAR model can be represented in a state space form, with the transition matrix

$$F = \begin{pmatrix} \Phi_1 & \Phi_2 & \cdots & \Phi_{p-1} & \Phi_p \\ I & 0 & \cdots & 0 & 0 \\ 0 & I & \cdots & 0 & 0 \\ \cdots & \cdots & \ddots & \cdots & \cdots \\ 0 & 0 & \cdots & I & 0 \end{pmatrix}.$$

The state space coefficients are $A = F$, $B = \Sigma^{1/2}(I,0)'$, $C = (I,0)$, $D = 0$. It is customary to initialize states using the stationary distribution, provided that all eigenvalues of F lie within the unit circle. The eigenvalues are determined by solving $|F - \lambda I| = 0$, resulting in the characteristic equation

$$\left| I\lambda^p - \Phi_1\lambda^{p-1} - \Phi_2\lambda^{p-2} - \cdots - \Phi_p \right| = 0.$$

The stationary variance, $E(S_t S_t') = \bar{\Sigma}$, is characterized by the Lyapunov equation

$$\bar{\Sigma} = F\bar{\Sigma}F' + BB'.$$

When the state dimension is high, solving the Lyapunov equation via an explicit formula or Schur decomposition becomes computationally intensive. Alternative methods, such as diffuse, large variance, or deterministic initialization, may be considered to mitigate computational challenges.

8.3.2 Linear Regression Form

Conditioning on the first p observations as presample data, the VAR model can be interpreted as a set of linear regressions with lagged variables as predictors, aligning with the SUR model discussed in Sections 1.8 and 4.1. Regarding our timing convention, let $Y_0, Y_{-1}, \ldots, Y_{-p+1}$ denote the presample, and $Y_1, \ldots, Y_T$ represent the estimation sample.

The regression form of the VAR model reads

$$Y_t' = X_t B + \varepsilon_t',$$

where

$$X_t = \left(Y_{t-1}', \ldots, Y_{t-p}'\right),$$
$$B = (\Phi_1, \ldots, \Phi_p)'.$$

By stacking observations for $t = 1, \ldots, T$, the VAR model is represented in a matrix form

$$Y = XB + \varepsilon,$$

where

$$Y = \begin{pmatrix} Y_1' \\ \cdots \\ Y_T' \end{pmatrix}, X = \begin{pmatrix} X_1 \\ \cdots \\ X_T \end{pmatrix}.$$

In classical inference, the SUR model is typically estimated by Generalized Least Squares (GLS) in the presence of a non-diagonal covariance matrix. However, the VAR model can be estimated by OLS for each equation:

$$\hat{B} = \left(\sum_{t=1}^{T} X_t' X_t\right)^{-1} \sum_{t=1}^{T} X_t' Y_t' = (X'X)^{-1} X'Y,$$

despite correlated disturbances across equations. The rationale is that the VAR is a special case of the SUR with symmetric predictors (i.e., the same set of predictors X_t are shared across equations), and the GLS with symmetric predictors coincides with the OLS. Moreover, Exercise 8.5 establishes that the OLS also maximizes the conditional likelihood given the presample data.

In Bayesian inference, the parameters B and Σ are assigned a prior distribution. Under a diffuse prior, the posterior mean aligns with the OLS estimator (refer to Proposition 1.10). It may be tempting to specify a conjugate NIW prior, as the posterior is analytically tractable. According to Proposition 1.9, the prior $NIW(U, \Lambda, \Omega, v)$ leads to the posterior $NIW(\bar{U}, \bar{\Lambda}, \bar{\Omega}, \bar{v})$, where

$$\bar{U} = (\Lambda + X'X)^{-1}(\Lambda U + X'Y),$$
$$\bar{\Lambda} = \Lambda + X'X,$$
$$\bar{\Omega} = \Omega + Y'Y + U'\Lambda U - \bar{U}'\bar{\Lambda}\bar{U},$$
$$\bar{v} = T + v.$$

An advantage of the conjugate NIW prior is that Bayesian estimators can be replicated by frequentist approaches, if the prior information is interpreted as pseudo data (see Proposition 1.11). Combining the conjugate prior with the likelihood is equivalent to concatenating the pseudo and genuine data, which can be implemented using the NIW summation, as explained in Section 1.8. For further reading, refer to Sims and Zha (1998) for prior elicitation methods by dummy observations.

8.3.3 Shrinkage in Large Bayesian VAR

When a high-dimensional VAR model is confronted with macroeconomic data sampled at monthly or quarterly frequencies, fitting a richly parameterized VAR model to such limited data raises the concern on overfitting. Regularization through shrinkage mitigates the overfitting risk and is crucial for successful out-of-sample forecasting.

From a Bayesian standpoint, shrinkage is accomplished by specifying a tight prior distribution with a mean of zero and a small variance, thereby steering the posterior distribution toward zero unless there is substantial evidence in the data to suggest otherwise. The Minnesota prior, introduced by Doan et al. (1984) and Litterman (1986), and later generalized by Kadiyala and Karlsson (1997), is designed to shrink a high-dimensional and high-order VAR process into a set of univariate random walk or autoregressive processes. Shrinkage is indispensable in large Bayesian VAR analyses (Banbura et al., 2010; Koop, 2013; Carriero et al., 2016; Chan, 2020; Cimadomo et al., 2022).

Let Φ_{ijk} denote the (i,j) element of the kth lag VAR coefficient Φ_k. The Minnesota-style shrinkage refers to a normal prior with mean and variance:

$$E(\Phi_{ijk}) = \begin{cases} 1 & i = j, k = 1 \\ 0 & otherwise \end{cases},$$

$$Var(\Phi_{ijk}) = \begin{cases} \frac{v_o}{k^d} & i = j \\ \frac{v_c}{k^d}\frac{\sigma_i^2}{\sigma_j^2} & i \neq j \end{cases},$$

where v_o and v_c represent the first-order variances of self-lags (i.e., diagonal elements in Φ_k) and cross-lags (i.e., off-diagonal elements), and d indicates the speed of tightness decay in higher-order lags. The ratio σ_i^2/σ_j^2 reflects variable scales.

As Bayesian computational technologies continue to evolve, the models and techniques employed by researchers and practitioners advance accordingly. The original Minnesota prior, as formulated by Litterman (1986), concentrates on normally distributed VAR coefficients, while the residual covariance matrix Σ is fixed (or even restricted as a diagonal matrix), leading to a normal posterior. Why not treat Σ as a random variable with an inverse Wishart prior? A plausible explanation lies in the limitations of Bayesian computational technologies during the 1980s. Before the advent of MCMC, non-conjugate Bayesian inference faced significant challenges. The Minnesota-style shrinkage is not fully compatible with the conjugate NIW prior (see Section 1.8) unless one imposes the constraint $v_o = v_c$, implying a symmetric treatment of self-lags and cross-lags across all equations. Under the conjugate NIW prior, it is possible to shrink the higher-order autoregressive terms, but stronger shrinkage cross-lags relative to self-lags is not feasible (see Exercise 8.6).

The Minnesota-style shrinkage is compatible with a semi-conjugate prior: $\Sigma \sim IW(\Omega, v)$ and $\beta | \Sigma \sim N(\mu, V)$, avoiding the Kronecker structure inherent in the conjugate prior $NIW(U, \Lambda, \Omega, v)$: $\beta | \Sigma \sim N\left(vec(U), \Sigma \otimes \Lambda^{-1}\right)$, where $\beta = vec(B)$ represents the vectorized coefficients. Bayesian VAR under the semi-conjugate prior is typically estimated by the Gibbs sampler, which was popularized in the 1990s. Within this computational context, Kadiyala and Karlsson (1997) discuss numerical methods for full Bayesian VAR analysis with an inverse Wishart or diffuse prior for the covariance matrix Σ.

Specifically, under the semi-conjugate prior, the posterior conditional distributions are normal and inverse Wishart (see Exercise 8.7):

$$\beta | \Sigma, Y \sim N(\bar{\mu}, \bar{V}),$$
$$\Sigma | \beta, Y \sim IW(\bar{\Omega}, \bar{v}),$$

where

$$\bar{\mu} = \left(V^{-1} + \Sigma^{-1} \otimes \sum_{t=1}^{T} X_t' X_t\right)^{-1} \left[V^{-1}\mu + vec\left(\sum_{t=1}^{T} X_t' Y_t' \Sigma^{-1}\right)\right],$$

$$\bar{V} = \left(V^{-1} + \Sigma^{-1} \otimes \sum_{t=1}^{T} X_t' X_t\right)^{-1},$$

$$\bar{\Omega} = \Omega + \sum_{t=1}^{T} (Y_t' - X_t B)' (Y_t' - X_t B),$$

$$\bar{v} = T + v.$$

In summary, the conjugate and semi-conjugate priors have their pros and cons. The conjugate prior offers computational convenience for large Bayesian VAR, as the posterior distribution can be analytically derived through NIW summation operations (see Section 1.8). Additionally, conjugate priors can be elicited via pseudo-data (Sims and Zha, 1998). By concatenating pseudo and genuine data, Bayesian estimation can be performed using standard statistical software designed for classical inference. However, the conjugate prior imposes a constrained Kronecker covariance structure for symmetric treatment of each VAR equation, complicating shrinkage toward univariate autoregressive processes. Conversely, the semi-conjugate prior fully supports Minnesota-style shrinkage, albeit requiring computationally intensive MCMC methods, such as the Gibbs sampler, for posterior inference in large Bayesian VAR.

The Kronecker structure inherent in the NIW conjugate prior necessitates symmetric treatment of self-lags and lags of other variables. This problem can be addressed by triangularization (similar to the approach used in Section 5.4), enabling the equation-by-equation estimation approach proposed by Carriero et al. (2019). Triangularization results in a recursive representation of VAR equations, allowing for stochastic volatility and time-varying parameters incorporated into large Bayesian VAR models (Chan, 2023).

With advancements in Bayesian computation, prior specification in large Bayesian VAR remains an active research area. Rather than determining the hyperparameters of the Minnesota prior through heuristic rules (e.g., maximizing out-of-sample forecast performance), Giannone et al. (2015) discuss the optimal choice of prior tightness by estimating hyperparameters through hierarchical modeling. The SSVS hierarchical prior (see Section 3.4) can be employed for searching VAR restrictions in a data-driven manner (George et al., 2008). Huber and Feldkircher (2019) adopt an adaptive shrinkage scheme via the normal-gamma prior (Griffin and Brown, 2010), akin to adaptive shrinkage by lasso discussed in Section 3.3. Chan (2021) designs a novel prior that captures the best features of both the adaptive hierarchical and Minnesota priors. Additionally, Korobilis and Pettenuzzo (2019) propose a simulation-free method for adaptive hierarchical priors.

8.3.4 Forecast

The VAR model is renowned for its empirical success in macroeconomic forecasting with multivariate time series data. Under suitable degree of shrinkage, a richly parameterized VAR can approximate the complex interdependencies between variables over time.

Out-of-sample forecasting using VAR is straightforward if parameter uncertainty is ignored. Denote

$$\hat{Y}_{T+h} = E\left(Y_{T+h} \left| Y_{1:T}, \beta, \Sigma \right.\right), h \geq 1.$$

The one-step forecast is given by

$$\hat{Y}_{T+1} = \sum_{j=1}^{p} \Phi_j Y_{T+1-j}.$$

By the law of iterated expectations, multiple-step forecasts can be recursively computed:

$$\hat{Y}_{T+h} = \sum_{j=1}^{p} \Phi_j \hat{Y}_{T+1-j},$$

where $\hat{Y}_t = Y_t, t \leq T$.

However, incorporating parameter uncertainty is crucial for credible intervals of predictions. A surprising challenge arises in that the posterior predictive distribution $p\left(Y_{T+h} \left| Y_{1:T} \right.\right)$ is not analytically tractable for horizons $h > 1$, even under the NIW conjugate prior (see Exercise 8.8). This contrasts with standard linear regression, where the predictive distribution follows the t distribution under the conjugate NIG or NIW prior. The key distinction is that predictors in linear regression are fixed over the forecasting horizon, whereas in VAR forecasts, the predictors themselves are subject to prediction. Despite these challenges, predictive distributions can be approximated using direct or indirect Monte Carlo methods.

First, using direct Monte Carlo, the mean of the posterior predictive distribution can be simulated as follows:

$$E\left(Y_{T+h} \left| Y_{1:T} \right.\right) = E\left[E\left(Y_{T+h} \left| Y_{1:T}, \beta, \Sigma \right.\right) \left| Y_{1:T} \right.\right] = \int \hat{Y}_{T+h} p\left(\beta, \Sigma \left| Y_{1:T} \right.\right) d\beta d\Sigma.$$

Under the conjugate prior, posterior draws of β, Σ can be sampled from the NIW distribution. Under the semi-conjugate prior, posterior draws can be obtained by the Gibbs sampler. Let $\beta^{(i)}, \Sigma^{(i)}$ be the ith posterior draw sampled from $p(\beta, \Sigma | Y_{1:T})$, and denote the corresponding point forecast by $\hat{Y}_{T+h}^{(i)}$. The Monte Carlo approximation of $E(Y_{T+h} | Y_{1:T})$ is given by $\frac{1}{n}\sum_{i=1}^{n} \hat{Y}_{T+h}^{(i)}$.

Second, indirect Monte Carlo involves a Markov chain whose stationary distribution is the posterior predictive distribution $p(Y_{T+h} | Y_{1:T})$. The out-of-sample forecast can be interpreted as a missing data problem, where the missing values $Y_{T+1:T+h}$ can be imputed by the Gibbs sampler with data augmentation. To generate draws from the joint distribution $p(Y_{T+1:T+h}, \beta, \Sigma | Y_{1:T})$, the Gibbs sampler cycles through the posterior conditional distributions. On the one hand, $p(Y_{T+1:T+h} | Y_{1:T}, \beta, \Sigma)$ is managed within the state space framework, where the Kalman filter and the simulation smoother provide posterior draws of $Y_{T+1:T+h}$. On the other hand, $p(\beta, \Sigma | Y_{1:T+h})$ is addressed within Bayesian linear regression framework, where the NIW summation operator concatenates the original and augmented data (i.e., $Y_{1:T}$ and $Y_{T+1:T+h}$) for the posterior distribution of model parameters.

8.3.5 Impulse Response, Variance Decomposition and Historical Decomposition

Beyond forecasting, VAR models are instrumental in policy analysis, particularly in studying the dynamic effects of monetary policy on macroeconomic variables. For instance, an expansionary monetary shock, such as an unexpected decrease in the federal funds rate, can be analyzed through its impacts on outputs, prices, and unemployment rates. This analysis is captured by the impulse response function (IRF).

However, the IRF cannot be directly computed from a reduced-form VAR because the innovation term lacks structural interpretation, necessitating identification constraints derived from economic theory, rather than statistical assumptions. For short-run constraints, see Bagliano and Favero (1998) and Eichenbaum and Evans (1995), among others. For long-run constraints, see Blanchard and Quah (1989) and King et al. (1991), among others. For sign restrictions, refer to Uhlig (2005), Canova and Nicolo (2002) and Antolin-Diaz and Rubio-Ramirez (2018).

In a structural VAR, contemporaneous interactions between endogenous variables are explicitly modeled:

$$B_0 Y_t = \sum_{k=1}^{p} B_k Y_{t-k} + u_t,$$

where the disturbance term u_t is interpreted as the structural shock: the ith element in u_t corresponds to a shock to the ith variable in Y_t. For simplicity, assume $E(u_t u_t') = I$. Due to endogeneity, direct estimation of structural parameters $B_0, B_1, \ldots, B_p$ is infeasible. Instead, the structural VAR is converted to the reduced form by inverting the contemporaneous matrix B_0:

$$Y_t = \sum_{k=1}^{p} \Phi_k Y_{t-k} + \varepsilon_t,$$

where $\Phi_k = B_0^{-1} B_k$ and $\varepsilon_t = B_0^{-1} u_t$. The innovation ε_t is a linear combination of the structural shocks in u_t, and the covariance matrix satisfies

$$\Sigma = B_0^{-1} B_0^{-1\prime}.$$

To recover structural parameters, at least $\frac{m(m-1)}{2}$ identification constraints are needed, given that B_0 contains m^2 unknown parameters, while Σ provides $\frac{m(m+1)}{2}$ unique elements. The recursive ordering, assuming B_0 is triangular, is a popular identification scheme. This approach characterizes contemporaneous effects by a one-way causal chain among ordered variables. Given these short-run constraints, the matrix B_0 can be computed using the Cholesky factor of the covariance matrix Σ, allowing recovery of the remaining structural parameters as $B_k = B_0\Phi_k$, $k = 1, \ldots, p$.

Conditioning on the structural parameters, the IRF can be derived through the moving-average representation of the structural VAR model:

$$Y_t = C_0 u_t + C_1 u_{t-1} + C_2 u_{t-2} + \cdots,$$

where $C_t = \sum_{k=1}^{p} \Phi_k C_{t-k}$, with the initial values $C_0 = B_0^{-1}$, $C_{-1} = C_{-2} = \cdots = 0$.

The IRF is defined as the partial effect of the jth shock to the ith response variable in k periods. In the linear VAR model, the IRF is given by the (i, j) element of the moving-average coefficient C_k:

$$\frac{\partial Y_{t+k,i}}{\partial u_{t,j}} = C_{k,ij}.$$

Closely related to the IRF, the forecast error variance decomposition (FEVD) attributes the volatility of observations to structural shocks. The FEVD measures relative importance of each shock in affecting the forecast error variance of observations.

Consider forecasting Y_{t+h}, given observations up to period t. Since future shocks are unobservable, the moving average form indicates that the forecast error is given by

$$E = C_0 u_{t+h} + C_1 u_{t+h-1} + \cdots + C_{h-1} u_{t+1},$$

with the forecast error variance

$$Var(E) = \sum_{k=0}^{h-1} C_k C_k'.$$

The jth shock to the ith response variable is given by

$$E_{ij} = C_{0,ij} u_{t+h,j} + C_{1,ij} u_{t+h-1,j} + \cdots + C_{h-1,ij} u_{t+1,j},$$

with the variance

$$Var(E_{ij}) = \sum_{k=0}^{h-1} C_{k,ij}^2.$$

The FEVD, defined as $\frac{Var(E_{ij})}{\sum_{j=1}^{m} Var(E_{ij})}$, quantifies the importance of the jth shock in the forecast error variance of the ith response variable.

In the context of VAR models, the analysis of structural shocks is crucial for understanding the dynamics of economic systems. While the FEVD provides a forward-looking perspective by attributing future forecast errors to structural shocks, the historical decomposition (HD) offers a backward-looking approach, revealing the contribution of realized shocks from difference sources to current observations.

The goal of HD is to perform a counterfactual decomposition:

$$Y_t = \mu + \sum_{j=1}^{m} Y_{jt}^*,$$

where Y_{jt}^* represents a counterfactual time series that indicates what Y_t would have looked like if the jth shock was the sole determinant throughout history, in the absence of other shocks. Two primary methods are available for computing HD.

The first method is based on the moving average representation of the structural VAR model, which indicates that actual observations are formulated by the weighted sum of historical structural shocks:

$$Y_t = \mu + C_0 u_t + C_1 u_{t-1} + C_2 u_{t-2} + \cdots .$$

In contrast with the IRF and FEVD, where the intercept term of the VAR model is irrelevant, the unconditional mean μ must be included in HD, ensuring that the sum of the counterfactual time series matches the original data.

Let Y^*_{ijt} be the ith element in the counterfactual series Y^*_{jt}:

$$Y^*_{ijt} = C_{0,ij} u_{t,j} + C_{1,ij} u_{t-1,j} + C_{2,ij} u_{t-2,j} + \cdots .$$

As the moving average coefficients $C_k, k = 0, 1, 2, \ldots$ diminish with $k \to \infty$, contributions from initial values and distant past shocks can be ignored, allowing for series truncation when C_k becomes sufficiently small.

The second HD method involves simulating simulate the counterfactual series Y^*_{jt} using the smoothed shocks:

$$Y^*_{jt} = \sum_{k=1}^{p} \Phi_k Y_{j,t-k} + B_0^{-1} u^*_{jt}.$$

Given the estimated coefficients $\Phi_1, \ldots, \Phi_p$ and residuals $\hat{\varepsilon}_t$, the smoothed structural shocks are represented by $\hat{u}_t = B_0 \hat{\varepsilon}_t$. The counterfactual shocks u^*_{jt} are constructed by retaining the jth element in $\hat{u}_t$ (setting other elements to zero). The counterfactual time series Y^*_{jt} can be simulated based on the counterfactual shocks.

A notable distinction between IRF/FEVD and HD is that the former depends solely on VAR parameters, while the latter relies on both data and parameters. From a Bayesian perspective, dependency on data does not pose a problem when simulating posterior distributions and constructing Bayesian p-values (refer to Section 1.7). Bayesian inference is advantageous in quantifying the uncertainty associated with estimated IRF/FEVD. Compared to the complexity of using the Delta method for asymptotic confidence intervals, Bayesian credible intervals can be straightforwardly computed using posterior samples.

A typical workflow for Bayesian computation of IRF, FEVD and HD involves the following steps:

1. Propose a reduced-form VAR model.
2. Specify a proper prior with shrinkage on VAR coefficients.
3. Obtain samples from the posterior distribution of VAR coefficients.
4. Impose identification constraints and recover structural VAR parameters.
5. Apply IRF, FEVD and HD formulae.
6. Repeat steps 3 - 5 for each posterior sample to obtain credible intervals.
7. Plot or tabulate the estimated IRF, FEVD and HD.

8.3.6 Scenario Analysis by Conditional Forecast

While IRF-based policy analysis is not feasible in a reduced-form VAR due to the lack of identification for structural shocks, the reduced-form model effectively captures autocorrelations and cross-correlations between multiple variables over time. These correlations are fundamental to conditional forecasting, enabling counterfactual scenario analysis.

Conditional forecasting in VAR models involves predicting future values of variables based on user-specified scenarios. A scenario refers to predetermined values for some endogenous variables within the forecast horizon. The conditional forecast provides an educated guess of future values for the remaining variables in the VAR model. Conditioning variables may include fiscal or monetary policy instruments, where known values reflect a combination of shocks that generate the specified scenario. For further insights on conditional forecasts in dynamic multivariate models, see Waggoner and Zha (1999).

In Bayesian VAR models, conditional forecasts are handled similarly to missing data imputation (Rubin, 1996; Murray, 2018). Let $Y_{1:T}$ be the estimation sample, and $Y^*_{T+1:T+h}$ be a partially specified matrix, in which numerical values represent the user-specified scenarios and missing values represent the unknown variables to be predicted. The objective is to obtain the posterior predictive distribution $p\left(Y_{T+1:T+h} \left| Y_{1:T}, Y^*_{T+1:T+h}\right.\right)$, incorporating parameter uncertainty by integrating out unknown parameters β, Σ in the VAR model.

It is desirable to contrast the forecast results with and without conditioning variables:

$$E\left(Y_{T+1:T+h} \left| Y_{1:T}, Y^*_{T+1:T+h}\right.\right) - E\left(Y_{T+1:T+h} \left| Y_{1:T}\right.\right),$$

where the former is the scenario point estimator and latter is the baseline point estimator of future variables.

The conditional forecast can be handled by the simulation smoother for state estimation and the conjugate NIW sampler for parameter estimation. The target distribution

$$p\left(Y_{T+1:T+h}, \beta, \Sigma \left| Y_{1:T}, Y^*_{T+1:T+h}\right.\right)$$

can be simulated by the Gibbs sampler, cycling through the posterior conditional distributions $p\left(Y_{T+1:T+h} \left| Y_{1:T}, Y^*_{T+1:T+h}, \beta, \Sigma\right.\right)$ by the simulation smoother and $p\left(\beta, \Sigma \left| Y_{1:T+h}\right.\right)$ by conjugate linear regressions that include the imputed values in the estimation sample.

Some remarks on the limitations of scenario analysis are noteworthy. Association does not imply causality. The conditional forecast leverages autocorrelations and cross-correlations for predictions based on a user-specified scenario, which differs from a treatment variable that can be manipulated by the experimenter. For instance, predicting GDP conditional on a scenario of increasing interest rates over the next few years does not necessarily indicate the causal effect of contractionary monetary policy (by setting interest rates) on the real economy. Instead, it may reflect a systematic monetary policy response to heightened economic activity. The reduced-form VAR model excels in out-of-sample forecasting, including conditional forecasts based on specific scenarios, but it is not equivalent to causal analysis that addresses the what-if question: how does a variable change in response to the intervention of another variable, all other conditions being equal?

8.4 Dynamic Factor Models

In a data-rich environment, data analysts may encounter extensive datasets characterized by a vast array of highly correlated variables. A pertinent example can be found in financial markets, where Treasury bills and bonds of varying maturities are actively traded. Similarly, statistical agencies disseminate numerous macroeconomic indicators, many of which demonstrate pronounced co-movement patterns, such as various price indices. Additionally, in the context of international economics, researchers have access to panel data of multiple countries over years. To address the data-rich challenge, it is possible to employ a high-dimensional VAR model that incorporates all these variables through Bayesian shrinkage (Banbura et al., 2010) or coefficient restrictions (Canova and Ciccarelli, 2009).

As a compelling alternative to the high-dimensional VAR, the dynamic factor model (DFM) posits that a multitude of observed time series can be explained by a limited number of latent factors that evolve dynamically. This approach is grounded in solid statistical foundations established by influential works such as Forni et al. (2000), Bai and Ng (2002), Stock and Watson (2002). These foundational studies have paved the way for DFM to become an active research area.

8.4.1 Dynamic Factor and Factor-Augmented Vector Autoregressions

The DFM is naturally expressed in the state space form:

$$S_t = AS_{t-1} + u_t,$$

$$Y_t = CS_t + \varepsilon_t,$$

where S_t represents m-dimensional latent factors and Y_t denotes d-dimensional observed series ($m < d$). The covariance matrices for the disturbances u_t and ε_t are given by Σ_u and Σ_ε, respectively. In practice, it is common to restrict Σ_ε as a diagonal matrix, since the common factors are presumed to encapsulate the inter-series correlations, while ε_t accounts for idiosyncratic variability unique to each series.

The $d \times m$ matrix C is referred to as the factor loading matrix. It is desirable to leverage domain-specific knowledge to impose a particular structure on this matrix. Otherwise, an identification problem arises in the unrestricted matrices A and C. To illustrate the problem of rotational indeterminacy, consider rotating the factors by an arbitrary $m \times m$ invertible matrix H:

$$HS_t = \left(HAH^{-1}\right)\left(HS_{t-1}\right) + Hu_t,$$

$$Y_t = CH^{-1}\left(HS_t\right) + \varepsilon_t,$$

where HS_t can be interpreted as the rotated factor, with the loading matrix CH^{-1}. Since the likelihood function is invariant under factor rotation, the model is only identified up to such linear transformation, allowing us to estimate at most the subspace spanned by the factors. To achieve identification of both the factors and the factor loading, constraints must be imposed.

From the perspective of state space modeling, the factors S_t serve as the state variables, and the Kalman filter and smoother can be employed for state estimation. Moreover, the Kalman filter provides the likelihood function in the prediction error decomposition form, allowing for maximum likelihood estimation of the model. Nevertheless, due to the plethora of parameters in the transition equation, the DFM is commonly approached through Bayesian simulation, specifically employing the Gibbs sampler. When the matrices A and Σ_u are unrestricted, the Gibbs sampler can cycle through the full posterior conditionals $p\left(S\,|Y, A, C, \Sigma_u, \Sigma_\varepsilon\right)$, $p\left(A, \Sigma_u\,|Y, S, C, \Sigma_\varepsilon\right)$ and $p\left(C, \Sigma_\varepsilon\,|Y, S, A, \Sigma_u\right)$. Posterior simulation for S can be handled by the simulation smoother of the state space model. Closed-form posterior conditionals for A, Σ_u are available under the conjugate NIW prior, akin to Bayesian VAR models. Posterior sampling for C, Σ_ε is problem-specific, potentially necessitating a Metropolis component if the posterior conditional distribution lacks a recognizable form.

The factor-augmented vector autoregression (FAVAR), introduced by Bernanke et al. (2005), closely aligns with the DFM. In FAVAR, a vast array of informational variables Y_t is assumed to be driven by a few latent factors S_t and some observed pervasive variables X_t. They jointly follow a VAR process:

$$\begin{pmatrix} S_t \\ X_t \end{pmatrix} = \begin{pmatrix} A_{11} & A_{12} \\ A_{21} & A_{22} \end{pmatrix} \begin{pmatrix} S_{t-1} \\ X_{t-1} \end{pmatrix} + \begin{pmatrix} u_{1t} \\ u_{2t} \end{pmatrix},$$

$$Y_t = CS_t + BX_t + \varepsilon_t.$$

The FAVAR can be viewed as a specialized form of the DFM, wherein both S_t and X_t are treated as factors, with the latter being perfectly observed. A trivial identity $X_t = X_t$ can be incorporated into the observation equation:

$$\begin{pmatrix} Y_t \\ X_t \end{pmatrix} = \begin{pmatrix} C & B \\ 0 & I \end{pmatrix} \begin{pmatrix} S_t \\ X_t \end{pmatrix} + \begin{pmatrix} \varepsilon_t \\ 0 \end{pmatrix}.$$

Rotational indeterminacy remains relevant to the FAVAR model. Let Q and R be $m \times m$ and $m \times k$ invertible matrices, where m and k are dimensions of S_t and X_t, respectively. The observation equation can be expressed as

$$Y_t = CQ\left(Q^{-1}S_t - Q^{-1}RX_t\right) + (B + CR)\,X_t + \varepsilon_t.$$

The factors can be rotated as

$$\begin{pmatrix} \tilde{S}_t \\ \tilde{X}_t \end{pmatrix} = \begin{pmatrix} Q^{-1} & -Q^{-1}R \\ 0 & I \end{pmatrix} \begin{pmatrix} S_t \\ X_t \end{pmatrix},$$

and the rotated factors still follow a VAR process

$$\begin{pmatrix} \tilde{S}_t \\ \tilde{X}_t \end{pmatrix} = \Gamma \begin{pmatrix} A_{11} & A_{12} \\ A_{21} & A_{22} \end{pmatrix} \Gamma^{-1} \begin{pmatrix} \tilde{S}_{t-1} \\ \tilde{X}_{t-1} \end{pmatrix} + \Gamma \begin{pmatrix} u_{1t} \\ u_{2t} \end{pmatrix},$$

where $\Gamma = \begin{pmatrix} Q^{-1} & -Q^{-1}R \\ 0 & I \end{pmatrix}$. Consequently, $m\,(m+k)$ constraints are needed for exact identification of the FAVAR model. For example, the upper $m \times m$ submatrix in C could be restricted as an identity matrix and the $m \times k$ submatrix in B as a zero matrix.

The Gibbs sampler for the FAVAR is similar to that for the DFM. The latent factors can be sampled by the simulation smoother of the state space model, and the remaining parameters can be sampled by Bayesian VAR or linear regressions in conjugate forms, or by Metropolis-within-Gibbs samplers in the absence of conjugacy.

8.4.2 Probabilistic Principal Component Analysis

Tipping and Bishop (1999) provide insightful analysis on a specialized form of DFM and FAVAR, where the factors are assumed to be independent and follow standard normal distributions. Their key insight lies in deriving an analytical solution for the maximum likelihood estimator, achieved through singular value decomposition (SVD) of the sample covariance matrix of regression residuals.

The probabilistic principal component analysis (PPCA) addresses a specific configuration of the factor model:

$$Y_t = CS_t + BX_t + \sigma\varepsilon_t,$$

where the high-dimensional observations Y_t and the low-dimensional factors S_t are $d \times 1$ and $m \times 1$ vectors, respectively. The latent factors are assumed to be independent and identically distributed as standard normal variates. From the perspective of the DFM in the state space form, the transition matrix is trivially a zero matrix. The disturbances ε_t are also independent and identically distributed as standard normal variates. PPCA represents an isotropic error model, characterized by a scalar variance σ^2 for all series in Y_t.

PPCA accommodates observed factors X_t in the model specification. Assume that S_t and X_t are independent across all leads and lags, implying that PPCA is special case of the FAVAR. Consequently, X_t can be treated as exogenous regressors without specifying the data generating process. For instance, $X_t = 1$ serves as an intercept term to accommodate non-zero means of observations.

In summary, the key assumptions of PPCA involve two normal distributions:

$$S_t \sim N(0, I),$$

$$Y_t | S_t, X_t \sim N(CS_t + BX_t, \sigma^2 I).$$

Despite the stringent constraints on factors, the PPCA model remains susceptible to rotational indeterminacy, which poses an identification challenge. Let V be a $m \times m$ orthonormal matrix such that $VV' = V'V = I$. The model can be rotated as

$$Y_t = (CV')(VS_t) + BX_t + \sigma\varepsilon_t.$$

The rotated factors VS_t continue to follow standard normal distributions because $E(VS_tS_t'V') = I$. The rotated matrix CV' and the original loading C are observationally equivalent. Consequently, the maximum likelihood estimator of C is determined only up to an orthonormal rotational matrix.

PPCA sheds light on state and parameter estimation of state space models. First, while the Kalman filter provides a recursive formula for state estimation, PPCA offers an explicit posterior state distribution. Second, the Kalman filter computes the likelihood function in the prediction error decomposition form, typically necessitating numerical methods for maximum likelihood estimation. Remarkably, PPCA furnishes an analytical solution for the maximum likelihood estimator via SVD: $C = UDV'$, where V corresponds to the orthonormal matrix that causes rotational indeterminacy. A thorough examination of PPCA before delving into DFM or FAVAR enhances our comprehension of factor models.

Conditional on the parameters C, B, σ^2, the state estimation problem can be addressed through analyzing the joint distribution of S_t and Y_t, as a linear combination of S_t and ε_t:

$$\begin{pmatrix} S_t \\ Y_t \end{pmatrix} = \begin{pmatrix} 0 \\ BX_t \end{pmatrix} + \begin{pmatrix} I & 0 \\ C & \sigma I \end{pmatrix} \begin{pmatrix} S_t \\ \varepsilon_t \end{pmatrix}.$$

Given that S_t and ε_t follow the standard normal distributions, their linear combination maintains multivariate normal

$$\begin{pmatrix} S_t \\ Y_t \end{pmatrix} \sim N\left[\begin{pmatrix} 0 \\ BX_t \end{pmatrix}, \begin{pmatrix} I & C' \\ C & \Omega \end{pmatrix}\right],$$

$$\Omega = CC' + \sigma^2 I_d.$$

The posterior distribution of the state variable $p(S_t | Y_t, C, B, \sigma^2)$ remains normal with mean and variance given by

$$E(S_t | Y_t, C, B, \sigma^2) = C'\Omega^{-1}(Y_t - BX_t),$$

$$Var(S_t | Y_t, C, B, \sigma^2) = I_m - C'\Omega^{-1}C,$$

mirroring the state estimation results obtained via the Kalman filter. The Kalman filter remains appealing for computational efficiency, as it allows for univariate treatment of multivariate series, thereby circumventing the inversion of a large $d \times d$ matrix Ω. Alternatively, the equation $Y_t = CS_t + BX_t + \sigma\varepsilon_t$ can be envisioned as a Bayesian linear regression for each period $t = 1, \ldots, T$. The response variable is $(Y_t - BX_t)/\sigma$, the predictors are C/σ, and the coefficients are S_t with the prior $N(0, I)$. There are d observations for the period-t regression. The posterior mean and variance can be derived by inverting a $m \times m$ matrix instead:

$$E(S_t | Y_t, C, B, \sigma^2) = (C'C + \sigma^2 I_m)^{-1} C'(Y_t - BX_t),$$

$$Var\left(S_t \left| Y_t, C, B, \sigma^2\right.\right) = \sigma^2 \left(C'C + \sigma^2 I_m\right)^{-1}.$$

The Kalman filter also provides the likelihood function by integrating out the states. In the context of PPCA, the marginal distribution of Y_t is $N(BX_t, \Omega)$, $t = 1, \ldots, T$. Consequently, the log-likelihood takes the form

$$\ln L\left(C, B, \sigma^2\right) = -\frac{Td}{2}\ln 2\pi - \frac{T}{2}\ln|\Omega| - \frac{1}{2}\sum_{t=1}^{T}(Y_t - BX_t)'\,\Omega^{-1}(Y_t - BX_t).$$

The likelihood function corresponds to the multivariate normal density. If the covariance matrix were unrestricted, the maximum likelihood estimators for B and Ω would align with the OLS estimator and the sample covariance matrix of regression residuals:

$$\hat{B}' = \left(\sum_{t=1}^{T} X_t X_t'\right)^{-1} \sum_{t=1}^{T} X_t Y_t',$$

$$\hat{\Omega} = \frac{1}{T}\sum_{t=1}^{T}\left(Y_t - \hat{B}X_t\right)\left(Y_t - \hat{B}X_t\right)'.$$

However, the covariance matrix in PPCA exhibits a distinctive structure: $\Omega = CC' + \sigma^2 I_d$, where C is a $d \times m$ matrix and σ^2 is a scalar, implying that $dm+1$ parameters dictate a $d \times d$ covariance matrix. It is intuitive that the maximum likelihood estimators of C and σ^2 should be derived from the sample covariance matrix $\hat{\Omega}$. Proposition 8.1 substantiates this intuition, asserting that the maximum likelihood estimator of C is constructed from m largest eigenvalues and eigenvectors of $\hat{\Omega}$, while that of σ^2 accommodates the remaining $d-m$ smallest eigenvalues, reflecting the variance loss due to dimension reduction. This closed-form solution is a remarkable application of SVD that reveals the internal structure of the factor loading matrix. Proposition 8.1 also elucidates the source of rotational indeterminacy, as $CC' = CV'VC'$ for an arbitrary orthonormal matrix V.

Proposition 8.1 (PPCA) *For the factor model $Y_t = CS_t + BX_t + \sigma\varepsilon_t$, where S_t and ε_t are independent and identically distributed as standard normal variates, the maximum likelihood estimator of $\hat{C}, \hat{B}, \hat{\sigma}^2$ can be obtained by the following algorithm:*

1 Regress Y_t on X_t for OLS estimator and residual covariance matrix:

$$\hat{B}' = \left(\sum_{t=1}^{T} X_t X_t'\right)^{-1} \sum_{t=1}^{T} X_t Y_t',$$

$$\hat{\Omega} = \frac{1}{T}\sum_{t=1}^{T}\left(Y_t - \hat{B}X_t\right)\left(Y_t - \hat{B}X_t\right)'.$$

2 Compute eigenvalues $\lambda_1 \geq \ldots \geq \lambda_d$ and eigenvectors $u_1, \ldots, u_d$ of $\hat{\Omega}$.

3 Set $\hat{\sigma}^2 = \frac{1}{d-m}\sum_{j=m+1}^{d}\lambda_j$.

4 Construct economy-size SVD: $\hat{C} = UDV'$, where

4.1 left singular vectors are $U = (u_1, \ldots, u_m)$,

4.2 singular values are $D = diag\left(\sqrt{\lambda_1 - \hat{\sigma}^2}, \ldots, \sqrt{\lambda_m - \hat{\sigma}^2}\right)$,

4.3 right singular vectors V are arbitrary $m \times m$ orthonormal matrix.

Traditional principal component analysis can be viewed as a limiting case of PPCA as $\sigma^2 \to 0$. By treating σ^2 as a user-specified hyperparameter approaching zero, without estimating σ^2 in step 3 of Proposition 8.1, the maximum likelihood estimator of the factor loading matrix remains the largest eigenvalues and eigenvectors: $\hat{C} = UDV'$, where

$U = (u_1, \ldots, u_m)$, $D = diag\left(\sqrt{\lambda_1}, \ldots, \sqrt{\lambda_m}\right)$. Taking the estimated factors as the posterior mean of the multivariate normal distribution:

$$\left(\hat{C}'\hat{C}\right)^{-1} \hat{C}'\left(Y_t - \hat{\mu}\right) = VD^{-1}U'\left(Y_t - \hat{\mu}\right),$$

where $\hat{\mu} = \hat{B}X_t$, the estimated factor model is

$$\hat{Y}_t = (UDV')\left[VD^{-1}U'\left(Y_t - \hat{\mu}\right)\right] + \hat{\mu} = UU'\left(Y_t - \hat{\mu}\right) + \hat{\mu}.$$

It is instructive to compare PPCA with the standard principal component algorithm:

1. Center the data as $Y_t - \hat{\mu}$.
2. Compute sample covariance matrix $\hat{\Omega} = \frac{1}{T}\sum_{t=1}^{T}\left(Y_t - \hat{\mu}\right)\left(Y_t - \hat{\mu}\right)'$.
3. Identify largest eigenvalues of $\hat{\Omega}$ as $\lambda_1 \geq \ldots \geq \lambda_m$, with eigenvectors $U = (u_1, \ldots, u_m)$.
4. Construct m factors (principal components) as $U'\left(Y_t - \hat{\mu}\right)$.
5. Set factor loading matrix as U.

In standard principal component analysis, the principal components are orthogonal with decreasing variances (i.e., the largest eigenvalues $\lambda_1, \ldots, \lambda_m$), because

$$\frac{1}{T}\sum_{t=1}^{T} U'\left(Y_t - \hat{\mu}\right)\left(Y_t - \hat{\mu}\right)' U = DD.$$

In contrast, the PPCA employs standardized factors: they are orthogonal, but the variances are uniformly one. The $m \times m$ invertible matrix VD^{-1} is used for standardizing the principal components, which reflects rotational indeterminacy inherent in the factor model.

8.4.3 An application of Yield Curve Forecast

A renowned example of applying DFMs to the macroeconomy and yield curve is presented in Diebold et al. (2006). Let $Y_t(\tau)$ denotes the interest rate of a bond with maturity τ at time t. The maturities of U.S. Treasury bills and bonds traded on the secondary market span from 1 month to 30 years. According to the Nelson and Siegel (1987) yield curve model, bonds of different maturities are interconnected by

$$Y_t(\tau) = S_{1t} + S_{2t}\left(\frac{1 - e^{-\lambda\tau}}{\lambda\tau}\right) + S_{3t}\left(\frac{1 - e^{-\lambda\tau}}{\lambda\tau} - e^{-\lambda\tau}\right),$$

where S_{1t}, S_{2t} and S_{3t} are interpreted as the level, slope, and curvature factors. The demeaned factors are assumed to follow a VAR process:

$$\begin{pmatrix} S_{1t} \\ S_{2t} \\ S_{3t} \end{pmatrix} = A \begin{pmatrix} S_{1,t-1} \\ S_{2,t-1} \\ S_{3,t-1} \end{pmatrix} + u_t.$$

The transition matrix A and the covariance matrix of u_t are unrestricted and treated as unknown parameters, while the factor loading exhibits a tight structure specified by the Nelson and Siegel (1987) model. The factor loading matrix is determined by a single unknown parameter λ such that

$$C = \begin{pmatrix} 1 & \frac{1-e^{-\lambda\tau_1}}{\lambda\tau_1} & \frac{1-e^{-\lambda\tau_1}}{\lambda\tau_1} - e^{-\lambda\tau_1} \\ \cdots & \ddots & \cdots \\ 1 & \frac{1-e^{-\lambda\tau_d}}{\lambda\tau_d} & \frac{1-e^{-\lambda\tau_d}}{\lambda\tau_d} - e^{-\lambda\tau_d} \end{pmatrix},$$

where $\tau_1, \ldots, \tau_d$ represent the maturities of various bonds.

Diebold et al. (2006) extend the yields-only model to a more comprehensive yields-macro version by incorporating three macroeconomic variables into the state vector. This extended model follows a 6-dimensional VAR(1) transition, capturing both yields-to-macro and macro-to-yields relationships. Notably, the macro factors themselves are observed and included in the observation vector, resulting in an expanded factor loading matrix represented in a block diagonal form:

$$\begin{pmatrix} C & 0_{d\times 3} \\ 0_{3\times 3} & I_{3\times 3} \end{pmatrix}.$$

Diebold et al. (2006) estimate unknown parameters using maximum likelihood, computed via the Kalman filter in the prediction error decomposition form. Bayesian estimation can be approached in two ways. First, the MH sampler evaluates the posterior density, which is proportional to the prior multiplied by the likelihood obtained through the Kalman filter. Second, the Gibbs sampler with data augmentation cycles through states and unknown parameters. Posterior simulation of states is executed using the simulation smoother of the state space model, and posterior simulation of parameters in the transition equation is managed by the Bayesian VAR. Additionally, a Metropolis-within-Gibbs step can be employed for simulating the scalar parameter λ.

An alternative to likelihood-based inference is a two-step approach (Diebold and Li, 2006). Initially, conditional on λ, factors can be estimated through linear regressions with cross-sectional yields data. Subsequently, the estimated factors are utilized for VAR regressions, facilitating yield curve forecasting by predicting factor movements. This two-step approach is also applicable to other factor models. In the first step, principal components are extracted by averaging cross-sectional data to construct factors. In the second step, a time series model is fitted to estimate the temporal correlations of factors. Interested readers may refer to Stock and Watson (2002), Bai (2003) and Forni et al. (2000) for further insights.

8.4.4 An application of Nowcasting

In a data-rich economy, the DFM is a powerful tool for nowcasting, which involves real-time forecasting of key economic variables, such as Gross Domestic Product (GDP), whenever high-frequency data of relevant economic indicators are published. An empirical model may encompass hundreds of mixed-frequency macroeconomic variables. To illustrate, consider a simplified model where the unemployment rate, Consumer Price Index (CPI), and GDP are driven by common factors:

$$\begin{aligned} S_t &= AS_{t-1} + u_t, \\ \tilde{Y}_t &= CS_t + \varepsilon_t, \end{aligned}$$

where $\tilde{Y}_t = \left(\tilde{Y}_{1t}, \tilde{Y}_{2t}, \tilde{Y}_{3t}\right)'$ represents the latent monthly data. While the unemployment rate and CPI are available monthly, GDP data are released quarterly. These are representative data types addressed by mixed frequency VAR analysis (Eraker et al., 2015; Schorfheide and Song, 2015). The actual observations are linear functions of latent variables:

$$\begin{aligned} Y_{1t} &= \tilde{Y}_{1t}, t = 1, 2, \ldots \\ Y_{2t} &= \tilde{Y}_{2t}, t = 1, 2, \ldots \\ Y_{3t} &= \sum_{j=0}^{2} \tilde{Y}_{3,t-j}, t = 3, 6, \ldots \end{aligned}$$

By incorporating lagged variables into the state vector, the model can be expressed in the state space form (see Exercise 8.9). The Kalman filter is implemented with the univariate treatment of multivariate series technique, allowing the filter to process one data point at a time. On the first day of the month t, data is available up to month $t-1$, conditional on which the Kalman filter forecasts GDP. On the first Friday of the month, the Bureau of Labor Statistics releases unemployment data in the Employment Situation report. Subsequently, data Y_{1t} is added, and the Kalman filter updates the GDP forecast, which remains the last known estimate until mid-month when CPI data is released. The GDP forecast is updated again by running the Kalman filter with the new data Y_{2t}. If the current month coincides with the end of the quarter, Y_{3t} becomes available for another update. Otherwise, the Kalman filter awaits the first Friday of the next month for an update. In essence, nowcasting is performed by the Kalman filter whenever a new data point is released.

The model requires periodic maintenance every few months. The model parameters are re-estimated using a rolling window of observations from recent years, a process that can be executed overnight. Apart from maintenance days, the Kalman filter operates on a pre-fitted model, minimizing computational demands. This efficiency makes it feasible to generate real-time forecasts immediately after data release. For a comprehensive workflow of nowcasting, refer to Giannone et al. (2008).

8.5 Time-varying-parameter Regressions

The ever-evolving economic environments over time raise concerns about parameter stability in statistical models (Stock and Watson, 1996; Paye and Timmermann, 2006), particularly when analyzing time series data that span several decades. Structural breaks can result in biased estimates and unreliable forecasts if not properly addressed. Various methodologies have been developed to tackle the issue of parameter instability, each offering distinct advantages.

First, rolling-window regressions involve estimating models over a moving subsample, such as a ten-year window, which is incrementally shifted forward in time. For instance, Swanson (1998) examines the predictability of money stock for fluctuations in real output using rolling-window and increasing-window estimation techniques. As a simple and robust approach for detecting structural breaks in parameter estimates, rolling-window regressions provide insights into the evolving relationships when the time series is sufficiently long, allowing for comparison of estimation results among many windows. A practical challenge is that estimation results can be sensitive to the choice of window size. Inoue et al. (2017) develop a window selection method by minimizing the quadratic loss function in forecasting.

Second, regime switching models accommodate structural breaks by allowing coefficients to switch between regimes according to transition probabilities (i.e., Markov switching) or thresholds of specific variables (i.e., threshold switching). This method can flexibility capture abrupt changes in economic relationships, a feature particularly useful in modeling large shocks and policy shifts. The pioneering study of Hamilton (1989) suggests periodic shift between positive and negative growth rate of the real output using a Markov switching autoregressive model. Sims and Zha (2006) fit models that accommodate Markov switching in coefficients and variances to examine potential regimes in monetary policy.

Third, time-varying-parameter (TVP) models allow parameters to evolve continuously over time, often modeled by an autoregressive or random walk process. This approach is particularly useful for capturing gradual changes in economic relationships. While the random coefficient models have been used in statistics for decades (Swamy, 1970; Hsiao,

1975), the foundational works of Cogley and Sargent (2005) and Primiceri (2005) inspired a growing interest in TVP models in empirical macroeconomic studies such as Cogley and Sbordone (2008), Del Negro and Otrok (2007), Koop and Korobilis (2013), Belmonte et al. (2014), Geraci and Gnabo (2018), Chan (2023), among others.

8.5.1 Model Specification

For illustrative purposes, consider a univariate linear regression model with time-varying parameters:

$$Y_t = X_t\beta_t + \sigma\varepsilon_t,$$
$$\beta_t = \beta_{t-1} + \eta u_t,$$

where ε_t and u_t follow standard normal distributions.

The TVP regression is naturally expressed in the state space form, with the state variable β_t following a random walk process. Alternatively, a persistent and stationary $AR(1)$ process may be considered, typically with the autocorrelation coefficient approaching one, as the relationship between X_t and Y_t is expected to be relatively stable in a short time interval. In the observation equation, the regressor X_t serves as the loading of the state variable. It is time-varying, but the loading is observed at each time point.

The initial state β_0 is a parameter of primary interest, as it represents the time-invariant coefficient in a regression without TVP (i.e., setting $\eta = 0$). More insights are gained by standardizing the state variable:

$$\tilde{\beta}_t = \frac{\beta_t - \beta_0}{\eta},$$

leading to a reparametrized the TVP regression:

$$Y_t = X_t\beta_0 + X_t\eta\tilde{\beta}_t + \sigma\varepsilon_t,$$
$$\tilde{\beta}_t = \tilde{\beta}_{t-1} + u_t,$$

where the standardized state variable $\tilde{\beta}_t$ is initialized at the origin by construction: $\tilde{\beta}_0 = 0$.

The reparametrized state space model features a regression component $X_t\beta_0$ in the observation equation. The maximum likelihood estimator of β_0 has an analytical solution conditional on σ, η. Consequently, β_0 can be concentrated out of the likelihood function, thereby simplifying numerical optimization. Refer to Section 6.5 for estimating regression effects in the state space model.

In the reparametrized model, the parameter η serves as the state loading coefficient, rather than a noise variance term, which is traditionally specified with an inverse gamma prior $\eta^2 \sim IG(a_0, b_0)$. The density function

$$p(\eta^2) \propto (\eta^2)^{-a_0-1} e^{-b_0\eta^{-2}}.$$

reveals a significant drawback the inverse gamma prior, which pushes the variance away from zero:

$$\lim_{\eta^2 \to 0} p(\eta^2) = 0.$$

However, the limiting case $\eta = 0$ is crucial for TVP analysis, as the model reduces to a time-invariant linear regression without TVP. The advantages of interpreting η as the state loading coefficient are twofold: avoiding the nonnegative constraint and allowing shrinkage toward time-invariant regressions via a symmetric prior distribution centered at zero. Fruhwirth-Schnatter and Wagner (2010) propose a normal prior for the loading coefficient: $\eta \sim N(0, \psi)$, implying that $\eta^2 \sim \Gamma\left(\frac{1}{2}, 2\psi\right)$, which is distinct from the traditional inverse gamma prior for the variance term.

8.5.2 Bayesian Inference

Posterior simulation of the TVP regression can be handled by the Gibbs sampler, which cycles through the full posterior conditional distributions. On the one hand, $p\left(\tilde{\beta}\left|Y,\beta_0,\eta,\sigma^2\right.\right)$ is managed within the state space framework, where the simulation smoother generates posterior samples for the sequence $\tilde{\beta}$ (omitting the time subscript for brevity). On the other hand, $p\left(\beta_0,\eta,\sigma^2\left|Y,\tilde{\beta}\right.\right)$ is handled within the framework of Bayesian linear regressions:

$$Y_t = \begin{pmatrix} X_t & X_t\tilde{\beta}_t \end{pmatrix}\begin{pmatrix} \beta_0 \\ \eta \end{pmatrix} + \sigma\varepsilon_t.$$

Since η serves as a coefficient in the linear regression, Bayesian shrinkage and variable selection techniques discussed in Chapter 3 are applicable, allowing the data to determine whether to use TVP or time-invariant regressions. This is particularly useful when a model contains many TVP variables, making overfitting a concern. Shrinkage toward time-invariant coefficients or selecting a subset of TVP variables is likely to improve out-of-sample forecasts.

Following Belmonte et al. (2014), we consider lasso shrinkage through a hierarchical prior: $\eta\left|\psi\right. \sim N(0,\psi)$ and $\psi \sim \exp\left(2\lambda^{-1}\right)$. The scale mixture prior can be handled by the Gibbs sampler, in which $p\left(\tilde{\beta}\left|Y,\beta_0,\eta,\sigma^2,\psi\right.\right)$ is managed by the state simulation smoother, and $p\left(\beta_0,\eta,\sigma^2\left|Y,\tilde{\beta},\psi\right.\right)$ is handled by Bayesian linear regressions. Additionally, the latent variables are sampled from the posterior conditional distribution

$$p\left(\psi^{-1}\left|Y,\tilde{\beta},\beta_0,\eta,\sigma^2\right.\right) \propto \psi^{-\frac{1}{2}}\exp\left(-\frac{\lambda^2}{2}\psi - \frac{\eta}{2}\psi^{-1}\right),$$

which is recognized as the inverse Gaussian distribution with parameters $\left|\frac{\lambda}{\eta}\right|$ and λ^2.

8.5.3 Multivariate Extension

In empirical studies, TVP is often deployed within the VAR framework, hence the name TVP-VAR in Primiceri (2005) and Cogley and Sargent (2005). The multivariate extension follows the same concept as the univariate TVP regression, but the notation becomes more complex. The TVP-VAR model takes the form:

$$Y_t = \sum_{j=1}^{p}\Phi_{jt}Y_{t-j} + \varepsilon_t,$$

$$\beta_t = \beta_{t-1} + \Pi u_t,$$

where $mp^2 \times 1$ vector $\beta_t = vec\left[(\Phi_1,\ldots,\Phi_p)'\right]$ represents the time-varying VAR coefficients. Due to high dimensionality, assume that each time-varying coefficient follows a univariate random walk process without interactions, implying a diagonal matrix

$$\Pi = diag\left(\eta_1,\ldots,\eta_{mp^2}\right).$$

As in the reduce-form VAR, contemporaneous effects are embodied in the covariance matrix $E\left(\varepsilon_t\varepsilon_t'\right) = \Sigma$.

Conditional on the presample data, the VAR can be represented in the linear regression form:

$$Y_t = X_t\beta_t + \varepsilon_t,$$

where $X_t = I_m \otimes \left(Y'_{t-1}, \ldots, Y'_{t-p}\right)$.

Akin to the univariate case, the initial state β_0 can be separated out of the drifting terms by defining the standardized states

$$\tilde{\beta}_t = \Pi^{-1}\left(\beta_t - \beta_0\right),$$

which reparametrizes the model as

$$Y_t = X_t\beta_0 + X_t\Pi\tilde{\beta}_t + \varepsilon_t,$$

$$\tilde{\beta}_t = \tilde{\beta}_{t-1} + u_t.$$

Bayesian estimation of the TVP-VAR is typically implemented via the Gibbs sampler, which cycles through the full posterior conditional distributions. On the one hand, states are sampled from $p\left(\tilde{\beta}\,|Y, \beta_0, \Pi, \Sigma\right)$ via the simulation smoother of the state space model. On the other hand, the parameters are sampled from $p\left(\beta_0, \Pi, \Sigma\,\middle|Y, \tilde{\beta}\right)$ using Bayesian linear regressions

$$Y_t = \left[\begin{array}{cc} X_t & X_t \cdot diag\left(\tilde{\beta}_t\right) \end{array}\right]\left[\begin{array}{c} \beta_0 \\ diag\left(\Pi\right) \end{array}\right] + \varepsilon_t,$$

which leverages the identity $\Pi\tilde{\beta}_t = diag\left(\tilde{\beta}_t\right) \cdot diag\left(\Pi\right)$, in which $diag\left(\tilde{\beta}_t\right)$ refers to expanding the vector $\tilde{\beta}_t$ as a diagonal matrix, while $diag\left(\Pi\right)$ refers to extracting diagonal elements from the diagonal matrix Π. See Exercise 8.10 for the functional forms of the posterior conditional distributions, under the semi-conjugate normal and inverse Wishart priors.

8.5.4 Approximated Filtering for Large TVP-VAR

For large TVP-VAR models, drawing posterior samples using the Gibbs sampler can be computationally intensive, as it requires alternating between the state simulation smoother and Bayesian linear regressions thousands of times for both state and parameter updates. Koop and Korobilis (2013) propose a one-pass approximated filter for state estimation, eliminating dependency on model parameters.

Consider the state space form of the TVP-VAR model

$$Y_t = X_t\beta_t + \varepsilon_t,$$

$$\beta_t = \beta_{t-1} + \Pi u_t,$$

where the initial state is assigned a diffuse distribution: $\beta_0 \sim N\left(0, \kappa I\right)$. Note that β_0 is handled by state estimation instead of parameter estimation. The Kalman filter proceeds with a set of recursive equations:

$$\begin{aligned}
\beta_{t|t-1} &= \beta_{t-1|t-1}, \\
Y_{t|t-1} &= X_t\beta_{t|t-1}, \\
P_{t|t-1} &= P_{t-1|t-1} + \Pi\Pi', \\
O_{t|t-1} &= X_tP_{t|t-1}X_t' + \Sigma, \\
L_{t|t-1} &= P_{t|t-1}X_t', \\
\beta_{t|t} &= \beta_{t|t-1} + L_{t|t-1}O_{t|t-1}^{-1}\left(Y_t - Y_{t|t-1}\right), \\
P_{t|t} &= P_{t|t-1} - L_{t|t-1}O_{t|t-1}^{-1}L_{t|t-1}'.
\end{aligned}$$

For large TVP-VAR models, it is too computationally expensive to run the Kalman filter thousands of times, conditional on the posterior draws of Π and Σ obtained by Bayesian linear regressions. It is desirable to replace the unknown parameters Π and Σ by some low-cost alternatives. Koop and Korobilis (2013) propose a one-pass filter by approximating the predictive variance as

$$P_{t|t-1} = \frac{1}{\alpha} P_{t-1|t-1},$$

where $\alpha \in (0,1)$ serves as a forgetting factor. For quarterly macroeconomic data, $\alpha = 0.99$ may be a reasonable choice. Additionally, the unknown parameter Σ is approximated by a weighted average

$$\hat{\Sigma}_t = w\hat{\Sigma}_{t-1} + (1-w)\,\hat{\varepsilon}_t\hat{\varepsilon}_t',$$

where $\hat{\varepsilon}_t = Y_t - X_t\beta_{t|t-1}$ is the prediction error and w is the weight. Since the unknown parameters are eliminated, a single pass of the approximated filter suffices for the state estimation.

8.6 Panel Data Analysis

Panel data, characterized by observations on multiple cross-sectional units over several time periods, offers a data-rich environment for econometric analysis. A typical panel dataset involves two dimensions: n cross-sectional units (e.g., individuals, firms, countries) and T periods. The response variable Y_{it} and predictors X_{it} are indexed for the ith unit in period t, where $i = 1, \ldots, n$ and $t = 1, \ldots, T$.

In a balanced panel, every unit has the same number of observations. However, real-world data often exhibit unbalanced panels, where some units are observed on fewer occasions, resulting in missing values (e.g., NaN entries in Y_{it} and/or X_{it}). Such datasets can be effectively managed within the state space framework using the Kalman filter, which handles missing observations by bypassing the update step during forward filtering.

While the accumulation of observations over both individuals and periods seems appealing, the true strength of panel data regression is not only a larger sample size in a pooled regression, but also lies in addressing unobserved heterogeneity in the regression:

$$Y_{it} = \alpha_i + X_{it}\beta + \varepsilon_{it},$$

where the latent variable α_i represents the unobserved heterogeneity specific to the ith unit. It captures time-invariant omitting variables for the regression. If the random variable α_i is uncorrelated with X_{it}, it is termed random effects. If the unobserved confounding α_i can be correlated with X_{it} in an arbitrary manner, it is termed fixed effects. It is crucial to note that the term fixed effects might refer to the coefficients β that are fixed for all i and t in the generalized mixed effects literature (particularly in medical studies). However, in econometric parlance the fixed effects refer to the whole vector $(\alpha_1, \ldots, \alpha_n)$ as unknown parameters estimated via dummy variable regressions. Tolerance on omitting variables is a remarkable feature of the panel data analysis. Endogenous predictors can hinder causal inference in cross-sectional regressions, but panel data regressions can address endogeneity if the unobserved confounding remains time-invariant.

Panel data analysis can be a standalone subject warranting an entire book (e.g., Baltagi (2013), Hsiao (2014), Arellano (2003), among others), yet we explore panel data regressions within the state space framework for two primary reasons. First, the Kalman filter routinely provides state estimation on the unobserved effects and the likelihood function for parameter estimation on the unknown parameters. Unbalanced panel data do not need special treatment, as the Kalman filter automatically handles missing data appropriately. Second, random and fixed effects can be managed within the unified framework. Fixed effects have a diffuse prior, while random effects have a proper prior, with a hierarchical structure in Bayesian models. In the state space model, the unobserved effects can be easily extended to random coefficients and time-varying parameters that follow random walk or autoregressive processes.

8.6.1 Random Effects

Consider the panel data regression with random effects

$$Y_{it} = \alpha_i + X_{it}\beta + \varepsilon_{it},$$

where the latent variable α_i follows $N\left(0, \sigma_\alpha^2\right)$ and is independent of X_{it}, implying that observing X_{it} does not change the distribution of the unobserved effects. Provided that an intercept term is included in the regressors, the mean of α_i can be normalized to zero. Similar to standard linear regressions, the idiosyncratic error ε_{it} follows $N\left(0, \sigma_\varepsilon^2\right)$ and is independent of α_i and X_{it}.

It is possible to assimilate the random effects into error terms, resulting in an error component model:

$$Y_{it} = X_{it}\beta + \tilde{\varepsilon}_{it},$$

where the composite errors $\tilde{\varepsilon}_{it} = \varepsilon_{it} + \alpha_i$ are correlated. The correlation pattern can be discerned by stacking T observations for each unit i:

$$Y_i = X_i\beta + \tilde{\varepsilon}_i.$$

The composite errors $\tilde{\varepsilon}_i = (\tilde{\varepsilon}_{i1}, \ldots, \tilde{\varepsilon}_{iT})$ have a non-diagonal covariance matrix:

$$E\left(\tilde{\varepsilon}_i\tilde{\varepsilon}_i'\right) = \sigma_\alpha^2\iota\iota' + \sigma_\varepsilon^2 I_T \equiv \Omega,$$

where ι represents a vector of ones.

In classical econometrics, regressions with correlated errors are typically addressed via generalized least squares (GLS), which involves three steps: 1) decomposing the inverse covariance matrix as $\Omega^{-1} = P'P$, 2) transforming the data as $\tilde{Y}_i = PY_i$ and $\tilde{X}_i = PX_i$, and 3) applying ordinary linear regressions to the transformed data

$$\hat{\beta} = \left(\sum_{i=1}^{n}\tilde{X}_i'\tilde{X}_i\right)^{-1}\sum_{i=1}^{n}\tilde{X}_i'\tilde{Y}_i = \left(\sum_{i=1}^{n}X_i'\Omega^{-1}X_i\right)^{-1}\sum_{i=1}^{n}X_i'\Omega^{-1}Y_i.$$

As an alternative to the Cholesky or eigenvalue decomposition, the specific structure in the covariance matrix allows for a closed-form matrix square root (as a symmetric matrix):

$$P = \sigma_\varepsilon^{-1}\left(I_T - \frac{\delta}{T}\iota\iota'\right),$$

where δ serves as a discount factor on the group mean:

$$\delta = 1 - \sqrt{\frac{\sigma_\varepsilon^2}{\sigma_\varepsilon^2 + T\sigma_\alpha^2}}.$$

Consequently, the coefficients β in the random effects model are estimated by GLS, which is implemented by standard regressions with quasi-within-transformed data:

$$\tilde{X}_{it} = X_{it} - \delta \frac{1}{T} \sum_{j=1}^{T} X_{ij},$$

$$\tilde{Y}_{it} = Y_{it} - \delta \frac{1}{T} \sum_{j=1}^{T} Y_{ij}.$$

However, two unresolved issues remain. First, the variances $\sigma_\alpha^2, \sigma_\varepsilon^2$ are crucial in GLS, yet they are unknown parameters that need to be estimated. Second, the random effects α_i have been assimilated into the composite error terms, but estimating the random effects remains of interest. Although classical solutions are available (e.g., feasible GLS), the state space model offers a simple and effective solution: all unknown variables (including unobserved effects α_i, regression coefficients β, and variance terms $\sigma_\alpha^2, \sigma_\varepsilon^2$) are jointly estimated by full-information maximum likelihood.

In contrast with batch processing methods such as GLS where all data points are utilized together, the Kalman filter features stream processing by scanning one data point at a time. Panel data have two dimensions: $i = 1, \ldots, n$ and $t = 1, \ldots, T$. By converting the subscript (i, t) to the linear index $\tau = (i-1)T + t$, two-dimensional data $Y_{11}, \ldots, Y_{1T}, \ldots, Y_{n1}, \ldots, Y_{nT}$ can be re-indexed as $Y_1, \ldots, Y_T, \ldots, Y_{(n-1)T+1}, \ldots, Y_{nT}$. For an unbalanced panel dataset, some elements in Y_τ can be labeled as missing values, and the Kalman filter routinely addresses missing data as appropriate. In the state space model, the unobserved effect α_i serves as the state variable. Under the linear index $\tau = 1, \ldots, nT$, the coefficients in the state equation $S_\tau = A_\tau S_{\tau-1} + B_\tau u_\tau$ and the measurement equation $Y_\tau = X_\tau \beta + C_\tau S_\tau + D_\tau \varepsilon_\tau$ are specified as follows:

$A_\tau = 0$, $B_\tau = \sigma_\alpha$, for $\tau = 1, T+1, \ldots, (n-1)T+1$,

$A_\tau = 1$, $B_\tau = 0$, for $\tau = 2, \ldots, T, \ldots, (n-1)T+2, \ldots, nT$,

$C_\tau = 1$, $D_\tau = \sigma_\varepsilon$, for all $t = 1, \ldots, nT$.

The state transition can be described as a periodic renew-and-hold process: the state variable is renewed every T periods by innovations from $N\left(0, \sigma_\alpha^2\right)$, and the state holds constant in the subsequent $T-1$ periods until the next renewal.

Joint estimation of unobserved effects, regression coefficients, and variance terms is achieved by likelihood concentration, which involves three steps.

First, conditional on $\beta, \sigma_\alpha^2, \sigma_\varepsilon^2$, the Kalman filter offers state estimation for the unobserved effects α_i.

Second, conditional on $\sigma_\alpha^2, \sigma_\varepsilon^2$, the profile maximum likelihood estimator of β has a closed-form solution, akin to the regression effects in the state space model (refer to Section 6.5). For the panel data regression, the estimator aligns with the GLS implemented by quasi-within-transformation:

$$\hat{\beta} = \left(\sum_{i=1}^{n} \sum_{t=1}^{T} \tilde{X}_{it}' \tilde{X}_{it} \right)^{-1} \sum_{i=1}^{n} \sum_{t=1}^{T} \tilde{X}_{it}' \tilde{Y}_{it},$$

$$\tilde{X}_{it} = X_{it} - \delta \frac{1}{T} \sum_{j=1}^{T} X_{ij},$$

$$\tilde{Y}_{it} = Y_{it} - \delta \frac{1}{T} \sum_{j=1}^{T} Y_{ij}.$$

Third, as α_i and β have been integrated out of the likelihood function, only two variance parameters $\sigma_\alpha^2, \sigma_\varepsilon^2$ remain for numerical optimization. The objective function is the (marginalized) likelihood, readily available as a by-product of the Kalman filter in the prediction error decomposition form.

8.6.2 Fixed Effects

The fixed effects model allows for arbitrary dependence between the unobserved effects α_i and the observed predictors X_{it}. If the unobserved effects are incorporated into the disturbance term, as in the error component model $Y_{it} = X_{it}\beta + \tilde{\varepsilon}_{it}$, where $\tilde{\varepsilon}_{it} = \varepsilon_{it} + \alpha_i$, an endogeneity problem arises due to the correlation between the regressors and the composite errors. In the program evaluation literature, heterogeneity may manifest as unobserved personal ability, which is often correlated with observable personal characteristics. This unobserved confounding poses a significant challenge for causal inference.

In classical econometrics, fixed effects refer to the outcome vector $\alpha = (\alpha_1, \ldots, \alpha_n)'$ as unknown parameters, which are coefficients of the dummy variable matrix D in the vector/matrix form of the panel data regression:

$$Y = X\beta + D\alpha + \varepsilon,$$

where the regression data matrices X and Y are constructed by stacking nT observations. The fixed effects estimator is defined by least squares dummy variables (LSDV) approach, which involves the joint estimation of β and α by the OLS:

$$\begin{pmatrix} \hat{\beta} \\ \hat{\alpha} \end{pmatrix} = \begin{pmatrix} X'X & X'D \\ D'X & D'D \end{pmatrix}^{-1} \begin{pmatrix} X' \\ D' \end{pmatrix} Y.$$

In practice, the LSDV method may become infeasible when n is large. Proposition 8.2 offers a more efficient method for computing the same estimator via two-step regressions. The first step involves projecting out α by regressing (Y, X) on D, resulting in residuals $\left(\tilde{Y}, \tilde{X}\right)$. The second step involves recovering $\hat{\beta}$ by regressing $\tilde{Y}$ on $\tilde{X}$. The residuals from the second step are identical to those of the original system. Proposition 8.2 also indicates that the transformed system suffices for statistical inference of β, as the covariance matrix of the LSDV estimator can be computed using the transformed data $\tilde{X}$ without retrieving the original data X and D. The only adjustment involves the loss of n degrees of freedom in the regression.

Proposition 8.2 (Frisch-Waugh-Lovell) *Consider the classical linear regression $Y = X\beta + D\alpha + \varepsilon$, where (X, D) is a $nT \times (k+n)$ full-rank matrix. Let $\hat{\beta}$ and $\hat{\alpha}$ be the OLS estimator.*

1. $\hat{\beta}$ can be obtained by OLS on the transformed data:

$$\hat{\beta} = \left(\tilde{X}'\tilde{X}\right)^{-1} \tilde{X}'\tilde{Y},$$

where $\tilde{X} = PX$ and $\tilde{Y} = PY$, and the idempotent matrix P is defined as

$$P = I - D\left(D'D\right)^{-1} D'.$$

2. $\hat{\alpha}$ is the solution to linear equations:

$$D\hat{\alpha} = \left(Y - X\hat{\beta}\right) - \left(\tilde{Y} - \tilde{X}\hat{\beta}\right).$$

3. Residuals of the projected system are the same as residuals of the original system:

$$\tilde{Y} - \tilde{X}\hat{\beta} = Y - X\hat{\beta} - D\hat{\alpha}.$$

4. Transformed data suffices for computing the covariance matrix of $\hat{\beta}$:

$$Var\left(\hat{\beta}\right) = \hat{\sigma}_\varepsilon^2 \left(\tilde{X}'\tilde{X}\right)^{-1},$$

$$\hat{\sigma}_\varepsilon^2 = \frac{1}{nT - n - k}\left(\tilde{Y} - \tilde{X}\hat{\beta}\right)'\left(\tilde{Y} - \tilde{X}\hat{\beta}\right).$$

The two-step regression strategy is applicable to any linear regressions with arbitrarily partitioned regressors, as Proposition 8.2 does not require a specific form of the matrix D. However, in the fixed effects model, the matrix D consists of dummy variables. Regressing (Y, X) on the dummy matrix D is numerically equivalent to within-transformation. In practice, neither the dummy variable matrix D nor the projection matrix P is explicitly created. The mathematical expressions $\tilde{X} = PX$ and $\tilde{Y} = PY$ can be implemented by subtracting the group mean from the original data:

$$\tilde{X}_{it} = X_{it} - \frac{1}{T}\sum_{j=1}^{T} X_{ij},$$

$$\tilde{Y}_{it} = Y_{it} - \frac{1}{T}\sum_{j=1}^{T} Y_{ij}.$$

The intuition is that within-transformation effectively removes heterogeneity through differencing, allowing for pooled regression on homogeneous data.

Recall that random effects are addressed by quasi-within-transformation under a discount factor

$$\delta = 1 - \sqrt{\frac{\sigma_\varepsilon^2}{\sigma_\varepsilon^2 + T\sigma_\alpha^2}}.$$

Consider four limiting cases:

1. If $\sigma_\alpha^2 \to \infty$, then $\delta \to 1$, which provides the fixed effects estimator.
2. If $\sigma_\varepsilon^2 \to 0$, then $\delta \to 1$, which provides the fixed effects estimator.
3. If $\sigma_\alpha^2 \to 0$, then $\delta \to 0$, which provides the pooled OLS estimator.
4. If $\sigma_\varepsilon^2 \to \infty$, then $\delta \to 0$, which provides the pooled OLS estimator.

The first case indicates that fixed effects can be viewed as the limiting case of random effects under a diffuse prior such that $\sigma_\alpha^2 \to \infty$. Consequently, the state space form for random effects is also applicable to fixed effects. This is not surprising, as the state space theory suggests that a time-invariant state initialized by a diffuse prior is akin to a fixed parameter.

In summary, the principal distinction between fixed and random effects is whether the unobserved heterogeneity can be correlated with the regressors. If correlation exists, we have fixed effects, where $\alpha_1, \ldots, \alpha_n$ can be treated as unknown parameters or diffuse states with $\sigma_\alpha^2 \to \infty$. In practice, statistical software designed for random effects can be repurposed for fixed effects by specifying a large value for σ_α^2 as a proxy for infinite variance.

8.6.3 Two-way and Multi-way Fixed Effects

In labor economics and other fields, researchers often aim to control for fixed effects across multiple dimensions: individuals, firms, regions, years, etc. As an extension of one-way fixed effects with unit-specific latent variables α_i, $i = 1, \ldots, n$, two-way fixed effects incorporate time-specific latent variables δ_t, $t = 1, \ldots, T$ in the regression:

$$Y_{it} = \alpha_i + \delta_t + X_{it}\beta + \varepsilon_{it}.$$

Similarly, a three-way fixed effects model can be specified as

$$Y_{ijt} = \alpha_i + \gamma_j + \delta_t + X_{ijt}\beta + \varepsilon_{ijt}.$$

Unbalanced panel data can be represented by missing values in Y_{ijt} and/or X_{ijt}.

In classical econometrics, fixed effects refer to the outcome vectors $\alpha = (\alpha_1, \ldots, \alpha_n)'$, $\gamma = (\gamma_1, \ldots, \gamma_J)'$, $\delta = (\delta_1, \ldots, \delta_T)'$ as unknown parameters associated with dummy variable matrices D_α, D_γ, D_δ in the matrix-form regression:

$$Y = D_\alpha \alpha + D_\gamma \gamma + D_\delta \delta + X\beta + \varepsilon,$$

or in a compact form:

$$Y = \chi\theta + \varepsilon,$$

where $\chi = (D_\alpha, D_\gamma, D_\delta, X)$ and $\theta = (\alpha', \gamma', \delta', \beta')'$. The multi-way fixed effects estimator is defined by the OLS with dummy variables (i.e., the LSDV estimator):

$$\hat{\theta} = (\chi'\chi)^{-1} \chi' Y.$$

The fixed effects estimator $\hat{\theta}$ may not be directly computable for large and high-dimensional datasets, as the matrix χ involves multitudinous dummy variables and may not fit into computer memory, complicating matrix multiplication and inversion.

For balanced panel data in a two-way fixed effects model, two-way within-transformation may be applied:

$$\tilde{X}_{it} = X_{it} - \frac{1}{T}\sum_{t=1}^{T} X_{it} - \frac{1}{n}\sum_{i=1}^{n} X_{it} + \frac{1}{nT}\sum_{i=1}^{n}\sum_{t=1}^{T} X_{it},$$

$$\tilde{Y}_{it} = Y_{it} - \frac{1}{T}\sum_{t=1}^{T} Y_{it} - \frac{1}{n}\sum_{i=1}^{n} Y_{it} + \frac{1}{nT}\sum_{i=1}^{n}\sum_{t=1}^{T} Y_{it}.$$

The two-way within-transformation can also be framed as the OLS regression of Y_{it} on X_{it}, $\frac{1}{T}\sum_{t=1}^{T} X_{it}$ and $\frac{1}{n}\sum_{i=1}^{n} X_{it}$.

For unbalanced panel data in a multi-way fixed effects model, Proposition 8.2 may be applied to eliminate one of the fixed effects. For instance, α can be projected out by regressing $(Y, D_\gamma, D_\delta, X)$ on D_α, which is numerically equivalent to within-transformation by groups $i = 1, \ldots, n$, without explicitly creating the dummy variable matrix D_α. However, the transformed D_γ, D_δ are no longer dummy variable matrices. To further project out γ and δ by Proposition 8.2, large and dense matrix operations pose a computational challenge.

To estimate multi-way fixed effects with unbalanced panel data, an elegant approach is alternating projections (Gaure, 2013). Define the projection matrices

$$P_k = I - D_k \left(D_k' D_k\right)^{-1} D_k', k \in \{\alpha, \gamma, \delta\}.$$

The idempotent matrices P_α, P_γ, P_δ serve as within-transformation by groups i, j, t, respectively. Although an iteration of alternating projections takes the mathematical form: $P_\delta P_\gamma P_\alpha (Y, X)$, it does not require creating large-size projection matrices for multiplication. Instead, it is implemented by within-transformation to the data (Y, X) by groups $i = 1, \ldots, n$, followed by groups $j = 1, \ldots, J$, and then groups $t = 1, \ldots, T$. Alternating projections cycle through all groups in multiple iterations until the transformed data converge, a process justified by projection theory where convergence holds in the strong operator topology (Gaure, 2013). The alternating projections performs effectively for large datasets, rendering the LSDV estimator feasible for in-memory computation.

8.6.4 Gibbs Samper for Random and Fixed Effects

In Bayesian literature, random effects are analyzed within the hierarchical (multilevel) longitudinal model, in which the random effects have hierarchical priors. The Gibbs sampler iteratively cycles through the full posterior conditional distributions of the unobserved effects and regression coefficients. A notable advantage of the Gibbs sampler is its adaptability to multi-way unobserved effects and limited dependent variable models.

We consider a two-way random effects model:

$$Y_{it} = \alpha_i + \delta_t + X_{it}\beta + \varepsilon_{it}.$$

This model exhibits a hierarchical structure: α_i follows $N\left(0, \sigma_\alpha^2\right)$, and σ_α^2 is assigned an inverse gamma prior for conjugate analysis. A distinctive feature of the hierarchical model is that the parameter σ_α^2 has no impact on the likelihood function conditional on α_i. Consequently, the posterior conditional distribution of σ_α^2 depends on α_i, but not on Y_{it}. Similarly, a hierarchical prior is specified for the time-specific effects: δ_t follows $N\left(0, \sigma_\delta^2\right)$ with an inverse gamma prior for σ_δ^2.

The Gibbs sampler, as proposed by Gelfand and Smith (1990), cycles through full posterior conditional distributions and can be implemented using a series of Bayesian linear regressions.

1. Updating β and σ_ε^2: Conditional on random effects, the model simplifies to a pooled regression, with $Y_{it} - \alpha_i - \delta_t$ as the response variable and X_{it} as predictors. Under the normal and inverse gamma priors for β and σ_ε^2, the posterior conditional distribution retains the normal and inverse gamma form, as in semi-conjugate Bayesian linear regressions.
2. Updating $\alpha_i, i = 1, \ldots, n$: This involves n separate linear regressions with a known disturbance variance (i.e., σ_ε^2). For each ith regression, the response variable is $Y_{it} - \delta_t - X_{it}\beta$, and the sole predictor is an intercept. Under a normal prior $N\left(0, \sigma_\alpha^2\right)$, T observations are available for updating the normally distributed α_i.
3. Updating δ_t, $t = 1, \ldots, T$: This involves T separate linear regressions with a known disturbance variance (i.e., σ_ε^2). For each tth regression, the response variable is $Y_{it} - \alpha_i - X_{it}\beta$, and the sole predictor is an intercept. Under a normal prior $N\left(0, \sigma_\delta^2\right)$, n observations are available for updating the normally distributed δ_t.
4. Updating σ_α^2: This corresponds to a trivial linear regression without regressors. The data are α_i, $i = 1, \ldots, n$. Under an inverse gamma prior for σ_α^2, the posterior conditional distribution remains inverse gamma.
5. Updating σ_δ^2: This corresponds to a trivial linear regression without regressors. The data are δ_t, $t = 1, \ldots, T$. Under an inverse gamma prior for σ_δ^2, the posterior conditional distribution remains inverse gamma.

The fundamental assumption of random effects is that α_i and δ_t are uncorrelated with X_{it}, implying that observing X_{it} does not alter the distribution of the unobserved effects. Fixed effects relax this assumption by allowing for arbitrary correlations between unobserved effects and observed regressors. In classical econometrics, fixed effects refer to the whole vector $\alpha = (\alpha_1, \ldots, \alpha_n)'$ and $\delta = (\delta_1, \ldots, \delta_T)'$ as unknown parameters (coefficients) associated with the dummy variable matrices D_α and D_δ in the matrix form of the panel data regression:

$$Y = D_\alpha \alpha + D_\delta \delta + X\beta + \varepsilon,$$

or in a compact form:

$$Y = \chi\theta + \varepsilon,$$

where $\chi = (D_\alpha, D_\delta, X)$ and $\theta = (\alpha', \delta', \beta')'$. The classical fixed effects estimator $\hat{\theta}$ is defined by the OLS with dummy variables (i.e., the LSDV estimator):

$$\hat{\theta} = (\chi'\chi)^{-1} \chi' Y.$$

A computational challenge arises as $\hat{\theta}$ may not be directly computable using the OLS formula, due to the huge dummy variable matrix χ. For large and high-dimensional fixed effects models, memory constraints may pose significant issues.

The Gibbs sampler offers a Bayesian solution to this computational problem. The core idea is to specify $\hat{\theta}$ as the mean of the target distribution, and then construct a Markov chain that has the target distribution as its stationary distribution. Provided that the Gibbs sampler avoids large matrices in memory, the MCMC samples can approximate the fixed effects estimator.

The Gibbs sampler for the fixed effects model is a variant of the Gibbs sampler for the random effects model. The steps for updating $\beta, \sigma_\varepsilon^2, \alpha, \delta$ remain unchanged, except for the diffuse specification of σ_α^2 and σ_δ^2, which are not estimated from the data.

Under a normal prior $N(0, V_\beta)$ for β and an inverse gamma prior $IG(a, b)$ for σ_ε^2, the target distribution for the Gibbs sampler takes the form

$$p\left(\theta, \sigma_\varepsilon^2 \left| Y\right.\right) \propto \left(\sigma_\varepsilon^2\right)^{-\left(a + \frac{nT}{2} + 1\right)} e^{-\sigma_\varepsilon^{-2} b - \frac{1}{2}\sigma_\varepsilon^{-2}(Y - \chi\theta)'(Y - \chi\theta) - \frac{1}{2}\theta'\Lambda\theta},$$

where $\Lambda = diag\left(V_\alpha^{-1}, V_\delta^{-1}, V_\beta^{-1}\right)$, $V_\alpha = diag\left(\sigma_\alpha^2, \ldots, \sigma_\alpha^2\right)$, $V_\delta = diag\left(\sigma_\delta^2, \ldots, \sigma_\delta^2\right)$.

The posterior conditional distribution $p\left(\theta \left| Y, \sigma_\varepsilon^2\right.\right)$ is normal with mean

$$E\left(\theta \left| Y, \sigma_\varepsilon^2\right.\right) = \left(\Lambda + \sigma_\varepsilon^{-2}\chi'\chi\right)^{-1} \sigma_\varepsilon^{-2}\chi' Y.$$

Under the diffuse prior specification $\Lambda = 0$, implying that $\sigma_\alpha^2 \to \infty$, $\sigma_\delta^2 \to \infty$ and $V_\beta \to \infty$, the posterior conditional mean of θ aligns with the classical LSDV estimator:

$$E\left(\theta \left| Y, \sigma_\varepsilon^2\right.\right) = \hat{\theta}.$$

As the posterior conditional mean does not depend on σ_ε^2, the posterior (unconditional) mean remains the LSDV estimator:

$$E\left(\theta \left| Y\right.\right) = \hat{\theta}.$$

The Gibbs sampler for the fixed effects cycles through the posterior conditional distributions of $\beta, \sigma_\varepsilon^2, \alpha, \delta$, mirroring steps 1, 2 and 3 for the random effects (with steps 4 and 5 skipped due to the diffuse variance specification). The Gibbs sampler avoids explicitly creating dummy variable matrices D_α, D_δ, by updating α_i, $i = 1, \ldots, n$ and δ_t, $t = 1, \ldots, T$

one at a time, thus circumventing memory constraints. The sample mean of MCMC draws converges to the classical LSDV estimator.

The Gibbs sampler facilitates the estimation of fixed and random effects within a unified framework. The primary distinction lies in the prior specification: fixed effects have a diffuse prior, whereas random effects have a hierarchical prior. While fixed effects estimation updates the distribution of parameters, random effects estimation updates the distribution of hyperparameters. Conditional on σ_α^2 and σ_δ^2, there is no difference between fixed and random effects in terms of updating model parameters.

Another strength of the Gibbs sampler is its straightforward extension to nonlinear and multivariate models via data augmentation techniques. For example, binary response variables can be addressed using the Gibbs sampler of Albert and Chib (1993).

8.6.5 Marginalization and Big Data Algorithm

Chib and Carlin (1999) propose a block sampler for the panel data regression

$$Y_{it} = \alpha_i + X_{it}\beta + \varepsilon_{it},$$

where the variables $\beta, \alpha_1, \ldots, \alpha_n$ are jointly updated in one block through marginal-conditional decomposition

$$p\left(\beta, \alpha \left| Y, \sigma_\alpha^2, \sigma_\varepsilon^2\right.\right) = p\left(\beta \left| Y, \sigma_\alpha^2, \sigma_\varepsilon^2\right.\right) \prod_{i=1}^{n} p\left(\alpha_i \left| Y, \beta, \sigma_\alpha^2, \sigma_\varepsilon^2\right.\right).$$

Similar to the Gibbs sampler, $p\left(\alpha_i \left| Y, \beta, \sigma_\alpha^2, \sigma_\varepsilon^2\right.\right)$ is handled by normal linear regressions, with $Y_{it} - X_{it}\beta$ as the response variable and a column of ones as the predictor. The posterior conditional distribution of α_i maintains normality.

The key to the block method is sampling from the marginalized distribution $p\left(\beta \left| Y, \sigma_\alpha^2, \sigma_\varepsilon^2\right.\right)$, as a Bayesian counterpart of GLS. Marginalization over α_i yields a non-diagonal covariance matrix $I_n \otimes \Omega_T$, where $\Omega_T = \sigma_\alpha^2 \iota\iota' + \sigma_\varepsilon^2 I_T$. In practice, the non-diagonal covariance matrix can be efficiently handled by quasi-within transformation. Alternatively, in an augmented state space model that includes β in the state vector, the Kalman filter effectively marginalizes heterogeneity and provides the posterior distribution of β. The appeal of the marginalized sampler lies in improving MCMC chain convergence and reducing sample autocorrelations.

Chib and Carlin (1999) integrate out α_i for block sampling, while an alternative marginalization scheme integrates out all parameters except α_i. The marginalized MCMC sampler has the target distribution

$$p\left(\alpha \left| Y\right.\right) \propto p\left(\alpha\right) p\left(Y \left| \alpha\right.\right),$$

where $\alpha = (\alpha_1, \ldots, \alpha_n)$. Parameters are partitioned in two groups, in which σ_α^2 is marginalized for evaluating $p(\alpha)$, and $\beta, \sigma_\varepsilon^2$ are marginalized for evaluating $p\left(Y \left| \alpha\right.\right)$, resulting in double marginalization:

$$p\left(\alpha\right) = \frac{p\left(\sigma_\alpha^2\right) p\left(\alpha \left| \sigma_\alpha^2\right.\right)}{p\left(\sigma_\alpha^2 \left| \alpha\right.\right)},$$

$$p\left(Y \left| \alpha\right.\right) = \frac{p\left(\beta, \sigma_\varepsilon^2 \left| \alpha\right.\right) p\left(Y \left| \beta, \sigma_\varepsilon^2, \alpha\right.\right)}{p\left(\beta, \sigma_\varepsilon^2 \left| Y, \alpha\right.\right)}.$$

Under an inverse gamma conjugate specification, $p\left(\sigma_\alpha^2\right)$ and $p\left(\sigma_\alpha^2 \left| \alpha\right.\right)$ follow $IG\left(a_z, b_z\right)$ and $IG\left(\bar{a}_z, \bar{b}_z\right)$, respectively. Under a NIG conjugate specification, $p\left(\beta, \sigma_\varepsilon^2 \left| \alpha\right.\right)$ and

$p\left(\beta, \sigma_{\varepsilon}^{2} | Y, \alpha\right)$ follow $NIG(\mu, \Lambda, a, b)$ and $NIG\left(\bar{\mu}, \bar{\Lambda}, \bar{a}, \bar{b}\right)$, respectively. Refer to Chapter 1 for the posterior forms for the upper-bar variables. Conjugate specifications allow for evaluating the marginalized density in a surprisingly simple form:

$$p(\alpha) \propto \left(\bar{b}_{z}\right)^{-\bar{a}_{z}},$$

$$p(Y|\alpha) \propto \left(\bar{b}\right)^{-\bar{a}}.$$

The essence of double marginalized subsampling is to integrate out all parameters, leaving a vector of latent variables for subsampling. As a big data algorithm, only one data point is accessed at a time for rank-one update of the cumulative statistics that determine $p(\alpha)$ and $p(Y|\alpha)$.

8.7 From GARCH to Stochastic Volatility

Forecasting asset returns based on past prices presents significant challenges, primarily due to the principles underlying efficient market hypotheses, which suggest that all available information is already incorporated in asset prices, leaving no arbitrage opportunities Fama (1970). However, it is feasible to forecast the conditional variance of returns, a critical component in financial econometrics. Volatility clustering, a phenomenon where large (small) changes in asset prices are followed by large (small) changes, is a characteristic feature of financial markets. Clustering is often observed through strong autocorrelations in squared returns.

The Generalized Autoregressive Conditional Heteroscedasticity (GARCH) model, introduced by Engle (1982) and extended by Bollerslev (1986), has become a cornerstone in volatility forecasting. The GARCH model can be viewed as a special case of the state space model, wherein the transition equation is deterministic. In this context, the conditional variance, acting as the state variable, is determined by historical data. By incorporating additive noise into the transition equation, the GARCH model extends to the stochastic volatility (SV) model. Despite the challenges in estimating states and parameters in SV models, empirical studies suggest that SV models often outperform GARCH in volatility forecasting (Kim et al., 1998; Fleming and Kirby, 2003).

8.7.1 GARCH

Linear regression serves as a conditional mean model of the response variable:

$$Y_t = X_t\beta + \varepsilon_t,$$

where X_t is exogenous or predetermined (as a function of $Y_{1:t-1}$), and ε_t follows a white noise process that satisfies $E(\varepsilon_t) = 0$, $E\left(\varepsilon_t^2\right) = \sigma^2$ and $E(\varepsilon_t\varepsilon_{t-j}) = 0$, $j \geq 1$. A white noise process is characterized by the absence of autocorrelations, which does not necessarily imply statistical independence. The GARCH model exemplifies this concept. While maintaining the conditional mean specification and the white noise assumption, the model elaborates on the data-generating process of the white noise:

$$\varepsilon_t = \sigma_t e_t,$$

$$\sigma_t^2 = f\left(Y_{1:t-1}\right),$$

where the disturbances e_t are independent and identically distributed as standard normal variates, and e_t is independent of $Y_{1:t-1}$. The conditional variance σ_t^2 is assumed to be a deterministic function $f(\cdot)$ of past observations. By the conditional variance, we mean

$$Var\left(Y_t \left| Y_{1:t-1}\right.\right) = E\left(\varepsilon_t^2 \left| Y_{1:t-1}\right.\right) = \sigma_t^2.$$

The most prevalent conditional variance model is GARCH(1,1), where the function $f(\cdot)$ is specified as

$$\sigma_t^2 = c + \phi\sigma_{t-1}^2 + \psi\varepsilon_{t-1}^2,$$

with some initial values of σ_0^2 and ε_0^2. The conditional variance σ_t^2 is a deterministic function of $Y_{1:t-1}$, since $\varepsilon_{t-1}^2 = (Y_t - X_t\beta)^2$.

Proposition 8.3 elucidates several intriguing properties of the GARCH(1,1) model. First, the white noise process $\{\varepsilon_t\}$ is uncorrelated but not independent, as $\{\varepsilon_t^2\}$ follows an ARMA process. Second, the white noise process can be homoscedastic (i.e., time-invariant unconditional variances) while being conditionally heteroscedastic (i.e., time-varying conditional variances determined by past observations). Third, the likelihood function is a product of univariate normal densities, yet the joint distribution is not multivariate normal.

Proposition 8.3 (GARCH process) *Consider a GARCH(1,1) model:*

$$Y_t = X_t\beta + \varepsilon_t,$$

$$\varepsilon_t = \sigma_t e_t,$$
$$\sigma_t^2 = c + \phi\sigma_{t-1}^2 + \psi\varepsilon_{t-1}^2,$$

where the disturbances e_t are independent and identically distributed as standard normal variates, and e_t is independent of $Y_{1:t-1}$. Parameters are subject to constraints:

$$c > 0, \phi \geq 0, \psi > 0, \phi + \psi < 1.$$

GARCH(1,1) exhibits the following statistical properties:

1. *If the process is initialized by $E\left(\sigma_0^2\right) = E\left(\varepsilon_0^2\right) = \frac{c}{1-\phi-\psi}$, then $\{\varepsilon_t\}$ follows a white noise process: $E\left(\varepsilon_t\right) = 0$, $E\left(\varepsilon_t^2\right) = \frac{c}{1-\phi-\psi}$ and $E\left(\varepsilon_t\varepsilon_{t-j}\right) = 0$, for all $t \geq 1, j \geq 1$.*
2. *If initialized arbitrarily, the process still has $E\left(\varepsilon_t\right) = 0$ and $E\left(\varepsilon_t\varepsilon_{t-j}\right) = 0$, and the unconditional variance converges to the stationary variance:*

$$\lim_{t\to\infty} E\left(\varepsilon_t^2\right) = \frac{c}{1-\phi-\psi}.$$

3. *The squared sequence $\{\varepsilon_t^2\}$ follows an ARMA(1,1) process:*

$$\varepsilon_t^2 = c + (\phi+\psi)\,\varepsilon_{t-1}^2 + v_t - \phi v_{t-1},$$

where $v_t = \varepsilon_t^2 - \sigma_t^2$ with $E\left(v_t\right) = 0$, $E\left(v_t^2\right) = 2E\left(\sigma_t^4\right)$ and $E\left(v_t v_{t-j}\right) = 0$.

4. *If parameters are subject to a constraint: $3\psi^2 + \phi^2 + 2\phi\psi < 1$, then*

$$\lim_{t\to\infty} E\left(\sigma_t^4\right) = \frac{c^2\left(1+\phi+\psi\right)}{\left(1-\phi-\psi\right)\left(1-3\psi^2-\phi^2-2\phi\psi\right)}.$$

5. *The likelihood function takes the form*

$$p\left(Y_{1:T}\right)=\prod_{t=1}^{T}\frac{1}{\sqrt{2\pi}\sigma_t}e^{-\frac{(Y_t-X\beta)^2}{2\sigma_t^2}},$$

but it is not a multivariate normal density.

The popularity of the GARCH model is bolstered by its predictive prowess in financial volatility. Given observations $Y_{1:T}$, the focus is on the h-step-ahead volatility forecast. The identity $\varepsilon_t=\sigma_t e_t$ indicates that

$$E\left(\varepsilon_{T+h}^2\left|Y_{1:T}\right.\right)=E\left(\sigma_{T+h}^2\left|Y_{1:T}\right.\right), h\geq 1.$$

Additionally, the GARCH(1,1) dynamics $\sigma_t^2=c+\phi\sigma_{t-1}^2+\psi\varepsilon_{t-1}^2$ indicates that

$$E\left(\sigma_{T+h}^2\left|Y_{1:T}\right.\right)=c+\left(\phi+\psi\right)E\left(\sigma_{T+h-1}^2\left|Y_{1:T}\right.\right),$$

which provides a recursive formula for the h-step-ahead volatility forecast. As the forecasting horizon extends, the conditional variance forecast converges to the unconditional variance of the GARCH(1,1) model:

$$\lim_{h\to\infty}E\left(\varepsilon_{T+h}^2\left|Y_{1:T}\right.\right)=\lim_{h\to\infty}E\left(\sigma_{T+h}^2\left|Y_{1:T}\right.\right)=\frac{c}{1-\phi-\psi}.$$

The GARCH(1,1) model captures volatility clustering, yet other financial market characteristics also merit attention. For instance, markets can exhibit extreme volatility during precipitous price declines. Volatility asymmetry can be incorporated into GARCH-family models through more flexible specifications of conditional variances. For example, the Exponential GARCH, introduced by Nelson (1991), models the logarithm of the conditional variance process with a leverage term:

$$\ln\sigma_t^2=c+\phi\ln\sigma_{t-1}^2+\alpha\left|\frac{\varepsilon_{t-1}}{\sigma_{t-1}}\right|+\delta\frac{\varepsilon_{t-1}}{\sigma_{t-1}}.$$

Similarly, the threshold Glosten-Jagannathan-Runkle GARCH (Glosten et al., 1993) uses an indicator function $I\left(\varepsilon_{t-1}<0\right)$ to capture asymmetry in volatility clustering. Regardless of the functional form of the conditional variance in the GARCH-family model, it remains a deterministic function of past observations. Consequently, the likelihood function resembles that in Proposition 8.3. Maximum likelihood is the standard estimation method employed by most statistical software, although Bayesian inference is feasible via the importance sampling (Geweke, 1989b) or the MH sampler (Nakatsuma, 2000; Vrontos et al., 2000).

8.7.2 Stochastic Volatility

The GARCH model is conceptually akin to a state space model, with the conditional variance as the state variable. However, the transition equation is trivial in that it is a deterministic function of past observations, obviating the need for the Kalman filter in state estimation and likelihood evaluation. Extending the GARCH model by incorporating additive noise into the transition equation is an appealing prospect. Researchers have noted that time-varying volatility is a salient feature of financial and macroeconomic data (see Cogley and Sargent (2005), Justiniano and Primiceri (2008), Diebold et al. (2017), among others). The extension from GARCH to SV is conceptually straightforward but computationally challenging, as the SV likelihood function is not analytically tractable, complicating state filtering and smoothing. Consequently, computationally intensive Monte Carlo methods are used for approximating the posterior distributions of states and parameters (Jacquier et al., 1994; Kim et al., 1998; Chib et al., 2002).

8.7.2.1 Model Specification

Consider a stylized SV model that parallels the GARCH(1,1) model specified in Proposition 8.3 (see also Section 4.2 for an alternative parameterization of the SV model):

$$Y_t = X_t\beta + \varepsilon_t,$$

$$\varepsilon_t = \sigma_t e_t,$$

$$\ln\sigma_t^2 = c + \phi\ln\sigma_{t-1}^2 + \delta u_t,$$

where the disturbances e_t and u_t are independent and identically distributed as standard normal variates. Modeling $\ln\sigma_t^2$ instead of σ_t^2 ensures the positive conditional variance. Notably, ε_t still follows a white noise process because

$$E(\varepsilon_t) = E(\sigma_t)E(e_t) = 0,$$

$$E(\varepsilon_t^2) = E(\sigma_t^2)E(e_t^2) = e^{\frac{c}{1-\phi}+\frac{\delta^2}{2(1-\phi^2)}},$$

$$E(\varepsilon_t\varepsilon_{t-j}) = E(e_t)E(\sigma_t\sigma_{t-j}e_{t-j}) = 0.$$

The fundamental distinction between SV and GARCH is that the conditional variance σ_t^2 in the GARCH model is a deterministic function of past observations. However, the volatility in the SV model is a latent random variable. Our objective is to compute the posterior distribution of σ_t^2 conditional on past observations.

It is feasible to replace the constant c by a function of lagged observations $f(Y_{1:t-1})$ in the SV model specification, which does not increase the complexity of the nonlinear state space model. By nonlinearity, we refer to a nonlinear function of the state variables. Functions of lagged observations do not complicate nonlinear filtering.

Whether a state space model is nonlinear depends on the definition of the state variable. A straightforward approach is to designate σ_t^2 as the state variable and Y_t as the measurement variable. The transition and observation equations are nonlinear functions of the state variable. An alternative specification involves reparametrizing $Z_t = \ln\sigma_t^2$ as the state variable and $\tilde{Y}_t = \ln(Y_t - X_t\beta)^2$ as the measurement variable, conditional on the parameter β. This approach offers linearized transition and observation equations:

$$Z_t = c + \phi Z_{t-1} + \delta u_t,$$

$$\tilde{Y}_t = Z_t + \ln e_t^2.$$

Although linearity is restored, the model departs from normality. The noise term $\ln e_t^2$ corresponds to the logarithm of a $\chi^2(1)$ distributed variate. This distribution is skewed, with mean $E(\ln e_t^2) = -1.2704$ and variance $Var(\ln e_t^2) = \frac{\pi^2}{2}$.

8.7.2.2 Nonlinear filtering

As the SV model represents a prominent application of nonlinear state space models, many simulation-based nonlinear filtering techniques discussed in Chapter 7 are applicable to volatility estimation. For instance, the bootstrap particle filter updates particles via the transition equation: $p(Z_t|Z_{t-1}) = \phi(Z_t; c + \phi Z_{t-1}, \delta^2)$, with incremental weights computed from the observation density $p(Y_t|Z_t) = \phi(Y_t; X_t\beta, e^{Z_t})$. The exponential function e^{Z_t} is likely to yield a large variance in the observation equation, which effectively mitigates the risk of weight degeneracy in the forward filtering. Reasonable dispersion of the weights on particles ensures accurate approximation for the filtering distributions of the stochastic volatility.

The performance of the particle filter hinges on the proposal distributions. As an alternative to the bootstrap filter, the auxiliary particle filter (Pitt and Shephard, 1999) discussed in Section 7.5 may be employed. By incorporating information in Y_t into particle generation, the auxiliary filter can effectively approximate the filtering distributions.

In the forward recursion of the particle filter, the sample average of the unnormalized weights provides an unbiased estimator of the likelihood function. Consequently, the particle marginal MCMC methods (Andrieu et al., 2010) discussed in Section 7.7 can be implemented for estimating unknown parameters in the SV model.

8.7.2.3 Single-move and Multiple-move Samplers

Although the general-purpose particle filters can manage nonlinear state space models, the unique structure of the SV model allows for designing customized MCMC samplers for Bayesian inference. The unknown variables in the SV model include dynamic states $\{Z_t\}$ and static parameters c, ϕ, δ, β. The Gibbs sampler cycles through the full posterior conditional distributions. On the one hand, conditional on the state sequence, the SV model is simplified to two linear regressions: the transition equation becomes an autoregression that can be managed within the framework of conjugate Bayesian linear regressions (refer to Chapter 1), and the observation equation resembles a normal regression with heteroskedasticity (refer to Section 4.1). See also Exercise 8.11 for strategies of imposing parameter constraints. On the other hand, conditional on the model parameters, the posterior state sequence can be generated using single-move or multiple-move samplers.

Jacquier et al. (1994) propose a single-move sampler that updates the state sequence one element at a time. Consider sampling Z_t conditional on the remaining states

$$Z_{-t} = (Z_1, \ldots, Z_{t-1}, Z_{t+1}, \ldots, Z_T).$$

The posterior conditional density takes the form:

$$p(Z_t | Y_{1:T}, Z_{-t}, c, \phi, \delta, \beta) \propto p(Z_t | Z_{t-1}, c, \phi, \delta)\, p(Z_{t+1} | Z_t, c, \phi, \delta)\, p(Y_t | Z_t, \beta),$$

where

$$p(Z_t | Z_{t-1}, c, \phi, \delta) \propto e^{-\frac{1}{2}\delta^{-2}(Z_t - c - \phi Z_{t-1})^2},$$

$$p(Z_{t+1} | Z_t, c, \phi, \delta) \propto e^{-\frac{1}{2}\delta^{-2}(Z_{t+1} - c - \phi Z_t)^2},$$

$$p(Y_t | Z_t, \beta) \propto e^{-\frac{1}{2}Z_t - \frac{1}{2}(Y_t - X_t\beta)^2 e^{-Z_t}}.$$

Although the posterior density does not belong to a common distribution, it can be handled using the MH sampler or rejection sampling (see Exercise 8.12 for a proposal).

A multiple-move sampler updates the entire state sequence by simulation smoothing. Following Kim et al. (1998), the distribution $\ln \chi^2(1)$ is approximated by a finite Gaussian mixture, which are sufficiently flexible to yield similarly shaped density functions under different mixture specifications. Table 8.1 presents a 7-regime Gaussian mixture that provides a satisfactory approximation.

Let $R_t \in \{1, \ldots, 7\}$ be a latent variable indicating the regime. For instance, $R_t = i$ indicates the regime $N(\mu_i, v_i^2)$ with weight q_i. The Gibbs sampler with data augmentation cycles through the posterior conditional distributions. On the one hand, conditional on the regime sequence $\{R_t\}$, a Gaussian linear state space model emerges, allowing for the simulation smoother to generate a path of the state sequence $\{Z_t\}$. Refer to Section 6.4 for alternative algorithms of the simulation smoothers (e.g., De Jong (1989) and Durbin and Koopman (2002)). On the other hand, conditional on the state sequence, the discrete-valued regimes can be updated by sampling from categorical distributions:

TABLE 8.1
Gaussian mixture approximation of skewed error distribution in stochastic volatility model.

Regime	1	2	3	4	5	6	7
Mean	−7.4564	−5.3531	−3.7979	−1.9447	−0.9792	0.0895	1.0347
Std	3.1897	1.7258	1.3631	1.2514	0.9196	0.7447	0.5568
Weight	0.0278	0.0452	0.0808	0.2438	0.1922	0.2928	0.1174

$$p(R_t = i\,|Y_t, Z_t) \propto q_i v_i^{-1} e^{-\frac{1}{2}v_i^{-2}\left(\tilde{Y}_t - Z_t - \mu_i\right)^2}.$$

The Gaussian mixture approximation of $\ln \chi^2(1)$ outperforms a crude approximator by $N\left(-1.27, \frac{\pi^2}{2}\right)$ suggested by Harvey et al. (1994). Despite the precision of Gaussian mixture approximations, minor errors may persist, as noted by Kim et al. (1998) who recommend a weighting step post-MCMC sampling for correction.

8.7.2.4 Double Marginalized Subsampling

The stochastic volatility model is well-suited for double marginalization, which simultaneously integrates out the parameters c, ϕ, δ in the transition equation and β in the observation equation, facilitating subsampling of the state sequence $\{Z_t\}$. Double marginalization is particularly advantageous for big data applications.

Denote $W_t = (1, Z_{t-1})$ and $\alpha = (c, \phi)'$, enabling the model to be succinctly expressed in regression form:

$$Y_t = X_t\beta + e^{\frac{1}{2}Z_t}e_t,$$
$$Z_t = W_t\alpha + \delta u_t.$$

A close-form marginal likelihood requires conditionally conjugate priors. The prior of β can be multivariate normal with mean μ_β and precision Λ_β (covariance matrix Λ_β^{-1}). The prior of α, δ^2 can be $NIG(\mu_\alpha, \Lambda_\alpha, a, b)$. The posterior conditional distributions are analytically tractable: $p(\beta\,|Y, Z)$ remains multivariate normal with mean $\bar{\mu}_\beta$ and precision $\bar{\Lambda}_\beta$, and $p(\alpha, \delta^2\,|Y, Z)$ follows $NIG(\bar{\mu}_\alpha, \bar{\Lambda}_\alpha, \bar{a}, \bar{b})$, where

$$\bar{\mu}_\beta = \bar{\Lambda}_\beta^{-1}\left(\Lambda_\beta\mu_\beta + \sum_{t=1}^{T} e^{-Z_t}X_t'Y_t\right),$$

$$\bar{\Lambda}_\beta = \Lambda_\beta + \sum_{t=1}^{T} e^{-Z_t}X_t'X_t,$$

$$\bar{\mu}_\alpha = \bar{\Lambda}_\alpha^{-1}\left(\Lambda_\alpha\mu_\alpha + \sum_{t=1}^{T} W_t'Z_t\right),$$

$$\bar{\Lambda}_\alpha = \Lambda_\alpha + \sum_{t=1}^{T} W_t'W_t,$$

$$\bar{a} = a + \frac{T}{2},$$

$$\bar{b} = b + \frac{1}{2}\sum_{t=1}^{T} Z_t^2 + \frac{1}{2}\mu_\alpha'\Lambda_\alpha\mu_\alpha - \frac{1}{2}\bar{\mu}_\alpha'\bar{\Lambda}_\alpha\bar{\mu}_\alpha.$$

The target density for the marginalized MH sampler can be evaluated by Bayes' rule:

$$p(Z|Y) = p(Z)\,p(Y|Z),$$

where

$$p(Z) = \left(\bar{b}\right)^{-\bar{a}} \left|\bar{\Lambda}_\alpha\right|^{-\frac{1}{2}},$$

$$p(Y|Z) = \left|\bar{\Lambda}_\beta\right|^{-\frac{1}{2}} e^{-\frac{1}{2}\sum_{t=1}^{T} Z_t - \frac{1}{2}\sum_{t=1}^{T} e^{-Z_t} Y_t^2 + \frac{1}{2}\bar{\mu}_\beta' \bar{\Lambda}_\beta \bar{\mu}_\beta}.$$

These expressions are fully determined by a set of cumulative statistics:

$$S_1(Z) = \sum_{t=1}^{T} e^{-Z_t} X_t' X_t,$$

$$S_2(Z) = \sum_{t=1}^{T} e^{-Z_t} X_t' Y_t,$$

$$S_3(Z) = \sum_{t=1}^{T} e^{-Z_t} Y_t^2,$$

$$S_4(Z) = \sum_{t=1}^{T} Z_t,$$

$$S_5(Z) = \sum_{t=1}^{T} W_t' W_t,$$

$$S_6(Z) = \sum_{t=1}^{T} W_t' Z_t,$$

$$S_7(Z) = \sum_{t=1}^{T} Z_t^2.$$

The random-walk MH sampler for double marginalized subsampling serves as a big data algorithm, allowing parameters β, α, δ^2 to be updated with each single observation, rather than sweeping through the entire dataset. Additionally, marginalization functions as a blocking (grouping) strategy that reduces correlations of MCMC draws and enhances the mixing of the Markov chain.

8.8 Linear DSGE Models

Introduced by Kydland and Prescott (1982) and Long and Plosser (1983), Dynamic Stochastic General Equilibrium (DSGE) models are grounded in solid microeconomic foundations, as their dynamic equations are derived from the optimality conditions of rational households, firms, and policymakers. Despite their theoretical coherence, small-scale DSGE models have historically struggled to capture all features of the business cycle and may not perform as well as reduced-form VAR models in out-of-sample forecasts. However, this trend has shifted in recent years. Smets and Wouters (2007) have demonstrated that medium-scale DSGE models possess excellent forecasting power and can compete with Bayesian VAR models in out-of-sample prediction. Accordingly, larger-scale DSGE models have gained popularity in central banks for macroeconomic forecasting and policy analysis.

Parameter estimation is a central theme in econometrics, but an econometrician cannot estimate parameters without formulating and solving a DSGE model. Consequently, a significant portion of this section is dedicated to solution techniques. The essential steps of DSGE modeling involve specification, solution, and estimation.

First, a DSGE model is specified through dynamic stochastic optimization problems for forward-looking households, firms, and policymakers, incorporating equilibrium conditions for market clearing and the law of motion for exogenous shocks.

Second, although analytical solutions to these optimization problems are seldom available, various approximation methods exist. A popular approach is to log-linearize the first-order conditions and other equations that characterize the general equilibrium. The linearized equations can be expressed as linear rational expectations models. Unstable eigenvalues impose linear constraints that illuminate the policy function (i.e., control variables as a function of state variables) and the co-movement between expectational errors and exogenous shocks. Solving the model involves replacing unstable eigenvalues with zeros and endogenizing expectational errors as functions of exogenous shocks.

Third, a solved DSGE model is in the state space form, where unknown parameters can be estimated via maximum likelihood or Bayesian methods. The latter is more prevalent due to the availability of expert opinions on the prior distribution of model parameters. For linear state space models, the Kalman filter provides the log-likelihood, which is combined with prior information to evaluate the posterior density. MCMC methods, such as the MH sampler, are popular for posterior inference. Recent years have also seen the application of sequential Monte Carlo methods for DSGE models (Fernandez-Villaverde and Rubio-Ramirez, 2007; Herbst and Schorfheide, 2014; Cai et al., 2021).

8.8.1 An Introductory Example

Before delving into general solution and estimation techniques for DSGE models, it is instructive to gain intuition by solving a stylized optimization problem:

$$V(k_0, a_0) = \max_{c_t, k_{t+1}} E \sum_{t=1}^{\infty} \beta^t \ln c_t,$$

where consumption c_t and capital k_t satisfy the resource constraint

$$c_t + k_{t+1} = A a_t k_t^{\alpha},$$

and the productivity shock $\{a_t\}$ follows an exogenously evolving process

$$\ln a_t = \phi \ln a_{t-1} + \sigma \varepsilon_t.$$

A guess-and-verify strategy proceeds by hypothesizing the value function

$$V(k_0, a_0) = B_0 + B_1 \ln k_0 + B_2 \ln a_0,$$

and validating the Bellman equation

$$V(k_t, a_t) = \max_{c_t, k_{t+1}} \ln c_t + \beta E_t \left[V(k_{t+1}, a_{t+1}) \right],$$

with the coefficients $B_1 = \frac{\alpha}{1-\alpha\beta}$ and $B_2 = \frac{1}{(1-\alpha\beta)(1-\beta\phi)}$. The optimal policy functions are given by

$$c_t = (1 - \alpha\beta) A a_t k_t^{\alpha},$$

$$k_{t+1} = \alpha\beta A a_t k_t^{\alpha}.$$

The solved DSGE model exhibits a state space form, with the transition equation

$$\begin{pmatrix} \ln a_{t+1} \\ \ln k_{t+1} \\ \ln c_{t+1} \end{pmatrix} = \begin{pmatrix} 0 \\ \ln \alpha\beta A \\ \bar{c} \end{pmatrix} + \begin{pmatrix} \phi & 0 & 0 \\ 1 & \alpha & 0 \\ \alpha+\phi & \alpha^2 & 0 \end{pmatrix} \begin{pmatrix} \ln a_t \\ \ln k_t \\ \ln c_t \end{pmatrix} + \begin{pmatrix} \sigma \\ 0 \\ \sigma \end{pmatrix} \varepsilon_{t+1},$$

where $\bar{c} = \ln(1-\alpha\beta) A + \alpha \ln \alpha\beta A$. Given observations of the output $Y_t = A a_t k_t^{\alpha}$, the measurement equation is

$$\ln Y_t = \ln A + \begin{pmatrix} 1 & \alpha & 0 \end{pmatrix} \begin{pmatrix} \ln a_t \\ \ln k_t \\ \ln c_t \end{pmatrix}.$$

From an econometric perspective, the solved DSGE model represents a linear state space model with unknown parameters $\alpha, \beta, \phi, \sigma, A$. The Kalman filter provides the log-likelihood, enabling parameter estimation via maximum likelihood or Bayesian MCMC samplers.

The toy model has an exact analytical solution, but most DSGE models do not, necessitating an approximated solver that can be generalized to more complex dynamic models. The idea is to cast the log-linearized equilibrium conditions as linear rational expectations models, solved by analyzing unstable eigenvalues.

At a high level, n equations characterizing the general equilibrium are required to solve for n endogenous variables. As the equations are dynamic, transversality conditions are needed in addition to the first-order conditions. In the toy model, three equations (i.e., the Euler equation, the resource constraint, and the exogenous shock process) can solve for three unknowns a_t, k_t, c_t such that

$$c_t^{-1} = E_t\left(\beta c_{t+1}^{-1} A a_{t+1} \alpha k_{t+1}^{\alpha-1}\right),$$

$$c_t + k_{t+1} = A a_t k_t^{\alpha},$$

$$\ln a_t = \phi \ln a_{t-1} + \sigma\varepsilon_t.$$

Given the steady states $a^* = 1$, $k^* = (\alpha\beta A)^{\frac{1}{1-\alpha}}$, $c^* = AK^{*\alpha} - k^*$, log-linearizing the model around the steady states yields a set of linear equations

$$\hat{c}_{t+1} = \hat{c}_t + (\alpha-1)\hat{k}_{t+1} + \hat{a}_{t+1} + \eta_{t+1},$$

$$\hat{k}_{t+1} = \beta^{-1}\hat{k}_t + \left(1-\alpha^{-1}\beta^{-1}\right)\hat{c}_t + \alpha^{-1}\beta^{-1}\hat{a}_t,$$

$$\hat{a}_{t+1} = \phi\hat{a}_t + \sigma\varepsilon_t,$$

where $\hat{a}_t = \ln a_t - \ln a^*$, $\hat{k}_t = \ln k_t - \ln k^*$ and $\hat{c}_t = \ln c_t - \ln c^*$ represent percentage deviations from the steady states. The expectation operator $E_t(\cdot)$ is replaced by the expectational error η_{t+1}. The corresponding matrix form is

$$\Gamma_0 x_{t+1} = \Gamma_1 x_t + \Psi\varepsilon_{t+1} + \Pi\eta_{t+1},$$

where

$$\Gamma_0 = \begin{pmatrix} 1 & 0 & 0 \\ 0 & 1 & 0 \\ -1 & 1-\alpha & 1 \end{pmatrix}, \Gamma_1 = \begin{pmatrix} \phi & 0 & 0 \\ \frac{1}{\alpha\beta} & \frac{1}{\beta} & 1-\frac{1}{\alpha\beta} \\ 0 & 0 & 1 \end{pmatrix},$$

$$x_t = \begin{pmatrix} \hat{a}_t \\ \hat{k}_t \\ \hat{c}_t \end{pmatrix}, \Psi = \begin{pmatrix} \sigma \\ 0 \\ \sigma \end{pmatrix}, \Pi = \begin{pmatrix} 0 \\ 0 \\ 1 \end{pmatrix}.$$

As the matrix Γ_0 is invertible, the equation can be simplified as

$$x_{t+1} = \Gamma_0^{-1}\Gamma_1 x_t + \Gamma_0^{-1}\Psi\varepsilon_{t+1} + \Gamma_0^{-1}\Pi\eta_{t+1},$$

where

$$\Gamma_0^{-1}\Gamma_1 = \begin{pmatrix} \phi & 0 & 0 \\ \frac{1}{\alpha\beta} & \frac{1}{\beta} & 1-\frac{1}{\alpha\beta} \\ \frac{\alpha-1}{\alpha\beta}+\phi & \frac{\alpha-1}{\beta} & \alpha-\frac{1}{\beta}+\frac{1}{\alpha\beta} \end{pmatrix}.$$

The stability of the difference equation system is determined by the eigenvalues of the transition matrix $\Gamma_0^{-1}\Gamma_1$, which can be diagonalized as

$$\Gamma_0^{-1}\Gamma_1 = VDV^{-1},$$

where $D = diag\,(\lambda_1, \lambda_2, \lambda_3)$ and $V = (v_1, v_2, v_3)$ consist of eigenvalues and eigenvectors.:

$$\lambda_1 = \phi, \lambda_2 = \alpha, \lambda_3 = \alpha^{-1}\beta^{-1},$$

$$v_1 = \begin{pmatrix} \frac{\phi-\alpha}{\phi} \\ \frac{1}{\phi} \\ 1 \end{pmatrix}, v_2 = \begin{pmatrix} 0 \\ \alpha^{-1} \\ 1 \end{pmatrix}, v_3 = \begin{pmatrix} 0 \\ \frac{1-\alpha\beta}{\alpha-1} \\ 1 \end{pmatrix},$$

$$V^{-1} = \begin{pmatrix} \frac{\phi}{\phi-\alpha} & 0 & 0 \\ \frac{\alpha(\alpha-\phi+\alpha\beta\phi-1)}{(\phi-\alpha)(1-\alpha^2\beta)} & \frac{\alpha(1-\alpha)}{1-\alpha^2\beta} & \frac{\alpha(1-\alpha\beta)}{1-\alpha^2\beta} \\ -\frac{1-\alpha}{1-\alpha^2\beta} & -\frac{\alpha(1-\alpha)}{1-\alpha^2\beta} & \frac{1-\alpha}{1-\alpha^2\beta} \end{pmatrix}.$$

Reparametrizing $w_t = V^{-1}x_t$ yields a diagonalized system:

$$w_{t+1} = Dw_t + V^{-1}\Gamma_0^{-1}\Psi\varepsilon_{t+1} + V^{-1}\Gamma_0^{-1}\Pi\eta_{t+1}.$$

The unstable eigenvalue $|\lambda_3| > 1$ potentially induces divergence of the third element of w_t (denoted by w_{3t}), violating the transversality condition. However, saddle-path stability can be achieved by manipulating the control variable to prevent the explosive path by fixing $w_{3t} = 0, \forall t \geq 0$.

On the one hand, the unstable eigenvalue imposes a constraint on the endogenous variables, from which the policy function can be derived. Specifically, w_{3t} is a linear combination of $\hat{a}_t$, $\hat{k}_t$ and $\hat{c}_t$, with the weights given by the third row of V^{-1}:

$$-\frac{1-\alpha}{1-\alpha^2\beta}\hat{a}_t - \frac{\alpha\,(1-\alpha)}{1-\alpha^2\beta}\hat{k}_t + \frac{1-\alpha}{1-\alpha^2\beta}\hat{c}_t = 0,$$

resulting in the linear policy function

$$\hat{c}_t = \alpha\hat{k}_t + \hat{a}_t,$$

which serves as a log-linearized version of the optimal policy function with the level variables:

$$c_t = (1-\alpha\beta)\,Aa_t k_t^{\alpha}.$$

On the other hand, the unstable eigenvalue imposes a constraint on the expectational error and the exogenous shock, such that the former adjusts to offset the latter. Specifically, $w_{3t} = 0$ indicates a linear relation between ε_{t+1} and η_{t+1} such that the third row of $V^{-1}\Gamma_0^{-1}\Psi\varepsilon_{t+1} + V^{-1}\Gamma_0^{-1}\Pi\eta_{t+1}$ must be zero. The solved expectational error is $\eta_{t+1} = 0$. This is a special case of the general result that η_{t+1} is endogenized by a linear function of ε_{t+1}, but not necessarily zero in general.

In summary, by analyzing unstable eigenvalues, we derive the policy function $\hat{c}_t = \alpha\hat{k}_t + \hat{a}_t$ and resolve the expectational error $\eta_{t+1} = 0$, yielding the linearized model solution

$$\begin{pmatrix} \hat{a}_{t+1} \\ \hat{k}_{t+1} \\ \hat{c}_{t+1} \end{pmatrix} = \begin{pmatrix} \phi & 0 & 0 \\ 1 & \alpha & 0 \\ \alpha+\phi & \alpha^2 & 0 \end{pmatrix} \begin{pmatrix} \hat{a}_t \\ \hat{k}_t \\ \hat{c}_t \end{pmatrix} + \begin{pmatrix} \sigma \\ 0 \\ \sigma \end{pmatrix} \varepsilon_{t+1}.$$

The solved DSGE model takes the state space form, in which the econometric-oriented state variables are $\hat{a}_t$, $\hat{k}_t$ and $\hat{c}_t$, while the macroeconomic-oriented state variables are $\hat{a}_t$ and $\hat{k}_t$.

Compared to the analytical solution, log-linearization removes all constant terms in the transition and measurement equations, consequently omitting parameters β and A. Nevertheless, the log-linearization technique is applicable to complex DSGE models lacking closed-form solutions, and eigenvalue analysis can be generalized to solve a variety of linear rational expectations models.

8.8.2 Solving Linear Rational Expectations Models

Linearized DSGE equations can be expressed as linear rational expectations models (Sims, 2002):

$$\Gamma_0 x_{t+1} = \Gamma_1 x_t + C + \Psi\varepsilon_{t+1} + \Pi\eta_{t+1},$$

where the exogenous shock ε_{t+1} and the endogenous expectational error η_{t+1} are assumed to satisfy $E_t(\varepsilon_{t+1}) = 0$ and $E_t(\eta_{t+1}) = 0$.

The asymptotic behavior of the dynamic system is determined by the eigenvalues of the coefficient matrices. If Γ_0 is invertible, the eigenvalues of $\Gamma_0^{-1}\Gamma_1$ can be computed as $\Gamma_0^{-1}\Gamma_1 v_i = \lambda_i v_i$, $i = 1, \ldots, m$, or in the matrix form

$$\Gamma_0^{-1}\Gamma_1 V = VD,$$

where $D = diag(\lambda_1, \ldots, \lambda_m)$ and $V = (v_1, \ldots, v_m)$.

If Γ_0 is not invertible, the generalized eigenvalues of Γ_1 and Γ_0 can be computed as $\Gamma_1 v_i = \lambda_i \Gamma_0 v_i$, or in the matrix form

$$\Gamma_1 V = \Gamma_0 VD.$$

Reparametrizing $w_t = V^{-1}x_t$ leads to a diagonalized system:

$$w_{t+1} = Dw_t + (\Gamma_0 V)^{-1}(C + \Psi\varepsilon_{t+1} + \Pi\eta_{t+1}).$$

Without loss of generality, the diagonal elements of D (i.e., eigenvalues) are sorted in ascending order. The ith element of w_t is considered unstable if the corresponding eigenvalue exceeds a threshold (e.g., $|\lambda_i| > 1$). Variables are partitioned into stable and unstable blocks:

$$\begin{pmatrix} w_{1,t+1} \\ w_{2,t+1} \end{pmatrix} = \begin{pmatrix} D_1 & 0 \\ 0 & D_2 \end{pmatrix} \begin{pmatrix} w_{1t} \\ w_{2t} \end{pmatrix} + \begin{pmatrix} Q_1 \\ Q_2 \end{pmatrix} (C + \Psi\varepsilon_{t+1} + \Pi\eta_{t+1}),$$

where D_1 and D_2 correspond to the stable and unstable eigenvalues, respectively. Additionally, $(\Gamma_0 V)^{-1} = \begin{pmatrix} Q_1 \\ Q_2 \end{pmatrix}$. The unstable variables can be solved using backward recursion:

$$w_{2t} = D_2^{-1}w_{2,t+1} - D_2^{-1}Q_2(C + \Psi\varepsilon_{t+1} + \Pi\eta_{t+1}).$$

By the law of iterated expectations, the solution is given by

$$w_{2t} = -\sum_{\tau=1}^{\infty} D_2^{-\tau} Q_2 C = (I - D_2)^{-1} Q_2 C \equiv \bar{w}_2, \forall t \geq 0.$$

On the one hand, the fixed w_{2t} imposes linear constraints on the endogenous variables, due to $w_t = V^{-1}x_t$. On the other hand, the fixed w_{2t} imposes linear constraints on the expectational error and exogenous shocks. The constant $w_{2t} = \bar{w}_2$ implies that

$$Q_2 (\Psi \varepsilon_{t+1} + \Pi \eta_{t+1}) = 0.$$

The equation indicates co-movement between the expectational error and exogenous shocks, but whether the former can fully offset the latter depends on the rank conditions of $Q_2\Pi$ and $Q_2\Psi$. To solve a linear equation $Ax = b$, where $A = Q_2\Pi$ and $b = Q_2\Psi$, three scenarios arise: no solution, a unique solution, or indeterminate solutions. A solution exists if and only if the column space of $Q_2\Psi$ is contained within that of $Q_2\Pi$. With some abuse of notation, we denote the solution as

$$\eta_{t+1} = -(Q_2\Pi)^{-1} Q_2 \Psi \varepsilon_{t+1},$$

where $(Q_2\Pi)^{-1} Q_2\Psi$ should be interpreted as solving the linear equation $Ax = b$, where $A = Q_2\Pi$ and $b = Q_2\Psi$. Typically, software generates: 1) the solution if unique, 2) one of the solutions if indeterminate, and 3) the ordinary least squares estimator if no solution exists. It follows that

$$Q_1 (\Psi \varepsilon_{t+1} + \Pi \eta_{t+1}) = \left[Q_1\Psi - Q_1\Pi (Q_2\Pi)^{-1} Q_2\Psi\right] \varepsilon_{t+1}.$$

Even if η_{t+1} is not uniquely determined by ε_{t+1}, it is still possible that $Q_1\Pi\eta_{t+1}$ is uniquely determined. Our notation $Q_1\Pi (Q_2\Pi)^{-1}$ can also be interpreted as solving the linear equation $Q_1\Pi = \Phi Q_2\Pi$ for the matrix Φ. Given the existence of a solution (i.e., the equation $Q_2 (\Psi\varepsilon_{t+1} + \Pi\eta_{t+1}) = 0$ is satisfied), the solution is unique if and only if the row space of $Q_1\Pi$ is contained within that of $Q_2\Pi$.

Under the constraints $w_{2t} = \bar{w}_2, \forall t \geq 1$ and $\eta_{t+1} = -(Q_2\Pi)^{-1} Q_2\Psi\varepsilon_{t+1}$, the solution of the linear rational expectations models is given by

$$\begin{pmatrix} w_{1,t+1} \\ w_{2,t+1} \end{pmatrix} = \begin{pmatrix} D_1 & 0 \\ 0 & 0 \end{pmatrix} \begin{pmatrix} w_{1t} \\ w_{2t} \end{pmatrix} + \begin{pmatrix} Q_1 C \\ \bar{w}_2 \end{pmatrix} + \begin{pmatrix} Q_1\Psi - Q_1\Pi (Q_2\Pi)^{-1} Q_2\Psi \\ 0 \end{pmatrix} \varepsilon_{t+1},$$

along with the initial condition at $t = 0$, namely $w_{20} = \bar{w}_2$. Given the reparameterization $w_t = V^{-1}x_t$, the solution can also be expressed as

$$x_{t+1} = V \begin{pmatrix} D_1 & 0 \\ 0 & 0 \end{pmatrix} V^{-1} x_t + V \begin{pmatrix} Q_1 C \\ \bar{w}_2 \end{pmatrix} + V \begin{pmatrix} Q_1\Psi - Q_1\Pi (Q_2\Pi)^{-1} Q_2\Psi \\ 0 \end{pmatrix} \varepsilon_{t+1}.$$

In summary, the solution has two features. First, unstable eigenvalues are replaced by zeros. The original transition matrix is $\Gamma_0^{-1}\Gamma_1 = V \begin{pmatrix} D_1 & 0 \\ 0 & D_2 \end{pmatrix} V^{-1}$, while the solved transition matrix is $V \begin{pmatrix} D_1 & 0 \\ 0 & 0 \end{pmatrix} V^{-1}$. Second, unstable eigenvalues necessitate co-movement of the expectational error and the exogenous disturbance. In the solved model, the former is endogenized as a function of the latter.

To circumvent matrix inversion issues with V^{-1} and $(\Gamma_0 V)^{-1}$, Sims (2002) develop an approach using the QZ decomposition, which gives access to intermediate results in computing generalized eigenvalues. Let

$$\Gamma_0 = Q'\Lambda Z',$$

$$\Gamma_1 = Q'\Omega Z',$$

where Q, Z are orthonormal matrices and Λ, Ω are upper-triangular matrices, of which the ratio of diagonal elements are the generalized eigenvalues. Partition variables by the stable and unstable eigenvalues:

$$\begin{pmatrix} \Lambda_{11} & \Lambda_{12} \\ 0 & \Lambda_{22} \end{pmatrix} \begin{pmatrix} w_{1,t+1} \\ w_{2,t+1} \end{pmatrix} = \begin{pmatrix} \Omega_{11} & \Omega_{12} \\ 0 & \Omega_{22} \end{pmatrix} \begin{pmatrix} w_{1t} \\ w_{2t} \end{pmatrix} + \begin{pmatrix} Q_1 \\ Q_2 \end{pmatrix} (C + \Psi\varepsilon_{t+1} + \Pi\eta_{t+1}),$$

where $\begin{pmatrix} w_{1t} \\ w_{2t} \end{pmatrix} = \begin{pmatrix} Z_1'x_t \\ Z_2'x_t \end{pmatrix}$. The unstable eigenvalues require a fixed w_{2t} such that

$$w_{2t} = \bar{w}_2 = (\Lambda_{22} - \Omega_{22})^{-1} Q_2C, \forall t \geq 0,$$

implying a linear constraint on the endogenous variables $Z_2'x_t = \bar{w}_2$, and co-movement between ε_{t+1} and η_{t+1} such that

$$Q_2 (\Psi\varepsilon_{t+1} + \Pi\eta_{t+1}) = 0.$$

Sims (2002) shows that the solution exists if and only if the column space of $Q_2\Psi$ is contained within that of $Q_2\Pi$, and the solution is unique if and only if the row space of $Q_1\Pi$ is contained within that of $Q_2\Pi$. When the conditions are satisfied, a matrix Φ satisfies $Q_1\Pi = \Phi Q_2\Pi$. The solution is represented in the state space form:

$$x_{t+1} = \Theta_1 x_t + \Theta_c + \Theta_0\varepsilon_{t+1},$$

along with the initial condition $Z_2'x_0 = \bar{w}_2$, where

$$\Theta_1 = Z_1\Lambda_{11}^{-1} \begin{pmatrix} \Omega_{11} & \Omega_{12} - \Phi\Omega_{22} \end{pmatrix} Z',$$

$$\Theta_0 = Z \begin{bmatrix} \Lambda_{11}^{-1} (Q_1 - \Phi Q_2) \\ 0 \end{bmatrix} \Psi,$$

$$\Theta_c = Z \begin{bmatrix} \Lambda_{11}^{-1} & -\Lambda_{11}^{-1} (\Lambda_{12} - \Phi\Lambda_{22}) \\ 0 & I \end{bmatrix} \begin{bmatrix} Q_1 - \Phi Q_2 \\ (\Lambda_{22} - \Omega_{22})^{-1} Q_2 \end{bmatrix} C.$$

In addition to the solver of Sims (2002), alternative techniques of solving linear rational expectations models are proposed in the literature, including Blanchard and Kahn (1980), King and Watson (1998), Klein (1999), among others.

8.8.3 Parameter Estimation

A solved DSGE model is represented in the state space form, where coefficient matrices A, B, C, D are tightly parameterized by a vector of structural parameters θ. In linear state space models, the likelihood function can be evaluated using the Kalman filter, allowing for numerical maximum likelihood estimation. However, Bayesian methods are prevalent for formal parameter estimation in DSGE models for the following reasons.

First, model complexity: DSGE models employed by central banks for economic forecasting and policy analysis are typically medium- or large-scale models that support complex dynamics. Small and stylized DSGE models struggle to capture the salient features of the real-world economy and are not suitable for empirical data confrontation. The seminal work of Smets and Wouters (2007), featuring 21 equations and 36 parameters, demonstrates the exceptional performance of DSGE models that can compete with Bayesian VAR in out-of-sample prediction. Various extensions have been developed since then, expanding both the state and parameter spaces. For a large econometric model with many parameters, the numerical maximum likelihood often struggles to find the maximum of the complex objective function. The problem can be mitigated by Bayesian regularization via tight priors.

Second, data availability: Economic data for fitting DSGE models are typically sampled at monthly or quarterly frequencies. Even with half a century of data, only several hundred observations are available. Concerns about structural breaks in the economy further reduce the sample size used in empirical studies. In the absence of regularization, the risk of overfitting looms large in a high-dimensional model estimated with small data.

Third, parameter interpretability: Parameters in DSGE models are interpretable. Economists possess expert knowledge on typical values or plausible ranges of structural parameters, facilitating prior elicitation (Garthwaite et al., 2005). Expert opinions, microeconomic studies, foreign and past data, and other sources of non-sample information are informative about the distributions of structural parameters in DSGE models. It is not unusual for tight or even dogmatic priors to be adopted in empirical works. For instance, Smets and Wouters (2007) specify the intertemporal elasticity of substitution parameter by a normal prior with mean 1.5 and standard deviation 0.375.

In most papers, the prior distributions and hyperparameters are clearly documented in a table. Hyperparameters are tuned to reflect typical parameter values (prior mean) and reasonable uncertainty (prior variance). Typically, a bounded variable, say $\beta \in (0, 1)$, is assigned a Beta or uniform prior. A positive-valued variable, say $\sigma > 0$, is assigned a gamma, inverse gamma, log normal or inverse Gaussian prior. A real-valued variable is assigned a normal prior.

As the structural parameters enter the dynamic equations in a specific form required by the model solution, conjugacy is generally unavailable. The posterior density, proportional to the prior times the likelihood, does not belong to a common distribution family. The MH sampler is popular for posterior inference. The canonical workflow for posterior sampling of DSGE models involves the following steps: 1) search for the posterior mode by numerical optimization, 2) compute the Hessian evaluated at the posterior mode by numerical derivatives, 3) make proposals from a multivariate normal or t distribution with the covariance matrix proportional to the inverse Hessian, and 4) follow the standard MH sampling algorithm to accept or reject proposals.

Let $p(\theta | Y) \propto p(\theta) p(Y | \theta)$ be the posterior density evaluated by the Kalman filter. Numerical optimization techniques (such as quasi-Newton algorithms discussed in Section 5.8) are employed to search for the mode of the posterior density:

$$\theta^* = \arg\max_\theta \ln p(\theta | Y).$$

The Hessian matrix evaluated at the posterior mode

$$H(\theta^*) = \frac{\partial \ln p(\theta^* | Y)}{\partial\theta^* \partial\theta^{*\prime}}$$

is often computed by finite differencing or automatic differentiation. However, the numerical Hessian is computationally expensive and inaccurate. It is possible to recycle the approximated Hessian matrix used in quasi-Newton algorithms. In most cases, the approximated Hessian is close to the true value, provided that the number of iterations is large.

The second-order Taylor approximation of the log posterior

$$p(\theta | Y) \approx \exp\left[\ln p(\theta^* | Y) + \frac{1}{2}(\theta - \theta^*)' H(\theta^*)(\theta - \theta^*)\right]$$

indicates that the posterior can be approximated by a multivariate normal distribution with mean θ^* and covariance matrix $-H^{-1}(\theta^*)$, which serves as a reasonable proposal for the MH sampler in posterior inference. In practice, the proposal covariance matrix can be rescaled as $-cH^{-1}(\theta^*)$, where c is a tuning parameter. A smaller c often provides a higher acceptance rate of the proposal draws. A desirable acceptance rate can be roughly 1/4 (Gelman et al., 1997).

The standard MH sampler faces considerable challenges in handling high-dimensional target distributions. When all parameters are updated simultaneously, proposals are typically accepted with low probabilities, resulting in slow mixing of the Markov chain. Chib and Ramamurthy (2010) develop a tailored MH sampler well-suited for DSGE applications. The high-dimensional parameters are divided into random blocks to increase the probability of accepting a proposal that updates a random subset of parameters.

Herbst and Schorfheide (2014) comment that the efficacy of the MH algorithm declines as the complexity of DSGE models increases, motivating alternative estimation methods via sequential Monte Carlo. The particle filter for static models, introduced by Chopin (2002), is applicable to DSGE parameter estimation. Additionally, sequential Monte Carlo methods are well-suited to parameter learning as new data become available, as illustrated by Cai et al. (2021) in the context of online estimation of DSGE models. Refer to Section 7.8 for parameter learning algorithms by particle filtering.

8.9 Nonlinear DSGE Models

Linearization significantly simplifies the solution and estimation of DSGE models, yet solving and estimating the original model is of interest when nonlinearity plays a crucial role (e.g., avoiding certainty equivalence).

The perturbation method, widely used for solving nonlinear dynamic systems in physics, can also approximate the policy functions of DSGE models. The second-order perturbation method provides a policy function in the state space form with quadratic terms. Nonlinear DSGE models are typically estimated using particle MCMC methods (Andrieu and Roberts, 2009; Andrieu et al., 2010). Interestingly, these methods were applied in empirical DSGE studies (Fernandez-Villaverde and Rubio-Ramirez, 2007) before rigorous theoretical foundations were established.

8.9.1 An Example of Perturbation

The perturbation approach approximates the solution of a nonlinear equation through a Taylor expansion around a specific point, at which it is feasible to solve the equation. To gain intuition about the perturbation technique, consider a scalar-valued differentiable equation

$$f(x,\sigma) = 0,$$

where x is the unknown variable and σ is the perturbation parameter. Denote the solution as

$$x = g(\sigma),$$

but the functional form of g is unknown, as the equation cannot be solved generally. Suppose the equation can be solved in a special case when $\sigma = 0$. Solving the equation $f(x,0) = 0$, yields $g(0)$, around which the Taylor expansion is performed:

$$g(\sigma) = g(0) + g'(0)\sigma + \frac{1}{2}g''(0)\sigma^2 + \cdots .$$

The implicit function theorem is fundamental to the perturbation method, allowing for computing the derivatives $g'(0)$, $g''(0)$, etc. by differentiating f. Define the value function

$$F(\sigma) = f(g(\sigma), \sigma),$$

which must be zero for all σ. Consequently, the derivatives of F equal to zero as well:

$$F'(\sigma) = F''(\sigma) = 0.$$

Let $f_x(x, \sigma)$ and $f_\sigma(x, \sigma)$ denote the partial derivatives of $f(x, \sigma)$. They satisfy the equation

$$F'(\sigma) = f_x(g(\sigma), \sigma) g'(\sigma) + f_\sigma(g(\sigma), \sigma) = 0.$$

By evaluating the equation at $\sigma = 0$, the first derivatives of g is recovered by the first partial derivatives of f:

$$g'(0) = -\frac{f_\sigma(g(0), 0)}{f_x(g(0), 0)}.$$

Similarly, the second derivatives of g can be recovered by the second partial derivatives of f. Let $f_{xx}, f_{x\sigma}, f_{\sigma x}, f_{\sigma\sigma}$ denote the second partial derivatives of f, evaluated at $\sigma = 0$ (input arguments of functions are suppressed for brevity). The solution to the equation

$$F''(0) = f_{xx} g'(0) g'(0) + f_{x\sigma} g'(0) + f_x g''(0) + f_{\sigma x} g'(0) + f_{\sigma\sigma} = 0.$$

is given by

$$g''(0) = -f_x^{-1} \left[f_{xx} g'(0) g'(0) + f_{x\sigma} g'(0) + f_{\sigma x} g'(0) + f_{\sigma\sigma} \right].$$

Perturbation via higher-order Taylor approximation follows a similar process. The kth order derivatives of g is computed recursively by the kth order partial derivatives of f, based on previously computed lower-order derivatives of g.

Interested readers can refer to Chapters 13 and 14 in Judd (1998) for more applications of the perturbation methods.

8.9.2 First-order Perturbance for DSGE

The equilibrium equations of DSGE models can be presented in a generic form:

$$E_t \left[f(Y_{t-1}, Y_t, Y_{t+1}, u_t) \right] = 0,$$

where Y_t is a $n \times 1$ vector of endogenous variables, and u_t is a $m \times 1$ vector of exogenous shocks such that $u_t = \sigma \varepsilon_t$ and $\varepsilon_t \sim N(0, \Sigma)$, with the perturbation parameter $\sigma > 0$ in the stochastic setting and $\sigma = 0$ in the deterministic model. Assume that the function $f : \Re^n \times \Re^n \times \Re^n \times \Re^m \to \Re^n$ is differentiable. There are n equations to solve for n unknown variables.

Regarding the timing convention, a variable decided at period t is stamped by a time subscript t. The expectation operator E_t takes the conditional mean of future random variables with a time script $t + 1$. For example, u_{t+1} is random, but u_t is known.

The solution of the DSGE model is characterized by the policy function (also known as the decision rule):

$$Y_t = g(Y_{t-1}, u_t, \sigma).$$

For a dynamic stochastic model, a closed-form solution is generally unavailable. However, it is possible to solve for the steady state such that $\sigma = 0$, $u_t = 0$, $Y_t = \bar{Y}$ and

$$f(\bar{Y}, \bar{Y}, \bar{Y}, 0) = 0,$$

$$\bar{Y} = g\left(\bar{Y}, 0, 0\right).$$

Although the functional form of g is unknown, it can be evaluated at the steady state. The first-order perturbation method focuses on the first-order Taylor approximation of g around the steady state:

$$Y_t = \bar{Y} + g_y\left(Y_t - \bar{Y}\right) + g_u u_t + g_\sigma \sigma,$$

where g_y, g_u, g_σ denote the three partial derivatives of $g\left(Y_{t-1}, u_t, \sigma\right)$. The input arguments of those functions are suppressed for brevity. When g_y, g_u, g_σ are evaluated at the steady state, they are treated as $n \times n$, $n \times m$ and $n \times 1$ unknown coefficients to be solved. The perturbation method recovers them by differentiation of the model equations. Let f_-, f_0, f_+, f_u denote the four partial derivatives (i.e., the Jacobian matrices) of $f\left(Y_{t-1}, Y_t, Y_{t+1}, u_t\right)$. Again, these are functions per se, but when f_-, f_0, f_+, f_u are evaluated at the steady state, they become $n \times n$, $n \times n$, $n \times n$, $n \times m$ known matrices (as f can be evaluated, and its Jacobian can be computed).

Substitute the policy function in the model equations and define

$$F\left(Y_{t-1}, u_t, \sigma\right) = f\left\{Y_{t-1}, g\left(Y_{t-1}, u_t, \sigma\right), g\left[g\left(Y_{t-1}, u_t, \sigma\right), \sigma\varepsilon_{t+1}, \sigma\right], u_t\right\}.$$

Due to the equation $E_t\left[F\left(Y_{t-1}, u_t, \sigma\right)\right] = 0, \forall Y_{t-1}, u_t, \sigma$, the expected derivatives of F with respect to Y_{t-1}, u_t, σ equal zero, facilitating the solution for g_y, g_u, g_σ at the steady state.

First, g_σ is recovered by differentiating F with respect to σ, and evaluating derivatives at the steady state:

$$\left(f_0 + f_+ g_y + f_+\right) g_\sigma + E_t\left(f_+ g_u \varepsilon_{t+1}\right) = 0.$$

Given $E_t\left(\varepsilon_{t+1}\right) = 0$, we have a homogenous linear equation for g_σ, leading to an important result $g_\sigma = 0$. Certainty equivalence is the main feature of the first-order perturbation method, indicating that the decision maker does not account for σ in the policy function.

Second, g_u is recovered by differentiating F with respect to u_t, and evaluating derivatives at the steady state:

$$f_0 g_u + f_+ g_y g_u + f_u = 0.$$

Given g_y (to be solved in the next step), the solution of the linear equation is

$$g_u = -\left(f_0 + f_+ g_y\right)^{-1} f_u.$$

Third, g_y is recovered by differentiating F with respect to Y_{t-1}, and evaluating derivatives at the steady state:

$$f_- + f_0 g_y + f_+ g_y g_y = 0.$$

This is a matrix-form quadratic equation. Our objective is to solve for the $n \times n$ matrix g_y given three $n \times n$ matrices f_-, f_0, f_+. The solution can be found by analyzing the generalized eigenvalues of two auxiliary matrices:

$$A = \begin{pmatrix} -f_0 & -f_- \\ I & 0 \end{pmatrix},$$

$$B = \begin{pmatrix} f_+ & 0 \\ 0 & I \end{pmatrix}.$$

Let λ, x be a generalized eigenvalue and eigenvector of A, B such that

$$Ax = \lambda Bx.$$

Notably, λ, x would be the standard eigenvalue and eigenvector of $B^{-1}A$, if B were invertible (but it is often not invertible due to the sparsity of f_+). Nevertheless, the generalized eigenvalues and eigenvectors exist and can be computed by software, such as, `eig(A,B)` in MATLAB.

As the matrices A and B are padded with zeros and identity matrices at the bottom rows, the eigenvectors exhibit a special structure, visible when representing $Ax = \lambda Bx$ in 2×2 blocks:

$$\begin{pmatrix} -f_0 & -f_- \\ I & 0 \end{pmatrix} \begin{pmatrix} u \\ v \end{pmatrix} = \lambda \begin{pmatrix} f_+ & 0 \\ 0 & I \end{pmatrix} \begin{pmatrix} u \\ v \end{pmatrix}.$$

The bottom rows indicate that $u = \lambda v$, implying $x = \begin{pmatrix} \lambda v \\ v \end{pmatrix}$. The top rows provide an equation:

$$\lambda^2 f_+ v + \lambda f_0 v + f_- v = 0.$$

The size of A is $2n \times 2n$, yielding $2n$ generalized eigenvalues in total. Sort them in ascending order by absolute values and choose n smallest eigenvalues for constructing a diagonal matrix $\Lambda = diag(\lambda_1, \ldots, \lambda_n)$, with the corresponding (bottom rows of) eigenvectors $V = (v_1, \ldots, v_n)$. It follows that

$$f_+ V \Lambda \Lambda + f_0 V \Lambda + f_- V = 0.$$

If the matrix V is invertible, the solution of the equation $f_- + f_0 g_y + f_+ g_y g_y = 0$ is

$$g_y = V \Lambda V^{-1}.$$

By construction, the eigenvalues of g_y are $\lambda_1, \ldots, \lambda_n$. If all eigenvalues lie within the unit circle, a non-explosive solution is achieved. In practice, some eigenvalues may be zero, and some columns of g_y may be zero, as a subset of endogenous variables are (economic) state variables that dictate the policy function.

If the matrix V is not invertible, a workaround involves the QZ decomposition for A and B such that

$$A = Q\tilde{A}Z,$$
$$B = Q\tilde{B}Z,$$

where $\tilde{A}, \tilde{B}$ are upper triangular and Q, Z are unitary. The equation $Ax = \lambda Bx$ can be represented as $Q\tilde{A}Zx = \lambda Q\tilde{B}Zx$, implying $\tilde{A}\tilde{x} = \lambda \tilde{B}\tilde{x}$, where $\tilde{x} = Zx$. Therefore, $\tilde{A}, \tilde{B}$ and A, B have the same set of generalized eigenvalues, and eigenvectors are linear transformation by Z. The QZ decomposition is an intermediate step for computing generalized eigenvalues: $\left|\tilde{A} - \lambda \tilde{B}\right| = 0$. Because $\tilde{A}, \tilde{B}$ are upper triangular, the generalized eigenvalues are simply the ratios of diagonal elements of $\tilde{A}, \tilde{B}$, and the $2n \times 2n$ matrix of concatenated eigenvectors is upper triangular. By reordering eigenvalues in QZ factorization, the selected n eigenvalues can be placed in the leading (upper left) diagonal blocks. For those selected eigenvalues, the equation $\tilde{x} = Zx$ exhibits the following pattern 2×2 blocks:

$$\begin{pmatrix} \tilde{x}_1 \\ 0 \end{pmatrix} = \begin{pmatrix} Z_{11} & Z_{12} \\ Z_{21} & Z_{22} \end{pmatrix} \begin{pmatrix} \lambda v \\ v \end{pmatrix},$$

resulting in the equation: $Z_{21}\lambda v + Z_{22} v = 0$, that is, $Z_{21} V \Lambda + Z_{22} V = 0$. The QZ solution of the equation $f_- + f_0 g_y + f_+ g_y g_y = 0$ is

$$g_y = -Z_{21}^{-1} Z_{22}.$$

8.9.3 Second-order Perturbation for DSGE

The second-order perturbation method approximates the policy function by the second-order Taylor expansion around the steady state. Recall that the first-order terms g_y, g_u, g_σ in the policy function are derived from the Jacobian of the model equations f_-, f_0, f_+, f_u. Similarly, the second-order terms $g_{yy}, g_{yu}, g_{uu}, g_{\sigma\sigma}, g_{\sigma y}, g_{\sigma u}$ in the policy function are obtained from the Hessian of the model equations. Fortunately, all second-order terms can be resolved using linear equations, avoiding matrix-form quadratic equations and explosive eigenvalues. However, the notation becomes more complex, requiring careful application of the chain rule of differentiation, particularly in the presence of multiple chains. For pedagogical purposes, a two-step process is recommended. First, assume all variables are scalars to ensure all chains in differentiation are properly addressed. Second, reshape matrices and three-dimensional arrays to facilitate multivariate extension.

Let g_{yy} denote the second-order derivative of the policy function $g(Y_{t-1}, u_t, \sigma)$ with respect to Y_{t-1} twice, evaluated at the steady state. The perturbation method solves g_{yy} by differentiating F twice, equivalent to differentiating both sides of the first-order equation:

$$f_- + f_0 g_y + f_+ g_y g_y = 0,$$

where f_-, f_0, f_+, g_y are functions of Y_{t-1}. In particular, $g_y g_y$ refers to a verbose expression

$$g_y \left[g\left(Y_{t-1}, u_t, \sigma\right), u_{t+1}, \sigma\right] g_y\left(Y_{t-1}, u_t, \sigma\right).$$

Let h_-, h_0, h_+ be partial derivatives of f_-, f_0, f_+ with respect to Y_{t-1}, evaluated at the steady state:

$$\begin{aligned} h_- &= f_{--} + f_{-0} g_y + f_{-+} g_y g_y, \\ h_0 &= f_{0-} + f_{00} g_y + f_{0+} g_y g_y, \\ h_+ &= f_{+-} + f_{+0} g_y + f_{++} g_y g_y, \end{aligned}$$

where $f_{--}, f_{-0}, f_{-+}, f_{0-}, f_{00}, f_{0+}, f_{+-}, f_{+0}, f_{++}$ represent the second-order partial derivatives (i.e., Hessian) of the model equations f, evaluated at the steady state. It follows that differentiating F twice yields the equation:

$$h_- + h_0 g_y + h_+ g_y g_y + \left(f_0 g_{yy} + f_+ g_{yy} g_y g_y + f_+ g_y g_{yy}\right) = 0.$$

As a linear equation with respect to g_{yy}, the prototype solution for the univariate case is given by

$$g_{yy} = -\left(f_0 + f_+ g_y g_y + f_+ g_y\right)^{-1}\left(h_- + h_0 g_y + h_+ g_y g_y\right).$$

As an application of the chain rule of differentiation, the main challenge lies in tracking multiple chains. Once a prototype solution for scalars is established, a multivariate extension can be made, where g_{yy} becomes a $n \times n \times n$ array. Some authors prefer tensor notations for three-dimensional arrays. Alternatively, we adhere to matrix notation by vectorizing the equation:

$$vec\left(f_-\right) + vec\left(f_0 g_y\right) + vec\left(f_+ g_y g_y\right) = 0.$$

The formula $vec(ACB) = (B' \otimes A)\, vec(C)$ facilitates differentiating a $n^2 \times 1$ vector $vec(C)$ with respect to a $1 \times n$ vector, yielding the derivative as a $n^2 \times n$ matrix. The multivariate version of the linear equation for g_{yy} (reshaped as a $n^2 \times n$ matrix) exhibits the Sylvester form:

$$\begin{aligned} \left(g_y' \otimes f_+\right) g_{yy} g_y + \left[I \otimes \left(f_0 + f_+ g_y\right)\right] g_{yy} + c &= 0, \\ c = h_- + \left(g_y' \otimes I\right) h_0 + \left(g_y' g_y' \otimes I\right) h_+ &. \end{aligned}$$

Exercise 8.14 discusses numerical methods for solving the Sylvester equation, which is inherently linear and analytically tractable, though different solvers vary in computational complexity.

Similarly, let g_{yu} denote the second-order derivative of the policy function $g(Y_{t-1}, u_t, \sigma)$ with respect to Y_{t-1}, u_t, evaluated at the steady state. Let H_-, H_0, H_+, H_u represent partial derivatives of f_-, f_0, f_+, f_u with respect to u_t, evaluated at the steady state:

$$H_- = f_{-0}g_u + f_{-+}g_y g_u + f_{-u},$$

$$H_0 = f_{00}g_u + f_{0+}g_y g_u + f_{0u},$$

$$H_+ = f_{+0}g_u + f_{++}g_y g_u + f_{+u},$$

$$H_u = f_{u0}g_u + f_{u+}g_y g_u + f_{uu}.$$

The linear equation for solving g_{yu} is given by

$$(I \otimes f_0)\, g_{yu} + (I \otimes f_+ g_y)\, g_{yu} + d = 0,$$

$$d = H_- + (g_y' \otimes I)\, H_0 + (g_y' g_y' \otimes I)\, H_+ + (g_y' \otimes f_+)\, g_{yy} g_u.$$

Similarly, the linear equation for solving g_{uu} is given by

$$(I \otimes f_0)\, g_{uu} + (I \otimes f_+ g_y)\, g_{uu} + k = 0,$$

$$k = H_u + (g_u' \otimes I)\, H_0 + (g_u' g_y' \otimes I)\, H_+ + (g_u' \otimes f_+)\, g_{yy} g_u.$$

Last but not least, let $g_{\sigma y}, g_{\sigma u}, g_{\sigma\sigma}$ denote the second-order derivative of the policy function $g(Y_{t-1}, u_t, \sigma)$ with respect to (σ, Y_{t-1}), (σ, u_t), (σ, σ), evaluated at the steady state. The coefficients $g_{\sigma y}, g_{\sigma u}, g_{\sigma\sigma}$ are solved by differentiating both sides of the first-order equation:

$$(f_0 + f_+ g_y + f_+)\, g_\sigma + E_t\left(f_+ g_u \varepsilon_{t+1}\right) = 0.$$

The steady state $g_\sigma = 0$ and $E_t(\varepsilon_{t+1}) = 0$ result in $g_{\sigma y} = 0$ and $g_{\sigma u} = 0$. However, differentiation with respect σ generates quadratic terms of ε_{t+1}, necessitating the covariance matrix Σ in the equation for $g_{\sigma\sigma}$. The prototype solution for the univariate case is

$$g_{\sigma\sigma} = -\left(f_0 + f_+ g_y + f_+\right)^{-1}\left(f_{++} g_u g_u + f_+ g_{uu}\right)\Sigma.$$

For the multivariate case, $g_{\sigma\sigma}$ becomes a $n \times 1$ vector such that

$$g_{\sigma\sigma} = -\left(f_0 + f_+ g_y + f_+\right)^{-1} b,$$

where the ith element of b equals

$$tr\left[f_{++}^{(i)} g_u \Sigma g_u'\right] + \sum_{j=1}^{n} f_+^{(i,j)} \sum_{k=1}^{m} g_{uu}^{(j,k)} \Sigma^{(:,k)}.$$

In summary, the main feature of the second-order perturbation solver is $g_{\sigma y} = 0$ and $g_{\sigma u} = 0$, but $g_{\sigma\sigma} \neq 0$, which avoids certainty equivalence present in the first-order perturbation. The second-order coefficients, including $g_{yy}, g_{yu}, g_{uu}, g_{\sigma\sigma}$, are solved from linear equations, leveraging the Hessian of model equations evaluated at the steady state.

More discussions on the perturbation methods and other nonlinear solution methods for DSGE can be found in Collard and Juillard (2001), Schmitt-Grohe and Uribe (2004), Aruoba et al. (2006), among others.

8.9.4 Parameter Estimation

A typical DSGE workflow involves solving and estimating the model. If the equilibrium equations are (log) linearized or the policy function is approximated using the first-order perturbation method, the solved DSGE model exhibits a linear state space form for econometric analysis. Alternatively, when using the second-order perturbation method, the policy function includes quadratic terms, resulting in a nonlinear state space model. Additionally, richer nonlinear features, such as stochastic volatility, can be incorporated into the model. A recent study by Diebold et al. (2017) demonstrates real-time forecast evaluation of a New Keynesian DSGE model with stochastic volatility in output growth, government spending, and technology shocks. Nonlinear filtering, smoothing, and parameter estimation techniques discussed in Chapter 7 are applicable to the econometric analysis of nonlinear DSGE models.

As an alternative to informal calibration by matching model-implied and sample moments, likelihood-based inference is preferable for formal parameter estimation in the state space framework. An econometric version of the nonlinear DSGE model includes a collection of transition and measurement equations

$$Y_t = g\left(Y_{t-1}, u_t, \theta\right),$$

$$Z_t = CY_t + D\varepsilon_t,$$

where Y_t is a $n \times 1$ vector of endogenous variables, and u_t is a $m \times 1$ vector of exogenous shocks. Applying the second-order perturbation solver, the approximated policy function g includes quadratic terms of Y_{t-1} and u_t. The unknown parameters in the policy function are estimated by macroeconomic data Z_t, such as output, consumption, investment, wage, price, interest rate, etc. In the observation equation, the coefficient C is typically a selection matrix, selecting a subset of Y_t as the observations. Assuming data are observed without errors, the observation variances are zero: $D = 0$, but the number of observed series should be less or equal to the number of exogenous shock variables to prevent stochastic singularity. If measurement errors are allowed, the matrix D is usually specified as a diagonal matrix.

An intuitive approach to parameter estimation is using the particle filter to evaluate the likelihood function

$$p\left(Z_{1:T} \left| \theta\right.\right) = \int \prod_{t=1}^{T} p\left(Y_t \left| Y_{t-1}, \theta\right.\right) p\left(Z_t \left| Y_t\right.\right) dY_{1:T}.$$

With the simulated likelihood function, unknown parameters are estimated using maximum likelihood or the MH sampler for the posterior distribution of the parameters θ. This likelihood-based inference approach was proposed by Fernandez-Villaverde and Rubio-Ramirez (2007), predating the theoretical development of the pseudo-marginal method (Andrieu and Roberts, 2009) and the particle-marginal method (Andrieu et al., 2010).

The pseudo-marginal and particle-marginal algorithms are discussed in Section 7.7. The intractable likelihood function is substituted by an unbiased estimator obtained from the mean of unnormalized particle weights. See Flury and Shephard (2011) for an intuitive explanation of Bayesian inference based on a simulated likelihood. Additionally, parameter learning by SMC^2 (Chopin et al., 2013) addresses intractability of the likelihood function using nested particle filter for each particle of the parameters. Refer to Section 7.8 for a detailed description of the algorithm.

8.10 Software and Computational Considerations

When the number of parameters is large, parameter estimation in state space models becomes computationally intensive, necessitating caution regarding numerical issues. We consider two examples.

The first example is the yields-macro model of Diebold et al. (2006), which includes 6 state variables (level, slope, and curvature factors, along with three macroeconomic indicators) and 17 observed series (yields of various maturities). The number of unknown parameters totals 81. Diebold et al. (2006) remark that parameter estimation presents a "challenging, but not insurmountable, numerical task". Challenges include specifying initial parameter values for numerical optimization and tuning the MCMC sampler with a reasonable proposal density for Bayesian inference. A prudent strategy is to first estimate a simpler model and then expand it with additional parameters. For instance, one might begin with ordinary linear regressions of the yields-only model under a fixed $\lambda = 0.0609$ in the observation equation. These results can serve as starting values for the numerical maximum likelihood estimation of 36 parameters in the yields-only model. The subsequent step involves extending the yields-only model to the yields-macro model and numerically maximizing the likelihood function with respect to 81 parameters. Finally, the Hessian matrix is calculated, and the inverse of the negative Hessian is used as the proposal covariance matrix for the MH sampler.

The second example involves Bayesian estimation of DSGE models using the MH sampler. Due to tight priors on model parameters, it is often preferable to numerically maximize the posterior density rather than the likelihood function. However, computing the Hessian poses a significant challenge. The numerical Hessian obtained through double finite differencing is notoriously inaccurate. If the objective function were the log likelihood, the outer product of gradients could approximate the inverse Hessian. Unfortunately, this option is unavailable for posterior density maximization. An alternative is to extract the approximated Hessian from the optimization program. For quasi-Newton algorithms, the optimization program accumulates Hessian information through changes in gradients. While it does not replicate the true Hessian, it approximately captures the curvature. In most cases, the approximation closely aligns with the true value after hundreds of iterations. An advantage of this method is that the approximated Hessian is always negative definite by construction, whereas the numerical Hessian obtained by finite differencing often loses definiteness. Another challenge is the inversion of the Hessian matrix, which is computationally demanding, numerically unstable, and prone to ill-conditioning. Fortunately, the numerical optimization program may track the inverse Hessian approximation (rather than the Hessian itself) at each iteration. This provides exactly what is needed for proposing a covariance matrix for the MH sampler, without an explicit matrix inversion.

8.11 Applications

In this section, we consider applications of Bayesian state space models by real-world data.

8.11.1 Conditional Forecast by Bayesian VAR and Factor Models

In Chapter 1, we considered unconditional forecasts using a large Bayesian VAR with 38 macroeconomic variables of the U.S. economy. In this study, we utilize the same model for conditional forecasts and compare the outcomes with factor models (DFM and FAVAR), which are fitted to the identical dataset.

The estimation window spans from 1975Q1 to 2021Q2. During the forecast validation period from 2021Q3 to 2023Q3, we simulate posterior predictive distributions conditional on three variables pertinent to monetary policy targets: the Consumer Price Index, Federal Funds Rate, and Reserves of Depository Institutions. Given observations of these conditional variables, we project the remaining 35 endogenous variables in the model. The forecasted values are then compared with actual realizations to validate the model.

The conditional forecasts by VAR, DFM, and FAVAR models can be implemented within the state space framework using the simulation smoother. Observations within the estimation window are augmented by the conditional variables in the forecast period. The variables for prediction are treated as missing values, and the Kalman smoother provides Monte Carlo imputation of the missing data.

The model specification of Bayesian VAR is the same as that in Chapter 1. In the DFM model, we assume that 6 latent factors S_t follow VAR(1) process and they determine 38 macroeconomic variables Y_t via the observation equation:

$$Y_t = CS_t + \varepsilon_t,$$

where the 38×6 factor loading matrix C is unknown, but parameter estimation of the factor loading has rotational indeterminacy. A workaround is to normalize the loading by a 6×6 invertible matrix, say an identity matrix for the first 6 factors. However, the forecasted Y_t is invariant to rotation. It is possible to ignore rotational indeterminacy and only track the identified variables in the Gibbs sampler. Like the DFM, we assume 6 factors in the FAVAR, but the first three factors are specified as the consumer price index, federal funds rate and reserves of depository institutions. The three observed factors and three latent factors jointly follow VAR(1) process and they determine the remaining macroeconomic variables in Y_t. The first three series in Y_t are the real GDP, the unemployment rate, and the manufacturing capacity utilization. A reasonable identification constraint is to assume that the first three rows of the factor loading take the form

$$\begin{pmatrix} 0 & 0 & 0 & 1 & 0 & 0 \\ 0 & 0 & 0 & 0 & 1 & 0 \\ 0 & 0 & 0 & 0 & 0 & 1 \end{pmatrix}.$$

An interpretation of the constraint is that the monetary shocks have no contemporaneous effect on the real economy, and the three latent factors are closely related to real economic activity. The forecasting results are invariant to the identification constraints.

Figure 8.3 illustrates the VAR conditional forecasts of GDP, consumption, investment, price, unemployment, manufacturing capacity, treasury yield, money stock, and stock price within the forecast interval. The forecasted figures align reasonably well with actual observations for GDP, consumption, unemployment, capacity, yield, and money. Nonetheless, our prediction does not capture the downward trajectory of investment and the fluctuations in the stock market.

Figures 8.4 and 8.5 depict the conditional forecasts by the DFM and the FAVAR, respectively. For most variables, the overall shape of the predictive curves under the VAR and factor models exhibit similarity, indicating that the joint consideration of multiple models provides a robust assessment of the conditional forecasts for the macroeconomy. Although the factor models offer a more concise depiction of the macroeconomy, their forecasting

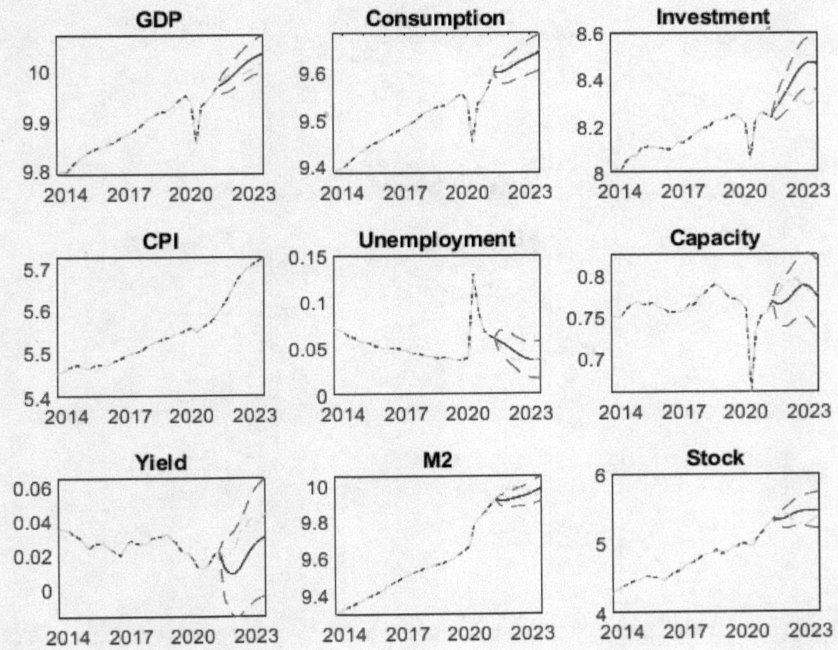

FIGURE 8.3
Conditional forecast by Bayesian VAR. Solid lines are posterior predictive means and dashed lines are 90% credible intervals. Dash-dot lines show actual observations.

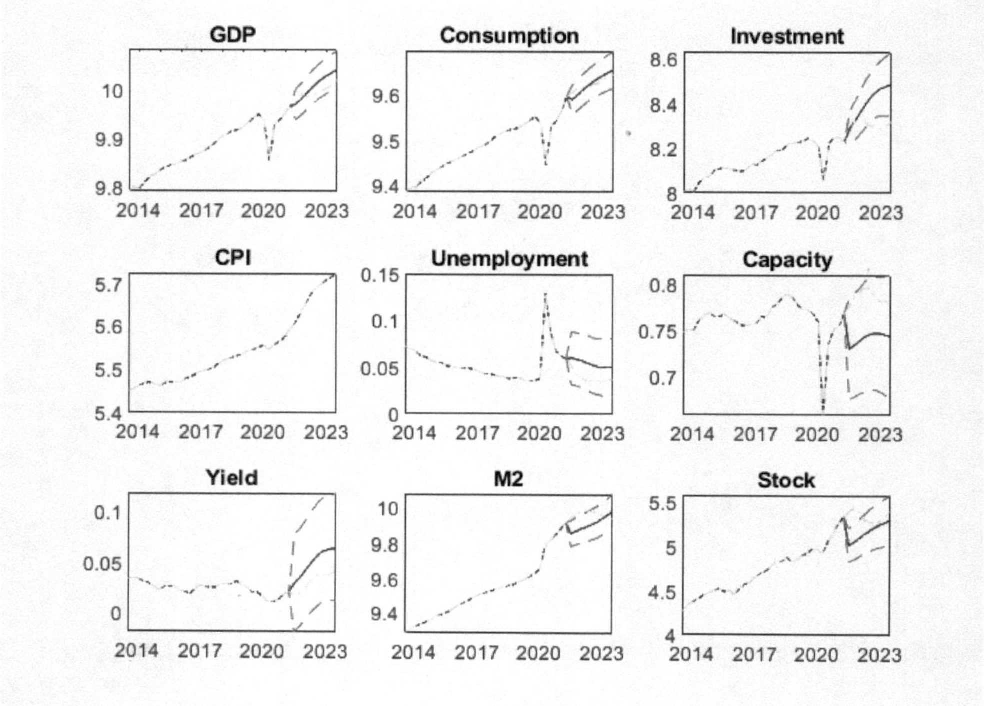

FIGURE 8.4
Conditional forecast by dynamic factor models. Solid lines are posterior predictive means and dashed lines are 90% credible intervals. Dash-dot lines show actual observations.

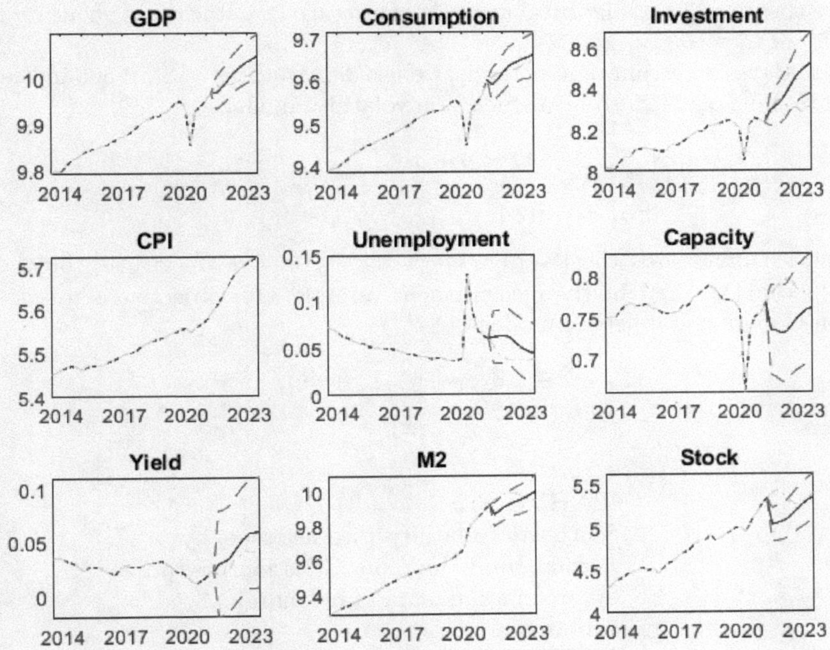

FIGURE 8.5
Conditional forecast by factor-augmented VAR. Solid lines are posterior predictive means and dashed lines are 90% credible intervals. Dash-dot lines show actual observations.

prowess seems marginally inferior to the VAR model. After all, the VAR model has endured criticisms and stood the test of time, continuing as the most favored atheoretic model for macroeconomic forecasting.

8.11.2 Stochastic Volatility with Leverage Effect

It is widely recognized that the conditional variances of asset returns exhibit predictable patterns, whereas the conditional means of returns may not. Forecasting volatility is instrumental for risk management, portfolio optimization, and the pricing of financial derivatives. A salient feature of financial time series is the heightened volatility during price declines. The escalation in the debt-to-equity ratio renders the firm riskier, thereby amplifying the expected volatility. This phenomenon is referred to as the financial leverage effect. For a comprehensive theoretical and empirical analysis of leverage in a stochastic volatility model, see Yu (2005) and Omori et al. (2007).

To illustrate Bayesian estimation methods, we use daily returns of S&P 500 indices from January 2012 to January 2020 to fit a stochastic volatility model:

$$Y_t = \mu + \sigma_t e_t,$$

$$\ln \sigma_t^2 = c + \phi \ln \sigma_{t-1}^2 + \gamma Y_{t-1} + \delta u_t.$$

The model accommodates the leverage effect if $\gamma < 0$. Let $Z_t = \ln \sigma_t^2$ be the state variable and $\tilde{Y}_t = \ln (Y_t - \mu)^2$ be the measurement variable. Our objective is to estimate a linear, yet non-Gaussian state space model:

$$Z_t = c + \phi Z_{t-1} + \gamma Y_{t-1} + \delta u_t,$$

$$\tilde{Y}_t = Z_t + \ln e_t^2.$$

TABLE 8.2
Stochastic volatility with leverage, by normal approximation, Gaussian mixture approximation and double marginalized subsampling.

	Normal	Mixture	Marginal
Intercept	0.040	0.065	0.067
	(0.021)	(0.012)	(0.012)
Constant	−0.010	−0.030	−0.029
	(0.005)	(0.008)	(0.008)
AR1	0.972	0.939	0.939
	(0.005)	(0.008)	(0.008)
Leverage	−0.215	−0.207	−0.196
	(0.019)	(0.022)	(0.021)
Variance	0.032	0.102	0.100
	(0.008)	(0.016)	(0.014)

It may be tempting to fit a Gaussian linear model by approximating $\ln e_t^2$ by a normal distribution with matched mean and variance: $N\left(-1.27, \frac{\pi^2}{2}\right)$. The results are shown in the first column of Table 8.2. At least it provides a reasonable starting value for numerical optimization and MCMC simulation.

However, the noise term $\ln e_t^2$ follows the logarithm of $\chi^2(1)$ distribution, exhibiting skewness with a stretched left tail. It is desirable to approximate the skewed distribution by a finite Gaussian mixture shown in Table 8.1. State space models with finite mixture distributions can be handled by the Gibbs sampler with data augmentation. The sampler cycles through the latent variables of the regime indicators, along with updating states and parameters in MCMC iterations. The simulation results are summarized in the second column of Table 8.2.

As an alternative to distributional approximation, double marginalized subsampling has the exact posterior distribution as the MCMC target distribution. With all model parameters $c, \phi, \gamma, \delta, \mu$ integrated out, the target density $p(Z \mid Y)$ can be evaluated by rank-one update of cumulative statistics.

The empirical results are presented in Table 8.2. The estimated ϕ is 0.94, indicating significant volatility clustering. Past returns are informative about expected future volatility. The estimated γ is about -0.2, providing solid evidence of the leverage effect. The volatility of returns tends to increase as the price plummets.

8.11.3 Random Effects in College Scorecard

The College Scorecard project by the U.S. Department of Education facilitates informed enrollment choices for prospective postsecondary students. It provides institution-level data for comparing costs (e.g., tuition and student loans) and outcomes (e.g., employment rate and earnings after graduation) of higher education. Debt and earning data in specific fields of study within institutions are also available.

The College Scorecard boasts high-quality earning data sourced from W-2 tax forms of students who receive federal financial aid. While not all students rely on financial aid for education, national statistics indicate that over half of undergraduate students utilize federal financial aid, suggesting that the College Scorecard dataset provides valuable insights into the returns of schooling.

Our empirical analysis focuses on the field-of-study College Scorecard data spanning from 2016 to 2020. The principal variable is the earnings of graduates one year after earning their credentials. Due to privacy considerations, some earnings data points are suppressed, resulting in the publicly available data comprising approximately 221,000 observations across more than 5,200 institutions and 370 fields.

We consider a hierarchical linear regression:

$$Y_{ijkt} = \alpha_i + \beta_j + \gamma_k + \delta_t + \sigma\varepsilon_{ijkt},$$

where 1) α_i represents institutional random effects distributed as $N\left(0, \sigma_\alpha^2\right)$, 2) β_j provides field-specific random effects with the distribution $N\left(\mu_\beta, \sigma_\beta^2\right)$, 3) γ_k corresponds to different levels of credentials, including Undergraduate Diplomas, Associate's Degrees, Bachelor's Degrees, Post-Baccalaureate Certificates, Master's Degrees, Doctoral Degrees, First Professional Degrees and Graduate Professional Certificates, and 4) δ_t captures the time effect. As possible values in γ_k and δ_t are moderate, we treat them as fixed effects coded as dummy variables. A column in each set of dummy variables is dropped, so that μ_β reflects the regression intercept. The response variable Y_{ijkt} is the median post-completion earning (in thousand dollar) for the kth degree holders from the jth field at the ith university in the tth year.

As for the prior specification, the variance terms $(\sigma_\alpha^2, \sigma_\beta^2, \sigma^2)$ have inverse-gamma priors, while the intercept μ_β and the coefficients of dummy variables in γ_k and δ_t have normal priors. The Gibbs sampler cycles through the full posterior conditional distributions, which can be implemented by Bayesian linear regressions. For example, to sample α_i conditional on the remaining parameters, we run a linear regression by data points in the ith university. The response variable is $Y_{ijkt} - \beta_j - \gamma_k - \delta_t$, and the regressor consists of a column of ones.

Similarly, β_j is updated by a linear regression with the response variable $Y_{ijkt} - \alpha_i - \gamma_k - \delta_t$ by data points in the jth field.

TABLE 8.3
Posterior means and standard deviations in College Scorecard earning data regressions (in thousand dollar).

	Mean	Std
Undergraduate	−31.34	0.25
Associate	−28.31	0.24
Bachelor	−20.8	0.22
Post-baccalaureate	−6.79	0.62
Master	−3.20	0.22
Doctor	14.24	0.27
FirstProfessional	11.10	0.37
$\delta_t = 2018$	2.58	0.06
$\delta_t = 2019$	1.84	0.06
$\delta_t = 2020$	4.13	0.06
μ_β	56.99	0.90
σ_β^2	268.97	20.08
σ_α^2	54.58	1.32
σ^2	92.16	0.28

Table 8.3 summarizes the posterior means and standard deviations for regression coefficients and noise variances, revealing significant returns to education: higher degrees correlate with increased earnings. For instance, a Ph.D. holder can earn an additional $35,000 compared to a bachelor's degree recipient, who earns roughly $10,000 more than an individual with an undergraduate certificate.

Figure 8.6 presents summary statistics of the field-specific random effects, aggregated at the 2-digit Classification of Instructional Programs code level (with a 2-digit code for display purposes and a 4-digit code utilized in regressions). The box chart illustrates the median, upper (75th percentile) and lower (25th percentile) quartiles of the income distribution, alongside the minimum and maximum values excluding outliers. Notably, graduates in science and engineering fields tend to earn significantly more than their peers from other disciplines.

The regression outcomes also encompass income distributions by institution. Due to space constraints, Table 8.4 displays the wage premiums of the largest universities, measured by the number of credentials conferred during the sample periods. This table offers a quantitative evaluation of graduate outcomes from prestigious universities. A comparison between Figure 8.6 and Table 8.4 highlights that variations in earnings across fields within a single institution are more pronounced than those aggregated across multiple institutions. These empirical findings suggest that for prospective students, selecting an appropriate field of study is paramount over the choice of university.

8.11.4 Liquor Price Elasticity in Hierarchical Linear Model

Price elasticity measures the percentage change in the quantity demanded of a good in response to a one percent change in its price. Given that controlled experiments may be cost-prohibitive, we estimate the price elasticities of distilled spirits using observational data: specifically, all liquor transactions in Iowa from 2012 to 2018.

TABLE 8.4
Posterior means and standard deviations of institutions-specific random effects in College Scorecard earning data regressions.

	Mean	Std		Mean	Std
U. Phoenix Utah	4.27	0.22	U. Georgia	5.04	0.50
Thunderbird	5.78	0.29	California State U. Long Beach	6.13	0.52
Pennsylvania State U.	7.79	0.22	Capella U.	4.39	0.55
Western Governors U.	13.17	0.95	U. Arizona	5.15	0.46
Grand Canyon U.	3.96	0.66	Boston U.	10.56	0.51
Ivy Tech Community College	1.65	0.67	California State U. Fullerton	5.84	0.58
Liberty U.	0.51	0.53	Purdue U. Main Campus	9.14	0.52
U. Washington Seattle	12.95	0.36	U. Houston	6.38	0.54
Southern New Hampshire U.	4.25	0.61	Ultimate Medical Academy	-12.77	1.26
Ohio State U.	6.33	0.37	Georgia State U.	1.47	0.54
Rutgers U. Camden	8.33	0.36	Ohio U. Chillico Campus	3.23	0.35
U. Central Florida	1.75	0.56	U. Cincinnati Main Campus	4.91	0.47
U. Southern California	15.29	0.47	Kent State U. at Ashtabula	-0.76	0.39
Texas A&M U.	8.77	0.45	U. California Davis	7.70	0.57
Florida International U.	0.91	0.54	U. North Texas	3.42	0.54
U. Florida	3.64	0.41	Lone Star College System	1.43	0.96
New York U.	14.58	0.49	U. California Irvine	5.63	0.62
U. South Florida	1.95	0.41	U. South Carolina Columbia	3.62	0.54
U. Texas at Austin	10.67	0.49	U. Alabama	3.82	0.60
U. Maryland	13.39	0.75	Temple U.	5.52	0.51
U. Michigan Ann Arbor	10.75	0.49	U. North Carolina	7.91	0.60
Chamberlain U. Illinois	4.42	1.12	George Mason U.	12.07	0.51
Walden U.	4.07	0.70	U. California San Diego	8.18	0.62
Miami Dade College	1.19	1.05	Northeastern U.	14.52	0.47
U. Texas at Arlington	5.05	0.61	Texas State U.	2.51	0.52
Florida State U.	1.86	0.48	San Diego State U.	5.65	0.57
U. Minnesota Twin Cities	9.24	0.46	Valencia College	0.97	1.13
U. California Los Angeles	10.88	0.54	U. Missouri Columbia	4.52	0.54
Michigan State U.	7.04	0.46	George Washington U.	14.96	0.52
Purdue U. Global Mason City	0.06	0.24	Strayer U.	1.42	0.32
U. Illinois Urbana Champaign	9.34	0.51	U. Massachusetts Amherst	7.79	0.58
American Public U. System	10.51	0.63	U. Pittsburgh Bradford	2.80	0.38
Ashford U.	2.65	0.71	Virginia Tech U.	8.68	0.57
U. California Berkeley	18.60	0.60	North Carolina State U.	6.61	0.52
U. Wisconsin Madison	9.40	0.50	Northern Arizona U.	4.25	0.55
Columbia U.	20.82	0.57	Iowa State U.	7.65	0.55
Indiana U. Bloomington	5.75	0.57	Columbia Southern U.	11.20	1.03
DeVry College of New York	-0.77	0.29	U. Colorado Boulder	6.86	0.63
U. Maryland College Park	11.88	0.58	Johns Hopkins U.	20.89	0.58
California State U. Northridge	5.43	0.57	Texas Tech U.	6.38	0.54

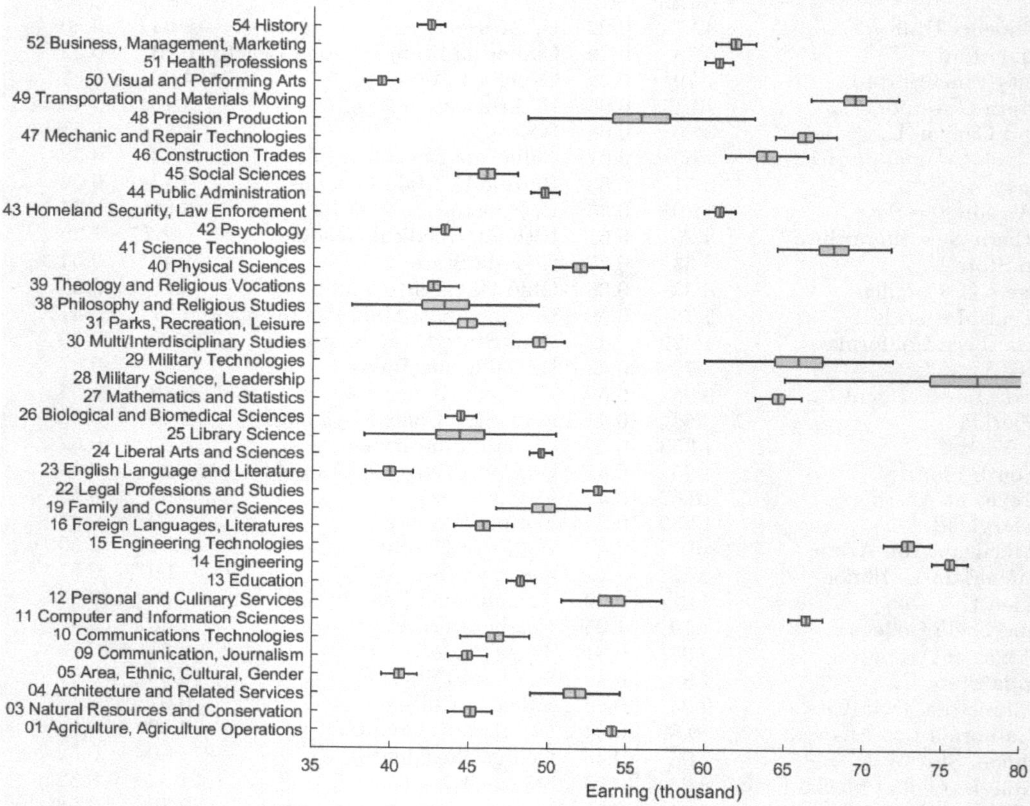

FIGURE 8.6
Posterior distributions of field-specific random effects in College Scorecard regressions. The box chart displays median, lower and upper quartiles, and non-outlier minimum and maximum earnings.

Iowa operates as an alcoholic beverage control state, mandating that private retailers procure spirits from the Alcoholic Beverages Division prior to retail sale. The authority routinely publishes all transactions, forming an extensive open dataset with over 15 million records. Each record corresponds to an invoice detailing the transaction date, store, price, and quantity of a brand-named item (e.g., Crown Royal Canadian whisky), categorized accordingly (e.g., Canadian whisky). The dataset encompasses over 8,000 items across 105 categories.

We preprocess the dataset by selecting 1331 top-selling items, which account for about 88% of the total market share. For each item, we aggregate transaction records to weekly sales volumes. The weekly prices are weighted average of transaction prices. The price and quantities are denoted by p_{it} and q_{it}, which are modeled by a panel data regression:

$$\ln q_{it} = z_i + \beta_0 \ln p_{it} + x_{it}\beta + \varepsilon_{it},$$

where z_i is the unobserved heterogeneity of the ith item. To partial out effects other than own-price changes, we add the weighted average price of other brand-named liquor items in the same category as the ith item, the coefficient of which reflects cross-price elasticities. We also include month and year dummies variables, as the liquor consumption patterns exhibit seasonality due to holidays, school breaks and sport events as well as the time trend.

TABLE 8.5
Posterior means and standard deviations in price elasticity regressions.

	Mean	Std
Intercept	6.035	0.050
Price	−0.714	0.016
CrossPrice	0.123	0.009
Feb	0.058	0.005
Mar	0.080	0.005
Apr	0.144	0.005
May	0.151	0.005
Jun	0.196	0.005
Jul	0.122	0.005
Aug	0.163	0.005
Sep	0.071	0.005
Oct	0.152	0.005
Nov	0.069	0.005
Dec	0.237	0.005
Year2013	−0.048	0.004
Year2014	−0.083	0.004
Year2015	−0.068	0.004
Year2016	−0.049	0.004
Year2017	−0.070	0.004
Year2018	−0.077	0.004
σ^2	0.376	0.001
σ_z^2	1.518	0.059

Empirical findings presented in Table 8.5 indicate that the own-price elasticity is -0.71, suggesting that liquor is an inelastic commodity, potentially due to alcohol addicting. A 1% increase in price leads to a 0.7% decrease in sales volume, thereby elevating sales revenue. The cross-price elasticity for other liquor items within the same category stands at 0.12. This positive value confirms that different brands within a category are considered substitute goods. Regarding seasonal consumption patterns, December, the holiday season, sees liquor sales surge by 24% compared to January. June ranks as the second-highest month for liquor consumption, with sales climbing 20% above January levels.

8.12 Bibliographic Notes

Traditionally, time series are analyzed using ARMA models (Box and Jenkins, 1976). Durbin and Koopman (2012) advocate the state space framework as an alternative to ARMA modeling. A key advantage of state space models is that missing data are handled in a principled manner. Early studies of Jones (1980), Ansley and Kohn (1983) and Harvey and Pierse (1984) demonstrate state space representations of the ARMA process subject to missing data and temporal aggregation, which is closely related to mixed frequency studies in Zadrozny (1988), Mariano and Murasawa (2003, 2010), Schorfheide and Song (2015), among others. Mixed frequency data can also be handled using the MIDAS regression introduced by Ghysels et al. (2006) and Ghysels et al. (2007).

In macroeconomic time series analysis, the debate between structural and data-driven models is longstanding. Keynesian models of simultaneous equations with aggregated variables were prevalent prior to the 1980s. Lucas (1976) criticized these models for neglecting the forward-looking behaviors of decision-makers, who, as rational economic agents, could anticipate and respond to monetary and fiscal policies. Sims (1980) contended that the identifying assumptions in structural equations were not credible and introduced the VAR, which has become a cornerstone model for macroeconomic forecasting. The VAR extensions such as dynamic factor models (Diebold et al., 2006) and TVP-VAR (Cogley and Sargent, 2005) are also in the toolkit of econometricians. In contrast to the data-driven VAR, the DSGE model has a strong degree of theoretical coherence, although a small-scale DSGE model may not capture all features of the business cycle and may not match the out-of-sample forecasting capabilities of reduced-form VAR models. Kydland and Prescott (1982) presented a seminal DSGE model that changed the empirical research of macroeconomics. Real business cycles (Long and Plosser, 1983) can be viewed as a precursor to DSGE modelling. Smets and Wouters (2007) showcased the impressive forecasting power of a larger DSGE model that rivals VAR models.

In panel data analysis, Wooldridge (2010) clarifies the distinctions between fixed and random effects models and discusses classical estimation strategies. In Bayesian literature, panel data are typically studied under hierarchical longitudinal models with random coefficients. The Gibbs sampler, as described by Gelfand and Smith (1990), cycles through the full posterior conditionals, while Chib and Carlin (1999) propose a block sampler that efficiently handles random effects.

In financial time series analysis, GARCH and SV are two primary frameworks for volatility prediction. Engle (1982) and Bollerslev (1986) pioneered the GARCH model, which has since been extended in various forms, such as Exponential GARCH (Nelson, 1991) and Glosten-Jagannathan-Runkle GARCH (Glosten et al., 1993). All GARCH variants share the characteristic that the conditional variance is predetermined as a function of past observations. Conversely, in SV models, the conditional volatility is treated as random. Even in its simplest form, the SV model poses estimation challenges due to its intractable likelihood function. Fleming and Kirby (2003) suggest that the preference for GARCH over SV models is largely due to computational convenience. To address these challenges, various approximation methods have been proposed for state and parameter estimation within Bayesian frameworks. Jacquier et al. (1994) analyze the SV model using a local neighbor sampler, while Kim et al. (1998) approximate the log chi-square distribution with a finite mixture of Gaussian distributions. This approach is appealing because data augmentation restores normality, allowing for the application of the Kalman filter/smoother, conditional on the latent variable indicating the class of the mixture distribution. See also Chib et al. (2002) and Omori et al. (2007) for refined MCMC methods for efficient computation of SV models.

Recent years witness a surge of interest in modeling large shocks such as the pandemic and regional conflicts within the statistical framework of VAR and stochastic volatility. See, for example, Lenza and Primiceri (2022), Bobeica and Hartwig (2023), Antolin-Diaz et al. (2024), Carriero et al. (2024), among others.

8.13 Proof of Propositions

Proposition 8.1 (PPCA)

The closed-form maximum likelihood estimator is derived from step-by-step concentration of the likelihood function:

$$\ln L\left(C, B, \sigma^2\right) = -\frac{Td}{2}\ln 2\pi - \frac{T}{2}\ln|\Omega| - \frac{1}{2}\sum_{t=1}^{T}\left(Y_t - BX_t\right)'\Omega^{-1}\left(Y_t - BX_t\right).$$

First, taking derivatives with respect to B, we obtain the maximum likelihood estimator. The OLS estimator is the maximizer, because of the same set of regressors for all response variables in Y_t.

$$\hat{B}' = \left(\sum_{t=1}^{T} X_t X_t'\right)^{-1}\sum_{t=1}^{T} X_t Y_t'.$$

We replace B by $\hat{B}$ in the likelihood function, which is concentrated as

$$\ln L\left(C, \sigma^2\right) = -\frac{Td}{2}\ln 2\pi - \frac{T}{2}\ln|\Omega| - \frac{1}{2}tr\left(\hat{\Omega}\Omega^{-1}\right),$$

where the sample covariance matrix is

$$\hat{\Omega} = \frac{1}{T}\sum_{t=1}^{T}\left(Y_t - \hat{B}X_t\right)\left(Y_t - \hat{B}X_t\right)'.$$

Second, the constrained covariance matrix is $\Omega = CC' + \sigma^2 I_d$, so its maximum likelihood estimator is not $\hat{\Omega}$. To optimize C, σ^2, we first hold σ^2 as fixed and take derivatives with respect to C by the chain rule, so we have the first-order condition:

$$\left(-\frac{T}{2}\Omega^{-1} - \frac{T}{2}\Omega^{-1}\hat{\Omega}\Omega^{-1}\right)2C = 0,$$

which indicates that the maximum likelihood estimator of C must satisfy the equation:

$$\hat{\Omega}\left(CC' + \sigma^2 I_d\right)^{-1}C = C.$$

In other words, the maximum likelihood estimator of C should be derived from the sample covariance matrix $\hat{\Omega}$ by manipulating the first-order condition.

The $d \times m$ matrix C has the economy-size and full-size SVD representations:

$$C = UDV' = \begin{pmatrix} U & U_0 \end{pmatrix} \begin{pmatrix} D \\ 0 \end{pmatrix} V',$$

where the $d \times m$ matrix U consists of orthonormal vectors, and the $d \times (d-m)$ matrix U_0 serves as the orthogonal complements. The $m \times m$ diagonal matrix D consists of singular values and V is a $m \times m$ orthonormal matrix. Because V disappears in the product CC', it has no effect on the likelihood function, hence an unidentified matrix.

By the full-size SVD representation, we can simplify the matrix inversion:

$$\begin{aligned}(CC' + \sigma^2 I_d)^{-1} &= \begin{pmatrix} U & U_0 \end{pmatrix} \begin{bmatrix} (DD+\sigma^2 I_m)^{-1} & 0 \\ 0 & \sigma^{-2} I_{d-m} \end{bmatrix} \begin{pmatrix} U' \\ U_0' \end{pmatrix} \\ &= U\left(DD+\sigma^2 I_m\right)^{-1} U' + \sigma^{-2} U_0 U_0'.\end{aligned}$$

The orthonormal matrix satisfies $U'U = I$ and $U_0'U = 0$. It follows that the first-order condition in the SVD form is

$$\hat{\Omega} U = U\left(DD + \sigma^2 I_m\right).$$

We inspect the first-order condition column by column. Let u_j be the jth column of U, and s_j be the jth diagonal element of D. We have

$$\hat{\Omega} u_j = \left(s_j^2 + \sigma^2\right) u_j,$$

which indicates that u_j is an eigenvector of $\hat{\Omega}$, and $\lambda_j \equiv s_j^2 + \sigma^2$ is the corresponding eigenvalue of $\hat{\Omega}$. Therefore, s_j takes the form

$$s_j = \sqrt{\lambda_j - \sigma^2}.$$

We have shown that the first-order condition requires U and D derived from the eigenvalues and eigenvectors of the sample covariance matrix $\hat{\Omega}$. The remaining task is to select m out of d eigenvalues to maximize the likelihood function. We need to show the largest eigenvalues are the optimizers.

Third, we plug $U = (u_1, \ldots, u_m)$ and $D = diag\left(\sqrt{\lambda_1 - \sigma^2}, \ldots, \sqrt{\lambda_m - \sigma^2}\right)$ back to the likelihood function for further concentration. Also, we fill in $U_0 = (u_{m+1}, \ldots, u_d)$ to fulfill an orthonormal matrix.

As the population and sample covariance matrices have the decomposition

$$\Omega = \begin{pmatrix} U & U_0 \end{pmatrix} \begin{bmatrix} DD + \sigma^2 I_m & 0 \\ 0 & \sigma^2 I_{d-m} \end{bmatrix} \begin{pmatrix} U' \\ U_0' \end{pmatrix},$$

$$\hat{\Omega} = \begin{pmatrix} U & U_0 \end{pmatrix} \begin{bmatrix} diag(\lambda_1, \ldots, \lambda_m) & 0 \\ 0 & diag(\lambda_{m+1}, \ldots, \lambda_d) \end{bmatrix} \begin{pmatrix} U' \\ U_0' \end{pmatrix},$$

the determinant term in the likelihood function is given by

$$\ln|\Omega| = \sum_{j=1}^{m} \ln \lambda_j + (d-m)\ln\sigma^2,$$

and the trace term in the likelihood function is given by

$$tr\left(\hat{\Omega}\Omega^{-1}\right) = tr\begin{bmatrix} I_m & 0 \\ 0 & \sigma^{-2} diag(\lambda_{m+1}, \ldots, \lambda_d) \end{bmatrix} = m + \sigma^{-2} \sum_{j=m+1}^{d} \lambda_j.$$

The concentrated likelihood function takes the form

$$\ln L = -\frac{T}{2}\left[d\ln 2\pi + \sum_{j=1}^{m}\ln\lambda_j + (d-m)\ln\sigma^2 + m + \sigma^{-2}\sum_{j=m+1}^{d}\lambda_j\right].$$

Taking derivatives with respect to σ^2, *we obtain the maximum likelihood estimator*

$$\hat{\sigma}^2 = \frac{1}{d-m}\sum_{j=m+1}^{d}\lambda_j.$$

Fourth, replacing σ^2 *by* $\hat{\sigma}^2$, *in the likelihood function, we further concentrate the likelihood:*

$$\ln L = -\frac{T}{2}\left[\sum_{j=1}^{m}\ln\lambda_j + (d-m)\ln\frac{1}{d-m}\sum_{j=m+1}^{d}\lambda_j\right] + constant.$$

We can only optimize the order of eigenvalues to maximize the likelihood function. Due to Jensen's inequality and concavity of logarithm, we have

$$\frac{1}{d-m}\sum_{j=m+1}^{d}\ln\lambda_j \leq \ln\frac{1}{d-m}\sum_{j=m+1}^{d}\lambda_j,$$

and the inequality is smaller if $\lambda_{m+1},\ldots,\lambda_d$ *are closer to each other. As the total* $\sum_{j=1}^{d}\ln\lambda_j$ *has been fixed, we need to minimize*

$$\ln\frac{1}{d-m}\sum_{j=m+1}^{d}\lambda_j - \frac{1}{d-m}\sum_{j=m+1}^{d}\ln\lambda_j.$$

The optimal $\lambda_{m+1},\ldots,\lambda_d$ *must be consecutive in a row of the ordered eigenvalues. As we require* $s_j = \sqrt{\lambda_j - \hat{\sigma}^2} \geq 0$, $\forall j \leq m$ *and* $\hat{\sigma}^2$ *is the average of* $\lambda_{m+1},\ldots,\lambda_d$, *we must select* $\lambda_{m+1},\ldots,\lambda_d$ *as the smallest eigenvalues. In other words,* $\lambda_1,\ldots,\lambda_m$ *are the largest eigenvalues.*

In summary, we concentrate the likelihood function in four steps, and we obtain the closed-form maximum likelihood estimator for the unknown parameters B, σ^2, C *such that*

$$\hat{B}' = \left(\sum_{t=1}^{T}X_tX_t'\right)^{-1}\sum_{t=1}^{T}X_tY_t',$$

$$\hat{\sigma}^2 = \frac{1}{d-m}\sum_{j=m+1}^{d}\lambda_j,$$

$$\hat{C} = UDV' = (u_1,\ldots,u_m)\begin{pmatrix}\sqrt{\lambda_1-\hat{\sigma}^2} & & \\ & \ddots & \\ & & \sqrt{\lambda_m-\hat{\sigma}^2}\end{pmatrix}V',$$

where $\lambda_1,\ldots,\lambda_d$ *are the sorted eigenvalues of the sample covariance matrix* $\hat{\Omega}$, *with the corresponding orthonormal eigenvectors* $u_1,\ldots,u_d$. *The arbitrary orthonormal matrix* V *has no effect on the likelihood function, which reflects rotational indeterminacy in the factor model.*

Lastly, if σ^2 is treated as a known parameter and we let $\sigma^2 \to 0$, the concentrated likelihood is dominated by the term $-\frac{T}{2}\sigma^{-2}\sum_{j=m+1}^{d}\lambda_j$. In that case, the largest eigenvalues still maximize the likelihood function.

Proposition 8.2 (Frisch-Waugh-Lovell)
1. The OLS estimator is the solution to the normal equations:

$$\begin{pmatrix} X'X & X'D \\ D'X & D'D \end{pmatrix}\begin{pmatrix} \hat{\beta} \\ \hat{\alpha} \end{pmatrix} = \begin{pmatrix} X' \\ D' \end{pmatrix} Y,$$

from which we first solve for $\hat{\alpha}$ (as a function of $\hat{\beta}$):

$$\hat{\alpha} = (D'D)^{-1}\left(D'Y - D'X\hat{\beta}\right).$$

Plugging $\hat{\alpha}$ into $X'X\hat{\beta} + X'D\hat{\alpha} = X'Y$, we obtain the normal equation for $\hat{\beta}$ alone:

$$X'PX\hat{\beta} = X'PY,$$

where $P = I - D(D'D)^{-1}D'$ is an idempotent matrix: $P = P' = P^2$. It follows that

$$\hat{\beta} = \left(\tilde{X}'\tilde{X}\right)^{-1}\tilde{X}'\tilde{Y},$$

where $\tilde{X} = PX$ and $\tilde{Y} = PY$.
2. Consider the residuals:

$$Y - X\hat{\beta} - D\hat{\alpha} = Y - X\hat{\beta} - D(D'D)^{-1}\left(D'Y - D'X\hat{\beta}\right) = PY - PX\hat{\beta}.$$

It follows that $\hat{\alpha}$ is the solution to linear equations:

$$D\hat{\alpha} = \left(Y - X\hat{\beta}\right) - \left(\tilde{Y} - \tilde{X}\hat{\beta}\right).$$

3. The above equation indicates that residuals of the projected system are the same as residuals of the original system

$$\tilde{Y} - \tilde{X}\hat{\beta} = Y - X\hat{\beta} - D\hat{\alpha}.$$

4. Consider inverting the 2×2 block matrix $\begin{pmatrix} X'X & X'D \\ D'X & D'D \end{pmatrix}$. The upper-left block of the inverted matrix takes the form:

$$\left[X'X - X'D(D'D)^{-1}D'X\right]^{-1},$$

which equals to $(X'PX)^{-1}$, or $\left(\tilde{X}'\tilde{X}\right)^{-1}$. Also, an estimator for σ_ε^2 is obtained by the mean squared residuals (which are the same for the transformed and original systems):

$$\hat{\sigma}_\varepsilon^2 = \frac{1}{nT - n - k}\left(\tilde{Y} - \tilde{X}\hat{\beta}\right)'\left(\tilde{Y} - \tilde{X}\hat{\beta}\right),$$

where the degrees of freedom are adjusted to reflect n regressors that are projected out.

Proposition 8.3 (GARCH process)

1. Because e_t is independent to σ_t (which is a function of $Y_{1:t-1}$), the equation $\varepsilon_t = \sigma_t e_t$ indicates that

$$E(\varepsilon_t) = E(\sigma_t) E(e_t) = 0,$$
$$E(\varepsilon_t^2) = E(\sigma_t^2) E(e_t^2) = E(\sigma_t^2),$$
$$E(\varepsilon_t \varepsilon_{t-j}) = E(e_t) E(\sigma_t \sigma_{t-j} e_{t-j}) = 0.$$

In particular, the identity $E(\varepsilon_t^2) = E(\sigma_t^2)$, $\forall t \geq 1$ will be used extensively throughout the proof.

Whether $E(\sigma_t^2)$ is time-varying depends on our initialization method. Suppose that we initialize σ_0^2 and ε_0^2 such that $E(\sigma_0^2) = E(\varepsilon_0^2) = \frac{c}{1-\phi-\psi}$. By induction, if $E(\sigma_{t-1}^2) = \frac{c}{1-\phi-\psi}$, then

$$E(\sigma_t^2) = c + \phi E(\sigma_{t-1}^2) + \psi E(\varepsilon_{t-1}^2) = \frac{c}{1-\phi-\psi}.$$

In summary, $\{\varepsilon_t\}$ follows a white noise process with the stationary variance $\frac{c}{1-\phi-\psi}$ for all periods, if the process is initialized by the stationary variance.

2. Now we consider the process initialized by an arbitrary variance. The identity $E(\varepsilon_t^2) = E(\sigma_t^2), \forall t \geq 1$ still holds. Taking the expectation of $\sigma_t^2 = c + \phi\sigma_{t-1}^2 + \psi\varepsilon_{t-1}^2$, we have

$$E(\sigma_t^2) = c + (\phi + \psi) E(\sigma_{t-1}^2).$$

The parameter constraint $\phi + \psi < 1$ indicates a contraction mapping, and we have

$$E(\sigma_t^2) = (\phi+\psi)^t E(\sigma_0^2) + c \sum_{\tau=0}^{t-1} (\phi+\psi)^\tau.$$

The asymptotic variance is given by

$$\lim_{t\to\infty} E(\varepsilon_t^2) = \lim_{t\to\infty} E(\sigma_t^2) = \frac{c}{1-\phi-\psi}.$$

In summary, the unconditional variance converges to the stationary variance under an arbitrary initialization scheme, provided that the GARCH parameters are subject to the constraint $\phi + \psi < 1$.

3. We have the identity:

$$\begin{aligned}
\varepsilon_t^2 &= \sigma_t^2 + v_t \\
&= c + \phi\sigma_{t-1}^2 + \psi\varepsilon_{t-1}^2 + v_t \\
&= c + (\phi+\psi)\varepsilon_{t-1}^2 + v_t - \phi v_{t-1}.
\end{aligned}$$

It remains to show $\{v_t\}$ is a white noise process. By construction,

$$v_t = \varepsilon_t^2 - \sigma_t^2 = \sigma_t^2 (e_t^2 - 1).$$

Because e_t is independent to σ_t^2, we have $E(v_t) = 0$, $E(v_t^2) = 2E(\sigma_t^4)$ and $E(v_t v_{t-j}) = 0$.

4. By the equation $\varepsilon_t = \sigma_t e_t$, we have

$$E(\varepsilon_t^4) = E(\sigma_t^4) E(e_t^4) = 3E(\sigma_t^4).$$

Taking the expectation of the squared equation $\sigma_t^2 = c + \phi\sigma_{t-1}^2 + \psi\varepsilon_{t-1}^2$, we have

$$E(\sigma_t^4) = c^2 \frac{1+\phi+\psi}{1-\phi-\psi} + (3\psi^2 + \phi^2 + 2\phi\psi) E(\sigma_{t-1}^4).$$

The parameter constraint $3\psi^2 + \phi^2 + 2\phi\psi < 1$ indicates a contraction mapping. Asymptotically, we have

$$\lim_{t\to\infty} E\left(\sigma_t^4\right) = \frac{c^2\left(1+\phi+\psi\right)}{\left(1-\phi-\psi\right)\left(1-3\psi^2-\phi^2-2\phi\psi\right)}.$$

In summary, the squared sequence $\{\varepsilon_t^2\}$ follows an ARMA(1,1) process, in which the noise term $\{v_t\}$ is a zero-mean and uncorrelated sequence, with the asymptotic variance converging to a constant.

5. By the marginal-conditional decomposition, the likelihood function is given by

$$p\left(Y_{1:T}\right) = \prod_{t=1}^{T} p\left(Y_t \left|Y_{1:t-1}\right.\right) = \prod_{t=1}^{T} \frac{1}{\sqrt{2\pi}\sigma_t} e^{-\frac{\left(Y_t - X_t\beta\right)^2}{2\sigma_t^2}},$$

where $p\left(Y_t \left|Y_{1:t-1}\right.\right)$ is a univariate normal density, with the mean $X_t\beta$ and the variance σ_t^2. However, $p\left(Y_{1:T}\right)$ is not a multivariate normal density. We show the proposition by contradiction. Suppose that $Y_{1:T}$ follows a multivariate normal distribution in the form:

$$\left(\begin{array}{c} Y_T \\ Y_{1:T-1} \end{array}\right) \sim N\left[\left(\begin{array}{c} \mu_1 \\ \mu_2 \end{array}\right), \left(\begin{array}{cc} \Sigma_{11} & \Sigma_{12} \\ \Sigma_{21} & \Sigma_{22} \end{array}\right)\right],$$

then we have

$$Var\left(Y_T \left|Y_{1:T-1}\right.\right) = \Sigma_{11} - \Sigma_{12}\Sigma_{22}^{-1}\Sigma_{21}.$$

It indicates that the conditional variance of a multivariate normal distribution is not a function of the conditioning variables $Y_{1:T-1}$. However, in the GARCH model, the conditional variance is a function of $Y_{1:T-1}$:

$$Var\left(Y_T \left|Y_{1:T-1}\right.\right) = \sigma_T^2 = f\left(Y_{1:T-1}\right).$$

We conclude that the GARCH likelihood function is not a multivariate normal density.

8.14 Exercises

Exercise 8.1 (ARMA representation) *Let $\{Z_t\}$ be an $AR(p)$ series*

$$Z_t = \sum_{j=1}^{p} \phi_j Z_{t-j} + \sigma\varepsilon_t.$$

Consider weighted sum of the lagged AR series

$$Y_t = Z_t + \sum_{j=1}^{q} \psi_j Z_{t-j}.$$

Show that $\{Y_t\}$ follows the $ARMA(p,q)$ process

$$Y_t = \sum_{j=1}^{p} \phi_j Y_{t-j} + \sigma\varepsilon_t + \sigma\sum_{j=1}^{q} \psi_j \varepsilon_{t-j}.$$

Exercise 8.2 (Companion matrix) *The companion matrix is crucial for the ARMA analysis, as its eigenvalues determine the stationarity of the time series. A $p \times p$ companion matrix takes the form*

$$F = \begin{pmatrix} \phi_1 & \phi_2 & \cdots & \phi_{p-1} & \phi_p \\ 1 & 0 & \cdots & 0 & 0 \\ 0 & 1 & \cdots & 0 & 0 \\ \cdots & \cdots & \ddots & \cdots & \cdots \\ 0 & 0 & \cdots & 1 & 0 \end{pmatrix}.$$

By definition, the eigenvalues satisfy

$$|F - \lambda I| = 0.$$

1. *Use the definition of the determinant operator to show that*

$$|F - \lambda I| = (\phi_1 - \lambda)\,\lambda^{p-1} + \phi_2\lambda^{p-2} + \phi_3\lambda^{p-3} + \cdots + \phi_p.$$

2. *Show that the eigenvalues are solutions to the characteristic equation*

$$\lambda^p - \phi_1\lambda^{p-1} - \phi_2\lambda^{p-2} - \cdots - \phi_p = 0.$$

Exercise 8.3 (Lyapunov equation) *Consider the Lyapunov equation*

$$\Gamma_0 = A\Gamma_0 A' + BB'.$$

Show that it has a positive definite solution Γ_0 such that

$$vec\,(\Gamma_0) = (I - A \otimes A)^{-1}\, vec\,(BB'),$$

if all eigenvalues of A are in the unit circle.

Exercise 8.4 (Seasonal Effects) *Suppose that we have k seasons, and we create dummies d_{it}, $i = 1, \ldots, k$, with the identity $\sum_{i=1}^{k} d_{it} = 1$, $\forall t$. Also, we include an intercept term in the regression:*

$$Y_t = \gamma_0 + \sum_{i=1}^{k} \gamma_i d_{it} + \sigma_\varepsilon \varepsilon_t.$$

Because of perfect multicollinearity (the rank of $k+1$ regressors is at most k), we must take some normalization method. Show that normalization by $\sum_{i=1}^{k} \gamma_i = 0$ is equivalent to removing one dummy variable.

Exercise 8.5 (Conditional likelihood of VAR) *We write the VAR model*

$$Y_t = \sum_{j=1}^{p} \Phi_j Y_{t-j} + \varepsilon_t$$

in the linear regression form

$$Y_t' = X_t B + \varepsilon_t',$$

where $X_t = \left(Y'_{t-1}, \ldots, Y'_{t-p}\right)$ *and* $B = (\Phi_1, \ldots, \Phi_p)'$. *We reserve the first p observations as the presample data, conditional on which we run the linear regression.*

1. Show that the conditional likelihood function takes the form

$$p(Y|B,\Sigma) \propto |\Sigma|^{-\frac{T}{2}} \exp\left[-\frac{1}{2}\sum_{t=1}^{T}(Y'_t - X_t B)\,\Sigma^{-1}(Y'_t - X_t B)'\right].$$

2. Show that an alternative form of the conditional likelihood function is

$$p(Y|B,\Sigma) \propto |\Sigma|^{-\frac{T}{2}} \exp\left[-\frac{1}{2}tr\sum_{t=1}^{T}(Y'_t - X_t B)(Y'_t - X_t B)'\,\Sigma^{-1}\right],$$

where $tr(\cdot)$ *is the trace of a matrix (the sum of diagonal elements). Is the alternative form reminiscent of the NIW distribution?*

3. Show that the conditional likelihood is maximized by the least squares estimator

$$\hat{B} = \left(\sum_{t=1}^{T} X'_t X_t\right)^{-1} \sum_{t=1}^{T} X'_t Y'_t.$$

How is $\hat{B}$ *related to Bayesian linear regressions?*

Exercise 8.6 (Minnesota prior) *The Minnesota prior imposes shrinkage on the VAR coefficients such that a high-dimensional and high-order VAR process reduces to a collection of one-dimensional and low-order AR processes, such as the random walks. To fix ideas, we consider a bivariate VAR(2) process:*

$$Y_{1t} = \phi_{11,1}Y_{1,t-1} + \phi_{12,1}Y_{2,t-1} + \phi_{11,2}Y_{1,t-2} + \phi_{12,2}Y_{2,t-2} + \varepsilon_{1t},$$

$$Y_{2t} = \phi_{21,1}Y_{1,t-1} + \phi_{22,1}Y_{2,t-1} + \phi_{21,2}Y_{1,t-2} + \phi_{22,2}Y_{2,t-2} + \varepsilon_{2t}.$$

Our goal is shrinkage toward the univariate random walks:

$$Y_{1t} = Y_{1,t-1} + \varepsilon_{1t},$$

$$Y_{2t} = Y_{2,t-1} + \varepsilon_{2t}.$$

To do so, we wish to impose tight priors on the VAR coefficients

$$B = \begin{pmatrix} \phi_{11,1} & \phi_{12,1} & \phi_{11,2} & \phi_{12,2} \\ \phi_{21,1} & \phi_{22,1} & \phi_{21,2} & \phi_{22,2} \end{pmatrix}',$$

or the vectorized version

$$\beta = \begin{pmatrix} \phi_{11,1} & \phi_{12,1} & \phi_{11,2} & \phi_{12,2} & \phi_{21,1} & \phi_{22,1} & \phi_{21,2} & \phi_{22,2} \end{pmatrix}'.$$

1. Consider the semi-conjugate prior: $\Sigma \sim IW(\Omega, v)$ *and* $\beta|\Sigma \sim N(\mu, V)$, *where the hyperparameters* μ *and* V *are specified as*

$$\mu = \begin{pmatrix} 1 & 0 & 0 & 0 & 0 & 1 & 0 & 0 \end{pmatrix}',$$

$$V = diag\begin{pmatrix} v_o & v_c & \frac{v_o}{2^d} & \frac{v_c}{2^d} & v_c & v_o & \frac{v_c}{2^d} & \frac{v_o}{2^d} \end{pmatrix}.$$

Discuss the effects of tuning parameters v_o, v_c *and* d *on the strength of shrinkage.*

2. Consider the conjugate prior: $\Sigma \sim IW(\Omega, v)$ *and* $\beta\,|\Sigma \sim N\left(\mu, \Sigma \otimes \Lambda^{-1}\right)$, *where the hyperparameters* μ *and* Λ^{-1} *are specified as*

$$\mu = \begin{pmatrix} 1 & 0 & 0 & 0 & 0 & 1 & 0 & 0 \end{pmatrix}',$$

$$\Lambda^{-1} = diag\begin{pmatrix} v_o & v_o & \frac{v_o}{2^d} & \frac{v_o}{2^d} \end{pmatrix}.$$

Why do we impose the constraint $v_o = v_c$ *and only use the tuning parameters* v_o *and* d *for controlling the strength of shrinkage? Is it possible to shrink the higher-order autoregressive terms? Is it possible to shrink cross-lags against self-lags?*

Exercise 8.7 (Semi-conjugate prior) *Consider Bayesian VAR under the semi-conjugate prior such that* $\Sigma \sim IW(\Omega, v)$ *and* $\beta\,|\Sigma \sim N(\mu, V)$:

$$p(\Sigma) \propto |\Sigma|^{-\frac{v+m+1}{2}} \exp\left[-\frac{1}{2} tr\left(\Omega\Sigma^{-1}\right)\right],$$

$$p(\beta\,|\Sigma) \propto \exp\left[-\frac{1}{2}(\beta-\mu)' V^{-1}(\beta-\mu)\right].$$

The likelihood function is given by

$$p(Y\,|B,\Sigma) \propto |\Sigma|^{-\frac{T}{2}} \exp\left[-\frac{1}{2}\sum_{t=1}^{T}(Y_t' - X_t B)\,\Sigma^{-1}(Y_t' - X_t B)'\right].$$

or in the alternative form:

$$p(Y\,|B,\Sigma) \propto |\Sigma|^{-\frac{T}{2}} \exp\left[-\frac{1}{2} tr \sum_{t=1}^{T}(Y_t' - X_t B)'(Y_t' - X_t B)\,\Sigma^{-1}\right],$$

where β *and* B *represent the same set of variables, namely* $\beta = vec(B)$.

1. Show that the posterior conditional distribution of Σ *is given by* $\Sigma\,|\beta, Y \sim IW(\bar{\Omega}, \bar{v})$, *where*

$$\bar{\Omega} = \Omega + \sum_{t=1}^{T}(Y_t' - X_t B)'(Y_t' - X_t B),$$

$$\bar{v} = T + v.$$

2. To derive the posterior conditional distribution of β, *we need some algebra. We start from a quadratic term in the likelihood function. Show the following identity (Hint: consider the trace operator).*

$$\sum_{t=1}^{T}(Y_t' - X_t B)\,\Sigma^{-1}(Y_t' - X_t B)'$$

$$= \sum_{t=1}^{T} Y_t'\Sigma^{-1}Y_t - 2\beta' vec\left(\sum_{t=1}^{T} X_t' Y_t' \Sigma^{-1}\right) + \beta'\left(\Sigma^{-1} \otimes \sum_{t=1}^{T} X_t' X_t\right)\beta.$$

In the prior, we also have a quadratic term:

$$(\beta-\mu)' V^{-1}(\beta-\mu) = \beta' V^{-1}\beta - 2\beta' V^{-1}\mu + \mu' V^{-1}\mu.$$

By completing the square, show that the posterior conditional distribution takes the form: $\beta\,|\Sigma, Y \sim N(\bar{\mu}, \bar{V})$, *where*

$$\bar{\mu} = \left(V^{-1} + \Sigma^{-1} \otimes \sum_{t=1}^{T} X_t' X_t\right)^{-1} \left[V^{-1}\mu + vec\left(\sum_{t=1}^{T} X_t' Y_t' \Sigma^{-1}\right)\right],$$

$$\bar{V} = \left(V^{-1} + \Sigma^{-1} \otimes \sum_{t=1}^{T} X_t' X_t\right)^{-1}.$$

3. An alternative representation of the posterior distribution can be obtained by rewriting $Y_t' = X_t B + \varepsilon_t'$ *as*

$$Y_t = \chi_t \beta + \varepsilon_t,$$

where $\chi_t = diag(X_t, \ldots, X_t)$ *is a block diagonal matrix in which the predictors replicate themselves* m *times. The likelihood function is given by*

$$p(Y\,|\beta, \Sigma) \propto |\Sigma|^{-\frac{T}{2}} \exp\left[-\frac{1}{2}\sum_{t=1}^{T} (Y_t - \chi_t\beta)' \Sigma^{-1} (Y_t - \chi_t\beta)\right].$$

To complete the square, we note that
$(\beta - \mu)' V^{-1} (\beta - \mu) + \sum_{t=1}^{T} (Y_t - \chi_t\beta)' \Sigma^{-1} (Y_t - \chi_t\beta)$
$= (\beta - \bar{\mu})' \bar{V}^{-1} (\beta - \bar{\mu}) - \bar{\mu}' \bar{V}^{-1} \bar{\mu} + \mu V^{-1} \mu + \sum_{t=1}^{T} Y_t' \Sigma^{-1} Y_t.$
where

$$\bar{\mu} = \left(V^{-1} + \sum_{t=1}^{T} \chi_t' \Sigma^{-1} \chi_t\right)^{-1} \left(V^{-1}\mu + \sum_{t=1}^{T} \chi_t' \Sigma^{-1} Y_t\right),$$

$$\bar{V} = \left(V^{-1} + \sum_{t=1}^{T} \chi_t' \Sigma^{-1} \chi_t\right)^{-1}.$$

Show that the posterior conditional distributions are given by $\beta\,|\Sigma, Y \sim N(\bar{\mu}, \bar{V})$. *Are the results conformable to those we derived previously?*

Exercise 8.8 (One-step-ahead forecast) *The predictive distribution* $p(Y_{T+h}\,|Y)$ *in the VAR model is not analytically tractable, but there is an exception. Under the conjugate NIW prior, the one-step-ahead predictive distribution has a closed form.*

Consider the VAR model $Y_t' = X_t B + \varepsilon_t'$ *and the conjugate prior* $NIW(U, \Lambda, \Omega, v)$ *for* B, Σ. *The posterior distribution is* $NIW(\bar{U}, \bar{\Lambda}, \bar{\Omega}, \bar{v})$, *where*

$$\bar{U} = (\Lambda + X'X)^{-1} (\Lambda U + X'Y),$$

$$\bar{\Lambda} = \Lambda + X'X,$$

$$\bar{\Omega} = \Omega + Y'Y + U'\Lambda U - \bar{U}'\bar{\Lambda}\bar{U},$$

$$\bar{v} = T + v.$$

For the one-step-ahead forecast, the predictors $X_{T+1} = (Y_T', \ldots, Y_{T-p+1}')$ *are predetermined conditional on the estimation sample.*

1. Consider conditional forecast under a fixed covariance matrix Σ. *Show that* $p(Y_{T+h}\,|Y, \Sigma)$ *is normally distributed with mean and variance given by*

$$E(Y_{T+h}\,|Y, \Sigma) = \bar{U}' X_{T+1}',$$

$$Var\left(Y_{T+h}\left|Y,\Sigma\right.\right)=\left(X_{T+1}\bar{\Lambda}^{-1}X'_{T+1}+1\right)\Sigma.$$

2. However, the covariance matrix Σ itself is a random variable, with the posterior distribution $\Sigma\left|Y\right.\sim IW\left(\bar{\Omega},\bar{v}\right)$. Therefore, $p\left(Y_{T+h}\left|Y\right.\right)$ can be viewed as a scale mixture of normal distributions, and the marginal distribution is a multivariate t distribution. Show that the mean and variance are given by

$$E\left(Y_{T+h}\left|Y\right.\right)=\bar{U}'X'_{T+1},$$

$$Var\left(Y_{T+h}\left|Y\right.\right)=\frac{1}{\bar{v}-m+1}\left(X_{T+1}\bar{\Lambda}^{-1}X'_{T+1}+1\right)\bar{\Omega}.$$

Exercise 8.9 (Mixed frequency dynamic factor model) *Consider a mixed frequency dynamic factor model, in which the observations include two monthly series and a quarterly series:*

$$S_t=AS_{t-1}+u_t,$$
$$\tilde{Y}_t=CS_t+\varepsilon_t,$$

where $\tilde{Y}_t=\left(\tilde{Y}_{1t},\tilde{Y}_{2t},\tilde{Y}_{3t}\right)'$ represents the latent monthly data, while the actual data satisfy

$$Y_{1t}=\tilde{Y}_{1t},t=1,2,\ldots$$
$$Y_{2t}=\tilde{Y}_{2t},t=1,2,\ldots$$
$$Y_{3t}=\sum_{j=0}^{2}\tilde{Y}_{3,t-j},t=3,6,\ldots$$

How to write the model in the standard state space form for the Kalman filtering?

Exercise 8.10 (TVP-VAR posterior conditionals) *Consider the TVP-VAR model in the standardized state space form:*

$$Y_t=X_t\beta_0+X_t\Pi\tilde{\beta}_t+\varepsilon_t,$$
$$\tilde{\beta}_t=\tilde{\beta}_{t-1}+u_t,$$

where $E\left(\varepsilon_t\varepsilon'_t\right)=\Sigma$ and $E\left(u_tu'_t\right)=I$.

1. Show that the diagonal matrix satisfies the identity:

$$\Pi\tilde{\beta}_t=diag\left(\tilde{\beta}_t\right)\cdot diag\left(\Pi\right),$$

in which $diag\left(\tilde{\beta}_t\right)$ refers to expanding the vector $\tilde{\beta}_t$ as a diagonal matrix, while $diag\left(\Pi\right)$ refers to extracting diagonal elements from the diagonal matrix Π. Explain why the identity is useful for the posterior simulation.

2. By the simulation smoother, we obtain posterior draws of $\tilde{\beta}_t$, conditional on which the model reduces to a multiple-equation linear regression:

$$Y_t=Z_tb+\varepsilon_t,$$

where

$$Z_t=\left[\begin{array}{cc}X_t & X_t\cdot diag\left(\tilde{\beta}_t\right)\end{array}\right],$$

$$b = \begin{bmatrix} \beta_0 \\ diag\,(\Pi) \end{bmatrix}.$$

We specify a semi-conjugate prior: $\Sigma \sim IW(\Omega, v)$ *and* $b|\Sigma \sim N(\mu, V)$. *Show that the posterior conditional distributions remain normal and inverse Wishart distributions:*

$$b\left|Y, \tilde{\beta}, \Sigma \sim N(\bar{\mu}, \bar{V}),\right.$$

$$\Sigma\left|Y, \tilde{\beta}, b \sim IW(\bar{\Omega}, \bar{v}),\right.$$

where

$$\bar{\mu} = \left(V^{-1} + \sum_{t=1}^{T} Z_t'\Sigma^{-1}Z_t\right)^{-1} \left(V^{-1}\mu + \sum_{t=1}^{T} Z_t'\Sigma^{-1}Y_t\right),$$

$$\bar{V} = \left(V^{-1} + \sum_{t=1}^{T} Z_t'\Sigma^{-1}Z_t\right)^{-1},$$

$$\bar{\Omega} = \Omega + \sum_{t=1}^{T} (Y_t - Z_t b)(Y_t - Z_t b)',$$

$$\bar{v} = v + T.$$

3. Is it possible to specify a conjugate NIW prior like $\Sigma \sim IW(\Omega, v)$ *and* $b|\Sigma \sim N(\mu, \Sigma \otimes \Lambda^{-1})$*?*

Exercise 8.11 (Transition equation of stochastic volatility) *Consider a stochastic volatility model*

$$Z_t = c + \phi Z_{t-1} + \delta u_t,$$
$$\tilde{Y}_t = Z_t + \ln e_t^2.$$

Conditional on the state sequence $\{Z_t\}$, *we are interested in the posterior conditional distributions of* c, ϕ, δ^2.

1. Suppose that we specify the conjugate prior: $c, \phi, \delta^2 \sim NIG(\mu, \Lambda, a, b)$. *Show that the posterior conditional distribution is* $NIG(\bar{\mu}, \bar{\Lambda}, \bar{a}, \bar{b})$, *where*

$$\bar{\mu} = \left(\Lambda + \sum_{t=1}^{T} W_t'W_t\right)^{-1} \left(\Lambda\mu + \sum_{t=1}^{T} W_t'Z_t\right),$$

$$\bar{\Lambda} = \Lambda + \sum_{t=1}^{T} W_t'W_t,$$

$$\bar{a} = a + \frac{T}{2},$$

$$\bar{b} = b + \frac{1}{2}\sum_{t=1}^{T} Z_t^2 + \frac{1}{2}\mu'\Lambda\mu - \frac{1}{2}\bar{\mu}'\bar{\Lambda}\bar{\mu},$$

$$W_t = (1, Z_{t-1}).$$

2. In many real-world financial datasets, the volatility is highly persistent, which implies that the coefficient ϕ *is close to one. However, it is not supposed to exceed one, because* Z

would diverge in an explosive system. We wish to impose a constraint $\phi < 1$ by the prior. Consider a Gibbs sampler that generates draws of c, ϕ, δ^2 conditional on Z, and generates draws of Z conditional on c, ϕ, δ^2. John and Mary implement the Gibbs sampler differently. John generates δ^2 from $IG(\bar{a}, \bar{b})$ and then generates a proposal c, ϕ from $N(\bar{\mu}, \delta^2\bar{\Lambda}^{-1})$. If $\phi > 1$, he rejects the proposal and generates another draw from the normal distribution until the proposal is accepted. Mary generates δ^2 from $IG(\bar{a}, \bar{b})$ and then generates a proposal c, ϕ from $N(\bar{\mu}, \delta^2\bar{\Lambda}^{-1})$. If $\phi > 1$, she rejects c, ϕ, δ^2, and generates another IG draw and normal draw until the proposal is accepted. Who is correct? What is the exact prior specification?

Exercise 8.12 (Rejection sampling of stochastic volatility) *Consider a stochastic volatility model*

$$Z_t = c + \phi Z_{t-1} + \delta u_t,$$
$$Y_t = X_t\beta + e^{\frac{1}{2}Z_t} e_t,$$

where e_t and u_t are independent standard normal disturbances. The single-move Gibbs sampler generates Z_t from the posterior conditional distribution

$$p(Z_t | Y_{1:T}, Z_{-t}, c, \phi, \delta, \beta) \propto p(Z_t | Z_{t-1}, c, \phi, \delta)\, p(Z_{t+1} | Z_t, c, \phi, \delta)\, p(Y_t | Z_t, \beta),$$

where

$$p(Z_t | Z_{t-1}, c, \phi, \delta) \propto e^{-\frac{1}{2}\delta^{-2}(Z_t - c - \phi Z_{t-1})^2},$$
$$p(Z_{t+1} | Z_t, c, \phi, \delta) \propto e^{-\frac{1}{2}\delta^{-2}(Z_{t+1} - c - \phi Z_t)^2},$$
$$p(Y_t | Z_t, \beta) \propto e^{-\frac{1}{2}Z_t - \frac{1}{2}(Y_t - X_t\beta)^2 e^{-Z_t}}.$$

1. The target density is the product of three terms, among which the first two are determined by the transition equation, and the third one depends on the observation equation. Suppose that we ignore the observation equation and only utilize the transition equation. Show that

$$p(Z_t | Y_{1:T}, Z_{t-1}, Z_{t+1}, c, \phi, \delta) \propto p(Z_t | Z_{t-1}, c, \phi, \delta)\, p(Z_{t+1} | Z_t, c, \phi, \delta)$$

is the density of the normal distribution $N(Z_t^*, \delta^{*2})$, where

$$Z_t^* = \frac{(1-\phi)c + \phi(Z_{t-1} + Z_{t+1})}{1+\phi^2},$$

$$\delta^{*2} = \frac{\delta^2}{1+\phi^2}.$$

2. Show that the log density of the target distribution $\ln p(Z_t | Y_{1:T}, Z_{-t}, c, \phi, \delta, \beta)$ is globally concave, so that the adaptive rejection sampling method is applicable.

3. To implement the rejection sampling, we propose an easy-to-sample distribution and calculate the upper bound of the target and proposal distributions. Note that

$$e^{-Z_t} \geq e^{-Z_t^*} - e^{-Z_t^*}(Z_t - Z_t^*).$$

We use this inequality to construct our proposal:

$$\tilde{p}(Z_t | Y_{1:T}, Z_{-t}, c, \phi, \delta, \beta) \propto p(Z_t | Z_{t-1}, c, \phi, \delta)\, p(Z_{t+1} | Z_t, c, \phi, \delta)\, \tilde{p}(Y_t | Z_t, \beta),$$
$$\tilde{p}(Y_t | Z_t, \beta) \propto e^{-\frac{1}{2}Z_t - \frac{1}{2}(Y_t - X_t\beta)^2\left[e^{-Z_t^*} - e^{-Z_t^*}(Z_t - Z_t^*)\right]}.$$

The only difference between the target and proposal distributions is that the term e^{-Z_t} *in the target distribution is replaced by the linear term* $e^{-Z_t^*} - e^{-Z_t^*}(Z_t - Z_t^*)$ *for the proposal. Show that the proposal is a normal distribution:* $N\left(Z_t^{**}, \delta^{**2}\right)$, *where*

$$Z_t^{**} = Z_t^* + \frac{1}{2}\delta^{*2}\left[(Y_t - X_t\beta)^2 e^{-Z_t^*} - 1\right],$$

$$\delta^{**2} = \frac{\delta^2}{1+\phi^2}.$$

Show that the ratio of the target and proposal distributions has the upper bound

$$\frac{p\left(Z_t\left|Y_{1:T}, Z_{-t}, c, \phi, \delta, \beta\right.\right)}{\tilde{p}\left(Z_t\left|Y_{1:T}, Z_{-t}, c, \phi, \delta, \beta\right.\right)} \leq 1.$$

Exercise 8.13 (PPCA Prior) *The factors in PPCA are assumed to follow the standard normal distributions. In this exercise we consider how the state and parameter estimation results change if we specify a different prior variance, say* $S_t \sim N(0, \kappa I)$. *The observation equation is still given by*

$$Y_t = CS_t + BX_t + \sigma\varepsilon_t.$$

1. Given the model parameters C, B, σ^2, *how does the state estimation in the form of the posterior conditional distribution* $p\left(S_t\left|Y_t, C, B, \sigma^2\right.\right)$ *change with the prior variance* κ?

2. How does the maximum likelihood estimator of C, B, σ^2 *change with the prior variance* κ?

Exercise 8.14 (Sylvester equation) *For the second-order perturbation solver of nonlinear DSGE models, we encounter the Sylvester equation:*

$$AXC + DXE = B,$$

where the coefficients A, B, C, D, E *are* $m \times m$, $m \times n$, $n \times n$, $m \times m$, $n \times n$ *matrices, and our aim is to solve the* $m \times n$ *matrix* X.

1. Show that the Sylvester equation is a linear system with respect to X. *(Hint: consider vectorization by the formula* $vec(AXC) = (C' \otimes A)\, vec(X)$.*)*

2. Does the Sylvester equation has a closed-form solution? If yes, show the solution and discuss its applicability to real-world DSGE models.

3. Consider QZ decomposition for A, D *such that*

$$A = Q\tilde{A}Z,$$

$$B = Q\tilde{B}Z,$$

where $\tilde{A}, \tilde{B}$ *are upper-triangular and* Q, Z *are unitary. Similarly, QZ decomposition for* C, E *takes the form*

$$C = U\tilde{C}V,$$

$$E = U\tilde{E}V,$$

where $\tilde{C}, \tilde{E}$ *are upper-triangular and* U, V *are unitary. Define* $\tilde{X} = ZXU$. *The Sylvester equation is reduced to an upper-triangular system:*

$$\tilde{A}\tilde{X}\tilde{C} + \tilde{D}\tilde{X}\tilde{E} = Q'BV'.$$

Is it possible to solve the upper-triangular system recursively?

4. Suppose that D and E are invertible, is there an alternative way to solve the Sylvester equation?

Bibliography

Aitkin, M. (1987). Modelling variance heterogeneity in normal regression using GLIM. *Journal of the Royal Statistical Society Series C Applied Statistics 36*(3), 332–339. 150

Albert, A. and J. A. Anderson (1984). On the existence of maximum likelihood estimates in logistic regression models. *Biometrika 71*, 1–10. 203

Albert, J. H. and S. Chib (1993). Bayesian analysis of binary and polychotomous response data. *Journal of the American Statistical Association 88*, 669–679. 197, 246, 391

Amemiya, T. (1985). *Advanced econometrics.* Cambridge: Harvard University Press. 246, 247

Anderson, S. and R. Newell (2003). Simplified marginal effects in discrete choice models. *Economics Letters 81*, 321–326. 196

Andrews, D. and C. Mallows (1974). Scale mixtures of normal distributions. *Journal of the Royal Statistical Society, Series B 36*, 99–102. 201, 205, 234, 246

Andrieu, C., A. Doucet, and R. Holenstein (2010). Particle Markov chain Monte Carlo methods. *Journal of the Royal Statistical Society, Series B 72*, 269–342. xix, 67, 68, 297, 328, 329, 333, 338, 396, 406, 412

Andrieu, C. and G. O. Roberts (2009). The pseudo-marginal approach for efficient Monte Carlo computations. *The Annals of Statistics 37*(2), 697–725. xix, 70, 84, 97, 160, 297, 328, 333, 338, 406, 412

Ansley, C. F. and R. Kohn (1983). Exact likelihood of vector autoregressive-moving average process with missing or aggregated data. *Biometrika 70*(1), 275–278. 261, 289, 424

Antolin-Diaz, J., T. Drechsel, and I. Petrella (2024). Advances in nowcasting economic activity: The role of heterogeneous dynamics and fat tails. *Journal of Econometrics 238*(2), 221–232. 425

Antolin-Diaz, J. and J. Rubio-Ramirez (2018). Narrative sign restrictions for SVARs. *American Economic Review 108*(10), 2802–2829. 369

Antoniak, C. E. (1974). Mixtures of Dirichlet processes with applications to Bayesian nonparametric problems. *The Annals of Statistics 2*(6), 1152–1174. 170

Arellano, M. (2003). *Panel Data Econometrics.* Oxford: Oxford University Press. 383

Arlot, S. and A. Celisse (2010). A survey of cross-validation procedures for model selection. *Statistics Survey 4*, 40–79. 15

Aruoba, S. B., J. Fernandez-Villaverde, and J. Rubio-Ramirez (2006). Comparing solution methods for dynamic equilibrium economies. *Journal of Economic Dynamics and Control 30*, 2477–2508. 411

Ascari, G., G. Fagiolo, and A. Roventini (2015). Fat-tail distributions and business-cycle models. *Macroeconomic Dynamics 19*(2), 465–476. 153

Avery, R., K. Brevoort, and G. Canner (2007). Opportunities and issues in using HMDA data. *Journal of Real Estate Research 29*(4), 351–380. 28

Azzalini, A. and A. Dalla Valle (1996). The multivariate skew-normal distribution. *Biometrika 83*(4), 715–726. 161

Bagliano, F. C. and C. Favero (1998). Measuring monetary policy with VAR models: an evaluation. *European Economic Review 42*(6), 1069–1112. 369

Bai, J. (2003). Inferential theory for factor models of large dimensions. *Econometrica 71*, 135–171. 378

Bai, J. and S. Ng (2002). Determining the number of factors in approximate factor models. *Econometrica 70*, 191–221. 373

Balke, N. (2000). Credit and economic activity: Credit regimes and nonlinear propagation of shocks. *Review of Economics and Statistics 82*(2), 344–349. 364

Baltagi, B. H. (2013). *Econometric Analysis of Panel Data.* Chichester: Wiley. 383

Banbura, M., D. Giannone, and L. Reichlin (2010). Large Bayesian vector auto regressions. *Journal of Applied Econometrics 25*(1), 71–92. 31, 364, 366, 372

Bang, H. and J. M. Robins (2005). Doubly robust estimation in missing data and causal inference models. *Biometrics 61*, 962–972. 209

Banterle, M., C. Crazian, A. Lee, and C. Robert (2019). Accelerating Metropolis-Hastings algorithms by delayed acceptance. *Foundations of Data Science 1*, 103–128. 81, 82, 97

Bardenet, R., A. Doucet, and C. Holmes (2017). On Markov chain Monte Carlo methods for tall data. *Journal of Machine Learning Research 18*, 1–43. 73, 76, 83, 87, 97

Bartlett, P., P. Long, G. Lugosi, and A. Tsigler (2020). Benign overfitting in linear regression. *Proceedings of the National Academy of Sciences 117*(48), 30063–30070. 122, 138

Baum, L., T. Petrie, G. Soules, and N. Weiss (1970). A maximization technique occurring in the statistical analysis of probabilistic functions of Markov chains. *Annals of Mathematical Statistics 41*(1), 164–171. 308

Bedrick, E., R. Christensen, and W. Johnson (2000). Bayesian accelerated failure time analysis with application to veterinary epidemiology. *Statistics in Medicine 19*(2), 221–237. 235

Belmonte, M., G. Koop, and D. Korobolis (2014). Hierarchical shrinkage in time-varying parameter models. *Journal of Forecasting 33*, 80–94. 123, 380, 381

Berger, A. L., V. J. D. Pietra, and S. A. D. Pietra (1996). A maximum entropy approach to natural language processing. *Computational Linguistics 22*(1), 39–71. 207

Berger, J. (1985). *Statistical Decision Theory and Bayesian Analysis.* New York: Springer. 3

Bernanke, B., J. Boivin, and P. S. Eliasz (2005). Measuring the effects of monetary policy: A factor-augmented vector autoregressive (FAVAR) approach. *The Quarterly Journal of Economics 120*(1), 387–422. 364, 373

Bernardo, J. (1979). Reference posterior distributions for Bayesian inference. *Journal of the Royal Statistical Society, Series B 41*(2), 113–147. 3

Bernardo, J. and A. Smith (2000). *Bayesian Theory.* Chichester: John Wiley and Sons. 20, 36

Bierkens, J., P. Fearnhead, and G. Roberts (2019). The zig-zag process and super-efficient sampling for Bayesian analysis of big data. *The Annals of Statistics 47*, 1288–1320. 97

Bishop, C. M. (2006). *Pattern Recognition and Machine Learning.* New York: Springer. 207

Blanchard, O. and C. Kahn (1980). The solution of linear difference models under rational expectations. *Econometrica 48*(5), 1305–1311. 404

Blanchard, O. and D. Quah (1989). The dynamic effects of demand and supply disturbances. *American Economic Review 79*, 655–673. 369

Bobeica, E. and B. Hartwig (2023). The COVID-19 shock and challenges for inflation modelling. *International Journal of Forecasting 39*(1), 519–539. 425

Bollerslev, T. (1986). Generalized autoregressive conditional heteroskedasticity. *Journal of Econometrics 31*, 307–327. 392, 424

Bolstad, W. (2010). *Understanding Computational Bayesian Statistics.* Hoboken: John Wiley & Sons. 247

Bolstad, W. and J. Curran (2017). *Introduction to Bayesian Statistics (Third Edition).* Hoboken: Wiley. 36

Box, G. and G. M. Jenkins (1976). *Time Series Analysis: Forecasting and Control.* San Francisco: Holden Day. 355, 357, 424

Briers, M., A. Doucet, and S. Maskell (2010). Smoothing algorithms for state-space models. *Annals of Institute of Statistical Mathematics 62*, 61–89. 289, 327, 337

Brockwell, A. E. (2006). Parallel Markov chain Monte Carlo simulation by pre-fetching. *Journal of Computational and Graphical Statistics 15*(1), 246–261. 91

Buchinsky, M. (1994). Changes in the U.S. wage structure 1963-1987: Application of quantile regression. *Econometrica 62*(2), 405–458. 171

Cai, M., M. Del Negro, E. Herbst, E. Matlin, R. Sarfati, and F. Schorfheide (2021). Online estimation of DSGE models. *Journal of Applied Econometrics 24*, 33–58. 399, 406

Cameron, A. and P. Trivedi (1986). Econometric models based on count data: comparisons and applications of some estimators and tests. *Journal of Applied Econometrics 1*(1), 29–54. 226

Cameron, A. C. and P. K. Trivedi (2005). *Microeconometrics: Methods and Applications.* New York: Cambridge University Press. 246

Cameron, A. C. and P. K. Trivedi (2013). *Regression Analysis of Count Data.* New York: Cambridge University Press. 225

Canova, F. and M. Ciccarelli (2009). Estimating multicountry VAR models. *International Economic Review 50*(3), 929–959. 372

Canova, F. and G. D. Nicolo (2002). Monetary disturbances matter for business fluctuations in the G-7. *Journal of Monetary Economics 49*, 1131–1159. 369

Cappe, O., E. Moulines, and T. Ryden (2004). *Hidden Markov Models.* New York: Springer. 307

Carlin, B. P. and S. Chib (1995). Bayesian model choice via Markov chain Monte Carlo methods. *Journal of the Royal Statistical Society. Series B 57*(3), 473–484. 69, 137

Carpenter, B., A. Gelman, M. D. Hoffman, D. Lee, B. Goodrich, M. Betancourt, M. Brubaker, J. Guo, P. Li, and A. Riddell (2017). Stan: A probabilistic programming language. *Journal of Statistical Software 76*(1), 1–32. 67

Carpenter, J., P. Clifford, and P. Fearnhead (1999). An improved particle filter for non-linear problems. *IEE Proceedings Part F Radar and Sonar Navigation 146*(1), 2–7. 311

Carriero, A., T. Clark, and M. Marcellino (2016). Common drifting volatility in large Bayesian VARs. *Journal of Business & Economic Statistics 34*(3), 375–390. 366

Carriero, A., T. Clark, and M. Marcellino (2019). Large Bayesian vector autoregressions with stochastic volatility and non-conjugate priors. *Journal of Econometrics 212*(1), 137–154. 31, 368

Carriero, A., T. E. Clark, M. Marcellino, and E. Mertens (2024). Addressing COVID-19 outliers in BVARs with stochastic volatility. *Review of Economics and Statistics 106*(5), 1403–1417. 364, 425

Carriero, A., G. Kapetanios, and M. Marcellino (2012). Forecasting government bond yields with large Bayesian vector autoregressions. *Journal of Banking and Finance 36*(7), 2026–2047. 31

Carter, C. K. and R. Kohn (1994). On Gibbs sampling for state space models. *Biometrika 81*(3), 541–553. 186, 272, 289

Casella, G. and E. I. George (1992). Explaining the Gibbs sampler. *The American Statistician 46*(3), 167–174. 58

Celeux, G., M. Hurn, and C. P. Robert (2000). Computational and inferential difficulties with mixture posterior distributions. *Journal of the American Statistical Association 95*(451), 957–970. 186

Celeux, G., K. Kamary, G. Malsiner-Walli, J.-M. Marin, and C. P. Robert (2019). Computational solutions for Bayesian inference in mixture models. In S. Fruhwirth-Schnatter, G. Celeux, and C. P. Robert (Eds.), *Handbook of Mixture Analysis.* New York: Chapman and Hall. 186

Cepeda, E. and D. Gamerman (2001). Bayesian modeling of variance heterogeneity in normal regression models. *Brazilian Journal of Probability and Statistics 14*(1), 207–221. 150

Chan, J. (2020). Large Bayesian VARs: A flexible Kronecker error covariance structure. *Journal of Business & Economic Statistics 38*(1), 68–79. 31, 366

Chan, J. (2021). Minnesota-type adaptive hierarchical priors for large Bayesian VARs. *International Journal of Forecasting 37*(3), 1212–1226. 368

Chan, J. (2023). Large hybrid time-varying parameter VARs. *Journal of Business & Economic Statistics 41*(3), 890–905. 368, 380

Chan, J., G. Koop, R. Leon-Gonzalez, and R. W. Strachan (2012). Time varying dimension models. *Journal of Business & Economic Statistics 30*(3), 358–367. 261

Chan, J., G. Koop, D. J. Poirier, and J. L. Tobias (2019). *Bayesian Econometric Methods (Second Edition)*. Cambridge: Cambridge University Press. 36

Chen, M., D. Dey, and Q. Shao (1999). A new skewed link model for dichotomous quantal response data. *Journal of the American Statistical Association 94*(448), 1172–1186. 161

Chen, M. and Q. Shao (1999). Monte Carlo estimation of Bayesian credible and HPD intervals. *Journal of Computational and Graphical Statistics 8*, 69–92. 14

Chen, R. and J. Liu (2000). Mixture Kalman filters. *Journal of the Royal Statistical Society. Series B 62*(3), 493–508. 317

Chib, S. (1992). Bayesian inference in the Tobit censored regression model. *Journal of Econometrics 51*, 79–99. 247

Chib, S. (1995). Marginal likelihood from the Gibbs output. *Journal of the American Statistical Association 90*(432), 1313–1321. 180

Chib, S. (1996). Calculating posterior distributions and modal estimates in Markov mixture models. *Journal of Econometrics 75*(1), 79–97. 309

Chib, S. and B. P. Carlin (1999). On MCMC sampling in hierarchical longitudinal models. *Statistics and Computing 9*, 17–26. 391, 424

Chib, S. and E. Greenberg (1994). Bayes inference in regression models with ARMA(p,q) errors. *Journal of Econometrics 64*, 183–206. 150, 356

Chib, S. and E. Greenberg (1995). Understanding the Metropolis-Hastings algorithm. *American Statistician 49*, 327–335. 97

Chib, S. and E. Greenberg (1998). Analysis of multivariate Probit models. *Biometrika 85*(2), 347–361. 200

Chib, S., E. Greenberg, and I. Jeliazkov (2009). Estimation of semiparametric models in the presence of endogeneity and sample selection. *Journal of Computational and Graphical Statistics 18*, 321–348. 218

Chib, S. and I. Jeliazkov (2001). Marginal likelihood from the Metropolis-Hastings output. *Journal of the American Statistical Association 96*(453), 270–281. 180

Chib, S., F. Nardari, and N. Shephard (2002). Markov chain Monte Carlo methods for stochastic volatility models. *Journal of Econometrics 108*(2), 281–316. 394, 425

Chib, S. and S. Ramamurthy (2010). Tailored randomized-block MCMC methods with application to DSGE models. *Journal of Econometrics 155*, 19–38. 63, 406

Chib, S. and S. Ramamurthy (2014). DSGE models with Student-t errors. *Econometric Reviews 33*, 152–171. 298, 337

Chopin, N. (2002). A sequential particle filter method for static models. *Biometrika 89*, 539–551. xix, 259, 297, 331, 332, 333, 406

Chopin, N. (2004). Central limit theorem for sequential Monte Carlo methods and its application to Bayesian inference. *The Annals of Statistics 32*(6), 2385–2411. 337

Chopin, N., P. Jacob, and O. Papaspiliopoulos (2013). SMC2: An efficient algorithm for sequential analysis of state space models. *Journal of the Royal Statistical Society: Series B 75*, 397–426. 297, 333, 338, 412

Christensen, R., W. Johnson, A. Branscum, and T. Hanson (2011). *Bayesian Ideas and Data Analysis: An Introduction for Scientists and Statisticians.* Boca Raton: CRC Press. 36, 247

Cimadomo, J., D. Giannone, M. Lenza, F. Monti, and A. Sokol (2022). Nowcasting with large Bayesian vector autoregressions. *Journal of Econometrics 231*(2), 500–519. 31, 366

Cogley, T. and T. J. Sargent (2005). Drift and volatilities: Monetary policies and outcomes in the post WWII U.S. *Review of Economic Dynamics 8*(2), 262–302. 335, 364, 380, 381, 394, 424

Cogley, T. and A. Sbordone (2008). Trend inflation, indexation, and inflation persistence in the new Keynesian Phillips curve. *American Economic Review 98*(5), 2101–2126. 380

Collard, F. and M. Juillard (2001). Accuracy of stochastic perturbation methods: The case of asset pricing models. *Journal of Economic Dynamics and Control 25*, 979–999. 411

Collett, D. (2015). *Modelling Survival Data in Medical Research.* New York: Chapman and Hall/CRC. 228

Congdon, P. (2014). *Applied Bayesian Modellings.* Hoboken: John Wiley & Sons. 247

Cook, R., A. Gelman, and D. Rubin (2006). Validation of software for Bayesian models using posterior quantiles. *Journal of Computational and Graphical Statistics 15*(3), 675–692. 89, 334

Cowles, M. (1996). Accelerating Monte Carlo Markov chain convergence for cumulative-link generalized linear models. *Statistics and Computing 6*, 101–111. 200

Cowles, M. and B. Carlin (1996). Markov chain Monte Carlo convergence diagnostics: a comparative review. *Journal of the American Statistical Association 91*(434), 883–904. 56

Cox, D. (1972). Regression models and life tables. *Journal of the Royal Statistical Society, Series B 34*, 187–220. 233

Cross, J., C. Hou, and A. Poon (2020). Macroeconomic forecasting with large Bayesian VARs: Global-local priors and the illusion of sparsity. *International Journal of Forecasting 36*(3), 899–915. 31

Curdia, V., M. DelNegro, and D. L. Greenwald (2014). Rare shocks, great recessions. *Journal of Applied Econometrics 29*, 1031–1052. 298

Damen, S. and E. Buyst (2017). Mortgage shoppers: How much do they save. *Real Estate Economics 45*(4), 898–929. 28

De Jong, P. (1988). The likelihood for a state-space model. *Biometrika 75*, 165–169. 285

De Jong, P. (1989). Smoothing and interpolation with the state-space model. *Journal of the American Statistical Association 84*(408), 1085–1088. 272, 289, 396

De Jong, P. (1991). The diffuse Kalman filter. *The Annals of Statistics 19*(2), 1073–1083. 283, 289

Dean, J. and S. Ghemawat (2008). MapReduce: Simplified data processing on large clusters. *Communications of the ACM 51*(1), 107–113. 73

Del Moral, P., A. Doucet, and A. Jasra (2006). Sequential Monte Carlo samplers. *Journal of the Royal Statistical Society, Series B 68*(3), 411–436. 311

Del Moral, P., A. Doucet, and S. Singh (2010). Forward smoothing using sequential Monte Carlo. *Technical Report at arXiv 1012.5390*, 1–26. 337

Del Negro, M. and C. Otrok (2007). 99 luftballons: Monetary policy and the house price boom across U.S. states. *Journal of Monetary Economics 54*(7), 1962–1985. 380

Del Negro, M. and F. Schorfheide (2004). Priors from general equilibrium models for VARs. *International Economic Review 45*, 643–673. 364

Del Negro, M., F. Schorfheide, F. Smets, and R. Wouters (2007). On the fit of new Keynesian models. *Journal of Business & Economic Statistics 25*, 123–143. 364

Deligiannidis, G., A. Doucet, and M. Pitt (2018). The correlated pseudo-marginal method. *Journal of the Royal Statistical Society: Series B 80*(5), 839–870. 71, 87, 330, 338

Dellaportas, P. and A. F. M. Smith (1993). Bayesian inference for generalized linear and proportional hazards models via Gibbs sampling. *Journal of the Royal Statistical Society Series C Applied Statistics 42*(3), 443–459. 202, 206, 233, 246, 247

Dempster, A. P., M. R. Selwyn, and B. J. Weeks (1983). Combining historical and randomized controls for assessing trends in proportions. *Journal of the American Statistical Association 78*, 221–227. 202

Diaconis, P. and D. Ylvisaker (1979). Conjugate priors for exponential families. *The Annals of Statistics 7*, 269–281. 20, 36

Diebold, F. X. and C. Li (2006). Forecasting the term structure of government bond yields. *Journal of Econometrics 130*(2), 337–364. 378

Diebold, F. X., G. D. Rudebusch, and S. Aruoba (2006). The macroeconomy and the yield curve: a dynamic latent factor approach. *Journal of Econometrics 131*(1), 309–338. 274, 377, 378, 413, 424

Diebold, F. X., F. Schorfheide, and M. Shin (2017). Real-time forecast evaluation of DSGE models with stochastic volatility. *Journal of Econometrics 201*, 322–332. 394, 412

Diebolt, J. and C. P. Robert (1994). Estimation of finite mixture distributions through Bayesian sampling. *Journal of the Royal Statistical Society, Series B 56*, 363–375. 165

Doan, T., R. Litterman, and C. Sims (1984). Forecasting and conditional projection using realistic prior distributions. *Econometric Reviews 3*(1), 1–100. 366

Douc, R., A. Garivier, E. Moulines, and J. Olsson (2011). Sequential Monte Carlo smoothing for general state space hidden Markov models. *The Annals of Applied Probability 21*, 2109–2145. 338

Douc, R., E. Moulines, and J. Olsson (2009). Optimality of the auxiliary particle filter. *Probability and Mathematical Statistics 29*(1), 1–28. 318

Doucet, A., S. J. Godsill, and C. Andrieu (2000). On sequential Monte Carlo sampling methods for Bayesian filtering. *Statistics and Computing 10*, 197–208. 315, 337

Doucet, A., N. Gordon, and V. Krishnamurthy (2001). Particle filters for state estimation of jump Markov linear systems. *IEEE Transactions on Signal Processing 49*(3), 613–624. 317

Doucet, A. and A. M. Johansen (2009). *A Tutorial on Particle Filtering and Smoothing: Fifteen Years Later.* Oxford: Oxford University Press. 337

Duane, S., A. Kennedy, B. Pendleton, and D. Roweth (1987). Hybrid Monte Carlo. *Physics Letters B. 195*(2), 216–222. 63

Duchateau, L. and P. Janssen (2008). *The Frailty Model.* New York: Springer. 235

Durbin, J. and S. J. Koopman (2002). A simple and efficient simulation smoother for state space time series analysis. *Biometrika 89*(3), 603–615. 186, 272, 289, 293, 396

Durbin, J. and S. J. Koopman (2012). *Time Series Analysis by State Space Methods: Second Edition.* Oxford: Oxford University Press. 357, 362, 424

Eberly, L. E. and G. Casella (2003). Estimating Bayesian credible intervals. *Journal of Statistical Planning and Inference 112*, 115–132. 14

Efron, B., T. Hastie, I. Johnstone, and R. Tibshirani (2004). Least angle regression. *The Annals of Statistics 32*(2), 407–451. 125

Eichenbaum, M. and C. L. Evans (1995). Some empirical evidence on the effects of shocks to monetary policy on exchange rate. *Quarterly Journal of Economics 110*, 975–1009. 369

Ellahie, A. and G. Ricco (2017). Government purchases reloaded: Informational insufficiency and heterogeneity in fiscal VARs. *Journal of Monetary Economics 90*, 13–27. 31

Engle, R. (1982). Autoregressive conditional heteroskedasticity with estimates of the variance of United Kingdom inflation. *Econometrica 50*(1), 987–1007. 392, 424

Eraker, B., C. Chiu, A. Foerster, T. Kim, and H. Seoane (2015). Bayesian mixed frequency VARs. *Journal of Financial Econometrics 13*(3), 689–721. 378

Escobar, M. D. and M. West (1995). Bayesian density estimation and inference using mixtures. *Journal of the American Statistical Association 90*(430), 577–588. 170

Fagiolo, G., M. Napoletano, and A. Roventini (2008). Are output growth-rate distributions fat-tailed? some evidence from OECD countries. *Journal of Applied Econometrics 23*(5), 639–669. 153

Fama, E. F. (1970). Efficient capital markets: A review of theory and empirical work. *Journal of Finance 25*(2), 383–417. 392

Fama, E. F. and K. R. French (1993). Common risk factors in the returns on stocks and bonds. *Journal of Financial Economics 33*, 3–56. 34

Fama, E. F. and K. R. French (2015). A five-factor asset pricing model. *Journal of Financial Economics 116*, 1–22. 34

Farrar, D. E. and R. R. Glauber (1967). Multicollinearity in regression analysis: The problem revisited. *The Review of Economics and Statistics 49*(1), 92–107. 8

Ferguson, T. S. (1973). A Bayesian analysis of some nonparametric problems. *The Annals of Statistics 1*(2), 209–230. 170

Fernandez, C. and M. Steel (1998). On Bayesian modeling of fat tails and skewness. *Journal of the American Statistical Association 93*, 359–371. 163, 164

Fernandez-Villaverde, J. and J. F. Rubio-Ramirez (2007). Estimating macroeconomic models: a likelihood approach. *Review of Economic Studies 74*(1), 1059–1087. 399, 406, 412

Ferrari, S. and F. Cribari-Neto (2004). Beta regression for modelling rates and proportions. *Journal of Applied Statistics 31*(7), 799–815. 237, 238

Firth, D. (1993). Bias reduction of maximum likelihood estimates. *Biometrika 80*, 27–38. 203

Flegal, J. M. and G. L. Jones (2010). Batch means and spectral variance estimators in Markov chain Monte Carlo. *The Annals of Statistics 38*(2), 1034–1070. 57

Fleming, J. and C. Kirby (2003). A closer look at the relation between GARCH and stochastic autoregressive volatility. *Journal of Financial Econometrics 1*(3), 365–419. 392, 424

Flury, T. and N. Shephard (2011). Bayesian inference based only on a simulated likelihood. *Econometric Theory 27*, 933–956. xix, 70, 330, 412

Forni, M., M. Hallin, M. Lippi, and L. Reichlin (2000). The generalized dynamic-factor model: identification and estimation. *Review of Economics and Statistics 82*, 540–554. 373, 378

Friedman, J., R. Tibshirani, and T. Hastie (2010). Regularization paths for generalized linear models via coordinate descent. *Journal of Statistical Software 33*(1), 1–21. 125

Fruhwirth-Schnatter, S. (2001). Markov chain Monte Carlo estimation of classical and dynamic switching and mixture models. *Journal of the American Statistical Association 96*(453), 194–209. 168, 192

Fruhwirth-Schnatter, S. (2006). *Finite Mixture and Markov Switching Models*. New York: Springer. 165, 307

Fruhwirth-Schnatter, S. and R. Fruhwirth (2007). Auxiliary mixture sampling with applications to logistic models. *Computational Statistics and Data Analysis 51*(7), 3509–3528. 202, 206, 246

Fruhwirth-Schnatter, S. and H. Wagner (2010). Stochastic model specification search for Gaussian and partially non-gaussian state space models. *Journal of Econometrics 154*, 85–100. 380

Fulop, A. and J. Li (2013). Efficient learning via simulation: A marginalized resample-move approach. *Journal of Econometrics 176*, 146–161. 333, 338

Gamerman, D. (1997). Efficient sampling from the posterior distribution in generalized linear mixed models. *Statistics and Computing 7*, 57–68. 202

Garthwaite, P., J. Kadane, and A. O'Hagan (2005). Statistical methods for eliciting probability distributions. *Journal of the American Statistical Association 100*(470), 680–701. 4, 405

Gaure, S. (2013). OLS with multiple high dimensional category variables. *Computational Statistics and Data Analysis 66*, 8–18. 388, 389

Gelfand, A. E. and D. K. Dey (1994). Bayesian model choice: Asymptotics and exact calculations. *Journal of the Royal Statistical Society (Series B) 56*, 501–514. 180, 183

Gelfand, A. E. and S. K. Sahu (1999). Identifiability, improper priors, and Gibbs sampling for generalized linear models. *Journal of the American Statistical Association 94*, 247–253. 247

Gelfand, A. E. and A. F. Smith (1990). Sampling-based approaches to calculating marginal densities. *Journal of the American Statistical Association 85*, 398–409. 58, 97, 202, 389, 424

Gelman, A., J. Carlin, H. Stern, D. Dunson, A. Vehtari, and D. Rubin (2014). *Bayesian Data Analysis (Third Edition)*. Boca Raton: CRC Press. 20, 36

Gelman, A., W. R. Gilks, and G. O. Roberts (1997). Weak convergence and optimal scaling of random walk Metropolis algorithms. *Annals of Applied Probability 7*(1), 110–120. 55, 203, 226, 406

Gelman, A., X. Meng, and H. Stern (1996). Posterior predictive assessment of model fitness via realized discrepancies. *Statistica Sinica 6*, 733–807. 15, 178

Gelman, A. and D. Rubin (1992). Inference from iterative simulation using multiple sequences. *Statistical Science 7*(4), 457–511. 56

Geman, S. and D. Geman (1984). Stochastic relaxation, Gibbs distributions, and the Bayesian restoration of images. *IEEE Transactions on Pattern Analysis and Machine Intelligence 6*, 721–741. 58, 97

George, E., D. Sun, and S. Ni (2008). Bayesian stochastic search for VAR model restrictions. *Journal of Econometrics 142*, 553–580. 128, 368

George, E. I. and D. P. Foster (2000). Calibration and empirical Bayes variable selection. *Biometrika 87*(4), 731–747. 137

George, E. I. and R. E. McCulloch (1993). Variable selection via Gibbs sampling. *Journal of the American Statistical Association 88*(423), 881–889. 115, 127, 137

George, E. I. and R. E. Mcculloch (1997). Approaches for Bayesian variable selection. *Statistica Sinica 1*, 339–374. 127, 137

Geraci, M. and J. Gnabo (2018). Measuring interconnectedness between financial institutions with Bayesian time-varying vector autoregressions. *Journal of Financial and Quantitative Analysis 53*(3), 1371–1390. 380

Geweke, J. (1989a). Bayesian inference in econometric models using Monte Carlo integration. *Econometrica 57*, 1317–1339. 310

Geweke, J. (1989b). Exact predictive density for linear models with ARCH disturbances. *Journal of Econometrics 40*(1), 63–86. 394

Geweke, J. (1991). Efficient simulation from the multivariate normal and Student-t distributions subject to linear constraints and the evaluation of constraint probabilities. In *Computer Sciences and Statistics Proceedings of the 23d Symposium on the Interface*, pp. 571–578. 197

Geweke, J. (1993). Bayesian treatment of the independent Student-t linear model. *Journal of Applied Econometrics 8*, 19–40. 155, 157, 298

Geweke, J. (1996). *Variable Selection and Model Comparison in Regression.* Oxford: Oxford University Press. 130

Geweke, J. (1999). Using simulation methods for Bayesian econometric models: Inference, development, and communication. *Econometric Reviews 18*, 1–126. 180

Geweke, J. (2004). Getting it right: Joint distribution tests of posterior simulators. *Journal of the American Statistical Association 99*, 799–804. 89, 334

Geweke, J. (2005). *Contemporary Bayesian Econometrics and Statistics.* Hoboken: John Wiley & Sons. 36, 186

Geweke, J. (2007). Interpretation and inference in mixture models: Simple MCMC works. *Computational Statistics & Data Analysis 51*(7), 3529–3550. 186

Geweke, J., M. Keane, and D. Runkle (1994). Alternative computational approaches to inference in the multinomial Probit model. *Review of Economics and Statistics 76*(4), 609–632. 200

Geyer, C. (1992). Practical Markov chain Monte Carlo. *Statistical Science 7*(4), 473–483. 57

Ghosh, J. and J. P. Reiter (2013). Secure Bayesian model averaging for horizontally partitioned data. *Statistics and Computing 23*(3), 311–322. 7

Ghysels, E. (2016). Macroeconomics and the reality of mixed frequency data. *Journal of Econometrics 193*(2), 294–314. 364

Ghysels, E., P. Santa-Clara, and R. Valkanov (2006). Predicting volatility: getting the most out of return data sampled at different frequencies. *Journal of Econometrics 131*(1), 59–95. 424

Ghysels, E., A. Sinko, and R. Valkanov (2007). MIDAS regressions: Further results and new directions. *Econometric Reviews 26*(1), 53–90. 424

Giannone, D., M. Lenza, and G. E. Primiceri (2015). Prior selection for vector autoregressions. *Review of Economics and Statistics 97*(2), 436–451. 368

Giannone, D., L. Reichlin, and D. Small (2008). Nowcasting: The real-time informational content of macroeconomic data. *Journal of Monetary Economics 55*(4), 665–676. 379

Gilks, W. R. and C. Berzuini (2001). Following a moving target - Monte Carlo inference for dynamic Bayesian models. *Journal of the Royal Statistical Society. Series B 63*, 127–146. xix, 297, 331

Gilks, W. R. and P. Wild (1992). Adaptive rejection sampling for Gibbs sampling. *Journal of the Royal Statistical Society Series C Applied Statistics 41*(2), 337–348. 206, 227, 233

Glosten, L. R., R. Jagannathan, and D. E. Runkle (1993). On the relation between the expected value and the volatility of the nominal excess return on stocks. *The Journal of Finance 48*(5), 1779–1801. 394, 424

Godsill, S. J. (2001). On the relationship between Markov chain Monte Carlo methods for model uncertainty. *Journal of Computational and Graphical Statistics 10*(2), 230–248. 69, 137

Gordon, N., D. Salmond, and A. Smith (1993). Novel approach to nonlinear/non-gaussian Bayesian state estimation. *Radar and Signal Processing, IEEE Proceedings 140*(2), 107–113. 309, 314, 337

Green, P. J. (1995). Reversible jump Markov chain Monte Carlo computation and Bayesian model determination. *Biometrika 82*, 711–732. 70, 137

Greenberg, E. (2013). *Introduction to Bayesian Econometrics.* Cambridge: Cambridge University Press. 36, 186

Greene, W. H. (2008). *Econometric Analysis, Sixth Edition.* New Jersey: Prentice Hall. 246, 335

Griffin, J. and P. Brown (2010). Inference with normal-gamma prior distributions in regression problems. *Bayesian Analysis 5*(1), 171–188. 368

Haario, H., E. Saksman, and J. Tamminen (2001). An adaptive Metropolis algorithm. *Bernoulli 7*(2), 223–242. 55, 97

Haldane, J. (1932). A note on inverse probability. *Mathematical Proceedings of the Cambridge Philosophical Society 28*(1), 55–61. 23

Hamilton, J. (1994). *Time Series Analysis.* New Jersey: Princeton University Press. 355

Hamilton, J. D. (1989). A new approach to the economic analysis of nonstationary time series and the business cycle. *Econometrica 57*(2), 357–384. 379

Hans, C. (2009). Bayesian lasso regression. *Biometrika 96*(4), 835–845. 123, 137

Harvey, A. C. (1989). *Forecasting, Structural Time Series Models and the Kalman Filter.* Cambridge: Cambridge University Press. 357

Harvey, A. C. and J. Durbin (1986). The effects of seat belt legislation on British road casualties: a case study in structural time series modelling. *Journal of Royal Statistical Society Series A 149*, 187–227. 259, 357

Harvey, A. C. and R. G. Pierse (1984). Estimating missing observations in economic time series. *Journal of the American Statistical Association 79*(385), 125–131. 289, 424

Harvey, A. C., E. Ruiz, and N. Shephard (1994). Multivariate stochastic variance models. *Review of Economic Studies 61*(2), 247–264. 166, 397

Hastie, T., A. Montanari, S. Rosset, and R. J. Tibshirani (2022). Surprises in high-dimensional ridgeless least squares interpolation. *The Annals of Statistics 50*(2), 949–986. 122, 138

Hastie, T., R. Tibshirani, and J. Friedman (2009). *The Elements of Statistical Learning: Data Mining, Inference, and Prediction.* New York: Springer. 8

Hastings, W. K. (1970). Monte Carlo sampling methods using Markov chains and their applications. *Biometrika 57*, 97–109. 97

Heckman, J. J. (1979). Sample selection bias as a specification error. *Econometrica 47*, 153–161. 218, 245

Heinze, G. and M. Schemper (2002). A solution to the problem of separation in logistic regression. *Statistics in Medicine 21*, 2409–2419. 203

Herbst, E. and F. Schorfheide (2014). Sequential Monte Carlo sampling for DSGE models. *Journal of Applied Econometrics 29*, 1073–1098. 399, 406

Herbst, E. and F. Schorfheide (2019). Tempered particle filtering. *Journal of Econometrics 210*(1), 26–44. 315, 351

Herbst, E. P. (2015). Using the Chandrasekhar recursions for likelihood evaluation of DSGE models. *Computational Economics 45*(4), 693–705. 268

Hilbe, J. M. (2011). *Negative Binomial Regression.* New York: Cambridge University Press. 226

Hirano, K., G. Imbens, and G. Ridder (2003). Efficient estimation of average treatment effects using the estimated propensity score. *Econometrica 71*, 1161–1189. 209

Hoerl, A. E. and R. W. Kennard (1970). Ridge regression: Biased estimation for nonorthogonal problems. *Technometrics 12*(1), 55–67. 7, 115, 118

Hoffman, M. and A. Gelman (2014). The No-U-Turn sampler: Adaptively setting path lengths in Hamiltonian Monte Carlo. *Journal of Machine Learning Research 15*(1), 1593–1623. 67

Holmes, C. C. and L. Held (2006). Bayesian auxiliary variable models for binary and multinomial regression. *Bayesian Analysis 1*, 146–168. 197, 202, 205, 206, 235, 246

Holtz-Eakin, D., W. Newey, and H. S. Rosen (1988). Estimating vector autoregressions with panel data. *Econometrica 56*, 1371–1395. 364

Horvitz, D. G. and D. J. Thompson (1952). A generalization of sampling without replacement from a finite universe. *Journal of the American Statistical Association 47*, 663–685. 209

Hsiao, C. (1975). Some estimation methods for a random coefficients model. *Econometrica 43*, 305–325. 379

Hsiao, C. (2014). *Analysis of Panel Data.* New York: Cambridge University Press. 383

Huber, F. and M. Feldkircher (2019). Adaptive shrinkage in Bayesian vector autoregressive models. *Journal of Business & Economic Statistics 37*(1), 27–39. 31, 368

Ibrahim, J. G., M. Chen, and D. Sinha (2001). *Bayesian Survival Analysis.* New York: Springer. 228

Imbens, G. and J. Wooldridge (2009). Recent developments in the econometrics of program evaluation. *Journal of Economic Literature 47*, 5–86. 209

Imbens, G. W. (2000). The role of the propensity score in estimating dose-response functions. *Biometrika 87*, 706–710. 209

Inoue, A., L. Jin, and B. Rossi (2017). Rolling window selection for out-of-sample forecasting with time-varying parameters. *Journal of econometrics 196*(1), 55–67. 379

Ishwaran, H. and M. Zarepour (2002). Dirichlet prior sieves in finite normal mixtures. *Statistica Sinica 12*, 941–963. 171

Jacob, P., C. Robert, and M. Smith (2011). Using parallel computation to improve independent Metropolis-Hastings based estimation. *Journal of Computational and Graphic Statistics 20*, 616–635. 91

Jacob, P. and A. Thiery (2015). On non-negative unbiased estimators. *The Annals of Statistics 43*(2), 769–784. 84

Jacquier, E., N. G. Polson, and P. E. Rossi (1994). Bayesian analysis of stochastic volatility models. *Journal of Business & Economic Statistics 20*, 69–87. 186, 394, 396, 424

Jacquier, E., N. G. Polson, and P. E. Rossi (2004). Bayesian analysis of stochastic volatility models with fat-tails and correlated errors. *Journal of Econometrics 122*, 185–212. 186

Jasra, A., C. C. Holmes, and D. A. Stephens (2005). Markov chain Monte Carlo methods and the label switching problem in Bayesian mixture modeling. *Statistical Science 20*(1), 50–67. 186

Jeffreys, H. (1946). An invariant form for the prior probability in estimation problems. *Proceedings of the Royal Statistical Society of London, Series A 186*, 453–461. xviii, 3, 20, 22, 24, 36

Johansen, A. and A. Doucet (2008). A note on auxiliary particle filters. *Statistics and Probability Letters 78*(12), 1498–1504. 318, 337

Johansen, S. (1991). Estimation and hypothesis testing of cointegration vectors in Gaussian vector autoregressive models. *Econometrica 59*(6), 1551–1580. 364

Jones, G. L., M. Haran, B. S. Caffo, and R. Neath (2006). Fixed-width output analysis for Markov chain Monte Carlo. *Journal of the American Statistical Association 101*, 1537–1547. 57

Jones, R. H. (1980). Maximum likelihood fitting of ARMA models to time series with missing observations. *Technometrics 22*(3), 389–395. 289, 424

Judd, K. (1998). *Numerical Methods in Economics*. Cambridge: MIT Press. 407

Julier, S. and J. Uhlmann (2004). Unscented filtering and nonlinear estimation. *Proceedings of the IEEE 92*(3), 401–422. 304, 306, 316, 337

Jungbacker, B., S. J. Koopman, and M. van der Wel (2011). Maximum likelihood estimation for dynamic factor models with missing data. *Journal of Economic Dynamics and Control 35*(8), 1358–1368. 261

Justiniano, A. and G. E. Primiceri (2008). The time-varying volatility of macroeconomic fluctuations. *American Economic Review 98*, 604–641. 394

Kadane, J. B. and N. A. Lazar (2004). Methods and criteria for model selection. *Journal of the American Statistical Association 99*(465), 279–290. 11

Kadiyala, K. R. and S. Karlsson (1997). Numerical methods for estimation and inference in Bayesian VAR models. *Journal of Applied Econometrics 12*(2), 99–132. 366, 367

Kalbfleisch, J. D. and R. L. Prentice (2002). *The Statistical Analysis of Failure Time Data.* Chichester: Wiley. 228

Kalman, R. E. (1960). A new approach to linear filtering and prediction problems. *Journal of Basic Engineering 82*, 35–45. 186, 259, 289, 337

Kantas, N., A. Doucet, S. Singh, J. Maciejowski, and N. Chopin (2015). On particle methods for parameter estimation in state-space models. *Statistical Science 30*(3), 328–351. 338

Kass, R. and A. Raftery (1995). Bayes factors. *Journal of the American Statistical Association 90*(430), 773–795. 12

Kass, R. and L. Wasserman (1996). The selection of prior distributions by formal rules. *Journal of the American Statistical Association 91*, 1343–1370. 36

Kastner, G. and F. Huber (2020). Sparse Bayesian vector autoregressions in huge dimensions. *Journal of Forecasting 39*(7), 1142–1165. 31

Keane, M. (1992). A note on identification in the multinomial Probit model. *Journal of Business & Economic Statistics 10*(2), 193–200. 200

Kim, S., N. Shephard, and S. Chib (1998). Stochastic volatility: Likelihood inference and comparison with ARCH models. *Review of Economic Studies 65*(3), 361–93. 166, 186, 206, 392, 394, 396, 397, 425

King, R., R. Plosser, J. Stock, and W. Watson (1991). Stochastic trends and economic fluctuation. *American Economic Review 81*(4), 819–840. 369

King, R. and M. Watson (1998). The solution of singular linear difference systems under rational expectations. *International Economic Review 39*(4), 1015–1026. 404

Kirkpatrick, S., C. Gelett, and M. Vecchi (1983). Optimization by simulated annealing. *Science 220*, 671–680. 97

Kitagawa, G. (1996). Monte Carlo filter and smoother for non-Gaussian non-linear state space models. *Journal of Computational and Graphical Statistics 5*(1), 1–25. 311

Klein, P. (1999). Using the generalized Schur form to solve a multivariate linear rational expectations model. *Journal of Economic Dynamics and Control 24*(10), 1405–1423. 404

Kobak, D., J. Lomond, and B. Sanchez (2020). The optimal ridge penalty for real-world high-dimensional data can be zero or negative due to the implicit ridge regularization. *Journal of Machine Learning Research 21*(1), 6863–6878. 122, 138

Koenker, R. (2005). *Quantile Regression.* Cambridge: Cambridge University Press. 172

Koenker, R. and G. Bassett (1978). Regression quantiles. *Econometrica 46*(1), 33–50. 171

Koop, G. (2003). *Bayesian Econometrics.* Hoboken: Wiley. 36, 186

Koop, G. (2013). Forecasting with medium and large Bayesian VARs. *Journal of Applied Econometrics 28*(2), 177–203. 31, 366

Koop, G. and D. Korobilis (2013). Large time-varying parameter VARs. *Journal of Econometrics 177*(2), 185–198. 380, 382, 383

Koop, G., D. J. Poirier, and J. L. Tobias (2007). *Bayesian Econometric Methods.* Cambridge: Cambridge University Press. 36, 186

Koop, G. and S. M. Potter (2011). Time varying VARs with inequality restrictions. *Journal of Economic Dynamics and Control 35*(7), 1126–1138. 335

Koopman, S. J. (1993). Disturbance smoother for state space models. *Biometrika 80*(1), 117–126. 259

Koopman, S. J. (1997). Exact initial Kalman filtering and smoothing for nonstationary time series models. *Journal of the American Statistical Association 92*, 1630–1638. 283, 289

Koopman, S. J. and N. Shephard (1992). Exact score for time series models in state space form. *Biometrika 79*(4), 823–826. 273

Korattikara, A., Y. Chen, and M. Welling (2014). Austerity in MCMC land: Cutting the Metropolis-Hastings budget. *Proceedings of the 31st International Conference on Machine Learning 1*, 181–189. 97

Korobilis, D. and D. Pettenuzzo (2019). Adaptive hierarchical priors for high-dimensional vector autoregressions. *Journal of Econometrics 212*(1), 241–271. 368

Kozumi, H. and G. Kobayashi (2011). Gibbs sampling methods for Bayesian quantile regression. *Journal of Statistical Computation and Simulation 81*(11), 1565–1578. 175, 178

Kuo, L. and B. Mallick (1998). Variable selection for regression models. *Sankhyau: The Indian Journal of Statistics, Series B 60*(1), 65–81. 137

Kydland, F. E. and E. C. Prescott (1982). The time-varying volatility of macroeconomic fluctuations. *Econometrica 50*, 1345–1370. 398, 424

Lancaster, T. (2004). *An Introduction to Modern Bayesian Econometrics*. Malden: Blackwell. 36, 186

Lenza, M. and G. E. Primiceri (2022). How to estimate a vector autoregression after March 2020. *Journal of Applied Econometrics 37*(4), 688–699. 364, 425

Li, M. (2007). Bayesian proportional hazard analysis of the timing of high school dropout decisions. *Econometric Reviews 25*(5), 529–556. 234

Li, M. and J. L. Tobias (2009). Bayesian methods in microeconometrics. In J. Geweke, G. Koop, and H. Van Dijk (Eds.), *The Oxford Handbook of Bayesian Econometrics*. Oxford: Oxford University Press. 150, 186, 247

Li, Q. and N. Lin (2010). The Bayesian elastic net. *Bayesian Analysis 5*(1), 151–170. 123, 137

Lindley, D. and A. Smith (1972). Bayes estimates for the linear model. *Journal of the Royal Statistical Society, Series B 34*, 1–41. 36

Litterman, R. B. (1986). Forecasting with Bayesian vector autoregressions - five years of experience. *Journal of Business & Economic Statistics 4*(1), 25–38. 31, 364, 366, 367

Liu, J. and R. Chen (1998). Sequential Monte Carlo computation for dynamic systems. *Journal of the American Statistical Association 93*(443), 1032–1044. 311, 337

Liu, J., F. Liang, and W. Wong (2000). The multiple-try method and local optimization in Metropolis sampling. *Journal of the American Statistical Association 95*, 121–134. 67, 91

Liu, J., W. Wong, and A. Kong (1994). Covariance structure of the Gibbs sampler with applications to the comparisons of estimators and augmentation schemes. *Biometrika 81*, 27–40. 60

Liu, J. and Y. Wu (1999). Parameter expansion for data augmentation. *Journal of the American Statistical Association 94*, 1264–1274. 113

Long, J. B. and C. I. Plosser (1983). Real business cycles. *Journal of Political Economy 91*(1), 39–69. 398, 424

Love, I. and L. Zicchino (2006). Financial development and dynamic investment behavior: Evidence from panel VAR. *Quarterly Review of Economics and Finance 46*, 190–210. 364

Lucas, R. (1976). *Econometric Policy Evaluation: A Critique*. New York: Elsevier. 424

Lutkepohl, H. (2005). *New Introduction to Multiple Time Series Analysis*. Berlin: Springer. 364

Lykou, A. and I. Ntzoufras (2013). On Bayesian lasso variable selection and the specification of the shrinkage parameter. *Statistics and Computing 23*, 361–390. 123

Maclaurin, D. and R. Adams (2014). Firefly Monte Carlo: Exact MCMC with subsets of data. *Uncertainty in Artificial Intelligence - Proceedings of the 30th Conference, UAI 2014 1*, 4289–4295. 83, 97

Madigan, D., J. York, and D. Allard (1995). Bayesian graphical models for discrete data. *International Statistical Review 63*(2), 215–232. 115, 130, 137

Manski, C. and D. McFadden (1981). *Structural Analysis of Discrete Data with Econometric Applications*. Cambridge: The MIT Press. 246

Mariano, R. S. and Y. Murasawa (2003). A new coincident index of business cycles based on monthly and quarterly series. *Journal of Applied Econometrics 18*(4), 427–443. 289, 424

Mariano, R. S. and Y. Murasawa (2010). A coincident index, common factors, and monthly real GDP. *Oxford Bulletin of Economics and Statistics 72*(1), 27–46. 289, 424

Marin, J.-M., K. Mengersen, and C. P. Robert (2005). Bayesian modelling and inference on mixtures of distributions. In D. Dey and C. Rao (Eds.), *Bayesian Thinking*, Volume 25 of *Handbook of Statistics*, pp. 459–507. New York: Elsevier Science Publishers. 168

McCullagh, P. and J. A. Nelder (1989). *Generalized Linear Models, Second Edition*. New York: Chapman and Hall/CRC. 247

McCulloch, R. E. and P. Rossi (1994). An exact likelihood analysis of the multinomial Probit model. *Journal of Econometrics 64*, 207–240. 221, 247

McLachlan, G. J. and D. Peel (2000). *Finite Mixture Models*. Chichester: Wiley. 165

Metropolis, N., A. W. Rosenbluth, M. N. Rosenbluth, A. H. Teller, and E. Teller (1953). Equation of state calculations by fast computing machines. *The Journal of Chemical Physics 21*, 1087–1092. 97

Minsker, S., S. Srivastava, L. Lin, and D. Dunson (2017). Robust and scalable Bayes via a median of subset posterior measures. *Journal of Machine Learning Research 18*, 1–40. 73

Munnell, A., G. Tootell, L. Browne, and J. McEneaney (1996). Mortgage lending in Boston: Interpreting HMDA data. *American Economic Review 86*(1), 25–53. 28

Murray, J. (2018). Multiple imputation: A review of practical and theoretical findings. *Statistical Science 33*(2), 142–159. 372

Nakatsuma, T. (2000). Bayesian analysis of ARMA-GARCH models: A Markov chain sampling approach. *Journal of Econometrics 95*(1), 57–69. 394

Neal, R. (2000). Markov chain sampling methods for Dirichlet process mixture models. *Journal of Computational and Graphical Statistics 9*(2), 249–265. 170

Neal, R. (2003). Slice sampling. *The Annals of Statistics 31*(3), 705–767. 62

Neal, R. (2011). *MCMC Using Hamiltonian Dynamics.* New York: Chapman and Hall. 63

Neiswanger, W., C. Wang, and E. P. Xing (2014). Asymptotically exact, embarrassingly parallel MCMC. In *Proceedings of the Thirtieth Conference on Uncertainty in Artificial Intelligence*, pp. 623–632. Arlington: AUAI Press. 73, 76

Nelder, J. A. and R. W. M. Wedderburn (1972). Generalized linear models. *Journal of the Royal Statistical Society, Series A 135*(3), 370–384. 247

Nelson, D. (1991). Conditional heteroscedasticity in asset returns: A new approach. *Econometrica 59*(2), 347–370. 394, 424

Nelson, R. C. and A. F. Siegel (1987). Parsimonious modeling of yield curves. *Journal of Business 60*(4), 473–489. 377

Nemeth, C. and C. Sherlock (2018). Merging MCMC subposteriors through Gaussian-process approximations. *Bayesian Analysis 13*(2), 507–530. 73, 77, 78

Nobile, A. (2000). Comment: Bayesian multinomial Probit models with a normalization constraint. *Journal of Econometrics 99*, 335–345. 221, 224

Olsen, R. (1978). A note on the uniqueness of the maximum likelihood estimator in the Tobit model. *Econometrica 36*, 1211–1215. 246

Omori, Y., S. Chib, N. Shephard, and J. Nakajima (2007). Stochastic volatility with leverage: Fast and efficient likelihood inference. *Journal of Econometrics 140*(2), 425–449. 186, 418, 425

Park, T. and G. Casella (2008). The Bayesian lasso. *Journal of the American Statistical Association 103*(482), 681–686. 123, 124, 137, 157

Paye, B. and A. Timmermann (2006). Instability of return prediction models. *Journal of Empirical Finance 13*(3), 274–315. 34, 379

Percy, D. (1992). Prediction for seemingly unrelated regressions. *Journal of the Royal Statistical Society, Series B 54*(1), 243–252. 147

Petris, G., S. Petrone, and P. Campagnoli (2009). *Dynamic Linear Models with R.* New York: Springer. 260

Pitt, M. K. and N. Shephard (1999). Filtering via simulation. auxiliary particle filters. *Journal of the American Statistical Association 94*, 590–599. 318, 337, 396

Poirier, D. (1995). *Intermediate Statistics and Econometrics: A Comparative Approach.* Cambridge: The MIT Press. 36

Polson, N., J. Scott, and J. Windle (2013). Bayesian inference for logistic models using Polya-Gamma latent variables. *Journal of the American Statistical Association 108*(504), 1339–1349. 202, 206, 246

Primiceri, G. E. (2005). Time varying structural vector autoregressions and monetary policy. *Review of Economic Studies 72*(3), 821–852. 335, 364, 380, 381

Qian, H. (2014). A flexible state space model and its applications. *Journal of Time Series Analysis 35*(2), 79–88. 261

Qian, H. (2016). A computationally efficient method for vector autoregression with mixed frequency data. *Journal of Econometrics 193*(2), 433–437. 364

Qian, H. (2018). Big data Bayesian linear regression and variable selection by normal-inverse-gamma summation. *Bayesian Analysis 13*(4), 1011–1035. 36, 134, 135, 136

Qian, H. (2019). Inequality constrained state space models. *Journal of Business & Economic Statistics 37*(2), 350–362. 335

Qian, H. (2022). Bayesian inference in common microeconometric models with massive datasets by double marginalized subsampling. *Journal of Business & Economic Statistics 40*(4), 1484–1497. 186

Quiroz, M., R. Kohn, M. Villani, and M. Tran (2019). Speeding up MCMC by efficient data subsampling. *Journal of the American Statistical Association 114*, 831–843. 85, 86, 87, 97

Raftery, A. E., D. Madigan, and J. A. Hoeting (1997). Bayesian model averaging for linear regression models. *Journal of the American Statistical Association 92*(437), 179–191. 11, 130, 137

Rasmussen, C. and C. Williams (2006). *Gaussian Processes for Machine Learning.* Cambridge: The MIT Press. 78

Rhee, C. and P. Glynn (2015). Unbiased estimation with square root convergence for SDE models. *Operations Research 63*(5), 1026–1043. 84

Richardson, S. and P. J. Green (1997). On Bayesian analysis of mixtures with an unknown number of components. *Journal of the Royal Statistical Society, Series B 59*, 731–792. 165

Robbins, H. and S. Monro (1951). A stochastic approximation method. *The Annals of Mathematical Statistics 22*(3), 400–407. 87, 240

Robert, C. and G. Casella (2004). *Monte Carlo Statistical Methods.* New York: Springer Science Business Media. 56, 97

Roberts, G. and J. Rosenthal (2001). Optimal scaling for various Metropolis-Hastings algorithms. *Statistial Science 16*(4), 351–367. 97

Roberts, G. O., J. S. Rosenthal, and Y. F. Atchade (2007). Coupling and ergodicity of adaptive Markov chain Monte Carlo algorithms. *Journal of Applied Probability 44*(2), 458–475. 55

Rosenbaum, P. R. and D. B. Rubin (1983). The central role of the propensity score in observational studies for causal effects. *Biometrika 70*, 41–55. 208, 209, 210

Rubin, D. B. (1974). Estimating causal effects of treatments in randomized and nonrandomized studies. *Journal of Educational Psychology 66*, 688–701. 208

Rubin, D. B. (1976). Inference and missing data. *Biometrika 63*, 581–592. 259

Rubin, D. B. (1978). Bayesian inference for causal effects: the role of randomization. *The Annals of Statistics 6*, 34–58. 208

Rubin, D. B. (1996). Multiple imputation after 18+ years (with discussion). *Journal of the American Statistical Association 91*, 473–520. 372

Schmitt-Grohe, S. and M. Uribe (2004). Solving dynamic general equilibrium models using a second-order approximation to the policy function. *Journal of Economic Dynamics and Control 28*, 755–775. 411

Schon, T., F. Gustafsson, and P. Nordlund (2005). Marginalized particle filters for mixed linear/nonlinear state-space models. *IEEE Transactions on Signal Processing 53*(7), 2279–2289. 317

Schorfheide, F. and D. Song (2015). Real-time forecasting with a mixed-frequency VAR. *Journal of Business & Economic Statistics 33*(3), 366–380. 259, 364, 378, 424

Scott, S., A. Blocker, F. Bonassi, H. Chipman, E. George, and R. McCulloch (2016). Bayes and big data: The consensus Monte Carlo algorithm. *International Journal of Management Science and Engineering Management 11*, 78–88. 73, 75, 76, 96, 97

Scott, S. L. (2017). Comparing consensus Monte Carlo strategies for distributed Bayesian computation. *Brazilian Journal of Probability and Statistics 31*, 668–685. 75

Shao, J. (1993). Linear model selection by cross-validation. *Journal of the American Statistical Association 88*(422), 486–494. 15

Sims, C. (1980). Macroeconomics and reality. *Econometrica 48*, 1–48. 30, 364, 424

Sims, C. (2002). Solving linear rational expectations models. *Computational Economics 20*(1), 1–20. 402, 403, 404

Sims, C., D. Waggoner, and T. Zha (2008). Methods for inference in large multiple-equation Markov-switching models. *Journal of Econometrics 146*(2), 255–274. 364

Sims, C. and T. Zha (1998). Bayesian methods for dynamic multivariate models. *International Economic Review 39*, 949–968. 17, 366, 367

Sims, C. and T. Zha (2006). Were there regime switches in U.S. monetary policy. *American Economic Review 91*(1), 54–81. 364, 379

Smets, F. and R. Wouters (2007). Shocks and frictions in US business cycles: A Bayesian DSGE approach. *American Economic Review 97*, 586–608. xvii, 274, 398, 404, 405, 424

Smith, A. F. M. and G. O. Roberts (1993). Bayesian computation via the Gibbs sampler and related Markov chain Monte Carlo methods. *Journal of the Royal Statistical Society, Series B 55*(1), 3–23. 58

Smith, M. and R. Kohn (1996). Nonparametric regression using Bayesian variable selection. *Journal of Econometrics 75*(2), 317–343. 130

Srivastava, S., C. Li, and D. Dunson (2018). Scalable Bayes via barycenter in Wasserstein space. *Journal of Machine Learning Research 19*, 1–35. 73

Stephens, M. (2000). Dealing with label switching in mixture models. *Journal of the Royal Statistical Society. Series B (Statistical Methodology) 62*(4), 795–809. 186

Stock, J. and M. Watson (1996). Evidence on structural instability in macroeconomic time series relations. *Journal of Business & Economic Statistics 14*, 11–30. 34, 379

Stock, J. and M. Watson (2002). Macroeconomic forecasting using diffusion indexes. *Journal of Business & Economic Statistics 20*(2), 147–62. 364, 373, 378

Stock, J. and M. Watson (2007). Why has U.S. inflation become harder to forecast. *Journal of Money, Credit and Banking 39*(1), 3–33. 335

Strachan, R. W. and B. Inder (2004). Bayesian analysis of the error correction model. *Journal of Econometrics 123*(2), 307–325. 364

Swamy, P. (1970). Efficient inference in a random coefficient regression model. *Econometrica 38*(2), 311–323. 379

Swanson, N. (1998). Money and output viewed through a rolling window. *Journal of Monetary Economics 41*, 455–474. 8, 34, 379

Tan, Z. (2010). Bounded, efficient and doubly robust estimation with inverse weighting. *Biometrika 97*, 661–682. 209

Tanner, M. A. and W. H. Wong (1987). The calculation of posterior distributions by data augmentation. *Journal of the American Statistical Association 82*, 528–550. 58, 97, 259

Teh, Y. W., H. A. Thiery, and S. J. Vollmer (2016). Consistency and fluctuations for stochastic gradient Langevin dynamics. *Journal of Machine Learning Research 17*, 1–33. 88, 97

Tibshirani, R. (1996). Regression shrinkage and selection via the lasso. *Journal of the Royal Statistical Society B 58*, 267–288. 115, 122, 137

Tierney, L. (1994). Markov chains for exploring posterior distributions. *The Annals of Statistics 22*, 1701–1728. 56, 97

Tipping, M. E. and C. M. Bishop (1999). Probabilistic principal component analysis. *Journal of the Royal Statistical Society, Series B 61*(3), 611–622. 374

Titterington, D. M., A. F. M. Smith, and U. E. Makov (1985). *Statistical Analysis of Finite Mixture Distributions.* Chichester: Wiley. 165

Tobin, J. (1958). Estimation of relationships for limited dependent variables. *Econometrica 26*, 24–36. 210, 225, 246

Train, K. (2003). *Discrete Choice Methods with Simulation.* New York: Cambridge University Press. 206, 207, 246

Tsay, R. S. (1998). Testing and modeling multivariate threshold models. *Journal of the American Statistical Association 93*, 1188–1202. 364

Uhlig, H. (2005). What are the effects of monetary policy on output? results from an agnostic identification procedure. *Journal of Monetary Economics 52*, 381–419. 369

van Hasselt, M. (2011). Bayesian inference in a sample selection model. *Journal of Econometrics 165*(2), 221–232. 218

Vrontos, I. D., P. Dellaportas, and D. Politis (2000). Full Bayesian inference for GARCH and EGARCH models. *Journal of Business & Economic Statistics 18*(2), 187–198. 394

Waggoner, D. F. and T. Zha (1999). Conditional forecasts in dynamic multivariate models. *Review of Economics and Statistics 81*(4), 639–651. 259, 372

Wang, X. and D. Dunson (2013). Parallelizing MCMC via Weierstrass sampler. *arXiv preprint arXiv:1312.4605*. 73

Wang, X., F. Guo, K. Heller, and D. Dunson (2015). Parallelizing MCMC with random partition trees. *Advances in Neural Information Processing Systems*, 451–459. 73

Wasserman, L. (2000). Bayesian model selection and model averaging. *Journal of Mathematical Psychology 44*, 92–107. 11

Wienke, A. (2011). *Frailty Models in Survival Analysis.* New York: Chapman and Hall/CRC. 235

Winkelmann, R. (2008). *Econometric Analysis of Count Data.* New York: Springer. 225

Wooldridge, J. (2010). *Econometric Analysis of Cross Section and Panel Data, Second Edition.* Cambridge: The MIT Press. 246, 424

Yao, W. and B. G. Lindsay (2009). Bayesian mixture labeling by highest posterior density. *Journal of the American Statistical Association 104*(486), 758–767. 186

Yu, J. (2005). On leverage in a stochastic volatility model. *Journal of Econometrics 127*(2), 165–178. 418

Yu, K. and R. A. Moyeed (2001). Bayesian quantile regression. *Statistics and Probability Letters 54*(4), 437–447. 175

Zadrozny, P. (1988). Gaussian likelihood of continuous-time ARMAX models when data are stocks and flows at different frequencies. *Econometric Theory 4*(1), 108–124. 289, 424

Zeger, S. L. and M. R. Karim (1991). Generalized linear models with random effects; a Gibbs sampling approach. *Journal of the American Statistical Association 86*(413), 79–86. 202, 206, 247

Zellner, A. and R. Peter (1984). Bayesian analysis of dichotomous quantal response models. *Journal of Econometrics 25*(3), 365–393. 202

Zhang, C., S. Bengio, M. Hardt, B. Recht, and O. Vinyals (2021). Understanding deep learning (still) requires rethinking generalization. *The Annals of Statistics 64*(3), 107–115. 122

Zhang, J. and A. Lawson (2011). Bayesian parametric accelerated failure time spatial model and its application to prostate cancer. *Journal of Applied Statistics 38*(3), 591–603. 235

Zorn, C. (2005). A solution to separation in binary response models. *Political Analysis 13*, 157–170. 203

Zou, D., J. Wu, V. Braverman, Q. Gu, and S. Kakade (2024). Benign overfitting of constant-stepsize SGD for linear regression. *Journal of Machine Learning Research 24*(1), 15502–15559. 138

Zou, H. and T. Hastie (2005). Regularization and variable selection via the elastic net. *Journal of the Royal Statistical Society, Series B 67*, 301–320. 137

Index

FSC
www.fsc.org
MIX
Papier | Fördert
gute Waldnutzung
FSC® C083411

Kolibri 360 GmbH
Ferdinand-Jühlke-Straße 7
99095 Erfurt, Deutschland
produktsicherheit@kolibri360.de